Advances in Intelligent Systems and Computing

Volume 1057

The series "Advances in Intelligent Systems and Computing" contains publications on theory, applications, and design methods of Intelligent Systems and Intelligent Computing. Virtually all disciplines such as engineering, natural sciences, computer and information science, ICT, economics, business, e-commerce, environment, healthcare, life science are covered. The list of topics spans all the areas of modern intelligent systems and computing such as: computational intelligence, soft computing including neural networks, fuzzy systems, evolutionary computing and the fusion of these paradigms, social intelligence, ambient intelligence, computational neuroscience, artificial life, virtual worlds and society, cognitive science and systems, Perception and Vision, DNA and immune based systems, self-organizing and adaptive systems, e-Learning and teaching, human-centered and human-centric computing, recommender systems, intelligent control, robotics and mechatronics including human-machine teaming, knowledge-based paradigms, learning paradigms, machine ethics, intelligent data analysis, knowledge management, intelligent agents, intelligent decision making and support, intelligent network security, trust management, interactive entertainment, Web intelligence and multimedia.

The publications within "Advances in Intelligent Systems and Computing" are primarily proceedings of important conferences, symposia and congresses. They cover significant recent developments in the field, both of a foundational and applicable character. An important characteristic feature of the series is the short publication time and world-wide distribution. This permits a rapid and broad dissemination of research results.

** Indexing: The books of this series are submitted to ISI Proceedings, EI-Compendex, DBLP, SCOPUS, Google Scholar and Springerlink **

More information about this series at http://www.springer.com/series/11156

Kedar Nath Das · Jagdish Chand Bansal ·
Kusum Deep · Atulya K. Nagar ·
Ponnambalam Pathipooranam ·
Rani Chinnappa Naidu
Editors

Soft Computing for Problem Solving

SocProS 2018, Volume 2

 Springer

Editors
Kedar Nath Das
Department of Mathematics
National Institute of Technology Silchar
Silchar, Assam, India

Kusum Deep
Department of Mathematics
Indian Institute of Technology Roorkee
Roorkee, Uttarakhand, India

Ponnambalam Pathipooranam
School of Electrical Engineering
VIT University
Vellore, Tamil Nadu, India

Jagdish Chand Bansal
Department of Mathematics
South Asian University
New Delhi, Delhi, India

Atulya K. Nagar
Department of Mathematics
Faculty of Science
Liverpool Hope University
Liverpool, UK

Rani Chinnappa Naidu
School of Electrical Engineering
VIT University
Vellore, Tamil Nadu, India

ISSN 2194-5357 ISSN 2194-5365 (electronic)
Advances in Intelligent Systems and Computing
ISBN 978-981-15-0183-8 ISBN 978-981-15-0184-5 (eBook)
https://doi.org/10.1007/978-981-15-0184-5

This Springer imprint is published by the registered company Springer Nature Singapore Pte Ltd.
The registered company address is: 152 Beach Road, #21-01/04 Gateway East, Singapore 189721, Singapore

Preface

SocProS stands for *Soft Computing for Problem Solving*. It is an Eight years old series of International Conferences held annually under the joint collaboration among a group of faculty members from the institutes of repute like NIT Silchar, IIT Roorkee, South Asian University Delhi, Liverpool Hope University, UK and VIT Vellore.

For the first time, SocProS was held at IE(I), RLC, Roorkee, India during Dec 20-22, 2011, with General Chairs as Prof Kusum Deep, Indian Institute of Technology Roorkee and Prof Atulya K. Nagar, Liverpool Hope University, UK. The second SocProS was held at JKLU, Jaipur, India during Dec 28–20, 2012. Similarly, the third SocProS was held at the Greater Noida Extension Centre of IIT Roorkee during December 26–28, 2013, fourth SocProS was held at NIT Silchar, Assam during December 27–29, 2014, Fifth SocProS was held at Saharanpur Campus of IIT Roorkee, during December 18–20, 2015, Sixth SocProS was held at Thapar University, Patiala, Punjab, during December 23–24, 2016, Seventh SocProS was held at IIT Bhubaneswar, Odisha, During December 23–24, 2017, Now the name 'SocProS' became a brand name which has already established its benchmark in last eight years through its successful milestones every time in attracting many participants from all over the world like UK, US, Korea, France, Dubai, South Africa etc.

This time, the Eighth SocProS has been held at VIT Vellore, India during Dec 17–19, 2018. Like earlier SocProS conferences, the focus of SocProS 2018 lies in Soft Computing and its applications to solve real life problems occurring in different domains in the field of medical and health care, supply chain management, signal processing and multimedia, industrial optimization, image processing, cryptanalysis etc. SocProS 2018 attracted a wide spectrum of thought-provoking research papers on various aspects of Soft Computing with umpteen applications, theories and techniques. A total 176 quality research papers are selected for publication in the form of proceedings in its Volume 1 and Volume 2.

We are sure that the research findings in the novel papers contained in this proceeding will be much fruitful and may inspire more and more researchers to work in the field of *soft computing*. The topics that are presented in this proceedings are Fuzzy logic & Fuzzy controller, Artificial Neural Network, Face Recognition & Classification, Feature Extraction, Machine learning, Reinforcement learning, Deep learning, Supervised learning, Different optimization techniques like Spider-Monkey Optimization, Particle Swarm Optimization, Meta heuristic Optimization, Artificial Bee Colony Optimization, Walk Grey Wolf Optimization, Algorithms like Flower Pollination Algorithm, Parallel Random Forest Algorithm, C-mode Clustering Algorithm, Crow Search Algorithm, Genetic Algorithm, Artificial Bee Colony Algorithm, Adaptive Multi-Swarm Bat Algorithm etc. Therefore this proceeding must provide an excellent platform to explore the assorted soft computing techniques to the readers.

The editors would like to express their sincere gratitude to its Patron, Plenary Speakers, Invited Speakers, Reviewers, Programme Committee Members, International Advisory Committee, and Local Organizing Committee; without whose support the quality and standards of the Conference could not be maintained. Special thanks to Springer and its team for this valuable publication.

Over and above, we would like to express our deepest sense of gratitude to 'VIT Vellore' for hosting this conference. Also, sincere thanks to all sponsors of SocProS' 2018.

Silchar, India	Kedar Nath Das
New Delhi, India	Jagdish Chand Bansal
Roorkee, India	Kusum Deep
Liverpool, UK	Atulya K. Nagar
Vellore, India	Ponnambalam Pathipooranam
Vellore, India	Rani Chinnappa Naidu

About This Book

The proceedings of SocProS 2018 will serve as an academic bonanza for scientists and researchers working in the field of Soft Computing. This book contains theoretical as well as practical aspects using fuzzy logic, neural networks, evolutionary algorithms, swarm intelligence algorithms, etc. with many applications under the umbrella of 'Soft Computing'. This book is beneficial for the young as well as experienced researchers dealing across complex and intricate real world problems for which finding a solution by traditional methods is a difficult task.

The different application areas covered in the proceedings are: Image Processing, Cryptanalysis, Industrial Optimization, Supply Chain Management, Newly Proposed Nature Inspired Algorithms, Signal Processing, Problems related to Medical and Health Care, Networking Optimization Problems etc. This will surely helpfully for the researchers/scientists working in similar fields of optimization.

Contents

About the Editors

Dr. Kedar Nath Das is an Assistant Professor at the Department of Mathematics, National Institute of Technology, Silchar, Assam, India. Over the past 10 years, he has made substantial contributions to research on soft computing, and has published several research papers in prominent national and international journals. His chief area of interest is in evolutionary and bio-inspired algorithms for optimization.

Dr. Jagdish Chand Bansal is an Associate Professor at the South Asian University, New Delhi, India and visiting research fellow at Liverpool Hope University, Liverpool, UK. He has an excellent academic record and is a leading researcher in the field of swarm intelligence. Further, he has published numerous research papers in respected international and national journals.

Prof. Kusum Deep is a Professor at the Department of Mathematics, Indian Institute of Technology Roorkee, India. Over the past 25 years, her research has made her a central international figure in the areas of nature-inspired optimization techniques, genetic algorithms and particle swarm optimization.

Prof. Atulya K. Nagar holds the Foundation Chair as Professor of Mathematical Sciences and is Dean of the Faculty of Science at Liverpool Hope University, UK. Prof. Nagar is an internationally respected scholar working at the cutting edge of theoretical computer science, applied mathematical analysis, operations research, and systems engineering. He received a prestigious Commonwealth Fellowship for pursuing his doctorate (DPhil) in Applied Non-Linear Mathematics, which he earned from the University of York (UK) in 1996; and he holds BSc (Hons.), MSc, and MPhil (with Distinction) from the MDS University of Ajmer, India.

Prof. Ponnambalam Pathipooranam is an Associate Professor at the School of Electrical Engineering, VIT University, India. His areas of research interests are Multilevel Converters, Fuzzy controller for multilevel converters, MPC controllers, Thermoelectric Generators for Solar Photo voltaic cells areas in which he is actively publishing. He is having 15 years of teaching experience.

Prof. Rani Chinnappa Naidu received the B.Eng. and M.Tech. degrees from VIT University, Vellore, India, and Ph.D. degree from Northumbria University, Newcastle upon Tyne, UK., all in Electrical Engineering. After that, she joined as a Postdoctoral Researcher in Northumbria Photovoltaic Applications Centre, Northumbria University, UK. She is currently an Associate Professor at VIT University. She is an Senior member in IEEE. She leads an appreciable number of research groups and projects in the areas such as solar photovoltaic, wind energy, power generation dispatch, power system optimization, and artificial intelligence techniques.

Artificial Neural Network-Based Smart Energy Meter Monitoring and Control Using Global System for Mobile Communication Module

P. Ashwini Kumari and P. Geethanjali

Abstract This paper presents smart and optimal way of allocating power to the utility using global system for mobile communication module-based remote automatic energy meter reading system. The designed device is installed with the energy meter at consumer premises. A smart communication is established between service provider and consumer using GSM module which is capable of calculating the energy consumed at different tariff and time. An artificial neural network using back-propagation approach is employed to obtain optimal allocation of service provider to meet the objective function. The novel idea of smart energy metering not only reduces the cost of energy consumption but also helps in proper repayments, optimal usage of power based on time of day tariff, and theft control with higher reliability and greater flexibility. A smart real-time prototype of the automatic energy reading system was built to demonstrate the effectiveness and efficiency of automatic meter reading, billing, and notification through the use of global system for mobile communication network.

Keywords Arduino microcontroller · Artificial intelligence (AI) · Global system for mobile communication (GSM) · GSM modem · Current transformer · Artificial neural network (ANN) · Back-propagation algorithm (BPA) · Independent power producer (IPP)

1 Introduction

Energy, being a strategic commodity, plays a significant role in economic development of a country. Energy consumption in efficient way has become a major issue

P. Ashwini Kumari · P. Geethanjali (✉)
VIT University, Vellore, India
e-mail: pgeethanjali@vit.ac.in

P. Ashwini Kumari
e-mail: kanchanapu@gmail.com

P. Ashwini Kumari
REVA University, Bengaluru, India

© Springer Nature Singapore Pte Ltd. 2020
K. N. Das et al. (eds.), *Soft Computing for Problem Solving*,
Advances in Intelligent Systems and Computing 1057,
https://doi.org/10.1007/978-981-15-0184-5_1

1

in current scenario. Consumer electricity usage has been increased without actual knowledge of his requirement which sometimes increases the risk of power theft. The service provider could not predict his own power consumption of particular location at any given time. The usage pattern of the consumer effectively judges the load changes. There is need for accurate measurement of energy consumption by using smart techniques [1]. An automated energy metering system reduces the human intervention and avoids the error produced due to manual billing. Advances in soft computing gives significant thrust to automatize smart cities. It also paves the way for increasing the operating efficiency by resolving issues with respect to wrong meter readings and theft which is auto self-corrected by adopting AI techniques. To combat power theft and fraud which represent economic losses, researchers are focusing on IoT, big data, and machine learning. The learning algorithms can not only learn and anticipate large amount of data but also diagnose unusual behavior in big data sets. Intelligent metering includes progressed metering innovation, which includes setting clever meters that measure vitality utilization, remotely switch on the supply to clients and even control the most extreme power utilization conceivable, in this manner perusing, preparing, and sending messages to clients [2]. An automatic remote meter reading system based on GSM is presented in this paper. This paper is useful to obtain meter reading when desired, so meter readers do not need to visit each customer for the consumed energy data collection and to distribute the bills [3]. Microcontroller can be used to monitor and record the meter readings. In case of a customer defaulter, no need to send a person of utility to cut-off the customer connection. Service provider can continue or stop the supply of power by sending the Short Message Service (SMS). GSM is employed for transferring the readings recorded by the energy meter. An IoT-enabled measurement of power line metrics and power consumption for monitoring is proposed as per IoT protocols. Android-based Web service feature is used to extract and respond to the messages which are sent through server [4]. Need-based energy consumption with suitable decision criteria to incorporate the shortest transmission path as per IEEE 21451 standard is presented in [5]. The measurements were made with the help of current and voltage transducers by considering the effect of harmonics using FFT to enhance the quality of power. Power theft being a major issue could be classified into two ways as mentioned in [6, 7]. By comparing energy usage with substation loads and by incorporating prediction based on machine learning anomalies present in the grid could be easily detected. The success rate of Advanced Energy Metering Infrastructure (AEMI) is determined by data analytics.

The key challenges in smart metering include:

- Ability to work with large data
- Privacy and security constraints
- Adoption of unsupervised learning techniques
- Effective control strategies and optimized latency.

2 Description

The block diagram of proposed work and its hardware implementation is shown in Fig. 1.

2.1 Independent Power Producers

These are the service providers who generate the power and sell it to utility with different tariff schemes. The service provider bids the cost as per the time of the day.

Data acquisition system receives the information from IPP and the large data will be dumped to cloud. Whenever required, the data can be retrieved for the processing. A wired or wireless secure cryptographic standard is employed in [8] for secured communication. In this paper, the communication between the energy meter and the user is established through microcontroller. GSM module consists of modem

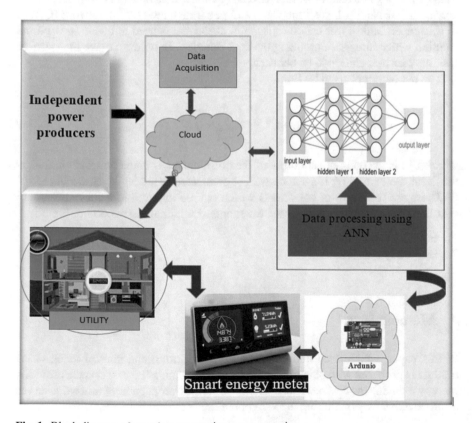

Fig. 1 Block diagram of complete automatic energy metering

assembly, interfacing units like USB and RS-232 and power supply circuits. GPRS is a communication protocol used in GSM technology [9]. IPP should provide SIM card with an IMEI number for ease of identification.

3 Artificial Neural Network

There are various intelligent techniques available in the literature which proposed a feasible solution to deal with nonlinear and unpredictable problems. One of such framework is artificial neural network which mimics the behavior of human brain. It has neurons, synapse, and dendrites. The neural network architecture includes perceptron, convolutional neural networks (CNNs) [10], which uses back-propagation with many layers incorporating feed-forward network, recurrent neural networks (RNNs) [11] that work by predicting the next term of the input data in a sequential manner, which are basically suited for supervised learning framework long-/short-term memory, gated recurrent units (GRUs) [12], Hopfield networks [13], deep belief networks (DBNs) [14], auto-encoders, and generative adversarial network (GAN). In this paper, multi-layer back-propagation algorithm is used to obtain the optimal solution which aims at minimizing the cost function. In order to make the learning fast, the bias α is chosen to be between -1 and 1.

The net output is given by Eq. 1

$$Y_{\text{net}} = \alpha + \sum_{i=1}^{n}(x_i w_i) \tag{1}$$

where Y_{net} is the optimal net output of the output layer, α is the bias function, X_i is the input vector, and W_i is synaptic weight connected to hidden layer.

There are two types of constraints which can be imposed to achieve minimum cost function which are defined over the range of minimum to maximum limits:

1. Equality constraints
2. Inequality constraints.

4 Methodology

In this paper, ANN-based automatic energy monitoring and control using GSM module is proposed with objective of optimally allocating the best service provider, thereby reducing the cost per unit consumption. Back-propagation framework serves as an effective way of altering the weights in feed-forward network.

Steps to compute optimum cost by lambda iteration method:

Step: 1   Read the data from all the service providers, and store it in cloud.
Step: 2   Feed the input to ANN framework.
Step: 3   Set learning with weighing factor and bias.
Step: 4   Check for the convergence; if not converged, change and update the bias with previous values.
Step: 5   Run various iterations until optimal cost function is achieved.
Step: 6   Stop.

These optimum values of cost function are fed to Arduino which sends SMS to utility using GSM about the best service provider, and it also informs the account balance along with next recharging period. If the customer fails to pay the amount on due date, the power will be disconnected automatically by controller. There are two modes of operation, namely postpaid mode and prepaid mode. In first mode, when the supply is given to microcontroller and if postpaid mode is selected, the customer after every 30 days receives a SMS from electricity board with the details of energy consumed and its tariff. If payment is not done within the stipulated time, the controller is enabled to automatically turn off the supply without visiting the consumer premises.

5   Hardware Implementation and Flowchart

The complete flowchart depicting the flow of the process is given in Fig. 2, and the complete hardware implementation with load and energy meter is as shown in Fig. 3.

The messages received by the user indicate the amount payable and the due date before which the bill has to be paid. The customer also receives the intimation of power supply connection and disconnection due to low balance.

6   Conclusion and Future Work

This paper presents the design of ANN-based, simple, low-cost, GSM-based automated energy monitoring system for automating billing and managing the collected data globally. The proposed system overcomes the drawbacks associated with traditional way of reading data. The service provider would be able to track and monitor the energy usage at faster rate and can optimally plan the load schedule with optimal cost.

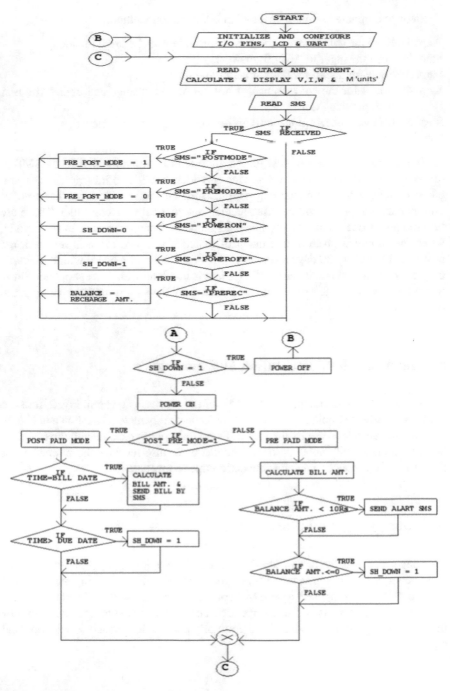

Fig. 2 Flowchart depicting the logic of the program

Fig. 3 Complete hardware implementation

The proposed work eases the process of accessing information, reducing power thefts, flexible online payments, and bill generation via SMS and reduction of manpower intervention. The work can further be taken a head by incorporating evolutionary computing techniques to check the best solution at reduced cost infrastructure. The overall paper aims at achieving optimal demand side management so that economic load dispatching is accompanied with ANN algorithm with its smart features.

References

1. Ashna, K., George, S.N.: GSM based automatic energy meter reading system with instant billing. In: 2013 International Multi-conference on Automation, Computing, Communication, Control and Compressed Sensing (iMac4s), pp. 65–72. IEEE, 22 Mar 2013
2. Zou, L., Chu, S., Guo, B.: The design of prepayment polyphase smart electricity meter system. In: 2010 International Conference on Intelligent Computing and Integrated Systems (ICISS), pp. 430–432. IEEE, 22 Oct 2010
3. Kurkute, S.R., Girase, G., Patil, P.: Automatic energy meter reading system using GSM technology. Telecommun. Syst. **4**(3) (2016)
4. Kim, D.Y., Kim, Y.C.: In: IEEE International Conference on Smart Energy Grid Engineering (SEGE) (2015)
5. Morello, R., et al.: A smart power meter to monitor energy flow in smart grids: the role of advanced sensing and IoT in the electric grid of the future. IEEE Sens. J. **17**(23), 7828–7837 (2017)
6. Arif, A., Al-Hussain, M., Al-Mutairi, N., Al-Ammar, E., Khan, Y., Malik, N.: Experimental study and design of smart energy meter for the smart grid. In: 2013 International Renewable and Sustainable Energy Conference (IRSEC), pp. 515–520. IEEE, 7 Mar 2013
7. Malhotra, P.R., Seethalakshmi, R.: Automatic meter reading and theft control system by using GSM. Int. J. Eng. Technol. (IJET) **5**(2), 2013

8. Das, V.V.: Wireless communication system for energy meter reading. In: 2009 International Conference on Advances in Recent Technologies in Communication and Computing (ART-Com'09), pp. 896–898. IEEE, Oct 2009

9. Silviya, E.M., Vinodhini, K.M., Salai Thillai Thilagam, J.: GSM based automatic energy meter system with instant billing. Int. J. Adv. Res. Electr. Electron. Instrum. Eng. **3**, 44–49 (2014)

10. Kim, Y.: Convolutional neural networks for sentence classification. arXiv preprint. arXiv:1408. 5882, 25 Aug 2014

11. Lipton, Z.C., Berkowitz, J., Elkan C.: A critical review of recurrent neural networks for sequence learning. arXiv preprint. arXiv:1506.00019, 29 May 2015

12. Chung, J., Gulcehre, C., Cho, K., Bengio, Y.: Gated feedback recurrent neural networks. In: International Conference on Machine Learning, pp. 2067–2075, 1 Jun 2015

13. Wen, U.P., Lan, K.M., Shih, H.S.: A review of Hopfield neural networks for solving mathematical programming problems. Eur. J. Oper. Res. **198**(3), 675–687 (2009)

14. Längkvist, M., Karlsson, L., Loutfi, A.: A review of unsupervised feature learning and deep learning for time-series modeling. Pattern Recogn. Lett. **42**, 11–24 (2014)

Efficient Analysis of User Reviews and Community-Contributed Photographs for Reputation Generation

V. Subramaniyaswamy, Logesh Ravi and V. Indragandhi

Abstract People can share their thoughts and opinions on any entities through the Internet. Normally, the attitude of the preferences of human can be predicted which are expressed in natural languages. Using sentimental mining method, the readership predictions are made on online reviews of locations. The reviews have been useful for the travelers to gain knowledge about the information of various locations and shortlist the best that is needed for them. In this paper, we categorize the locations based on the reviews and community-contributed photographs with the help of yelp and Tripadvisor datasets. In the proposed approach, opinions are filtered to eliminate unrelated ones through opinion pertinence calculation, and later grouped into a number of fused principal opinion sets. Based on the experiments conducted on large-scale datasets, the proposed approach is found to be useful for the user to make a decision.

Keywords Sentiment analysis · Prediction · Decision making · Opinion mining · Reputation generation · Data mining

1 Introduction

The recent developments in web technologies have generated massive user data in the form of reviews and comments. Users tend to share their opinions on the items in the websites [9]. The user-generated comments and reviews are more helpful to customers and merchants and the opinion mining enhances the sales and user satisfaction [8]. Customers can make purchase decisions, and the merchants can

V. Subramaniyaswamy (✉)
School of Computing, SASTRA Deemed University, Thanjavur 613401, India
e-mail: vsubramaniyaswamy@gmail.com

L. Ravi
Sri Ramachandra faculty of Engineering and Technology, Sri Ramachandra Institute of Higher Education and Research, Chennai, India

V. Indragandhi
School of Electrical Engineering, Vellore Institute of Technology, Vellore 632014, India

© Springer Nature Singapore Pte Ltd. 2020
K. N. Das et al. (eds.), *Soft Computing for Problem Solving*,
Advances in Intelligent Systems and Computing 1057,
https://doi.org/10.1007/978-981-15-0184-5_2

9

make use of the reviews to improve their products [2–4]. The processing of the reviews is more complicated and understanding the meaning in the users' context. Similar to digital items, locations and tourist spots can also have reviews. People around the world may visit different places on vacations to begin a new life, to start a business, on a pilgrimage, and so on [6]. The people will get the information about the location using social media. A traveler who wishes to travel around the world heavily depends upon these reviews to select the places he has to visit. Different people have different tastes, and their preferences about a place can differ. To suit everyone's needs, in this paper, we suggest a system which using the datasets taken from online travel websites, displays information about a location.

The main challenge faced is deciding upon the places to visit, trying to gain knowledge about the place and coming to a conclusion. Here, we accumulate the information and identify the appropriate pieces of data that cohesively form a part of the larger whole. Given any input of data, we satisfy them giving them all the appropriate information about the places so that they could find one of their comforts. We take the user preference into account and also find out the common shared interests to travel their plans that satisfy the users.

The major contributions of this paper can be summarized as:

- We propose a system that incorporates datasets taken from travel websites.
- We demonstrate techniques used to group those data.
- We develop a system that analyzes and displays information based on user queries.
- We display the location and Google map and present the Wikipedia article about the location.

This paper is structured as follows. The next section presents an outline of related works. In Sect. 3, we specify the research challenges and we explain about our proposed approach. In Sect. 4, we present our experimental results and discussions. Finally, Sect. 5 concludes the article with summary of work with future work directions.

2   Related Works

An important problem with reviews is their degree of relevance to our current scenario. Just because the forum is open, anyone can write anything. The less relevant and irrelevant reviews can be labeled as noisy. They can also be called as review-spam. Such reviews have to be removed before considering them for our project. The degree of relevance of a review to the relevant forum is called review pertinence [9]. Extraction of opinion target experimented on open-domain contents like news. Reviews were also helpful in prediction of political events based on people's opinion. Kim and Hovy [5] used a designed approach to name the one who holds the opinion and the target. They developed an election prediction system for this purpose.

Shapiro [7] stated that in calculation of reputed scores, time is the main constraint. Reputation systems are becoming popular among internet users, nowadays, because we can measure the quality of a product using metrics provided by these systems. These systems play a vital role in e-commerce and review websites. Abdel-Hafez. et al. [1] introduced a function that generates scores for item reputation. Micro-blogs are used by various people ranging from normal people to popular persons. They contain many opinions which are helpful and can be processed. To summarize opinions, Zhou et al. [11] adapted the traditional aspect-based summarization framework to manipulate opinions contained in micro-blogs.

One of the important media to express opinion is blogs. People write blogs about anything and everything. This information can also be processed based on our needs. By extracting the bloggers' information, we can also group people based on their interests. To improve the ability to retrieve, Weng and Zhao [10] determine the relations between a blogger and another and give information about the interfaces of query. Unlike ratings and scores, natural texts contained in reviews are complex to analyze [8]. Processing texts such as reviews help the system at intelligent by evaluating the current sentimental context of the user by making better predictions.

3 Proposed Methodology

In this section, we present the case scenario to understand the research problem and describe the proposed approach. Mr. Shiva is on an executive visit in Dubai. He is extremely preoccupied and can devote a weekend for visiting nearby places. Dubai is a place known for its buildings and hangout places. Shiva wants to roam around Dubai and makes some beautiful memories. Being new to the place, he does not know anything about tourist locations. He wants to know about famous spots to visit and also some additional information about the location. To save time, he also wants to locate the place in map so that he can travel without any chaos.

In order to suggest these, we allow him to explore his affinities based on his own choice and point of interest. We suggest the places which match his points of interests. For example, he could see the places which are all nearer to him and also he could explore the details and make up his mind whether to visit the place or not.

3.1 System Architecture

The descriptions about a place given by the user with or without ratings on the target item as a dataset help us to recommend places on their point of interest. Information collection and preprocessing is done in order to avoid raw datasets. Raw datasets will not account into efficiency. The overall efficiency will be let down if we have data taken from the raw datasets. Filtering of the data collected is done in order to increase the efficiency and also gather only related data. Thus, this will reduce the possibilities of the irrelevant data being present in the system. Filtering is a must because having the data which are not needed not only results in time lag but also the

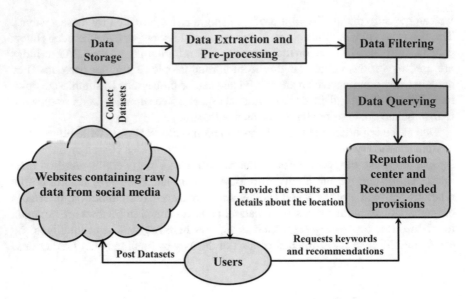

Fig. 1 Proposed system architecture

inconsistency of the system as it might wrongly retrieve the data which are similar. The data are grouped after the filtering based on the user's request and given back. Reputations are generated based on the popularity of the locations. This is executed to increase the user's trust toward the software. The user should find the software to be useful and easy so that it would be a great reach. Other locations related to the keyword are also recommended based on their point of interest. The detailed architecture is shown in Fig. 1.

3.2 Proposed Architecture

We collect the dynamic data from the website with the help of crawler. The HTML parser is exploited to analyze the dataset. The preprocessing of data is helpful in achieving effective results. The first step in preprocessing is tokenization, where all the words in the dataset are divided into tokens. The preprocessed dataset is used for further calculations. An important step in mining approach is data preprocessing. If there are so many irrelevancy and redundancy present along with dirty data, then gaining sufficient information about that is a tedious task. Preparation of data and filtering of data will take a decent time of processing. Hence data cleansing is done.

It is the method of detection, removal and correction of the misread, not precise, or damaged records and so on from an entity, datasets, or database center, and this refers to the identification of the incomplete, imperfect, and unmatched data and therefore correcting it by replacement, modification, or deletion of the misread or corrupt data.

Post-cleaning, the dataset should be consistent along with other datasets in our system. Sometimes, the user could have entered the data wrong. This might also lead to inconsistencies which have been detected or removed. Also the inconsistencies may be caused by corruption in the storage system, or by the way it is defined in the dictionary in different stores. Data cleansing is different from data validation because, in the latter, it means that the date have been removed from the system while entering rather than being performed on groups of data.

3.3 Data Filtering

Data filtering is also an important task. It deals with the reduction of the content having noise or errors from processed data. Noise indicates the important features in the data and will result in the limitation of their usefulness in practice. Hence, it must be a very important task to filter the data. Many techniques are developed to filter the processed data.

To increase efficiency, this method eliminates the unrelated opinions on the target entities. This approach targets on the related and relevant opinions which could be used later. The data are usually collected from multiple websites. This data which had been preprocessed and filtered are sent as input for further processing in the next stage.

3.4 Data Querying

Data querying involves the retrieval of a subset of the existing data as specified by the user. It is simply the process of expecting a specific answer to a specific question. The queries here have a pattern and they follow the pattern. This type of coding is done through a language called structured query language—SQL.

For retrieval of data, we used various techniques in the central database. The data set imported in the system is thoroughly searched for the queried information and required data are retrieved using the following algorithm.

Algorithm 1. Data retrieval

Input: Database containing datasets, user query
Output: Queried information
MatrixConst(run by ADL using key k)
for $x = 1$ to d do

set l as highest rank in queries choosing Dict[x]

```
for y = 1 to r do
    if y < r - 1 then
        A[x, y] = Ek(1)
    Else
        A[x, y] = Ek(0)
```

vary a and b such that survival rate is 1
Filtering (from cloud)
for each file Fy from the cloud do
for $x = 1$ to d do

$k = y \bmod r;$
$ey = cy ^\wedge Fy;$

map(cy, ey) a times to a buffer of size b

4   Experimental Results

We collected datasets from many travel websites like Tourpedia and Tripadvisor. These datasets are in .csv format that contains much information about locations and tourist spots. They are preprocessed, filtered, and used based on our requirements. We can import them as shown in Fig. 2, in our system and discard whenever there is a change in real-time information.

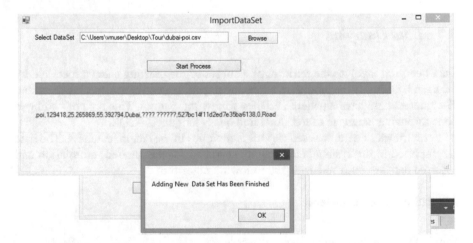

Fig. 2 Importing dataset

4.1 Sentiment Analysis Based on Category

In this module, the user can able to search the particular location. The location which was going to be searched is to be placed in the dataset. The sentiment analysis which is based on the category can be efficient in order to categorize only the particular place. The sentiment analysis which is based on the query then generates the map. The location map and the information about the location are displayed. Figure 3 presents the sentiment analysis by category.

Before using sentiment analysis, we processed the datasets and reputation score must be generated for all reviews. Using these reputations, we can identify relevant reviews and even sort them based on the scores so that highly relevant reviews are easily accessible to users in search of that information. In order to find the reputation score, we have to consider opinion sets x. S_x is the number of similar opinions in the major opinion set x which indicates its popularity. The reputation of an entity A can be generated as follows.

$$\text{Rep}(A) = \frac{1}{x} \sum_{x=1}^{x} \frac{\theta(N_x) \cdot V_x \cdot S_x}{N_x \cdot N_x} \tag{1}$$

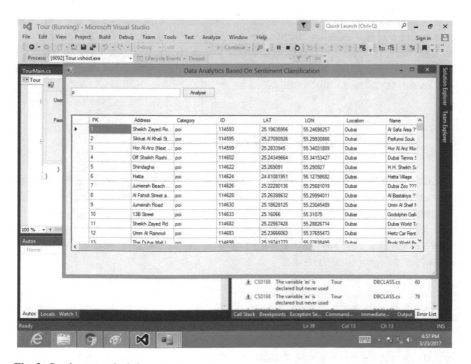

Fig. 3 Sentiment analysis by category

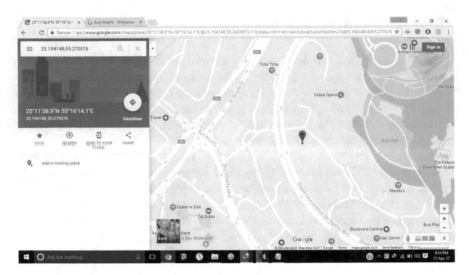

Fig. 4 Locating the user query in Google maps

4.2 Sentiment Analysis Based on Place

In this module, the user can able to search the particular location. The user can now able to analyze the location based on place. The sentiment analysis which is based on the giving the place will search only the place. The sentiment analysis which is based on the query then generates the map. The location which is given by the user will be displayed as a map as portrayed in Fig. 4.

4.3 Sentiment Analysis Based on Keyword

In this module, the user can able to search the location by giving the dataset. The user can now able to analyze the location based on keyword. The sentiment analysis which is based on giving the keyword will extract the database. Based on the keyword, the admin can view the whole location information. The admin now able to view the search result based on keyword. Figure 5 depicts the information article presentation module of the proposed system.

4.4 Recommendations Based on Image Features and Tags

In this module, the community-contributed photographs are collected from the Flickr provided by tourists. Along with the images, tags associated with the photographs

Fig. 5 Wikipedia article about the location

are also crawled to make personalized recommendations. The image tag information is considered as the metadata that influences the recommendation generation process with the geo-tagged information. The initial dataset from the Flickr is further processed for filtering based on the photograph content and tags associated with them. The tag information also includes the image categories and the landmark associated with it. Some common tags of photographs in a user's travel include "mytrip", "travel", "tour", and "myvacation". Though these tags have no specific information on the photograph category, it can be considered as noise and be further removed. Then, we employ TF-IDF to analyze the importance of the tags associated with the image. The noisy tags are removed and other important data associated with the user-contributed images are used for the clustering process. The photographs are clustered based on computed clustering attributes such as number of users and content weight for each user. Along with the image clustering, the data associated with the photographs are also clustered for the indexing purpose. Figure 6 represents the sample photograph

Fig. 6 An example image for the recommendation processing

Sunset

Vacation

Beach

Nature

Clouds

Blue

Wave

Fig. 7   Processed relevant images for the sample image based on image features

and its tags used for the recommendation processing. Figure 7 portrays the similar relevant photographs retrieved by the proposed approach.

5   Conclusion

The presented research work will help the travelers to travel through the different location without guide. This may provide the travelers to visit different locations and display the exact location through the map. The dataset is imported and discarded depend upon the admin decision. This project is user-friendly to the user since the exact name and exact location are not inserted as a query by the user. The sentiment analysis which is based on category, place, and keyword displayed the map and the information about the location. As the future work, we can consider that the user's GPS is tracked down and hence if he/she searches a query, the locations near to their places can be displayed first. Also, datasets of the locations unavailable can be created and imported into the database. The portal can overlook the interests of the user and hence can recommend based on the user's interests.

Acknowledgements The authors are grateful to Science and Engineering Research Board (SERB), Department of Science & Technology, New Delhi, for the financial support (No. YSS/2014/000718/ES).

References

1. Abdel-Hafez, A., Xu, Y.: An accurate rating aggregation method for generating item reputation. In: IEEE International Conference on Data Science and Advanced Analytics (DSAA), pp. 1–8 (2015)
2. Ahluwalia, R.: Examination of psychological processes underlying resistance to persuasion. J. Consum. Res. **27**(2), 217–232 (2000)

3. Angst, C.M., Agarwal, R.: Adoption of electronic health records in the presence of privacy concerns: the elaboration likelihood model and individual persuasion. MIS Q. **33**(2), 339–370 (2009)
4. Aral, S.: The Problem With online ratings (2013). http://sloanreview.mit.edu/article/the-problem-with-online-ratings-2/2013
5. Kim, S.M., Hovy, E.: Extracting opinions, opinion holders, and topics expressed in online news media text. In: Proceedings of the Workshop on Sentiment and Subjectivity in Text, pp. 1–8. Association for Computational Linguistics (2006)
6. Logesh, R., Subramaniyaswamy, V., Malathi, D., Sivaramakrishnan, N., Vijayakumar, V.: Enhancing recommendation stability of collaborative filtering recommender system through bio-inspired clustering ensemble method. In: Neural Computing and Applications (2019)
7. Shapiro, C.: Consumer information, product quality, and seller reputation. Bell J. Econ. **13**, 20–35 (1982)
8. Shri, J.M.R., Subramaniyaswamy, V.: An effective approach to rank reviews based on relevance by weighting method. Indian J. Sci. Technol. **8**(11) (2015)
9. Wang, J.Z., Yan, Z., Yang, L.T., Huang, B.X.: An approach to rank reviews by fusing and mining opinions based on review pertinence. Inf. Fusion **23**, 3–15 (2015)
10. Weng, Y., Zhao, L.: A blogger reputation evaluation model based on opinion analysis. In: IEEE, Asia-Pacific Services Computing Conference (APSCC), pp. 27–34 (2010)
11. Zhou, X., Wan, X., Xiao, J.: CMiner: opinion extraction and summarization for Chinese microblogs. IEEE Trans. Knowl. Data Eng. **28**(7), 1650–1663 (2016)

Optimization of Parking Lot Area in Smart Cities Using Game Theory

R. Padma Priya and Saumya Bakshi

Abstract The freedom of movement for people in urban cities in the twenty-first century is curtailed due to traffic congestion and parking problems. It has brought out the need for development of smart cities through incorporation of Internet of Things and other technologies in the planning of cities. Chennai, a city in India with a population size of 9.1 million and a vehicle pool of more than 3.7 million units, is in dire need of intelligent solutions to its traffic and transportation issues. Sustainable and practical city planning not only requires considering increased air pollution, longer travel duration, accidents, but also with frustration among travelers. Hence, smart city planning measures are incomplete if parking space issues are ignored. In this paper, we address the parking space problem and how they can be tackled in smart cities in the interest of both government and travelers. Based on the Stackelberg game model, our proposed scheme considers a game between the public authority and the travelers. The government with the help of its policies *can influence* the decisions taken by the travelers. In this study, we work upon proposing two utility function—one for traveler and one for government. The utility function of the government aims to maximize the public transit usage while maintaining flow of travelers to urban city centers in Chennai. City center refers to an urban center such as a shopping mall, heritage site, or government building. The utility function of an individual traveler aims to minimize travel duration, cost of travel, and other inconveniences. A trade-off demand of the players is sought using variable parking space. A small version of the game is envisioned for urban reality-based scenarios like strong center, weak center, weak transit system, strong transit system, respectively. Consequently, on the basis of the results, optimized parking space allotment is obtained and we infer that reducing parking space works more promisingly in a strong urban center scenario. The solution provided in our paper can be collaborated with existing smart parking

R. Padma Priya (✉) · S. Bakshi
School of Computer Science and Engineering, Vellore Institute of Technology, Vellore,
Tamil Nadu, India
e-mail: padmapriya.r@vit.ac.in

S. Bakshi
e-mail: saumya.bakshi2014@vit.ac.in

© Springer Nature Singapore Pte Ltd. 2020
K. N. Das et al. (eds.), *Soft Computing for Problem Solving*,
Advances in Intelligent Systems and Computing 1057,
https://doi.org/10.1007/978-981-15-0184-5_3

systems for optimal results. Conclusions inferred can be applied by the government towards development of practical, efficient, and sustainable parking schemes.

Keywords Stackelberg game · Parking space optimization · Maximization of public transit · Urban transportation · Smart city

1 Introduction

There is an overall rise in the demand for transport in Indian cities because of population rise as a result of migration from smaller cities. Due to this, many urban areas are bothered by rise in traffic congestion leading to increased air pollution, longer travel duration, accidents but also with frustration among travelers. These issues cannot be solved just by constructing new roads or widening the existing ones. Till now, the main focus of studies has been congestion, environmental damage, and safety of travelers as described in [1]. Slowly, metropolitan authorities are considering parking measures as a means to bring about changes in the roadways. They are now considering parking as a part of their transport congestion management policies.

Urban life in this modern generation of living faces quite a lot of difficulties with regards to commuting from one place to another. In the city of Chennai, India the vehicle pool stands more than 3.7 million units [2] and the vehicle density per kilometer of road is as high as 2093 as of 2016. The carbon emissions of the city are the fifth-highest from 54 South Asian cities as in [3]. This is a wakeup call for the city to take drastic measures to upgrade its transportation system to a sustainable one. Parking system is a cardinal element in the development of smart cities, and hence, solution of parking challenges takes priority. With steeply rising traffic and ever-increasing demand for parking space, creative and efficient solutions are required.

Game theory is a means of modeling real-life scenarios to bring out effective solutions. It presents innumerable applications in the study of transportation problems [4]. Noncooperative game theory is used to describe the games in transportation. Stackelberg games between the government or authority and the travelers can be used for decision making as discussed in [5]. Previously, the Stackelberg-Logit model has been used for evaluation of an intelligent transportation system as ascertained by authors in [6–8]. The Logit model is extensively used in transportation problems to ascertain the probability of a mode choice. The aim is to use a similar modeling for deciding the amount of parking space that should be allotted by the government for a city center.

In Sect. 2, we present a literature survey of the need for parking lot optimization and the measures taken up till now. The problem is proposed to be solved using the mode choice model for travelers namely the multinomial Logit model.

2 Literature Survey

As a developing country, India is seeing huge increase in car ownership coupled with unplanned traffic management and land area use. This leads to increased traffic, higher fuel usage, and more inconvenience on the roads. Due to financial and political reasons, it is not easy to bring out radical laws to combat transportation issues. In [9], the authors have suggested using measures such as hiking parking fees, restricting parking supply, betterment of public transit system, and increasing fuel taxes to deal with the increased traffic on roads. The development of the Delhi metro is an example of increase in public transport system. The project was taken very seriously and it managed to create a positive image in the minds of the travelers in terms of coverage and connectivity as outlined in [10]. Our work aims to promote usage of public transit systems as a daily means of reaching a destination. In India, according to the authors in [11], the focus has mostly been on the betterment of the public transport system and not on traffic management measures. Parking policy is seen as a chief measure to manage traffic congestion. Alternative measures to on-street parking can help decongest roads in India. Kolkata has been taken as a case study in [12] and being one of the four big metropolitan cities in India, traffic congestion is one of its major issues. Through this paper, we aim to promote implementation of smart cities in India. Angeldiou [13] explains the definition of smart cities in the age of urbanization. The author explicates the need for balance between the technology push of smart products and solutions for cities and demand pull of addressing the issues of sustainability and efficiency. The research route of our paper treads on the path of smart city development, keeping the need for balance in mind. Zanella et al. [14] discuss smart solutions in context of the Padova Smart City project in Italy. Some of the solutions outlined were waste management system using smart garbage containers, traffic congestion monitoring via air quality and noise monitoring, and smart parking systems using sensors. Hence, smart city planning measures are incomplete if parking space issues are ignored. The solution provided in our paper can be collaborated with existing smart parking systems for optimal results.

The parking supply problem revolves around the choice of mode of travel by commuters. Some models cater to choosing between types of parking or choosing the parking site as outlined in [15]. They usually do not take into account the effect of policy change on the traveler behavior. Morrall and Bolger [16] describe the importance of parking policies in the urban transportation public transit domain as a whole affecting ridership and public transit usage. They have found a strong connection between public transit usage and downtown parking during peak period. They elaborate on an inverse relationship between the commuters choosing public transit and the number of available parking stalls. Chow [17] proposes an electronics coupon system for public transit fares decided by a third party. He shows the effect of this can have in a monopolistic, an oligopolistic and a government-controlled environment. The Stackelberg game between the government authorities and the private firms decides the ticket price and helps in formulation of policies. The model described in our paper considers the effect of government policy change on the decision made by travelers with respect to parking. The authors in [18] have used game theory principles to determine the parking area allotment for an urban center in a city. The government or the public welfare authority aims to maximize the public transit ridership. In India, public transit mainly consists of bus services and metro/local train services. The utility of a single traveler is obtained via surveys and observations from real-life scenarios. The authors in [19] take into account some of the factors on which travel mode choice utility depends such as age, car, distance, income, time taken, and price. In this paper, the utilities are assumed. They can be modified by giving input of the values derived on the basis of real disaggregate data.

3 Methodology

3.1 Definition and Assumptions of the Game

Stackelberg games are games wherein the players are classified into leaders and followers. The leader makes the first move, anticipating the reaction of the followers. In our paper, we have considered the public authority or government as the leader and a composite body of all travelers as the followers. In the transportation scenario, the Stackelberg games are noncooperative which means there is no external body to enforce cooperation and each player aims to maximize his/her own objective. The strategies in our game are not the alternatives faced by individual travelers. Each strategy of a player aggregates to a choice distribution of all travelers. Hence, the objective function of the travelers does not reflect each individual's will but their propensity to reach certain equilibrium amongst the choices wherein each traveler tries to maximize his/her individual gain as. An individual traveler's objective is to reduce the duration of travel, the travel cost and face minimum inconvenience while traveling. The government, on the other hand, aims to increase ridership in public transit systems. Hence, preference coefficients are used in the designing of utility function of the government. Through the game formed above, we aim to ascertain the proper amount of parking supply. Parking supply can be measured in terms of amount of parking spaces, the time required to find a parking slot or parking charge. We assume parking supply to refer to number of parking spaces in this research and a variable x to describe it. Each value of the variable represents a possible solution for government in the given transportation setting. This variable influences the behavior of the traveler, since the utility function of an individual traveler contains the variable x. Thus, it can be used as a policy tool by the government.

In the Stackelberg game defined between government and the travelers, the former resolves to change the amount of parking space taking into account the choice distribution of the travelers and the latter change their behavior based on the decision taken by the government. Among the many route choice models, the Logit model is used. Such a model is behavioral. It is derived from data obtained via surveys and observations. The multinomial Logit model is one of the simplest choice models

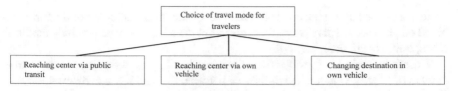

Fig. 1 Travel mode choices of travelers in multinomial Logit model

used in the transportation scenario as shown in Eq. (4). Its drawback is the property of independence of irrelevant attributes. It states that the relative odds between any two outcomes are independent of the number and nature of items being considered simultaneously. The choices of strategies for our game are divided into three alternatives. For every origin-destination (OD) combination, a traveler chooses one of the following three alternatives as displayed in Fig. 1—city center by transit (CT), city center by own vehicle (CV), and other destination by own vehicle (OV). They leave out the case of shared ride taken by travelers since that makes modeling more complex. City center refers to an urban center such as a shopping mall, heritage site, or government building.

3.2 Objective Function

The objective function for the government is created keeping in mind the travelers' proneness and not the traveler's will. It is considered to be linear to represent the basic scenario excluding variations such as car-pooling facility and traffic conditions. It is written using preference coefficients S_i where maximum preference is given to public transit (CT) followed by center by own vehicle (CA) and lastly changing destination using own vehicle (OA). The parking variable $x = 0$ indicates completely removing the parking area and $x = 1$ indicates no change in the parking area.

$$\max_V U_L = \sum_{i=1}^{N} (S_i X_i) \tag{1}$$

where

X_i: The actual rate of travelers choosing alternative i for traveling
N: The number of alternatives available for travel
S_i: Preference coefficient for each mode of transit decided by the government

The utility of an individual traveler is given as:

$$U_i(x) = c + bx \tag{2}$$

c and b are constants which are determined according to the behavior of the travelers based on preference surveys or other form of evidential data. c can include the time taken, traffic coefficient, etc.

The minimization of the objective of the travelers aggregated using values from a number of trips A, which is the difference between the actual travel distribution and that according to the Logit choice model is given as:

$$\min U_F = \sum_{i=1}^{N} \left(X_i - \left(\frac{1}{A} \right) \sum_{a=1}^{A} p_i^a \right)^2 \tag{3}$$

where

A: Number of trip samples
p_i^a: Probability of choosing alternative i in trip a by traveler

The multinomial Logit model probability using (2) is given as:

$$p_i = e^{(U(x))} / \sum_{i=1}^{N} e^{(U(x))} \tag{4}$$

3.3 Notations

The notations used in the paper are as follows:

Notation	Denotes
A	Number of trip samples
a	Trip index
N	Number of travel mode choices
I	Alternative travel mode index
a_i, b_i	Utility function parameters
CT	Abbreviation for reaching center via public transit
CV	Abbreviation for reaching center via own vehicle
OV	Abbreviation for reaching another destination via own vehicle
S_i	The preference coefficient of government for alternative i
X_i	The rate of travelers selecting ith traveling mode alternative
x	The parking variable to be determined by the government
U_i	Utility of single traveler choosing alternative i
U_L	The objective function of the government
U_F	The objective function of the travelers

3.4 Algorithms

Figure 2 describes the algorithm for solving the Stackelberg game between the public authority and the travelers. The probability choice distributions of the travelers are plotted according to the Logit model which takes the utilities of individual travelers as input. The variables in the game are the choice distribution of the travelers, X_i, and the parking space decision variable x. The X_i values are put in the government's objective function. The value of x for which the function is the maximum is considered the optimum value for that transportation setting. The government must reduce the parking area by the optimum rate x so as to ensure that more people are diverted to public transit services instead of changing their destination altogether. Change of destination would cause loss to the center. The center could be a mall or a government office.

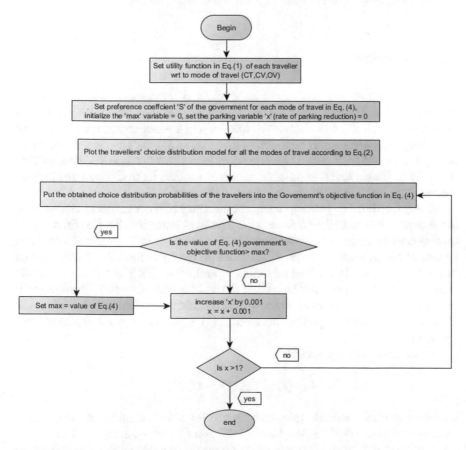

Fig. 2 Algorithm to determine optimal rate of parking reduction 'x'

4   Results and Discussions

The formulations are tested for four different scenarios—strong center, weak center, strong transit system, and weak transit system. A lot of different combinations of scenarios are possible. Out of them, four basic scenarios are discussed here with the help of which understanding of the other scenarios becomes plausible. It is assumed that the constants in the utility function of travelers are constant though they are dynamic in real life. These assumptions will change in accordance with the data used to model them for real-life conditions. Change in the choice of the travelers will affect factors such as traffic on the road and time taken. The preference coefficients for the government objective function are taken as $S_{CT} = 1$, $S_{CV} = 0.5$, and $S_{OV} = -0.5$. For the graphical representation, the values of the variable x are taken between 0 and 1. At $x = 0$, indicates that the parking space is completely removed. At $x = 1$, indicates that the parking space is not reduced at all. The number of trips, A, is assumed as 1.

4.1   Scenarios

A city is composed of a plethora of buildings and centers each with its own significance and ability to attract travelers. Changing parking space rules in different centers may draw different reactions by travelers. In a strong city center, the result of reducing parking space may be different than that in a weak city center. The punctuality and condition of public transit systems also affect the travelers' decisions on using them. We explore four possible scenarios for the city of Chennai and explore the results of modifying the parking space in them.

Strong Center: For a strong center, the tendency of the travelers to change their destination is less since the urban center is a point of attraction. People will come by their own automobile and by public transit. The center of India Post in Chennai can be considered as a strong center. It is situated on the Anna Salai Road which is one of the busiest roads in Chennai according to authors in [3]. The number of people changing their destination will be less. As the parking space is reduced, travelers will prefer to come via bus services or metro services rather than change their destination. Hence, when no parking policies are enforced, i.e., the value of $x = 1$, the utilities are assumed as:

For multinomial Logit model

$$U_{CV}(1) > U_{CT}(1) > U_{OV}(1).$$

Based on the above, suitable values for the utilities of individual travelers have been identified as $U_{CV} = -14 + 4x$, $U_{CT} = -12$ and $U_{OV} = -13$. Figure 3 shows the distribution of the travelers in terms of their mode of transportation and the objective function of the government. The utilities of the individual traveler are in the order as mentioned above for multinomial Logit model. U_{CV} is assumed to be constant

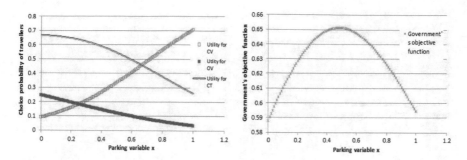

Fig. 3 Changes in choice of travelers as parking space changes for India Post, Anna Salai Road (left). Changes in the value of government's objective function as the parking space are reduced (right). Positive effects of reducing parking space seen till $x = 0.48$

for this scenario. The objective function of the government is an inverted U shaped curve. The Stackelberg equilibrium is at the maximum point of the curve which is at $x = 0.48$. A decrease in the parking space leads to increase in the distribution of travelers choosing the government's alternative. But as the parking space becomes lesser, the rate of increase of travelers via the government's alternative decreases. Hence, negative effects of the parking restrictions can be seen at this point.

Weak Center: In the case of a weak center, the areas of attraction lie outside the center. A weak center in the city of Chennai could be shopping malls which are close by. For example, the Express Avenue mall on Mount Road and Spencer Plaza in Anna Salai are close by. If a commuter does not find parking space in one mall, he/she will go the other mall since both centers are similar in their functionality. In such a scenario, the utilities assumed are as follows:

For multinomial Logit model

$$U_{OV}(1) > U_{CV}(1) > U_{CT}(1).$$

The individual utilities have been identified as $U_{CV} = -14 + 4x$, $U_{CT} = -12$ and $U_{OV} = -9.5$. Figure 4 shows that the rate of travelers choosing to change their destination increases as the parking restrictions are increased. Reducing x does not have any positive effects of moving people towards using the public transit system. Hence, the objective function of the government in Fig. 4 is at its maximum when no changes are made to the parking space. The Stackelberg equilibrium is at $x = 1$ for the assumed scenario.

Weak Transit System: The initial state of the transit services such as bus and local train services is varied over regions. The connectivity, timeliness, cleanliness, and other factors affect the number of people willing to use it. Weak transit systems can also be attributed to absence of bus stops, improper location of bus stops, and poor connectivity among different locations of the city. In Chennai, the metro system is being made. It still requires a lot of work to be put into it in terms of connectivity and user-friendliness. Hence presently, it can be considered as a weak transit system. For weak services, the utilities assumed are as follows:

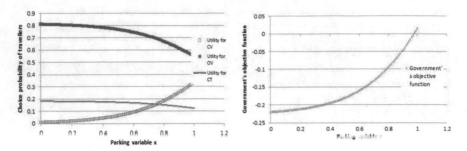

Fig. 4 Changes in choice of travelers as parking space changes for Express Avenue (left). Changes in the value of government's objective function as the parking space are reduced (right).Reducing parking space has no effect on the number of people choosing public transit

For multinomial Logit model

$$U_{CV}(1) > U_{OV}(1) > U_{CT}(1).$$

The individual utilities have been identified as $U_{CV} = -14 + 4x$, $U_{CT} = -12$ and $U_{OV} = -11.5$. Figure 5 shows the changes in the rate of travelers as the parking restrictions are increased. The reduction in parking space does not have any positive effect to lead the travelers to the desired choice of the government. Hence, the Stackelberg equilibrium is at $x = 1$ for the changes in the objective function of the government.

Strong Transit System: In the case of strong transit system, people will prefer it to other modes of travel choice. In the city of Chennai, the bus services are rampantly used by the public for commuting. It is a daily means of travel for many, especially the poorer sections of the society due to its cheap pricing. For example, the Chennai Mofussil Bus Terminus also known as CMBT bus stand, located in Koyembedu is the largest bus terminus in Asia. It can handle over 2000 buses a day and is an important entry–exit point to the city. It is available as a suitable option for commuters going to

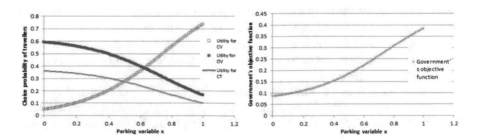

Fig. 5 Changes in the value of government's objective function as the parking space is reduced in a weak transit system scenario (right). The reduction in parking space does not have any positive effect to lead the travelers to the desired choice of the government

and from Koyembedu area. Hence, the utilities, with relation to such a transit system, will be assumed as follows:

For multinomial Logit model

$$U_{CT}(1) > U_{CV}(1) > U_{OV}(1).$$

The individual utilities have been identified as $U_{CV} = -14 + 4x$, $U_{CT} = -9$ and $U_{OV} = -11.5$. For a strong transit network, the distribution of travelers is shown as in Fig. 6. The maximum optimum of the objective function of the government is at $x = 0$. Removing all parking space leads to maximum ridership for the transit system. This happens in cases where reducing the parking space has no negative effects on the utility of the traveler traveling via the transit system. Such a case is only theoretical, since increased ridership will have a burden on the transit network whether it is a bus network or train.

The government's objective function graph can take various shapes. The maximum point on the curve gives the Stackelberg equilibrium value for the game. The inverted U shaped curve indicates that while the transit system is strong, the parking restrictions lead to a negative effect on U_{CT}. The increasing V shaped curve as x nears 0, indicates that the center or transit system is strong and there are no side effects of increasing parking restrictions. Such a case is only theoretical. The solution is always $x = 0$. The decreasing V shaped curve as x nears 0, indicates that the center or transit system is weak and reducing parking area only worsens the condition. The solution is always $x = 1$. For U shaped curves, the center or transport is weak but reducing the parking has a positive effect. In such a case, the government can either choose $x = 0$ or $x = 1$. Hence, decision has to be based on external factors. For the four scenarios discussed in the paper, the values of the parking variable are shown in Table 1.

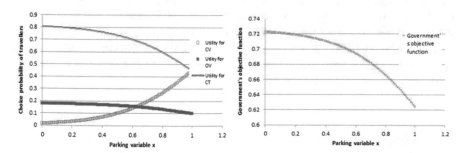

Fig. 6 Changes in choice of travelers as parking space changes for a Chennai Mofussil Bus Terminus or CMBT (left). Changes in the value of government's objective function as the parking space are reduced for the strong transit system scenario (right). Removing all parking space leads to maximum ridership for the transit system

Table 1 Equilibrium value of parking variable x in different scenarios

Scenarios	Equilibrium value of x
Strong center	0.48
Weak center	1
Strong transit	0
Weak transit	1

5 Conclusion

Subsequent to globalization, the scale of urbanization and vehicular growth has been massive in the country of India, which can be best seen in its metropolitan cities. Availability of reduced space in the cities has led to an increase in the demand for parking areas especially in urban city centers. Swelling traffic congestion, traveler frustration, deferred travel time, and severe environmental impact are only adding fuel to the fire. In such times, increase in the usage of public transit can facilitate decompression of the roadways in metropolitan cities. In Chennai, increasing population is creating difficulty in traffic conditions of the city. To counter those, some measures suggested previously have been improving road conditions, building more flyovers, providing connectivity between mass transit facilities and fast and effective metro system. The aim of the paper is to put forward a method, apart from the ones already suggested for regulating parking supply. We focus on optimization of the parking spaces by the government or concerned public authority in city centers and obtain optimal results for four scenarios with respect to the city of Chennai. It is necessary for a populated country like India to look into measures to improve traffic management besides improving and growing the public transit system. The government authorities can take data through surveys and model the parameters in the equations to get valuable results. The parking variable values can be used effectively by the government during policy and decision making. Future work can be done by modulating the equations for a specific city/region or scenario. Also, better choice models can be selected to give more accurate results. We are working on using nested Logit model to give segregation between the public transit systems which might be more beneficial for the government.

References

1. Marsden, G.: The evidence base for parking policies—a review. Transp. Policy **13**(6), 447–457 (2006)
2. Transport in Chennai—Wikipedia, the free encyclopedia [Online]. Available: https://en.wikipedia.org/wiki/Transport_in_Chennai
3. Oliver, J.: Chennai—alarming vehicular traffic and the City' S stand among global pioneers of sustainability. Int. J. Manage. Appl. Sci. **2**, 190–194 (2016)
4. Zhang, H., Su, Y., Peng, L., Yao, D.: A review of game theory applications in transportation analysis. In: 2010 International Conference on Computer and Information Application,

pp. 152–157 (2010)
5. Hollander, Y.: "The applicability of non-cooperative game theory in transport system analysis. TRB 2006 Annu. Meet. **33**(5), 481–496 (2006)
6. Lin, N., Wang, C.: A stackelberg game and improved logit model based traffic guidance model. In: Proceedings—2013 International Conference on Mechatronic Sciences, Electric Engineering and Computer, MEC 2013, pp. 2382–2386 (2013)
7. Wen, C.-H., Koppelman, F.S.: The generalized nested logit model. Transp. Res. Part B Methodol. **35**(7), 627–641 (2001)
8. Shang, B., Zhang, X.N.: Study of travel mode choice behavior based on nested logit model. Appl. Mech. Mater. **253**, 1345–1350 (2013)
9. Pucher, J., Korattyswaropam, N., Mittal, N., Ittyerah, N.: Urban transport crisis in India. Transp. Policy **12**(3), 185–198 (2005)
10. Siemiatycki, M.: Message in a metro: building urban rail infrastructure and image in Delhi, India. Int. J. Urban Reg. Res. **30**(2), 277–292 (2006)
11. Alam, M., Ahmed, F.: Urban transport systems and congestion: a case study of indian cities. Transp. Commun. Bull. Asia Pacific **82**, 33–43 (2013)
12. Chowdhury, I.R.: Traffic congestion and environmental quality: a case study of Kolkata City. Int. J. Humanit. Soc. Sci. Invent. **4**(7), 20–28 (2015)
13. Angelidou, M.: Smart cities: a conjuncture of four forces. Cities **47**, 95–106 (2015)
14. Zanella, A., Bui, N., Castellani, A., Vangelista, L., Zorzi, M.: Internet of things for smart cities. IEEE Internet Things J. **1**(1), 22–32 (2014)
15. Still, B., Simmonds, D.: Parking restraint policy and urban vitality. Transp. Rev. **20**(3), 291–316 (2000)
16. Froeb, L., Tschantz, S., Crooke, P.: Bertrand competition with capacity constraints: mergers among parking lots. J. Econom. **113**(1), 49–67 (2003)
17. Chow, J.Y.J.: Policy analysis of third party electronic coupons for public transit fares. Transp. Res. Part A Policy Pract. **66**, 238–250 (2014)
18. Qin, H., Gao, J., Zhang, G., Chen, Y., Wu, S.: Nested logit model formation to analyze airport parking behavior based on stated preference survey studies. J. Air Transp. Manag. **58**, 164–175 (2017)
19. Tanaka, S., Ohno, S., Nakamura, F.: Analysis on drivers parking lot choice behaviors in expressway rest area. Transp. Res. Procedia **25**, 1342–1351 (2017)

Toward Automatic Scheduling Algorithm with Hash-Based Priority Selection Strategy

Xiaonan Ji and Kun Ma

Abstract Not only the organizations or groups but also the laboratory or the store is in demand of a system with automatic scheduling algorithm. Current automatic scheduling with computer is time consuming. With the goal to develop an innovative system which can increase the productivity, we finally design an advanced algorithm based on priority and hash map. It firstly provides a method called linear programming (LP) for the problem. Then, we interpret the similar backtracking approaches and compared with another two algorithms. We rebuild their process and design and propose a more efficient and simpler algorithm based on priority and hash. After the development of such a web system, it is proved to be a simple, efficient, and easy-implement method to solve the problem.

Keywords Scheduling algorithm · Duty table · Priority selection strategy · Duty management

1 Introduction

In some student self-managing organization, students need to go to the office on duty. For example, the student manager is responsible for handling daily shift work. Unfortunately, some students might switch the work time due to the change of classes and other school affairs. Therefore, the maintenance work of the duty table (also called duty management) is time and energy consuming. Actually, it is a heavy work full of several challenges that we have to consider while arranging the weekly schedule [1]. It is more frequent in the case of holidays and examination seasons. Therefore, we attempt to propose an automatic scheduling algorithm with hash-based priority selection strategy. Currently, there are some approaches to automatic

X. Ji · K. Ma (✉)
Shandong Provincial Key Laboratory of Network Based Intelligent Computing,
University of Jinan, Jinan 250022, China
e-mail: ise_mak@ujn.edu.cn, kun.ma.cn@ieee.org

© Springer Nature Singapore Pte Ltd. 2020
K. N. Das et al. (eds.), *Soft Computing for Problem Solving*,
Advances in Intelligent Systems and Computing 1057,
https://doi.org/10.1007/978-981-15-0184-5_4

scheduling such as linear programming and backtracking [1, 2]. However, it caused too much time and might produce incorrect results.

In our previous experience, we should arrange over 40 crews for shift work weekly. Especially, there are a lot of course adjustments at the beginning and the end of the semester, which might change the scheduling table of work duty frequently. Not only the student organization but also the management of a team requires such a system using automatic scheduling algorithm. Therefore, we designed a simple automatic scheduling algorithm with hash-based priority selection strategy to build a system for such organizations. The contrition of our method is double hash-based priority ranking. The first ranking is based on crew number, and the second is based on ranking value. Our method has these features: simple and efficient to process, easy to implement to solve the scheduling problem in small/medium group. The time complexity is related to ranking algorithm, not involved in crew number.

The remainder of the paper is organized as follows. The current work of scheduling algorithm is discussed in Sect. 2. In Sect. 4, our automatic scheduling algorithm with hash-based priority selection strategy is proposed. The processing and algorithm are discussed. In Sect. 5, our automatic scheduling experiments show that our approaches are effective and efficient. Brief conclusions and future research directions are outlined in the last section.

2   Related Work

An facility maintenance management framework based on building information modeling is proposed to implement automatic scheduling of maintenance work orders [1, 2]. Another scheduling method is based on genetic algorithm [3]. The model described the optimization of warp beam looming schedule in weaving process to reduce the schedulers' labor. Compared with current methods, our method is simple and efficient based on hash-based priority selection.

Next, two categories of scheduling algorithm are analyzed.

2.1   Linear Programming

Linear programming (LP) algorithm is a method to achieve the best outcome [4]. LP works by minimizing, maximizing, or optimizing an objective function given a set of constraints. Compared to several scheduling methods, LP has two main advantages [5]. The first advantage is that LP always produces an optimal solution given the correct formulation. Another advantage is that there are many open-source tools in the software market. There are a disadvantage as shown in Table 1.

Table 1 Example of the disadvantages of linear programming

	A	B
Part 1	Available	Unavailable
Part 2	Unavailable	Available

In this case, there are some problems difficult to solve. If we arrange sequentially A to B, Part 1 to Part 2 in order, we will get A which is on duty at Part 1 and B at Part 2. But if we arrange B, then A, we will get nothing.

2.2 Backtracking

The schedule table of duty work is like a Sudoku game table [6]. Scheduling with backtracking approaches is more complex than in a typical backtracking solution to Sudoku game [7, 8]. The zero-one integer program given by models described the project scheduling problem, where the objective is to minimize project duration (completion time or make-span) [7]. The disadvantage of this method is that this method might come back to the beginning if the process gets unsuitable result at the last step. Compare with LP, backtracking cannot always get a correct result. Besides, the computing is time consuming on large schedule table.

3 Scheduling Requirements

There is a list of requirements of scheduling system on duty management, which is shown in Table 2. 4 operating crews (students with duty) are selected everyday. Each student is required to accept two duty tasks for the sake of fairness weekly.

Table 2 Weekly schedule table

Duty part	Monday	Tuesday	Wednesday	Thursday	Friday
Part 1 8:00–9:40					
Part 2 10:20–12:00					
Part 3 14:00–15:40					
Part 4 16:20–18:00					

4 Automatic Scheduling Algorithm with Hash-Based Priority Selection Strategy

An efficient automatic scheduling algorithm with hash-based priority selection strategy is proposed. It is a simple one which can be easy to implement in many applications. There are some variables of the formulation as shown in Table 3. w means number of days on duty weekly, s means number of parts of one day, n means number of duty one person weekly, N means no more than N duty one person a week, and m_i means the crew in the ith box of schedule table.

$$z = s \left(\sum_{w=1}^{5} \sum_{s=1}^{4} \sum m_i \right) \bigcap \sum_{n=1}^{N} (s (R (m_i))) \tag{1}$$

Objective Function (OF) is shown in Eq. 1.

– S(x) represents sorting function.
– R(x) represents ranking function.

It requires free time table of each person, and w, s, n you defined. The first step of our algorithm is to make a hash map for each size of crews in the ith box of schedule table, which is shown in Fig. 1. Then, it is sorted from small to large in order. This selection strategy is exactly how we mostly write the schedule.

Table 3 Variables table

Symbol	Definition
w	Number of days on duty weekly
s	Number of parts of one day
n	Number of duty one person weekly
N	No more than N duty one person a week
m_i	The crew in the ith box of schedule table

	Monday	Tuesday	Wednesday	Thursday	Friday
Part 1 8:00–9:40			One Box		
Part 2 10:20–12:00					
Part 3 14:00–15:40					
Part 4 16:20–18:00					

Fig. 1 Scheduling table of work duty

Table 4 Initial ranking

Name	AA	BB	CC	$\ldots$
Rank	1	0	2	0

Next, we make a hash map of rank for each person (see Table 4). According to this hash map, we sort all the crews from small to large by ranking value in each box of schedule. Initial ranking is 0. The next and last step is to arrange the first N staff on duty.

The ranking value represents its availability. The smaller the ranking value is, the more available it is. For example, the ranking value is changed to 2 if a crew is arranged twice.

As shown in Table 4, crew AA, BB, CC have been arranged once, 0 time and twice. Then, BB should be considered to be arranged next time. It will output only one explicit result when all the process is done with the hash-based priority selection strategy algorithm. Another different scheduling result is got if and only if the initial value is different. The flow chart of our automatic scheduling algorithm with hash-based priority selection strategy is shown in Fig. 2.

5 Experiments

Comparing with linear programming and backtracking algorithm, we refactor the scheduling process and design an efficient and simple algorithm. Given ranges of w, s, and n, it is possible to have up to 5 (days) $\times$ 4 (parts) $\times$ 2 = 40 times computing for arranging one week duty table. It is independent of crew numbers. This is the contribution of our improved method. The time complexity of our method depends on selected sorting algorithm. In our experiment, we choose the Quicksort algorithm with time complexity $O(nlogn)$. Considering the process of building hash map, the time complexity of our algorithm is approximately equal to $O(n + nlogn)$. The environment of this experiment is a standalone machine with an Intel Core(R) i5-8500 CPU and 8GB memory.

We tested the executing time of scheduling duty table of 40–480 crews with increment 40 crews. Figure 3 shows our scheduling result. Each time point is the average result of 10 times scheduling. The scheduling time is shown with the "test data." The result of backtracking algorithm with time complexity $y = x^2$ is the same as the green line. The scheduling time is a little higher than the yellow line and is almost the same as the blue line. The experiment results show our automatic scheduling is effective. Our method obviously has an absolute advantage in speed than backtracking algorithm, and it can output only one result. Sequential computation and invariant ranking cause this result. This is stable to generate final results. The same initial value will generate the same result.

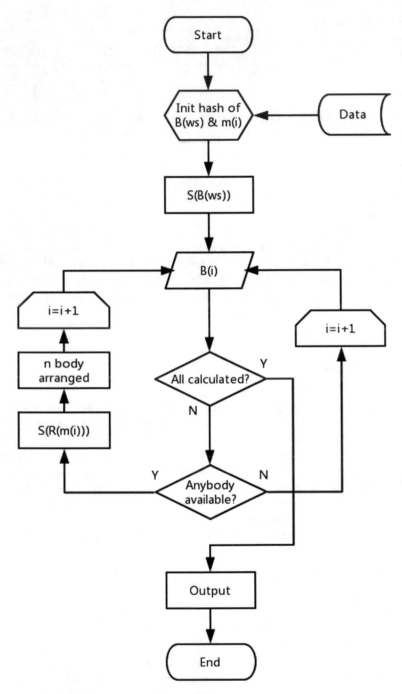

Fig. 2 Flow chart of our automatic scheduling algorithm with hash-based priority selection strategy

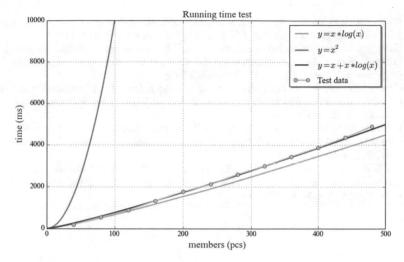

Fig. 3 Running time test

6   Conclusions

In this paper, we design an efficient and simple automatic scheduling algorithm based on priority and hash. With the idea of sequential processing, we complete the development of an improved algorithm based on priority and hash. It has many features and advantages:

- Simple and efficient process.
- Easy to implement.
- Provide a solution to scheduling problem on small/medium organization/group.
- Low latency of scheduling, dependent of sorting algorithm, and independent of staff numbers.

Acknowledgements This work was supported by the National Natural Science Foundation of China (61772231), the Shandong Provincial Natural Science Foundation (ZR2017MF025), the Shandong Provincial Key R&D Program of China (2018CXGC0706), the Project of Shandong Province Higher Educational Science and Technology Program (J16LN13), the Science and Technology Program of University of Jinan (XKY1734 & XKY1828), and the Project of Shandong Provincial Social Science Program (18CHLJ39).

References

1. Chen, W., Chen, K., Cheng, J.C., Wang, Q., Gan, V.J.: Bim-based framework for automatic scheduling of facility maintenance work orders. Autom. Constr. **91**, 15–30 (2018)
2. Ma, K., Yang, B., Yu, Z.: Optimization of stream-based live data migration strategy in the cloud. Concurrency Comput. Practice Experience **30**(12), e4293 (2018)

3. Wang, J.A., Pan, R., Gao, W., Wang, H.: An automatic scheduling method for weaving enterprises based on genetic algorithm. J. Textile Inst. **106**(12), 1377–1387 (2015)
4. Dantzig, G.: Linear Programming and Extensions. Princeton University Press, Princeton (2016)
5. Zhang, Y., Sun, X., Wang, B.: Efficient algorithm for k-barrier coverage based on integer linear programming. China Commun. **13**(7), 16–23 (2016)
6. Geem, Z.W.: Harmony search algorithm for solving sudoku. In: International Conference on Knowledge-Based and Intelligent Information and Engineering Systems, pp. 371–378. Springer, Berlin (2007)
7. Patterson, J.H., Brian Talbot, F., Slowinski, R., Weglarz, J.: Computational experience with a backtracking algorithm for solving a general class of precedence and resource-constrained scheduling problems (1990)
8. Chaib, A., Bouchekara, H., Mehasni, R., Abido, M.: Optimal power flow with emission and non-smooth cost functions using backtracking search optimization algorithm. Int. J. Electr. Power Energy Syst. **81**, 64–77 (2016)

Prediction of Flashover Voltage on 11 kV Composite Insulator Using Kernel Support Vector Machine

K. Kumar, B. Vigneshwaran, K. Vishnu Priya
and T. Sreevidya Chidambara Vadivoo

Abstract This paper presents the forecasting of flashover mechanism on polymeric insulator using nonlinear support vector machine regression (NLSVMR). In transmission and distribution systems, insulator plays a vital role in steadiness of the power systems. Flashover of composite insulators makes the whole system into danger. For the past decades, owing to good hydrophobicity properties of composite insulators made a replacement of porcelain insulators in service. To analyze the performance of composite insulator for long-term process, it is necessary to predict their pollution performance behavior before the breakdown occurs. In this proposed work, 11 kV composite insulator with three straight shed with clevis end fitting (Type A) and three alternate sheds with ball end fitting (Type B) are taken into consideration for flashover prediction. For experimental analysis, solid layer method was used in artificial pollution test, in which coastal, desert, and industrial pollution layer is used for further analysis. Using even-rising method, the flashover voltage has been obtained for different equivalent salt deposit density (ESDD) values. Finally, radial basis function (RBF) SVMR shows better flashover prediction when compared to linear and polynomial NLSVMR.

Keywords Solid layer method · Flashover mechanism · ESDD · NLSVMR · RBF

1 Introduction

In an electrical power system, open-air insulators take part in a considerable role in maintaining the trustworthiness of the network. The dependability and stability of the electrical power systems and its equipment show a prominent role for the satisfactory performance of power systems. In recent years, extra high-voltage transmission is

K. Kumar (✉) · B. Vigneshwaran · K. Vishnu Priya · T. S. Chidambara Vadivoo
Department of EEE, National Engineering College, Kovilpatti, Tamilnadu 628503,
India
e-mail: kumarkathir92@gmail.com

B. Vigneshwaran
e-mail: bvigneshwar89@gmail.com

© Springer Nature Singapore Pte Ltd. 2020
K. N. Das et al. (eds.), *Soft Computing for Problem Solving*,
Advances in Intelligent Systems and Computing 1057,
https://doi.org/10.1007/978-981-15-0184-5_5

widely used in order to transmit the electrical energy from power stations to consumers. In order to have a steady support for the overhead transmission lines on poles or towers, the conductors must be properly insulated from poles. Such insulation is provided by the high-voltage insulators which prevent the flow of leakage current from tower to earth and holds a good insulation between the tower and conductors. Porcelain insulators are extensively used in electrical power transmission and distribution systems for a long period. Nowadays, polymer insulators are most likely to be preferred for the reason of their better-quality insulation performance, 90% weight reduction, reduced fracture, lesser installation costs and enhanced handling of shock loads, better contamination performance [1, 2].

The design of high-voltage insulators used for transmission purposes is a foremost concern that should able to bear up the system voltage as well as transient voltage occurred due to natural calamities. Different types of insulators when installed in the vicinity of industrial or coastal areas the airborne particles may settled down on the exterior part of the insulators and as a consequence contamination builds up step by step. These particles will not reduce the dielectric strength of insulator under the presence of dry atmospheric conditions. When smog or raindrop wets the surface of the insulator, the conductive layer will form and initiate the leakage current to flow on the exterior part of the insulator. Unless the natural clean-up and satisfactory maintenance, the conducting layer will affect the property of the insulator by surface flashover [3, 4].

In practice, the varieties of pollutants which settle down on the surface of the insulator are listed in Table 1 as follows.

Humidity in the form of fog, mist, drizzle, light rain on the insulator surface wets the contamination layer and dissolves the salts and any soluble materials to create a slight conducting layer on the insulator surface. The severity of the contamination layer can be characterized by measuring the equivalent salt deposit density in case of soluble sea salt and non-soluble deposit density in case of sand, cement, etc.

Hence, it is more important to analyze the performance of insulators under polluted conditions and by artificial pollution solid layer method according to IEC 61245 standard. The test has been carried out on composite insulators because of its following advantages.

Table 1 Typical source of pollution

Type of pollutant	Area of location
Salt (NaCl)	Coastal areas and salt industries
Cement	Cement industries, building sites, and rock quarries
Metallic	Mining and mineral processing industries
Coal	Coal mining, coal handling plants/thermal plants
Sand	Desert areas

- Lightweight
- Excellent hydrophobicity
- Low leakage current
- Excellent tracking and arc resistance
- Highly resistant to vandalism
- High mechanical strength
- Unbreakable.

2 Contamination on Polymeric Insulators

Any malfunction happening in the characteristics of polymeric insulators will have the end result of substantial loss of capital (millions of dollars), as there are abundant power sectors that depend upon the ease of use of an uninterrupted power supply. The key factors disturbing the polymeric insulator are categorized by mechanical factors, ecological circumstances, design and operation of electrical power systems. Owing to the above factors, the polymeric insulator surface can be damaged and degraded. Among the above factors, the electric field stress and contamination become the most significant factor to be investigated thoroughly. Pollution is the main problem which creates deterioration of insulators and harshly deteriorates the electrical properties being one of the major problems of malfunctioning of insulators. This contamination is the main key factor which creates the surface flashover in the insulators and transformer bushings. The insulator begins to fail when the contaminant which stays alive in the atmosphere that patch up in the surface of the insulator and merge with the moisture content present around the environment of insulators [5, 6].

The intensity and the category of contamination of an area are related to the types of contamination, as well as with climatic conditions of the place. The usual phase in which a flashover can come into view with the insulator is by means of contamination are explained as follows:

- Outdoor polymeric insulators are being subjected to a variety of working circumstances and climatic conditions.
- Pollution deposits in the exterior shell of the polymeric insulators will improve the probability of flashover.
- Owing to waterless dry situation, the polluted surfaces do not conduct, and thus contamination is of little significance in dry periods.
- With respect to light drizzle, smog or dew, the pollution on the exterior part dissolve which creates a conducting layer on the exterior part of the polymeric insulator and the operating transmission voltages on that insulator will initiates the leakage current.
- High current density close to the insulator part results in the heating and aeration of the contamination layer.

- An arc is promoted if the voltage stress from corner to corner of the dry band exceeds the withstand voltage capability. The extension of the arc across the insulator ultimately results in surface flashover.

3   Types of Contamination

In this proposed work, three different types of contamination like coastal [7] desert [8] and industrial pollution [9] with different conductivities are used. The solid layer method with artificial pollution was adopted to coat the contaminant on the surface of the specimens. The breakdown voltage of the specimens is varied due to the different conductivities between the surface of the specimen and contamination layer. Initially, a flashover is formed on the surface of the specimen at a particular instant, later the flashover extends to breakdown. Hence for long-time performance of the insulator, it is necessary to study and analyze the flashover characteristics of different specimens with different contaminants. During the artificial pollution test, the polymeric insulator is subjected to defined quantity of contaminant in milligrams/cm^2. The allocation of contaminant on the insulator is in nonuniform manner.

4   Experimental Setup and Methodology

The polymeric insulator of different shed profiles and their dimensions was shown in Table 2.

Here, the test was performed in an artificial fog chamber in the high-voltage laboratory. The 11 kV test specimen of type A and B was used for contamination testing. The specimen is vertically suspended in the artificial fog chamber and the high-voltage supply was given at the top portion of the insulator while the bottom portion is grounded. Before the test ensure that the specimen was carefully cleaned to remove the dirt and then dry naturally.

Table 2 Profile of test specimen

Parameter	Type A	Type B
Diameter (cm)	6.7	5.1
Height (cm)	31	38
Creepage distance (cm)	30	34
Dry arcing distance (cm)	12.5	21
Creepage factor	2.4	1.6
Profile factor	3.9	6.2
Form factor	1.4	2.1

In case of coastal pollution, the NaCl mixture along with keiselghur is taken on different ratios and mixed with distilled water in order to apply the contaminant on the surface of the polymeric insulator. After natural atmospheric drying for 5 h, the specimen is subjected to test condition by even-rising method. In this method, the series of surface flashover test was taken 4–5 times at same contaminant level simultaneously. The minimum flashover voltage was noted for each specimen. The average flashover voltage was calculated by the given expression,

$$U_{av} = \sum_{i=1}^{N} U_{fm}(i)/N \tag{1}$$

where $U_{fm}(i)$ is the minimum flashover voltage taken and N is total times of effective test. The standard deviation of the flashover voltage is given as,

$$\sigma(\%) = \frac{\sqrt{\sum_{i=1}^{N}(U_{fm}(i) - U_{av})^2}}{N} \times \frac{100\%}{U_{av}}. \tag{2}$$

After the test procedure is over, the salt contaminant was wiped out from the marked surface and dissolved in 20 ml of distilled water using stirrer as shown in figure. The conductivity of the solution was found by the expression,

$$ESDD = S_A \times V/A$$
$$\text{where } S_A = (5.7 \times \sigma_{20})^{1.03} \text{Kg/m}^3 \tag{3}$$

$$\sigma_{20} = \sigma\theta[1 - b(\theta - 20)] \tag{4}$$

where, θ = temperature of solution (°C), b is determined by using empirical formula is expressed as;

$$b = -3200 \times 10^{-8} \times \theta^3 + 1032 \times 10^{-5} \times \theta^2 - 8272 \times 10^{-4} \times \theta + 3544 \times 10^{-2}.$$

The distilled water-containing non-soluble pollutants like cement and sand are filtered out by using a grade XXX filter paper. The filter-paper-containing pollutants shall be dried and then weighted. The NSDD was measured using the formula

$$NSDD = 1000(W_f - W_i)/A \text{ mg/cm}^2 \tag{5}$$

where, W_f, is the weight of contaminant with filter paper under dehydrated conditions in milligrams. W_i, is the initial weight of filter paper under dry condition in milligrams, and A is the surface area of the insulator in cm^2 [3, 10]. The experimental setup was shown in Fig. 1.

Fig. 1 Experimental setup

5 Results and Discussion

Before conducting pollution test on the insulators, the test specimen was subjected to withstand test before pollution and the flashover voltage was measured. It was found that the flashover voltage of type A was 77 kV and type B was 91 kV.

5.1 Coastal Pollution Test Results of Type A and B

The artificial pollution based on solid layer method was used to contaminate the surface of the insulator [11]. The NaCl and the keiselghur were taken and made into a mixture in order to make the strong binding of the contaminant settled on the surface of the insulator. The test results of the specimen are tabulated in Tables 3 and 4 where the NaCl level is kept constant with varying keiselghur.

The above test results were plotted in graph which relates the pollution severity level and flashover voltage was shown in Fig. 2.

Table 3 Coastal pollution on type A

S. No.	NaCl (gm)	Keiselghur (gm)	U_{av} (kV)	ESDD (mg/cm^2)	σ (%)	E_{L} (kV/cm)	E_{H} (kV/cm)
1	30	1	57.35	0.083	1.9	1.91	4.58
2		2	54.18	0.101	1.06	1.80	4.33
3		3	48.7	0.120	2.2	1.62	3.89
4		4	44	0.143	0.67	1.46	3.52

Table 4 Coastal pollution on type B

S. No.	NaCl (gm)	Keiselghur (gm)	U_{av} (kV)	ESDD (mg/cm^2)	σ (%)	E_L (kV/cm)	E_H (kV/cm)
1	30	1	77.25	0.083	1.61	2.27	3.67
2		2	75	0.101	3.6	2.20	3.57
3		3	71.65	0.120	2.1	2.10	3.41
4		4	69.3	0.143	1.17	2.03	3.3

Fig. 2 Relationship between ESDD and U_{av} of coastal pollution

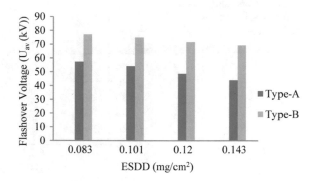

5.2 Desert Pollution Test Results of Type A and B

The desert atmosphere is characterized for sand storms and hurricanes that surround sand particles which are in movement to a high speed. Such particles will be settled down at high-speed wind on the surface of the insulator which causes material erosion. Here desert sand along with 20 ml distilled water is used to coat the contaminant. Tables 5 and 6 show the amount of sand taken and its individual NSDD values. The relationship between NSDD and average flashover voltage was shown in Fig. 3.

Flashover voltage on insulators depends on the profile of the insulator and the amount of pollution severity. From the test results, it is inferred that the average flashover voltage of type B is higher than type A. The standard deviation error was within 7% based on IEC 61245.

Table 5 Desert pollution on type A

S. No.	Sand (gm)	U_{av} (kV)	NSDD (mg/cm^2)	σ (%)	E_L (kV/cm)	E_H (kV/cm)
1	10	69.25	0.0125	1.6	2.3	5.54
2	20	56.8	0.025	1.21	1.89	4.5
3	30	53.75	0.035	1.12	1.7	4.3
4	40	50.64	0.04	2.91	1.68	4.05

Table 6 Desert pollution on type B

S. No.	Sand (gm)	U_{av} (kV)	NSDD (mg/cm^2)	σ (%)	E_L (kV/cm)	E_H (kV/cm)
1	10	80.6	0.0125	2.8	2.37	3.83
2	20	75.23	0.025	1.08	2.21	3.58
3	30	70.15	0.035	2.1	2.06	3.34
4	40	68.61	0.04	1.11	2.01	3.26

Fig. 3 Relationship between NSDD and U_{av} of desert pollution

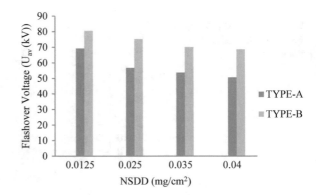

5.3 Industrial Pollution Test Results of Type A and B

In industrialized region, the polymeric insulators were primarily pretentious to the deposition of cement, smoke, and carbon dust on the surface of the insulator by force of wind. Here, the industrial contaminant used is cement which has maximum amount of gypsum that absorbs huge percentage of moisture. Pollution along with moisture may result in the formation of contamination layer and leading to flashover. It mainly deals with NSDD. The cement samples taken from cement industry were applied on the surface of the insulators. The cement ratios and flashover voltage of test results were tabulated in Tables 7 and 8. The relationship between NSDD and U_{av} of cement pollution was shown in Fig. 4.

E_L and E_H are the flashover along the creepage distance and flashover along the insulation distance. The differences in profile of the insulator make differences in flashover performances.

Table 7 Industrial pollution on type A

S. No.	Cement (gm)	U_{av} (kV)	NSDD (mg/cm^2)	σ (%)	E_L (kV/cm)	E_H (kV/cm)
1	10	62.73	0.01	1.9	2.09	5.01
2	20	59.8	0.02	2.6	1.9	4.78
3	30	53.4	0.03	3.8	1.78	4.27
4	40	48.17	0.04	2.5	1.6	3.85

Table 8 Industrial pollution on type B

S. No.	Cement (gm)	U_{av} (kV)	NSDD (mg/cm^2)	σ (%)	E_{L} (kV/cm)	E_{H} (kV/cm)
1	10	82.5	0.01	0.94	2.42	3.92
2	20	79.63	0.02	3	2.34	3.77
3	30	74.81	0.03	1.8	2.2	3.56
4	40	72.35	0.04	2.71	2.12	3.44

Fig. 4 Relationship between NSDD and U_{av} of industrial pollution

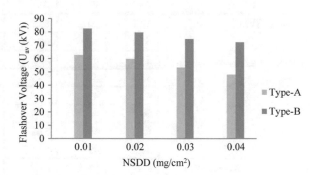

5.4   Support Vector Machine Regression Model

Supervised learning is a type of system having both input and desired output data. SVM is a type of supervised learning algorithm based on finding a hyperplane that can separate the cases of binary classes. This method is applied in various applications such as text classification and image categorization [12]. The main advantages of SVM over other algorithms like neural networks, K-nearest neighbor, random forests, and linear discriminant analysis were due to the ability of handling high-dimensional features and providing high classification accuracy. In practical circumstances, the system may operate under nonlinear categories. Thus, SVM hyperplane is obtained via a kernel function which can be linear, polynomial, or radial basis function (RBF).

In this work, linear SVM, polynomial kernel SVM, and RBF kernel SVM regression models are used for predicting the flashover voltage of polluted insulator. These regression models are trained and then used to predicting the flashover voltage. Prediction of high-dimensional data set is performed by SVM. Initially, train a SVM regression model using sample data stored in database. During training, SMO solver is used for regression model. Convergence of model is necessary for better accuracy [13–15].

Test inputs vectors are given to the regression model and compute the mean-squared error for the new model. To improve the accuracy of the prediction rate, instead of linear function kernel SVM regression function is used. In this proposed work, Gaussian or radial basis function (RBF) and polynomial kernel are used. The prediction rate for various kernel functions for Type A and Type B insulators is shown in Tables 9 and 10.

Table 9 Prediction of flashover voltage for type A insulator

	Types of pollution layer					
	Breakdown voltage (kV/cm)–AV					
	(Actual value)–PV (Predicted value)					
	Type A					
	Coastal pollution		Desert pollution		Industrial pollution	
	AV	PV	AV	PV	AV	PV
Linear SVM	57.1	50	69.2	65	62.7	55
	55.2	54	65.2	60	61.3	56
	54.3	60	58.8	55	59.8	51
	52.7	48	56.2	53	57.3	58
	48.7	45	55.8	58	54.5	56
	46.2	44	53.7	55	53.4	48
	45.2	49	51.2	45	52.1	48
	44.5	47	50.6	48	48.1	44
Polynomial Kernel SVM	57.1	55	69.2	67	62.7	65
	55.2	59	65.2	64	61.3	68
	54.3	57	58.8	60	59.8	55
	52.7	50	56.2	50	57.3	60
	48.7	45	55.8	59	54.5	51
	46.2	47	53.7	50	53.4	50
	45.2	49	51.2	50	52.1	50
	44.5	45	50.6	45	48.1	45
RBF Kernel SVM	57.1	56	69.2	68	62.7	63
	55.2	55	65.2	6	61.3	60
	54.3	53	58.8	57	59.8	58
	52.7	51	56.2	55	57.3	56
	48.7	48	55.8	54	54.5	53
	46.2	45	53.7	52	53.4	53
	45.2	44	51.2	50	52.1	50
	44.5	44	50.6	50	48.1	49

6 Conclusion

From the above test results and analysis, it is concluded that the nonuniform contamination distribution has an impact on the pollution flashover voltage of the polymeric insulator. The contamination flashover voltage is associated with the degree and type of contaminant, the dissimilarity in the pollution distribution on the pinnacle and bottom surface. The average flashover voltage decreases with increase in soluble and

Table 10 Prediction of flashover voltage for type B insulator

	Types of pollution layer					
	Coastal pollution		Desert pollution		Industrial pollution	
	Breakdown voltage (kV/cm)–AV					
	(Actual value)–PV (Predicted value)					
	Type B					
	AV	PV	AV	PV	AV	PV
Linear SVM	77.2	72	80.6	84	82.5	80
	76.5	71	78.2	79	81	78
	75	70	76.5	73	79.1	79.7
	74.5	72	75.2	72	78	77.1
	72.8	72.8	73.4	73	77.5	76
	71.5	64	72.1	74	76.2	74
	70.3	63	70.6	67	74.8	78
	69.3	65	68	67	72.3	70
Polynomial Kernel SVM	77.2	75	80.6	81	82.5	81
	76.5	73	78.2	77	81	79
	75	78	76.5	75.2	79.1	77
	74.5	72	75.2	75.9	78	76
	72.8	72.5	73.4	73.5	77.5	75
	71.5	68	72.1	70	76.2	73
	70.3	68.5	70.6	68.2	74.8	74.1
	69.3	67	68	68.5	72.3	73
RBF Kernel SVM	77.2	77	80.6	80.3	82.5	82
	76.5	76.5	78.2	78.9	81	80
	75	74	76.5	76.4	79.1	79.6
	74.5	74.1	75.2	75.1	78	77.5
	72.8	72.5	73.4	74	77.5	76.5
	71.5	71.4	72.1	71	76.2	75.5
	70.3	71	70.6	69	74.8	74
	69.3	69	68	68.2	72.3	72.8

non-soluble pollution deposition. On comparing the shed profile of the test specimens, the accumulation of pollution on type B with alternate sheds is decreased than the similar sheds. In this proposed work, composite insulator subjected to three different types of pollution layer, namely coastal, desert, and industrial pollution. The flashover voltage is predicted for the polluted insulator using linear and nonlinear SVM regression model. In general, the performance of the outdoor insulator is degraded due to pollution layer. Effective prediction of flashover voltage leads to avoid the damage of insulators. Therefore, it is suggested that RBF kernel SVM function shows better prediction rate when compared to the other techniques. The future research will be focused toward optimization of kernel function parameters.

Acknowledgements We acknowledge the Department of Electrical and Electronics Engineering, National Engineering College for the support by permitting to do this research work in high-voltage laboratory.

References

1. Mackevich, J., Minesh, Shah: Polymer outdoor insulating materials Part I: comparison of porcelain and polymer electrical insulation. IEEE Electr. Insul. Mag. **13**(3), 5–12 (1997)
2. Gubanski, S.M.: Modern outdoor insulation—concerns and challenges. IEEE Mag. Dielectr. Elect. Insul. **21**(6), 5–11 (2005)
3. Zhang, Zhijin, Liu, Xiaohuan, Jiang, Xingliang, Jianlin, Hu, Gao, David Wenzhong: Study on ac flashover performance for different types of porcelain and glass insulators with non-uniform pollution. IEEE Trans. Power Delivery **28**(3), 1691–1698 (2013)
4. Jiang, Xingliang, Wang, Shaohua, Zhang, Zhijin, Jianlin, Hu, Qin, Hu: Investigation of flashover voltage and non-uniform pollution correction coefficient of short samples of composite insulator intended for ±800 kV UHVDC. IEEE Trans. Dielectr. Electr. Insul. **17**(1), 71–80 (2010)
5. Douar, M. A., Mekhaldiand, A., Bouzidi, M. C.: Flashover process and frequency analysis of the leakage current on insulator model under non-uniform pollution conditions. IEEE Trans. Dielectr. Electr. Insul. **17**(4) (2010)
6. Belkheiri, M., Zegnini, B., Mahi, D.: Modeling the critical flashover voltage of high voltage insulators using artificial intelligence. J. Intell. Comput. Appl. (JICA) **2**(2), 137–154 (2009). ISSN0974-410X
7. Engelbrecht, Chris S., Hartings, Ralf, Tunell, Helena, Engstrom, Bjorn, Janssen, Harald, Hennings, Raimund: Pollution tests for coastal conditions on an 800-kV composite bushing. IEEE Trans. Power Delivery **18**(3), 953–959 (2003)
8. Akbar, M., Zedan, F. M.: Performance of HV transmission line insulators in desert conditions. III. Pollution measurements at a coastal site in the eastern region of Saudi Arabia. IEEE Trans. Power Delivery **6**(1) (1991)
9. Mei, Hongwei, Wang, Liming, Guan, Zhicheng, Mao, Yingke: Influence of sugar as a contaminant on outdoor insulation characteristics of insulators in a substation. IEEE Trans. Dielectr. Electr. Insul. **19**(4), 1318–1324 (2012)
10. Dong, Bingbing, Jiang, Xingliang, Jianlin, Hu, Shu, Lichun, Sun, Caixin: Effects of artificial polluting methods on AC flashover voltage of composite insulators. IEEE Trans. Dielectr. Electr. Insul. **19**(2), 714–722 (2012)
11. Xingliang, Jiang, Jihe, Yuan, LichunShu, Zhijin, Zhang, Jianlin, Hu, Feng, Mao: Comparison of DC pollution flashover performances of various types of porcelain, glass, and composite insulators. IEEE Trans. Power Delivery **23**(2) (2008)
12. Vapnik, V.: The Nature of Statistical Learning Theory. Springer, NJ (1995)
13. TunayGencoglu, Muhsin, Uyar, Murat: Prediction of flashover voltage of insulators using least squares support vector machines. Expert Syst. Appl. **36**, 10789–10798 (2009)
14. Sahli, Z., Mekhaldi, A., Boudissa, R., Boudrahem, S.: Prediction parameters of dimensioning of insulators under non-uniform contaminated conditions by multiple regression analysis. Electr. Power Syst. Res. **81**(4), 821–829 (2011)
15. Zegnini, B., Mahdjoubi, A. H., Belkheiri, M.: A least squares support vector machines (LS-SVM) approach for predicting critical flashover voltage of polluted insulators. In: 2011 Annual Report Conference on Electrical Insulation and Dielectric Phenomena. https://doi.org/10.1109/ceidp.2011.6232680

Analysis of Human Error of EHS in Healthcare Industry Using TISM

R. K. A. Bhalaji, S. Bathrinath, Chitrasen Samantra and S. Saravanasankar

Abstract Human error plays a key role in healthcare industry. The risk factors for human error involved in environmental health and safety section is extremely high when compared to other sections. The objective of this paper is to recognize the most influential factors for human error of EHS in healthcare industry. The risk factors are identified from the literature survey as well as inputs from industrial experts. The identified factors are analyzed using one of the soft computing tools known as total interpretive structural modeling (TISM). The model of TISM clearly shows the driving power and reliance power of the factor. A case empirical study is conducted in an Indian healthcare industry for verifying the suggested model. The outcomes of the paper will help industrial managers for implementing the framework of EHS and also will enhance the managerial excellence. Finally, to verity the TISM results, a statistical validation is done using covariance-based structural equation modeling (CBSEM).

Keywords Human error · EHS · TISM · CBSEM · India

1 Introduction

In healthcare industry, human error of EHS is a major issue because it affects the production as well as turnover [7] of the industry. The errors occurred in the working environment lead to wound or disease to the employees. These errors can negatively affect employee's health. Moreover, there are several risk factors involved due to human error in EHS and it is difficult to control. Hence, identifying the factor in working environment should be intended and sorted out to diminish the probability

R. K. A. Bhalaji · S. Bathrinath (✉) · S. Saravanasankar
Department of Mechanical Engineering, Kalasalingam Academy of Research and Education, Krishnankoil, Tamil Nadu, India
e-mail: bathri@gmail.com

C. Samantra
Department of Production Engineering, Parala Maharaja Engineering College, Berhampur, Odisha 761003, India

© Springer Nature Singapore Pte Ltd. 2020
K. N. Das et al. (eds.), *Soft Computing for Problem Solving*,
Advances in Intelligent Systems and Computing 1057,
https://doi.org/10.1007/978-981-15-0184-5_6

and effect of errors. In this paper, TISM method is applied to analyze the risk factors and also it suggests the model for controlling the risk factors. TISM method is useful for understanding the relationship between different risk factors. This method clearly indicates the risk factors and their levels. The practitioners and industrial managers can use the results identified for reducing the human error in EHS section. For the management excellence, human safety is also a key part in the healthcare industry. Further, the human error of EHS and the related information has been studied by many investigators [12, 15].

2 Relevant Literatures

Relevant literature is sorted into two subsections namely (a) risk factors involved in human error of EHS and (b) literature gap

(a) Risk factors involved in human error of EHS

Bowie et al. [4] analyzed the accidents happened in the healthcare industry due to the human error by using root cause analysis. The outcomes shown that lack of training for employees is the key reason for human error and they need to conduct safety training for employees per week. Flashpoler et al. [10] examined the man–machine interface (MMI) by using Delphi technique in healthcare sector. The results depicted that MMI faults are mainly due to operating error and by implementing/erecting advanced machines and tools workers health and safety can be improved. Charles et al. [5] explored employees in healthcare management after work-related exposure to worm disease. The findings of the study show that risk is mainly due to poor EHS plans. For lessening wounds and risk among employees in healthcare, they suggest a proper EHS plan. Redinger et al. [13] assessed the occupational health and safety (OHS) in three leading healthcare organizations. The consequences demonstrated that organization C is the most influential one in the process of OHS and they need to implement OHSAS standards. Rivers and Glover [14] analyzed the relationship between employees in the healthcare sector. The outcomes exhibited that poor communication between employees that lead to accidents in the sector because of stress in the working environment. Warren et al. [19] examined the system performance in healthcare industry based on the perspective of employee. From the above study, the results showed that working conditions, organizational climate and regulations are the main factors that are affecting system performance.

(b) Literature Gap

So far, there is no research paper related to human error of EHS by using soft computing tools like TISM method. In order to fulfill this gap, this paper analyzed the risk factors for human error of EHS using the suggested method. Based on the literature survey as well as experts inputs, risk factors are recognized as shown in Table 1.

Table 1 Factors affecting human error of EHS

S.No	Factors	Factors code	References
1	Adequacy of EHS systems	H1	[13]
2	Working conditions	H2	[17, 19]
3	Adequacy of organization support	H3	[6]
4	Availability of EHS procedures/plans	H4	[5, 8]
5	Number of simultaneous goals	H5	[14]
6	Adequacy of MMI/operational support	H6	[9, 16]
7	Adequacy of training/experience	H7	[4]

Table 2 Reachability matrix

Factors	H1	H2	H3	H4	H5	H6	H7
H1	1	1	0	1	0	0	0
H2	0	1	1	1	1	1	0
H3	1	1	1	0	1	0	0
H4	0	1	0	1	1	0	1
H5	0	0	0	0	1	1	1
H6	0	1	1	0	0	1	1
H7	0	1	1	0	1	1	1

3 Case Empirical Study

For verifying the suggested model, case empirical study is conducted in south Indian healthcare industry. More than 1000 employees are employed in the industry and over 50 crores of turnovers. Based on the reports gathered from the industries, employees faced a lot of problems in the EHS section because of human error for the past 10 years. Due to the risks involved in the industry, it affects the production rate as well as works are getting delayed. Risk assessment is needed for reducing the risks and implementation of EHS section. Therefore, to identify the risk factors due to human error in EHS are collected from the relevant literatures and also from the inputs from industrial experts. Members in the expert committees such as chief engineer, EHS manager and senior executive engineer are formed to give decisions for risk factors as per the TISM rule as indicated in Step 5. They will give the ratings in questionnaire form as mentioned in Appendix. They are selected based upon the experience and skill in decision making. Here to analyze the risk factors, a soft computing tool known as Total interpretive structural modelling (TISM) is suggested. Finally, the outcomes will help industrial mangers for reducing and eliminating the risks and also for the implementation of EHS framework.

3.1 Proposed Methodology

In this paper, Total Interpretive Structural Modelling (TISM) is used to show the relationship between factors with various leveled. Based on the procedure of Interpretive structural modeling (ISM), to build the method of TISM and it is created by Warfield [18]. TISM is an advanced version of ISM. This method is used for understanding the relationship between every risk factors. From the experts perspective, to recognize the relationship between every risk factors. To create a model for TISM of 'm' factors, experts should characterize the set of relationships '$m\,(m-1)$'. The procedure for TISM method is discussed below [2, 11].

Step 1: To define and recognize the risk factors

To define and recognize the risk factors are the initial phase of this procedure and show the relationships of every factor from the experts' view by using questionnaire and it is mentioned in Table 2.

Step 2: Define the relationship of contextual

To build up the structure for linking the factors, it is easy to characterize the relevant relationship among the factors.

Step 3: Interpretation relationship

To create a model more obviously and to limit comments from various clients. From the experts and shareholders, comments were requested to specify the two factors that are linked.

Step 4: Pairwise comparison

To get the opinions from experts on the relationship among every risk factors, a relationship of absence or presence matrix has been established by placing whichever "Yes (Y) or No (N)" for every link of a–b.

Step 5: To check transitivity and reachability matrix

To convert the pairwise comparison in reachability matrix form, the numeric '1' is allocated to 'Yes' and '0' for 'No'. Transitivity rule is used to check the matrix for accomplishing the final reachability matrix as given in Table 3.

Step 6: To partition the reachability matrix

Based on the reachability and antecedents sets, matrix of final reachability is divided into various levels for each factor by an iteration series called partition level. Table 3 shows the final level.

Step 7: To develop the diagraph

Table 3 Level matrix

Factors	Human errors	Levels
H1	Adequacy of EHS systems	IV
H2	Working conditions	I
H3	Adequacy of organization support	I
H4	Availability of EHS procedures/plans	III
H5	Number of simultaneous goals	I
H6	Adequacy of MMI/operational support	I
H7	Adequacy of training/experience	II

Based on the levels, the factors are graphically arranged and to provide the direct link according to the relationships appeared in reachability matrix. A diagraph is attained to show the relationships of transitive whose interpretation is significant.

Step 8: Matrix for interaction

Finally, the diagraph is converted into a matrix for binary interaction signifying to every interaction by the entry of 1. To interpret the entry 1 cell by choosing the related interpretation from the information base in an interpretive form as appeared in Table 4.

Step 9: Model of TISM

Figure 1 shows the TISM model in which all interpreted links are in between the side of the specific links and they are in written form. For any modification and theoretical differences, the TISM model is verified.

4   Result and Discussions

This paper used TISM method for analyzing the risk factors of human error in EHS in the healthcare industry and it is used for the purpose of reducing the eliminating the risks. We recognized seven crucial factors from the expert inputs as well as relevant literature. The result reveals that seven risk factors have divided into four levels based on the reliance power and driver power between risk factors. Detailed discussion about result of this paper is detailed below.

Level I. In this paper, the most extreme power was adequacy of EHS system and also it is a self-reliant factor. Therefore, industry needs to improve the framework of EHS by implementing environmental standards like ISO 14,000. These standards will improve the status of the industry as well as the employees perspective and also it will help to reduce the environmental effect. EHS standards will reduce the incidents due to environment, loss in finance, work-related exposure, ecological occasions, and human error.

Table 4 Interpretive matrix

Factors	Adequacy of EHS system	Working conditions	Adequacy of organization support	Availability of EHS procedures/plans	No. of simultaneous goals	Adequacy of MMI/OS	Adequacy of training/experience
Adequacy of EHS system	–	Impact to environment and controlling OHS risks	–	It impacts safety	–	–	–
Working conditions	–		It influences the quality of organizational support and affection unity	EHS plan will be regularly updated	Effect on working conditions on employee performance	Affects the performance of tasks	–
Adequacy of organization support	Importance to high standards in EHS	Extensive ranging source cutbacks and personal shortages influence working conditions	–	–	Number of projects a person can work	–	–
Availability of EHS procedures/plans	–	To implement plans for all personnel to report unsafe working conditions	–	–	To new achievements in EHS knowledge	–	Usefulness of appropriate health and safety training programs

(continued)

Table 4 (continued)

Factors	Adequacy of EHS system	Working conditions	Adequacy of organization support	Availability of EHS procedures/plans	No. of simultaneous goals	Adequacy of MMI/OS	Adequacy of training/experience
No. of simultaneous goals	–	–	–	–	–	Technological changes will affect MMI	Work situations assess the impact of training on set goals
Adequacy of MMI/OS	–	Poor MMI design and mental fatigue	Influenced by new system as well as advanced machines	–	–	–	Reduce time and costs for training
Adequacy of training/experience	–	Training people to recognize their working conditions	To conduct EHS training for all employees	–	Successful training is in meeting objectives	Mental set and multiple fault complexity affect MMI	

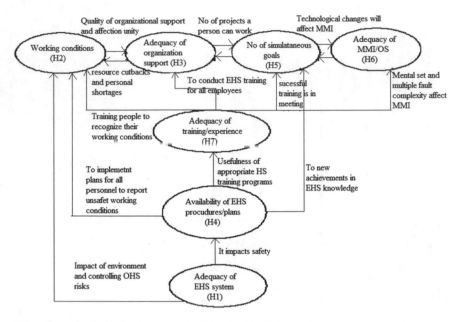

Fig. 1 TISM model for human error of EHS in healthcare industries

Level II. Availability of proper EHS plans is in the third level of this industry. Organization requires increasing the no. of EHS plans for control well-being and safety of employees as well as property. They need to conduct the training for employees to understand about the EHS plans and how to use it.

Level III. In the working environment, many risks are happened due to inexperienced and illiterate employees in the industry. They have less experience in this field and don't know about the risks. To solve this issue, conducting the drill for new employees consistently for clearly know about the safety in working environment.

Level IV. In many places, healthcare industry runs 24 h a day. It affects employee's health and safety, physical stress, responsibilities, and legal rights. Therefore to maintain happiness, motivation and communication between employees for reducing the stress is very essential. If workers feel good and safe, then the working conditions are good. Many employees believe the management for some basic and personal needs but many industries do not fulfill the needs of employees. Organization doesn't satisfy the workers because of providing low wages, working hours, safety and privileges. The main aim of the organization is to satisfy the needs of employees because if employees believed that organization, it will improve the organization's name. In many industries, managers set a target for employees for manufacturing the product within limited time. It mentally affects the worker's mind as well as reduces the performance of products. So, the management will give more time for workers; then, it provides better results. Work-related mishaps, illness, pressure and improper use of tools are mainly due to poor MMI. Therefore, the industries need to implement the significance of the MMI for the safety of employees.

Table 5 Covariance matrix

	H1	H2	H3	H4	H5	H6	H7
H1	3.974						
H2	1.057	0.642					
H3	1.429	0.468	1.797				
H4	1.387	−0.064	0.927	1.269			
H5	1.695	−0.216	−0.243	0.842	1.797		
H6	2.949	−0.730	−0.838	1.847	1.737	6.714	
H7	1.413	0.011	0.151	1.127	0.926	1.889	1.269

5 Statistical Validation

This paper used covariance-based structural equation modeling (CBSEM) for validating the TISM results. Structural equation modeling (SEM) that comprises of various types of mathematical models, algorithms and techniques in statistics for creating the systems of data construction and is in the shape of causative modeling [1]. CBSEM is used to assess the goodness of fit which concentrates on diminishing the variations between both observed and estimated covariance matrix and also it is termed as confirmatory factor analysis. For solving the problems, many researchers used CBSEM [3, 10]. Here, questionnaire is developed for obtaining the decision makers opinion working in the healthcare industry and based on the judgements they provide the rating from 0–4 scale (0-no importance to 4-very high importance scale). This ratings are used for analyzing the risk factors. The selected decision makers are more than 10 years' experience in healthcare industry and they are skilled and expertise in decision making. Then, the collected data are solved by using covariance-based structural equation modeling.

Finally, the CBSEM results revealed that H2, H3, H5, and H6 are in the top level and it is shown in Table 5. Both TISM and CBSEM outcomes are the similar one and they are 100% verified. Figure 2 displays the rank between these both methods and Table 6 shows that mean and standard deviation of every risk factor.

6 Conclusion and Scope for Future Research

For the better perceptions of EHS framework, this paper analyzed the factors for human error of EHS in the healthcare industry using a soft computing tool TISM method. For the importance of EHS framework, this paper will be helpful for the considered case industry. TISM method can be applied to perform analysis using inputs from industrial experts and it easily captures the expert's skill and experience efficiently. A TISM shows the relationship and also the logical connection between every

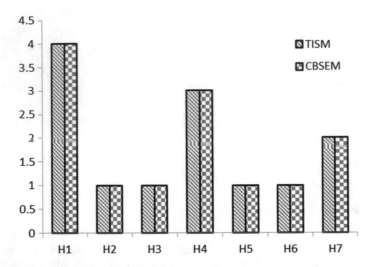

Fig. 2 Ranking for both TISM and CBSEM

Table 6 Mean and standard deviation

Factors	Mean	Standard deviation	Level
H1	2.128	0.802	IV
H2	0.856	1.994	I
H3	0.759	1.994	I
H4	1.203	1.127	III
H5	0.759	1.994	I
H6	1.466	1.994	I
H7	1.203	1.341	II

factors and the way of their relationship. The TISM model depicts the most influential factor and their interaction in improving environmental management system. Finally, results from TISM method are verified through statistical validation using covariance-based structural equation modeling (CBSEM). This method is valuable for the scholars and academicians for problem solving.

This study has some drawbacks as well that can be explored later. More factors have not been recognized. These can additionally be investigated and verified thoroughly. By using other tools like ISM and DEMATEL, interrelationships among the risk factors can be evaluated for future perspectives. For achieving the aims and targets of the industry, EHS manager used the TISM model to find out the most influential factors.

Appendix: Questionnaire

Note: For the review, a questionnaire was framed and provided to experts in the field to evaluate risk factors for human error involved in the EHS section in south Indian healthcare industry. The gathered data will be used for purposes of research only. Your comments will not be shared to a third party or social media. It will be kept confidential. Your input will help to bring about positive outcomes in this research, and hence, it is very important. My heartful thanks for your time and effort for providing a rating.

Please choose any one rating that you suitable for each item (refer Step 5 for TISM and segment 5 for CBSEM).

Factors	H1	H2	H3	H4	H5	H6	H7
H1							
H2							
H3							
H4							
H5							
H6							
H7							

Profile of the expert:

1. Name: ……………………………………..
2. Experience in the healthcare industry (in years): ……
3. Name of the organization: ………………………..
4. Current position in the organization:
5. Mobile No. & Email: …………………………..

Thank you very much for your time and effort in filling up this questionnaire.

References

1. Afthanorhan, W.M.A.B.W.: A comparison of partial least square structural equation modeling (PLS-SEM) and covariance based structural equation modeling (CB-SEM) for confirmatory factor analysis. Int. J. Eng. Sci. Innovative Technol. **2**(5), 198–205 (2013)
2. Agarwal, A., Vrat, P.: A TISM based bionic model of organizational excellence. Global J. Flex. Syst. Manage. **16**(4), 361–376 (2015)
3. Ali, F., Kim, W.G., Li, J., Cobanoglu, C.: A comparative study of covariance and partial least squares based structural equation modeling in hospitality and tourism research. Int. J. Contemp. Hospitality Manage. **30**(1), 416–435 (2018)

4. Bowie, P., Skinner, J., de Wet, C.: Training health care professionals in root cause analysis: a cross-sectional study of post-training experiences, benefits and attitudes. BMC Health Serv. Res. **13**(1), 50 (2013)
5. Charles, P.G., Grayson, M.L., Angus, P.W., Sasadeusz, J.J.: Management of healthcare workers after occupational exposure to hepatitis C virus. Med. J. Aust. **179**(3), 153–157 (2003)
6. Chow, C.W., Ganulin, D., Haddad, K., Williamson, J.: The balanced scorecard: a potent tool for energizing and focusing healthcare organization management. J. Healthc. Manag. **43**(3), 263–280 (1998)
7. Donaldson, M. S., Corrigan, J. M., & Kohn, L. T. (Eds.).: *To err is human: building a safer health system* (Vol. 6). National Academies Press, (2000)
8. Dünnebeil, S., Sunyaev, A., Blohm, I., Leimeister, J.M., Krcmar, H.: Determinants of physicians' technology acceptance for e-health in ambulatory care. Int. J. Med. Inform. **81**(11), 746–760 (2012)
9. Flaspöler, E., Hauke, A., Pappachan, P., Reinert, D., Bleyer, T., Henke, N., Beeck, R. O. D.: The Human-Machine Interface as an Emerging Risk. *EU-OSHA* (European Agency for Safety and Health at Work), Luxemburgo (2009)
10. Hair Jr., J.F., Babin, B.J., Krey, N.: Covariance-based structural equation modeling in the Journal of Advertising: Review and recommendations. J. Advertising **46**(1), 163–177 (2017)
11. Jayalakshmi, B., Pramod, V.R.: Total interpretive structural modeling (TISM) of the enablers of a flexible control system for industry. Glob. J. Flex. Syst. Manage. **16**(1), 63–85 (2015)
12. Leape, L.L., Woods, D.D., Hatlie, M.J., Kizer, K.W., Schroeder, S.A., Lundberg, G.D.: Promoting patient safety by preventing medical error. JAMA **280**(16), 1444–1447 (1998)
13. Redinger, C.F., Levine, S.P., Blotzer, M.J., Majewski, M.P.: Evaluation of an occupational health and safety management system performance measurement tool—III: measurement of initiation elements. AIHA J. **63**(1), 41–46 (2002)
14. Rivers, P.A., Glover, S.H.: Health care competition, strategic mission, and patient satisfaction: research model and propositions. J. Health Organ. Manage. **22**(6), 627–641 (2008)
15. Rogers, A.E., Hwang, W.T., Scott, L.D., Aiken, L.H., Dinges, D.F.: The working hours of hospital staff nurses and patient safety. Health Aff. **23**(4), 202–212 (2004)
16. Rooney, J.J., Heuvel, L.N.V., Lorenzo, D.K.: Reduce human error. Qual. Prog. **35**(9), 27–36 (2002)
17. Scotti, D.J., Driscoll, A.E., Harmon, J., Behson, S.J.: Links among high-performance work environment, service quality, and customer satisfaction: an extension to the healthcare sector. J. Healthcare Manage. **52**(2), 109–124 (2007)
18. Warfield, J.N.: On arranging elements of a hierarchy in graphic form. IEEE Trans. Syst. Man Cybern. **2**, 121–132 (1973)
19. Warren, N., Hodgson, M., Craig, T., Dyrenforth, S., Perlin, J., Murphy, F.: Employee working conditions and healthcare system performance: the Veterans Health Administration experience. J. Occup. Environ. Med. **49**(4), 417–429 (2007)

Congestion Control Using Fuzzy-Based Node Reliability and Rate Control

Bhawana Pillai and Rakesh Singhai

Abstract The mobile ad hoc network (MANET) wireless system offers a few favorable circumstances, including minimal effort, basic system support, and helpful administration scope. In these systems, congestion can happen anywhere in the route while information getting forwarded from source node to destination node. These nodes can introduce long delays, which affect the overall efficiency of a system. Congestion leads to high data loss, long postponement, and time misuse of asset. To avoid it, effective congestion control mechanism should be applied that makes best use the accessible system assets and keeps the load underneath the limit. This paper proposed a mechanism to control the congestion as well as bestow better performance with respect to network parameters under dynamic behavior. Existing on-demand routing protocol is modified with local route establishment process and fuzzy-based reliable node detection mechanism is used to resolve the collision. The result taken in different scenarios by varying number of nodes concludes that the proposed approach is effective as compared to existing congestion control implemented in on-demand routing protocol.

Keywords Mobile ad hoc network · AODV · Packet delivery ratio · Round-triptime · CTS · RTS · Collision · Congestion

1 Introduction

Ad hoc network is a collection of wireless links connecting wireless terminals and it is adaptive in nature, also characterized by self-organizing nodes. This is normally a decentralized network. This network is said to be an ad hoc because determination of nodes that are willing to forward the data is made dynamically. Due to lack of central controller, if any new terminal is detected, ad hoc network automatically inducts it whereas if any terminal moves out of network, the rest of the terminals routinely reconfigure themselves. Each wireless terminal in the ad hoc network is

B. Pillai (✉) · R. Singhai
Department of Electronic and Communication, U.I.T., RGPV, Bhopal, India
e-mail: bhawanapillai@gmail.com

© Springer Nature Singapore Pte Ltd. 2020   67
K. N. Das et al. (eds.), *Soft Computing for Problem Solving*,
Advances in Intelligent Systems and Computing 1057,
https://doi.org/10.1007/978-981-15-0184-5_7

operational with a transceiver, an antenna, and a power source. These wireless nodes are mobile and so this network is termed as mobile ad hoc network (MANET). Wireless medium is error prone with limited bandwidth and factors like multiple access, signal fading, noise, and interference causes significant throughput loss in MANET. In MANET, there is need of a routing procedure as there is no infrastructure support. Routing is also required to trace the destination node which might be out of range of a source node. In this case to implement routing, each terminal must be able to forward data to other nodes. Mobility of nodes leads to frequent changes in the network topology and also makes existing routing protocols inefficient for the dynamic topologies. Therefore, reactive routing algorithm Ad hoc on-demand Distance Vector (AODV) is specially developed for MANET. AODV works in two phases, viz. routing discovery and maintenance which consist of transmission of large number of data packets through relatively small number of nodes and hops. The main module of AODV is flooding the route discovery packets and reply to original source node to establish the route. As well as congestion or link error is propagated to source node increasing control overhead. There is no mechanism to find out trustworthy node so as to improve congestion mechanisms.

2 Related Work

Alwadiyeh et al. [1] presented two mechanisms least delay, interference-aware multipath routing protocol (LIMR) and shortest path, interference-aware multipath routing protocol (SIMR) to reduce the power of interference between the selected node disjoint multipath schemes.

Authors presented [2] more adaptive and more reliable congestion control protocol for ad hoc network using bypass-route selection. It is done by distributing traffic on multiple routes by applying traffic splitting function. The work has achieved reduction in packet drop and overhead. Liu Ban-teng et al. [3] presented node degree theory for congestion control and utilised node with less degree to forward the packets.

Mallalpur et al. [4] introduced the multipath load balancing technique for congestion control in MANET. To efficiently balance the load among the desirable paths, the link cost and the path cost parameters are effectively used. The literature [5–7] described that reactive protocols lay down the routes on-demand, need raised by source nodes in ad hoc manner. Whenever any node wants to open communication with another node, the routing protocol makes efforts to setup a route. This protocol floods the network with route request (RREQ) and route reply (RERP) messages. With RREQ, the route is discovered from source to target node and when target node sends RERP for the confirmation, the route has been established. It is on-demand means a route is established by AODV from a destination only on demand and it is capable of unicast as well as multicast routing. It keeps these routes as long as they are desired by the sources. To ensure newness of established routes, protocol applies sequence number generated on every RREQ. Every RREQ carries a time to live TTL value. This numeric value stores the count of hops, this message should be forwarded. Every node keeps two separate counters: a node sequence number and broadcast_id which in pair uniquely identifies a RREQ. RERR message is used to notify other nodes of the loss of the link. Each node keeps a precursor list containing IP address for each of its neighbors which are likely to use the link as forwarding hop. If a link crack occurs while the route is active, RERR message is propagated to the source node. After receiving the RERR, if still source node desires the route, it can re-initiate route discovery.

MANET faces challenge to maintain and allocate shared network resources—bandwidth of the link and queue on the terminal. When excessive packets are pending for transmission in the link or the queue, MANET is said to be congested. Existing congestion control methods are implemented by existing routing method on top of transmission control protocol traffic flow. Congestion resolving can be done by distributed coordination function (DCF) of TCP. But to overcome the limitations, reduction of control overhead and finding out reliable node in turn reliable route will benefit the overall packet transmission and control the congestion due to excessive control packets.

2.1 Problem Statement

AODV lacks an efficient route maintenance technique. The routing information is always obtained on demand, including for common traffic and information cannot be reuse. The messages can be misused for insider attacks including route disruption, route invasion, node isolation, and resource consumption.

AODV is designed to support the shortest hop-count metric. This metric favors long, low bandwidth links over short, high-bandwidth links. Existing mechanism of AODV supports congestion control based on traditional TCP mechanism which is not suitable and fruitful due to ad hoc nature of MANET and misleading the network performance. Similarly, it lacks in collision detection and avoidance control.

The aim of this work is to improve the congestion control for improving the performance in MANET by reducing the limitation through the enhancement of AODV routing protocol.

To apply local route discovery procedure, identifying the intermediate node to find an alternative path instead of routing RERR to source. To identify the reliable node, direct trust computation using network performance parameter based on number of packets transmitted and received by node. One of such measure can be PDR. To regularize the process, threshold can be determined. The objective is to determine formula for threshold calculation and PDR calculation. Whenever the reliability (packet delivery ratio) becomes lower from the fixed threshold value, a local route discovery procedure has to be on track to search other available path by the predecessor of that node instead of the sender which generally searches a new path only when the detected route has been destroyed due to node movement or route failure. Transmit the data packet by delaying them after certain time. To analyze and identify RTS/CTS mechanism and role of RTT to work on congestion avoidance.

The main objective is to improve congestion control to minimize end-to-end delay, buffer overflow, reduce control packets to improve PDR, and to improve throughput for the better performance of the network.

3 Proposed Congestion Control

In order to accommodate the limitations of existing AODV of excessive control overhead, following modification is suggested and implemented in AODV in three phases:

3.1 Steps for Local Route Establishment

If node faces the packet loss due to the congestion at node or forwarding path, then it reports to RERR to its just previous node, not the original source node, this can control RERR control message transmission to the source node. The previous neighbor node will start the route discovery by multicasting RREQ to only its neighbors. This node will send the RREQ packet to a group of node except the congested node and source node. This first step will try to reduce congestion due to flooding of RERR and RREQ. This converts the route discovery process to local route establishment [1].

Route Repair Algorithms Pseudocode

Initialization: M: mobile nodes, S: sender nodes, and R: receiver nodes
I: {i1, i2, ... ij, ij + 1, ... in − 1, in} intermediate nodes
L: {l1, l2, ... lk, lk + 1, ... Lm − 1, lm} i.e. l1 = i1 i2 link between nodes
Step time: {1, 2, ... 100} $\lambda = 0$ step change
While λ <=100 Do
Check route-table(S, ij, R)
If lk update && R! exist then Check path lk to lm
If (ij to ij+1 route break) then
Local-route-repair (lk, lk+2,R)
End if
End if
$\lambda = \lambda + 10$ End do
$s(t + 1) = s(t) + k$//if congestion is not detected $s(t + 1) = s(t) * l$//if congestion detected where s(t) data sending rate at time slot t k -additive increase parameters l(0<l<1) multiplicative decrease factor

3.2 Fuzzy-Based Reliable Route Establishment

To identify an alternate route to control further congestion, it is proposed to apply local route discovery by finding a reliable node for data transfer. To ensure the reliability of the node, trust computation can be done. There are direct and indirect trust computation methods [2] as summarized in Sect. 2. This work is based on direct trust computation based on network performance factor packet delivery ratio (PDR) [5]. PDR is a significant contributory factor while determining the congestion status of the network. Its calculation is stated in the next section. After receiving the RREQ packet, every node will reply with the RREP packet with its packet data ratio (PDR) value. Sender node receives RREP by a different neighbor and compares the received PDR values with the threshold value (this work has fixed it as 70% by averaging simulation runs). Thus, a reliable node and reliable route are discovered.

The trust value calculated is fuzzy classified as {High, Medium, Low} and packet preference is set as {0,1}. Set of fuzzy rules are prepared:

If PDR is high, preference is 1 then node is marked as reliable and route is marked as reliable
If PDR is medium, preference is 1 then node is marked as reliable and route is marked as reliable
If PDR is low, preference is 0 then node is marked as unreliable and route is marked as unreliable-trust computation is again done for other intermediate neighboring nodes.

3.3 Congestion Resolving Using TCP DCF

To take care of collision detection and avoidance control on this reliable route, the RTS/CTS mechanism is implemented on the route. To avoid collision of packets on a common channel, upon finding the probability of congestion, the sender node will delay transmission of the next packet by three times the round-trip time (RTT). This will delay the next packet until the previous packet will complete its transmission on the one-way path and acknowledgment will be received by the sender. This mechanism will control the congestion and avoid collision up to some extent

4 Simulation Experiment

Proposed modified TCP control flow is implemented and executed by changing number of nodes in simulation environments for number of runs to plot and tabulate results to analyze the work done.

4.1 Simulation Environment

The simulation is done by using widely used powerful network simulator NS-2 version 2.31 for mobile ad hoc networks [7]. AODV protocol is in-built part of NS-2 installation. The paper is implemented by modifying HELLO packet and routing table entry in NS2 with the help of configuration files. NS2 code is edited to implement congestion control of TCP. Tcl script file is written to specify node configurations parameters and other related NS commands. The output of simulation is trace file. Each and every event during the simulation updates this trace file. This file is used for analyzing performance parameters of MANET [7]. The random topologies and varying nodes are generated using setdest tool in NS2 for the simulations. Simulation configuration file chooses the source–destination terminals randomly. Pause time 0 means each node moves constantly throughout the simulation. The queue length is set to 50 packets to analyze congestion situation. Congestion control is observed for a 800 m * 800 m grid by varying number of nodes as 10, 25, 50, 75, and 100 with random movement.

Within the limit of 80–100 nodes it is medium network. The work is simulated for small to medium-size network by varying number of nodes. NS2 is implemented using combination of OTcl (for scripts describing the network topology) and C++ (The core of the simulator) and compatible whereas NS3 is not backward compatible with NS2. NS3 contains limited number of contributed codes as compared to NS2. So MANET and routing protocols can be studied and experimented in NS2 in better manner. In addition to this, NS2 employs with network animator (NAM), it is a Tcl-based animation system that produces a visual representation of the network described which helps in better implementation and simulation of proposed MANET work.

4.2 Performance Factors

This subsection provides the terminology of performance factors that are considered for comparing effective implementation of TCP congestion control with AODV routing presented in this paper. The two main factors analyzed are:

- Packet delivery ratio (PDR): PDR is analyzed by counting number of successfully received by the terminal. Simulation is done by varying number of nodes as 10, 25, 50, 75, and 100 nodes. Packets may queue up at the source and never enter the network in a good disciplines network on getting unstable due to congestion. This affects the network throughput.
- Normalized routing load (NRL): NRL is counted at destination terminal. Minimum NRL shows less network overhead and maximum channel utilization. Modified AODV with TCP congestion control periodically broadcast HELLO packet for early congestion detection and increases delivery of data packets by alternate route to destination.

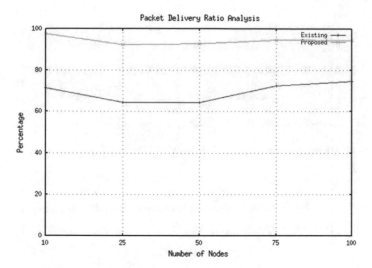

Fig. 1 Packet drop ratio analysis—PDR against no. of nodes

4.3 Packet Delivery Ratio

Graph is plotted for percentage PDR against number of nodes, shown in Fig. 1. When any intermediate node senses congestion with the help of HELLO message, it requests previous adjacent node to check for alternate route. Availability of alternate route avoids congestion and increase PDR.

The algorithm supports the threshold PDR value for nodes 70%. Reliable route finding in second module finds the nodes with PDR equal or more than 70% and this threshold is determined based on repetitive simulation for number of nodes and averaging of obtained PDR values. So after final simulation PDR achieved is more than 70% and because of reliable route and reliable nodes, congestion control and initial step of routing request from previous node instead of additional control packets always initiated by start node, PDR achieved is much higher that is around 90%.

4.4 Normalized Routing Load (NRL)

Figure 2 displays the graph of NRL simulated by varying number of nodes and shows significant improvement.

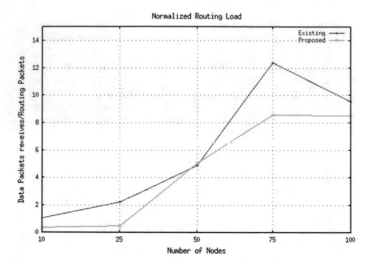

Fig. 2 NRL analysis—routing packets against no. of nodes

4.5 Throughput

Throughput of the network is calculated in Kbps in simulator. It is maximum 5.96 as provided in the attached table. (Fig. 3, and in Table 1)

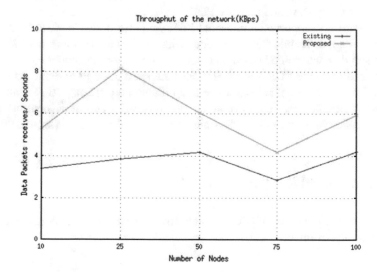

Fig. 3 Network throughput—data packets against no. of nodes

Table 1 Tabulated results of network throughput

No. of nodes	Throughput of the network (KBps)	
	Existing	Proposed
10	3.4	5.32
25	3.86	8.15
50	4.17	6.02
75	2.86	4.17
100	4.2	5.96

5 Conclusion

The proposed approach gives greater than 90% of PDR, the routing overhead is lowest while number of nodes is minimum and packet drop ratio is also reduced as compared to existing AODV routing. Local route repair methodology for route establishment process mechanism is used to control and resolve collision. Reliable route finding is done using fuzzy mechanism. This is resulted in improved packet delivery and other network parameters. The work can be extended to study congestion control with other reactive routing protocols.

References

1. Alwadiyeh, E.S., et al.: Interference-aware multipath routing protocols for mobile ad hoc networks. In: 22nd National Conference of the Australian Society for Operations Research, pp. 156–162. Adelaide, Australia, 1–6 Dec 2013
2. Vadivel, R., et al.: Adaptive reliable and congestion control routing protocol for MANET. Wirel. Netw. https://doi.org/10.1007/s11276-015-1137-3
3. Liu, B.T., et al.: Research on Congestion Control Routing Algorithm for Mobile Ad Hoc Networks Based on Node Degree Thoery. ISBN: 978-1-4244-6252-0/11/$26.00 ©2011 IEEE
4. Mallapur, S.V., Patil, S.R., Agarkhed, J.V.: Load Balancing Technique for Congestion Control Multipath Routing in Mobile Ad Hoc Networks. ISBN: 978-1-4799-8641-5/15/_c2015IEEE IEEE Elsevier (2015)
5. Seddik-Ghaleb, A., Ghamri-Doudane, Y., Senouci, S.M.: Effect of ad hoc routing protocols on TCP performance within MANETs. In: Proceedings 3rd Annual IEEE SECON 2006, vol. 3, pp. 866–873 (2006)
6. Kazi, S.H.: Congestion control in mobile ad-hoc networks (MANETs). Unpublished thesis, Brac University April 2011
7. Cuong Do D., et al.: Improving AODV protocol to avoid congested areas in mobile ad hoc networks. Indian J. Sci. Technol. 9(39) (2016). https://doi.org/10.17485/ijst/2016/v9i39/97552

Crowdsourcing Advent to Optimize Supply Chain Network in Rural India

Chandra Kant Upadhyay, Vijayshri Tiwari, Vineet Tiwari and B. Pandiya

Abstract The present scenario of India is highly urban-centric regarding business development. The growing knowledge and consciousness of brands in rural India attract attention in spite of tedious supply chain which has levels of hierarchies for suppliers, distributors and finally the consumers. The business requires a mechanism to track the goods at every level of the supply chain system. The report released by the Internet and Mobile Association of India (IAMAI) in 2017 points that out of the total 481 million Internet users, urban India has 295 million Internet users and the rest 186 million is rural. This is a huge opportunity to tap the rural market. The proposed model presents a framework of crowdsourcing applications for the optimization of the supply chain. The focus is on developing effective communication and mapping value of the goods produced and making available goods and services at low cost to the consumers. There is a potential for parcel delivery to villagers who are registered and willing to deliver for a nominal amount. Crowdsourcing approach can fasten the delivery time and lessen the cost too by the implementation of the algorithm. The trend analysis was done to understand the current demand. The gap between urban and rural markets can be reduced with this endeavor which involves Crowdshipping as a tool to optimize resources.

Keywords ICTs · E-commerce · Crowdsourcing · Rural · Supply chain · India

1 Introduction

The FMCG segment in India has a market scope in billions and ranks fourth in the Indian economy sector wise. It accounts for a revenue share of approximately 60%. The rural division is the largest contributor to the overall revenue generated by the FMCG sector in India and recorded a market size of around US$ 29.4 billion in

C. K. Upadhyay (✉) · V. Tiwari · V. Tiwari · B. Pandiya
Indian Institute of Information Technology Allahabad, Allahabad, UP, India
e-mail: ckupadhyay18@gmail.com

© Springer Nature Singapore Pte Ltd. 2020
K. N. Das et al. (eds.), *Soft Computing for Problem Solving*,
Advances in Intelligent Systems and Computing 1057,
https://doi.org/10.1007/978-981-15-0184-5_8

Fig. 1 Urban–rural Internet users in India. *Source* [14]

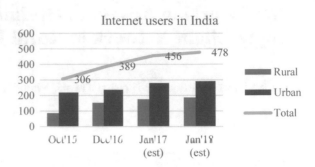

2016–2017. FMCG products amount for around 50% of total rural expenditure [14]. The understanding and demand for superior goods and services have risen in the rural market. With the rise in better distribution channels of FMCG companies, this demand can be met. The quick development of Information and Communication Technology with uses of the Internet in urban and rural has provided the impetus for the growth of E-commerce as shown in Fig. 1. The enterprise solutions are in the process of creating E-commerce to overlook traditional businesses. This provides the new rural settlement and promotes rural modernization. The rural market in India has amassed a huge 700 million consumers which are around 70% of the Indian population. Particularly in the FMCG segment, the personal care category accounts for 50% of the total market. An average citizen of India spends 8% of his income on this segment. It comprises basically of oral care, skin and hair care, cosmetics and toiletries, etc. The total rural income stands at US$ 572 billion which is expected to reach US$ 1.8 trillion by FY21 [14].

The growth of E-commerce recently generates more parcels and packets for distribution of products [9, 19, 30]. India is showing significant development in E-commerce since the behavior and preferences of customers are increasingly shifting from physical buying to online sales thus concreting the way for the success of the online shopping Web sites. The study reveals the influencing factors which can be both encouraging and discouraging for the rural consumers in the direction of online buying. The rationale of this study is to find out the potential of E-commerce in rural India and how the recent developments in the Information and Communication Technology can be profitable for exploring the entrepreneurial initiatives and promoting active social growth linking with a digital gap.

The trend analysis was done to understand the current demand, in which India is the second largest country for searched in fast-moving consumer goods depicted in Fig. 2.

Fig. 2 Google trend's search at global level for 2004 onward regarding fast-moving consumer goods

2 Literature Review

2.1 Rural Customer Behavior (RCB)

According to Nielsen [18], the various parameters used in the model include rural marketing mix or internal factors like (availability, affordability, acceptability and awareness), external factors, (like demographic factors, sociocultural factors, psychographic factors, economic factors and political factors) buyers buying behavior process, cognitive, affective and behavioral factors, pre-purchase, post-purchase factors and a feedback system. The rural marketing involves four crucial factors as shown in Fig. 3, namely awareness, affordability, acceptability and availability. There has to be aware of the goods or services which are targeted in the rural space along with the affordability because the income level in the rural areas is comparatively low. So, keeping in mind the purchasing power of the customers is very important. The products or services also need to be acceptable to the customers as they should be in synchronization with their culture and requirements. The gap in the requirement, i.e., demands and supply, is the key issue in availability and that can be targeted to enter the market.

Fig. 3 Rural marketing—four A's structure. *Source* [15]

2.2 An Initiative of Social Commerce

Social networking has gained an immense presence in recent times in the online space of the country as a promotion mix [16]. It has become an integral part of the customers and organizations that are trying to find new ways to use social media as a tool for marketing [24]. The initial usage was connectivity and linkage with friends and dear ones, but off late it has entered digital marketing strategy with Web sites and applications being the face of the organization. The target customers can be reached out in a maximum manner with this approach. Customer relationship management (CRM) has merged with social marketing to know more about prospective customers and make a business out of it. It also creates a brand and helps in launch and advertisement of new products along with getting feedback for the present ones.

2.3 Crowdsourced Logistics

Crowdsourcing is one of the emerging phenomena attracting prominent attention from both corporate business practitioners and scholars in recent years. It helps in facilitating the connectivity and collaboration of people in society and organizations [31]. The term crowdsourcing was initially given by Howe in Wired Magazine article. It is defined as the act of a company or institution taking a function once performed by employees and outsourcing it to an undefined network of people in the form of an open call [12]. In the logistics sector for shipping, additional market share for crowdshippers gain [2]. It can be applied in rural areas as a tool for community development as initiated in Namibia [26]. Table 1 depicts the year wise literature review according to author framework related to Crowdshipping, and Fig. 4 shows Google trend analysis for the years 2004–present (2018) using a keyword 'crowdsourcing' and 'shipping,' at worldwide; the below graph of crowdsourcing upward in nature due to demand increases in social media.

Table 1 Literature review regarding the concept of Crowdshipping

Authors	Year	Title	Technique	Methodology	Focus
Chen et al. [5]	(2014)	Taxi exp: a novel framework for Citywide package express shipping via taxi crowdsourcing	Optimization	Framework	Crowdsourcing logistics
Mladenow et al. [17]	(2015)	Crowdsourcing in logistics: concepts and applications using the social crowd	Optimization	Framework	Crowdsourcing logistics
Pan et al. [19]	(2015)	A crowdsourcing solution to collect E-commerce reverse flows in metropolitan areas	Optimization	Simulation	Crowdsourcing logistics
Weinelt [28]	(2016)	World economic forum white paper digital transformation of industries: logistics	Optimization	Framework	General, Crowdsourcing logistics
Xiao [29]	(2017)	Development of city logistics in china. In: Contemporary Logistics in China	Optimization	Framework	Last mile delivery

Source Author compilation (2018)

2.4 Cognitive Factors

The factors are directly associated with barriers regarding knowledge, skill and confidence which are of E-commerce usage. These problems are compounded by highly underdeveloped and unreliable services. For providing Internet at rural areas, the setup must be built up for mutually fixed and wireless technologies. In direction to use ICTs by the rural population, the concerns linking to nature of demand prospective revenue and logistic tools are addressed to confirm dependable provision [4].

Fig. 4 Google trend's search on 'crowdsourcing' and 'shipping' worldwide. *Source* Author compilation 2018

The connected research models of parcel pickup behavior are of [8] and shared city logistics [6]. The demography involved in sharing transport in urban logistics is age and education related to on-demand rides [13, 21] as common age group will have a greater chance for sharing the vehicle [25]. The attributes which are pertinent in the case of ride sourcing [1, 10] along with urban site and shipment are the usage of single sharers [7].

As discussed by [23], there is an imminent need to develop interconnected business models to reinvent delivery. The typology of five archetypal business models has been proposed by the researchers. The various issues which play a major role in delivery can be divided into two macro factors—operational and behavioral [20]. The operational performances include concerns of delivery rate, the security of delivery and variations in the performances. The other behavioral issues are choices of customers regarding the delivery location and type. The acceptance of crowd delivery can be dicey at times and the details of it need to be chalked out beforehand. Other customer preferences like time and location should also be kept in mind. The attributes which should be present in a crowdsourcing plan were discussed by Hughes and Cohen [13] for developing a new interconnected business model to find a novel way of improving delivery by the feedback of online customers. The feedbacks can be summarized for a better crowdsourcing platform.

The attributes or variables which are the factors of crowdsourcing like a driver, type of vehicle and delivery rate and time have been discussed in conditions in Table 2. The two conditions are of different types in each factors portraying the two different possible ways of delivery.

Table 2 Last mile delivery attributes

Variables/attributes	Condition 1	Condition 2 Route
Driver	Professional/occasional	Full time/part time optimization
Type of vehicle	Two wheeler	Four wheeler
Delivery rate	Bid	Fixed
Pickup condition	Driver/owner decision	Third-party decision
Delivery time	Days/hour	Week/days
Rating	Star	Feedback
Daily/weekly travel	Km	Meter

Source Author compilation (2018)

2.5 Threats in the Rural E-commerce Supply Chain Network

The positives which are present in the attractive opportunity also carry along with them some serious threats and challenges. The logistics sector in India has several complexities and problems which have the latent power to decrease the business opportunity. The problems include major inefficient transportation, poor condition, and infrastructure of a storage facility, recent tax structure, i.e., Goods and Service Taxes (GST) is quite complex for many, obsolete technology usage and also the minimum rate of technology adoption according to [3]. The human factor which is inherently involved is the inadequate skills of the human resource involved in the process and their reluctant nature to get learning and training. The lack of awareness by the consumers in some rural areas is also a challenging issue. All these factors combined give a major jolt to the booming business expectations.

3 Key Drivers of E-commerce in India

High-speed Internet has brought a big boom in the E-commerce sector with more opportunities. The huge percentage of the population has a subscription to broadband Internet, and the number is increasing day by day. The rapid increase in the number of 3G Internet users with the recent beginning of 4G in India has promising hopes for the business. Other related factors are the explosive growth in the number of smartphone users which shows that more and more consumers can scroll the products and services in their hand. The mobile phone is no longer a device only meant for communication. It provides an amazing online shopping experience to the customers. The variety of product range offers a lot of options to the customers to choose from. They can sort the products or services according to the price, availability, specifications, etc. The standard of living has increased considerably with rising income levels, and this has paved the way for higher purchasing power. The competition between the rival

organizations also brings the price of the product or service to the least level as the
market economy is maintained mostly.

3.1 Local Service Providers

The local digital service providers, i.e., CSCs are the integral pillar for the e-
governance plan by Government of India. It was started in 2004 with a vision to
promote these centers as the face of the rural market. It is to be used as the front end
delivery spot for private players and which could employ rural citizens of India. The
CSC centers would not only the digital delivery point but act as a telecenter which
results financial or operational revenue generation [11]. The CSCs have a huge role
to play in this whole transaction. The Common Services Center (CSC) program is a
proposal of the Ministry of Electronics under the aegis of the Government of India.
CSCs are the point of contact for the delivery of electronic services to the rural sector
in India. They help in rural inclusion and financially contributing rural sector. Actually, CSCs are not just service delivery points in villages but are expected to position
as change agents. They can bring transformation by promoting rural entrepreneur-
ship. The rural potential can be uplifted, and employment can be generated for the
rural unemployed youths. They are the cornerstone of rural community participation.
The bottom-up approach provides a unique way in which the rural citizen can see a
better future by bringing in a social change. The inclusion of the rural scenario has
ushered in a new era of business with expanded boundaries for the E-commerce that
earlier had focused mainly on the urban sector. The Ministry of Telecommunications,
Government of India, has also introduced Wi-Fi Choupal to enhance the connectivity
of rural areas. The very objective and aim of the initiative are to connect Bharat-Net
infrastructure to provide broadband services to the Gram Panchayats in villages. The
number of Wi-Fi Choupal is expected to be around 5000 at the Gram Panchayat level
[22].

3.2 Postal Service Network in Rural India

The postal department has a role to play for the delivery of products from one
center to the other. Apart from posts, the services the department can include are the
delivery of goods and also suffice reverse logistics requirements. The reach of the
postal department is good in rural areas and is improving day by day. According to
the Ministry of Communication, GOI, there are 1,39,182 post offices in rural India
which accounts to around 90% of the total number [22]. This huge number opens
opportunities for more transactions and deliveries in the rural segment.

4 Methodology

4.1 Data Collection and Validation

The procedure of data collection initiated with framing pilot survey schedule for the rural respondents. The schedule is convenient for the villagers as some of them are illiterate or not interested to fill the forms themselves. It included open-ended questions which paved the way for the respondents to answer according to their wish and speed. The respondents were the villagers from the North Indian state of Uttar Pradesh, and the district was Allahabad. The urban CSCs visited were in Center 1 area in Allahabad city, and the rural one was in Center 2 Block in the district of Allahabad. Interviews were also conducted at the Common Service Centers (CSCs) to understand the demand and problems at the very place where the research is based. The names were kept anonymous and would remain confidential for any commercial use. Instead, only the responses will be used for research purpose. The outcome of the data collection and viewpoint provided evidence that in spite of individuals having transport vehicles of different types, they were not able to get the requisite and enough amount of business. They had the wish to work independently but lack of knowledge skills has held them back. They did not have the platform to proceed in the area of transportation (Table 3).

Table 3 Profile of rural and urban respondents

$N = 100$	Frequency	Urban	Rural
		Center 1	Center 2
Male	70	38	32
Female	30	22	08
Age 18–24 (Gen Z)	36	24	12
25–35 (Gen Y)	48	22	26
>36	16	09	07
No. of SIM card	04	02	02
Smartphone	92	52	40
Other (i.e., PAD, palmtop, laptop, etc.)	63	45	18
Crowd response	100	50	50
Transport type (vehicle ownership)	02	01	01
Family dependent	08	03	05
Family monthly income	₹5,000 to ₹10,000	12	28
	More than ₹10,000	38	22

Source Data collection by author (2018)

5 Proposed Framework

5.1 Crowdsourcing Applications for Optimization of Supply Chain

The crowdsourcing application helps to bridge the digital divide between urban and rural usage. The rural population is still largely untouched by the technical developments in the outer world. This concept of crowdsourcing which leads to Crowdshipping has to be explained to them, and training along with knowledge needs to be imparted for their usage.

In depicted Fig. 5, the framework has been proposed for the implementation of digital supply chain distribution networks in rural India at CSC centers. The framework describes how the rural customer can order the product and get it delivered by Common Service Center (CSC) to him with the help of a Crowdshipping platform. When the order or demand of the product is generated, the information is updated automatically at the CSC and received by the E-commerce enterprise. A Crowdshipping platform is present in the form of Information Technology which maps the processes on a daily basis. The platform binds the stakeholders involved in the process and with the use of real-time communication tools sends the message of the requirement of delivery to the postal department or the individuals interested in Crowdshipping the product. These users are registered individuals who ply between the routes for their daily work and can deliver the product for an extra income. They have social capital and trust within the area which is a must for such services. They

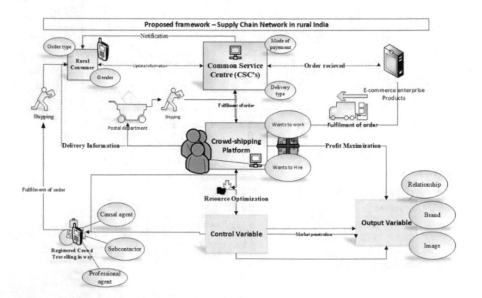

Fig. 5 Proposed framework for rural supply chain network in India. *Source* Author owns

can pick the product and deliver it to the customer who has ordered it initially. The information of delivery requirement is floated by the Crowdshipping platform. The ant colony algorithm is used for the optimization technique to save time and cost. The tracking of the delivery can be done by GIS maps, and this can help in understanding the exact location of the product. It also provides safety to the delivery. This system can help in reducing the cost of delivery and also the time because the transport is done in a convenient and usual manner without extra efforts for delivery [27]. The long-term impact of this quick and safe delivery is the relationship which will be developed between the customer and the E-commerce organization, and this image and goodwill will help the organization to grow in rural areas and have a big market overall. It also provides employment to rural youths.

6 Impacts

The direct impact of this research work is the implementation of the framework in the existing traditional model of transport for optimization. The impact can be seen in the extra income which can be gained by the rural citizens. The huge influx of rural populace to the urban cities for employment and other facilities can be reduced as such amenities, and living standard is expected to be delivered in the village itself. The number of rural entrepreneurs can increase with such idea of transport services and with the help of CSCs. The migration of labors for the delivery process to the urban market can be mitigated if the movement of E-commerce increases in the rural sector. The indirect and gradual impact will be the increase in sales of the E-commerce enterprises as their customer base has increased in the rural market. The better resource utilization in the rural area will give rise to productivity which in turn will bring income to the rural businesses.

7 Conclusions

The paper discusses the scope for growth in E-commerce enterprises at a rural market and how crowdsourcing can bring in changes with respect to time and cost. The rural sector is still unorganized in India, and it will take a gradual approach to bring in changes. With more brand awareness and consciousness, the demand is always going to high, and it can be met in smart ways crowdsourcing being one. The framework discussed describes overall how the demand for an order from a rural customer is met by crowdsourcing and fulfill by CSCs centre. The tracking feature will help in knowing the exact position of the delivery. The macro environment in reach of goods is captured by Chinese markets, and so with 'Make in India' Project the delivery and transport also should support it. Our aim is to show how this different way of delivery can help rural youth and customers. Poor resource utilization is one of the big reasons why rural India is suffering from the problems of poverty and unemployment. The literacy rate, healthcare and infrastructure can all be improved if more and more business is developed in the region. The economy of an urban unit depends on the productivity and the volume of sales according to the demand it generates. It already has the advantage of low-cost labor which only needs to be trained for the specific business needs. The weaknesses like low connectivity and transport can be met gradually by self-sufficiency. Thus, this framework can be implemented for a better rural upliftment as the future of our country lies in villages and cities combined.

8   Future Work

This area of crowdsourcing is huge and has immense potential to grow. The planning and design of a Crowdshipping software architecture for implementation with entity relationship diagram and data flow diagram need to be developed. The ant colony algorithm is applied in a manner to optimize the path. With the help of the systematic and comprehensive literature review, the present status of research in this area can be understood, and future researches can be benefitted from it. The flowchart of the supply chain can be revised for better and more economically viable solutions. Based on the framework and parameters, further data collection can be done to get the response from grassroot level which will pave the way for further empirical data analysis and interpretation. The present theoretical research work opens new vistas in this area of crowdsourcing as opportunities can be chalked out gradually for a better rural world. The proposed framework can be enhanced and personalized according to the industry and region.

Acknowledgements Apart from few academicians we contacted in Indian Institute of Management, Lucknow (IIM-L), for academic and research guidance, this theoretical work was possible only because of the extensive and timely support of the CSCs and the staff present there. The knowledge and the real-life ground realities explained to us by them helped us to understand the problems and opportunities at the grassroot level. The coordination with them provided us with valuable insights and has made the project possible.

References

1. Agatz, N., Erera, A., Savelsbergh, M., Wang, X.: Optimization for dynamic ride-sharing: a review. Eur. J. Oper. Res. **223**(2), 295–303 (2012)
2. Ashe, D.: Can couriers mobilize crowdsources for package delivery? daily crowdsource. http://dailycrowdsource.com/20-resources/projects/287-can-couriers-mobilize-crowdsourcers-forpackage-delivery (2017)
3. Kearney, A.T.: CSCMP study: supply chain. Trends & Implications for India (2014)
4. Bhatnagar, S.: Enhancing telecom access in rural India: some options. In: Emerging Market Forum, pp. 28–29. Asia Pacific Center, Stanford (2000)
5. Chen, C., Zhang, D., Wang, L., Ma, X., Han, X., Sha, E.: Taxi exp: a novel framework for city-wide package express shipping via taxi crowd sourcing. In: 2014 IEEE 11th International

Conference on Ubiquitous Intelligence and Computing and 2014 IEEE 11th International Conference on Autonomic and Trusted Computing and 2014 IEEE 14th International Conference on Scalable Computing and Communications and Its Associated Workshops, pp. 244–251 (2014)

6. Chowdhury, M.S.: Managing the freight deliveries in Manhattan: opportunities for collaborative city logistics measures. In: International Conference on Transportation and Development 2016, pp. 162–170 (2016)
7. Clewlow, R.R.: Carsharing and sustainable travel behavior: results from the San Francisco Bay Area. Transp. Policy **51**, 158–164 (2016)
8. Collins, A.T.: Behavioural influences on the environmental impact of collection/delivery points. In: Green Logistics and Transportation, pp. 15–34 (2015)
9. Dylan, A.: Can Couriers Mobilize Crowdsources for Package Delivery? Daily Crowdsource (2017)
10. Furuhata, M., Dessouky, M., Ordóñez, F., Brunet, M.E., Wang, X., Koenig, S.: Ridesharing: the state-of-the-art and future directions. Transp. Res. Part B: Methodol. **57**, 28–46 (2013)
11. Harris, R.W., Kumar, A., Balaji, V.: Sustainable telecentres? two cases from India. The Communication Initiative. http://www.com~nit.co~st2OO3/sld-7727.html (2007)
12. Howe, J.: The rise of crowdsourcing. Wired Mag. **14**(6), 1–4 (2006)
13. Hughes, S., Cohen, D.: Can online consumers contribute to drug knowledge? a mixed-methods comparison of consumer-generated and professionally controlled psychotropic medication information on the internet. J. Med. Internet Res. **13**(3) (2011)
14. IBEF Report: Fast moving consumer goods (FMCG). Retrieved from https://www.ibef.org (2017)
15. Kotler, P., Keller, K.L., Koshy, A., Jha, M.: Marketing Management: A South Asian Perspective, 13th edn. Pearson Education, New Delhi (2009)
16. Mangold, W.G., Faulds, D.J.: Social media: the new hybrid element of the promotion mix. Bus. Horiz. **52**(4), 357–365 (2009)
17. Mladenow, A., Bauer, C., Strauss, C.: Crowdsourcing in logistics: concepts and applications using the social crowd. In: Proceedings of the 17th International Conference on Information Integration and Web-based Applications & Services, p. 30 (2015)
18. Nielsen white paper report: emerging consumer demand: rise of the small town in India (2012)
19. Pan, S., Chen, C., Zhong, R.Y: A crowdsourcing solution to collect E-commerce reverse flows in metropolitan areas. In: 15th IFAC Symposium on Information Control Problems in Manufacturing INCOM (2012)
20. Punel, A., Stathopoulos, A.: Modeling the acceptability of crowdsourced goods deliveries: role of context and experience effects. Transp. Res. Part E: Logistics Transp. Rev. **105**, 18–38 (2017)
21. Rayle, L., Dai, D., Chan, N., Cervero, R., Shaheen, S.: Just a better taxi? a survey-based comparison of taxis, transit, and ride sourcing services in San Francisco. Transp. Policy **45**, 168–178 (2016)
22. Report on Wi-Fi Choupal. http://usof.gov.in/usof-cms/csc-wifi-choupal.jsp (2018)
23. Rougès, J.F., Montreuil, B.: Crowdsourcing delivery: new interconnected business models to reinvent delivery. In: 1st International Physical Internet Conference, pp. 1–19 (2014)
24. Scott, D.M. (ed.): The New Rules of Marketing and PR: How to Use Social Media, Online Video, Mobile Applications, Blogs, News Releases, and Viral Marketing to Reach Buyers Directly, Wiley (2015)
25. Shaheen, S.A., Cohen, A.P.: Growth in worldwide carsharing: an international comparison. Transp. Res. Rec. **1992**(1), 81–89 (2007)
26. Stanley, C., Winschiers-Theophilus, H., Onwordi, M., Kapuire, G.K.: Rural communities crowdsource technology development: a Namibian expedition. In: Proceedings of the Sixth International Conference on Information and Communications Technologies and Development: Notes, vol. 2, pp. 155–158 (2013)
27. Upadhyay, C.K., Pandiya, B., Tewari, V.: Economical branding approach for entrepreneurs to achieve sustainable development. In: ICSM Conference Proceedings, International Conference on Sustainable Management, IIM, Kashipur (2018)

28. Weinelt, B.: World Economic Forum White Paper Digital Transformation of Industries: Logistics. World Economic Forum (2016)
29. Xiao, J.: Development of city logistics in China. In: Contemporary Logistics in China. Springer, Singapore, pp. 139–162 (2017)
30. Yu, Y., Wang, X., Zhong, R.Y., Huang, G.Q.: E-commerce logistics in supply chain management: practice perspective. Proc. Cirp **52**, 179–185 (2016)
31. Zhao, Y., Zhu, Q.: Evaluation on crowdsourcing research: current status and future direction. Inf. Syst. Front. **16**(3), 417–434 (2014)

Role of Cloud Forensics in Cloud Computing

Shaik Khaja Mohiddin and Yalavarthi Suresh Babu

Abstract Some positive edges of cloud computing such scalability, elasticity, accessing the cloud from anywhere, and device-independent nature of cloud, has opened many doors for many intruders to carry out their mischief acts with respect to the data stored in the clouds. It is sometimes very difficult to trace out the exact fact what has happened with data, due to region effect, due to legal issues, and due to technical issues are the main and huge hurdles that exist in the present scenario; also there lies a main drawback that the present direct cloud forensics methods are not too much effective to overcome the problem. One among the buzzing technologies in today's techno world is cloud forensics, which has its own importance in many aspects due to exponential growth to data, to reduce maintenance cost, to provide security measures to the data, and these things make the individual persons to move and attract toward cloud concept, as cloud can go with these things easily where the client has to pay for the desired flavors of services which they want to utilize. As the huge and huge amount of data is being accumulated at a place due to which it sometimes compromise may be arose when cloud service provider do not take standard measures to overcome this.

Keywords Cloud forensic (CF) · Cloud service providers (CSP) · Cloud forensic tools (CFT)

S. K. Mohiddin (✉)
Department of CSE, Acharya Nagarjuna University, Namburu, Guntur,
Andhra Pradesh, India
e-mail: mail2mohiddin@gmail.com

Y. S. Babu
Deparment of CSE, JKC, Guntur, Andhra Pradesh, India
e-mail: Yalavarthi_S@yahoo.com

© Springer Nature Singapore Pte Ltd. 2020   91
K. N. Das et al. (eds.), *Soft Computing for Problem Solving*,
Advances in Intelligent Systems and Computing 1057,
https://doi.org/10.1007/978-981-15-0184-5_9

1   Introduction

The organization of this chapter is carried out in four sections starting with the introduction of forensics in cloud, broad challenges of cloud forensics which are divided into several sub categories, various tools which are used during the forensic investigation in the cloud which are a part of digital forensics, and importance of dedicated cloud forensic framework.

Introduction to forensics in cloud: cloud forensics is the combination of digital forensic and cloud computing, when certain mischief happens with the data stored in the cloud then cloud forensic is the scenario where one can get relevant information regarding what has been carried out with respect to the clients data which is stored in the cloud (Fig. 1; Table 1).

Steps involved in the cloud forensic process: Cloud forensic is carried out by the following steps:

- **Identification: Before** starting the forensic investigation, one has to exactly find out whether really some mischief has happened or not after conformation and identification then investigator can start his investigation.

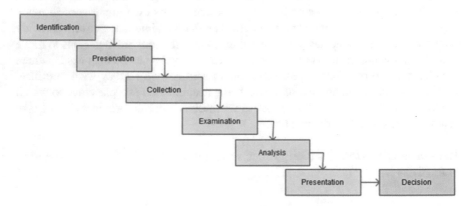

Fig. 1 Steps to be carried out during the forensic process in the cloud

Table 1 Showing the basic advantages and disadvantages among various clouds [1]

Nature\type of cloud	Public cloud	Private cloud	Hybrid cloud
Advantages	Easy to implement	Gives privilege for complete control	Highly cost effective
	Utilization efficiency is through server	Utilization efficiency is through server	Utilization efficiency is through server
Disadvantages	For long-term expensive	More upfront costs	It required complex management schemes
	Suitable for prolonged services	Suitable for prolonged services	A dedicated data center is required

- **Preservation**: The collected clues during the investigation play an important role during the investigation process so they should be preserved carefully for the future references.
- **Collection**: Whatever were the clues which are collected during the investigation should be preserved properly.
- **Examination**: All the collected clues during the forensic investigation in the cloud should be examined properly to get a conclusion.
- **Analysis**: The collected, examined data should be analyzed properly so that a conclusion is derived by the investigator on what might be happened with the data by the intruder.
- **Presentation**: As the collected information is in the form of technical information which has to present in the cyber court where some judge may not be aware of the technical terminology completely so the presentation plays a vital role.
- **Decision**: Depending on the analyzed and presented data before the court, decision is taken by the court which has to be followed strictly and the accused should be convicted.

2 Applying Three-Dimensional Concepts for Cloud Forensics

When major cloud service providers such as Google, Amazon, and Salesforce.com are compared among themselves, common aspects is noticed that these cloud tycoons have extended their cloud data centers throughout the world and they provide the services on the basis of cost effectiveness and service availability; data in the cloud centers are being replicated among other data centers which are located in various jurisdictions so that during an unexpected failure, they can have the backup of their relevant data. The way, in which these service providers deals with the customers during the forensics concepts differ from each other; the emergence of multi- tenancy and multi-jurisdiction strengthens to have them as a default setting for cloud forensic. When a problem is being encountered, these cloud service providers have their different approaches to overcome the same problem. When certain mischief is happened with respect to the data using cloud forensic, one can trace out the cause for the mischief and related information regarding the status of data. During the early developments of cloud forensics, it was assumed to be associated with three dimensions such as legal, technical, and organizational (Fig. 2; Table 2).

3 Cloud Forensic Challenges

Various challenges which are being faced in cloud forensics are mainly categorized into the following and again each main category has certain sub category depending

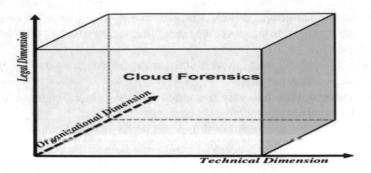

Fig. 2 Three dimensions of cloud forensic

Table 2 Three-dimensional cloud forensics and their associated parameters

Cloud forensic dimensions	Associated with parameters
Technical	• They compromise the tools and the techniques which are being carried out during forensic process in the cloud
Organizational	• **Investigators**: They carry out the examination when certain misconduct is carried out with the cloud data they should be an expertise person
	• **IT professionals**: They provide technical support to the investigators during the investigation by the cloud forensic expert
	• **Incident Handlers**: When certain incidents are happening with respect to data leakage in the cloud, breach of data, when cloud data effected by malicious codes they play a vital role in the above said situations
	• **Legal Advisors**: They deal with legally issued pertaining to the cloud such as multi-tenancy and jurisdictional issues so that forensic activities should not disturb the integrity of other's data stored in the cloud
	• **External Assistance**: External parities and CSP should be taken help during the forensic investigation process
Legal	• Here, development with respect to the SLA agreements between the CSP and clients along with certain regulations which assures that there is no breach carried out when the investigation on the data stored in the cloud is carried out [2, 3]

on the type of forensic challenge they may fall, they are being categorized in that manner. Out of the listed challenges, they are the general challenges which are being faced during forensic investigation and depending on the kind of attack on the data by the intruder these challenges may vary in certain methods. All the challenges are being faced and in some, only a few are being faced by the investigator but on an overall, the following are the general challenges which are nearly 65 which are explained as follows. Depending on the complexity, the data forensic process is carried out, i.e. attack done on what kind of data whether the data is confidential data, clients individual data, organizational data, or country-related data, in the cloud along with the forensic investigating data there is other data of other clients so care should be taken by the investigator not to disturb the integrity of others data.

CF main category	CF Sub category	Description of challenge
Architecture	General	• **Deletion in the cloud**: Client data may be some times deleted by the intruders, during his mischief with the data
		• **Recovering overwritten data**: sometimes intruder may be success to change the integrity of data, he may overwrite the data without leaving a clue, at that point, investigator has to face challenge to trace out the original data
		• **Interoperability issues among providers**: Same service provider at different locations of data centers in different countries may be imposing different rules as per the laws of the geo graphical location
		• **Single points of failure**: Even a single failure lead by the intruder during his act with the data may become a big clue for the investigator
		• **No single point of failures for criminals**: Sometimes criminals show their upper hand without leaving any evidence which can be a turning point to an investigator
		• **Detection of malicious act**: During his act, intruder may be successful some time to delete his traces which may be the clue for investigator to carry out the forensic process
		• **Criminals access to low-cost computing power**: Without expending much, intruders may carryout their required task to compromise the data in the cloud at low costs

(continued)

(continued)

CF main category	CF Sub category	Description of challenge
		• **Real-time investigation intelligence processes not possible**: One may not know at what time on which data at which data center of cloud an intruder may carry out his mischief act so it is difficult to trace out such a situation
		• **Malicious code may circumvent VM isolation methods**: To gain access to the clients data intruder in his act may compromise VM by running a malicious code, which is a positive thing for him to
	Multi-tenancy	• **Errors in cloud management portal configurations**: During the management or maintenance of cloud with respect to the customers data stored in the cloud, which gives a change for intruder to do his task easily
		• **Potential evidence segregation**: It may be difficult to collect the data in cloud where along with the effected client data related to other clients is also available, if investigation is carried out then it may sometimes lead to the integrity of others data stored in the same cloud
		• **Boundaries**: While accessing the data stored in the cloud there should be no boundaries regarding the data stored in the cloud
	Data segregation	• **Potential evidence segregation**: The key evidences which are collected during the investigation by the investigator should be preserved well

(continued)

(continued)

CF main category	CF Sub category	Description of challenge
	Provenance	• **Secure provenance**: Though different clouds have different data centers at different geo graphical locations but the same should be maintained from a single point of contact, i.e. from a single place only anything to be updated with respect to cloud either with respect to security or policies or any other should be carried out
		• **Data chain of custody**: Every data that is collected during the investigation should be related to the other
Data collection	General	• **Decreased access and data control**: Clients data which is stored in cloud should be accessed through a protected manner
		• **Chain of dependencies**: During the investigation, the clues which are being collected should be carried out in the order, i.e. there is a link between one step of evidence to the other if the collected evidence is the correct one then
		• **Locating evidence**: With respect to the mischief, proper and correct evidence should be traced out, which would be helpful to take necessary steps during the investigations and that evidence should be a clue to the next step
		• **Data location**: As replicated, data is available in the cloud data centers located at different places it is important to indentify at which location of data, intruder has done certain mischief
		• **Imaging and isolating data**: Every time, it is good evident to have the images of data so that one can easily identify that where the data that was stored by the client in the trusting cloud is being isolated or not

(continued)

(continued)

CF main category	CF Sub category	Description of challenge
		• **Data available for a limited time**: At certain point of time, during the cloud forensic process, it may take certain time to carryout examination of the mischief data but the time limit is also important because the mischief data may be lost after certain time, as designed by the intruder so investigator must tune his task to complete them within the stipulated time otherwise data will be lost and he has nothing to do then
		• **Locating storage media**: Investigator has to trace out first where the data is located among several available cloud data centers to carry out his crucial steps to move forward for investigation
		• **Evidence identification**: This is the crucial step that has to be carried out during the forensic process, i.e. to collect the required evidence which plays a vital role in the forensic process; if correct evidence is traced out then one can easily find the cause and trace the exact fact for the mischief of data done by intruder
		• **Dynamic storage**: When the situation demands and during certain natural disasters, a replica of the same copy of data should be maintained properly within the data centers of the cloud service providers
		• **Live forensics**: Certain point of time, investigator has to face the situation where he has to carry out his task on the live data where the intruder is doing his task
		• **Resource abstraction**: The resources which are necessary for the collection of evidence during the cloud forensic process should be collected fastly

(continued)

(continued)

CF main category	CF Sub category	Description of challenge
		• **Selective data acquisition**: Some times during the investigation, investigator has to deal with an organizational data which may be of huge size, at that time, it would be very beneficial to him if he successfully traces out the exact part of the data, where the intruders have played there task
		• **Cryptographic key management**: User credential along with the data should be stored in the cloud in a cryptographic manner and every time during the login this could protect some efforts of intruders
	Data integrity	• **Ambiguous trust boundaries**: CSP should provide full support to the investigator during investigation without any litigations
		• **Data Integrity and evidence preservation**: Integrity of data should be maintained in the cloud and collected evidence with respect to the data should be preserved
		• **Root of trust**: There should be a trust believe between CSP and client along with this a good trust between CSP and investigation agency or forensic investigator
Analysis	General	• **Evidence correlation**: Evidences which are collected during the forensic process should be co-related
		• **Reconstructing virtual storage**: Copies of the same data of client in the cloud should be stored in the virtual storage which would be helpful during the investigation by the investigator
	Metadata logs	• **Timestamp synchronization**: The time gap between the mischief of data and the time to carry out the steps to overcome is very important

(continued)

(continued)

CF main category	CF Sub category	Description of challenge
		• **Log format unification**: Different cloud service providers follow different log format for their client that should be unique so that forensic investigation can be carried out easily
	Metadata	• **Use of metadata**: Information related to client from SLA to log details and his log in credentials should be maintained properly
		• **Log capture**: The log in details of all the authenticated users must be maintained properly which gives a major key during investigations. It is one of the old concepts in cloud forensics, Sometimes with anti-forensics intruders may compromise with respect cloud centers and carry out their task easily, well that should not be given a chance to them
Incident first responders	General	• **Competence and trust worthiness**: Due to competence among various cloud service providers, there should be trustworthiness for cloud service providers and investigators which is not carried out properly
Legal	Contract/SLA	• **Missing terms in contract or SLA**: When a client enrolls in a particular cloud, he should go with the service level agreement and asks the cloud provider whether the standards are sufficient during an cyber attack on cloud
	General	• **Limited investigative power**: Due to reputation of the cloud, CSP does not support the investigators to a full-fledged manner to carry out the investigation they may be some times compromised by the intruders

(continued)

(continued)

CF main category	CF Sub category	Description of challenge
		• **Reliance on cloud providers**: Proper trust should be created by the cloud providers to their clients that they have the necessary steps which could be carried out during the investigation process, but that is not actually done by them
		• **Physical data location**: Original copy of data in the cloud which is replicated to other data centers of the same cloud should be kept in a protected manner
		• **Port protection**: The port from where the data is being accessed
		• **Transfer protocol: standard**: Protocols should be followed by every CSP while the client is dealing with his data within the cloud
		• **E-discovery**: An e-statement of client data and his details should be maintained properly which a copy should be sent to the client and a copy to cloud providers
	Jurisdiction	• **Lack of international agreements and Laws**: As cloud providers have extended their services a par from location so there should be an agreement between the countries in different jurisdictions so that if certain mischief happens with the data stored in other data center which is stored in other country or other place not related to the present place
		• **International cloud services**: While dealing with international cloud services, various cloud providers have extended their services to other places also
		• **Jurisdiction**: Depending on the location of cloud data centers, the jurisdiction falls in that area only

(continued)

(continued)

CF main category	CF Sub category	Description of challenge
		• **International communication**: The data stored in different data centers at other geographical location should be well communicated
	Privacy	• **Confidentiality**: As each cloud user/client is given user credential which should not be disclosed to others
	Ethical	• **Reputation fate**: During the investigation, cloud providers may think that if other clients know that certain mischief has carried out in the cloud which they are using they may withdraw their accounts
Role management	Identity management	• **Identifying account owner**: Owner authentication is needed every time during the usage of the cloud and also it is important to gather the information about the cloud accounts which a client has in different clouds and every time he has to change his log credentials
	General	• **Fictitious identities**: Anything which is suspected should be noticed properly
		• **Decounting user credentials and Physical locations**: Every time periodically user's credentials should be changed with a prior message to the client
		• **Authentication and access control**: Only authenticated users should be given access to their data and periodically they should maintain the changing of their user credentials
Standards	General	• **Testability, validation and scientific principles not addressed**: In most of the cloud providers during their SLA with the client they would not reveal the standards which they follow to recover or to overcome mischief

(continued)

(continued)

CF main category	CF Sub category	Description of challenge
	No single process	• **Lack of standard processes and models**: There are lagging in standard process which has to be framed
Training	General	• **Limited knowledge of logs and records**: A particular record for the logs in cloud by the users should be maintained
	Qualification and certification	• **Cloud training for investigators**: Lack of skilled investigators is a major problem because proper trained and experienced investigator may solve a critical situation in easy manner

4 Vulnerabilities Traced Cloud

During attacks on cloud by the intruders, they can be traced out using various forensic analysis tools which are being used in the cloud during any kind of cyber attack [4]. As shown below, both Tables 3 and 4 are obtained. Various common vulnerabilities which are traced out in cloud are discussed as follows (Fig. 3).

5 Threats in Cloud Computing

There are certain commonly observed threats in cloud computing are discussed in the Table 4.

6 Conclusion and Future Work

In the present scenario, cloud forensics is carried out with respect to digital forensic and cloud computing, every time when certain forensic related tasks has to be carried out in the cloud then one has to relive on digital forensics, where digital forensics is assumed to be a part of cloud forensics, where digital forensic methods are applied

Table 3 Common vulnerabilities observed in cloud

S. no.	Vulnerabilities in	Cause	Which service is affected
1	Virtual network	When virtual bridges are shared among several virtual machine [5]	IaaS
2	Virtual machine image	When VM images are placed in public repositories [6]. Due to dormant nature of VM images they are not patched [7]	IaaS
3	Hypervisors	Due to complexity of code at hypervisor [8]	IaaS
4	Virtual machine	Due to de-allocation and allocation of resources to VMs in an unrestricted manner [9]. Due to migration of VM in an uncontrolled manner [7, 10] by compromising VM IP address [11]	IaaS
5	Data	Sometimes data cannot be deleted completely [12–14]	SaaS, PaaS, IaaS
6	Resource allocation which is not limited	Leads to overbooking due to unequal modeling of resources [14]	SaaS, PaaS, IaaS

with certain tools which gives a sound results with respect to digital method. Cloud forensic requires its own dedicated framework because if there exists a dedicated framework of cloud forensic, investigator can tailor the needs of his requirements as they are all ready being added in the tools of cloud forensics framework but if it is the case of carrying out cloud forensic using digital forensics then it requires certain skills to tailor digital forensics with respect to the cloud. There is also a huge requirement of skills for the investigators who are dealing with cloud forensic concepts. With respect to jurisdictions, there should be common laws that should be imposed without giving barriers and a lot has to be carried out with respect to cyber laws and cyber judicial content because there is requirement to present the cloud crime scene in front of judicial persons who may or may not have a sound technical knowledge. Beside these there should be a full co-operation of CSP to the investigators during the investigation. As there is a huge data segregation is carried out in the cloud so more standards have to be increased in the cloud with respect to forensic perspective also, and it should be done mandatory for forensic as services along with the regular services offered by the cloud referring to the various cyber crime reports collected through various resources points to the fact that one has to increase the concept of forensic readiness either the cloud provides the services to an organizations with respect to single or multiple users, and data with respect to any

Table 4 Common threats observed in cloud

S. no.	Threat	Description	Which service is affected
1	Denial of service	Resources made busy and are unavailable to real user during need	IaaS
2	Data leakage	Happens when data goes to the hands of unauthorized persons [11, 15]	SaaS, PaaS, IaaS
3	Data scavenging	Though data is deleted, still it remains in the devices which the hacker takes the advantage [15, 16]	SaaS, PaaS, IaaS
4	Account and service hijacking	Due to weak credential attacker gains access to users data [17]	SaaS, PaaS, IaaS
5	Customer data manipulation	Data is manipulated during users attack on Web application with the help of server application [18]	
6	VM hopping	Here, one VM gains access over the other VM [19]	IaaS
7	Malicious VM creation	Intruder after gaining access to a VM, he creates an image of VM and injects Trojan horse in it [13, 20]	

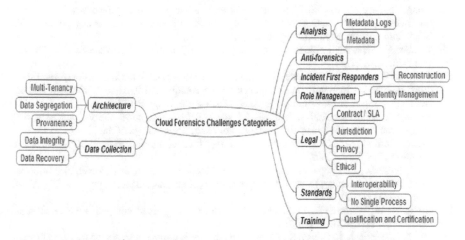

Fig. 3 NIST mind map for cloud computing

perspective should be given the same importance. One has to overcome these defects before the question in forensic heads toward more complexity levels, and trust of people no more pay attention toward cloud providers.

References

1. Hu, et al.: Advantages and disadvantages of different cloud types (2011)
2. Broadhurst, R.: Developments in the global law enforcement of cyber crime. Policing Int. J. Police Strat. Manage. **29**(2), 408–433 (2006)
3. Liles, S., Rogers, M., Hoebich, M.: A survey of the legal issues facing digital forensic experts. In: Peterson, G., Shenoi, S. (eds.) Advances in Digital Forensics V. Springer, Heidelberg, Germany, pp. 267–276 (2009)
4. Khaja Mohiddin, S., Yalavarthi, S.B., Kondragunta, V.: An analytical comparative approach of cloud forensic tools during cyber attacks in cloud. In: Bansal, J., Das, K., Nagar, A., Deep, K., Ojha, A. (eds.) Soft Computing for Problem Solving. Advances in Intelligent Systems and Computing, vol. 817. Springer, Singapore (2019)
5. Wu, H., Ding, Y., Winer, C., Yao, L.: Network security for virtual machine in cloud computing. In: 5th International Conference on Computer Sciences and Convergence Information Technology (ICCIT), pp. 18–21. IEEE Computer Society, Washington, DC, USA (2010)
6. Morsy, M.A., Grundy, J., Müller, I.: An analysis of the cloud computing security problem. In: Proceedings of APSEC 2010 Cloud Workshop. APSEC, Sydney, Australia (2010)
7. Garfinkel, T., Rosenblum, M.: When virtual is harder than real: security challenges in virtual machine based computing environments. In: Proceedings of the 10th Conference on Hot Topics in Operating Systems, Santa Fe, NM, vol. 10, pp. 227–229. USENIX Association Berkeley, CA, USA (2005)
8. Wang, Z., Jiang, X.: HyperSafe: a lightweight approach to provide lifetime hypervisor control-flow integrity. In: Proceedings of the IEEE Symposium on Security and Privacy, pp. 380–395. IEEE Computer Society, Washington, DC, USA (2010)
9. Winkler, V.: Securing the cloud: cloud computer security techniques and tactics. Elsevier Inc., Waltham, MA (2011)
10. Dawoud, W., Takouna, I., Meinel, C.: Infrastructure as a service security: challenges and solutions. In: The 7th International Conference on Informatics and Systems (INFOS), Potsdam, Germany, pp. 1–8. IEEE Computer Society, Washington, DC, USA (2010)
11. Ristenpart, T., Tromer, E., Shacham, H., Savage, S.: Hey, you, get off of my cloud: exploring information leakage in third-party compute clouds. In: Proceedings of the 16th ACM Conference on Computer and Communications Security, Chicago, Illinois, USA, pp. 199–212. ACM, New York, NY, USA (2009)
12. Townsend, M.: Managing a security program in a cloud computing environment. In: Information Security Curriculum Development Conference, Kennesaw, Georgia, pp. 128–133. ACM, New York, NY, USA (2009)
13. Grobauer, B., Walloschek, T., Stocker, E.: Understanding cloud computing vulnerabilities. IEEE Secur. Priv. **9**(2), 50–57 (2011)
14. Ertaul, L., Singhal, S., Gökay, S.: Security challenges in cloud computing. In: Proceedings of the 2010 International conference on Security and Management SAM'10, pp. 36–42. CSREA Press, Las Vegas, USA (2010)
15. ENISA: Cloud Computing: benefits, risks and recommendations for information security (2009)
16. Jansen, W.A.: Cloud hooks: security and privacy issues in cloud computing. In: Proceedings of the 44th Hawaii international conference on system sciences, Koloa, Kauai, HI, pp. 1–10. IEEE Computer Society, Washington, DC, USA (2011)

17. Jasti, A., Shah, P., Nagaraj, R., Pendse, R.: Security in multi-tenancy cloud. In: IEEE International Carnahan Conference on Security Technology (ICCST), KS, USA, pp. 35–41. IEEE Computer Society, Washington, DC, USA (2010)
18. Cloud Security Alliance: Top threats to cloud computing V1.0. Available: https://cloudsecurityalliance.org/research/top-threats (2010)
19. https://www.owasp.org/index.php/Category:OWASP_Top_Ten_Project (2010)
20. Hashizume, K., Rosado, D.G., Fernández-Medina, E., et al.: J. Internet Serv. Appl. **4**, 5 (2013). https://doi.org/10.1186/1869-0238-4-5

Hydromagnetic Squeeze Film in a Longitudinally Rough Conducting Conical Plates

Jatinkumar V. Adeshara, M. B. Prajapati, G. M. Deheri and R. M. Patel

Abstract This article wishes to study the presentation of the longitudinally rough and hydromagnetically conducting conical plates. Here, both the plates are chosen to be conducting electrically while an electrically conducting lubricant fills the clearance space between the plates. A transverse magnetic field is applied. Christensen and Tonder's used stochastic averaging process regarding roughness, the associated stochastically averaged Reynolds' type equation is resolved. This gives pressure and consequently, the load taking capacity. It is indicated by the graphical results that an augmented performance is registered by the bearing system in comparison with traditional fluid-based lubrication. The results demonstrate that longitudinal roughness is more helpful as compared to transverse roughness. Here, the role of standard deviation remains crucial even if suitable numerical values of the angle of the cone are considered. Thus, this investigation gives ample measures for bettering the bearing performance with appropriate choice of hydromagnetization and conductivity parameters.

Keywords Conical plates · Squeeze film with hydromagnetization · Longitudinal roughness · Electrical conductivity · Load bearing capacity

J. V. Adeshara · M. B. Prajapati
Department of Mathematics, Hemchandracharya North Gujarat University,
Patan, Gujarat 384 265, India
e-mail: adesharajatin01@gmail.com

G. M. Deheri · R. M. Patel (✉)
Department of Mathematics, Sardar Patel University, Vallabh Vidyanagar, Gujarat 388 120, India
e-mail: rmpatel2711@gmail.com

R. M. Patel
Gujarat Arts and Science College, Ellis bridge, Ahmedabad, Gujarat 380 006, India

© Springer Nature Singapore Pte Ltd. 2020   109
K. N. Das et al. (eds.), *Soft Computing for Problem Solving*,
Advances in Intelligent Systems and Computing 1057,
https://doi.org/10.1007/978-981-15-0184-5_10

Nomenclature

x, y Cartesian coordinates
H Lubricant film thickness
$\dot{h}$ Squeeze film velocity
μ Viscosity
B_0 Applied (transverse) magnetic field between both plates
s Lubricant conductivity
M $B_0 h \left(\frac{s}{\mu}\right)^{1/2}$ = Hartmann number
ω Cone's Vertical angle
p Lubricant pressure
w Load bearing capacity
σ^* Standard deviation in non-dimensional form (σ/h)
α^* Non-dimensional variance (α/h)
ε^* Skewness in non-dimensional form (ε/h^3)
P Dimensionless pressure
W Dimensionless load bearing capacity.

1 Introduction

One comes across various theoretical and experimental investigations of hydromagnetic pressurization [11, 12]. The behavior of hydromagnetic squeeze films between two non-porous surfaces was discussed by Shukla and Prasad [24] and studied the effect of surface conductivity on the bearing performance. In Patel and Gupta [17] applied Morgan–Cameron calculation and simplified the investigated the hydromagnetic squeeze films between two plates for different shapes.

Many investigations deal with the surface roughness effect [5–9, 26]. Stochastic averaging model of Christensen and Tonder [6–8] was applied in several studies [4, 13, 20, 25]. Transverse roughness pattern in the presence of a magnetic fluid has been the matter of investigation in a number of articles. Patel and Deheri [18] analyzed the behavior of annular plates on surface roughness for magnetic fluid-based squeeze film. A magnetohydrodynamic squeeze film characteristic between curved annular plates was studied by Lin et al. [15]. Performance of squeeze film formed by magnetic fluid on transversely rough slider bearing was done by Deheri et al. [10]. Patel and Deheri [19] discussed the effect of ferrofluid-based squeeze film on porous conical plates. Hydromagnetic squeeze film was studied by Vadher et al. [27] on truncated conical plates. Lin et al. [16] studied effect of fluid inertia forces on squeeze film with ferromagnetic model on conical plates.

Generalized form of longitudinal surface roughness was introduced by Andharia et al. [4] on hydrodynamic lubricated slider bearing. Roughness effect on the behavior of squeeze film on spherical bearing was analyzed by Andharia and Deheri [1].

Andharia and Daheri [2] discussed the effect on the ferrofluid-based squeeze film between conical plates having longitudinal surface roughness, which was extended by them [3] for elliptical plates. Shimpi and Deheri [23] extended the analysis of Andharia and Deheri [2] to consider the mutual effect of slip and deformation in truncated conical plates. It was observed that longitudinal roughness came to help the ferrofluid lubrication to counter the deformation effect when slip was moderate. The longitudinal roughness pattern has been a little bit sober as indicated by Lin [14] for magnetic fluid lubricated journal bearing. Vadher et al. [28] recorded an adverse roughness effect on the behavior of hydromagnetic squeeze film among accompanying conical plates with porosity.

Since somewhat a smaller number of discussions regarding squeeze film in conical plates, it was deemed proper to launch an investigation on the hydromagnetic squeeze film performance in longitudinally rough-conducting conical plates.

2 Analysis

Figure 1 indicates the squeezing film geometry and configuration for present study.

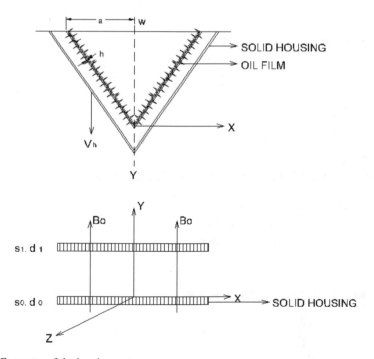

Fig. 1 Geometry of the bearing system

The upper plate moves along its normal toward a non-porous fixed plate. Both plates are electrically conducted and lubricant fills the space between two plates. A standardized transverse magnetic field is used among the conical plates.

The surfaces are assumed to be longitudinally rough. The lubricated film thickness is taken from Christensen and Tonder [6–8].

Under the standard considerations of hydromagnetic lubrication, the customized Reynolds' type equation governing the hydromagnetic flow for non-porous plates is given by Prajapati [21]

$$\frac{1}{x}\frac{d}{dx}\left(x\frac{dp}{dx}\right) = \frac{\mu\dot{h}}{\left[\frac{2}{M^3}\left[\frac{M}{2} - \tanh\frac{M}{2}\right]\right]\left[\frac{\varnothing_0+\varnothing_1+1}{\varnothing_0+\varnothing_1+\left(\tanh\frac{M}{2}\right)/\frac{M}{2}}\right]} \tag{1}$$

here ϕ_0 and ϕ_1 are permeability parameters of lower and upper surfaces.

Using averaging process as discussed in Christensen and Tonder [6–8] Reynolds' expression for squeeze film pressure turns out to be

$$\frac{1}{x}\frac{d}{dx}\left(x\frac{dp}{dx}\right) = \frac{\mu\dot{h}h^{-3}\left[1 - \alpha h^{-1} + 6h^{-2}(\sigma^2 + \alpha^2) - 10h^{-3}(\varepsilon + 3\sigma^2\alpha + \alpha^3)\right]}{\left[\frac{2}{M^3}\left[\frac{M}{2} - \tanh\frac{M}{2}\right]\right]\left[\frac{\varnothing_0+\varnothing_1+1}{\varnothing_0+\varnothing_1+\left(\tanh\frac{M}{2}\right)/\frac{M}{2}}\right]} \tag{2}$$

Applying Reynolds' type boundary conditions

$$p(a\operatorname{cosec}\omega) = 0$$

and

$$\frac{dp}{dx} = 0 \quad \text{at } x = 0 \tag{3}$$

on Eq. (1) one gets the squeeze film pressure distribution as

$$p = \frac{\mu\dot{h}h^{-3}\left[1 - \alpha h^{-1} + 6h^{-2}(\sigma^2 + \alpha^2) - 10h^{-3}(\varepsilon + 3\sigma^2\alpha + \alpha^3)\right]\left(\frac{x^2 - a^2\operatorname{cosec}^2\omega}{4}\right)}{\left[\frac{2}{M^3}\left[\frac{M}{2} - \tanh\frac{M}{2}\right]\right]\left[\frac{\phi_0+\phi_1+1}{\phi_0+\phi_1+\left(\tanh\frac{M}{2}\right)/\frac{M}{2}}\right]} \tag{4}$$

Use of the following dimensionless terms

$$\sigma^* = \frac{\sigma}{h} \quad \alpha^* = \frac{\alpha}{h} \quad \varepsilon^* = \frac{\varepsilon}{h^3}$$

leads to the pressure in non-dimensional form as

$$p = \frac{ph^3}{\mu \dot{h} \pi a^2 \mathrm{cosec}\,\omega} = \frac{\left[1 - 3\alpha^* + 6\left(\sigma^{*2} + \alpha^{*2}\right) - 10\left(\varepsilon^* + 3\sigma^{*2}\alpha^* + \alpha^{*3}\right)\right]\mathrm{cosec}\,\omega \left(1 - \frac{x^2 \sin^2 \omega}{a^2}\right)}{8\pi \left[\frac{2}{M^3}\left[\frac{M}{2} - \tanh \frac{M}{2}\right]\right]\left[\frac{\phi_0 + \phi_1 + 1}{\phi_0 + \phi_1 + \left(\tanh \frac{M}{2}\right)/\frac{M}{2}}\right]} \tag{5}$$

Then, the distribution of load bearing capacity of the system in dimensionless form is calculated as

$$
\begin{aligned}
W &= -\frac{wh^3}{\mu \pi^2 a^4 \mathrm{cosec}^2 \omega \dot{h}} \\
&= \frac{\left[1 - 3\alpha^* + 6\left(\sigma^{*2} + \alpha^{*2}\right) - 10\left(\varepsilon^* + 3\sigma^{*2}\alpha^* + \alpha^{*3}\right)\right]\mathrm{cosec}^2 \omega}{8\pi \left[\frac{2}{M^3}\left[\frac{M}{2} - \tanh \frac{M}{2}\right]\right]\left[\frac{\phi_0 + \phi_1 + 1}{\phi_0 + \phi_1 + \frac{\left(\tanh \frac{M}{2}\right)}{\frac{M}{2}}}\right]}
\end{aligned} \tag{6}
$$

where

$$W = 2\pi \int\limits_0^{a\,\mathrm{cosec}\,\omega} px\,\mathrm{d}x$$

3 Results and Discussions

The study of Prakash and Vij [22] can be reduced for smooth bearing cases, setting the parameter of magnetization is to be zero when non-conducting plates are in place. The effect of conductivities on load carrying capacity is decided by

$$\frac{1}{\frac{(\phi_0 + \phi_1 + 1)}{\phi_0 + \phi_1 + (\tanh \frac{M}{2})/\frac{M}{2}}}$$

When values (large) of M are involved above factor tends to

$$\frac{1}{\frac{(\phi_0 + \phi_1 + 1)}{\phi_0 + \phi_1}}$$

Therefore, the load bearing capacity gets increased with increasing values of $\phi_0 + \phi_1$.

The fact that the load carrying capacity gets enhanced by the M for various values of $\phi_0 + \phi_1$, σ^*, α^*, ε^* and ω, is reflected from Figs. 2, 3, 4, 5, and 6.

Possibly, this happens because of the fact that the magnetization increases the viscosity of the lubricant. The effect of conductivity is to increase the load bearing capacity which is found in Figs. 7, 8, 9, and 10. Initially, the increase in load bearing capacity is more.

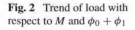

Fig. 2 Trend of load with respect to M and $\phi_0 + \phi_1$

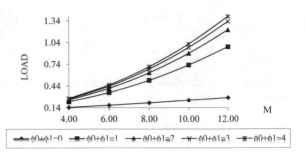

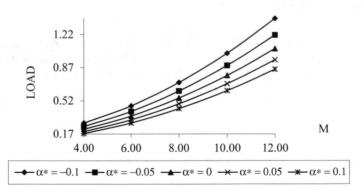

Fig. 3 Distribution of load with respect to M and σ^*

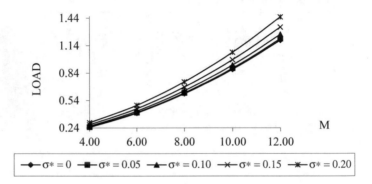

Fig. 4 Profile of load bearing capacity with regard to M and α^*

Interestingly, standard deviation (σ^*) helps in increasing the load bearing capacity which is unlikely in the case of transverse roughness. Figures 11, 12, 13, 14, 15, and 16 establish that the (−ve) skewed roughness augments the load as in the case of variance (−ve). These trends reverse for positively skewed roughness and variance (+ve).

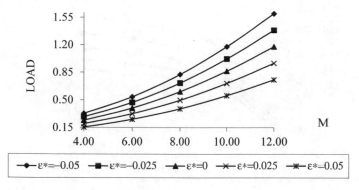

Fig. 5 Variation of load carrying capacity with respect to M and ε^*

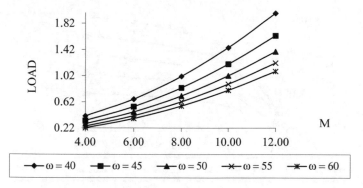

Fig. 6 Distribution of load with respect to M and ω

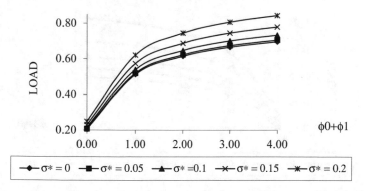

Fig. 7 Profile of load bearing capacity with respect to $\phi_0 + \phi_1$ and σ^*

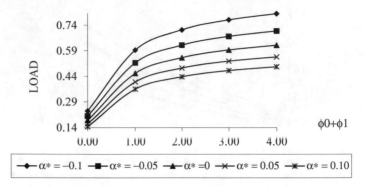

Fig. 8 Variation of load carrying capacity with respect to $\phi_0 + \phi_1$ and α^*

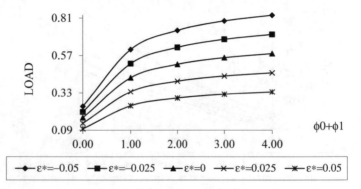

Fig. 9 Distribution of load with respect to $\phi_0 + \phi_1$ and ε^*

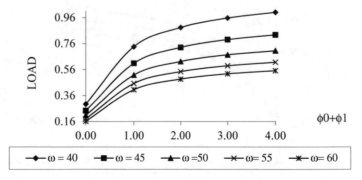

Fig. 10 Profile of load bearing capacity with regard to $\phi_0 + \phi_1$ and ω

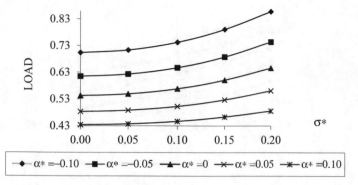

Fig. 11 Variation of load carrying capacity with respect to σ^* and α^*

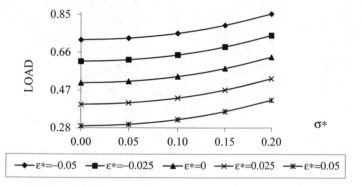

Fig. 12 Distribution of load with respect to σ^* and ε^*

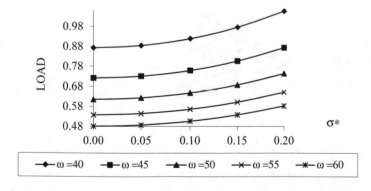

Fig. 13 Profile of load bearing capacity with regard to σ^* and ω

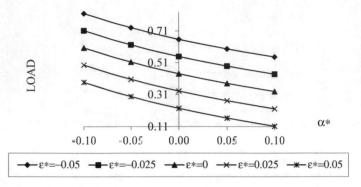

Fig. 14 Variation of load carrying capacity with respect to α^* and ε^*

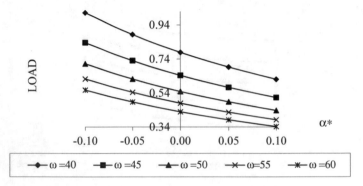

Fig. 15 Distribution of load with respect to α^* and ω

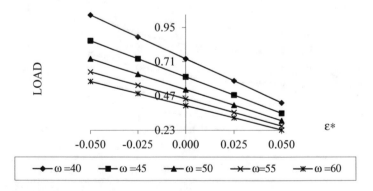

Fig. 16 Profile of load bearing capacity with regard to ε^* and ω

A close observation at some of the figures suggests that the collective effect of magnetization parameter and electrical permeability can help in compensating the unfavorable effect of positively skewed roughness and variance (+ve).

4 Conclusions

It is easily observed that the longitudinal roughness remains more favorable as compared to the transverse roughness case. In fact, the standard deviation makes all the difference. Hence, the roughness needs to be considered carefully while designing the bearing system. A close observation indicates that some amount of load is supported by the system even if no flow occurs. This does not occur in the case of traditional fluid-based system.

Acknowledgements Fruitful comments and constructive suggestions for improving the quality of the article are acknowledged with thanks.

References

1. Andharia, P.I., Deheri, G.M.: Effect of longitudinal surface roughness on the behaviour of squeeze film in a spherical bearing. Int. J. Appl. Mech. Eng. **6**(4), 885–897 (2001)
2. Andharia, P.I., Deheri, G.M.: Longitudinal roughness effect on magnetic fluid based squeeze film between conical plates. Ind. Lubr. Tribol. **62**(5), 285–291 (2010)
3. Andharia, P.I., Deheri, G.M.: Performance of magnetic fluid-based squeeze film between longitudinally rough elliptical plates. ISRN Tribol., 6. Article ID 482604 (2013)
4. Andharia, P.I., Gupta, J.L., Deheri, G.M.: Effect of longitudinal surface roughness on hydrodynamic lubrication of slider bearings. In: Proceedings of Tenth International Conference on Surface Modification Technologies, pp. 872–880. The Institute on Materials, Singapore (1997)
5. Burton, R.A.: Effects of two-dimensional, sinusoidal roughness on the load support Characteristics of a lubricant film. ASME J. Basic Eng. **85**, 258–264 (1963)
6. Christensen, H., Tonder, K.: Tribology of rough surfaces, Stochastic models of hydrodynamic lubrication. In: SINTEF, Section for Machine Dynamics in Tribology. Technical University of Norway, Trondheim, Norway, Report No. 10/69-18 (1969)
7. Christensen, H., Tonder, K.: Tribology of rough surfaces, parametric study and comparison of lubrication models. In: SINTEF, Section for Machine Dynamics in Tribology. Technical University of Norway, Trondheinm, Norway, Report No. 22/69-18 (1969)
8. Christensen, H., Tonder, K.: The hydrodynamic lubrication of rough bearing surfaces of finite width. In: ASME-ASLE Lubrication Conference Cincinnati. Ohio, Lub-7, 12–15 Oct 1970
9. Davis, M.G.: The generation of pressure between rough lubricated moving deformable surfaces. Lubr. Eng. **19**, 246 (1963)
10. Deheri, G.M., Andharia, P.I., Patel, R.M.: Transversely rough slider bearings with squeeze film formed by a magnetic fluid. Int. J. Appl. Mech. Eng. **10**(1), 53–76 (2005)
11. Dodge, F.T., Osterle, J.F., Rouleau, W.T.: Magnetohydrodynamic squeeze film bearings. J. Basic Eng. Trans. ASME **87**, 805–809 (1965)
12. Elco, R.A., Huges, W.F.: Magnetohydrodynamic pressurization in liquid metal lubrication. Wear **5**, 198–207 (1962)

13. Gupta, J.L., Deheri, G.M.: Effect of roughness on the behaviour of squeeze film in a spherical bearing. Tribol. Trans. **39**, 99–102 (1996)
14. Lin, J.R.: Longitudinal surface roughness effects in magnetic fluid lubricated journal bearing. J. Mar. Sci. Technol. **24**(4), 711–716 (2016)
15. Lin, J.R., Lu, R.F., Liao, W.H.: Analysis of magneto-hydrodynamic squeeze film characteristics between curved annular plates. Ind. Lubr. Tribol. **56**, 300–305 (2004)
16. Lin, J.R., Lin, M.C., Hung, T.C., Wang, P.Y.: Effects of fluid inertia forces on the squeeze film characteristics of conical plates ferromagnetic fluid model. Lubr. Sci. **5**(7), 429–439 (2013)
17. Patel, K.C., Gupta, J.L.: Behaviour of hydromagnetic squeeze film between porous plates. Wear **56**, 327–339 (1979)
18. Patel, R.M., Deheri, G.M.: Magnetic fluid based squeeze film behavior between annular plates and surface roughness effect. In: International Tribology Conference, pp. 631–638. Rome, Italy (2004)
19. Patel, R.M., Deheri, G.M.: Magnetic fluid-based squeeze film between porous conical plates. Ind. Lubr. Tribol. **59**(3), 143–147 (2007)
20. Prajapati, B.L.: Squeeze film behaviour between rotating porous circular plates with concentric circular pocket: surface roughness and elastic deformation effects. Wear **152**, 301–307 (1992)
21. Prajapati, B.L.: On certain theoretical studies in hydrodynamic and electromagnetohydrodynamic lubrication. Ph.D. thesis. S.P. University, Vallabh Vidyanagar, Gujarat, India (1995)
22. Prakash, J., Vij, S.K.: Load capacity and time height relations for squeeze film between porous plates. Wear **24**, 309–322 (1973)
23. Shimpi, M.E., Deheri, G.M.: Combine effect of bearing deformation and longitudinal roughness on the performance of a ferrofluid based squeeze film together with velocity slip in truncated conical plates. Imperial J. Interdiscip. Res. **2**(8), 1423–1430 (2016)
24. Shukla, J.B., Prasad, R.: Hydromagnetic squeeze films between two conducting surfaces. J. Basic Eng. Trans. ASME **87**, 818–822 (1965)
25. Ting, L.L.: Engagement behavior of lubricated porous annular disks part I: squeeze film phase, surface roughness and elastic deformation effects. Wear **34**, 159–182 (1975)
26. Tonder, K.C.: Surface distributed waviness and roughness. In: First World Conference in Industrial Tribology, p. 128, New Delhi, A3 (1972)
27. Vadher, P.A., Deheri, G.M., Patel, R.M.: A study on the performance of hydromagnetic squeeze film between two conducting truncated conical plates. Jordan J. Mech. Ind. Eng. **2**(2), 85–92 (2008)
28. Vadher, P.A., Deheri, G.M., Patel, R.M.: Performance of hydromagnetic squeeze films between conducting porous rough conical plates. Meccanica **45**, 767–783 (2010)

Performance of a Hydromagnetic Squeeze Film Between Longitudinally Rough Conducting Triangular Plates

Hardik P. Patel, G. M. Deheri and R. M. Patel

Abstract This study discusses the effect of longitudinal roughness on the performance of a hydromagnetic squeeze film in conducting triangular plates. A stochastic random variable characterizes the longitudinal roughness of the bearing surface. The associated Reynolds' equation is recourse to the stochastically averaging method of Christensen–Tonder, solving the Reynolds' equation with Reynolds' boundary conditions; the pressure is obtained which gives load profile as well. Unlike the transverse roughness case, here, it is found that the load bearing capacity increases due to the standard deviation related to roughness. This situation further improves with the involvement of negatively skewed roughness and variance ($-$ve). The effect of magnetization and conductivity elevates the situation further.

Keywords Triangular plates · Hydromagnetic lubrication · Reynolds' expression · Longitudinal roughness · Load bearing capacity

Nomenclature:

a Length of the sides
h Film thickness
$\dot{h}$ Squeeze film velocity

H. P. Patel
Department of Humanity and Science, L. J. Institute of Engineering and Technology, Ahmedabad, Gujarat, India
e-mail: hardikanny82@gmail.com

G. M. Deheri
Department of Mathematics, Sardar Patel University, Vallabh Vidyanagar, Gujarat 388120, India

R. M. Patel (✉)
Department of Mathematics, Gujarat Arts and Science College, Ahmedabad, Gujarat 380006, India
e-mail: rmpatel2711@gmail.com

© Springer Nature Singapore Pte Ltd. 2020
K. N. Das et al. (eds.), *Soft Computing for Problem Solving*,
Advances in Intelligent Systems and Computing 1057,
https://doi.org/10.1007/978-981-15-0184-5_11

121

μ Viscosity
B_0 Standardized transverse magnetic field incorporated between the plates
s Electrical conductivity of the lubricant
M $B_0 h \left(\frac{s}{\mu} \right)^{\frac{1}{2}}$ = Hartmann number
p Lubricant pressure
w Load carrying capacity
σ^* Non-dimensional standard deviation (σ/h)
α^* Non dimensional variance (α/h)
ε^* Non-dimensional skewness (ε/h^3)
P Dimensionless pressure
W Dimensionless load bearing capacity.

1 Introduction

Liquid metals (Mercury and Sodium) filled in between two conducting plates support heavy load by applying suitable magnetic field. The effect of external magnetic field on electromagnetic pressurization and corresponding load has been studied and scrutinized.

In liquid–metal lubrication, Elco and Huges [10] studied magnetohydrodynamic pressure. The behavior of magnetohydrodynamic squeeze films was analyzed by Kuzma [11]. The hydromagnetic theory for squeeze films in the presence of a transverse magnetic field was investigated by Shukla [20] to perform lubricants between two non-conductive non-porous surfaces. Patel and Gupta [14] used the approximation of Morgan–Cameron and simplified the analysis of a number of geometric shapes including parallel plates. Although, increasing the conductivity of the plate results in improved performance for circular plates [16]. All the investigations mentioned above established that the squeeze film enhanced due to magnetization. Besides, the conductivities of the plates play a key role in boosting the performance characteristics.

For both types of surface roughness, Christensen–Tonder [7–9] suggested a comprehensive analysis. The effect of surface roughness on squeeze film performance has been analyzed in Prajapati [15, 17], Andharia et al. [5, 6], and Patel et al. [13]. These studies underlined that the performance of squeeze film was significantly influenced by transverse surface roughness, mostly the influence being adverse. However, the negatively skewed roughness remained favorable from design point of view.

Effect of roughness on the behavior of squeeze film in a spherical bearing was analyzed by Andharia and Deheri [2]. Andharia and Daheri [3] discussed the effect on the ferrofluid-based squeeze film between conical plates having longitudinal surface roughness, which was extended by them [4] for elliptical plates. Shimpi and Deheri [19] extended the analysis of Andharia and Deheri [3] to consider the mutual effect of slip and deformation in truncated conical plates. It was observed that longitudinal roughness came to help the ferrofluid lubrication to counter the deformation effect when slip was moderate. The longitudinal roughness pattern has been a little bit sober as indicated by Lin [12] for magnetic fluid lubricated journal bearing. The effect of longitudinal surface roughness for hydromagnetic circular step bearing was discussed by Adeshara et al. [1]. All these investigations cleared that the effect of deviation (σ) remained crucial from bearing performance point of view. In addition as compared to transverse roughness, the longitudinal roughness proved to be more conducive from industry point of view.

Roughness effect on conducting porous triangular plates with hydromagnetic squeeze film was discussed by Vadher et al. [21]. The transverse surface roughness turned in somewhat negative influence on the squeeze film behavior.

Therefore, this article aims to analyze the effect of longitudinal roughness on the structure of Vadher et al. [21].

2 Analysis

Figure 1 shows the squeezing film geometry for present study.

The upper plate moves along its normal toward a non-porous fixed lower plate. Conducting lubricant is filled in between plates, which are electrically conducting. A standardized magnetic field is applied to triangular plates.

Roughness seems random in character and does not execute any pattern. The complexity of the geometrical structure affected by the randomness and multiple roughness scales. The bearings are considered to be longitudinally rough. Squeeze film thickness of the lubricant is chosen from Christensen and Tonder [7–9]. With usual suppositions of hydromagnetic lubrication, the improved Reynolds' type equation governing the hydromagnetic flow for non-porous plates is given by Prajapati [16]

$$\frac{\partial^2 p}{\partial x^2} + \frac{\partial^2 p}{\partial z^2} = \frac{\mu \dot{h}}{\left[\frac{2}{M^3}\left(\frac{M}{2} - \tanh \frac{M}{2}\right)\right]\left[\frac{\phi_0 + \phi_1 + 1}{\phi_0 + \phi_1 + (\tanh \frac{M}{2})/\frac{M}{2}}\right]} \tag{1}$$

here ϕ_0 and ϕ_1 are permeability parameters of lower and upper surfaces.

Applying averaging process as discussed in Christensen and Tonder [7–9] and Vadher et. al. [21], Reynolds' type equation for squeeze film pressure turns out to be

$$\frac{\partial^2 p}{\partial x^2} + \frac{\partial^2 p}{\partial z^2} = \frac{\mu \dot{h} h^{-3}\left[1 - 3\alpha h^{-1} + 6h^{-2}(\sigma^2 + \alpha^2) - 10h^{-3}(\varepsilon + 3\sigma^2\alpha + \alpha^3)\right]}{\left[\frac{2}{M^3}\left(\frac{M}{2} - \tanh \frac{M}{2}\right)\right]\left[\frac{\phi_0 + \phi_1 + 1}{\phi_0 + \phi_1 + (\tanh \frac{M}{2})/\frac{M}{2}}\right]} \tag{2}$$

Using Reynolds' type boundary conditions

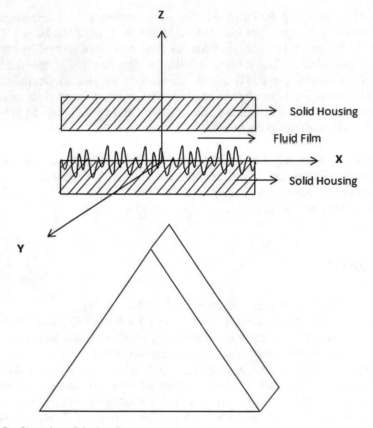

Fig. 1 Configuration of the bearing system

$$p(x, z) = 0$$

wherein,

$$(x - a)\left(x - \sqrt{3}z + 2a\right)\left(z + \sqrt{3}z + 2a\right) = 0 \tag{3}$$

on Eq. (2), one gets the squeeze film pressure distribution

$$p = \frac{\mu \dot{h} h^{-3}\left[1 - 3\alpha h^{-1} + 6h^{-2}\left(\sigma^2 + \alpha^2\right) - 10h^{-3}\left(\varepsilon + 3\sigma^2\alpha + \alpha^3\right)\right](x - a)\left(x - \sqrt{3}z + 2a\right)\left(x + \sqrt{3}z + 2a\right)}{\left[\frac{2}{M^3}\left(\frac{M}{2} - \tanh\frac{M}{2}\right)\right]\left[\frac{\phi_0 + \phi_1 + 1}{\phi_0 + \phi_1 + \left(\tanh\frac{M}{2}\right)/\frac{M}{2}}\right]} \tag{4}$$

Use of the following non-dimensional parameters

$$\sigma^* = \frac{\sigma}{h}, \quad \alpha^* = \frac{\alpha}{h}, \quad \varepsilon^* = \frac{\varepsilon}{h^3}$$

gives dimensionless pressure as

$$P = -\frac{ph^3}{\mu h 3\sqrt{3}a^2} = \frac{\left[1 - 3\alpha^* + 6\left(\sigma^{*2} + \alpha^{*2}\right) - 10\left(\varepsilon^* + 3\sigma^{*2}\alpha^* + \alpha^{*3}\right)\right]\left(1 - \frac{x}{a}\right)\left(1 - \frac{\sqrt{3}z}{2a} + \frac{x}{2a}\right)\left(1 + \frac{\sqrt{3}z}{2a} + \frac{x}{2a}\right)}{9\sqrt{3}\left[\frac{2}{M^3}\left(\frac{M}{2} - \tanh\frac{M}{2}\right)\right]\left[\frac{\phi_0 + \phi_1 + 1}{\phi_0 + \phi_1 + \left(\tanh\frac{M}{2}\right)/\frac{M}{2}}\right]}$$

$$(5)$$

The load capacity is given by

$$w = \int\limits_{-2a}^{a} \int\limits_{-\frac{x+2a}{\sqrt{3}}}^{\frac{x+2a}{\sqrt{3}}} p\,dx\,dz$$

is calculated in non-dimensional form as

$$W = -\frac{wh^3}{27\mu h a^4} = \frac{\left[1 - 3\alpha^* + 6\left(\sigma^{*2} + \alpha^{*2}\right) - 10\left(\varepsilon^* + 3\sigma^{*2}\alpha^* + \alpha^{*3}\right)\right]}{20\sqrt{3}\left[\frac{2}{M^3}\left(\frac{M}{2} - \tanh\frac{M}{2}\right)\right]\left[\frac{\phi_0 + \phi_1 + 1}{\phi_0 + \phi_1 + \left(\tanh\frac{M}{2}\right)/\frac{M}{2}}\right]} \quad (6)$$

3   Results and Discussion

It is clear that for smooth bearing surfaces, this becomes a performance of hydromag-netic squeeze film in triangular plates again for smooth bearing surfaces the results of Prakash and Vij [18] are obtained for $M \to 0$. In addition, the results of Patel and Gupta [14] are recovered when $\phi_0 + \phi_1 = 0$.

The conductivity effect on load taking capacity is decided by the factor

$$\frac{1}{\frac{(\phi_0 + \phi_1 + 1)}{\phi_0 + \phi_1 + \left(\tanh\frac{M}{2}\right)/\frac{M}{2}}}$$

which turns to

$$\frac{1}{\frac{(\phi_0 + \phi_1 + 1)}{\phi_0 + \phi_1}}$$

For higher values of M. Therefore, an increase in $\phi_0 + \phi_1 = 0$ would lead to increase load bearing capacity. Further, the plate conductivity increases load bearing capacity.

The profile of load presented in Figs. 2, 3, 4, and 5 suggests that the load bearing capacity sharply increases due to hydromagnetic parameter M.

The effect of plate conductivities gives in Figs. 6, 7, and 8 makes it clear that

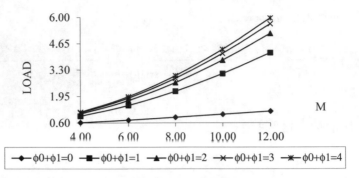

Fig. 2 Variation of load carrying capacity with respect to M and $\phi_0 + \phi_1$

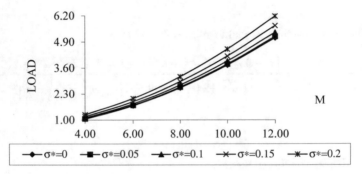

Fig. 3 Distribution of load for M and σ^*

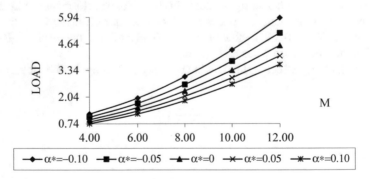

Fig. 4 Profile of load bearing capacity with regards to M and α^*

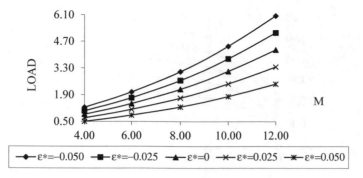

Fig. 5 Variation of load carrying capacity with respect to M and ε^*

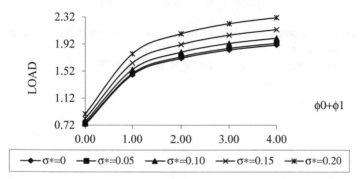

Fig. 6 Profile of load taking capacity with respect to $\phi_0 + \phi_1$ and σ^*

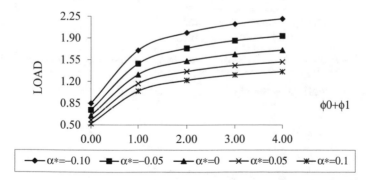

Fig. 7 Distribution of load for $\phi_0 + \phi_1$ and α^*

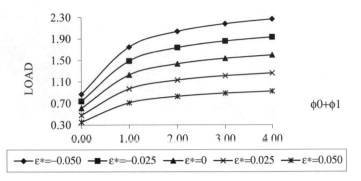

Fig. 8 Profile of load bearing capacity with regard to $\phi_0 + \phi_1$ and ε^*

the load carrying capacity enhances with increase in the conductivity. Of course, the increase is more in [0, 1].

The standard deviation σ^* associated with roughness lifts the load carrying capacity is manifest in Figs. 9 and 10.

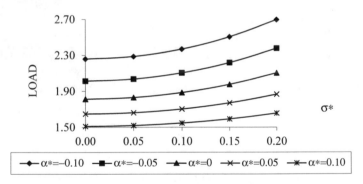

Fig. 9 Profile of load bearing capacity with regard to σ^* and α^*

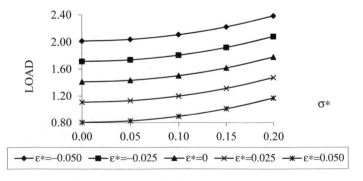

Fig. 10 Profile of load carrying capacity with respect to σ^* and ε^*

For longitudinal roughness, the deviation increases the load carrying capacity which is completely opposite to the nature of transverse roughness. The load bearing capacity gets augmented for skewed (negatively) roughness and variance ($-$ve). These trends reverse when positively skewed roughness and variance (+ve) occur.

It is observed that the bearing suffers when the plates are assumed electrically non-conducting as compared to the hydromagnetic squeeze film in conducting triangular plates. Probably, this happens to fringing phenomena when the plates are conducting.

4 Conclusion

From application point of view, the longitudinal roughness remains more favorable as compared to the transversely rough bearing system. It is appealing to note that the effect of standard deviation appears to be more crucial, although the initial effect is somewhat less so this paper emphasizes that the roughness can be properly addressed while the bearing system is being designed. Here, there is a possibility that the adverse effect of roughness can be completely overcome due to the positive effect of hydromagnetization with suitable choice of plate conductivities.

Acknowledgements Fruitful comments and constructive suggestions for improving the quality of the article are acknowledged with thanks.

References

1. Adeshara, J.V., Prajapati, M.B., Deheri, G.M., Patel, R.M.: A study of hydromagnetic longitudinal rough circular step bearing. Adv. Tribol., 7. Article ID 3981087 (2018)
2. Andharia, P.I., Deheri, G.M.: Effect of longitudinal surface roughness on the behaviour of squeeze film in a spherical bearing. Int. J. Appl. Mech. Eng. 6(4), 885–897 (2001)
3. Andharia, P.I., Deheri, G.M.: Longitudinal roughness effect on magnetic fluid based squeeze film between conical plates. Ind. Lubr. Tribol. 62(5), 285–291 (2010)
4. Andharia, P.I., Deheri, G.M.: Performance of magnetic fluid based squeeze film between longitudinally rough elliptical plates. ISRN Tribol., 6. Article ID482604 (2013)
5. Andharia, P.I., Gupta, J.L., Deheri, G.M.: Effect of longitudinal surface roughness on hydrodynamic lubrication of slider bearings. In: Proceedings of the 10th International Conference on Surface Modification Technologies, pp. 872–880. The Institute of Materials, London (1997)
6. Andharia, P.I., Gupta, J.L., Deheri, G.M.: Effect of transverse surface roughness on the behavior of squeeze film in a spherical bearings. J. Appl. Mech. Eng. 4, 19–24 (1999)
7. Christensen, H., Tonder, K.C.: The hydrodynamic lubrication of rough bearing surface of finite width. In: ASME-ASLE Lubrication Conference, Paper No. 70–Lub-7 (1970)
8. Christensen, H., Tonder, K.C.: Tribology of rough surfaces: parametric study and comparison of lubrication models. SINTEF Report, no. 22/69–18 (1969)
9. Christensen, H., Tonder, K.C.: Tribology of rough surfaces: stochastic models of hydrodynamic lubrication. SINTEF Report, no. 10/69–18 (1969)
10. Elco, R.A., Huges, W.F.: Magnetohydrodynamic pressurization in liquid metal lubrication. Wear 5, 198–207 (1962)

11. Kuzma, D.C.: Magnetohydrodynamic squeeze films. J. Basic Eng. Trans. ASME **86**, 441–444 (1964)
12. Lin, J.R.: Longitudinal surface roughness effects in magnetic fluid lubricated journal bearing. J. Mar. Sci. Technol. **24**(4), 711–716 (2016)
13. Patel, H.P., Deheri, G.M., Patel, R.M.: Combined effect of magnetism and roughness on a ferrofluid squeeze film in porous truncated conical plates: Effect of variable boundary conditions. Italian J. Pure Appil. Math. **39**, 107–119 (2018)
14. Patel, K.C., Gupta, J.L.: Behavior of hydromagnetic squeeze film between porous plates. Wear **56**, 327–339 (1979)
15. Prajapati, B.L.: Behavior of squeeze film rotating porous circular plates: surface roughness and elastic deformation effects. Pure Appl. Math. Sci. **33**(1–2), 27–36 (1991)
16. Prajapati, B.L.: On certain theoretical studies in hydrodynamic and electromagnetohydrodynamic lubrication. Ph.D. thesis. S.P. University, Vallabh Vidyanagar, Gujarat, India (1995)
17. Prajapati, B.L.: Squeeze film behavior between rotating porous circular plates with a concentric circular pocket: surface roughness and elastic deformation effects. Wear **152**(2), 301–307 (1992)
18. Prakash, J., Vij, S.K.: Load capacity and time height relations for squeeze film between porous plates. Wear **24**, 309–322 (1973)
19. Shimpi, M.E., Deheri, G.M.: Combined effect of bearing deformation and longitudinal roughness on the performance of a ferrofluid based squeeze film together with velocity slip in truncated conical plates. Imperial J. Interdiscip. Res. **2**(6), 1423–1430 (2016)
20. Shukla, J.B.: Hydromagnetic theory of squeeze films. Trans. ASME **87**, 142–147 (1965)
21. Vadher, P., Daheri, G.M., Patel, R.M.: Hydrpmagnetic squeeze film Between conducting porous transversely rough triangular plates. J. Eng. Ann. Fac. Eng. Hunedora **6**(1), 155–168 (2008)

GPU Computing for Compute-Intensive Scientific Calculation

Sandhya Parasnath Dubey, M. Sathish Kumar and S. Balaji

Abstract GPU has emerged as a platform that off-loads computation intensive work from CPU and performs numerical computations in less time. One such mathematical operation is matrix multiplication. Matrix is one of the fundamental mathematical objects used in the scientific calculation, with applicability in various fields such as computer graphics, analysis of electrical circuits, computer networks, DNA sequence comparison, protein structure prediction, etc. This work presents a comparative analysis of scalar matrix multiplication in three modes, namely: (i) sequential programming in C language (ii) parallel implementations using OpenCL, and (iii) MPI. The testbed comprises of input matrices ranging from small size of 100×100 to a higher size of $12,800 \times 12,800$. We observe that parallel execution in OpenCL outperforms MPI and sequential C for higher dimensional matrices. In contrast, sequential C outperforms both MPI and OpenCL for small dimension matrices. Besides, we analyze that OpenCL program has attained a speedup of $9\times$. Therefore, we conclude that parallel execution of code is more efficient for data of computationally large sizes and hence provides a potentially useful solution to address NP-complete problems.

Keywords HPC · GPU · Matrix multiplication · OpenCL · MPI

S. P. Dubey (✉)
Department of Computer Applications, Manipal Institute of Technology,
Manipal Academy of Higher Education, Manipal 576104, India
e-mail: sandhyadubey24@yahoo.co.in

M. S. Kumar
Deparment of Electronic and Communication, Manipal Institute of Technology,
Manipal Academy of Higher Education, Manipal 576104, India
e-mail: sathish.kumar@manipal.edu

S. Balaji
Department of Bio-Technology, Manipal Institute of Technology,
Manipal Academy of Higher Education, Manipal 576104, India
e-mail: biobalaji@gmail.com

© Springer Nature Singapore Pte Ltd. 2020
K. N. Das et al. (eds.), *Soft Computing for Problem Solving*,
Advances in Intelligent Systems and Computing 1057,
https://doi.org/10.1007/978-981-15-0184-5_12

131

1 Introduction

A graphics processing unit (GPU) has developed from graphics cards into a platform for high-performance computing (HPC) and perhaps, the most important development in HPC for many years [1]. GPU emerged as an accelerator and work along with the central processing unit (CPU) to accelerate the compute-expensive problems [2, 3]. Protein folding, drug discovery, and data analysis are few research areas in the field of bioinformatics which required huge computation power [4].

Even though there is a plenty research going on in understanding the protein folding and subsequent prediction of the protein structure from last six decades, still it is considered as a holy grail and the difference between the known sequences and the known structures are widened [5]. In order to deal with complexity and computational cost, hydrophobic-polar (HP) model is the most widely used model for coarse-level structure prediction [6]. Although the HP model has reduced the complexity, still it is a *NP-complete* problem where it needs huge computation power and time [7]. One of the reported study state that to model a protein structure through the HP model, it requires 12 hour time for one simulation. The aforementioned study was carried out at two different grid system namely Victorian Partnership for Advanced Computing (VPAC) and Monash Campus Grid (MCG) with protein sequences of the maximum length of 100 [8]. The working hypothesis of this paper is stated as follows: "the time and cost involved in such simulations can be significantly reduced by the application of GPU integrated system."

This work presents the use of GPU computing for such problems where a pilot study has been done for scalar matrix multiplication. Matrix is one of the fundamental mathematical objects in scientific computation, with applicability in various fields such as computer graphics, design, and analysis of electrical circuits, computer network, DNA sequence comparison, protein structure prediction, etc. GPU has emerged as a platform to off-load computation intensive work from CPU to GPU and perform the numerical computation efficiently with much less time. One such mathematical operation is scalar matrix multiplication. This work presents the comparative analysis of scalar matrix multiplication using conventional sequential programming (in C), parallel implementation using OpenCL and MPI. The testbed comprises of matrix dimensions ranging from 100×100 to $12,800 \times 12,800$.

It has been observed that for matrices of small sizes, sequential execution takes less time. However, as the size of the matrix increases, sequential execution becomes inefficient with an exponential increase of time. On the other hand, parallel execution has shown the reverse relation.

The remainder of the paper is organized as follows: Sect. 2 provides the brief on GPU, OpenCL, and MPI followed up with the next section comprising result and discussion. Section 4 briefs the conclusion of the study and the future work.

2 Methodology

2.1 Graphics Processing Unit

Graphics processing unit (GPU) is one of the most important developments for high-performance computing (HPC). It has been developed from graphics cards into a platform for high-performance computing and perhaps, the most important development in HPC for many years. GPUs serve as power vitality productive server farms in government laboratories, colleges, ventures, and small and medium organizations around the globe [9]. CUDA and OpenCL are two important frameworks that facilitate developers to map the sequential programs on GPU for parallel execution. CUDA is a proprietary application programming interface (API) from NVIDIA, whereas OpenCL is a non-proprietary framework that is supported by many GPU manufacturers. This work is developed with OpenCL.

2.2 Open Computing Language

Open Computing Language (OpenCL), is developed as a non-proprietary framework. It is an arrangement of establishment-level APIs intended to extract the fundamental equipment and to give a system (Fig. 1) for building parallel applications. In light of the prominent ISO C99 standard normally known as the C language, OpenCL can be seen as a variation, in this way, making it a broadly worthy. OpenCL has two main components first, set of model and second, software stack where prior specify OpenCL developmental flow and how it handles data and tasks whereas later shows the flow of OpenCL program development using the OpenCL libraries.

OpenCL program starts with host program which identified available computation devices and defines the context to use them [10]. Figure 1 represents two OpenCL devices, a CPU and a GPU. However, there is no restriction on the number of devices and of their type.

Next is command queues, here, we are having two command queues, an in-order command queue, and an out-of-order command queue. GPU works over the former command queue whereas CPU uses an out-of-order command queue. The host program at that point characterizes a program protest that is accumulated to produce portions for both OpenCL gadgets (the CPU and the GPU). Memory objects required by the program are characterized through the host program and further maps these memory objects onto the contentions of the kernels. Finally, the host program enqueues commands to the command queues to execute the kernels. The kernel is a function which consumes most of the computation time and hence, it is offloaded on the GPU. On completion of the computation, it returns the result to the host program.

Figure 2 depicts the kernel function used in OpenCL implementation and corresponding scalar matrix multiplication function used in MPI and C implementation.

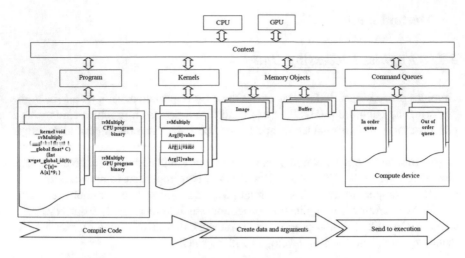

Fig. 1 OpenCL operational flow. Figure adapted from Khronos group OpenCL framework, 2010

/* execute on GPU for matrix multiplication */
__kernel void svMultiply
(__global float* inputA,
__global float* outputC)
{ int wA=10; int wB=10;
int row_el=get_global_id(0);
outputC[row_el]=inputA[row_el]
*984;
}
	/* perform scalar matrix multiplication */
	/* Use in sequential and MPI program */
	For (i=0; i<n; i++) {
	For (j=0; j<n; j++) {
	Matrix[i] [j] =
	scalar*matrix[i][j];
	}

Fig. 2 Matrix multiplication function

2.3   Message-Passing Interface

The message-passing interface (MPI) has been developed to exploit the computation power of numerous processors. MPI is a set of functions and is a library of functions and macros that can be utilized as a part of C, FORTRAN, and C++ programs. As its name suggests, MPI is planned for use in programs that adventure the presence of numerous processors by message-passing. Accordingly, it is one of the primary measures for programming parallel processors, and the first depends on message-passing. Further details on MPI and its implementation are available in [11]. This work has used two important MPI functions *viz.* "MPI_Scatter" and "MPI_Gather."

These two library functions are used to distribute data to be processed on the multiple nodes where the multiplication function (computation) performs and resultant values are written back to the master node, respectively.

MPI_Scatter: MPI_Scatter distributes the elements of an array among processes. If the array to be distributed contains n elements, and there are p processes, then the first n/p sent to process 0, the next n/p to process 1, and so on using the following syntax.

If the communicator *comm* contains processes of *comm_sz*, then MPI_Scatter divides *send_buf_p* referred data into *comm_sz* parts where the first part handle by process 0, the second by process 1, the third by process 2, and so on. For instance, in this work, we have used a block distribution. In this, Process 0 has read in all of an *n-component* vector into *send_buf_p*. Then, process 0 will get the first *local_n* = *n/comm_sz* components, similarly process 1 will get the next *local_n* components, and so on. Each process should pass its local vector as the *recv_buf_p* argument and the *recv_count* argument should be *local_n*. Both *send_type* and *recv_type* should be MPI_DOUBLE and *src_proc* should be 0. Perhaps surprisingly, *send_count* should also be *local_n-send_count* is the amount of data going to each process; it is not the amount of data in the memory referred to by *send_buf_p*.

MPI_Gather: MPI_Gather is the "inverse operation" to MPI_Scatter. If each process stores a subarray containing m elements, MPI_Gather will collect all of the elements onto process 0, and then process 0 can print all of the components. The data available on process 0 is stored in the first block in *recv_buf_p*, similarly, the data stored in the memory of process 1 is stored in the second block referred to by *recv_buf_p*, and so on. *recv_count* is the number of data items received from each process, not the total number of data items received.

3 Result and Discussion

Performance analysis is done using a square matrix where dimension varies from 100×100 to $12,800 \times 12,800$. In every next iteration, matrix dimension is getting doubled, so we consider here seven-run to compare the execution time of kernel and program in sequential, parallel, and MPI execution. Table 1 shows the execution time of kernel function (function which perform computation on GPU) and Table 2 presents the execution time for complete program.

Figures 3 and 4 depict the relationship between computation time and various implementations for varying dimension matrices. Figures 3 and 4 show the line graph for execution time of kernel and complete program. It has been observed that for matrices of small sizes such 100×100, 200×200 sequential execution takes less time, whereas, for large-size matrices such as 1600×1600, 3200×3200, parallel execution takes less time than compared to sequential execution. Three different events considered for MPI implementation wherein first have only one processor, second with five, and third with ten processors. It has been observed that MPI with one processor performs similar to a sequential C program whereas when the number

Table 1 Execution time of kernel

Matrix dimension		400	800	1600	3200	6400
Sequential Exec. C (sec.)		0	0	0	31	140
		0	0	15	31	141
		0	0	16	31	141
OpenCL Exec. (sec.)		0.065	0.288	0.905	3.52	14.4
		0.063	0.288	0.9	3.5	14.4
		0.063	0.288	0.898	3.66	14.4
MPI Exec. (sec.)	P-1	0.55	2.28	9.18	37.7	146
		0.572	2.27	9.15	36.6	151
		0.609	2.28	9.15	36.9	149
	P-5	0.143	0.465	2.39	9.65	34.2
		0.143	0.458	2.31	9.66	34.2
		0.143	0.465	2.3	9.65	34.2
	P-10	0.056	0.233	1.51	4.77	19.2
		0.056	0.267	1.18	4.77	19
		0.056	0.252	1.51	4.8	19

Table 2 Execution time of complete program

Matrix dimension		400	800	1600	3200	6400	12,800
Sequential Exec. C (sec.)		0	0.016	0.031	0.093	0.359	1.45
		0	0.015	0.031	0.109	0.359	1.4
		0	0.016	0.047	0.125	0.343	1.49
OpenCL Exec. (sec.)		0.125	0.125	0.156	0.218	0.468	1.1
		0.125	0.125	0.14	0.234	0.499	1.18
		0.125	0.125	0.171	0.203	0.517	1.1
MPI Exec. (sec.)	P-1	0.001	0.007	0.03	0.118	0.466	1.2
		0.001	0.007	0.029	0.117	0.472	1.25
		0.001	0.007	0.029	0.118	0.469	1.22
	P-5	0.002	0.007	0.025	0.095	0.373	1.19
		0.002	0.007	0.022	0.095	0.357	1.19
		0.002	0.007	0.025	0.095	0.373	1.17
	P-10	0.004	0.009	0.027	0.102	0.48	1.2
		0.004	0.009	0.03	0.108	0.394	1.29
		0.005	0.01	0.03	0.103	0.394	1.26

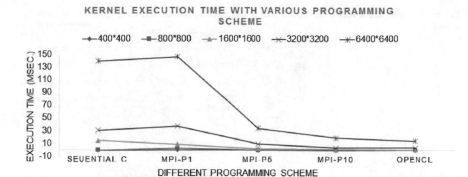

Fig. 3 Line graph of kernel execution time with various programming scheme for different matrix dimension

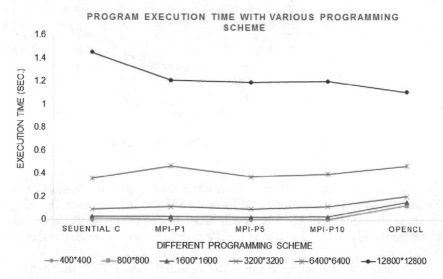

Fig. 4 Line graph for program execution with various programming scheme for different matrix dimension

Table 3 Speedup attain with OpenCL program

Matrix dimension	Speedup of kernel	Speedup of program
400 × 400	0	0
800 × 800	0	0.128
1600 × 1600	0	0.198
3200 × 3200	8.85	0.458
6400 × 6400	10	0.767
12,800 × 12,800	9.9	1.313

Fig. 5 Speedup graph for kernel

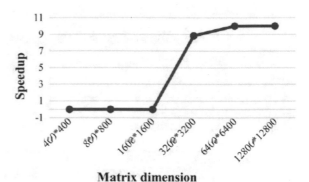

Matrix dimension

Fig. 6 Speedup graph for complete program

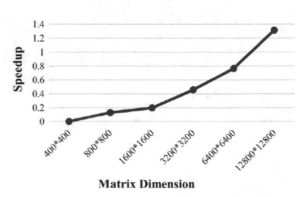

Matrix Dimension

of processor increases as in the case of ten processors, it performs analogous to GPU implementation. Furthermore, MPI has an added advantage over OpenCL, as it can make use of available coprocessor hence reduce the economic burden of new resources. Moreover, it is also relatively easier to port the existing codes to coprocessors.

In the context of pure multiplication function, there is a huge difference in time taken by the sequential and parallel implementation (Figs. 3 and 4) with the speedup of $10\times$ (Table 3). Whereas, the complete program attended the speedup of $1\times$ (Fig. 5). Even though this speedup is low but when considered for higher-dimension matrices, it has shown good efficiency. Moreover, when the application has many such functions like matrix multiplication in the problem it needs to done by mean of multiple kernels where each kernel execute in parallel on different data set.

Hence, the resultant performance and speedup attain an efficient point (Fig. 6) at speedup of 10. Once optimum speedup attended, it becomes constant. This represents the ideal case where it efficiently uses resources.

This study reveals that for matrix scalar multiplication, it is best to use GPU when the matrices size is equal or greater than 3200×3200.

4 Conclusion

We compare the performance of MPI and OpenCL program with sequential C for scalar matrix multiplication for a dense matrix. We conclude that sequential program performs well for small-size matrix, whereas MPI and OpenCL give better results than a sequential program for higher-dimension matrix. Similar is the case of the PSP problem which addresses with evolutionary algorithms [12]. An evolutionary algorithm mainly comprised of four steps, namely initial population generation, fitness calculation of generated individuals, selection process (selection of individuals for crossover and mutation), and followed by crossover and/or mutation. Dependency between the solution obtained through evolutionary algorithms is limited; hence, it presents an ideal situation for parallelism. Multiple optimal solutions can be obtained at the same time from the different core of the GPU. Whereas obtained solutions can be further subjected to the optimization by either crossover and/or mutation. Evolutionary algorithms are one of the most data-intensive task wherein single iteration, it deals with many thousands of data element and in every iteration, it needs to process each data. Hence, with this case study, we are proposing the use of GPU computing to be used for bioinformatics problem which is mostly *NP-complete* problem.

Although, this work is restricted to the separate use of OpenCL and MPI, this work can be extending to develop a hybrid process using MPI and OpenCL, where MPI library functions distribute and collect the data on various node (GPU enabled system) of network, where each node powered with GPUs and use OpenCL to harness the GPU computing power, this may result in a significant increase in the efficiency as simultaneously it executes on multiple nodes. However, it results in excess time to distribute and collect data on multiple nodes but the resultant improvement in time will nullify this overhead. Matrix multiplication can be used to define the mutation operator of evolutionary programming.

The future work will include optimizing the use of CPU and GPU together. Our final goal is to develop a parallel framework for the evolutionary algorithms to address the protein folding and protein structure prediction problem.

References

1. Hwu, W.W.: GPU Computing Gems, vol. 2. Elsevier (2012)
2. Li, J.: GPU Algorithms for Bioinformatics. University of Florida (2014)
3. de Oliveira, F.B., Davendra, D., Guimaraes, F.G.: Multi-objective differential evolution on the GPU with C-CUDA. In: Snasel, V., Abraham, A., Corchado, E. (eds.) Soft Computing Models in Industrial and Environmental Applications. Advances in Intelligent Systems and Computing, vol. 188. Springer, Berlin, Heidelberg (2013)
4. Pevsner, J.: Bioinformatics and Functional Genomics. Wiley (2015)
5. Rigden, D.J. (ed.): From Protein Structure to Function with Bioinformatics. Springer, Berlin (2009)
6. Dill, K.A., Bromberg, S., Yue, K., Chan, H.S., Ftebig, K.M., Yee, D.P., Thomas, P.D.: Principles of protein folding—a perspective from simple exact models. Protein Sci. 4(4), 561–602 (1995)

7. Berger, B., Leighton, T.: Protein folding in the hydrophobic-hydrophilic (HP) model is NP-complete. J. Comput. Biol. **5**(1), 27–40 (1998)
8. Islam, M.K., Chetty, M.: Clustered memetic algorithm with local heuristics for ab initio protein structure prediction. IEEE Trans. Evol. Comput. **17**(4), 558–576 (2013)
9. Cormen, T.H., Leiserson, C.E., Rivest, R.L., Stein, C.: Introduction to Algorithms. MIT Press, Leiserson (2009)
10. Gaster, B.R., Howes, L., Kaeli, D., Mistry, P., Schaa, D.: Heterogeneous Computing with OpenCL. Morgan Kaufmann (2012)
11. Quinn, M.J.: Parallel Programming in C with MPI and OpenMP. McGraw-Hill (2003)
12. Dubey, S.P.N., Kini, N.G., Balaji, S., Kumar, M.S.: A review of protein structure prediction using lattice model. Crit. Rev. Biomed. Eng. **46**(2), 147–162 (2018)

A Fuzzy-Controlled High Voltage Gain DC–DC Converter for Renewable Applications

T. Arunkumari and V. Indragandhi

Abstract In this manuscript, a non-isolated converter with high static voltage gain is presented. The designed converter has the feature of stable frequency and output even though disturbance occurs. It also achieves extreme voltage conversion, good efficiency, low voltage stress and less switching loss. The voltage doubler technique is implemented in the designed converter. With reduced duty cycle, the high voltage conversion is achieved. The proposed single-switch converter is controlled by fuzzy-controlled technique. The functioning process of the converter below continuous conduction mode (CCM) is explained. The input source of 30 V is stepped up to 400 V. The simulative analysis of the proposed converter is complete with MATLAB and Simulink.

Keywords Continuous conduction mode · Duty cycle · Fuzzy controller · High voltage conversion · Less switching loss

1 Introduction

The usage of non-conventional resources such as photovoltaic (PV) is being raised sequentially, due to global warming and the demand for fossil fuels. PV schemes utilize the solar source to generate electrical energy. On the other hand, the produced DC output voltage through PV scheme is too short. To handle this downside, power electronics switching converters develop a significant unit of non-conventional resources [1–3]. Hence, various DC–DC mode converters through large conversion gain aimed at PV schemes are designed. Predictable boost mode converter was utilized for huge voltage conversion with extreme duty ratio. Due to occurrence of the high losses in all components and the voltage conversion static limit of converter are restricted [4, 5].

T. Arunkumari · V. Indragandhi (✉)
School of Electrical Engineering, Vellore Institute of Technology, Vellore 632014, India
e-mail: arunindra08@gmail.com

T. Arunkumari
e-mail: aak.thi90@gmail.com

© Springer Nature Singapore Pte Ltd. 2020   141
K. N. Das et al. (eds.), *Soft Computing for Problem Solving*,
Advances in Intelligent Systems and Computing 1057,
https://doi.org/10.1007/978-981-15-0184-5_13

To exceed the limits of non-isolated boost converter, several converter models, for example, the interleaved-type combined with boost mode converter design, Soft-Switching type Boost mode converters (SSB), couple-inductive schemes [6–8], voltage multiplier(VM) technique converter models, [9–12] had been designed which are capable to deliver extreme voltage gain limit compared to traditional boost mode converters. Also, it is perceptible that an additional scheme of this model relies on the mixture of such topography which is designed in current centuries [13–17]. Amongst these designs, all have their own merits and demerits. For case, SSB functions in zero-voltage switching (ZVS) design which paves towards the decrement of ON and OFF fatalities and simultaneously results in growing efficiency. The key drawback of SSB is the difficulty of the circuit, since exploiting huge quantity of components together with ON and OFF switches and inductors.

Many techniques have been implemented for high voltage conversion technique such as inductive coupling method, switched capacitor method, voltage doubler method and voltage multiplier techniques. All the above methods have advantages and disadvantages. The common drawback in all those techniques is the size of the system is increased. Recently, a different VM converter topography is being depicted in [10–12]. A new VM scheme is presented in [10] which points as multilevel flying capacitor (MFC). A new scheme of VM has been presented in [11, 12]. It is essential to remind an input source current of designed models which is a limitation. Also, the static gain of this topography is stable and not able to be synchronized. Hence, these cannot be exploited in PV schemes due to their restraint to follow the maximum power point (MPP). Then, voltage lift technique has been proposed and it is implemented for boost converter [18–20]. In this paper, the voltage lift technique is implemented for zeta converter. In Fig. 1, the block diagram is depicted.

In this manuscript, a new single switch modified SEPIC converter with extreme static gain for PV solicitations is projected which can overwhelm the above-revealed limitations of the conventional converters. In Sect. 2, the operation of the designed converter mode in CCM is explained. Followed in Sect. 3, converter analysis is detailed. Section 4 deals with converter control fuzzy logic. Section 5 explains simulation results. Section 6 is followed with conclusion.

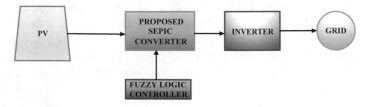

Fig. 1 Overview of conversion system

2 Operating Principle of the Designed Converter

The designed model is shown in Fig. 2. Limited suppositions are done to abridge the circuit model and process of the designed converter.

(1) The converter drives beneath CCM at the stable state circumstance.
(2) Semiconductor devices are deliberated as idyllic.
(3) The capacitor operated has more storage capacity, and they are expected to be constant. Hence, the input storage capacitor and the output storage capacitor are measured as equivalent.

The circuit consists of main switch S_1, inductors L_1, L_2, diodes D_1, D_2, D_3, D_4, D_5 and D_6, capacitors C_1, C_2, C_3, C_4, C_5 and output capacitor C_6. The voltage tripler (VT) circuit is combined with conventional SEPIC model to progress the static limit.

2.1 Continuous Conduction Mode Operation

The designed converter functions in two modes. The working functions are shown in Figs. 3 and 4.

Mode I [$t0 - t1$]: Once the single switch S_1 is switched ON, diodes D_2, D_4 and D_6 are active. Diodes D_1, D_3 and D_5 are OFF in condition. The input voltage V_{in} is conveyed to L_1, and $V_{C3} - V_{C2} - V_{C1}$ is transported to L_2. These inductors support

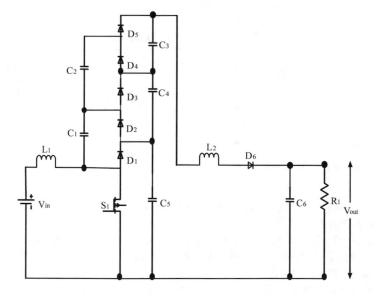

Fig. 2 Proposed single-switch DC–DC converter

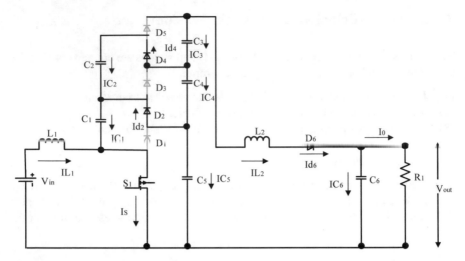

Fig. 3 Proposed DC–DC converters turn on mode

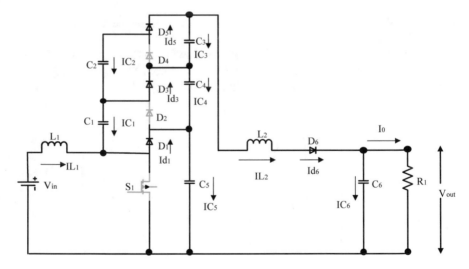

Fig. 4 Proposed DC–DC converters turn off mode

in stowing the energy. The storage element C_0 exonerates the essential vitality to the load for its function.

Mode II $[t1 - t2]$: When the terminal S_1, is switched OFF, the diodes such as D_2 and D_4 are in OFF state. The diodes D_1, D_3 and D_5 are in active state. The storage elements are electric by the inductance L_1 and L_2. The load obtains the energy by liquidating mode of the element. This condition stops when terminal is ON. The subsequent cycle endures. The key functioning graph is embodied in Fig. 5. The entire capacitive voltage is equivalent to the output capacitive.

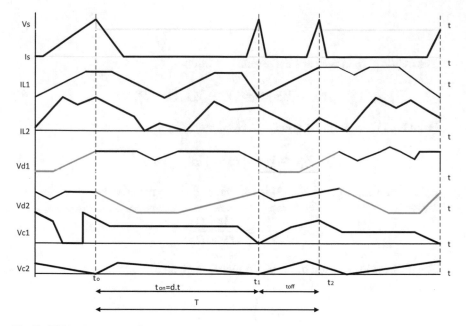

Fig. 5 CCM operation waveforms

$$V_0 = V_{C3} + V_{C4} + V_{C5} + V_{C6} \qquad (1)$$

3   Analysis of the Designed Converter

Here, the theoretical examination of the designed SEPIC converter is explained by following with design procedural of the converter proposed.

3.1   Static Gain and Switching Voltage Analysis

At steady condition, the inductor value is considered as null and the equation is termed as below, and hence in CCM operation, the inductor L_1 is given as

$$V_{in}d = V_{C0} - 3V_{in}(1 - D) \qquad (2)$$

Here, d is duty cycle and input voltage is V_{in}. On rearranging the values in Eq. (2), the capacitor C_0 value is obtained same as static gain of the converter.

$$V_{c0} = \frac{3V_{in}}{1 - D} \tag{3}$$

The inductor L_2 is zero at steady-state condition,

$$(V_{C0} - V_{C1})D = (V_0 - V_{C0})(1 - D) \tag{4}$$

From Eq. (1), capacitor 1 voltage is given by

$$V_{C1} = V_0 - V_{C2} - V_{C3} - V_{C0} \tag{5}$$

By substituting Eqs. (2) and (3) in (4), the static gain is derived as

$$\frac{V_0}{V_{in}} = 3\frac{(1 + D)}{(1 - D)} \tag{6}$$

The duty cycle equation of proposed model is given as

$$\frac{1 - D}{D} = 3\frac{V_{in}}{V_0} \tag{7}$$

The capacitor voltages in C_2, C_3 are given as

$$V_{C2} = \frac{3 \times V_{in}}{1 - D} \tag{8}$$

$$V_{C3} = \frac{3 \times V_{in}}{1 - D} \tag{9}$$

$$D = \frac{V_0 + V_d}{V_0 + 3V_{in} + V_d} \tag{10}$$

The attained duty cycle is 0.81.

The static gain is compared with other classical converters in Fig. 6, and from Fig. 7, it is observed that the modelled converter attains required static gain with reduced duty cycle $D = 0.814$. It is given that the switching voltage is equal to the totalling of insource and outsource as of boost converter. The switching energy of the proposed design is determined, and it is paralleled with conventional boost converter.

The proposed converters' step-up and step-down of voltage vary due to the duty cycle control. And also, it is dependent on elements in the circuit. Static gain and voltage gain of the proposed converter is depicted in Figs. 6 and 7.

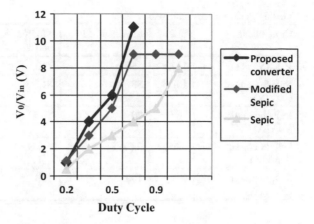

Fig. 6 Voltage gain analysis

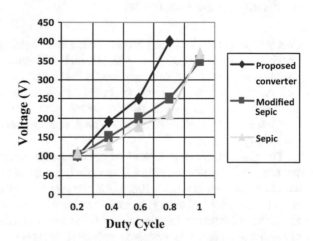

Fig. 7 Switching voltage of proposed converter

3.2   Comparison of Designed Converter with Other Converter

Here, the designed model is analysed with the conventional converters in the basis of formulas, performance, and output source voltage and duty cycle. Overall, the proposed converter shows better performance and it well suits for PV-based applications for different changes in temperature. The converter attains steady in a short period of time even though disturbances occur.

The enactment of the designed model is compared in Table 1. On comparing, it is noted that the static gain of the designed converter is high but its overall performance is good on comparing with other converters. The output power obtained is high. And the same way, the input current is 15 A. It is less compared to other converters.

Table 1 Performance comparisons

Components	SEPIC	Modified SEPIC	Modified SEPIC boost	Proposed converter
V_{out} (V)	147.5	150	150	400
Switch voltage (V_s) (V)	162.3	83.8	81.3	200
Inductor current (I_{in}) (A)	76	70	54.5	15
Output current (I_{out}) (A)	0.6572	0.6715	0.6699	0.9
Output power (P_{out}) (W)	97	100	100	400

4 Fuzzy Logic Controller

Fuzzy logic controller (FLC) is one of the most booming applications of fuzzy theory, implemented by Zadeh in 1965 [2]. Its best features are the usage of linguistic variables relatively than numerical contents. Linguistic variables are utilized as normal variables are such as small and large, which is represented as fuzzy sets. It is an addition of a crisp set which exhibits the membership or non-membership function. Fuzzy sets permit specific function which defines that an element may belong or may not belong to the set.

The FLC is utilized to control the zeta converter operation. This fuzzy is utilized for any other control operation. The main concept of fuzzy is the rules. The rules should be defined first according to the converter operation. Here in this design, the error in reference and change in error in delay of reference voltage are given as input to the fuzzy converter. The change in error is created by delay unit. The output source of the fuzzy organizer is compared with relational operator, and it is given to the gate of the MOSFET. The fuzzy block is depicted in Fig. 8.

FLC entails of three main blocks specifically fuzzification, FIS and defuzzification. In fuzzification, the data is converted to linguistic variable. The fuzzification method is of two types; they are Mamdani and Sugeno. Design of membership function for input error, input change in error and output duty is shown in Figs. 9, 10 and 11. Rule base and database are given to decision-making. The converter rules for fuzzy controller are depicted in Table 2, and surface plot of the input and output is represented in Fig. 12.

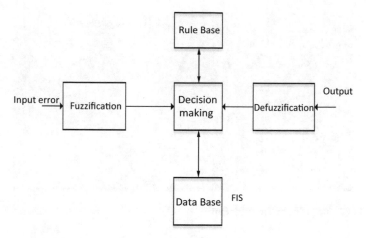

Fig. 8 Fuzzy block

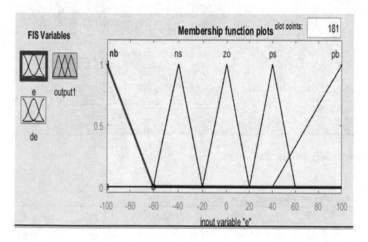

Fig. 9 Input error membership

5 Simulation Results

The designed converter is tested with MATLAB and Simulink. The 30 V of insource is enhanced up to 400 V. The key benefit of the designed converter is its single switch, condensed size, increased efficiency, less duty cycle, high voltage boost up with reduced ripple value. The output voltage source and source current are posted in Figs. 13 and 17. The ripple and peak overvalues are reduced. The enactment of the designed converter is well equalled to additional converters. The waveforms of other components are depicted from Figs. 14, 15, 16 and 17. The converter parameters are represented in Table 3.

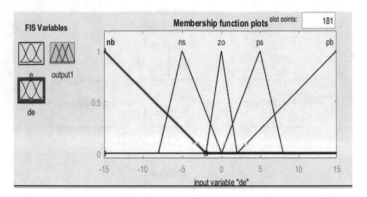

Fig. 10 Input change in error membership

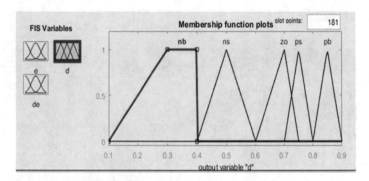

Fig. 11 Output duty membership

Table 2 Fuzzy rule for proposed converter

ERR	NB	NS	Z0	PS	PB
NB	nb	ns	ns	ns	ns
NS	ns	ns	ns	ns	ns
Z0	zo	zo	zo	zo	zo
PS	ps	ps	ps	ps	ps
PB	pb	pb	pb	pb	pb

6 Conclusion

In this manuscript, an extreme voltage static gain non-isolated converter is modelled. The CCM operation is explained in detail with the waveform. The mathematical expression for the proposed converter is derived and analysed with the simulation results. In this designed converter, the duty limit attained is 0.81 and the ripple input current is reduced to 0.113 A. The main advantages such as reduced size of

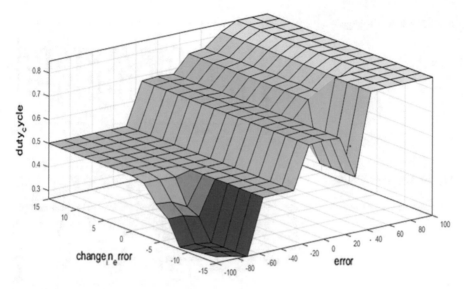

Fig. 12 Surface plot of fuzzy

Fig. 13 Output voltage

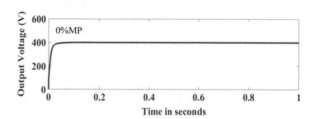

Fig. 14 Diode 1 voltage

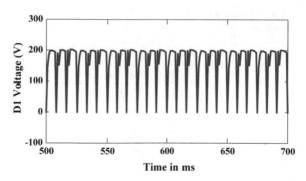

Fig. 15 Diode 2 voltage

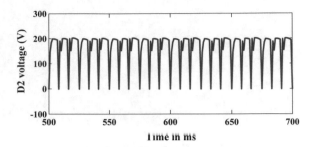

Fig. 16 Capacitor 2 voltage

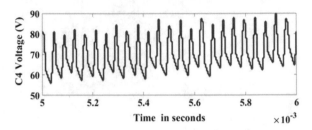

Fig. 17 Output current

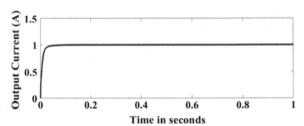

Table 3 Converter parameters

Components	Parameter
Input power	30 V
Output power	400 V
Input current	15 A
Output current	0.99 A
Inductor L_1, L_2	205e^{-6}, 180e^{-6} μH
Capacitor C_1, C_2	2.2e^{-6} μF
Capacitor C_0	40e^{-6} μF
Resistor	400 Ω

the converter, reduced number of switch and reduced ripple current are attained. This converter has high efficiency, while switching and conduction loss are less. The efficiency attained is stable on comparing to other converters. This converter is suitable for PV energy conversion, and it is also suitable to connect to the inverter source. This can be analysed through hardware results in future.

References

1. Hosseini, S.H., Alishah, R.S., Kurdkandi, N.V.: Design of a new extended zeta converter with high voltage gain for photovoltaic applications. In: 2015 9th International Conference on Power Electronics and ECCE Asia (ICPE-ECCE Asia), pp. 970–977. IEEE (2015)
2. Alishah, R.S., Nazarpour, D., Hosseini, S.H., Sabahi, M.: Design of new power electronic converter (PEC) for photovoltaic systems and investigation of switches control technique. In: Proceedings of the 28th Power System Conference (PSC), pp. 1–8 (2013)
3. Pan, C.T., Lai, C.M., Cheng, M.C.: A novel integrated single-phase inverter with auxiliary step-up circuit for low-voltage alternative energy source applications. IEEE Trans. Power Electron. 25(9), 2234–2241 (2010)
4. Huang, Y., Shen, M., Peng, F.Z., Wang, J.: Z-source inverter for residential photovoltaic systems. IEEE Trans. Power Electron. 21(6), 1776–1782 (2006)
5. Ye, Y.M., Cheng, K.W.E.: Quadratic boost converter with low buffer capacitor stress. IET Power Electron. 7(5), 1162–1170 (2013)
6. Abdel-Rahim, O., Orabi, M., Abdelkarim, E., Ahmed, M., Youssef, M.Z.: Switched inductor boost converter for PV applications. In: 2012 Twenty-Seventh Annual IEEE on Applied Power Electronics Conference and Exposition (APEC), pp. 2100–2106. IEEE (2012)
7. Chen, C., Wang, C., Hong, F.: Research of an interleaved boost converter with four interleaved boost convert cells. In: 2009. PrimeAsia 2009. Asia Pacific Conference on Postgraduate Research in Microelectronics & Electronics, pp. 396–399. IEEE (2009)
8. Park, S.H., Park, S.R., Yu, J.S., Jung, Y.C., Won, C.Y.: Analysis and design of a soft-switching boost converter with an HI-bridge auxiliary resonant circuit. IEEE Trans. Power Electron. 25(8), 2142–2149 (2010)
9. Li, W., Lv, X., Deng, Y., Liu, J., He, X.: A review of non-isolated high step-up DC/DC converters in renewable energy applications. In: Twenty-Fourth Annual IEEE Applied Power Electronics Conference and Exposition, 2009. APEC 2009, pp. 364–369. IEEE (2009)
10. Pan, Z., Zhang, F., Peng, F.Z.: Power losses and efficiency analysis of multilevel DC-DC converters. In: Twentieth Annual IEEE Applied Power Electronics Conference and Exposition, 2005. APEC 2005, vol. 3, pp. 1393–1398. IEEE (2005)
11. Qian, W., Cao, D., Cintrón-Rivera, J.G., Gebben, M., Wey, D., Peng, F.Z.: A switched-capacitor DC-DC converter with high voltage gain and reduced component rating and count. IEEE Trans. Ind. Appl. 48(4), 1397–1406 (2012)
12. Abutbul, O., Gherlitz, A., Berkovich, Y., Ioinovici, A.: Step-up switching-mode converter with high voltage gain using a switched-capacitor circuit. IEEE Trans. Circ. Syst. I Fundam. Theory Appl. 50(8), 1098–1102 (2003)
13. Arunkumari, T., Indragandhi, V.: An overview of high voltage conversion ratio DC-DC converter configurations used in DC micro-grid architectures. Renew. Sustain. Energy Rev. 77, 670–687 (2017)
14. Hsieh, Y.P., Chen, J.F., Liang, T.J., Yang, L.S.: Novel high step-up DC-DC converter with coupled-inductor and switched-capacitor techniques. IEEE Trans. Industr. Electron. 59(2), 998–1007 (2012)
15. Li, W., Li, W., Ma, M., Deng, Y., He, X.: A non-isolated high step-up converter with built-in transformer derived from its isolated counterpart. In: IECON 2010–36th Annual Conference on IEEE Industrial Electronics Society, pp. 3173-3178. IEEE (2010)

16. Laird, I., Lu, D.D.C., Agelidis, V.G.: High-gain switched-coupled-inductor boost converter. In: International Conference on Power Electronics and Drive Systems, 2009. PEDS 2009, pp. 423–428. IEEE (2009)

17. Zhao, Y., Li, W., Deng, Y., He, X.: Analysis, design, and experimentation of an isolated ZVT boost converter with coupled inductors. IEEE Trans. Power Electron. 26(2), 541–550 (2011)

18. Savakhande, V., Bhattar, C.L., Bhattar, P.L.: A voltage-lift DC-DC converter using modular voltage multiplier cell for photovoltaic application. In: 2017 International Conference on Circuit, Power and Computing Technologies (ICCPCT), pp. 1–7. IEEE (2017)

19. Savakhande, V.B., Bhattar, C.L., Bhattar, P.L.: Voltage-lift DC-DC converters for photovoltaic application—a review. In: 2017 International Conference on Data Management, Analytics and Innovation (ICDMAI), pp. 172–176. IEEE (2017)

20. Shahir, F.M., Babaei, E., Farsadi, M.: Voltage-lift technique based non-isolated boost DC-DC converter: analysis and design. IEEE Trans. Power Electron. 33(7), 5917–5926 (2017)

Comprehensive Study on Diabetic Retinopathy

R. S. Rajkumar, A. Grace Selvarani and S. Ranjithkumar

Abstract Diabetes is a chronic disease that is found common nowadays among the working age groups. Diabetes affects various organs of human. Diabetic retinopathy (DR) is a disease which affecting the human eye leading to vision impairment caused by diabetes mellitus. No medication is still available to cure DR but can be controlled. Hence, there exists lot of literatures in detecting the DR automatically. Comprehensive study on the various DR detection algorithms and their performance metrics has been discussed in this paper.

Keywords Diabetic retinopathy · Blood vessel segmentation · Microaneurysms · Exudates · Lesions

1 Introduction

Raise in the blood sugar level of a person will cause diabetes. This is on the grounds that body does not either emit a sufficient measure of the insulin or react to the insulin discharged. This affects the retina and damages its tiny veins. Retina forms the vital division of the vision as it senses the light and converts the light into signals, thereby sending the signal to the brain all the way through optic nerve.

Diabetes in people is continuously increasing due to their change in life style and additional factors [1]. The most significant reason for sightlessness in the western world is the DR [2]. DR leads to outflow of liquid or hemorrhage causing the damage to vision. As the age increases, the likelihood to prone to DR also increases. The four development stages of the DR in accordance with ETDRS [3] and ICDR are

1. Mild non-proliferative diabetic retinopathy(MiNPR)—small swellings happen in the tiny veins of the retina and liquid leakage may happen in the retina from the swelling.

R. S. Rajkumar (✉) · A. G. Selvarani · S. Ranjithkumar
Sri Ramakrishna Engineering College, Coimbatore, Tamil Nadu, India
e-mail: rajkumar.rs@srec.ac.in

© Springer Nature Singapore Pte Ltd. 2020
K. N. Das et al. (eds.), *Soft Computing for Problem Solving*,
Advances in Intelligent Systems and Computing 1057,
https://doi.org/10.1007/978-981-15-0184-5_14

155

2. Moderate non-proliferative DR(MoNPR)—veins in the retina swells and distorts as the DR progress. It influences the blood transportation in the retina and the presence of the retina changes.
3. Severe non-proliferative DR(SNPR)—at this stage, many veins are blocked and form the barrier for blood supply in retina. This will increase the growth factors of developing new veins.
4. Proliferative DR(PDR)—This is the advanced stage and new veins grow into the retina and vitreous liquid. The eye is packed up with vitreous liquid. The developed new vessels are more prone to leak and bleed as they are fragile in nature. This will direct to retinal disconnection, and the retinal disconnection causes permanent vision loss.

Development of DR is given in Fig. 1 [4]. Distinction between normal and DR

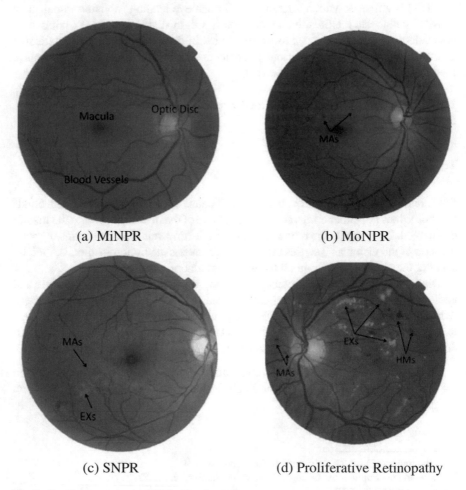

(a) MiNPR   (b) MoNPR

(c) SNPR   (d) Proliferative Retinopathy

Fig. 1 Development stages of diabetic retinopathy

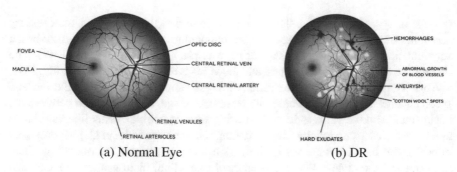

Fig. 2 Normal eye and DR eye

eye is given in Fig. 2 [5]. More than 300 million people are affected by DR according to World Health Organization. This eye disease comes in association with diabetes patients who suffer from it for the long time. Vision loss due to DR can be backed off if it is detected at the earlier stage and proper medication is taken. One challenge in predicting the DR at earlier stage is that it seldom shows its symptoms unless it is too late to begin the treatment. Still no treatment exists to cure DR, but it may be controlled if detected in advance. Hence, detecting DR at its earlier stage is very important.

Vascular abnormalities in the eye are used by the clinicians to recognize the disease. The equipments needed and experts to spot the DR are very few in the areas diabetic patients' population are high in local areas. Since the diabetics' population are increasing, infrastructure to detect DR has to be grown to put a stop to them from permanent vision loss.

2 Literature Survey on DR

Many numerous analysts had influenced different writing to study on the detection of DR. Among their works, different strategies for the recognition of DR incorporate veins detection and nearness of injuries like microaneurysms, exudates, and hemorrhages. Thus, the researchers work can be categorized into two ways, namely segmenting veins automatically [2, 6] and automation of lesion detection [1, 7].

2.1 Detection of DR Using Fuzzy Logic

Bajestani et al. [8] used type-2 fuzzy regression model to forecast the DR. Use of type-2 fuzzy model helped them to handle suspicions and imprecision in the tiny size of fundus images from large group of patients. It addressed vulnerability and dubiousness. Extra computational complexities were also leveraged by them with the use of type-2 fuzzy sets. This helps diabetic patient in avoiding unnecessary screening of their fundus image after the happening of the diabetics. Fuzzy-based segmentation—fuzzy C-means clustering was proposed by Kande et al. [9]. The divergence between the background and the veins is recognized using filtering. Non-uniformity illumination in color fundus images were corrected using the red and green channel intensity information in the retinal image. The distinction between the background and the veins is identified using filter matching. Veins like capillaries are not detected accurately by filter matching. Hence, Cinsdikici et al. [10] proposed hybrid algorithm of ant colony optimization technique with filter matching. They preprocessed the picture like green channel extraction, picture improvement, and square stamping process. The preprocessed images are then processed using ant colony optimization for exploring vessels with marked block, merging block, and binarization. The vessels are then detected using green channel normalization, logsig scaling, and Gaussian filtering. These ant colony optimizations are done in parallel with match filtering. The results of ant colony optimization technique and match filtering are then combined for DR detection.

2.2 Detection of DR Using Machine Learning Techniques

In addition to these unsupervised researches, there are also equivalent and popular researches by means of the educated datasets. Some of the techniques include ANN, KNN classifier [11], SVM [12].

Sinthanayothin et al. [11] used multilayer perceptron (MLP) neural network for DR discovery. The major features considered by them are optic disk, veins, and fovea. Optic disk of fundus picture was recognized as the pixels having high variation with the adjacent pixels. Fovea was identified by similar correlation with the characteristics of fovea like the dark near optic disk. They used standard deviation(SD) of Gaussian distribution for calculating the correlation. Veins are recognized using MLP, a category of ANN with three layers. It has 200 input nodes, 20 hidden nodes, 2 output nodes which was trained using backpropagation algorithm. Ricci et al. [12] constructed feature vector (features of veins and arteries in the retina) using SVM for notice the DR. They used unsupervised binary pixel classification for thresholding the fundus images and then supervised algorithm is used for constructing the vectors with two line separators along with the linear SVM. Niemeijer et al. [13] used KNN classifier for detecting the exudates and cotton wool spots. They used KNN classifier for detecting the bright lesions followed by linear discriminant analysis for differentiating the lesion types. Pixel order was done, and the pixels with high likelihood were assembled into cluster. In light of the attributes of cluster, each brilliant lesion cluster was named exudates or cotton fleece spot. Garcia et al. [14] distinguished exudates from the retinal image by using network classifiers, namely SVM, MLP, and RFBN. They preprocessed the image by normalizing the color, contrast, and brightness of the image. The exudates were then detected using segmentation algorithm from preprocessed image. Neural network classifiers were used to train the segmentation parameters. Islam et al. [15] used sack of words and SVM for automatically detecting DR. Features from the image are extracted, clustered using *k*-means algorithm, and the sack of words are formed. SVM classifier was used by them to classify the presence of diabetic retinopathy.

2.2.1 DR Detection Using CNN

Most of the current medical applications use DNN which are special category of ANN with more number of hidden layers. Deep neural networks increase their performance by adjusting their parameters. The presentation of convolution neural networks had pulled in numerous analysts particularly in the field of medicinal picture applications. Deep CNN [16], CNN with deep neural nets along with the dropouts and the activation functions like rectified linear units (ReLU) and modified ReLU has dramatically increased the DR detection with high accuracy.

Convolution neural network is a category of feedforward neural network. Back-propagation algorithm is used for training such networks. The fundus highlights are naturally extricated and the quick component extraction, and choice of the picture is controlled by CNN [17, 18]. It has feature extraction layer and selection layer. Convolution layer alias feature extraction layer uses various features for mapping with the image. In order to identify the unique feature of the image, the entire image is scanned with the sliding window of its respective feature map. Hence, each feature has a feature map. Feature map is a vector of weights applied on various positions of the input image vector. This minimizes the number of weights to be learned and updated compared to other neural network architectures. Pooling layer also called as sampling layer helps in selecting the features extracted. They also reduce the spatial resolution by reducing the number of parameters and reduce the considerable quantity of computation. Normally, pooling layer lies between two convolution layer.

Pratta et al. [19] developed CNN for identifying the features microaneurysms, exudates, and hemorrhages of the retina. They used data augmentation, dropout at dense layer and L2 normalization to reduce overfitting. Their CNN architecture consists of 10 deep convolution layers followed by three fully connected layers. They achieved 95% specificity, 75% accuracy, and 30% sensitivity on publicly available Kaggle dataset.

The presence of microaneurysms was used as the major factor for detecting the DR at early stage. In order to detect it by classifying into bright and lesser lesions, preprocessing like color normalization along with fuzzy C-means clustering was used by Osareh et al. [20]. They utilized fuzzy C-means clustering to detect the exudates and picked neural network for characterizing the fundus picture. Their neural net consists of 10 nodes in the input layer, 1 output neuron node, and 2–35 hidden neurons. Backpropagation and scaled conjugate gradient descent algorithms were used for training the network. Zhiguang Wang et al. [21] used global pooling layer and regression activation map instead of fully connected layer after the last convolution layer in CNN. It helped them in visualizing the region of interest which are more discriminative for the given regression value. Roychowdhury et al. [22] used machine learning technique for analysis of bright and red lesions by segmenting the optic disk and veins. They selected the most prevalent 30 features from 78 features by ranking the features using feature weights generated from AdaBoost [23]. Gaussian mixture model was used for spot the bright lesions and KNN for spot the red lesions. Exudates and cotton fleece spots are classified as bright lesions. Microaneurysms and hemorrhages are classified as red lesions. Gulshan et al. [7] used deep CNN for extracting various features from the fundus picture. They used pixel intensity for that. The underlying weights for the system are haphazardly allocated. The network was then trained by updating the weights associated with the feature through back-propagation algorithm. Stochastic gradient descent method was used by them for calculating updated weight of the network.

2.2.2 DR Detection by Other Methods

Labhade et al. [24] used various soft computing techniques like AdaBoost, gradient boost, SVM, Gaussian naive Bayes, and random forests for detecting DR. They used statistical texture analysis—statistical moments and GLCM for extracting fundus image features. Saranya Rubini et al. [25] detected the microaneurysms and hemorrhages by analyzing the image's Hessian matrix eigenvalues. They preprocessed fundus image and created Hessian matrix. The two eigenvalues for the matrix were then computed. The candidate regions were then identified by analyzing those two eigenvalues of the matrix, and feature analysis was performed. By using SVM classifier, the ending candidates were obtained and classified into microaneurysms and hemorrhages. Besides detecting the vessel like structure, biasing the eigenvalues helps them in identifying the sheet-like and blob-like structure [26, 27].

Comparisons of the above works are listed in Table 1.

3   Performance Measure Parameters

Generally, DR detection system is measured by means of three parameters, namely sensitivity, specificity, and accuracy. The four parameters used for determining specificity, sensitivity, and accuracy are

True Positive (TP): The images with DR are correctly identified as image having DR
True Negative (TN): The images with no DR are correctly identified as images with no DR
False Positive (FP): The images with no DR are wrongly identified as images with DR
False Negative (FN): The images with DR are wrongly identified as images with no DR
The specificity, sensitivity, and accuracy of the DR detection algorithm are given by

$$Specificity = TN/(TN + FP)$$

Table 1 Comparison of various DR detection methods

Researchers	Techniques used	Sensitivity	Specificity	Accuracy
Kande et al.	Fuzzy c-means clustering	–	–	0.8911 (DRIVE database) 0.896 (STARE database)
Cinsdikici et al.	Ant colony optimization + filter matching	–	–	0.9293
Osareh et al.	Fuzzy c-means clustering + neural network with scaled conjugate gradient descent (30 hidden neurons)	0.827	0.934	0.901
	Fuzzy c-means clustering + neural network with backpropagation (10 hidden neurons)	0.89	0.898	0.896
Roychowdhury et al.	Machine learning (AdaBoost, Gaussian mixture model and KNN)	0.80 (red lesions) 0.742 (bright lesions)	0.85 (red lesions) 0.98 (bright lesions)	–
Ricci et al.	Orthogonal lines SVM	–	–	0.9646 (STARE database) 0.9595 (DRIVE database)
Gulshan et al.	Deep CNN with Stochastic gradient descent	0.97	0.93	0.99
Niemeijer et al.	KNN classifier along with linear discriminant analysis	0.95 (bright lesions) 0.70 (exudates) 0.77 (cotton wool spots)	0.88 (bright lesions) 0.93 (exudates) 0.88 (cotton wool spots)	–
Garcia et al.	SVM	1	0.7778	0.9104
	MLP	1	0.9259	0.9701
	RBFN	1	0.8148	0.9254
Pratt et al.	CNN	0.30	0.95	0.75

Sensitivity = TP/(TP + FN)
Accuracy = (TP + TN)/(TP + TN + FP + FN).

In addition to the above measurements, precision is also used to measure the algorithm and is given by

Precision = TP/(TP + FP).

4 Conclusions

Vision loss in diabetic patients can be prevented by detecting the diabetic retinopathy. Ophthalmoscope helps the ophthalmologist to identify the DR stage by visualizing the features like veins, microaneurysms, exudates. This screening is regularly needed for the patients so that DR can be controlled on time. Manual screening by ophthalmologist is costlier, and hence, automated system to recognize DR is highly required. Several popular methods and algorithms for automatic detection of DR have been discussed in this paper. Among those deep neural networks outperform other methodologies in automatic detection of DR. In particular, CNN plays a major role in classification and segmentation of computer vision applications.

References

1. Abràmoff, M.D., Lou, Y., Erginay, A., Clarida, W., Amelon, R., Folk, J.C., Niemeijer, M.: Improved automated detection of DR on a publicly available dataset through integration of deep learning. Invest Ophthalmol Vis Sci **57**(13), 5200–5206 (2016)
2. Liskowski, P., Krawiec, K.: Segmenting retinal blood vessels with deep neural networks. IEEE Trans. Med. Imaging **35**(11), 2369–2380 (2016)
3. Early Treatment DR Study Research Group: Grading DR from stereoscopic color fundus photographs—an extension of the modified Airlie house classification: ETDRS report number 10. Ophthalmol. **98**(5), 786–806 (1991)
4. https://www.researchgate.net/figure/Stages-of-Diabetic-Retinopathy_fig1_280231772
5. Mateen, M., Wen,, J., Nasrullah, Song, S., Huang, Z.: Fundus image classification using VGG-19 architecture with PCA and SVD (2018). https://doi.org/10.3390/sym11010001
6. Jan, O., Kolar, R., Budai, A., Angelopoulou, E.: Retinal vessel segmentation by improved matched filtering: evaluation on a new high-resolution fundus image database. IET Image Process. **7**(4), 373–383 (2013)
7. Gulshan, V., Peng, L., Coram, M., Stumpe, M.C., Wu, D., Narayanaswamy, A., Venugopalan, S., Widner, K., Madams, T., Cuadros, J., Kim, R., Raman, R., Nelson, P.C., Mega, J.L., Webster, D.R.: Development and validation of a deep learning algorithm for detection of DR in retinal fundus photographs. JAMA **316**(22), 2402–2410 (2016)
8. Bajestani, N.S., Kamyad, A.V., Esfahani, E.N., Zare, A.: Prediction of retinopathy in diabetic patients using type-2 fuzzy regression model. Eur J Oper Res **264**, 859–869 (2017)
9. Kande, G.B., Subbaiah, P.V., Savithri, T.S.: Unsupervised fuzzy based vessel segmentation in pathological digital fundus images. J. Med. Syst. **34**(5), 849–858 (2010)

10. Cinsdikici, M.G., Aydın, D.: Detection of blood vessels in ophthalmoscope images using MF/ANT (matched filter/ant colony) algorithm. Comput. Methods Programs Biomed. **96**(2), 85–95 (2009)
11. Sinthanayothin, C., Boyce, J.F., Cook, H.L., Williamson, T.H.: Automated localisation of the optic disc, fovea and retinal blood vessels from digital colour fundus images. Br. J. Ophthalmol. **83**(8), 902–910 (1999)
12. Ricci, E., Perfetti, R.: Retinal blood vessel segmentation using line operators and support vector classification. IEEE Trans. Med. Imaging **26**(10), 1357–1365 (2007)
13. Niemeijer, M., van Ginneken, B., Russell, S.R., Suttorp-Schulten, M.S.A., Abràmoff, M.: Automated detection and differentiation of drusen, exudates and cotton–wool spots in digital color fundus photographs for DR Diagnosis. In: Proceedings of Image Visual Computing New Zealand, vol. 48, pp. 2260–2267 (2007)
14. Garcia, M., Sanchez, C.I., Lopez, M.I., Abasolo, D., Hornero, R.: Neural network based detection of hard exudates in retinal images. Comput. Methods Programs Biomed. **93**, 9–19 (2008)
15. Islam, M., Dinh, A.V., Wahid, K.A.: Automated DR detection using sack of words approach. J. Biomed. Sci. Eng. **10**, 86–96 (2017)
16. Bengio, Y., Lamblin, P., Popovici, D., Larochelle, H.: Greedy layer-wise training of deep networks. In: Advances in Neural Information Processing Systems, pp. 153–160 (2007)
17. Krizhevsky, A., Sutskever, I., Hinton, G.E.: Image net classification with deep convolutional neural networks. In: Advances in Neural Information Processing Systems, pp. 1097–1105 (2012)
18. LeCun, Y., Bengio, Y.: Convolutional networks for images, speech and time series. In: The Handbook of Brain Theory and Neural Networks, vol. 3361(10), 1995(1995)
19. Pratta, H., Coenen, F., Broadbent, D.M., Harding, S.P., Zhenga, Y.: Proc. Comput. Sci. **90**, 200–205 (2016)
20. Osareh, A., Mirmehdi, M., Thomas, B., Markham, R: Classification and localisation of diabetic-related eye disease. In: European Conference on Computer Vision. Springer, Berlin, Heidelberg, pp. 502–516 (2002)
21. Wang, Z., Yang, J.: DR Detection via Deep Convolutional Networks for Discriminative Localization and Visual Explanation (2017)
22. Roychowdhury, S., Koozekanani, D.D., Parhi, K.K.: Dream: DR analysis using machine learning. IEEE J. Biomed. Health Inf. **18**(5), 1717–1728 (2014)
23. Shen, L., Bai, L.: Abstract AdaBoost Gabor feature selection for classification. In: Proceedings of Image Visual Computing New Zealand, pp. 77–83(2004)
24. Labhade, J.D., Chouthmol, L.K., Deshmukh, S.: DR detection using soft computing techniques. In: IEEE: International Conference on Automatic Control and Dynamic Optimization Techniques (ICACDOT) (2016). https://doi.org/10.1109/icacdot.2016.7877573
25. Rubini, S.S., Kunthavai, A.: DR detection based on eigen values of the Hessian matrix. Proc. Comput. Sci. **47**, 311–318 (2015)
26. Frangi, A.F., Niessen, W.J., Vincken, K.L., Viergever, M.A.: Multiscale vessel enhancement filtering. In: Medical Image Computing and Computer–Assisted Intervention (MICCAI), vol. 1496. Lecture Notes in Computer Science, pp. 130–137 (1998). Online: http://www.tecn.upf.es/~afrangi/articles/miccai1998.pdf
27. Yu, Y., Zhao, H.: Enhancement filter for computer-aided detection of pulmonary nodules on thoracic CT images. In: Proceedings of 6th International Conference on Intelligent Systems Design and Applications, pp. 1200–1205 (2006)

A Critical Review on Federated Cloud Consumer Perspective of Maximum Resource Utilization for Optimal Price Using EM Algorithm

Pradeep Kumar V and Kolla Bhanu Prakash

Abstract Federated clouds have been a solution to some of the challenges of cloud computing like vendor lock-in and performance-related issues in terms of a wide range of resource utilization and pricing for cloud consumers. This paper provides much insight into the problems faced by cloud consumers while utilizing resources for particular price in relation to SLA violation, QoS awareness and cloud brokerage. A brief review of resource utilization with pricing in perspective of cloud consumers is presented, and a layered agent-based model was proposed for simulating federated cloud. To analyze maximum resource utilization on pricing option a MaxResourceUtility, an expected maximization (EM) algorithm was proposed to consider the influence of missing QoS factors while estimating it for resource utility. The results show that 5–10% increase in maximum resource utility and 10–20% decrease in pricing are observed by considering QoS factor response time while utilizing resources.

Keywords Cloud computing · Federated clouds · Resource maximization · Expected maximization (EM) · QoS factors · Cloud broker

1 Introduction

Cloud computing is a tremendous computing environment where different sets of resources are made available to cloud consumers on request. The service-level agreement (SLA) incorporates a contractual agreement between cloud providers and consumers in providing flexibility while utilizing resources. The cloud consumers were given less importance in terms of utilizing resources for specific prices. Many pricing models were proposed for trying to maximize the cloud provider profits but not looking for user satisfaction in utilization of resources.

P. K. V (✉) · K. B. Prakash
Department of Computer Science and Engineering, Koneru Lakshmaiah Education Foundation, Vaddeswaram,Vijayawada, Andhra Pradesh, India
e-mail: pradeepvadla@gmail.com

K. B. Prakash
e-mail: drkbp@kluniversity.in

© Springer Nature Singapore Pte Ltd. 2020
K. N. Das et al. (eds.), *Soft Computing for Problem Solving*,
Advances in Intelligent Systems and Computing 1057,
https://doi.org/10.1007/978-981-15-0184-5_15

Cloud service providers failed in providing vast resources with wide variations as demanded by cloud consumers due to increase of their utilization of cloud resources. Cloud consumers are suffering from SLA violations and vendor lock-in problems in order to shift to another provider who was flexible enough to meet their requirements in specification with provisioning resources. Resource pricing was a critical challenge for consumers in cloud computing because they never framed or considered their concerns in terms of implementation. Many pricing models as reviewed in [1] to state that revenue maximization for providers was given more priority but concerns of consumers were not taken into consideration in terms of their satisfaction for meeting their various ranges of demand for service utilization at particular instances.

Federated cloud was a collaborative cloud computing model where different sets of cloud service providers get collated to meet the demands of cloud consumers at any instance of their request. The federated cloud provider's collation formation would resolve much of vendor lock-in problems but SLA violation would become more complex to manage. The collation formation needs one more agreement among the collaborated providers named as federation-level agreement (FLA) whose information is not in awareness of cloud consumers. QoS (Quality of Service) violations occurred during the service utilization made by consumers in the federated cloud will lead to significant profit loss for consumers.

Resource management in federated cloud is a very critical task to analyze because much of the issues are related to over-provisioning and under-provisioning of resources and it was a big challenge for providers to manage these consumer requests when they are given to these collated providers. Liaqat et al. [2] reviewed that the federated resource management should work on with QoS-differentiated resource pricing and QoS-differentiated resource discovery to enhance the resource pricing and identification of relevant resources to consumers. In Sect. 2, detailed review on resource utilization and pricing with different factors like QoS and SLA violation is done. Section 3 gives proposed layered agent-based model. Section 3.1 discusses in detail about considered system model with EM algorithm, and Sect. 4 highlights the results. Section 5 gives the conclusion and future work.

2 Related Work

Auto-scaling of computational resources needs to meet the advanced reservation requests and on-demand requests of cloud consumers. It also needs to be analyzed for proactive and reactive approach by using machine learning algorithms and grade of service mechanism. In [3], a hybrid auto-scaling of computational resources was managed by intermediate enterprise represented by broker to a single-client enterprise which contains a set of cloud consumers. Instead of single consumer directly requesting for resources with cloud provider, this model of intermediated enterprise effectively manages all kinds of requests which may not be supported by cloud provider. Resource pricing is automatically managed by broker by buying them from cloud provider and providing them at reduced cost for consumers. This hybrid approach of auto-scaling resources was not applied for storage and bandwidth usage of consumers.

Trading of all cloud computing resources was done through virtual machine as an instance which manages to provide set of computational, storage and bandwidth resources on demand of cloud consumers. The cloud service broker (CSB) [4] was proposed to manage the performance variation in instances of the same type for heterogeneity clouds. The pricing issues related to scaling of applications instantly for certain time period varied depending on its performance for different cloud user requests. To address this price vs performance variation for cloud user-specified requirements, CSB runs a secondary market to offer an on-demand workload performance-specific assurance service. This price discovery mechanism needs to be analyzed for heterogeneous users for different requests per workload.

The wide variety of heterogeneous resource request for particular applications in public clouds is done effectively in hybrid clouds. The multi-agent system (MAS) [5] is analyzed for handling cost-efficient distribution of system between the resources for on/off premises. An optimal framework was proposed for performance evaluation process to optimize the deployment costs by providing variant options by partitioning resources to meet the increased demands of cloud users. The graph partitioning algorithm is used for hybrid cloud deployment cost management. The cloud broker needs to be incorporated in multi-agent systems to provide automatic adaptive deployment solutions to meet the demands of cloud consumers.

Cloud providers have a provision of gaining profits while delivering resources to cloud consumers. The adaptive resource provisioning of cloud providers is analyzed with different QoS factors and SLA violation while pricing resources. The reinforcement learning scheme in [6] considers QoS demands of cloud consumers while provisioning the resources by cloud providers. The request admission control policy influences the pricing of resources for consumers by collecting their QoS demands along with their requests, and any SLA violation was strictly monitored by providers for reducing their cost. In this approach, federated cloud setup was not tested for QoS constraints while provisioning resources by collated cloud providers.

The elastic nature of cloud computing needs an automated prediction suite [7] to predict the workload usage patterns of cloud consumers and try to provision resources according to their demands. The effective resource management is made possible if the time series prediction is done to auto-scale resources depending on future demands of cloud consumers. Different machine learning algorithms were theoretically analyzed for auto-scale of resources dynamically to meet the demands of cloud consumers. The risk minimization principle was evolved to handle over-provisioning of resources in order to sustain the utilization of resources by cloud consumers. The federated cloud setup needs to incorporate these time series prediction algorithms to predict resource demands of cloud consumers.

Cloud consumers need to be provided with an automatic decision-making tool in choosing cloud provider which is relevant to his request satisfying the specific constraints like reliability and capacity of availability of application to host in cloud. The multi-criteria decision making with stochastic model method [8] was used to analyze and rank the cloud infrastructures according to customer request considerations in terms of resource availability, downtime and cost. Federated cloud also needs some criteria of decision-making model while forming collation during outages of other collated providers and provides continuous services to cloud consumers. The cloud broker needs to gather all possible QoS constraints related to availability and reliability of cloud consumers and make available to cloud providers to choose for collaboration with other providers in making uninterrupted services available to cloud consumers.

Enterprises have a huge demand of using cloud computing resources to meet their goals and objectives of providing effective service to their customers. Each enterprise will get subscribed with cloud providers by specifying some limitations in terms of service usability. Cloud providers need to have a specific agreement called SLA to satisfy all the constraints specified by subscribed enterprises. The SLAs need to incorporate service-level objectives for more clear specifications of cloud subscribers to meet their huge demand of heterogeneous resources. A resource allocation plan (RAP) [9] for workload burstiness was used by cloud providers to predict the time-varied resource allocation to satisfy bursts and use service-level planning tool to frame the SLA along with specified service-level objectives to meet the demands of cloud subscribers. Federated clouds are also in need of such SLA plan to monitor the resource utilization of cloud consumers and predict burstiness workload to get serviced without any time delay.

Federated cloud resolving vendor lock-in problems for cloud consumers was made possible by implementing a cloud broker in discovering capabilities of different cloud providers for meeting the QoS requirements. The cloud brokering based on genetic algorithm QBrokerage [10] was analyzed for different QoS constraints to satisfy customer's preferences to run their applications at cloud provider. The cost models varied based on QoS-aware allocation of resources both at local in terms of storage and computational constraints and global in terms of network constraints. This QBrokerage was analyzed for only static applications; it needs to be implemented for elasticity of applications for satisfying cloud consumers.

Edge computing provisions resources for cloud consumers through network edge. The pricing of these geo-distributed edge nodes for resource allotment was analyzed to achieve market equilibrium approach [11]. The market equilibrium takes into considerations of cloud consumer's requests within their budget constraints and allocates resource bundles to provide service. This edge computing needs to be correlated to federated clouds where heterogeneous set of resources are provisioned through broker for forming collation and act as edge nodes to test for achieving market equilibrium.

Cloud computing made tremendous set of capabilities possible for cloud consumers in provisioning resources at different costs but much of these cost never incorporated the data center maintenance cost or network cost. The federated cloud resolves this issue of burdening the cloud consumers with these extra costs by forming collation with other sets of providers to provision resources collaboratively to meet the demands of cloud consumers. The coordinated resource provisioning mechanism [12] caused a trading cooperation between collaborated cloud providers to provide resources at affordable cost of cloud consumers. A two-phase coordinated resource reservation and provisioning were analyzed mathematically to provide resources at minimum cost of resource usage satisfying on-demand request instances. Cloud providers' uncertainty in fixing prices for change of demand of consumers instantly was not taken into considerations.

Cloud computing service model infrastructure as a service (IaaS) needs to provide a variant set of resources for instant request made by cloud consumers. An adaptive reward model was used in [13] to predict the vast utilization of resources within a specific constraint without any SLA violation. The Markov chain model [15] was used to access the service blocking and required number of resource instances for each request. This adaptive reward model has significance incorporating all probability of losing resources due to SLA violation, outages and downtime at cloud providers. Federated clouds are also in need of such reward model at cloud service broker so as to make easy selection of collated providers meeting the consumer's resource request.

Federated cloud brings the wide range of geo-distributed data centers available for cloud consumers at an instance, but this causes local data center cloud providers facing a significant loss in their profits earned in order to stabilize the resource utilization of consumers. A contract-based resource sharing model was proposed in [14], to enable cloud providers to form a contract with geo-distributed data centers in order to share resources to meet the demands of cloud consumers. This model needs to be analyzed with different QoS factors taken into consideration while forming contracts and then tries to ensure the provisioning of resources without any profit loss for cloud providers.

Detail of parameters covered by the implementation of above referred papers approaches are listed in Table 1. The list gives us clear understanding of need of a

Table 1 Performance metrics for resource utilization and pricing for cloud consumers in federated cloud

Referred paper	Scalability	Accuracy	Fairness	Availability	Fault tolerance
Biswas et al. [3]	Yes	No	No	Yes	Yes
O'Loughin et al. [4]	No	Yes	Yes	No	No
Alsarhan et al. [5]	Yes	No	Yes	No	Yes
Nkarvesh et al. [6]	Yes	Yes	No	Yes	Yes
Labba et al. [7]	Yes	Yes	No	Yes	Yes
Araujo et al. [8]	Yes	Yes	No	Yes	No
Youssef et al. [9]	Yes	No	No	Yes	Yes
Anasasi et al. [10]	Yes	No	No	Yes	Yes
Nguyen et al. [11]	Yes	Yes	No	Yes	No
Reddy et al. [12]	Yes	No	No	Yes	No
Chang et al. [13]	Yes	Yes	No	Yes	Yes
Xu et al. [14]	Yes	Yes	No	Yes	Yes

model to analyze the resource utilization and pricing through cloud consumer per-spective by taking different QoS factors into consideration while utilizing resources. Much of referred papers have neglected the fairness aspect in pricing distribution for cloud consumers. The earlier implementation approaches listed in Table 1 were not estimated missed QoS factors while anlayzing resource utilization made by cloud consumers.

3 Proposed Layered Agent-Based Model for Analyzing Federated Cloud

Layered agent-based model consists of layering the agents to simulate federated cloud nature for analyzing different aspects like maximizing resource usability and providing optimal price option for consumer agents by considering different QoS factors. The layered agent-based model considers four layers for analyzing commu-nication to satisfy consumers getting serviced for optimal price. The four layers in Fig. 1 are provider agent layer, collated provider agent layer, broker agent layer and cloud consumer agent layer. Simulating federated nature of clouds is possible by using this layered approach. Each layer has its own significance task to handle the functionality of federated nature to make it possible for analyzing resource utilization of cloud consumers for optimal service time.

Provider agent layer will form collation with other provider agents to make het-erogeneous resources available at any instant. Collation formation is a tedious task which involves federation-level agreement (FLA) to share resources for certain QoS constraints and price. Each cloud provider will initiate the collation formation by communicating this FLA in order to generate maximum profit by collaboration. The next layer collated provider agents will provide all set of collaborated resources for

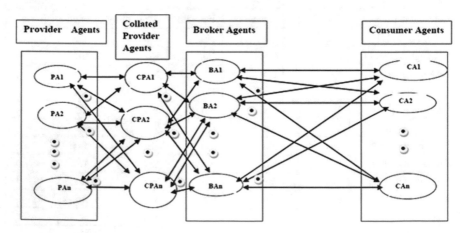

Fig. 1 Layered agent-based model

brokers to allocate for cloud consumers. The broker agent layer performs major task in generation of service classes for different sets of collated provider agent resources. This service class generation can be used for QoS-aware differential pricing for providing different prices for cloud consumers to meet their QoS demands along with resource requests. The cloud consumer agents will have a choice of selection of these service classes to get serviced his resource requests for an optimal price. The cloud consumers will estimate QoS factors for his next request of resources using EM algorithm.

3.1 System Model

The layered agent-based model provides scope for analyzing resource utilization and pricing options for federated cloud consumers using mathematical approach. Let us consider the provider agents $(PA_1, PA_2, \ldots, PA_n)$, collated provider agents $(CPA_1, CPA_2, \ldots, CPA_n)$, broker agents $(BA_1, BA_2, \ldots, BA_n)$ and consumer agents $(CA_1, CA_2, \ldots, CA_n)$. The resource request made by consumer agent was treated as vector of VM instances through which they specify the resources required $REQV(VM_1, VM_2, \ldots, VM_n)$. Each VM instance will have its specified resources to get serviced by broker agent. Along with resource request, vector of each VM instance will have equivalent QoS factors to meet while getting serviced. The QoS factor response time considered for each VM instance is $QOSV(RT_1, RT_2, \ldots, RT_n)$. The resource utilization is measured using $RUtil_\theta(i)$ where θ is the user instance for particular time period T and 'i' is the service class selection to utilize maximum resources of that VM instance by user θ.

$$\text{Tpri}_j = \text{Acepr}_i + \text{Sret}_i * \text{CVM}_i \tag{1}$$

where Tpri_j is the total pricing function of 'j' cloud consumer, $Acepr_i$ is the acceptable price by cloud consumer from broker service class-i for utilizing those VM instance resources, Sret_i is the service time of service class-i and CVM_i is the capacity of resources of service class-i VM instances for utilization.

$$\text{CVM}_i = \sum_{i=1}^{N} \text{Nr}_i * \Pr_i \tag{2}$$

where Nr_i is the number of resources of that VM instance of service class-i and Pr_i is the price associated with that resource of service class-i for that particular instance.

The resource utilization of 'j' cloud consumer is given by

$$\text{RUtil}_\theta(j) = Ut_{\text{Max}} - \text{Tpri}_j - \theta \, \text{QosFunc}(N_i, \text{CVM}_i) \tag{3}$$

U_{Max}	Maximum utilized resources made by θ users from service class-i
Tpri_j	Total pricing function of the jth user
$\text{QosFunc}(N_i, \text{CVM}_i)$	Congestion function which measures the influence of violation of that QoS factors while utilizing those resources for time period t.

Choosing a service class-i is possible maximizing this $\text{RUtil}_\theta(j)$ for particular time period t for jth user

$$i = \arg \text{Max}_{j \in \{1,2...m\}} \text{RUtil}_\theta(j)$$

or else not to select the service class-i

if $\text{RUtil}_\theta(j) < 0$.   (4)

$\text{QosFunc}(N_i, \text{CVM}_i)$ is a critical function of considering QoS factors affecting the resource utilization of that service class-i for a no. of users N_i while allocating its capacity of resources of VM instances CVM_i. Let us consider the QoS evaluation made for a particular service class by N users in a time slot T is E_i.
$E_i = \{e_{i,1}, e_{i,2}, \ldots \ldots, e_{i,N}\}$ where $e_{i,1}$ is the QoS evaluation made by first user of that service class-i.

Consider the mean value for QoS evaluations of VM instances.

$$\text{Esm}_i = \frac{1}{N} \sum_{k=1}^{N} e_{i,k}$$ (5)

$$\text{QoSFunc}(N_i, \text{CVM}_i) = \frac{1}{N_i} \sum_{i=1}^{N} e_{i,k} * \text{CVM}_i$$ (6)

Algorithm MaxResourceUtiltyEM For jth consumer of service class-i
   Input: REQV_j, QOSV_j, CVM_i, Acepr_i, Sret_i, E_i, Pr_i, Nr_i, Ut_{Max}
   Output: RUtil_θ for jth consumer

1. For each service class-i at broker for a particular time period t.
2. Compute CVM_i utilization of VM instances of service class-i using (2).
3. Compute total pricing of utilized resources Tpri_j of jth consumer using (1).

   Expected Step:

4. Compute the $\text{QosFunc}(N_i, \text{CVM}_i)$ by estimating the E_i on expected QoS evaluation values at 't' affecting the resource utilization using (6).

   Maximization Step:

5. Compute RUtil_θ for jth consumer by (3).
6. Choose the maximum $\text{RUtil}_\theta(j)$ for given QoS values of jth user.
7. If $\text{RUtil}_\theta(j) < 0$, reject that service class.
8. End for.

4   Results and Analysis

Let us consider the four service classes with set of resources as in Table 2 and Table 3 give prices of resources and service time. The following data is taken with reference to Amazon EC2 prices to analyze mathematically for the above algorithm. Figure 2a gives the comparison of maximum resource utility made by four consumer agents without QoS and with QoS, and Fig. 2b gives the comparison of prices paid by four consumer agents without QoS and with QoS parameters into consideration.

The results in Fig. 2a show 5–10% increase in resource utilization because of considering the QoS factor response time of service classes as evaluated by consumer agent. The results in Fig. 2b show 10–20% decrease in pricing of resources because of considering the QoS factor response time of service classes, and QoS value effect is measured by QoS evaluation function at cloud broker.

Figure 3(a) provides the estimated QoS for number of consumer agents which is computed with Eq. (6) during expected step. Figure 3(b) gives results of maximization step for the number of consumer agents using Eq. (3). Both graphs show that QoS function monotonically increases more no. of consumer agents.

Table 2 Service classes with resources

Service classes	Small			Medium			Large		
	CPU	RAM	HD	CPU	RAM	HD	CPU	RAM	HD
Class 1	4	4	20	10	8	120	20	16	200
Class 2	5	4	30	12	8	150	17	16	250
Class 3	2	4	40	15	8	110	25	16	240
Class 4	4	4	20	11	8	150	22	16	270

Table 3 Service classes with price and service time

Service classes	Prices in $			Service time (ms)
	Per CPU	Per GB	Per TB	
Class 1	0.12	0.10	0.05	0.23
Class 2	0.15	0.20	0.08	0.5
Class 3	0.17	0.08	0.02	0.25
Class 4	0.20	0.15	0.09	0.13

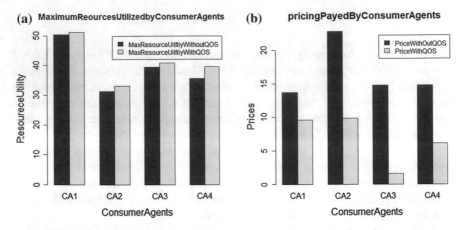

Fig. 2 **a** Comparison of maximum resource utility with QoS and without QoS. **b** Comparison of price without QoS and with QoS

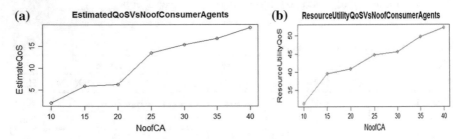

Fig. 3 **a** Estimated QoS vs. no. of consumer agents. **b** Resource utility QoS vs. no. of consumer agents

5 Conclusion and Future Work

The proposed model was analyzed to enable possibility of creating federated cloud setup for performing analytical study and to understand its functionality. The algorithm proposed is meant for analyzing the resource utilization and pricing of federated clouds in view of cloud consumers.

The result shows the comparison of resource utilization and pricing with QoS and non-QoS, and it clearly states that resource utilization increases and pricing of resources will decrease because of QoS evaluation function at cloud broker. More fine results need to be tracked by increasing no. of consumer agents and taking more QoS factors for QoS evaluation function.

Future scope of our work needed to enhance this model with varied factors to get the actual feel of working on federated clouds, perform in-depth analysis on auto-scaling of resources dynamically with variety of QoS factors and develop economic model to provide an optimal pricing option for consumers while utilizing resources.

References

1. Soni, A., Hasan, M.: Pricing schemes in cloud computing: a review. Int. J. Adv. Comput. Res. **7**(29), 60 (2017)
2. Liaqat, M., Chang, V., Gani, A., Ab Hamid, S.H., Toseef, M., Shoaib, U., Ali, R.L.: Federated cloud resource management: review and discussion. J. Netw. Comput. Appl. **77**, 87–105 (2017)
3. Biswas, A., Majumdar, S., Nandy, B., El-Haraki, A.: A hybrid auto-scaling technique for clouds processing applications with service level agreements. J. Cloud Comput. **6**(1), 29 (2017)
4. O'Loughlin, J., Gillam, L.: A performance brokerage for heterogeneous clouds. Future Gener. Comput. Syst. **87**, 831–845 (2017)
5. Alsarhan, A., Itradat, A., Al-Dubai, A.Y., Zomaya, A.Y., Min, G.: Adaptive resource allocation and provisioning in multi-service cloud environments. IEEE Trans. Parallel Distrib. Syst. **29**(1), 31–42 (2018)
6. Nikravesh, A.Y., Ajila, S.A., Lung, C.-H.: An autonomic prediction suite for cloud resource provisioning. J. Cloud Comput. **6**(1), 3 (2017)
7. Labba, C., Narjès Saoud, B.B., Dugdale, J.: A predictive approach for the efficient distribution of agent-based systems on a hybrid-cloud. Future Gener. Comput. Syst. **86**, 750–764 (2017)
8. Araujo, J., Maciel, P., Andrade, E., Callou, G., Alves, V., Cunha, P.: Decision making in cloud environments: an approach based on multiple-criteria decision analysis and stochastic models. J. Cloud Comput. **7**(1), 7 (2018)
9. Youssef, A.A., Krishnamurthy, D.: Burstiness-aware service level planning for enterprise application clouds. J. Cloud Comput. **6**(1), 17 (2017)
10. Anastasi, G.F., Carlini, E., Coppola, M., Dazzi, P.: QoS-aware genetic cloud brokering. Future Gener. Comput. Syst. **75**, 1–13 (2017)
11. Nguyen, D.T., Le, L.B., Bhargava, V.: Price-based resource allocation for edge computing: a market equilibrium approach. IEEE Trans. Cloud Comput.1-1 (2018)
12. Reddy, K.H., Kumar, G.M., Roy, D.S.: A novel coordinated resource provisioning approach for cooperative cloud market. J. Cloud Comput. **6**(1), 8 (2017)
13. Chang, B.J., Lee, Y.W., Liang, Y.H.: Reward-based Markov chain analysis adaptive global resource management for inter-cloud computing. Future Generation Comput. Syst. **79**, 588–603 (2017)
14. Xu, J., Palanisamy, B.: Optimized contract-based model for resource allocation in federated geo-distributed clouds. IEEE Trans. Serv. Comput. **1**, 1–11 (2018)
15. Prakash, K.B., Rangaswamy, D.: Content extraction of biological datasets using soft computing techniques. J. Med. Imaging Health Inf. **6**(4), 932–936 (2016)

Node Localization in Wireless Sensor Networks Using Multi-output Random Forest Regression

K. Madhumathi and T. Suresh

Abstract More advanced developments have been made in the field of wireless communications, and it has further accelerated the growth of compact and low power-consuming wireless sensor nodes. During communication, each source node estimates the shortest path to the destination node by using the location information. Location information also helps in securing the network in the prevention of intruders. Previously available sensor node localization methods in the literature such as radio signals, time of arrival (ToA), and time difference of arrival (TDoA) suffers from various drawbacks. Also, the usage of sophisticated devices like GPS to sense the location of the node increases the deployment cost and in parallel, the energy consumption is also increased. This paper aims at developing a model to predict the future location of a dynamic sensor node. The linear model is built using the historical location information of the respective node. The trained model is capable of predicting the X- and Y-coordinates of a node accurately. For each of the node, a separate model is built and their future locations are predicted. If a node has data packets to transmit to a sink node, it obtains the present and next location of the sink node from the base node.

Keywords Node localization · Linear regression · Time of Arrival · Time difference of arrival · Random forest regression

K. Madhumathi (✉)
Department of BCA, Anna Adarsh College for Women, Chennai, Tamil Nadu, India
e-mail: madhumathi@hotmail.com

T. Suresh
Department of Computer Science and Engineering, Annamalai University, Chidambaram, Tamil Nadu, India
e-mail: sureshaucse@gmail.com

© Springer Nature Singapore Pte Ltd. 2020
K. N. Das et al. (eds.), *Soft Computing for Problem Solving*,
Advances in Intelligent Systems and Computing 1057,
https://doi.org/10.1007/978-981-15-0184-5_16

177

1 Introduction

A wireless sensor network is composed of numerous homogeneous or heterogeneous nodes distributed spatially over a geographical region. The role of WSN in our daily life ranges from home automation systems to healthcare applications. The advancements in the area of microelectromechanical systems (MEMS), wireless transmission technology are the major cause for the advancements in the area of smart cities and smart vehicles. Based on the problem-specific requirements, the sensor nodes can be deployed either in deterministic or random fashion. The functioning of a node in WSN can be categorized into following stages, namely acquisition, operation, and communicating the acquired data. The various stages in the process of communication in WSN are shown in Fig. 1. Each of the sensors captures the spatiotemporal information from the environment and forwards the captured information to a gateway node or a cluster head node (if hierarchical clustering is adopted). These gateway or cluster head nodes again forward the information to the base station. The processing of the acquired information will be done in the base station to perform any of the following tasks such as fusion, location tracking. Location tracking helps in providing location-based services, network optimization, and environment characterization.

Wireless sensor networks are generally second hand to measure physical phenomena of environment, and in many of the applications, the sensed values are tagged with geo-spatial data, i.e., both time stamp and the geo-location of the sensor. The WSNs can be used in many areas where each sensing node receives the information from a certain location and transmits it to a base or control station. It is an essential task to find the location from where the sensors have gathered the respective information. Researches have contributed many node localization methods, and almost all of them require more computing and other resources. Due to advancements in GPS technology, localization has been made simple and due to which other approaches have become obsolete. The low-cost GPS receivers used in smartphones have high error in the location estimation. The price of the more advanced GPS devices ranges hundreds of dollars, and they consume nearly 50 mW of energy and find the location with an error of $+1$ m or -1 m if the weather is clear. This high cost and high-power requirement do not suit many WSN applications [1].

In literatures, many algorithms have been developed for node localization used to solve multiple problems in various domains. In general, hybrid approaches for node localization exhibit more reliable and accurate results [2]. The other approaches

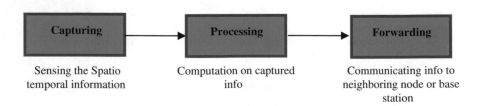

Fig. 1 Process flow in wireless sensor network

include distance or hop count-based methods which can be used for both connectivity-based and range-free algorithms. The methods proposed in the literatures have two drawbacks majorly: (i) The beacon broadcasting interval decides the accuracy of the location estimation and (ii) the variations in the radio propagation affects the accuracy of the localization [3, 4].

The location of a node can be assessed by tracking the communication between localized and un-localized node. And also, it is assessed by measuring the angular and spatial distance between the nodes. When an anchor node is used for predicting the location, then the following methods are adopted lateration, angulation, trilateration, multilateration, and triangulation.

- Lateration—node is localized based on estimate distance between nodes.
- Angulation—node is localized based on measure angle between nodes.
- Trilateration—node is localized based on estimate distance between three nodes. The position of unknown node is localized by using the three intersecting circles around the nodes.
- Multilateration—node is localized based on the position of more than three nodes.
- Triangulation—the location is assessed by measuring at least two angles of localized and un-localized node position. After measuring the angles of node, the trigonometric laws along with mathematical formulas of sine and cosine are calculated for attaining the position of a node.

Node location estimation methods can be categorized as anchor-based or anchor-free, centralized or distributed, using with or without GPS devices, fine or coarse-grained, static or dynamic sensor nodes, and based on with or without communication range. The cooperative node localization scheme is one of the successful methods which follow a simple and cost-effective distributed methodology. The cooperative node localization scheme can be further classified as range-based localization and range-free localization in wireless senor networks. The distance between two neighboring nodes is measured by using strength of the received signal, temporal methods based on time of arrival (ToA) [5, 6], time difference of arrival (TDoA) [7, 8], and angle of arrival (AoA) [9]. The location of the nodes can be estimated by using an optimization technique or by solving simple simultaneous equations for minimizing the error-rate in the prediction of node location [1, 10, 11].

This paper focuses on modeling a multi-output random forest regression for predicting the future location of a node based on the past trajectory data. The rest of the paper is organized as follows. Section 2 consists of a mathematical representation of the problem statement and the approach used for node localization. Section 3 describes the dataset used for modeling the location predictor. The methodology adopted for location prediction using regression technique is explained in Sects. 4 and 5 and gives the details about experiments conducted along with their results. Finally, Sect. 6 concludes the paper with a scope for future work.

2 Problem Statement and Multi-output Regression Approach

The aim of location estimation in WSN which is composed of total m sensor nodes is to find the location of n target nodes using the historical location coordinates of the respective nodes. In a two-dimensional location estimation problem, a total of $2n$ unknown coordinates,

$$\theta = \{\theta_x, \theta_y\}$$

where $\theta_x = \{x_1, x_2, \ldots, x_n\}$; $\theta_y = \{y_1, y_2, \ldots, y_n\}$ are to be calculated by using the available historical data coordinates $\{x_n + 1, \ldots, x_n + m\}$ and $\{y_n + 1, \ldots, y_n + m\}$.

The location estimation of the coordinates in the target nodes can be expressed mathematically for a given feature vectors $x_i \in \mathbb{R}^n$, $i = 1, 2, \ldots l$ and a vector of x–y coordinates $y \in \mathbb{R}^l$. Multi-output regression tree can be used to predict the x–y coordinates.

Multi-output regression tree has several advantages, and two of the advantages are most important among them when compared to using a separate regression tree for each of the dependent variable. In general, a single multi-target regression model is much lesser than the total size of the individual single target trees for all dependent variables. Also, a multi-target regression tree is more capable of identifying the dependencies among the various dependent variables.

Multi-output trees partition the given data training vector such that the samples with the similar X–Y coordinates are clustered together. Consider that the data at each tree node m be denoted by Q. For each candidate, node split $\theta = (j, t_m)$ containing a parameter j with a threshold t_m. Partition of the data into $Q_{\text{left}}(\theta)$ and $Q_{\text{right}}(\theta)$ subsets where $Q_{\text{left}}(\theta) = (x, y)|x_j <= t_m$ and $Q_{\text{right}}(\theta) = Q \backslash Q_{\text{left}}(\theta)$. An impurity function $H()$ is used to estimate the impurity at m.

$$G(Q, \theta) = \frac{n_{\text{left}}}{N_m} H(Q_{\text{left}}(\theta)) + \frac{n_{\text{right}}}{N_m} H\left(Q_{\text{right}}(\theta)\right) \tag{1}$$

$$\theta^* = \arg \min_{\theta} G(Q, \theta) \tag{2}$$

Continue the recursive procedure until the maximum permissible depth of the tree is reached, $N_m < \min$ samples or $N_m = 1$.

3 Dataset

A dataset of body-sensor traces collected during football matches was used in experiments and analysis. The dataset includes field position, speed, and heading which were accurately sampled at a rate of 20 Hz using 3D sport tracking system [12].

```
'timestamp','tag_id','x_pos','y_pos','heading','direction','energy','speed','total_distance'
...
'2013-11-03 18:30:00.000612',31278,34.2361,49.366,2.2578,1.94857,3672.22,1.60798,3719.61
'2013-11-03 18:30:00.004524',31890,45.386,49.8209,0.980335,1.26641,5614.29,2.80983,4190.53
'2013-11-03 18:30:00.013407',0918,74.5904,71.048,-0.961152,0,2.37406,0,0.285215
'2013-11-03 18:30:00.015759',109,60.2843,57.3384,2.19912,1.22228,4584.61,8.14452,4565.93
'2013-11-03 18:30:00.023466',909,45.0113,54.7307,2.23514,2.27993,4170.35,1.76589,4070.6
```

Fig. 2 Samples from 20 Hx XYZ traces

The body-sensor data was recorded using sensor belts wrapped to the waist of the football players, and the data was collected by placing static radio receivers fixed on poles or on the gallery roof of the stadium. The radio receivers had a 90° approximate field-of-view. The area covered by these sensors has overlapping zones, and this helps to offer high resistance to signal blocking and occlusions. The static radio receivers estimate the location of players in the field by vector-based computation of the collected signals from the sensors attached to the players' waist belt. Conventionally, the player's location on the field is projected using triangulation methods but using vector-based estimation the direct projection of the player location in the field became easier. The body sensors are equipped with an accelerometer for capturing the body movements as a three-dimensional data, a gyroscope, a sensor for monitoring pulse rate of heart and a compass. The accelerometer helps to collect additional data like distance covered under different running speed of the players [10]. The combination of gyroscope and compass tracks the actual heading of the player. The dataset contains both a raw and an interpolated dataset. The raw dataset is composed of the data as collected from the body sensors, and the processed dataset has been interpolated to match with the time stamps. In our experiments, raw dataset is used for evaluating the performance of the dataset. Each record in the file represents one set of values captured from one body sensor at time 't.' The sensor nodes are given a unique identifier termed as tag-id, and each football player has only one body sensor attached to his waist belt. A sample of the records is shown in Fig. 2, and Table 1 presents the description of the attributes in the record.

4 Methodology

The conventional regression problems have a single real-valued output representing a single-dependent variable; multivariate regression problems have $|\Upsilon|$ different outputs, each measuring one of $|\Upsilon|$ different concepts. There are two distinct methodologies for approaching a solution to solve a multivariate regression problem: either transforms the problem so that it can be solved using regular single-output regress or adapt the algorithm into directly handling multiple outputs. A set of multivariate regression trees (MRT) that have the ability to predict multiple dependent variables which are based on the least-squares split function. The objective of the least-squares function is to minimize the sum of squared errors among the child nodes and is represented as:

Table 1 Description of data attributes

time stamp (string)	Local Central European Time (CET) time encoded as ISO-8601 format string
tag id (int)	The sensor identifier
x pos (float)	Relative position in meters of the node in the field's x-direction
y pos (float)	Relative positions in meters of the node in the field's y-direction
heading (float)	Direction the node is facing in radians where 0 is the direction of the y-axis
direction (float)	Direction the node is traveling in radians where 0 is direction of the y-axis
energy (float)	Estimated energy consumption since the last sample. The value is based on step frequency as measured by the on-board accelerometer. The unit for this value is undefined and might be relative to each individual
speed (float)	Node speed in meters per second
total distance (float)	The number of meters travelled so far during the game

$$\varphi(s, t) = SS(t) - SS(t_L) - SS(t_R) \tag{3}$$

where $SS(t)$ is the sum of squared errors in node t, defined as

$$SS(t) = \sum_{i \in t} (y_i - \bar{y}(t))^2 \tag{4}$$

For accommodating multivariate output, a multivariate random forest is constructed by an ensemble of multivariate regression tree using a bootstrap resampling technique.

Performance analysis measure—the root-mean-square error of a regression model is the square root of the sum of squared errors (the difference between the actual and the predicted output) as measured on the validation dataset.

$$RMSE(h) = \sqrt{\frac{\sum_{(x,y) \in Z_{test}} (y - h(x))^2}{|Z_{test}|}} \tag{5}$$

5 Experiment and Results

This section presents the experiments and evaluation results that demonstrate the effectiveness of the regression-based approach for localizing the sensor nodes problem. The main overhead for this approach is the training procedure. The training procedure requires a huge collection of historical data of each of the sensor node to be tracked. The location data captured by the body sensors used for training and

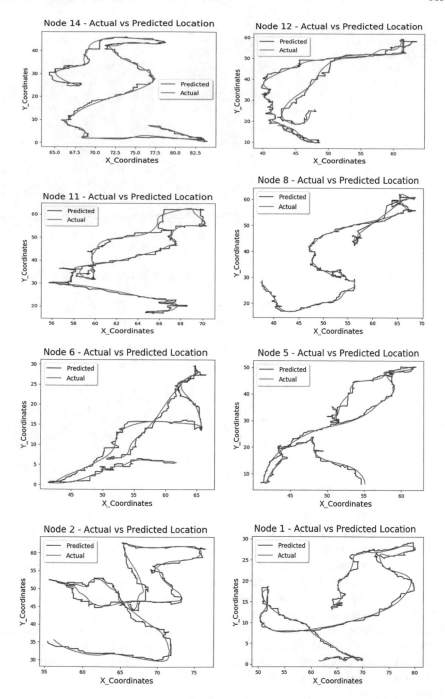

Fig. 3 Plot of actual and predicted locations

Fig. 4 Plot of RMSE values
for *X*-co-ordinate and
Y-co-ordinate predictions

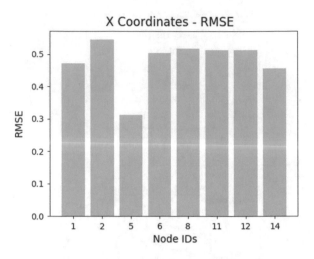

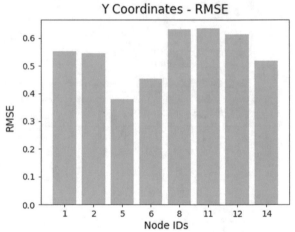

testing the prediction model. Using a lag function, the time series data captured from each of the sensor nodes is shifted (offsets) such that the future location values are aligned with the current location data. The lags are shifted 100 numbers of units, which helps to represent the position of the node after 400–500 ms from current position. As the problem of node localization is approached as regression problem in our experiments, the current *X*- and *Y*-coordinates along with other five attributes becomes the independent variables and the lagged two variables representing the future location of the nodes becomes the dependent variables. For each of the node, a multi-output random forest is constructed by fitting the model on the training data. The trained model can be able to predict the future location of a node based on the current location and other attributes sensed at the current time.

Initially, k-fold cross-validation is used for partitioning the training set into k mutually exclusive subsets. Then $k - 1$ partitions are used for training the models, and the performance of the model is evaluated using remaining partition. The optimal model is picked based on the overall performance from these k repetitions. The utilization of k different subsets instead of one random split training set increases the reliability of the model as it is based on more data. In the experiments, as each of the nodes adopts a random way-point mobility model, a separate regression model is trained for each of the node based on its historical trajectory data. The prediction accuracy of the models is estimated using unknown test cases. Figure 3 presents the comparison of actual and predicted locations of the node for the test dataset.

In general, fixing a threshold value for RMSE is difficult to establish. The unit of RMSE is the same as the unit of dependent variables which are expressed in meters in our dataset. The RMSE values of the results obtained from test dataset range between 0.3 and 0.6 m for both X- and Y-coordinates. A plot of RMSE in predictions for each of the node's X and Y coordinates is presented separately in Fig. 4.

6 Conclusion

In this paper, the node localization problem is approached as a multi-output regression problem and implemented using random forest algorithm. The trained multi-output regression model is capable of predicting the future location of the node in terms of X- and Y-coordinates with least error when compared to other methods in the literature. The method adopted in the experiments has an overhead of using training data for constructing the prediction model. Majority of the methods in the literature used anchor node-based node localization. In real-time applications, the WSNs are deployed in large-scale field/area. Under such situations, the anchor-based localization methods fail as they are not scalable and are designed for small areas. The method adopted in this paper is scalable in nature since it does not rely on the anchor nodes. As future work, the proposed node localization method must be integrated with a WSN routing algorithm to ensure the stability of the link before transmitting the data in the selected path.

References

1. Kulkarni, R.V., Venayagamoorthy, G.K., Cheng, M.X.: Bio-inspired node localization in wireless sensor networks. In: Proceedings of IEEE International Conference on Systems, Man and Cybernetics, pp. 205–210 (2009)
2. Zaidi, S., El Assaf, A., Affes, S., Kandil, N.: Accurate range-free localization in multi-hop wireless sensor networks. IEEE Trans. Commun. 64, 3886–3900 (2016)
3. Singh, M., Khilar, P.M.: An analytical geometric range free localization scheme based on mobile beacon points in wireless sensor network. Wirel. Netw. 22, 2537–2550 (2016)
4. Singh, M., Khilar, P.M.: Mobile beacon based range free localization method for wireless sensor networks. Wirel. Netw. 23, 1285–1300 (2017)
5. He, J., Yu, Y., Wang, Q.: RSS assisted ToA-based indoor geolocation. Int. J. Wireless Inf. Networks 20(2), 157–165 (2013)

6. Sun, S., Zhu, S., Ding, Z., Xu, B.: ToA-based source localization: a linearization approach adopting coordinate system translation. Int. J. Distrib. Sens. Netw. **2013**(5), 140–154 (2013)
7. Kaune, R.: Accuracy studies for TDoA and to a localization. In: International Conference on Information Fusion, pp. 408–415 (2012)
8. Giacometti, R., Baussard, A., Cornu, C., Khenchaf, A., Jahan, D., Quellec, J.M.: Accuracy studies for TDoA-AoA localization of emitters with a single sensor. In: IEEE Radar Conference, pp. 1–4 (2016)
9. Stoica, P., Sharman, K.: Maximum likelihood methods for direction-of-arrival estimation. IEEE Trans. Acoust. Speech Signal Proc. **38**(7), 1132–1143 (1990)
10. Oupakumar, A., Jacob, L.: Localization in wireless sensor networks using particle swarm optimization. In: Proceedings of IET International Conference on Wireless, Mobile and Multimedia Networks, pp. 227–230 (2008)
11. Kulkarni, R.V., Venayagamoorthy, G.K.: Particle swarm optimization in wireless-sensor networks: a brief survey. IEEE Trans. Syst. Man Cybern. Part C Appl. Rev. **41**, 262–267 (2011)
12. Pettersen, S.A, Johansen, D., Johansen, H., Berg-Johansen, V., Gaddam, V.R., Mortensen, A., Langseth, R., Griwodz, C., Stensland, H.K., Halvorsen, P.: Soccer video and player position dataset. In: International Conference on Multimedia Systems (MMSys), pp. 18–23. Singapore, Mar 2014

Ferro Fluid Based Squeeze Film in a Longitudinally Rough Surface Bearing of Infinite Width: A Comparative Study

Ankit S. Acharya, R. M. Patel and G. M. Deheri

Abstract A hydromagnetic squeeze film in a longitudinally rough parallel surface bearing has been discussed with the consideration of two different form of the magnitude of the associated magnetic field. In the light of the model of Christensen—Tonder regarding surface roughness, the load capacity is obtained after getting the pressure distribution by solving the stochastically averaged Reynolds' equation. The results presented here indicate that the magnetization offers a good amount of help in decreasing the unfavorable effect of roughness. In the case of the trigonometric form of magnitude it helps most. Besides, providing an additional degree of freedom, this article suggests some scopes for reducing the adverse effect of variance positive, skewness positive by magnetization and standard deviation in the case of negatively skewed roughness when negative variance is involved. Comparison of performance for both the forms of the magnitude informs that the trigonometric form of the magnitude presents a better picture to be used in the industry.

Keywords Squeeze film · Ferro fluid lubrication · Rough surface · Bearing of infinite width · Load bearing capacity

A. S. Acharya (✉)
Department of Applied Sciences and Humanity, L.J. Engineering College, Ahmedabad, India
e-mail: ankit.acharya@ljinstitutes.edu.in

R. M. Patel
Department of Mathematics, Gujarat Arts and Science College, Ahmedabad, Gujarat 380006, India
e-mail: rmpatel2711@gmail.com

G. M. Deheri
Department of Mathematics, Sardar Patel University, Vallabh Vidyanagar, Gujarat 388120, India
e-mail: gmdeheri@rediffmail.com

© Springer Nature Singapore Pte Ltd. 2020   187
K. N. Das et al. (eds.), *Soft Computing for Problem Solving*,
Advances in Intelligent Systems and Computing 1057,
https://doi.org/10.1007/978-981-15-0184-5_17

1 Introduction

The effect of longitudinal roughness has been analyzed for slider bearing [1] and for spherical bearing [2]. These investigations confirmed that the longitudinal roughness was a little better than transverse roughness pattern in terms of bearing design. The fact that the longitudinal roughness pattern improved the load-carrying capacity of a hydrodynamic short journal bearing when a couple stress fluid was present, assorted by Hsu et al. [3]. Andharia and Deheri [4, 5] discussed effect of longitudinal roughness on ferrofluid squeeze film between conical plates and truncated conical plates. These two discussions also submitted that the longitudinal roughness was of more help as compared to transverse roughness pattern.

Andharia and Deheri [6] considered the squeeze film in longitudinal rough elliptical plates with a ferrofluid lubrication. Panchal et al. [7] studied the influence of ferrofluid through a series of flow factors on the performance of a longitudinal rough finite slider bearing. Shimpi and Deheri [8] discussed the bearing deformation effect on a performance of the ferrofluid based squeeze film when longitudinal roughness occurred. Here, velocity slip effect was also investigated for truncated conical plates.

The ferrofluid has been fund to be extremely useful in many tribological problems occurring in bearing industry. Different properties and several flow models for ferrofluid have been discussed in Bhat [9]. The friction was found to be lowering at the moving plate of an infinitely long bearing is studied by Patel et al. [10] when a ferrofluid was employed for lubrication. The ferrofluid lubrication is investigated by Hsu et al. [11] on a short journal bearing. The friction coefficient was found to be modified by combined influence of magnetism and roughness. Acharya et al. [12] deliberated on behavior of a ferrofluid squeeze film infinitely long porous rough rectangular plates. Here, the magnitude of the external magnetic field was quite different from the usually adopted ones [9].

Wang et al. [13] conducted experimental and numerical studies on the surface roughness effect lubricated point contact. The roughness amplitude strongly influences the bearing performance here. In [14] Lin investigated the effect of longitudinal surface roughness on subcritical and super critical limit cycles of short journal bearings. It was established that for fixed bearing parameter the effects of longitudinal roughness structure extended the linear stability region. Surface roughness effect of slider bearing having hydrodynamic lubrication was studied by Andharia et al. [2]. The friction was found to be decreased depending on the roughness patterns.

In the literature, one comes across several forms of the magnitude of applied magnetic field. But little comparison has been made. Thus, here it has been proposed to compare the behavior of squeeze film considering different forms of magnitude for a longitudinally rough ferrofluid based squeeze film in parallel surface bearing of infinite width.

2 Analysis

The configuration and geometry of the bearing are shown in Fig. 1.

Adopting the stochastic averaging method of Christensen and Tonder [15–17] one is eventually led to the following governing Reynolds' equation for pressure distribution [18, 19].

$$\frac{\partial^2}{\partial x^2}\left(p - 0.5\mu_0\bar{\mu}H^2\right) = -\frac{12\mu\dot{h}}{m(h)^{-1}} \tag{1}$$

where μ_0 is the permeability of the free space, $\bar{\mu}$ is the magnetic susceptibility, μ is the fluid viscosity and

$$m(h) = \frac{1}{h^3}\left(1 - 3\alpha h^{-1} + 6h^{-2}(\sigma^2 + \alpha^2) - 10h^{-3}(\varepsilon + 3\sigma^2\alpha + \alpha^3)\right)$$

For more calibration considering the mean α, the standard deviation σ and skewness ε can be obtained from the deliberation of Christensen and Tonder [15–17].

For the comparative study, we have considered the following two distinct forms of magnitude.

Form I

$$H^2 = k\left(x^2 - \frac{l^2}{4}\right) \tag{2}$$

Form II

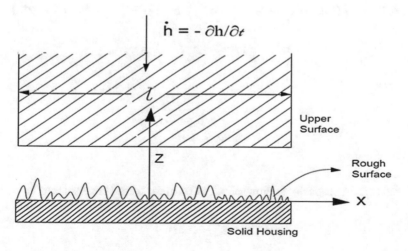

Fig. 1 Configuration of bearing system

$$H^2 = kl^2\left(\frac{3\pi x}{2l} + \frac{3\pi}{4}\right)\cos\left(\frac{3\pi x}{2l} - \frac{\pi}{4}\right) \tag{3}$$

where k is suitable constant.

Boundary conditions associated are:

$$p = 0 \quad \text{when } x = \pm\frac{l}{2} \tag{4}$$

Integrating Eq. (1) w.r.t. the above mentioned boundary conditions, we obtained the pressure distribution for the Form I

$$p_1 = 0.5\mu_0\overline{\mu}k\left(x^2 - \frac{l^2}{4}\right) + \frac{3}{2}\mu\dot{h}m(h)(l^2 - 4x^2) \tag{5}$$

and for the Form II, the pressure distribution is obtained as

$$p_2 = 0.5\mu_0\overline{\mu}kl^2\left(\frac{3\pi x}{2l} + \frac{3\pi}{4}\right)\cos\left(\frac{3\pi x}{2l} - \frac{\pi}{4}\right) + \frac{3}{2}\mu\dot{h}m(h)(l^2 - 4x^2) \tag{6}$$

It is easily observed that pressure distribution is parabolic in nature for the form I while this profile appears to be distorted in the form II.

The dimensionless quantities are

$$\overline{x} = \frac{x}{l}; \quad \mu^* = -\frac{k\mu_0\overline{\mu}h^3}{\mu\dot{h}}; \quad p = -\frac{ph^3}{\mu\dot{h}l^2}; \quad W = \frac{wh^3}{\mu\dot{h}l^3}$$

$$M(h) = m(h)h^3 = \left(1 - 3\alpha^* + 6\left(\sigma^{*2} + \alpha^{*2}\right) - 10\left(\varepsilon^* + 3\sigma^{*2}\alpha^* + \alpha^{*3}\right)\right)$$

$$\sigma^* = \frac{\sigma}{h}; \quad \alpha^* = \frac{\alpha}{h}; \quad \varepsilon^* = \frac{\varepsilon}{h^3}$$

and using

$$w = \int_{-\frac{l}{2}}^{\frac{l}{2}} p\,dx$$

the load carrying capacity in dimensionless form for form I comes out to be

$$W_1 = \frac{\mu^*}{12} + M(h) \tag{7}$$

while the load bearing capacity in non-dimensional form for the form II turns out to be

$$W_2 = \frac{\mu^*}{2}\left(1 + \frac{2}{3\pi}\right) + M(h) \tag{8}$$

3 Results and Discussion

Equations (7) and (8) confirm that the non-dimensional load carrying capacity gets raised by

$$\frac{\mu^*}{2} \approx 0.083\mu^*$$

and

$$\frac{\mu^*}{2}\left(1 + \frac{2}{3\pi}\right) \approx 1.212\mu^*$$

with regards to the conventional lubricant based bearing system. Further, the trigonometric form of the magnitude registers more load carrying capacity. The investigation of Prakash and Vij [20] for smooth bearing system for conventional fluid is obtained from this study.

The load taking capacity increases w.r.t. the increase in magnetization parameter, as linearity of expression is obtained for magnetization. Needless to say is that magnetization increased viscosity of fluid causing increased pressure and load-bearing capacity (cf. Figs. 2, 3, 4, 5, 6 and 7.

A noticeable fact is that the σ^* associated with roughness causes sharply increased load-bearing capacity. It can be seen from Figs. 8, 9, 10 and 11.

This trend of standard deviations is approximately opposite to the case of transverse roughness. An appealing scenario occurs in the case of skewed roughness ($-$ve)

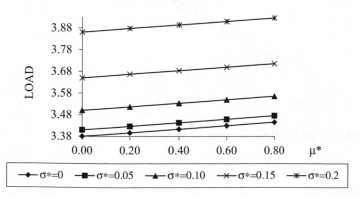

Fig. 2 Variation of load carrying capacity for μ^* and σ^*

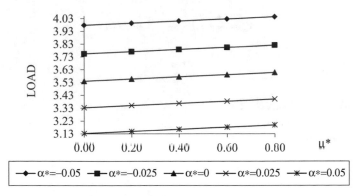

Fig. 3 Distribution of load-carrying capacity with respect to μ^* and α^*

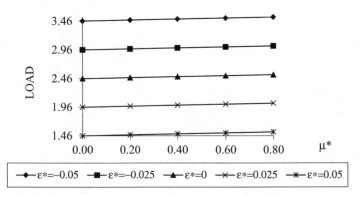

Fig. 4 Profile of load carrying capacity for μ^* and ε^*

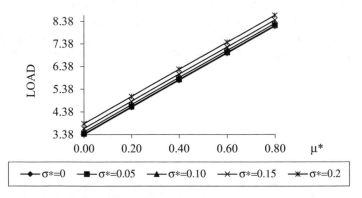

Fig. 5 Profile of load-carrying capacity with respect to μ^* and σ^*

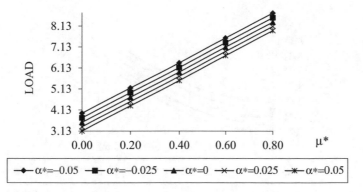

Fig. 6 Distribution of load-carrying capacity with respect to μ^* and α^*

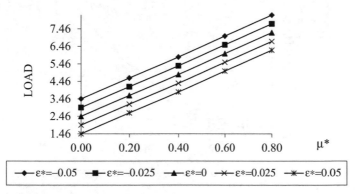

Fig. 7 Profile of load carrying capacity for μ^* and ε^*

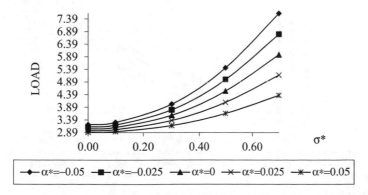

Fig. 8 Profile of load carrying capacity for σ^* and α^*

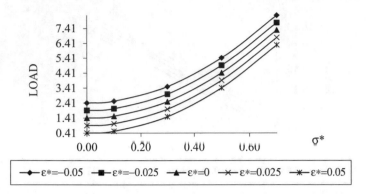

Fig. 9 Profile of load carrying capacity for σ^* and ε^*

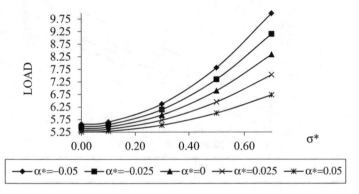

Fig. 10 Profile of load carrying capacity for σ^* and α^*

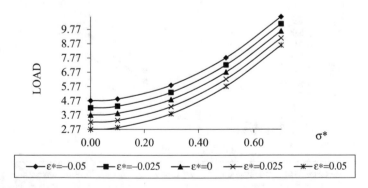

Fig. 11 Distribution of load carrying capacity for σ^* and ε^*

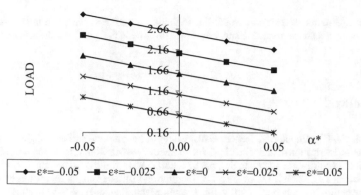

Fig. 12 Profile of load carrying capacity for α^* and ε^*

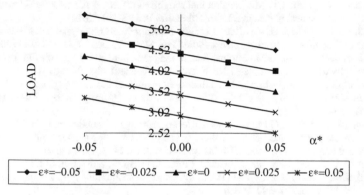

Fig. 13 Profile of load carrying capacity for α^* and ε^*

because the effect of roughness can be countered at least in case of trigonometric form of magnitude of magnetic field when variance ($-$ve) is in place which can be depicted from Figs. 12 and 13.

4 Conclusion

The trigonometric form of the magnitude registers a relatively enhanced performance in comparison with algebraic form of magnitude. In general effect of roughness, remains adverse. The absence of flow fails to defer the bearing system from sustaining a considerable amount of load which is rarely visited in the case of traditional lubrication. This investigation underlines that it is the standard deviation which makes the difference between the performances in the case of longitudinal and transverse roughness. Although a good picture emerges from this investigation, roughness needs to be given due respect while designing the bearing system.

Acknowledgements The authors would like to thank both the reviewers and the editor for their fruitful comments and constructive suggestions for improving the quality of the article.

References

1. Andharia, P.I.: Gupta, J.L., Deheri, G.M.: Effect of longitudinal surface roughness on hydro-dynamic lubrication of slider bearings. In: Proceedings of the 10th International Conference on Surface Modification Technologies, pp. 872–880. The Institute of Materials, London (1997)
2. Andharia, P.I., Deheri, G.M.: Effect of longitudinal surface roughness on the behaviour of squeeze film in a spherical bearing. Int. J. Appl. Mech. Eng. **6**(4), 885–897 (2001)
3. Hsu, C.-H., Lin, J.-R., Chiang, H.-L.: Combined effects of couple stresses and surface roughness on the lubrication of short journal bearing. Ind. Lubr. Tribol. **55**(5), 233–243 (2003)
4. Andharia, P.I., Deheri, G.M.: Longitudinal roughness effect on magnetic fluid based squeeze film between conical plates. Ind. Lubr. Tribol. **62**(5), 285–291 (2010)
5. Andharia, P.I., Deheri, G.M.: Effect of longitudinal roughness on magnetic fluid based squeeze film between truncated conical plates. Fluid Dyn. Mater. Process. **7**(1), 111–124 (2011)
6. Andharia, P.I., Deheri, G.M.: Performance of magnetic fluid based squeeze film between longitudinally rough elliptical plates. ISRN Tribol. 6. Article ID482604 (2013)
7. Panchal, G., Patel, H., Deheri, G.M.: Influence of magnetic fluid through a series of flow factors on the performance of a longitudinally rough finite slider bearing. Glob. J. Pure Appl. Math. **12**(1), 783–796 (2016)
8. Shimpi, M.E., Deheri, G.M.: Combined effect of bearing deformation and longitudinal roughness on the performance of a ferrofluid based squeeze film together with velocity slip in truncated conical plates. Imperial J. Interdisc. Res. **2**(6), 1423–1430 (2016)
9. Bhat, M.V.: Lubrication with a Magnetic Fluid. Team Spirit, India, Pvt. Ltd. (2003)
10. Patel, R.M., Deheri, G.M., Vadher, P.A.: Lubrication of an infinitely long bearing by a magnetic fluid. FDMP **6**(3), 277–290 (2010)
11. Hsu, T.C., Chen, J.H., Chiang, H.L., Chou, T.L.: Lubrication performance of short journal bearing considering the effects of surface roughness and magnetic field. Tribol. Int. **61**, 169–175 (2013)
12. Acharya, A.S., Patel, R.M., Deheri, G.M.: Ferro fluid squeeze film in infinitely rough rectangular plates. Int. J. Sci. Eng. Res. **6**(8), 2109–2120 (2015)
13. Wang, S., Hu, Y.-Z., Wang, W.-Z., Wang, H.: Effects of surface roughness on sliding friction in lubricated-point contacts: experimental and numerical studies. J. Tribol. **129**, 809–817 (2007)
14. Lin, J.-R.: The influences of longitudinal surface roughness on sub-critical and super-critical limit cycles of short journal bearings. Appl. Math. Model. **38**(1), 1–18 (2014)
15. Christensen, H., Tonder, K.C.: Tribology of rough surfaces: parametric study and comparison of lubrication models. SINTEF report, no. 22/69–18 (1969a)
16. Christensen, H., Tonder, K.C.: Tribology of rough surfaces: stochastic models of hydrodynamic lubrication. SINTEF report, no. 10/69–18 (1969b)
17. Christensen, H., Tonder, K.C.: The hydrodynamic lubrication of rough bearing surface of finite width. In: ASME-ASLE Lubrication Conference, Paper no. 70- Lub-7 (1970)
18. Hamrock, B.J.: Fundamentals of Fluid Film Lubrication. McGraw-Hill, New York (1994)
19. Prajapati, B.L.: On certain theoretical studies in hydrodynamic and electromagnetohydrodynamic lubrication, Ph.D. thesis, S.P. University. Vallabh Vidyanagar, Gujarat, India (1995)
20. Prakash, J., Vij, S.K.: Load capacity and time-height relations for squeeze film between porous plates. Wear **24**, 309–322 (1973)

Impact of Inertia Weight and Cognitive and Social Constants in Obtaining Best Mean Fitness Value for PSO

Yallapragada V. R. Naga Pawan⬚ and K. Bhanu Prakash⬚

Abstract The performance of an Inertia Weight-based Particle Swarm Optimization (IWPSO) is relied on parameters like inertia weight, cognitive, and social coefficients. This paper is investigating effective range of inertia weights and relationship between cognitive and social coefficients over different swarm sizes for selective iterations and dimensions in assessing performance of IWPSO. The experimental results show that the inertia weights in the range 0.1–0.8 and when cognitive coefficient is less than social coefficient, the results are more aspiring.

Keywords Particle swarm optimization · Inertia weight · Cognitive constant and social constant

1 Introduction

Particle Swarm Optimization (PSO) is a biologically inspired Swarm Intelligence (SI) technique where bird-like objects tend to move toward a goal. This technique is proposed by Kennedy and Eberhart [1, 2] used for solving global optimization problems. Due to its ease in implementation, it is used in a wide spectrum of domains as promising optimization technique due to faster convergence to an optimum.

The work of Kennedy and Eberhart [2] is considered as a Basic Particle Swarm Optimization (BPSO) technique. In BPSO, the velocities and positions of the particles are randomly initialized. The equations used BPSO are given as Eqs. (1) and (2).

$$v[i+1][d] = v[i][d] + constant1 * random() * (pBestx[i][d] - currentX[i][d])$$
$$+ constant2 * Random() * (pBestx[gBest][d] - currentX[i][d]) \tag{1}$$

$$currentX[i+1][d] = currentX[i][d] + v[i][d] \tag{2}$$

Y. V. R. Naga Pawan (✉) · K. B. Prakash
Department of CSE, Koneru Lakshmaiah Education Foundation, Vijayawada, India
e-mail: ynpawan@gmail.com

K. B. Prakash
e-mail: drkbp@kluniversity.in

© Springer Nature Singapore Pte Ltd. 2020
K. N. Das et al. (eds.), *Soft Computing for Problem Solving*,
Advances in Intelligent Systems and Computing 1057,
https://doi.org/10.1007/978-981-15-0184-5_18

where $d = 1, 2, 3, \ldots, n$, and $i = 1, 2, 3, \ldots N$ which represents dimension and particle index, respectively. Constant1 and constant2 are cognitive and social constant parameters which are set to 2 as mentioned in [1]. Random() and random() are two random numbers of the range [0, 1].

As particles tend to explore the search space hugely, the velocities of the particles are limited to the constant Vmax [3]. The particle velocity is adjusted using Eq. (3).

$$V[i + 1][d] = \begin{cases} v[i + 1][d] & if \ V[i + 1][d] < V\text{max}, d \\ V\text{max} & if \ V[i + 1][d] > V\text{max}, d \end{cases} \tag{3}$$

where Vmax, d is the maximum velocity in the d dimension.

Shi and Eberhart [4] introduced inertia weight to address the trade-off between local search and global search in the Eq. (1). The modified equation is given as Eq. (4).

$$v[i + 1][d] = w * v[i][d] + \text{constant1} * \text{random}() * (p\text{Best}x[i][d] - \text{current}X[i][d]$$
$$+ \text{constant2} * \text{Random}() * (P\text{Best} x[g\text{Best}][d] - \text{current}X[i][d] \tag{4}$$

The first part of Eq. (1) represents previous velocity, the second part is the cognition part of the particle, and the third part represents the cooperation among the particles [3]. The flowchart of BPSO is shown in Fig. 1. The flowchart has three parts, the first part is local, second part is based on vicinity, and the last part on global. The pseudocode of BPSO is shown in Fig. 2.

From [4], it is known that sum of c1 and c2 is 4. Using Eq. (4)

2   PSO Variants Based on Inertia Weight

Many researchers worked on Inertia Weight-based PSO, the variants are tabulated in Table 1.

3   Experiment and Results

Equations (2) and (4) as the base equations, three experiments are conducted to assess the impact of inertia weight (IW), constant1 (c1), cognitive constant, and constant2 (c2), social constant on IWPSO. In the experimentation, the inertia weight, swarm size, with c1 value is equal to (EQ) c2, i.e., c1 and c2 equal to 2, c1 less than (LT) c2, i.e., c1 = 1 and c2 = 3, and c1 greater than (GT) c2, i.e., c1 = 3 and c2 = 1, are considered. The IWPSO is implemented in R version 3.4.3 and RStudio 1.1.456. The details of the experiments are tabulated in Table 2.

For different inertia weights from 0.1 to 1.0 on swarm sizes (25, 50, 75, 100, 500) the time consumed in attaining mean fitness value is computed when c1 equal to c2

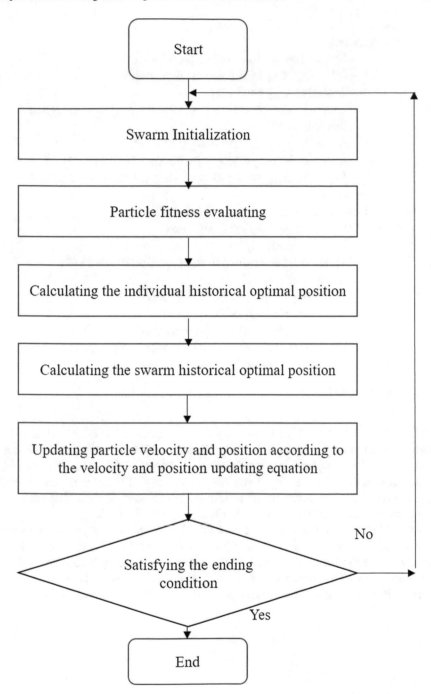

Fig. 1 Flowchart of basic PSO

Step 1:

Initialization

       For each particle, i, in the population

       Initialize X[i] with uniform distribution

       Initialize V[i] randomly.

       Evaluate the objective function of X[i] and assigned the value to fitness[i].

       Initialize $p_{phest}[i]$ with a copy of X[i].

       Initialize pbest_ftness[i] with a copy of fitness[i].

       Initialize p_{gbest} with index of the particle with the least fitness.

Step 2:

Repeat until stopping criterion is reached

       For each particle, i,:

Update V[i] and X[i] according to the equations (1) and (2)

       Evaluate fitness[i]

       If fitness[i] < pbest_fitness[i] then

              Pbest[i] = X[i]

              Pbest_fitness [i] = fitness[i]

       Update pgbest by the particle with current least fitness among the population

Fig. 2 Pseudocode for basic PSO

(c1 EQ c2), c1 less than c2 (c1 LT c2), and when c1 greater than c2 (c1 GT c2). The fitness function used is "modified Rastrigin function." The minimum and maximum search space for modified Rastrigin function is from $[-5.12, 5.12]$. The number of dimensions is 10 with clamped velocity, Vmax, to 4. The maximum number of iterations is 1000. The experimental results are tabulated in Table 3. The graphical representation of the results is shown in Figs. 3, 4, 5, 6, 7, and 8.

From Figs. 3, 4, 5, 6 and 7, it is found that the IWPSO converges to zero when inertia weight is from 0.1 to 0.8 in less than 1000 iterations. It is also observed that when c1 is greater than c2, the mean fitness value is zero for IWPSO in majority of the cases when inertia weights from 0.1 to 0.8. It is also observed that from Fig. 8 when c1 is greater than c2 in majority cases, the IWPSO is converged fastly.

Table 1 Inertia weight-based variants in PSO

Author	Method/approach	Findings
Shi and Eberhart [4]	• To control exploration and exploitation	• Converges to optimal solution fastly for certain inertia weights
Li and Gao [5]	• Linearly decreasing inertia weight PSO (LDIW PSO) [5] • Exponent decreasing inertia weight (EDW PSO) [5] • Exponent decreasing inertia weight and stochastic mutation (EDM PSO) [5]	• EDM PSO [5] comes out quickly from local optimum • EDM PSO [5] is faster convergence than LDIW PSO [5] and EDW PSO [5]
Chongpeng et al. [6]	• Decreasing inertia weight (DIW) [6]	• Best performance to balance global and local search
Shi and Eberhart [7]	• Fuzzy adaptive PSO • Linearly decreasing inertia weight (LDIW) [7]	• Fuzzy system tuning its inertia weight improves basic PSO
Zhang et al. [8]	• Processing in parallel • Random number inertia weight (RNW) [8]	• RNW [8]-based PSO has better results than LDIW [7] • RNW overcomes LDIW which lacks local search • Ability at early stage of run and global search ability at the end of run using linearly decreasing inertia weight method were overcome
Wei et al. [9]	• Dynamically changing inertia weight, DPSO • To escape from local optimum dimension mutation operator is designed	• DPSO is faster and better than LDW. • The proposed dimension mutation operator is helped from escaping local optimum
Pant et al. [10]	• GWPSO + ED, GWPSO + GD *and* GWPSO + UD [10]	• GWPSO + ED, GWPSO + GD *and* GWPSO + UD [10] *gave superior performance over BPSO*
Liu et al. [11]	• Multistart PSO • Improved PSO with linearly decreasing Inertia [11]	• Avoids premature convergence and enhance global search ability • Improved PSO has better performance with the linearly decreasing of inertia weight from 0.9 to 0.4 [11]

Table 2 Experimentation details

Experiment no.	Inertia weight	Swarm sizes	c1 and c2 relation
Experiment 1	0.1–1.0	25, 50, 75, 100, and 500	c1 EQ c2
Experiment 2	0.1–1.0	25, 50, 75, 100, and 500	c1 LT c2
Experiment 3	0.1–1.0	25, 50, 75, 100, and 500	c1 GT c2

Table 3 IWPSO performance based on inertia weight, versus swarm size when c1 EQ c2, c1 GT c2, and c1 LT c2

Inertia weight	Swarm size	Time in secs				Mean fitness value		
		C1 EQ c2 (1)	C1 LT c2 (2)	C1 GT c2 (3)	Best	C1 EQ c2 (1)	C1 LT c2 (2)	C1 GT c2 (3)
0.1	25	1.59	2.2	1.62	1	0	−0.99496	0
0.1	50	3.13	4.56	3.04	3	0	0	0
0.1	75	4.56	6.68	4.47	3	0	0	0
0.1	100	6.13	9.46	5.93	3	0	0	0
0.1	500	30.28	43.61	29.39	3	0	0	0
0.2	25	1.61	2.33	1.66	1	0	0	0
0.2	50	3.09	4.47	3.1	1	0	0	0
0.2	75	4.61	6.81	4.52	3	0	0	0
0.2	100	6.23	8.47	5.97	3	0	0	0
0.2	500	29.84	42.27	28.89	3	0	0	0
0.3	25	1.64	2.36	1.67	1	0	0	0
0.3	50	3.17	4.63	3.09	3	0	0	0
0.3	75	4.64	6.53	4.47	3	0	0	0
0.3	100	6.16	8.53	5.99	3	0	0	0
0.3	500	29.82	42.01	29.08	3	0	0	0
0.4	25	1.58	2.48	1.55	3	0	0	0
0.4	50	3.13	4.53	3.09	3	0	0	0
0.4	75	4.88	6.77	4.54	3	0	0	0
0.4	100	6.09	8.25	5.98	3	0	0	0
0.4	500	29.78	37.63	29.08	3	0	0	0
0.5	25	1.64	2.42	1.62	3	0	0	0
0.5	50	3.19	4.68	3.07	3	0	0	0
0.5	75	4.73	6.78	4.53	3	0	0	0
0.5	100	6.1	8.79	6	3	0	0	0
0.5	500	30.14	29.04	29.14	2	0	0	0
0.6	25	1.64	2.3	1.57	3	0	0	0
0.6	50	3.06	4.38	3.04	3	0	0	0
0.6	75	4.57	6.5	4.49	3	0	0	0
0.6	100	6.09	9.61	5.94	3	0	0	0
0.6	500	30.08	29	29.16	2	0	0	0
0.7	25	1.58	2.41	1.68	1	0	0	0
0.7	50	3.19	4.62	3.1	3	0	0	0
0.7	75	4.64	6.92	4.58	3	0	0	0
0.7	100	6.16	8.98	6.12	3	0	0	0
0.7	500	29.86	29.19	29.17	3	0	0	0
0.8	25	1.64	2.28	1.65	1	0	−2.79E−07	−4.71E−12
0.8	50	3.11	4.91	3.14	1	0	−3.39E−08	−5.27E−07
0.8	75	4.68	6.7	4.59	3	0	−4.09E−11	−1.13E−06
0.8	100	6.25	8.78	6.11	3	0	−6.94E−08	−2.58E−09
0.8	500	29.95	29.16	29.27	2	0	−3.16E−11	−9.24E−14
0.9	25	1.67	2.55	1.6	3	−0.00261	−0.08225	−0.01762
0.9	50	3.14	4.73	3.09	3	−0.00205	−0.01465	−0.05144

(continued)

Table 3 (continued)

Inertia weight	Swarm size	Time in secs				Mean fitness value		
		C1 EQ c2 (1)	C1 LT c2 (2)	C1 GT c2 (3)	Best	C1 EQ c2 (1)	C1 LT c2 (2)	C1 GT c2 (3)
0.9	75	4.77	6.74	4.56	3	−0.00086	−0.00137	−0.00116
0.9	100	6.19	9.22	6.1	3	−7.48E−05	−0.00066	−0.00694
0.9	500	30.06	29.26	29.53	2	−6.84E−06	−0.00045	−0.00019
1.0	25	1.69	2.52	1.81	1	−0.02662	−0.0053	−0.07417
1.0	50	3.26	4.63	3.13	3	−0.06247	−0.01791	−0.03282
1.0	75	4.73	7.11	4.53	3	−0.07446	−0.02678	−0.02032
1.0	100	6.09	9.47	6.06	3	−0.01756	−0.05412	−0.15429
1.0	500	30.34	29.41	29.31	3	−0.01387	−0.00083	−0.00846

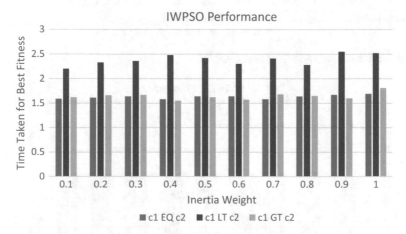

Fig. 3 Performance of IWPSO for swarm size 25

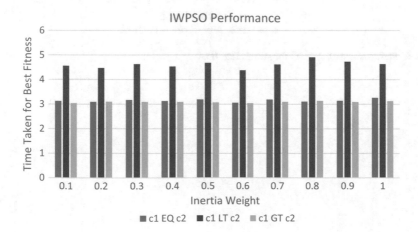

Fig. 4 Performance of IWPSO for swarm size 50

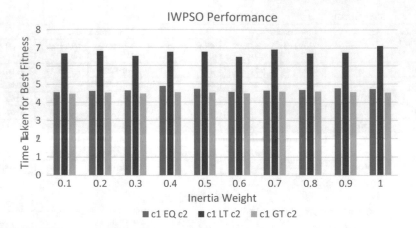

Fig. 5 Performance of IWPSO for swarm size 75

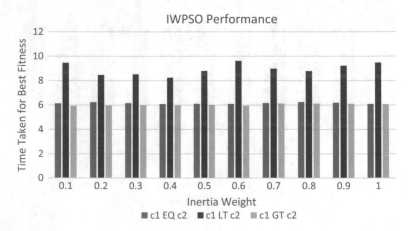

Fig. 6 Performance of IWPSO for swarm size 100

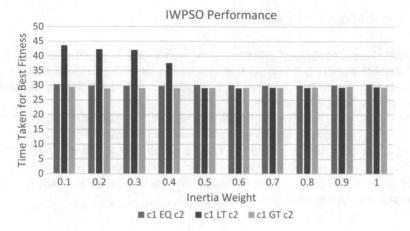

Fig. 7 Performance of IWPSO for swarm size 500

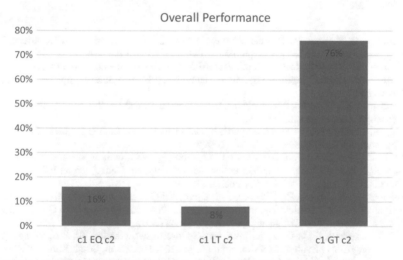

Fig. 8 Overall Performance of IWPSO for swarm sizes 25, 50, 75, 100 and 500

4   Conclusion and Future Work

IWPSO is used for identifying the impact of inertia weight, cognitive constant (c1), and social constant (c2). The experimental investigations show that for inertia weights ranging from 0.1 to 0.8 and when c1 is greater than c2, the faster convergence is observed for different swarm sizes.

In future, an exhaustive investigation is required to study the impact of inertia weights, cognitive constant, and social constant for various fitness functions on IWPSO and other variants of Inertia Weight-based Particle Swarm Optimization. The investigations are further extended to the complex domains like biological systems analysis [12] and unstructured multilingual web documents which have true potential of optimization.

References

1. Kennedy, J., Eberhart, R.: Particle warm optimization. In: Proceedings of IEEE International Conference on Neural Networks, pp. 1942–1948. Perth, Australia (1995)
2. Eberhart, R., Kennedy, J.: A new optimizer using particle swarm theory. In: Proceedings of the Sixth International Symposium on Micro Machine and Human Science, pp. 39–43. Nagoya, Japan (1995)
3. Eberhart, R.C., Simpson, P.K., Dobbins, R.W.: Computational Intelligence PC Tools, 1st edn. Academic Press Professional (1996)
4. Shi, Y., Eberhart, R.: A modified particle swarm optimizer. In: IEEE World Congress on Computational Intelligence, pp. 66–69 (2009)
5. Li, H.-R., Gao, Y.-L.: Particle swarm optimization algorithm with exponent decreasing inertia weight and stochastic mutation. In: Second International Conference on Information and Computing Science, pp. 66–69. Manchester (2009)
6. Chongpeng, H., Yuling, Z., Dingguo, J., Baoguo, X.: On some non-linear decreasing inertia weight strategies in particle swarm optimization. In: Proceedings of the 26th Chinese Control Conference, pp. 570–753. Zhangjiajie, Hunan, China (2007)
7. Shi, Y., Eberhart, R.C.: Fuzzy adaptive particle swarm optimization. In: Proceedings of the IEEE Congress on Evolutionary Computation, pp. 101–106. Seoul, South Korea (2001)
8. Zhang, L., Yu, H., Hu, S.: A new approach to improve particle swarm optimization. In: Proceedings of the 2003 International Conference on Genetic and Evolutionary Computation, pp. 134–139 (2003)
9. Wei, J., Wang, Y.: A dynamic particle swarm algorithm with Dimension Mutation. IJCSNS Int. J. Comput. Sci. Netw. Secur. **6**, 221–224 (2006)
10. Pant, M., Thangaraj, T., Singh, V.P.: Particle swarm optimization using gaussian inertia weight. In: International Conference on Computational Intelligence and Multimedia Applications, pp. 97–102. Sivakasi, Tamil Nadu (2007)
11. Liu, X., et al.: Particle swarm optimization with dynamic inertia weight and mutation. In: Third International Conference on Genetic and Evolutionary Computing, pp. 620–623. Guilin (2009)
12. Prakash, K.B., et al.: Content extraction of biological datasets using soft computing techniques. J. Med. Imaging Health Inform. **6**(5), 932–936 (2016)
13. Prakash, K.B., et al.: Information extraction in unstructured multilingual web documents. Indian J. Sci. Technol. **8**(16), 1–8 (2015)

Study of Squeeze Film in a Ferrofluid Lubricated Longitudinally Rough Rotating Plates

Hardik P. Patel, G. M. Deheri and R. M. Patel

Abstract This chapter discusses the performance of a squeeze film in longitudinally rough rotating circular plates in the presence of ferrofluid lubrication. The ferrofluid model of Neuringer—Rosensweig has been used. The roughness characterization has been adopted, taking the stochastic averaging model of Christensen–Tonder. The distribution of pressure is obtained solving the concerned Reynolds' type equation. This provides load taking capacity. The computed results are presented in graphical form, which establishes that the roughness (longitudinal) remains more favorable as compared to the transverse surface roughness. Undoubtedly, ferrofluid lubrication enhances the bearing performance, but it alone may not be sufficient to overcome the adverse effect of roughness and rotation.

Keywords Squeeze film · Rotation · Ferrofluid · Roughness · Load bearing capacity

Nomenclature

r	Radial coordinate
a	Circular plate's radius
h	Film thickness of lubricant
$\dot{h}$	Velocity of squeeze film
μ_0	Free space's permeability

H. P. Patel (✉)
Humanity and Science Department, L.J. Iinstitute of Engineering and Technology, Ahmedabad, Gujarat, India
e-mail: hardikanny82@gmail.com

G. M. Deheri
Mathematics Department, S. P. University, Vidyanagar, Gujarat 388120, India

R. M. Patel
Gujarat Arts and Science College, Ellis Bridge, Ahmedabad, Gujarat 380006, India
e-mail: rmpatel2711@gmail.com

© Springer Nature Singapore Pte Ltd. 2020
K. N. Das et al. (eds.), *Soft Computing for Problem Solving*,
Advances in Intelligent Systems and Computing 1057,
https://doi.org/10.1007/978-981-15-0184-5_19

207

$\bar{\mu}$ Magnetic susceptibility

μ Absolute viscosity of the lubricant

μ^* Magnetization parameter

σ Standard deviation of film thickness

α Mean of film thickness

ε Measure of symmetry of film thickness

p Lubricant pressure

P Pressure in the dimensionless form

σ^* Dimensionless standard deviation

α^* Non-dimensional mean

ε^* Dimensionless skewness

ρ Density of lubricant

S Non-dimensional rotational inertia

w Load taking capacity

W Load carrying capacity in the dimensionless form

Ω_u Upper plate's angular velocity

Ω_l Lower plate's angular velocity

Ω_r $= \Omega_u - \Omega_l$

Ω_f $= \Omega_l/\Omega_u =$ Rotation ratio.

1   Introduction

Wu [24] laid down the criterion for neglecting the effect of inertia. A similar method was adopted by Ting [20] simplifying the investigation of Wu [24]. Of course, only the lower disk's rotation was only considered here. Vora-Bhat [23] modified the above approach to consider a squeeze film performance where the curved porous upper plate and lower plate is an impermeable flat plate.

The adverse effect of roughness has been a matter of discussions in many investigations, wherein, a magnetic fluid has been used to overcome the negative effect induced by roughness. Ting [21], Prakash and Tiwari [17], Verma [22], Bhat and Deheri [4, 5], Prajapati [16], Gupta and Deheri [12], Andharia et al. [3] and Deheri et al. [10] discussed behavior of squeeze film between porous circular plates with porous matrix of variable thickness, and Lin et al. [13] analyzed the effect of non-Newtonian couple stresses on circular disks with ferrofluid lubrication. Effect of longitudinal roughness on ferrofluid lubricated journal bearing was studied by Lin [14]. In most of the investigations, it was found that the bearing was supporting certain amount of load even when the flow was absent and this load was more when longitudinal roughness was in place. Recently, ferrofluid based rotatory bearing has been treated in Ellahi et al. [11]. Here, homotopy analysis was adopted.

An outcome of roughness (longitudinal) was discussed by Andharia–Deheri [1] on the conical plates having squeeze film lubricated with ferrofluid. In this study, longitudinal roughness had a significant role in elevating the performance of the squeeze film. Andharia–Deheri [2] studied the magnetic fluid on a squeeze film in longitudinally rough elliptical plates. Shimpi and Deheri [19] extended the analysis of Andharia and Deheri [1] to consider the mutual effect of slip and deformation in truncated conical plates. It was observed that longitudinal roughness came to help the ferrofluid lubrication to counter the deformation effect when slip was moderate.

The effect of rotation of circular plates transversely rough and porous on the squeeze film having magnetic fluid lubrication was investigated by Patel et al. [15]. Therefore, it is appropriate to evaluate impact of longitudinal roughness pattern in the bearing configuration of the above article.

2 Analysis

The geometry and configuration of the bearing system given above have two circular plates having radius 'a.' The upper plate is moving in the direction of lower plate with squeeze velocity $\dot{h}(=dh/dt)$. The angular velocity of the upper plate is Ω_u, and Ω_l is angular velocity of lower plate (Fig. 1).

The longitudinal roughness of the bearing system is described by a stochastic variable. Invoking Christensen–Tonder's [7–9] analysis, the film thickness is taken.

Flow of the ferrofluid is axially symmetric which lies in between the plates having an oblique magnetic field $\bar{H} = (H(r) \cos \theta, 0, H(r) \sin \theta)$ is taken into consideration [6]. The magnitude of the magnetic field is very small at the center and at the boundary. Using the standard assumptions of hydromagnetic lubrication [1, 19]

Fig. 1 Configuration of the bearing

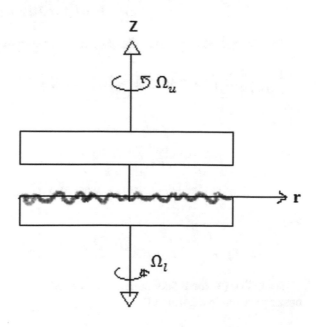

and stochastic averaging model of longitudinal roughness by Christensen–Tonder's [7–9], the associated Reynolds' type equation for the present bearing system comes out to be

$$\frac{1}{r}\frac{d}{dr}\left\{[m(h)]^{-1}r\frac{d}{dr}\left(p - 0.5\mu_0\bar{\mu}H^2\right)\right\} = 12\mu\dot{h} + 2\rho[m(h)]^{-1}$$

$$\left(\frac{3}{10}\Omega_r^2 + \Omega_r\Omega_l + \Omega_l^2\right) \tag{1}$$

where

$$H^2 = Kr(a - r); \quad 0 \leq r \leq a$$

and

$$m(h) = h^{-3}\left[1 - h^{-1}\alpha + 6h^{-2}(\sigma^2 + \alpha^2) - 10h^{-3}(\varepsilon + 3\sigma^2\alpha + \alpha^3)\right].$$

Magnetic field's inclination angle is governed by the partial differential equation [4]

$$\cot\theta\frac{\partial\theta}{\partial r} + \frac{\partial\theta}{\partial z} = \frac{1}{2(a - r)}. \tag{2}$$

In view of the Reynolds' boundary condition

$$p = 0 \text{ at } r = 0 \text{ and } p = 0 \text{ at } r = a. \tag{3}$$

The dimensionless pressure distribution gets determined by

$$\frac{1}{R}\frac{d}{dR}\left\{R\frac{d}{dR}[P - 0.5\mu^*R(1 - R)]\right\} = -12M(h) + \frac{S}{5}\left(3\Omega_f^2 + 4\Omega_f + 3\right) \tag{4}$$

where

$$P = -ph^3(\mu\dot{h}a^2)^{-1} \quad \mu^* = -\mu_0\bar{\mu}h^3(\mu\dot{h})^{-1} \quad R = a^{-1}r$$

$$\sigma^* = h^{-1}\sigma \qquad\qquad \alpha^* = h^{-1}\alpha \qquad\qquad \varepsilon^* = h^{-3}\varepsilon$$

$$\Omega_f = \Omega_l\Omega_u^{-1} \qquad\qquad W = -wh^3(\mu\dot{h}a^4)^{-1} \quad S = -\rho\Omega_u^2h^3(\mu\dot{h})^{-1}$$

and

$$M(h) = 1 - 3\alpha^* + \left(6\sigma^{*2} + 6\alpha^{*2}\right) - \left(10\varepsilon^* + 30\sigma^{*2}\alpha^* + 10\alpha^{*3}\right).$$

Using Eq. (3) (boundary conditions) solved Eq. (4), we get the pressure in the bearing system in the form of

$$P = \frac{\mu^*}{2} R(1 - R) + \left[12M(h) - \frac{S}{5}(3\Omega_f^2 + 4\Omega_f + 3) \right] \frac{1 - R^2}{4}. \tag{5}$$

In the dimensionless form, the LBC w is given by

$$W = \frac{\mu^*}{24} + \frac{3}{4} M(h) - \frac{S}{80}(3\Omega_f^2 + 4\Omega_f + 3). \tag{6}$$

3   Results and Discussion

Equation (6) indicates that the load carrying capacity gets increased by $\mu^*/24$ as compared to traditional lubricant based such system. The linearity of expression (6) with respect to μ^* says that an increase in magnetization would give increased load taking capacity is depicted in Figs. (2, 3, 4, 5 and 6).

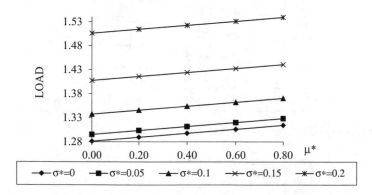

Fig. 2 Variation of load with respect to μ^* and σ^*

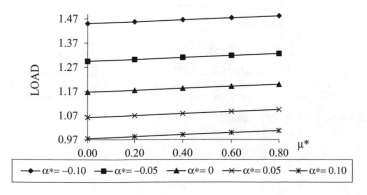

Fig. 3 Profile of load with regard to μ^* and α^*

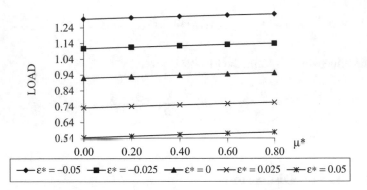

Fig. 4 Distribution of load for μ^* and ε^*

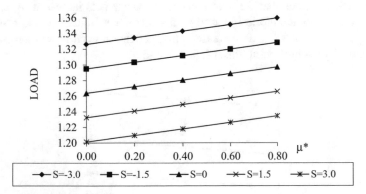

Fig. 5 Profile of load with regard to μ^* and S

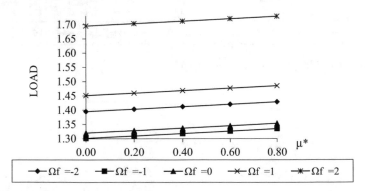

Fig. 6 Distribution of load for μ^* and Ω_f

For smooth bearing surfaces, this study is a study of circular plates in the absence of the rotation in magnetic fluid-based squeeze film. While assuming μ^* has value zero, one gets the investigation of [18].

For longitudinal roughness, the standard deviation increases the load distribution, which is entirely opposite to the nature of transverse roughness. This is exhibited in Figs. (7, 8, 9 and (10).

The variance $(-ve)$ increased the load capacity, while LBC falls due to variance $(+ve)$ (Figs. 11, 12 and 13).

The skewness follows the path of variance so far as load distribution is concerned which can be seen from Figs. (14 and 15).

The LBC is observed to assume a maximum value when the plates rotate in different directions $(-1 < \Omega_f < -0.5)$. But the maximum value is attained when Ω_f is approximately -0.67 (Fig. (16). Besides this optimum value occurs at $\Omega_f = 0$ when lower plate is taken non-rotating.

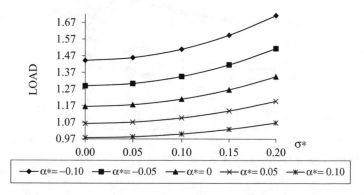

Fig. 7 Profile of load with regard to σ^* and α^*

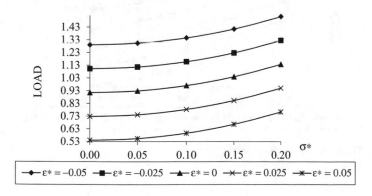

Fig. 8 Distribution of load for σ^* and ε^*

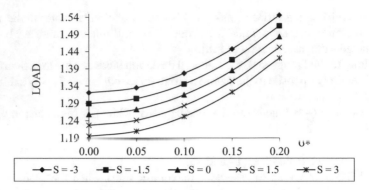

Fig. 9 Distribution of load for σ^* and S

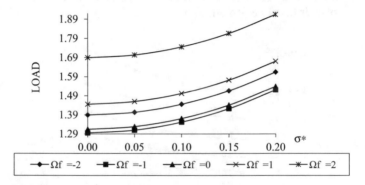

Fig. 10 Distribution of load for σ^* and Ω_f

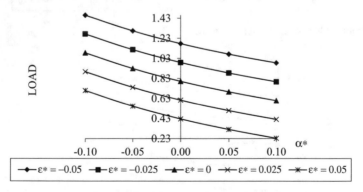

Fig. 11 Distribution of load for α^* and ε^*

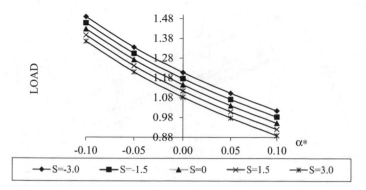

Fig. 12 Distribution of load for α^* and $\underline{S}$

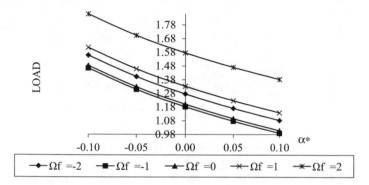

Fig. 13 Distribution of load for α^* and Ω_f

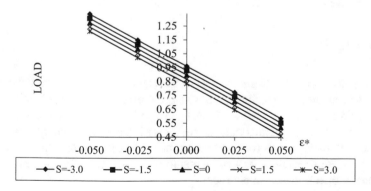

Fig. 14 Profile of load with regard to ε^* and S

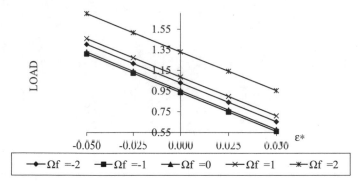

Fig. 15 Distribution of load for ε^* and Ω_f

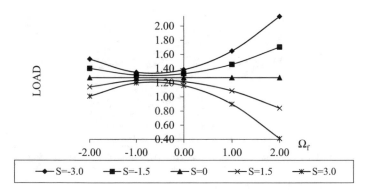

Fig. 16 Distribution of load for Ω_f and S

4 Conclusion

For this type of bearing system, the role of the standard deviation is crucial, even if the plates rotate in the opposite direction and the magnetic strength is in force. It is appealing to note that the trio of σ^*, ε^* ($-$ve) and α^* ($-$ve) tends to enhance the LBC. Therefore, longitudinal roughness remains a litter better for adoption in bearing design in comparison with transverse roughness. Lastly, the ferrofluid lubrication allows the bearing system to bear some quantity of load even if the flow is absent. In traditional lubrication, it is rarely seen.

Acknowledgements The authors would like to thank both the reviewers and the editor for their fruitful comments and constructive suggestions for an overall improvement of the quality of the research article.

References

1. Andharia, P.I., Deheri, G.M.: Longitudinal roughness effect on magnetic fluid based squeeze film between conical. Ind. Lubr. Tribol. **62**(5), 285–291 (2010)
2. Andharia, P.I., Deheri, G.M.: Performance of magnetic fluid based squeeze film between longitudinally rough elliptical plates. ISRN Tribol. ID482604, 6 (2013)
3. Andharia, P.I., Gupta, J.L., Deheri, G.M.: Effect of longitudinal surface roughness on hydrodynamic lubrication of slider bearings. In: Proceedings of the 10th International Conference on Surface Modification Technologies 1997, pp. 872–880. The Institute of Materials, London (1997)
4. Bhat, M.V., Deheri, G.M.: Magnetic fluid based squeeze film in curved porous circular disks. J. Magn. Magn. Mater. **127**, 159–162 (1993)
5. Bhat, M.V., Deheri, G.M.: Squeeze film behavior in porous annular disks. Lubr. Magn. Fluid Wear **151**, 123–128 (1991)
6. Bhat, M.V.: Lubrication with a Magnetic Fluid. Team Spirit (India) Pvt. Ltd., New Delhi (2003)
7. Christensen, H., Tonder, K.C.: The hydrodynamic lubrication of rough bearing surface of finite width. In: ASME-ASLE Lubrication Conference 1970, Paper no. 70-Lub-7 (1970)
8. Christensen, H., Tonder, K.C.: Tribology of Rough Surfaces: Parametric Study and Comparison of Lubrication Models. SINTEF Report, no. 22/69–18 (1969a)
9. Christensen, H., Tonder, K. C.: Tribology of Rough Surfaces: Stochastic Models of Hydrodynamic Lubrication. SINTEF Report, no. 10/69–18 (1969b)
10. Deheri, G.M., Patel, H.C., Patel, R.M.: Behavior of magnetic fluid based squeeze film between porous circular plates with porous matrix of variable thickness. Int. J. Fluid Mech. Res. USA **34**(6), 506–514 (2007)
11. Ellahi, R., Tariq, M.H., Hassan, M., Vafai, K.: On boundary layer nano-ferroliquid flow under the influence of low oscillating stretchable rotating disk. J. Mol. Liq. **229**, 339–345 (2017)
12. Gupta, J.L., Deheri, G.M.: Effect of roughness on the behavior of the squeeze film in a spherical bearing. Tribol. Trans. **167**, 173–179 (1996)
13. Lin, J.R., Lu, R.F., Lin, M.C., Wang, P.Y.: Squeeze film characteristics of parallel circular disks lubricated by ferrofluids with non-Newtonian couple stresses. Tribol. Int. **61**, 56–61 (2013)
14. Lin, J.R.: Longitudinal surface roughness effects in magnetic fluid lubricated journal bearing. J. Mar. Sci. Technol. **24**(4), 711–716 (2016)
15. Patel, H.C., Deheri, G.M., Patel, R.M.: Magnetic fluid-based squeeze film between porous rotating rough circular plates. Ind. Lubr. Tribol. **61**(3), 140–145 (2009)
16. Prajapati, B.L.: Squeeze film behavior between rotating porous circular plates with a concentric circular pocket: surface roughness and elastic deformation effects. Wear **152**(2), 301–307 (1992)
17. Prakash, J., Tiwari, K.: Roughness effect in porous circular squeeze-plates with arbitrary wall thickness. J. Lubr. Technol. **105**, 90 (1983)
18. Prakash, J., Vij, S.K.: Load capacity and time height relations for squeeze film between porous plates. Wear **24**, 309–322 (1973)
19. Shimpi, M.E., Deheri, G.M.: Combined effect of bearing deformation and longitudinal roughness on the performance of a ferrofluid based squeeze film together with velocity slip in truncated conical plates. Imperial J. Interdisc. Res. **2**(6), 1423–1430 (2016)
20. Ting, L.L.: A mathematical analog for determination of porous annular disks squeeze film behavior including the fluid inertia effect. J. Basic Eng. **94**(2), 417 (1972)
21. Ting, L.L.: Engagement behavior of lubricated porous annular disks part I: Squeeze film phase surface roughness and elastic deformation effects. Wear **34**, 159–182 (1975)
22. Verma, P.D.S.: Magnetic fluid based squeeze films. Int. J. Eng. Sci. **24**(3), 395 (1986)
23. Vora, K.H., Bhat, M.V.: The load capacity of a squeeze film between curved porous rotating plates. Wear **65**, 39 (1980)
24. Wu, H.: The squeeze film between rotating porous annular disks. Wear **18**, 461 (1971)

Study of Longitudinal Roughness on Hydromagnetic Squeeze Film Between Conducting Rotating Circular Plates

Jatinkumar V. Adeshara, M. B. Prajapati, G. M. Deheri and R. M. Patel

Abstract This investigation addresses the problem of squeeze film with electrical conduction between longitudinally rough surfaces and electrical lubricant in the existence of a transverse magnetic field for rotating circular plates. The surfaces are taken to be longitudinally rough in nature. In view of Christensen and Tonder's stochastic averaging method, the arbitrary irregularity of the bearing surfaces is modeled by a stochastic arbitrary inconstant with non-zero variance, skewness, and mean. The Reynolds' type equation for the distribution of pressure is stochastically averaged with regards to the arbitrary roughness constraint. A solution for SF pressure is obtained by using suitable Reynolds' type BC, which is further used to calculate the LBC. Based on the results obtained, the bearing is generally suffering due to longitudinal roughness. On the whole, the hydromagnetic effect characterized by the Hartmann number produces an increase in LCC as compared to the classical NL case. However, in the case of (−ve) roughness (skewed) in particular, the condition can be retrieved to some extent when (−ve) variance occurs by selecting the appropriate plate conductivity and standard deviation.

Keywords Load-bearing capacity · Circular plates · Longitudinal roughness · Rotation · Hydromagnetization

J. V. Adeshara · M. B. Prajapati
Mathematics Department, H. N. G. U, Patan, Gujarat 384265, India
e-mail: adesharajatin01@gmail.com

G. M. Deheri
Mathematics Department, S. P. U, Vallabh Vidyanagar, Gujarat 388120, India

R. M. Patel (✉)
15 Badrinath, Society, Ghodasar, Ahmedabad, Gujarat 380050, India
e-mail: rmpatel2711@gmail.com

© Springer Nature Singapore Pte Ltd. 2020   219
K. N. Das et al. (eds.), *Soft Computing for Problem Solving*,
Advances in Intelligent Systems and Computing 1057,
https://doi.org/10.1007/978-981-15-0184-5_20

Nomenclature

r	Radial coordinate
a	Plate's radius
$\dot{h}$	Velocity of squeeze film
B_0	Transverse magnetic field applied between the plates
h_0	Initial film thickness
h	Lubricant film thickness
s	Electrical conductivity of the lubricant
μ	Viscosity
M	$= B_0 h \left(\frac{s}{\mu}\right)^{1/2}$ = Hartmann number
h_0'	Lower plate's width surface
h_1'	Upper plate's width surface
s_0	Lower surface's electrical conductivity
s_1	Upper surface's electrical conductivity
$\phi_0(h)$	$= \frac{s_0 h_0'}{sh}$ = Lower surface's electrical permeability
$\phi_1(h)$	$= \frac{s_1 h_1'}{sh}$ = Upper surface's electrical permeability
ρ	Density of lubricant
Ω_u	Upper plate's angular velocity
Ω_l	Lower plate's angular velocity
Ω_r	$\Omega_u - \Omega_l$
Ω_f	Ω_l/Ω_u—Rotation ratio
S	$= -\frac{h^3 \rho \Omega_u^2}{\mu \dot{h}}$ = rotational inertia in non-dimensional form
p	Pressure of Lubricant
w	Load-carrying capacity
$\sigma*$	Non-dimensional standard deviation (σ/h)
$\alpha*$	Dimensionless variance (α/h)
$\varepsilon*$	Dimensionless skewness (ε/h^3)
P	Dimensionless pressure
W	Dimensionless load-carrying capacity
LCC	Load-carrying capacity
LBC	Load-bearing capacity
MHD	Magnetohydrodynamic
HL	Hydrodynamic lubrication
HSF	Hydromagnetic squeeze film
MF	Magnetic fluid
BC	Boundary conditions.

1 Introduction

Several analyses studied on the HL for plane metal bearing have discussed. Elco and Huges [14] analyzed MHD pressurization in lubrication of liquid metal, while Kuzma [16, 17] discussed the performance of magneto-hydrodynamic squeeze film. The behavior of hydromagnetic squeeze film between couple of conducting non-spongy surfaces has been investigated by Shukla and Prasad [24]. In this article, the effect of the plate conductivities on the behavior of the bearing system has been examined. Using approximation of M-C, Patel and Gupta [20] simplified the scrutiny for HSF lies between spongy annular disks with tangential velocity slip. The rough-ness is generated after having wear and run-in. The roughness effect was scrutinized by many researchers (Davis [13], Burton [8], Tonder [26], Tzeng and Saibel [27], Christensen and Tonder [10–12], Berthe and Godet [6]). Stochastic averaging process of Christensen and Tonder [10–12] suggests analysis for both roughnesses (longitu-dinal and transverse). This approach established the basis to study and analyze the outcome of surface irregularity in a many researchers [25], Prajapati [21], Gupta and Deheri [15], Andharia et al. [4].

Hydromagnetic squeeze film characteristics for different geometries have been discussed by Prajapati [21], Chou et al. [9]. Patel et al. [19] analyzed the MF based SF between porous rotating rough circular plates. The longitudinal roughness was subjected to study and analyze in Andharia and Deheri [1–3], Andharia et al. [5], Lin [18] and Shimpi and Deheri [23]. All these investigations have proposed that the longitudinal roughness offered a less adverse effect in comparison with the transverse roughness case.

Thus, in the present study, it has been recommended to examine the effect of rough-ness (longitudinal) on hydromagnetic squeeze film between conducting rotating cir-cular plates, wherein the standard deviation associated with longitudinal roughness plays significant role.

2 Analysis

Figure 1 shows the geometry and configuration of the bearing system.

The lower plate is considered to be rotating, while the upper plate moves to the lower plate. The clearance space of plates is electrically conducting and filled by lubricant.

The lubricant film is taken to be non-viscous, non-compressible, and the flow is laminar in nature. Prajapati [21] and Bhat [7] used hydromagnetic lubrication, modified Reynolds' type equation explaining the film pressure is obtained as

$$\frac{1}{r}\frac{\partial}{\partial r}\left[r\frac{\partial p}{\partial r}\right] = \frac{\mu \dot{h}}{AB} + 2\rho\left(\frac{3}{10}\Omega_r^2 + \Omega_r\Omega_l + \Omega_l^2\right) \tag{1}$$

Fig. 1 Geometry of the
bearing system

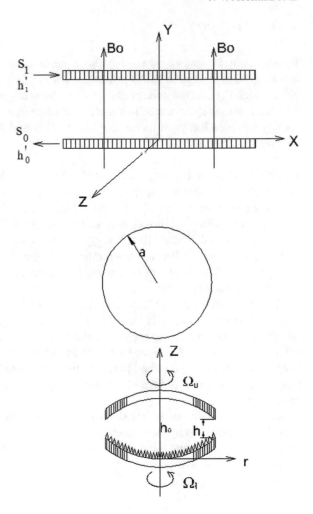

where

$$A = \frac{2}{M^3}\left[\frac{M}{2} - \tan h\frac{M}{2}\right], B = \left[\frac{\phi_0 + \phi_1 + 1}{\phi_0 + \phi_1 + (\tan h\frac{M}{2})/\frac{M}{2}}\right].$$

As a result with traditional theories of hydromagnetic lubrication and following the discussions of Andharia and Deheri [1–3], Christensen and Tonder [10–12], and Patel et al. [19], one arrives at the concerned stochastically averaged Reynolds' type equation resorting to the longitudinal roughness model as

$$\frac{1}{r}\frac{\partial}{\partial r}\left[r\frac{\partial p}{\partial r}\right] = \frac{\mu\dot{h}m(h)}{AB} + 2\rho\left(\frac{3}{10}\Omega_r^2 + \Omega_r\Omega_l + \Omega_l^2\right) \qquad (2)$$

where

$$m(h) = \frac{1}{h^3}\left[1 - \alpha h^{-1} + \left(6h^{-2}\sigma^2 + 6h^{-2}\alpha^2\right) - 10h^{-3}\left(\epsilon + 3\sigma^2\alpha + \alpha^3\right)\right]$$

with the use of Reynolds' BC
  $r = a$, then $p = 0$
  and

$$r = 0 \text{ then } \frac{\partial p}{\partial r} = 0. \tag{3}$$

Now, integrating Eq. (2) and applying BC (3), one gets the squeeze film pressure distribution in the form of

$$p = \left[\frac{\mu \dot{h} m(h)}{AB} + 2\rho\left(\frac{3}{10}\Omega_r^2 + \Omega_r\Omega_l + \Omega_l^2\right)\right]\left(\frac{r^2 - a^2}{4}\right). \tag{4}$$

Using the following non-dimensional terms in Eq. (4)

$$R = ra^{-1} \quad \sigma^* = h\sigma^{-1} \quad \alpha^* = \alpha h^{-1} \quad \varepsilon^* = \varepsilon h^{-3} \quad S = -h^3\rho\Omega_u^2\left(\mu\dot{h}\right)^{-1}$$

one derives the distribution of pressure in dimensionless as

$$P = ph^3(\mu\pi a^2\dot{h})^{-1} = \left(\frac{M(h)}{AB} - \frac{S}{5}\left(3\Omega_f^2 + 4\Omega_f + 3\right)\right)\frac{(1 - R^2)}{4\pi} \tag{5}$$

where

$$M(h) = m(h)h^3 = 1 - 3\alpha^* + 6\sigma^{*2} + 6\alpha^{*2} - \left(10\varepsilon^* + 30\sigma^{*2}\alpha^* + 10\alpha^{*3}\right).$$

In fact, the load is the internal pressure originated between the opposite surfaces due to the dynamic action. The LBC is

$$W = \int_0^a 2\pi r.p(r)dr$$

and is computed in dimensionless form as

$$W = -wh^3(\mu\pi a^4\dot{h})^{-1} = \frac{1}{8\pi}\left(\frac{M(h)}{AB} - \frac{S}{5}\left(3\Omega_f^2 + 4\Omega_f + 3\right)\right). \tag{6}$$

3   Results and Discussion

It is easily detected from Eqs. (5) and (6) that the distribution of pressure and the load depends on different parameters such as $M, \phi_0 + \phi_1, \sigma^*, \varepsilon^*, \alpha^*, S,$ and Ω_f. Setting the values of $\sigma^*, \alpha^*, \varepsilon^*, S,$ and Ω_f to be zero, the present study reduces to investigation of Shukla and Prasad [24] for a smooth non-spongy, irrotating conducting plate, while in limiting case taking $M \to 0$, this study trims down to the analysis of Prakash and Vij [22] for non-magnetic case. The results of Patel and Gupta [20] are found when electrical permeability of both the surfaces is considered to be zero.

It is clearly seen that as hydromagnetization parameter and electrical permeability of both the surfaces increase, the LBC increases for a fixed values of $\sigma^*, \alpha^*, \varepsilon^*, S,$ and Ω_f. The study of conductivity on the profile of pressure and distribution of LBC originates from the factor

$$\frac{\phi_0 + \phi_1 + \left(\tan h \frac{M}{2}\right)/\frac{M}{2}}{\phi_0 + \phi_1 + 1}.$$

This factor tends to

$$\frac{\phi_0 + \phi_1}{\phi_0 + \phi_1 + 1}$$

for higher values of M as $\tanh(M/2) \to 0$. As both these functions are increasing functions of $\phi_0 + \phi_1$, and from the mathematical expressions, one can notice that as $\phi_0 + \phi_1$ increases, the pressure and LCC increase as well.

The distribution of LBC with regards to the parameter of magnetization M for different values of $\phi_0 + \phi_1, \sigma^*, \alpha^*, \varepsilon^*, S,$ and Ω_f, respectively, given in Figs. (2, 3, 4, 5, 6 and 7).

A closed look at these figures depicts that the load W increases marginally with regards to the magnetization parameter M. Further, (−ve) skewed roughness helps to improve the bearing behavior, same as with (−ve) variance. Here, it is appealing

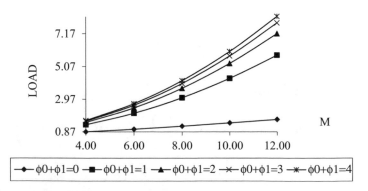

Fig. 2 Variation of load-carrying capacity with respect to M and $\phi_0 + \phi_1$

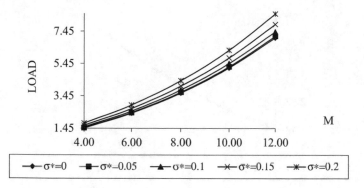

Fig. 3 Distribution of load for M and $\sigma*$

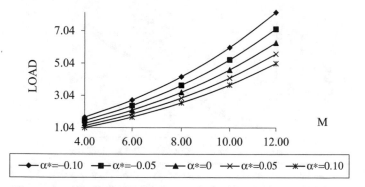

Fig. 4 Profile of load-bearing capacity with regards to M and $\alpha*$

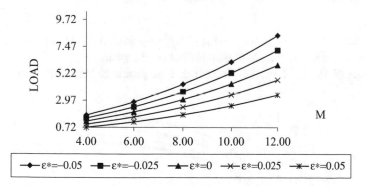

Fig. 5 Variation of load-carrying capacity with respect to M and $\varepsilon*$

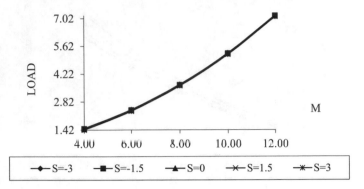

Fig. 6 Distribution of load for *M* and *S*

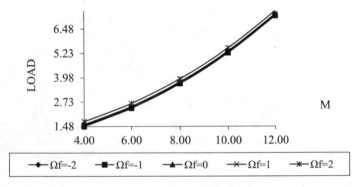

Fig. 7 Profile of load-bearing capacity with regards to *M* and Ω_f

to note that the increase in load induced by the variance ($-$ve) is sharper, and the variance (+) generally has an negative effect on the system.

Figures (8, 9, 10, 11 and 12) characterize the profile of load with regards to elec-

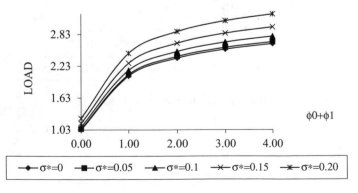

Fig. 8 Distribution of load with respect to $\phi_0 + \phi_1$ and $\sigma*$

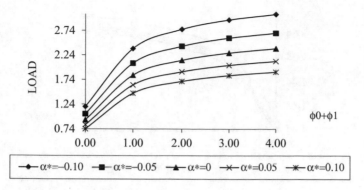

Fig. 9 Distribution of load for $\phi_0 + \phi_1$ and α^*

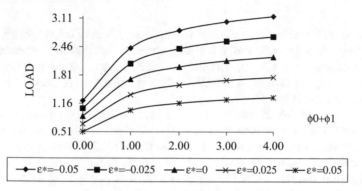

Fig. 10 Profile of load-bearing capacity with regards to $\phi_0 + \phi_1$ and ε^*

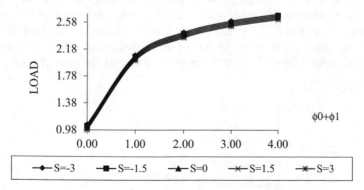

Fig. 11 Variation of load-carrying capacity with respect to $\phi_0 + \phi_1$ and S

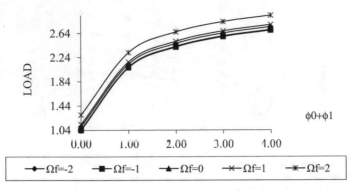

Fig. 12 Distribution of load for $\phi_0 + \phi_1$ and Ω_f

trical permeability of both the surfaces for different values of σ^*, α^*, S, ε^*, and Ω_f, respectively. From these figures, it is clearly observed that the LCC increases significantly with respect to $\phi_0 + \phi_1$. The standard deviation associated with longitudinal irregularity helps in increasing the LBC which is unlikely in the case of transverse surface roughness (Figs. 13, 14, 15 and 16).

Figures (17, 18, 19, 20 and 21) establish that the negatively skewed roughness enhances the load as in the case of variance (−ve). These trends reverse for positively skewed roughness and variance (+ve). The profile of LCC with respect to rotational inertia and rotation ratio is presented in Fig. (22). A maximum value of load is observed when the plates rotate in opposite directions ($-1 < \Omega_f < -0.5$). For $\Omega_f \approx -0.67$, the LBC attains its maximum value.

From these figures, one can conclude that the opposing effect induced by variance positive, positive skewness, and rotational inertia can be neutralized up to certain extent in the case of skewness(−ve), especially when the negative variance is involved. In general, the bearing suffers due to roughness. Given analysis offers plenty scopes for enhancing the bearing performance in the case of (−ve) skewed

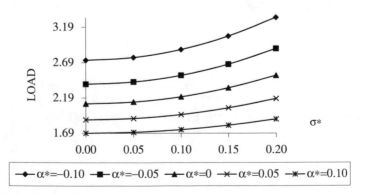

Fig. 13 Profile of load-bearing capacity with regards to σ^* and α^*

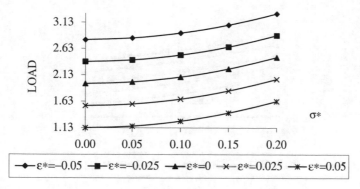

Fig. 14 Variation of load-carrying capacity with respect to σ^* and ε^*

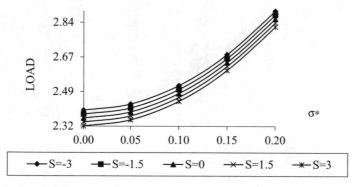

Fig. 15 Distribution of load for σ^* and S

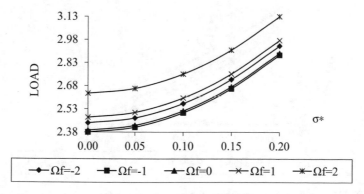

Fig. 16 Profile of load-bearing capacity with regards to σ^* and Ω_f

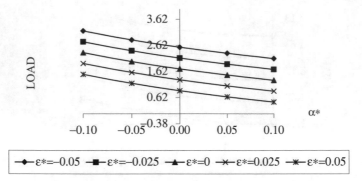

Fig. 17 Variation of load-carrying capacity with respect to α^* and ε^*

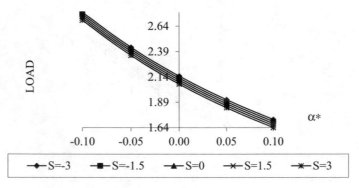

Fig. 18 Distribution of load for α^* and S

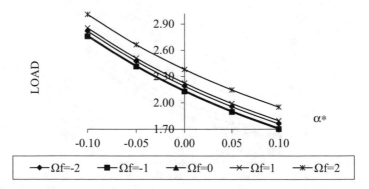

Fig. 19 Profile of load-bearing capacity with regards to α^* and Ω_f

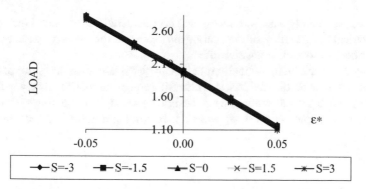

Fig. 20 Variation of load-carrying capacity with respect to ε^* and S

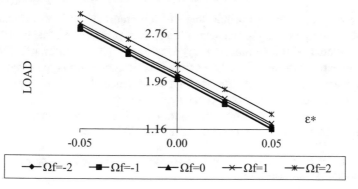

Fig. 21 Distribution of load for ε^* and Ω_f

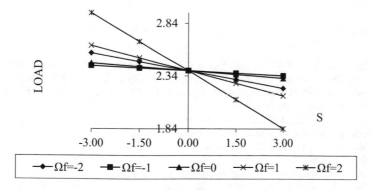

Fig. 22 Profile of load-bearing capacity with regards to S and Ω_f

roughness, especially when variance (−ve) is considered, and the combined effect of the electrical permeability of the bearing surfaces and the magnetization parameter (M) further improves this positive effect.

Using the case of longitudinally rough bearing system, machine's life period can be extended. Due to this analysis, it is clear that longitudinal irregularity parameters must be given due consideration; while designing such type of bearing systems, even if the suitable choice of $\phi_0 + \phi_1$ and (M), Hartmann number has been taken into account.

4 Conclusions

It is clearly seen from this article that the longitudinal roughness remains more effective as compared to the transverse roughness case. Indeed, the standard deviation associated with longitudinal roughness makes all the difference. Hence, the roughness essentials are to be taken care while designing this type of bearing system. Of course, in augmenting the bearing performance, the plate conductivities and the hydromagnetization may play crucial role. A close scrutiny indicates that some amount of load is supported by the system even if no flow occurs. This does not occur in the case of conventional fluid based bearing system.

Acknowledgements Comments and constructive suggestions for improving the overall quality of this article have been acknowledged.

References

1. Andharia, P.I., Deheri, G.M.: Effect of longitudinal surface roughness on the behaviour of squeeze film in a spherical bearing. Int. J. Appl. Mech. Eng. **6**(4), 885–897 (2001)
2. Andharia, P.I., Deheri, G.M.: Longitudinal roughness effect on magnetic fluid based squeeze film between conical plates. Ind. Lubr. Tribol. **62**(5), 285–291 (2010)
3. Andharia, P.I., Deheri, G.M.: Performance of magnetic fluid based squeeze film between longitudinally rough elliptical plates. ISRN Tribol. **ID482604**, 6 (2013)
4. Andharia, P.I., Gupta, J.L., Deheri, G.M.: Effect of longitudinal surface roughness on hydrodynamic lubrication of a slider bearings. In: Proceeding of The Tenth International Conference on surface modification Technologies, pp. 872–880 (1997)
5. Andharia, P.I., Gupta, J.L., Deheri, G.M.: Effect of transverse surface roughness on the behaviour of squeeze film in spherical bearing. J. Appl. Mech. Eng. **4**, 19–24 (1999)
6. Berthe, D., Godet, M.A.: A more general form of Reynolds' equation—application of rough surfaces. Wear **27**, 345–357 (1973)
7. Bhat, M.V.: Lubrication with a Magnetic Fluid. Team Spirit, India, Pvt. Ltd. (2003)
8. Burton, R.A.: Effect of two-dimensional sinusoidal roughness on the load support characteristics of a lubricant film. Trans. ASME J. Basic Eng. **85**, 258–264 (1963)
9. Chou, T.L., Lai, J.W., Lin, J.R.: Magneto-hydrodynamic squeeze film characteristics between a sphere and a plane surface. J. Mar. Sci. Technol. **11**(3), 174–178 (2003)

10. Christensen, H., Tonder, K.C.: Tribology of rough surfaces: a stochastic model of mixed lubrication. SINTEF, Report no. 18/70, 21 (1970)
11. Christensen, H., Tonder, K.C.: Tribology of rough surfaces: parametric study and comparison of lubrication models. SINTEF, Report no 22/69, 18 (1969)
12. Christensen, H., Tonder, K.C.: Tribology of rough surfaces: stochastic models of hydrodynamic lubrication. SINTEF, Report No. 10/69, 18 (1969)
13. Davis, M.G.: the generation of pressure between rough lubricated moving deformable surfaces. Lubr. Eng. **19**, 246 (1963)
14. Elco, R.A., Huges, W.E.: Magnetohydrodynamic pressurization in liquid metal lubrication. Wear **5**, 198–207 (1992)
15. Gupta, J.L., Deheri, G.M.: Effect of roughness on the behaviour of squeeze film in a spherical bearing. Tribol. Trans. **39**, 99–102 (1996)
16. Kuzma, D.C., Maki, E.M., Donnelly, R.J.: The magnehydrodynamic squeeze films. J. Fluid Mech. **19**, 395–400 (1964)
17. Kuzma, D.C.: Magnetohydrodynamic squeeze films. J. Basic Eng. Trans. ASME **86**, 441–444 (1964)
18. Lin, J.R.: Longitudinal surface roughness effects in magnetic fluid lubricated journal bearing. J. Mar. Sci. Technol. **24**(4), 711–716 (2016)
19. Patel, H., Deheri, G.M., Patel, R.M.: Magnetic fluid based squeeze film between porous rotating rough circular plates. Ind. Lubr. Tribol. **61**(3), 140–145 (2009)
20. Patel, K.C., Gupta, J.L.: Behaviour of hydromagnetic squeeze film between porous plates. Wear **56**, 327–339 (1979)
21. Prajapati, B.L.: On the certain theoretical studies in hydrodynamic and electromagnet hydrodynamic lubrication. Ph.D. Thesis, S. P. university, Vallabh Vidyanagar (1995)
22. Prakash, J., Vij, S.K.: Load carrying capacity and time height relation for squeeze film between porous plates. Wear, **24**, 309–332 (1973)
23. Shimpi, M.E., Deheri, G.M.: Combine effect of bearing deformation and longitudinal roughness on the performance of a ferrofluid based squeeze film together with velocity slip in truncated conical plates. Imperial J. Interdisc. Res. **2**(8), 1423–1430 (2016)
24. Shukla, J.B., Prasad, R.: Hydromagnetic squeeze films between two conducting surfaces. Transection. J. Basic Eng. **87**, 818–822 (1965)
25. Ting, L.L.: Engagement behavior of lubricated porous annular disks part I: squeeze film phase, surface roughness and elastic deformation effects. Wear **34**, 159–182 (1975)
26. Tonder, K.C.: Surface distributed waviness and roughness. In: The Firdst World Conference in Industrial Tribology, New Delhi, A3, 128 (1972)
27. Tzeng, S.T., Saibel, E.: Surface roughness effect on slider bearing lubrication. Transection ASME. J. Lubr. Technol. **10**, 334–338 (1967)

Squeeze Film in a Ferrofluid Lubricated Rough Conical Plates: Comparison of Porous Structures

R. M. Patel, Gunamani Deheri and Pragna A. Vadher

Abstract Present article is a comparison of the effects of porous structures on a Ferro fluid-based squeeze film in rough porous conical plates. The globular sphere model of Kozeny—Carman and capillary fissures model due to Irmay for porous structure have been considered for the investigation. The model of Christensen and Tonder has been imposed to calculate the effect of roughness. The distribution of pressure in the bearing is got by solving the concern equation of Reynolds'. Obtained results show that increasing values of the magnetization parameter result in increased load-carrying capacity. The effect of transverse roughness has been established to be adverse for both the structures. However, this effect is sharper for Irmay's model. Besides, the contrary effect of porosity and roughness can be reduced by the positive effect of magnetization at least in the case of globular sphere model due to Kozeny—Carman. The semi-vertical angle of the cone also provides support in minimizes the poor effect of negatively skewed roughness. This article offers the suggestion that the Kozeny—Carman model may be preferred as related to the model proposed by Irmay.

Keywords Conical plates · Magnetic fluid · Porous structure · Roughness · Squeeze film

1 Introduction

During the last few years, the use of magnetic particles (nano) in various engineering systems has been an intensive area of investigation all over the world. It is not surprising that magnetic nanoparticles have been found to be applied in biomedicine, such

R. M. Patel (✉)
Gujarat Arts and Science College, Ellis Bridge, Ahmedabad, Gujarat 380006, India
e-mail: rmpatel2711@gmail.com

G. Deheri
Mathematics Department, S. P. University, V.V. Nagar, Gujarat 388120, India

P. A. Vadher
Physics Department, Govt. Sci. College, Gandhinagar 382016, India

© Springer Nature Singapore Pte Ltd. 2020
K. N. Das et al. (eds.), *Soft Computing for Problem Solving*,
Advances in Intelligent Systems and Computing 1057,
https://doi.org/10.1007/978-981-15-0184-5_21

235

as magnetic separation, drug delivery, magnetic resonance image, and hyperthermia. A remarkable property of magnetic nanoparticle is their super para-magnetism. In addition, in magnetic gradient fields these ferrofluids exhibit increased viscosity. The improvement of magnetic lubricant over the conventional one is that the former can be retained at a chosen location by a magnetic field (external). This property alone remains sufficient for making use of these ferrofluids in almost all sealed systems such as food preparation machines.

Prakash and Vij [1] dealt with the investigation of load-carrying capacity and response time for squeeze film between plates. In this study different geometries including the conical shapes were discussed. Verma [2] investigated the influence of magnetic fluid lubricant on the behavior of a squeeze film. Here the load-carrying capacity was found to be increased. Subsequently, squeeze film performance in annular disks (porous) lubricated with magnetic field was researched by Bhat–Deheri [3]. The bearing performance was found to be significantly enhanced as compared to the case of lubricants. The discussion of Bhat–Deheri [3] confirmed a significant enhancement on the magnetic squeeze film in annular plates. This analysis was modified by Andharia et al. [4] for curved annular plates. Lin et al. [5] analyzed the performance of a magnetohydrodynamic squeeze film among curved annular plates. Patel and Deheri [6] obtained an augmented squeeze film performance under ferrofluid lubrication in conical plates by choosing suitably the angle of the cone (semi-vertical). The method adopted in the investigation of Patel–Deheri [6] was modified and established by Vadher et al. [7] to calculate the adverse result of transverse roughness for a fluid-based squeeze film between rough porous conical plates. Here it was presented that the raised load-carrying capacity received a further boost in the case of (−ve) skewed roughness.

The investigation of Andharia and Deheri [8] established the opposite rule of standard deviation with regards to performance of the squeeze film in conical plates. Lin [9] derived the equation of ferrofluid lubrication for cylindrical squeeze films considering fluid inertia forces with applications to disks. Recently, Andharia and Deheri [10] considered the enactment of a magnetic fluid based squeeze film between rough (longitudinally) plates. It has been noticed that the increased load-carrying capacity due to the magnetization got significantly increased owing to the collective effect of skewness (−ve) and standard deviation. Patel and Deheri [11] discussed the Sliomis model-based ferrofluid lubricated squeeze film in transversely rough rotating curved circular plates. Very recently, Patel and Deheri [12] developed the method adopted in above article to investigate the behavior of a magnetic squeeze film in rough curved annular plates considering assorted porous structures.

In this current article, it has been analyzed and examines the fluid lubrication of a squeeze film rough conical plate considering various porous structures.

2   Analysis

In Fig. 1, the geometry and configuration of the bearing system is given, which
involves of the upper plate moves towards to the lower. Developing the analysis
of Prakash and Vij [1], Patel and Deheri [6] obtained the equation for the pressure
considering ferrofluid lubrication:

$$\frac{1}{x}\frac{d}{dx}\left[x\frac{d}{dx}\left(p - 0.5\,\mu_0\bar{\mu}H^2\right)\right] = \frac{12\eta\dot{h}_0 \sin \omega}{h^3 \sin^3 \omega + 12\,\phi H} \tag{1}$$

where $H^2 = a^2\left(1 - \frac{x^2 \sin^2 \omega}{a^2}\right)$,

η is viscosity of fluid, susceptibility of magnetization is $\bar{\mu}$, μ_0 is free space's
permeability, taken from Bhat and Deheri [3].

The roughness is categorized by arbitrary variable with mean, skewness and variance following the discussion of Christensen and Tonder [13–15].

Adopting the stochastic averaging method of Christensen and Tonder [13–15]
for Eq. (1) one is eventually led to the following governing Reynolds' equation for
pressure distribution

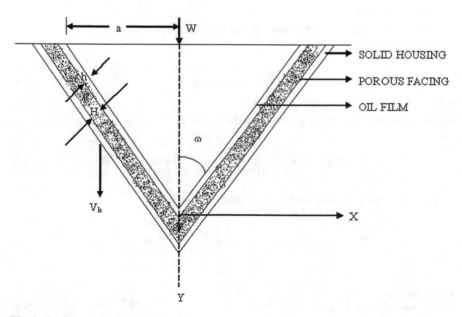

Fig. 1 Configuration of the bearing system

$$\frac{1}{x}\frac{d}{dx}\left\{x\frac{d}{dx}(p - 0.5\mu_0\bar{\mu}H^2)\right\} = \frac{12\eta\dot{h}_0\sin\omega}{g(h) + 12\psi l_1} \tag{2}$$

where

$$g(h) = h^3\sin^3\omega + 3\sigma^2 h\sin\omega + 3\alpha h^2\sin^2\omega + 3\alpha^2 h\sin\omega + 3\sigma^2\alpha + \alpha^3 + \varepsilon \tag{3}$$

and ψ is the porous structure of the porous region, l_1 is the porous facing thickness.

2.1   Case—I (A Globular Sphere Model Given in Fig. 2)

This model requires the porous substance to be packed by the globular spherical particles. The mean particle size is D_c. In view of the investigation of Kozeny and Carman [16] the permeability of the porous region was found to be

$$\psi = \frac{D_c^2 e_1^3}{180(1 - e_1)^2} \tag{4}$$

where e_1 is the porosity.

Integrate the Eq. (2) with regards to the given conditions
$p = 0; x = a\,\mathrm{cosec}\,\omega$

$$\frac{dp}{dx} = 0; x = 0 \tag{5}$$

results in the expression for pressure distribution as

$$p = 0.5\mu_0\bar{\mu}H^2 + \frac{3\eta\dot{h}_0}{g(h) + 12\psi l_1}(x^2 - a^2\mathrm{cosec}^2\omega) \tag{6}$$

Introducing the non–dimensional quantities

$$P = -\frac{ph_0^3}{\eta\dot{h}_0 a^2\mathrm{cosec}\,\omega}\quad \mu^* = -\frac{\mu_0\bar{\mu}h_0^3}{\eta\dot{h}_0}\quad \bar{x} = \frac{x}{a\,\mathrm{cosec}\,\omega}$$

Fig. 2 Structure model of porous sheets given by K-C

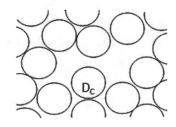

$$W = -\frac{wh_0^3}{\eta h_0 a^4 \mathrm{cosec}^2 \omega} \quad \bar{\psi} = \frac{D_c^2 l_1}{h_0^3} \bar{h} = h h_0^{-1}$$

$$\bar{h}_1 = h_1 h_0^{-1} \quad \bar{h}_2 = h_2 h_0^{-1} \quad \sigma^* = \sigma h_0^{-1} \quad \alpha^* = \alpha h_0^{-1}$$

$$\varepsilon^* = \varepsilon h_0^{-3}$$

and

$$g(\bar{h}) = \bar{h}^3 \sin^3 \omega + 3\sigma^{*2} \bar{h} \sin \omega + 3\alpha^* \bar{h}^2 \sin^2 \omega + 3\alpha^{*2} \bar{h} \sin \omega + 3\sigma^{*2} \alpha^* + \alpha^{*2} + \varepsilon^* \tag{7}$$

one arrives at the expressions for dimensionless pressure distribution

$$P = \frac{\mu^*}{2}(1 - \bar{x}^2) \sin \omega + \frac{3(1 - \bar{x}^2)}{g(\bar{h}) + \frac{\bar{\psi} e_1^3}{15(1 - e_1)^2}} \tag{8}$$

The dimensionless load carrying capacity is given by

$$W = \frac{\mu^*}{8} + \frac{3}{4}\left[\frac{\mathrm{cosec}\, \omega}{g(\bar{h}) + \frac{\bar{\psi} e_1^3}{15(1 - e_1)^2}}\right] \tag{9}$$

where in the load carrying capacity w is calculated from

$$w = \int_0^{\mathrm{acosec}\, \omega} x p(x) \mathrm{d}x.$$

Time Δt in non-dimensional form is given

$$\Delta \bar{t} = -W \int_{h_1}^{\bar{h}_2} \frac{1}{g(\bar{h}) + 12\bar{\psi}} \mathrm{d}\bar{h} \tag{10}$$

2.2 Case—II (Model of Capillary Fissures Model Is Represented in Fig. 3)

This model deals with three sets of reciprocally orthogonal fissures. Irmay [17] assumed that hydraulic gradient have no loss at the junction and originated the expression for the porous structure parameter as

Fig. 3 Structure model of
porous sheets given by Irmay

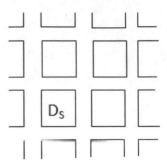

$$\psi = \frac{\left(1 - m^{2/3}\right)D_s^2}{12m} \tag{11}$$

where $m = 1 - e_1$, e_1 is the porosity.

In this case the non-dimensional pressure is found to be

$$P = \frac{\mu^*}{2}(1 - \bar{x}^2)\sin\omega + \frac{3(1 - \bar{x}^2)}{g(\bar{h}) + \frac{\bar{\psi}(1 - m^{2/3})}{m}} \tag{12}$$

Then the governing argument for dimensionless load carrying capacity is calculated as

$$W = \frac{\mu^*}{8} + \frac{3}{4}\left[\frac{1}{g(\bar{h}) + \frac{\bar{\psi}(1 - m^{2/3})}{m}}\right] \tag{13}$$

Lastly, the response time $\Delta \bar{t}$ as discussion earlier takes the following non-dimensional form

$$\Delta\bar{t} = -W \int_{\bar{h}_1}^{\bar{h}_2} \frac{1}{g(\bar{h}) + 12\bar{\psi}}d\bar{h} \tag{14}$$

3   Results and Discussions

Equation (13) establishes that the increase in load bearing capacity remains with regards to traditional lubrication. Indeed, magnetic fluid lubrication causes 5.63% increase in load carrying capacity (Figs. 4, 5, 6, 7, 8, 9, 10, 11, 12, 13, 14, 15, 16, 17, 18, 19, 20, 21, 22, 23, 24 and 25).

The graphical representations presented below indicate following:

Fig. 4 Variation of load-carrying capacity with respect to μ^* and ψ

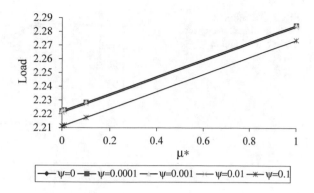

Fig. 5 Variation of load-carrying capacity with respect to μ^* and e_1

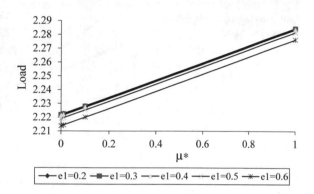

Fig. 6 Variation of load-carrying capacity with respect to σ^* and ψ

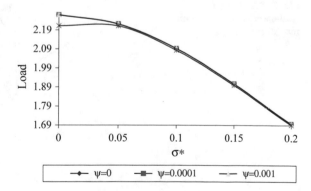

- The bearing suffers because of transversely rough surface.
- The negative influence of roughness is observed to be less when K. C. Model is adopted.
- Undoubtedly, magnetic fluid lubrication results in improved load profile.
- But, this positive effect of magnetization turns out to be more effective when one takes recourse K. C. Model

Fig. 7 Variation of load-carrying capacity with respect to σ^* and e_1

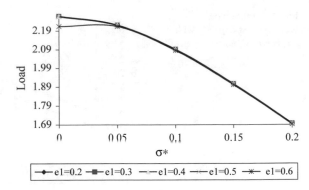

Fig. 8 Variation of load-carrying capacity with respect to α^* and ε^*

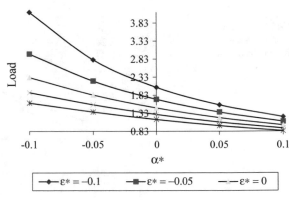

Fig. 9 Variation of load-carrying capacity with respect to α^* and ψ

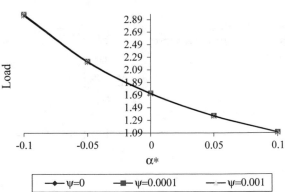

Fig. 10 Variation of
load-carrying capacity with
respect to α^* and ω

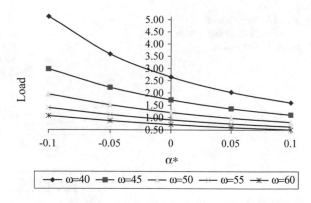

Fig. 11 Variation of
load-carrying capacity with
respect to ε^* and ψ

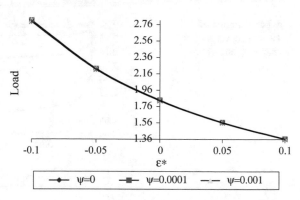

Fig. 12 Variation of
load-carrying capacity with
respect to ε^* and e_1

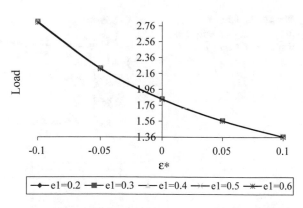

- The positive influence if negatively skewed roughness adds more to load-carrying capacity when Kozeny–Carman Model is in force.
- Even if, a suitable selection of semi-vertical angle of cone has been taken the role of Kozeny–Carman Model goes ahead of Irmay's Model with a view to roughness aspect only.

Fig. 13 Variation of
load-carrying capacity with
respect to ε^* and ω

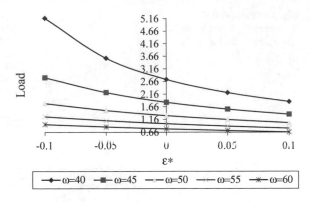

Fig. 14 Variation of
load-carrying capacity with
respect to ψ and e_1

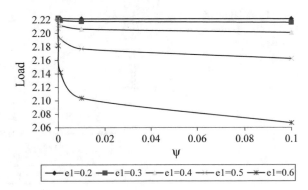

Fig. 15 Variation of
load-carrying capacity with
respect to μ^* and ψ

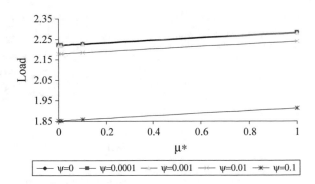

4 Conclusion

This investigation establishes that the K. C. Model may be a better choice for this
type of bearing systems. Further, this analysis indicates that even if suitable magnetic strength is in force. The roughness aspect must be carefully considered while
designing the bearing systems. For overall improved performance in the case of
Irmay's model, the angle may play a vital role. Besides, the bearing system supports

Fig. 16 Variation of
load-carrying capacity with
respect to μ^* and e_1

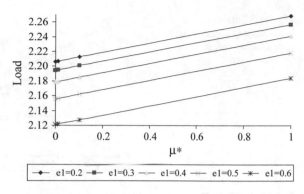

Fig. 17 Variation of
load-carrying capacity with
respect to σ^* and ψ

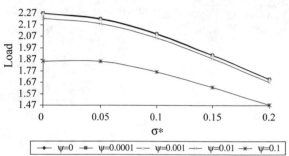

Fig. 18 Variation of
load-carrying capacity with
respect to σ^* and e_1

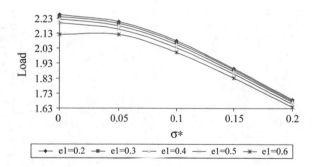

Fig. 19 Variation of
load-carrying capacity with
respect to α^* and ε^*

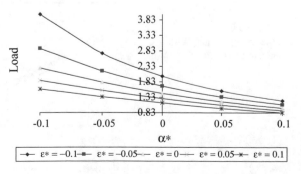

Fig. 20 Variation of
load-carrying capacity with
respect to α^* and ψ

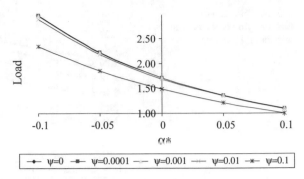

Fig. 21 Variation of
load-carrying capacity with
respect to α^* and ω

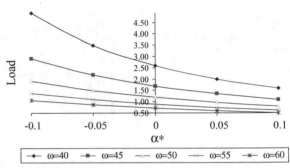

Fig. 22 Variation of
load-carrying capacity with
respect to ε^* and ψ

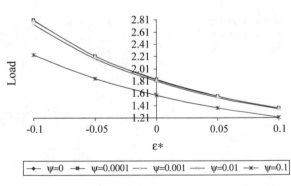

Fig. 23 Variation of
load-carrying capacity with
respect to ε^* and e_1

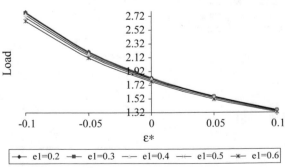

Fig. 24 Variation of
load-carrying capacity with
respect to ε^* and ω

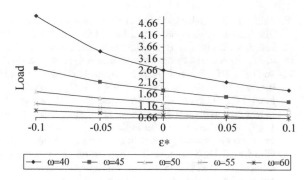

Fig. 25 Variation of
load-carrying capacity with
respect to ψ and e_1

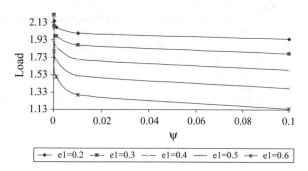

some amount of load in the absence of flow for both the models which never happens for conventional lubricants and this load is comparatively more in the case of Kozeny–Carman model.

Acknowledgements Constructive suggestions and positive remarks by the reviewers for the improvement of the presentation and quality of this research article are acknowledged with regards.

References

1. Prakash, J., Vij, S.K.: Load capacity and time height relations for squeeze film between porous plates. Wear **24**, 309–322 (1973)
2. Verma, P.D.S.: Magnetic fluid based squeeze films. Int. J. Eng. Sci. **24**(3), 395–401 (1986)
3. Bhat, M.V., Deheri, G.M.: Squeeze film behavior in porous annular disks lubricated with magnetic fluid. Wear. **151**, 123–128(1991)
4. Deheri, G.M., Patel, R.M., Abhangi, N.D.: Magnetic fluid-based squeeze film behavior between transversely rough curved annular plates. Ind. Lubr. Tribol. (2011)
5. Lin, J.-R., Lu, R.-F., Liao, W.-H.: Analysis of magneto-hydrodynamic squeeze film characteristics between curved annular plates. Ind. Lubr. Tribol., **56**(5), 300–305(2004)
6. Patel, R.M., Deheri, G.M.: Magnetic fluid based porous conical plates. Ind. Lubr. Tribol. **59**(3), 143–147 (2007)
7. Vadher, P.A., Deheri, G.M., Patel, R.M.: Performance of hydromagnetic squeeze films between conducting porous rough conical plates. Meccanica **45**(6), 767–783 (2010)

8. Andharia, P.I., Deheri, G.M.: Longitudinal roughness effect on magnetic fluid based squeeze film between conical plates. Ind. Lubr. Tribol. **62**(5), 285–291 (2010)

9. Lin, J.-R.: Derivation of ferrofluid lubrication equation of cylindrical squeeze films with convective fluid inertia forces and application to circular disks. Tribol. Int. **49**, 110–115 (2012)

10. Andharia, P.I., Deheri, G.M.: Performance of magnetic fluid based squeeze fil between longitudinally rough elliptical plates. ISRN Tribology, Article ID 482604, 6 p (2013)

11. Patel, J.R., Deheri, G.M.: Shliomis model based magnetic fluid lubrication of a squeeze film in rotating rough curved circular plates. Caribb. J. Sci. Technol. **1**, 138–150 (2013)

12. Patel, J.R., Deheri, G.M.: Theoretical study of Shliomis model based magnetic squeeze film in rough curved annular plates with assorted porous structures FME Transaction (2014)

13. Christensen, H., Tonder, K.: Tribology of rough surfaces. Stochastic models of hydrodynamic lubrication. SINTEF, Section for Machine Dynamics in Tribology, Technical University of Norway, Trondheim, Norway, Report No.10/69-18(1969a)

14. Christensen, H., Tonder, K.: Tribology of rough surfaces, parametric study and comparison of lubrication models. SINTEF, Section for Machine Dynamics in Tribology, Technical University of Norway, Trondheim, Norway, Report No.22/69-18(1969b)

15. Christensen, H., Tonder, K.: The hydrodynamic lubrication of rough bearing surfaces of finite width. ASME-ASLE Lubrication conference Cincinnati, Ohio, October 12–15, Lub-7 (1970)

16. Liu, Jun: Analysis of a porous elastic sheet damper with a magnetic fluid. J. Tribol. **131**, 0218011–0218015 (2009)

17. Irmay, S.: Flow of liquid through cracked media. Bull. Res. Counc., Isr., **5A**(1), 84 (1955)

IOT for Capturing Information and Providing Assessment Framework for Higher Educational Institutions—A Framework for Future Learning

Mayank Srivastava, Praneet Saurabh and Bhupendra Verma

Abstract Internet of Things (IoT) has been changing the way of operations for multiple segments like Industries, Health Care, and Manufacturing. It also holds a chance to change how Educational Institutions operates and enhance student learning experience. It has enormous opportunities for Educational Segment which will enhance the learning experiences for students, teachers and other stakeholders. The development of IOT Systems, devices, applications, and services are already in the consideration and process by the students and researchers. Therefore, this paper presents a framework to capture validated information of individual Higher Educational Institutions (HEI) through IOT devices to avail the assessment based platform to evaluate and enhance the educational experience. It also describes the processes to automate the survey of Educational Institutions and provide analytical report using IOT components and Machine Learning. To ease the understanding of different methods we provide a prototype with its practical implementations using common processes in a friendly manner.

Keywords Misuse · Cloud computing · Database · Education segment · Higher education institutions · Framework · Internet of things · Survey · Machine learning

1 Introduction

Internet of things, a number of sectors have implemented and predicted about the future of its implementation, and its potential impact and benefits in those sectors. There is an enormous number of physical devices across the globe which will be

M. Srivastava (✉)
Technocrats Institute of Technology Advance, Bhopal, MP 462021, India
e-mail: mayank.live@hotmail.com

P. Saurabh (✉) · B. Verma
Technocrats Institute of Technology, Bhopal, MP 462021, India
e-mail: praneetsaurabh@gmail.com

B. Verma
e-mail: bkverma3@gmail.com

embedded with sensors and are connected to utilize the network. Over the years, the Internet has connected the people to people then things to people and now things to things. This process of connectivity using sensors enables things to talk, transmit information and even pass on command to each other by processing data for decision-making system. This will create enormous opportunity for the people in more appropriate ways, and will also ensure the delivery of right information to right entity (machine or person), efficiently and effectively.

The widespread functionality of IOT has increased the data produced by its component, so as the data collection through different sensors has been widely increased. The quantity of data generated through applications, sensors, and its components are in large amounts of structured, unstructured and semi-structured data. This amount of data has resulted into big data as traditional database management system is not efficient in managing this massive data for storage, processing and analyzing [1, 2] The big data analytics has also become challenging as the quantity of data has increased [3]. Applications of Internet of Things have already been leveraged in the industrial segment and many others like health care, customer services, manufacturing, etc. Now, Educational Institution has an opportunity to lead the technical development and implement innovations models of IOT to enhance the capabilities, In some ways it already has come into the picture in this domain by connecting peers for providing education in their comfort zone [4], availing hassle-free education for disabled, increasing efficiency and providing security through RFID, chips, and sensors.

Internet of Learning Things are now into the picture and, the internet of things or everything (IOE)- the connection between people and things to process data into information has become the foundation for it. The process of interaction between objects allows individual to access unlimited information anytime and anywhere. The implementation of this in the Education Sector opens a new horizon of ideas and developments that have already been considered in the research segment. But its implementation for the benefit of the segment will enable the ocean of opportunities to expand. However, the questions are how the sensor data will be processed to take the intelligent decision by its own to produce the desired result. Also, the questions are how the raw data form will be transferred and would be converted into useful and meaningful data, i.e., information, which could be understandable to machine to interpret the information collected to create decision-making system [5]. We are in the era of data-driven processing and decisions, where the data analysis is a challenging task which would help to rethink teaching and learning on a global scale. We will understand its implementation to take data-driven decision in this segment. Education sector is transforming from knowledge transfer model to a self-directed, active, engaging and collaborative model. The sector uses technology as a catalyst to develop such model which helps students to increase their knowledge along with the skill development which is needed to succeed in the learning society [6]. The technologies have been integrated into the classrooms and learning to facilitate students through voice, video, and text-based collaborations, practices of learning through practical equipment like Arduino Microcontrollers, RFID sensors, cards, 3D printers, chips, etc. Now, the education segments have variety of technology-enabled solutions to enhance the teaching methodology. Also, as per Pew Research Centre, it has been observed that 95% of teens are using smartphones, and 45% of them are online constantly online [7]. So, these stats can be useful to bring change comprising the practices of smart learning methodologies.

Apparently, the quality assurance for these learning methods, capturing the information and analyzing the same to present among all are also important. The future of institutions is about how the Educational Institutions are implementing these changes to avail the quality assurance and its future implications. The future of universities is not only about using and applying trending technologies. It will depend upon the adaptability of the technology for the changing needs to develop the concurrent solution for the future and the knowledge. Across the education system, the inability to get the exact information about the available resources, infrastructure, and other details are quite difficult. There are ways to collect this information manually but this consumes lots of time, energy and resources to avail this. Even though, to provide the information in real-time would be quite a difficult task. In this paper, we propose a prototype to collect the information through IOT components which are being collected in the surveys which would help the stakeholders, i.e., accreditation bodies, students and parents in rank analysis, selection of the appropriate course and institutions through comparative analysis. IOT will help to collect the information through different sensors and will connect the seekers with the institutions who are based out of different locations of the country for the students and institutions as well. It will also help in collecting real-time information about the current scenario of the institutions for accreditation of the institutions to provide quality assurance for high-quality learning and peer to peer interaction.

In Sect. 2, we state more precisely about the architecture, processes, and motivation behind this framework. Section 3 introduces the conventional system which is used in the current scenario. Section 4 gives an overview about the proposed framework from a technical and research point of view and discusses the current requirements and its implications. Section 5, we shortly introduce the simulation process which was followed for use cases. Section 6 gives a conclusion on this.

2 Related Work

Internet today has connected all the things with the people to empower us and offer enhanced mobility in the current scenario and for the future as well. The communication between people has transformed gradually into things, eliminating all the interaction between people and things. The connectivity of the objects can be ascertained using sensors, physical objects and even perform intuitive action based on the analysis done while performing any task. In this way, IOT has connected people, things in the most desired manner to facilitate the delivery of relevant information with efficiency and high level of precision. The estimate of number of devices which will be connected, have gone up to over 30 billion and will form the requisite network of all the devices with the arrival of the Internet of Things in real-life scenarios.

The Internet of Things is changing the way students used to learn in classical old days. The education has been switched to the online learning methods, video lectures, youtube recordings, and skill-based learning. The Higher Educational Institutions are restricted to the confined knowledge and also does not rely on the fixed resource, with rapid change in environment, it must embrace the system with the latest technology to present as a learning tool to capture and provide platform with accelerated learning outcomes. Cisco describes the four prominent pillars of IOE in Education as People, Process, Data and Things [8]. IOT in Education Sector is at a very nascent stage, but some educational institutions are leading the way for the advanced development of it and also showing way to others that how IOT can be effectively used to advance the system and education people of all ages [9].

In current scenario, the information of any Higher Educational Systems are being collected through classical manual process in which physical movements are required to generate the statistical data of any particular institute which are called as various surveys and can be considered out-dated. The information is being collected through surveys which are being done through organizing surveys at certain interval of time. It does not provide the updated and latest information with reference to rapid change in the environment. In this technical era, the old information collected through manual processes would not benefit in efficient ways. So, the change in the system will be required which could give real-time information to make it useful in the upcoming time.

IOT devices in the different system can enable the HEI's to send the information at regular intervals of time which would help in the proper analysis to derive the assessment based framework to help all the stakeholders including accreditation body, institutes, parents and students. In order to accomplish this result, IoT components comprising sensors, base stations, and server will be implemented to add sense and communication medium to things in different areas of Educational Institutions. It will be equipped and assembled with sensors, transceivers, and microcontroller to communicate digitally based on a different set of rules (protocols) [10]. Those protocols can be also integrated for Auto Survey Methods for providing real-time reports.

The data-driven decisions can be more helpful in improving the quality of services of any institutions and also can enhance the success rate of students and institutes. Academic Institutions are now collecting real-time data through sensors to implement data-driven decisions [11]. The process of including sensors can enable the surveying body to generate automated reports which will be helpful to analyze the information gathered from the different resources and would facilitate in easy analysis and improvement of ways the student's learning. The process can change pattern and process of Educational Institutions and enhance its outcomes.

3 Conventional Systems

As of 2016, there were more than 799 universities which include central, state, deemed and private universities. There are more than 39,071 colleges in India which includes government and private degree colleges including women's colleges. Indian Higher Education System has rapidly increases and expanded its educational area. In last one decade from 2000–2001 to 2010–11, the number of educational institutions has rapidly increases in India [12]. More than 20,000 colleges and more than 8 million students have increases in this tenure [13]. India comprises the diverse Education System and is also one of the largest education systems. The expansion in different learning platforms and its diversification has enhanced the approach and opportunities to Higher Education. In parallel, the concern of quality and relevance of the Higher Education also arises with this. In order to provide a solution to these points, the Government of India had created National Policy on Education (NPE) in 1986 and the Programme of Action (POA) in 1992 which subsequently turned into National Assessment and Accreditation Council (NAAC) in 1994. It is an autonomous governed body of University Grant Commission (UGC). The main directive of this Accreditation Council is to avail quality assurance as an integral part of the operations of HEI (Higher Educational Institutions) [14]. NAAC criteria for assessment are as follows:

i.   Curricular Aspects.
ii. Teaching-Learning and Evaluation.
iii. Research, Innovations, and Extension.
iv. Infrastructure and Learning Resources.
v.   Student Support and Progression.
vi. Governance, Leadership, and Management.
vii. Institutional Values and Best Practices.

There are multiple sources and governing bodies for information on Higher Educational Institutions through which information is collected and processed in current scenario like from University Grant Commission (UGC), NAAC, Ministry of Human Resource and Development (MHRD), Higher Educational Institute Websites (HEI) Websites, All India Council for Technical Education (AICTE). But, it is quite difficult to find and define validated information with reference to the assessment made. The information available through the resources is quite old. The data is available in 2015 and even before that. So in the fast-moving era, we prefer to collect real-time information or say before time information. The time has changed and age of analytics has been trending. The world is moving on the track of data-driven world which suggest the proper path to follow on the basis of analytics. This signifies that the range of applications and opportunities has rapidly expanded and will continue to grow. Taking these aspects into consideration and with reference to the rapid advancement of technology, the main challenge would be to integrate this into the system for new capabilities, their operations, and strategies. There is a huge demand

and requirement for higher education institutions to digitize their content and activities, and adapt new methods to allow academic and researchers to work effectually in a digital environment [15].

4 Proposed Methodology

The IoT is not just a technology update and development within the industry but can lead to expanding the change to the whole society including higher education institutions. The IOT Technology will create impact on every part of society at some point in the coming future. Higher Educational Institutions commence to develop and leverage solutions such as Machine Learning, RFID (Radio Frequency Identification) and Cloud computing through IOT technologies which can be used to manage and analyze Big Data and can also help in making Data-Driven Decisions [16]. In the system proposed, the institute has been equipped with sensors, microcontrollers, learning management system where an individual institute has been considered as a single node and will transmit information to the base station. In similar way multiple institutes will be equipped with the similar components and multiple nodes have to be linked with the base station where the nodes will transmit information to the base station. The base station has to be linked with the cloud server where the data and information are being collected from different nodes. Furthermore, the data collected at the end of cloud storage process the information to present the analytical results to the end-user.

4.1 Objective

Following are the objectives of the proposed system:

i.   IOT can collect and store data.
ii. Smart Objects in the Institutions.
iii. Database to Store the Collected Data.

Architecture to process data into meaningful information to assist in decision-making system.

4.2 Expected Outcome

i.   Determine the basic information details of HEI.
ii. Collect the information related to infrastructure and available facilities.
iii. Measure the impact of information collected with reference to comparative analysis of other institutions.

iv. Architecture to collect and process the real-time information of Higher Educational.

Institutions with reference to integration with Education Management System.

4.3 Implementation in the Current System

It focuses on processing the information collected through the smart objects and online learning habits given in Fig. 1. The defined algorithm will collect the information through the smart intelligent agents. The data will be analyzed and processed to observe the habit and provide habit details without human intervention.

Our research methods include

Conducting Preliminary research study in order to understand the basic information available for Higher Educational Institutions (HEI). A system that will be able to link multiple HEIs with cloud to provide the comparative analysis using results of preliminary research studies. The research methods are divided into four categories.

i. A process to collect the information from the Higher Educational Institutions through sensors.
ii. Integration with online learning management system to learn behavior.
iii. The system to process the information collected and generates the analysis results and feedback without human intervention.
iv. Framework using the above two methods with integration of online learning management system to develop a prototype for Smart Educational Learning System.

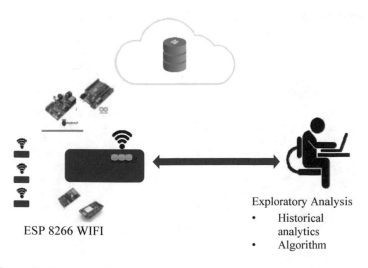

Fig. 1 Data collection in cloud through sensors

4.4   Methodology and Algorithm

The conventional system includes the manual process of gradual collection through visiting and collecting information from different institutions. It's the process which consumes more time and also does not produce real-time or updated information illustrated in Fig. 2.

The world is rapidly changing and so technology. In this technical era, the data and information of this much old-time cannot be validated over time and would not be useful in current scenario. For this, we need real-time information which could be helpful in validating the information and would also help in proper assessment of the institution. The developed prototype is based on automation and machine learning mechanism which includes the following processes. The process which will be followed:

i.   Heterogeneous sensors to collect the details.
ii.   Database to maintain the information.
iii. Intelligent agent to analyze the requirement of the user.
iv.   Provide the result and suggestion without human intervention.

The information collection and integration would be done in different formats. But the integration should have uniform view and it also provides a single view for the data arriving from different sources [17]. The produced data from the different sources of HEI can be categorized in three groups which shown in Fig. 2.

Structured data which will comprise the data stored in the database management system including both tuples and attributes. Semi-structured, such as web data like HTML, XML, and formatted data like JSON files unstructured data, such as images, audios, and images [18].

Algorithm-1

Step 1: Sensors Installation within a single institutions as multiple nodes.

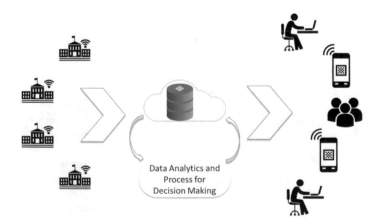

Fig. 2 Data analytics and processing for multiple HEI—The framework

Step 2: Connecting nodes with a central hub for single institution.
Step 3: Treating an institute as a node to send data to the cloud- (different performing sections and their current status)—Using Bevywise to trigger status through nodes.
Step 4: Treating an institute as a node to send data to the cloud- (different performing sections and their current status)—Using Bevywise to trigger status through nodes.
Step 5: Send data to the database on the cloud from the nodes as information.
Step 6: Data Processing and Analytics to provide the information to the end-user (as described in Fig. 2).
Step 7: Output: Analytical Report to the End User with respect to the triggered data.

4.5 Challenges and Requirements

The main challenge would be to adjust structures in semi-structured and unstructured data before integrating and analyzing these types of data [19]. There would be a requirement to address the risks factors which comprise about Trust, Identity, Privacy, Protection, Safety and Security (TIPSS factor) related to IOT.

5 Result and Analysis

The data transmitted by sensors consists of time-series values, which are sampled over a specified period and then transmitted to the gateway for further proceedings. There are several researches which have been processed in the domain IOT to investigate and access data through different sensors and devices. It is a challenge that how the data will be accessible and will be interpreted in a meaningful way and how actionable information can be extracted and processed to take the necessary actions.

5.1 Simulation Application

There are several IOT based simulation model available which facilitates to create simulation model for different connectivity and data collection. For example, Iotify, MATLAB, NetSim, Bevywise, AnSys, IBM Bluemix. The simulator software facilitates to design, create and test IOT based application without using real boards and gadgets [20].

Bevywise is a simulation application based on MQTT, which has been used in experimental practice to create simulation architecture for collection of information from multiple HEI presented in Fig. 3. Message Queuing Telemetry Transport (MQTT) is a machine to machine (M2 M) data transfer protocol. It is subscriber-based messaging protocol which uses message broker for its transmission given in

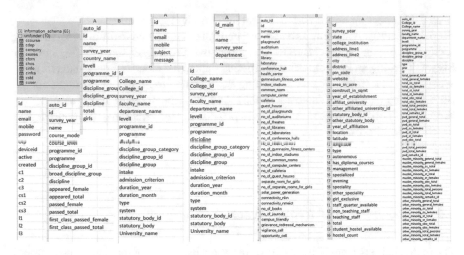

Fig. 3 Data structures and tables created to store the information collected from HEI

Fig. 4. It is used for establishing connections with remote (multiple) locations where the network bandwidth is limited and a "small code footprint" is required. It works on the top of the TCP/IP Protocol. Facebook, AWS (Amazon Web Services), Microsoft Azure uses MQTT as its protocol for telemetry messages [21–23].

```
C:\Bevywise\IotSimulator\Broker.exe
MQTTRoute for Bevywise IoT Simulator
Bevywise MQTTRoute - Trial Version - expires on Wed Jun 27 12:39:19 2018
Starting MQTT Broker at port - 1883
Client No:1 New connection request from 127.0.0.1 clientid is AC118
Client No:2 New connection request from 127.0.0.1 clientid is CM123AB
Client No:3 New connection request from 127.0.0.1 clientid is CO2_Sensor
Client No:4 New connection request from 127.0.0.1 clientid is DM4523
Client No:5 New connection request from 127.0.0.1 clientid is Door_Sensor
Client No:6 New connection request from 127.0.0.1 clientid is MSensor8967
MSensor8967-disconnected sucessfully.
Client No:7 New connection request from 127.0.0.1 clientid is MSensor8967
Welcome back! your connection resumed
Client No:7 New connection request from 127.0.0.1 clientid is PT432X
Client No:8 New connection request from 127.0.0.1 clientid is Sensor432
Client No:9 New connection request from 127.0.0.1 clientid is Smart_Meter
Client No:10 New connection request from 127.0.0.1 clientid is TSensor123
Client No:11 New connection request from 127.0.0.1 clientid is Ventilation_Senso
r
Client No:12 New connection request from 127.0.0.1 clientid is WSensor
TSensor123-disconnected sucessfully.
Door_Sensor-disconnected sucessfully.
CO2_Sensor-disconnected sucessfully.
WSensor-disconnected sucessfully.
Sensor432-disconnected sucessfully.
Smart_Meter-disconnected sucessfully.
```

Fig. 4 Data transmission through sensors in Bevywise simulator application using MQTT

Table 1 Changes in instruction of Education Institution

Conventional system (Current state)	Proposed system (Potential with IOT)
Out-dated information of very large gap with reference to current time	Real-Time Information with updated data
Information collection through manual process and survey	Information collection through sensors and automatically
One information at a time about any educational institutions	Multiple information of all the educational institutions
Linear and Static Information with very low control	Richer and interactive updated content
Basic information with minimum analysis and no comparative analysis available	Comparative analysis of multiple analysis among the institutions
Costly and long term process as collected manually through survey	Crowd Sourced information collected on cloud storage
Ad Hoc Decision Making	Data-Driven Decision Making

5.2 Comparative Result Analysis

As being very large number of institutions, it was not possible to collect details in real-time or the updated information through the conventional system. In the conventional system, the information was being collected through organizing surveys at certain intervals like 5 years or even more than that. In this span of time so many information and exact picture changes which cannot be helpful in exact analysis and taking any data-driven decisions. The information collected does not match with the real-time scenario which makes it useful for the predictions only. In current scenario, the information for educational institutions are referred to the survey details collected in 2011, 2013 and latest by 2015–2016, but these information are out-dated and in this advanced technical era, real-time information and before time information are demanded [24]. So, through the proposed system the information related to any educational institutions can be captured in real time which will give real time information collection and result analysis presented in Table 1.

6 Conclusions

The data collection process in the defined framework consumes very less time and can be used for automated processes and real-time report generation. Comprising the fact, IOT is considered as emerging paradigm to connect things in massive amount and collect distributed data from different nodes. The data collected from different resources can be converted into meaningful information and can be used for data-driven decisions as similar to Ranking and Analysis of Higher Educational Institutions. In current scenario, the IOT framework is not smart enough to understand the collected information and to learn, think and recognize cyber, physical and social world by itself which can be used for decision making. The technological advancement to the integrated object to behave as smart object with respect to significant learning thought the collected information may generate more revolutionary results.

This paradigm shift has been adopted by the Industries and other similar segment to get benefited. Now, Educational Institution needs to adopt this reform. The proposed framework will enable the HEI to provide its updated information regardless of its location. Also, the framework will process the information enabling data analytics in wireless IoT for multiple HEIs. Taking these aspects into consideration, the rate of data production has been increased in rapid pace over past years, and its implementation will also pass in parallel through this phase. It would lead to big data management, whereas the interaction between IoT and big data is currently at a stage where processing, transforming, and analyzing large amounts of data at a high frequency are necessary. Furthermore, potential key enables the proposed framework and integration into the Higher Educational Institutions and managing data collected from thousands of Institutions will be the key challenge.

References

1. Nasaruddin, F., Gani, A., Karim, A., Abaker, I., Hashem, T., Siddiqa, A., Yaqoob, I., Marjani, M.: Big IoT data analytics: architecture, opportunities, and open research challenges, IEEE. Access **5**, 5247–5261 (2017)
2. Puschmann, D., Barnaghi, P., Carrez, F., Ganz, F.: A practical evaluation of information processing and abstraction. Internet Things J. (2015)
3. Iera, A., Morabito, G., Atzori, L.: The internet of things: a survey. Comput. Netw. **54**(15), 2787–2805 (2010)
4. Rehman, U., Ghazal, S., Umar, I., Aldowah, H.: Internet of things in higher education: a study on future learning. J. Phys.: ICCSCM (2017)
5. Ahamed, B.B., Ramkumar, T., Shanmugasundaram, H.: Data integration progression in large data source using mapping affinity, advanced software engineering and its applications (ASEA). In: 7th International Conference (2014)
6. Zhang, X., Liu, J.: Data integration in fuzzy XML documents. Inf. Sci. **280**, 82–97 (2014)
7. Bernstein, P., Bertino, E., Davidson, S., Dayal, U., Agrawal, D.: Challenges and opportunities with big data, whitepaper, computing community consortium (2012)
8. Bernstein, P., Bertino, E., Davidson,S., Dayal,U., Agrawal, D.: Challenges and opportunities with big data, whitepaper, computing community consortium, (2012)
9. Mell, P., Grance, T.: The NIST definition of cloud computing. National Institute of Standards and Technology, U.S. Department of Commerce (2011)
10. Kanagavalli, R., Dr. Vagdevi, S.: A mixed homomorphic encryption scheme for secure data storage in cloud. In: IEEE International Advanced Computing Conference IACC2015 (2015)
11. Tebaa, M., Elhajii, S.: Secure cloud computing through Homomorphic Encryption. Int. J. Adv. Comput. Technol. **5**(16), 29–38 (2013)
12. Parmar, P.V.: Survey of various Homomorphic Encryption algorithms and schemes. Int. J. Comput. Appl. (0975–8887), **91**(8), 26–32 (2014)
13. Ogburn, M., Turner, C., Dahal, P.: Homomorphic Encryption in Complex Adaptive Systems, Publication 3, pp. 502–509. Elsevier, MD, Baltimore (2013)
14. Rivest, R., Shamir, A., Adleman, L.: A method for obtaining digital signatures and public key cryptosystems. Commun. ACM **21**(2), 120–126 (1978)

15. Song, X., Wang, Y.: Homomorphic cloud computing scheme based on hybrid homomorphic encryption. In: 3rd IEEE International Conference on Computer and Communications (2017)
16. Geetha, J.S., Amalarethinam, D.I.G.: ABCRNG—swarm intelligence in public key cryptography for random number generation. Intern. J. Fuzzy Mathematical Archive, **6**(2), 177–186 (2015)
17. Chean, T.L., Ponnusamy, V., Fati, S.M.: Authentication scheme using unique identification method with homomorphic encryption in mobile cloud computing. IEEE (2018)
18. Oppermann, A., Toro, F.G., Seifert, T., Seifert, J.P.: Secure cloud computing: communication protocol for multithreaded fully homomorphic encryption for remote data processing. Int. J. Commun. Syst. 1–26 (2017)
19. Das, D.: Secure cloud computing algorithm using homomorphic encryption and multi-party computation. IEEE (2018)
20. Ding, Y., Li, X.: Policy based on homomorphic encryption and retrieval scheme in cloud computing. In: IEEE International Conference on Computational Science and Engineering (CSE) and IEEE International Conference on Embedded and Ubiquitous Computing (EUC) (2017)
21. Anescu, G., Prisecaru, I.: NSC-PSO, a novel PSO variant without speeds and coefficients. In: 17th International Symposium on Symbolic and Numeric Algorithms for Scientific Computing (2016)
22. Abraham, A., Sharma, T.K., Pant, M.: Blend of local and global variant of PSO in ABC. IEEE (2013)
23. Tiwari, S., Mishra, K.K., Misra, A.K.: Test case generation for modified code using a variant of particle swarm optimization (PSO) Algorithm. In: 10th International Conference on Information Technology: New Generations (2013)
24. Singh, S., Shivangna, Mittal, E.: Range based wireless sensor node localization using PSO and BBO and its variants. In: International Conference on Communication Systems and Network Technologies (2013)

Fuzzy Logic Based Packet Dropping Detection Approach for Mobile Ad-Hoc Wireless Network

Sheevendra Singh, Isha Sharma, Praneet Saurabh and Ritu Prasad

Abstract Mobile ad hoc network (MANET) is a lively network that is self configuring in nature because of its mobility characteristics. MANET plays an important role in situations where fixed wired pre-defined backbone cannot be created or remains more costly. MANET brings ease and flexibility but at the same time it also introduces various limitations. This paper proposes Fuzzy logic based efficient packet dropping detection approach (FL-EPDDA) for MANET to overcome that challenges of malicious nodes in MANET. FL-EPDDA use fuzzy inference system (FIS) to identify malicious nodes in MANET. FIS uses three input parameter packet delivery ratio, packet forward and residual energy of node. FIS classifies the network nodes weather it is malicious or not with the help of these inputs. FL-EPDDA identifies the above discussed activities and it also discovers a trusted and secure path for secure data transmission. FL-EPDDA very efficiently classifies normal working mobile nodes and malicious nodes within MANET. Experimental are performed under parameters like packet delivery ratio, average throughput and packet drop rate with variation in malicious nodes in MANET. Comparative experimental results shows the proposed FL-EPDDA outperforms current state of art AODV under all the test conditions.

Keywords Fuzzy logic · FIS · Malicious node · MANET

S. Singh (✉) · I. Sharma · R. Prasad
Technocrats Institute of Technology (Advance), Bhopal, MP 462021, India
e-mail: shivdj.singh78@gmail.com

I. Sharma
e-mail: ishasharma0701@gmail.com

R. Prasad
e-mail: rit7ndm@gmail.com

P. Saurabh (✉)
Technocrats Institute of Technology, Bhopal, MP 462021, India
e-mail: praneetsaurabh@gmail.com

© Springer Nature Singapore Pte Ltd. 2020
K. N. Das et al. (eds.), *Soft Computing for Problem Solving*,
Advances in Intelligent Systems and Computing 1057,
https://doi.org/10.1007/978-981-15-0184-5_23

1 Introduction

Wireless technology is one of the most popular and accepted technologies in today's world. Data is transferred without using physical wires as it employs radio channels to achieve it. Again, wireless networks can be broadly divided into two types; Infrastructure based wireless networks and Infrastructure less wireless networks. Infrastructure based wireless network involves more infrastructure to facilitate communication among different users while Infrastructure less wireless network requires minimal infrastructure for making communication possible. Mobile adhoc networks (MANET) are infrastructure less wireless networks that requires minimal infrastructure [1]. MANET configuration is very dynamic and different wireless nodes (more than one nodes) can communications with each other without any central authority [2].

In recent years the use and its subsequent proliferation of mobile ad hoc network have grown tremendously since it did not require any fixed infrastructure, no configuration overhead and administrator [3]. Various different flexibilities make MANET an interesting option for communication [4]. All the nodes in MANET form a network and work both as a node and a router for forward packets [5]. Over the past few years, there has been a quick rise in this research domain due to its broad potential and applications within many scenarios, like gathering or compilation of data from military area, deploying in emergency or rescue situation etc. These all flexibilities introduces challenges like, self-configuration topology, congestion, energy consumption, variable capacity links and security, also called design challenges of MANET. MANET's dynamic nature many a times bring in malicious a node, that compromises the network efficiency and proves to be detrimental for the network.

In this work, a Fuzzy logic based efficient packet dropping detection approach (FL-EPDDA) based on fuzzy inference system (FIS) is introduced to efficiently detect undesired packet drop in MANET. FIS in this work is chosen to make control decision in order to improve performance so that it can overcome the limitations of the existing conventional solutions. FIS enables FL-EPPDDA to take decision and then generate output just like human mind. Scalar values of input in FIS are converted in fuzzified values and thereafter mathematical operations with different linguistic fuzzified values using membership function and fuzzy rules are performed. The decision making system then based on the rules and calculations spawns appropriate decision. In last the gathered fuzzified output is converted back into scalar values.

This paper is subdivided into different sections, Sect. 2 put forwards the related work, Sect. 3 presents the proposed work, Sect. 4 presents results and analysis and Sect. 5 concludes the paper.

2 Related Work

MANET is self organizing networks with no dedicated administrator and routers. Each node acts as a node and a router for forwarding a packet from source to destination. Routing can be divided into three groups, proactive routing protocols also called table driven approaches, reactive routing protocols overwhelmingly called on demand approaches and hybrid routing protocol (combination of both proactive and reactive approaches). Since there is no dedicated router so selection of routers and subsequent routes needs to happen in such a manner that probability of successful delivery of the packets from source to destination should be maximum [6]. In this context, AODV routing protocol is a reactive routing protocol that generates route identification process only when needed which reduces the workload of the network [7]. AODV router after receiving a request message checks its routing table, if route is found then AODV forwards the message to next node otherwise it saves message and begin a route request procedure to find route to update own routing table for future use [8]. AODV routing protocol search for shortest optimal path to exchange data between source and destination node but attacker nodes try to confuse source node to collect original data to destroy the linearity of the network [9]. Dynamic nature of MANET also introduces malicious nodes in MANET. Many different researches have discussed the detrimental effects of malicious nodes in MANET [5, 7]. Recently fuzzy based systems have grabbed eyeballs of researchers as it demonstrated problem solving traits for malicious nodes [7]. Some explores involving bio-inspired techniques [10, 11] can be looked upon to find more realistic solution in this domain [12, 13]. The next section introduces Fuzzy logic based efficient packet dropping detection approach (FL-EPDDA) for MANET to identify malicious nodes.

3 Proposed Method

MANET a group of wireless mobile nodes, working independently without any centralized infrastructure. Dynamic configuration of MANET allows nodes to enter and get connected in any network very easy and seamless. This stand true for nodes exiting a network also that results in more complexity. Due to the dynamic nature of MANET any unauthorized node (malicious node) can easily enter the network, participate in the communication, can absorb all the network traffic and drop all packets. A node can be considered as malicious node when packet maliciously dropped through neighboring nodes, congestion at the forwarding node, collision at the destination and less residual node energy. Therefore, Fuzzy logic based efficient packet dropping detection approach (FL-EPDDA) for MANET is introduced that use fuzzy inference system (FIS). In this scheme, each node of network collects all the information regarding packets and path creation in the form of trace file. On the basis of that trace following parameter values like packet received, packet sent, packet forwarded, packet dropped, packet receiving sending ratio, remaining energy are collected for each node. Out of these, three key parameters like packet receiving sending ratio (PRSN), packet forwarding (PFORN) and remaining energy (REN) as selected as input for fuzzy inference system. These inputs are used for finding packet dropping nodes from the network. Proposed FL-EPDDA has two major steps, MANET Monitoring module and Packet dropping identification module.

3.1 Monitoring Module

In this module, MANET creates and analyzes different nodes under conditions; detailed procedure is given in algorithm-1 given below:
   Algorithm-1

Step 1: Create Mobile Ad hoc Network: Initially, nodes are configured with random assignment of mobility and energy. Some connections of CBR-UDP are defined between source to destination nodes.

Step 2: Multiple sender (SN) and destination nodes (DN) define in network and AODV routing protocol configures to create route between them.

Step 3: In AODV routing, when sender node wants to communicate with the destination node then

- If destination node comes under the transmission range of source node then SN direct communicates with DN.
- Otherwise source node uses intermediate (peer) nodes (IN) for packet forwarding to DN.

Step 4: To create effective routing for packet transmission, source node initially broadcasts Route request packet (RREQ) to its intermediate nodes.

   (a) If (IN = = valid node)
        Use in routing process and SN can transmit data
        Otherwise
        IN may be malicious node and go to (Module II)

Step 5: If neighbor of intermediate node is destination node then data packets are forwarded to that node otherwise it will broadcast to next neighbor nodes.

Step 6: Finally destination node (DN) sends route reply packet (RREP) through INto SN and connection is established.

Step 7: Intermediate nodes are configured with random mobility therefore sometime communication link between source and destination can be changed. In this condition, intermediate nodes send route error packet to their sender node.

Step 8: Initially using previous steps simulation of MANET is done in analyzing module and analyzes performance of each node in the form of trace record. On the basis of that trace following parameter values collected for each node such as:

   a. Packet received, packet sent,

b. Packet forwarded and packet dropped,
c. Packet receiving sending ratio (PRSN),
d. Packet forwarding (PFORN) and
e. Remaining energy (REN).

Step 9: After analysis of trace record, performance of MANET is calculated in following terms

a. Packet delivery Ratio in variation of number of malicious nodes.
b. Throughput in variation of number of malicious nodes.
c. Packet loss Rate in variation of number of malicious nodes.

3.2 Packet Dropping Identification Module

It is the second module and in this module, fuzzy inference system (FIS) is defined for identifying malicious behavior of node in MANET. It includes three performance parameters of each node that act as an input, packet receiving sending ratio (PRSN), packet forwarding (PFORN) and remaining energy (REN) of node. These inputs are mapped in single output weather it is malicious node or normal node. FIS works like classifier that classifies the network nodes with the help of these inputs. Fuzzy inference system involves some basic steps like input variables, fuzzification process using membership function, define knowledge base or rule base for mapping input into output and defuzzification process. This module works in following manner:

Step 1: In proposed FIS, Gaussian membership function is adopted and for each input and output parameter this membership function used to define range (low, good and best) to categorize mobile nodes with the help of defined rules. The formula of gaussmf can be represented in Eq. (1):

$$\text{Gaussmf} = \mu_{Ai}(x) = \exp\left(-\frac{(c_i - x)^2}{2\sigma_i^2}\right) \tag{1}$$

where c_i represents midpoint ith fuzzy set, while A^i and σ_i are width can be also termed as standard deviation of the same ith fuzzy set.

Step 2: FIS rules are created by analyzing network in different environment of MANET. According to that, values of PRSN, PFORN and REN are calculated. Table 1 indicates the linguistic terms with membership values while Table 2 describes the rule base.

Step 3: For defuzzification, the centroid defuzzification technique is used given in Eq. (2).

$$x^* = \frac{\int \mu_i(x)x\mathrm{d}x}{\int \mu(x)\mathrm{d}x} \tag{2}$$

Table 1 Linguistic terms with membership values

Input variables	Membership function	Linguistic terms with their range
PRS_N	Gaussmf	0–0.5: Low 0.5–0.75: Good 0.75–1: Best
$PFOR_N$	Gaussmf	0–0.25: Low 0.25–0.5: Good 0.5–1: Best
RE_N	Gaussmf	0–0.4: Low 0.4–0.7: Good 0.7–1: Best
Normal node or malicious node	Gaussmf	0–0.25: Low (Malicious Node) 0.25–0.5: Good (Normal node) 0.5–1: Best (Best node)

Table 2 Rule base

PRS_N	RE_N		
$PFOR_N$	Low	Good	Best
Low	Low	Good	Good
Good	Best	Best	Best
Best	Best	Best	Best

where x^* represents output in defuzzified status; $\mu_i(x)$ denotes aggregated membership function and x signifies the variable for final output. The process of converting the fuzzy output is called defuzzification.

After defuzzification, final values are converted into crisp values corresponding to each node. When that calculated value of node come under the threshold value then that node is considered as malicious node. If (node_value <= 0 && node_value <= 0.4), then node will be considered "malicious node" and can be eliminated from the network as a precautionary measure otherwise node is proper and can use for routing process.

4   Simulation and Result Analysis

Simulation is used to design and develop different algorithms for MANET as it provides freedom from difficulty of analyzing and verifying any protocol for large scale systems. It provides flexible opportunity to test system with different topologies, mobility patterns along with several physical and network layer protocols. Simulation also gives an opportunity to understand how the changes impact the performance result of the network. There are well known simulators used for the simulation of MANET which are QualNet, NS-2, GloMoSim and OPNET, DIANEmu, GTNets,

J-Sim, Jane, NAB, OMNet ++ and SWANS. All the experiments in this work are carried in NS-2 simulator (NS-2.35).

4.1 Network Configuration for Experiments

- Channel Type: Wireless
- MAC type: Mac/802_11
- Max packet in Queue: 50 · Total Number of Mobile Nodes: 50
- Routing protocol: AODV
- Topography Dimensions: 1000 × 1000
- End Time of Simulation: 100 s
- Initial Node Energy: 300 J
- Traffic flow between nodes: UDP/CBR
- Packet Size: 512 kb · Data Rate: 8 kbps.

4.2 Result Determining Parameters

These are the parameters on which results will be calculated for both current state of the art and the proposed FL-EPDDA.

Packet Delivery Ratio: Packet Delivery Ratio (PDR) can be calculated by dividing the number of packet received by the destination from the number of packets initiated by the source as given in Eq. 3. If packet delivery ratio value is large then it indicates the better performance of the protocol.

$$\text{PDR} = \frac{\sum \text{Number of packet receive by destination}}{\sum \text{Number of packet sent from source}} \quad (3)$$

Throughput: Throughput is the number of successfully received packets in a unit time and it is represented in bps stated in Eq. 4. Throughput is calculated using awk script which processes the trace file and produces the result shown in equation.

$$\text{Throughput (kbps)} = \frac{\text{Packet received by node}}{\text{Data transmission period}} * 1024 \quad (4)$$

Table 3 Packet delivery ratio versus malicious node

Malicious nodes (%)	PDR AODV (%)	PDR FL-EPDDA (%)
0	88	94
2	38	69
4	30	59
6	26	43
8	22	37
10	13	31

Fig. 1 Packet delivery ratio versus malicious node

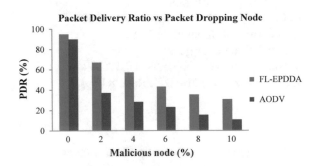

4.2.1 Effect of Malicious Node on Packet Delivery Ratio

This experiment is performed to calculate Packet delivery ratio (PDR) for both current state of art (AODV) and the proposed FL-EPDDA with the increase in malicious nodes percentage in the network from 0 to 10%.

Results presented in the Table 3 and Fig. 1 reflects that PDR decreases in both current state of art (AODV) and the proposed FL-EPDDA with increase in malicious nodes percentage in the network when it increases from 0 to 10%. But, rate of drop of PDR in current state of art AODV is much more higher as compared to the newly developed FL-EPDDA. This is due to the fact that AODV is designed for ideal scenario and it do not have the mechanism to deal with presence of malicious nodes in the MANET. This leads to lowering of packets delivery in the network. In comparison proposed FL-EPDDA gives better PDR even with increase in malicious nodes in the network due to the new incorporations like FIS. It makes FL-EPPDDA to take decision and then generate output just like human mind.

4.2.2 Effect of Malicious Node on Average Throughput

Objective of this experiment is to calculate average throughput for both current state of art (AODV) and the proposed FL-EPDDA with the increase in malicious nodes percentage in the network from 0% to 10%.

Table 4 Average throughput versus malicious node

Malicious nodes (%)	Throughput AODV (%)	Throughput FL-EPDDA (%)
0	86	97
2	35	46
4	21	39
6	19	27
8	11	20
10	6	10

Fig. 2 Average throughput versus malicious node

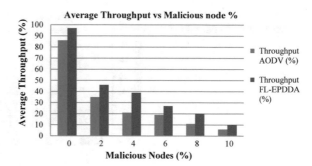

Above results given in Table 4 and Fig. 2 represents average throughput under AODV and proposed FL-EPDDA when percentage of malicious nodes increases in the network from 0 to 10%. Results show that as malicious nodes percentage increases from 0 to 10% in subsequent experiments then average throughput decreases. These experimental results also signify that malicious nodes in the network have detrimental effect in the network. With the experimental results it can be easily said that proposed FL-EPDDA with new incorporations deals with malicious node more effectively and demonstrates better results.

4.2.3 Effect of Malicious Node on Packet Loss Rate in MANET

This experiment is carried to measure the impact of Malicious nodes in Packet loss rate under both current state of art (AODV) and the proposed FL-EPDDA when malicious nodes percentage increases in the network from 0 to 10%.

In Table 5 and Fig. 3, it can be very easily observed that when percentage of malicious nodes increases in the network, proposed FL-EPDDA approach reports lower packet loss rate as compared to existing AODV. Even in normal scenario when malicious nodes percentage remains 0% in the network then also FL-EPDDA demonstrates lower packet loss rate as compared to AODV. Experimental results very clearly highlights the importance of new integrations and its impact in packet loss rate.

Table 5 Packet loss rate (%) versus malicious node

Malicious nodes (%)	Packet loss rate AODV (%)	Packet loss rate FL-EPDDA (%)
0	30	20
2	34	28
4	41	28
6	45	31
8	52	33
10	61	39

Fig. 3 Packet loss rate (%) versus malicious node (%)

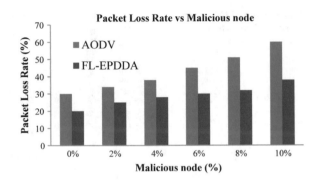

5 Conclusion

MANET is a group of wireless mobile nodes, which works independently without any centralized access point. This paper presented a Fuzzy logic based efficient packet dropping detection approach (FL-EPDDA) using fuzzy inference system (FIS) for MANET to overcome the challenges of malicious nodes in the MANET. Experimental results are evaluated under variation of malicious modes percentage from 0 to 10% in MANET. Experimental results reflected that FL-EPDDA reported higher packet delivery ratio, higher average throughput, and lower packet drop rate under all experiment scenarios. Experimental results put forward the fact that FL-EPDDA is more acquainted to deal with malicious nodes and reported better results as compared to current state of art.

References

1. Wu, L., Yu, R.: A Threshold-based method for selfish nodes detection in MANET. In: Computer Symposium (ICS), pp. 875–882 (2010)
2. Buttyn, L., Hubaux, J. P.: Enforcing service availability in mobile Ad-Hoc WANs. In: Proceedings of the 1st ACM International Symposium on Mobile Ad Hoc Networking and Computing, pp. 87–96 (2000)

3. Lal, C., Petroccia, R., Pelekanakis, K., Conti, M., Alves, J.: Toward the development of secure underwater acoustic networks. IEEE J. Ocean. Eng. **42**(4), 1075–1087 (2017)
4. Khan, S., Prasad, R., Saurabh, P., Verma, B.: Weight-Based Secure Approach for Identifying Selfishness Behavior of Node in MANET. In: Satapathy S., Tavares J., Bhateja V., Mohanty J. (eds.) Information and Decision Sciences. Advances in Intelligent Systems and Computing, vol. 701, pp 387–397 (2018)
5. Bisen, D., Sharma, S.: Fuzzy based detection of malicious activity for security assessment of MANET, Natl. Acad. Sci. Lett., 23–28 (2017)
6. Bisen, D., Sharma, S.: An energy-efficient routing approach for performance enhancement of MANET through adaptive neuro-fuzzy inference system. Int. J. Fuzzy Syst. 2693–2708 (2018)
7. Bisen, D., Sharma, S.: An enhanced performance through agent-based secure approach for mobile ad hoc networks. Int. J. Electron. 116–136 (2017)
8. Ravi, G., Kashwan, K.R.: A new routing protocol for energy efficient mobile applications for ad-hoc networks. Comput. Electr. Eng. **48**, 77–85 (2015)
9. The Network Simulator ns-2, Information Sciences Institute, USA. Viterbi School of Engineering, September. Retrieved from http://www.isi.eu/nsnam/ns/ (2017)
10. Saurabh, P., Verma, B.: An efficient proactive artificial immune system based anomaly detection and prevention system. Expert. Syst. Appl., Elsevier **60**, 311–320 (2016)
11. Saurabh, P., Verma, B.: Immunity inspired cooperative agent based security system. Int. Arab. J. Inf. Technol. **15**(2), 289–295 (2018)
12. Saurabh, P., Verma, B, Sharma, S.: An immunity inspired anomaly detection system: a general framework a general framework. In: 7th International conference on bio-inspired computing: theories and applications (BIC-TA 2012), AISC (vol. 202, pp. 417–428). Springer (2012)
13. Saurabh, P., Verma, B, Sharma, S.: Biologically Inspired Computer Security System: The Way Ahead, Recent Trends in Computer Networks and Distributed Systems Security, CCIS (vol. 335, pp. 474-484). Springer (2011)

Performance Analysis of Various Feature Sets for Malaria-Infected Erythrocyte Detection

Salam Shuleenda Devi, Ngangbam Herojit Singh and Rabul Hussain Laskar

Abstract Malaria being prevalent disease in urban areas, demands its accurate and fast diagnosis. Due to malaria infection in human being, the erythrocyte features got distorted. To diagnose these, various techniques have been developed, i.e., machine learning-based system, rapid diagnostic test, quantitative buffy coat, etc. In machine learning, the system performance depends on the feature set and classifier model. In this paper, the analysis of the importance of the feature set on malaria-infected erythrocyte classification has been performed. Further, a classifier model based on ANN-GA has been developed to classify the erythrocyte. The process consists of illumination correction, erythrocyte segmentation, feature extraction with or without feature selection techniques, and classification. Erythrocytes segmentation is done using image binarization with marker-controlled watershed segmentation. The six feature sets (morphological feature, texture and intensity feature) have been evaluated using various classifiers such as support vector machine (SVM), k-nearest neighbor (k-NN), and Naive Bayes to choose the better feature set. From the experimental results, it has been observed that the feature set f_6 (combination of morphological, texture and intensity feature ranked with ANOVA) outperforms other feature sets. Further, erythrocyte classification has been performed using ANN-GA with f_6 feature set. It may also conclude that the various features such as morphological feature, texture and intensity feature are equally important to detect the malaria-infected erythrocyte.

Keywords Malaria · Erythrocyte · Intensity and texture feature · Morphological feature · Feature selection

S. S. Devi (✉) · N. Herojit Singh
National Institute of Technology Mizoram, Aizawl, Mizoram, India
e-mail: shuleenda26@gmail.com

N. Herojit Singh
e-mail: herojitng@gmail.com

R. Hussain Laskar
National Institute of Technology Silchar, Silchar, Assam, India
e-mail: rabul18@yahoo.com

© Springer Nature Singapore Pte Ltd. 2020
K. N. Das et al. (eds.), *Soft Computing for Problem Solving*,
Advances in Intelligent Systems and Computing 1057,
https://doi.org/10.1007/978-981-15-0184-5_24

275

1 Introduction

Malaria is a parasitic epidemic disease caused by Plasmodium species which shows complex life cycles [1]. Different systems have been developed to diagnose the malaria, i.e., digital imaging system, machine learning-based system, quantitative buffy coat (QBC), rapid diagnosis tests (RDTs), manual microscopic examination of peripheral blood smear, etc. [1]. Various researches have been conducted to develop fast and rigorous system for malaria diagnosis [2–6]. With the advancement of digital imaging system and machine learning approach, different models have been proposed [7] to detect the infected erythrocyte from the microscopic images of the blood smears. In the machine learning approach, the system performance depends on the feature set, feature selection techniques, and classifiers. There are various features which have been used for erythrocyte classification, i.e., morphological features, texture and intensity features. The morphological features include area granulometry, regional extrema, Hu's moment, relative shape measurements, scale invariance, relative size, eccentricity, etc. [6–9]. Texture and intensity features are color histogram, color auto-correlogram, saturation histogram, green channel histogram, R-G channel histogram, local binary pattern, LBP-GLCM (co-occurrence of local binary pattern), gray-level run length matrix (GLRLM), etc. [10]. The feature selection techniques include analysis of variance (ANOVA), ANOVA with incremental feature selection, etc. Some of the classifiers used for malaria classification are SVM, Naive Bayes, k-NN, artificial neural network (ANN), hybrid, etc. [10]. The contributions of the paper are as follows:

a. Here, the feature set analysis has been done for infected erythrocyte detection.
b. The feature sets include morphological, texture and intensity feature set.
c. Moreover, the performance analysis has been done using various classifier models such as SVM, Naive Bayes, and k-NN for two different models (with and without feature selection techniques).

2 Proposed Method

Two different models have been developed for the analysis of the importance of features and feature selection techniques for malaria-infected erythrocyte classification. In the first model, feature selection technique has not been considered. The model I consists of erythrocyte extraction; feature extraction; and classification. In model II, the step consists of erythrocyte extraction; feature extraction; feature selection; and classification. The analysis model is shown in Fig. 1. In model I, the features set such as morphological feature (f_1), texture and intensity feature, (f_2), morphological, texture and intensity feature (f_3) are analyzed. For model II, the feature such as morphological feature, texture and intensity feature is ranked using ANOVA, and the feature sets are morphological feature with ANOVA ranking (f_4), texture and in-

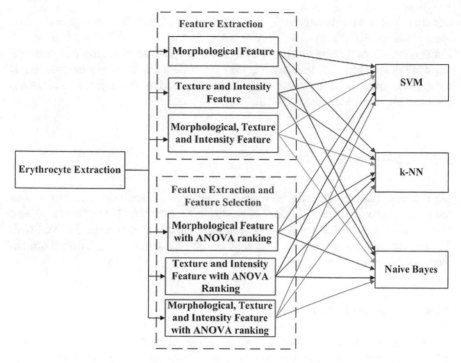

Fig. 1 System model

tensity feature with ANOVA ranking (f_5) , morphological, texture and intensity with ANOVA ranking (f_6). Further, the different feature sets are fed into three different classifiers such as SVM, k-NN and Naive Bayes.

2.1 Erythrocyte Extraction

Due to camera lighting effect and discrepancy during the thin smear preparation, the microscopic image quality becomes poor and non-uniform illumination occurs. So, the illumination correction is done using adapted gray world normalization method to improve the image quality. After preprocessing, the erythrocyte extraction from the background containing other blood components deals with different steps: image binarization, morphological filtering (hole filling and opening), calculation of area of the binary objects, thresholding based on the area to remove artifact, and clump splitting. The entire stained components are segmented from background using Otsu's thresholding technique [11]. Further, the unwanted stained components are separated from the erythrocytes by morphological filtering process. Moreover, some of the artifacts are separated from the erythrocytes using area thresholding. Here, the areas of all the binary objects obtained from the morphological filtered image are calculated and threshold (*Th*) has been set, i.e., *Th* = 0.3 of average area of the binary objects. The binary object in which area is less than the *Th* value is removed. Finally, the filtered binary image is undergone through marker-controlled watershed segmentation to segment the clump erythrocyte into an isolated erythrocyte. Here, the regional minima of the H-transform image are used as an internal marker of the marker-controlled watershed [10–12].

2.2 Feature Sets

Six different feature sets have been used to perform the analysis. The feature sets are formed by using morphological features (Hu's moment and shape features), and texture and intensity feature (GLCM, GLRLM, local binary pattern, LBP-GLCM, Kapur's entropy, Renyi entropy, Yeagers measure, Shannon entropy, fractal dimension, and histogram feature) [10, 11].

2.2.1   Morphological Features

Morphological features such as Hu's moment and shape features are used to characterize the erythrocyte for malaria detection. Geometric feature used to describe the shape includes form factor, minor axis area, perimeter, major axis, eccentricity, circularity, orientation, and roundness.

2.2.2   Prediction Error

It is a textural feature which provides the distortion information of the pixels homogeneity in an image.

2.2.3   GLCM

Gray-level co-occurrence matrix is a textural feature providing the information of how often gray-level combination occurs in an image.

2.2.4   GLRLM

The gray-level run length matrix describes the size of homogeneous runs for each gray level.

2.2.5 Local Binary Pattern

LBP is a powerful grayscale texture feature representing the local neighborhood for an image.

2.2.6 Fractal Dimension

Fractal dimension is used to identify the textural variation of an image.

2.2.7 Entropy

It measures the uncertainity associated with randomness. Here, entropy measure such as Shannon entropy, Renyi entropy, Havrda and Charvat entropy, and Kapur's entropy are used to define the feature vector for erythrocyte classification.

2.2.8 Histogram

Histogram, such as chrominance channel histogram, saturation channel histogram, green channel histogram, and R-G channel histogram, is used to represent the histogram feature to classify the infected erythrocyte.

2.2.9 LBP-GLCM

It represents the co-occurrence of binary pattern feature extractor based on local binary pattern and gray-level co-occurrence matrix.

2.3 Feature Extraction

In feature extraction, two different analyses have been done as shown in Fig. 1. Different features such as morphological, intensity and texture have been extracted and analyzed the importance of the features without feature selection techniques. Further, the importance of feature along with feature selection has also been done to discuss the importance of feature as well as selection technique.

2.4 Classification

Here, the classification of malaria-infected erythrocyte has been performed using SVM, k-NN, and Naive Bayes.

a. **SVM**: It is a supervised learning algorithm that scrutinizes the data for classification. It mapped the data into the high-dimensional feature spaces and classifies the data based on which side of the gap they fall [12, 13]. SVM with different kernel functions has been evaluated to set the better function for erythrocyte classification.

b. **k-NN**: k-nearest neighbors algorithm is a nonparametric classification [14] method where input consists of k closest training data in the feature space. The output is a class membership where data are classified by the majority voting of its neighbors. Here, various k value are evaluated for infected erythrocyte classification.

c. **Naive Bayes**: It is probabilistic classifier based on Bayes' theorem with nave independence assumptions between the features [15]. Various probabilistic functions have been evaluated for better classification result.

d. **ANN-GA**[16]: ANN is a biological inspired computational intelligence technique which requires setting and tuning of complex structure, whereas genetic algorithm helps to optimize the complex problem where the numbers of parameters are large. Here, GA is used to select the optimal number of features for erythrocyte classification from feature sets. In this process, feedforward neural network has been used with one hidden layer. Size of hidden neuron and number of features are optimized based on the GA.

3 Experimental Results and Discussion

The experimental study has been conducted using 1302 erythrocyte images (186 erythrocytes per life cycle of malaria infection). The database has been collected from Silchar Pathological Laboratories and contains seven different life cycles of *Plasmodium* species, i.e., *Plasmodium falciparum* (ring and schizont), *Plasmodium vivax* (ring, amoeboid ring, schizont, and gametocyte), and non-infected erythrocyte. The experimental results have been executed using MATLAB R2017b. The overall performance is calculated by averaging the classification accuracy of each subset of fourfold cross-validation process. Here, performance of the two models has been evaluated. In model I, three different feature sets (morphological features, intensity and texture features, and feature set form by their combinations) are fed into three different classifiers such as SVM, k-NN, and Naive Bayes. In model II, same three feature sets are ranked by using ANOVA ranking and fed into classifiers (SVM, k-NN, and Naive Bayes) and performances have been evaluated. Altogether, a total of six feature sets are used for the experimental purpose, i.e., f_1 = morphological feature, f_2 = texture and intensity feature, f_3 = morphological, texture and intensity , f_4 = morphological feature with ANOVA ranking, f_5 = texture and intensity feature with ANOVA ranking, and f_6 = morphological, texture and intensity with ANOVA ranking. For k-NN classifier, the performance of the different feature sets may rank as $f_6 > f_2 > f_3 > f_5 > (f_4, f_1)$. For Naive Bayes, the feature set may rank as $(f_2, f_5) > f_6 > f_3 > (f_4, f_1)$. In case of SVM, the feature sets are ranked as $f_5 >$

Table 1 Performance of different classifiers for various feature sets

Classifier	f_1	f_2	f_3	f_4	f_5	f_6
No.features	16	208	224	16	208	224
SVM(rbf kernel funcion)	59.07 ± 9.58	79.84 ± 1.57	76.84 ± 4.07	67.92 ± 4.93	89.69 ± 5.23	76.84 ± 4.04
Naive Bayes(kernel)	76.67 ± 6.23	87.22 ± 1.52	83.74 ± 4.23	76.67 ± 6.23	87.22 ± 4.65	83.75 ± 4.23
k-NN(k = 3)	87.00 ± 1.90	91.08 ± 2.77	90.61 ± 2.05	87.00 ± 1.90	89.69 ± 1.63	$\mathbf{91.62 \pm 2.04}$

Table 2 Comparative analysis of different classifiers

Classifier	Features set	Accuracy (%)
SVM	f_5	89.69 ± 5.23
Naive bayes	f_2, f_5	87.22 ± 4.65
k-NN	f_6	91.62 ± 2.04
ANN-GA	f_6 (103 nos. of features)	$\mathbf{92.01 \pm 1.56}$

$f_2 > (f_6, f_3) > f_4 > f_1$. From the complete analysis, it has been observed that the k-NN with f_6 feature set provides better result in comparison to other feature sets as well as classifiers as shown in Table 2. The overall performance of the best model (k-NN with f_6) is 91.62 ± 2.04 % . From the experimental analysis of various classifiers show that f_6 provides better results. So, f_6 is further used for ANN-GA model for erythrocyte classification. GA helps to choose the best optimal features from f_6 set. ANN-GA provides an accuracy of 92.01 ± 1.56 % with 103 no. of features (Table 1).

4 Conclusion

The experimental results of the various feature sets which consist of morphological feature, texture and intensity feature have been achieved using classifiers such as SVM, Naive Bayes, and k-NN. Here, the importance of different feature sets has been evaluated. From the analysis, it has been observed that the feature set f_6 (combination of morphological, intensity and texture feature with ANOVA ranking) provides the better performance in comparison to other feature sets with an accuracy of 91.62 ± 2.04%. Further, ANN-GA along with f_6 provides an accuracy of $\mathbf{92.01 \pm 1.56}$% with 103 number of features. Here, GA helps to optimize the number of features required for erythrocyte classification. So, it may also be concluded that the feature set as well as the feature selection technique play a vital role in malaria-infected erythrocyte detection. It has also been noticed that the features such as morphological, intensity and texture feature play a very important role and provide a significant improvement in erythrocyte classification.

Acknowledgements The research work has been done in the Speech and Image Processing Laboratory of NIT Silchar, Assam-788010. For malaria parasite identification and database collection, we would like to express our gratitude to Dr. S. A. Sheikh, Silchar Medical College and Hospital, Assam and Dr. A. Talukdar, Head of the Department, Pathology, Cachar Cancer Hospital and Research Centre, Assam.

References

1. Cuomo, M.J., Noel, L.B., White, D.B.: Diagnosing Medical Parasites: A Public Health Officers Guide to Assisting Laboratory and Medical Officers http://www.phsource.us/PH/PARA/ Diagnosing Medical Parasites (2012)
2. Di, Ruberto C., Dempster, A., Khan, S., Jarra, B.: Analysis of infected blood cell images using morphological operators. Image Vis. Comput. **20**(2), 133–146 (2002)
3. Nicholas, R.E., Charles, J.P., David, M.R., Adriano, G.D.: Automated image processing method for the diagnosis and classification of malaria on thin blood smears. Med Biol Eng Comput. **44**(5), 427–436 (2006)
4. Tek, F.B., Dempster, A.G., Kale, I.: Parasite detection and identification for automated thin blood film malaria diagnosis. Comput Vis Image Und. **114**(1), 21–32 (2010)
5. Diaz, G., Gonzalez, F.A., Romero, E.: A semi-automatic method for quantification and classification of erythrocytes infected with malaria parasites in microscopic images. J. Biomed. Inform. **42**(2), 296–307 (2009)
6. Springl, V.: Automatic Malaria Diagnosis Through Microscopic Imaging. Faculty of Electrical Engineering, Prague (2009)
7. Das, D.K., Ghosh, M., Pal, M., Maiti, A.K., Chakraborty, C.: Machine learning approach for automated screening of malaria parasite using light microscopic images. Micron **45**, 97–106 (2013)
8. Devi, S.S., Sheikh, S.A., Laskar, R.H.: Erythrocyte features for malaria parasite detection in microscopic images of thin blood smear: a review. Int. J. Interact. Multimed Artif. Intel. **4**(2), 35–39 (2016)
9. Devi, S.S., Kumar, R., Laskar, R.H.: Recent advances on erythrocyte image segmentation for biomedical applications. In: Fourth International Conference on Soft Computing for Problem Solving (pp. 353–359). Springer, India (2015)
10. Devi, S.S., Roy, A., Singha, J., Sheikh, S.A., Laskar, R.H.: Malaria infected erythrocyte classification based on a hybrid classifier using microscopic images of thin blood smear. Multimedia Tools Appl. (2016). https://doi.org/10.1007/s11042-016-4264-7
11. Otsu, N.: A threshold selection method from gray-level histograms. IEEE Trans. Sys. Man and Cyber **9**(1), 62–66 (1979)
12. Devi, S.S., Singha, J., Sharma, M., Laskar, R.H.: Erythrocyte segmentation for quantification in microscopic images of thin blood smears. J. Intell. Fuzzy Syst. **32**(4), 2847–2856 (2017)
13. Burges, C.J.C.: A tutorial on support vector machines for pattern recognition. Data Min. Knowl. Discov. **2**, 121–167 (1998)
14. Weinberger, K.Q., Saul, L.K.: Distance metric learning for large margin nearest neighbor classification. J. Mach. Learn. Res. **10**, 207–244 (2009)

15. Russell S, Norvig P (2003) Artificial Intelligence: A Modern Approach, 2nd edn. Prentice Hall. ISBN 978-0137903955
16. Ahmad, F., Mat-Isa, N.A., Hussain, Z., Boudville, R., Osman, M.K.: Genetic algorithm-artificial neural network (GA-ANN) hybrid intelligence for cancer diagnosis. In: 2nd International Conference on Computational Intelligence, Communication Systems and Networks, pp. 78–83 (2010)

A Demonstration on Initial Segmentation Step on Closed Digital Planar Curve

R. Mangayarkarasi and M. Vanitha

Abstract Polygonal approximation technique is used to represent boundary of a digital image, where the boundary is approximated using piece line segments to present the shape of the original boundary of a digital image. This paper uses the value of the turn angle between two line segments multiplied with the length of those line segments as a measure to detect good curvature points. The boundary acquisition technique produces boundary with many duplicate points. To expedite the process of obtaining final polygon, an initial segmentation step is mandatory. This paper demonstrates the contribution of initial segmentation by freeman chain code on the digital planar curves. The experimental results too support the same.

Keywords Digital image · Boundary · Segmented points · Chain code · Polygon

1 Introduction

Digital image processing is the field which deploys many mathematical techniques to process different types of images in order to extract meaningful information. One among them is feature extraction from a digital image. Polygonal approximation is a classical as well as robust method to extract sensible information about a boundary of a digital image. Due to this, it finds its application in the analysis of images, analysis of signals, pattern recognition and computer vision. In shape classification, the first step might be to extract the boundary of the image. To represent the boundary, many applications in the literature use polygonal/closed curve approximation techniques. These techniques use different methodology to make approximation one such methodology which is split-and-merge method [1]. And the other is metaheuristic methods that include various stochastic ideal estimations which are provided by using ant colony improvization, particle swarm improvization and genetic algorithm, and the one in [2] uses k means clustering. The best method which always gives tries to produce optimal output using PA techniques is constrained to provide optimality

R. Mangayarkarasi (✉) · M. Vanitha
SITE, Vellore Institute of Technology, Vellore, Tamil Nadu, India
e-mail: rmangayarkarasi@vit.ac.in

© Springer Nature Singapore Pte Ltd. 2020
K. N. Das et al. (eds.), *Soft Computing for Problem Solving*,
Advances in Intelligent Systems and Computing 1057,
https://doi.org/10.1007/978-981-15-0184-5_25

with a higher cost of computation in terms of storage and execution time. Some of these ideal approaches are based on dynamic programming methods [3] or use an A* algorithm and so on. These methods finds its usage in many real time applications such as vehicle number identification kumar et al. [4], Electro-oscillography (EOG) Bio signal processing by Semyonov [5]. The methods narrated so far use line segments for approximation. Apart from these methods, the technique in [6] investigates shape approximating algorithm which divides the given 2D binary image into a collection of arched or out curved polygon that represents that shape. It involves in the process of identifying a single-arch polygon initially and then used for further approximation. The initial component obtained by incrementing and selection of series of point from the native shape is identified on the basis of image boundary features that are obtained from the given image at various levels of scaling measures. This will be further added with some other approximated shapes of that same image contour which are based on the difference among the given shape and the initial component using some of the available operations for repairing specific concavities which are occurred by doing set difference operator. Finally, the divided components are made into a single combined shape which represents the approximated shape that depicts the ideal given image. The union operator is ideal for recombination of these separate polygons into whole, and also some minute details are included in this algorithm which gives advantage of using an algorithm that comes under split-and-merge approach in the field of the approximating polygons. In their calculation, a shape is disintegrated into raised parts of different shapes and sizes. Their approach is used to develop a curved shape segment that gives a decent estimate relating to the characteristic of shape. The shape segments reflect both the overall outline and certain nearby subtle elements of the comparing shape parts. A precise estimation for guaranteed shape can be built utilizing just a modest number of curved segments. The important use of morphological decay is compression and recreation. The authors [7] suggest a method to derive the initial segmented points which can be used as an alternative for freeman chain code [8]. Most of the closed curve approximation techniques use freeman chain to obtain initial set of points (ISP), whereas the method in [7] uses concavity tree to obtain ISP. And the technique uses ratio of integral square error and compression ratio as the termination criteria. This ISP technique can be used by any closed curve approximation to expedite the polygon generation process. The method in [9] obtains break points where integral square error is zero. The final set of dominant points are derived through iterative procedure, where the points deviation measure is greater than the threshold value. And the threshold value is also incremented by 0.5 after completing every iteration. The method [9] finds more redundant points in a single iteration and it made the technique to define the final set of dominant points quickly. The method in [10] presents a split–merge-based method to make closed curve approximation. In this paper [10], the authors discuss the redundant points on the digital boundary which slows down the polygon generation and proposes a technique which does not require any parameter to make output. The work presented in this paper demonstrates the use of removing redundant points before applying any measure to define the contribution of points on the digital boundary curve.

2 Analysed Framework

The method presented in this paper uses boundary extraction function in MATLAB 2017 to obtain the boundary. The resultant boundary has many duplicate vertices. To remove those vertices, the freeman chain code [8] is assigned to the every side of boundary as mentioned in Fig. 1.

The first set of redundant points are removed from the boundary using the below condition

If chaincode $(B_i) \neq (B_{i+1})$ Then
B_i is a dominant point
Else
B_i is not a dominant point

To demonstrate the importance of initial segmentation step, the method analysed in this paper uses the value of the turn angle between the consecutive line segments and multiplied with the length of two immediate line segments to refine the first set of dominant points as shown in Fig. 2. To produce the output polygon, the tested framework uses formulae described in Fig. 3 and deletes the vertex having least value for the metric suggested in this section. While continuously performing the deletion leads to deformation to the original shape, the amount of deformation can be measured using approximation error.

Fig. 1 Freeman's chain code

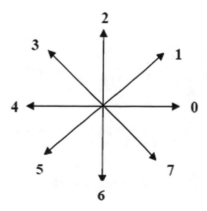

Fig. 2 Turn angle computation

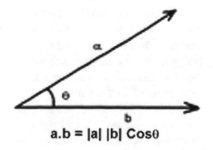

$$a.b = |a| \ |b| \ Cos\theta$$

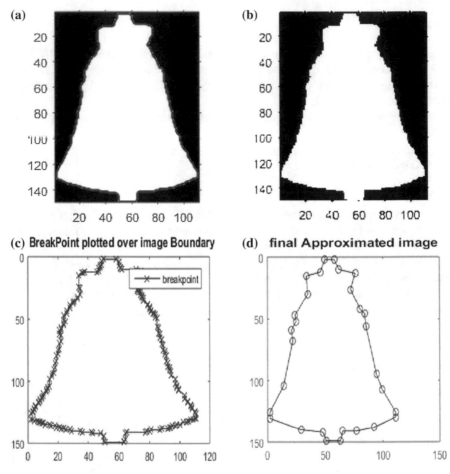

Fig. 3 **a** Input image; **b** binary image; **c** breakpoints over image; **d** polygon using only required number of points

3 Experimental Outcomes

To test the contribution of the preprocessing step, bell-7 image from MPEG-7 database has been used. After the binarization, the boundary of an image is extracted and processed using the metric mentioned. The input contour bell-7 consists of 408 vertices. In Table 1, we display the execution time by the method to produce output polygon both in the presence and absence of preprocessing steps for the bell-7, key and chromosome at varying number of line segments. Tables 1 and 2 tabulate the time incurred in the presence of preprocessing steps. By analysing the contents in Tables 1 and 2, it reveals the process of obtaining final output polygon which is faster because of the preprocessing step. Any PA technique execution time gets reduced provided if we eliminate unnecessary vertices.

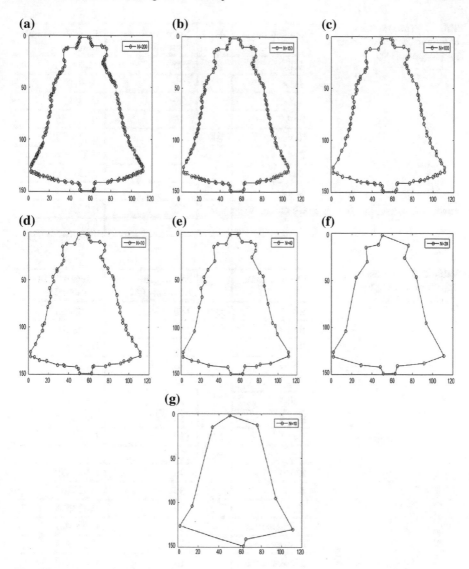

Fig. 4 Output curve at variant number of line segments (*m*) **a** *m* = 200, **b** *m* = 150 **c** *m* = 100, **d** *m* = 70, **e** *m* = 40, **f** *m* = 20, **g** *m* = 10

Table 1 Time taken to approximate in the absence of preprocessing step

Contour name	N	No. of line segments (m)	Execution time in seconds
Bell-7	408	150	0.102003
Bell-7		75	0.031015
Bell-7		39	0.03582
Bell-7		19	0.034755
Key	792	200	0.138757
Key		150	0.062389
Key		75	0.071400
Key		35	0.069614
Chromosome	60	35	0.0038431
		20	0.004178

Table 2 Time taken to approximate using the proposed method

Contour name	N	No. of line segments (m)	Execution time in seconds
Bell-7	193	180	0.003603
Bell-7		150	0.006216
Bell-7		100	0.012211
Bell-7		70	0.011910
Bell-7		40	0.014079
Bell-7		20	0.015475
Bell-7		10	0.016325
Key	283	100	0.018377
Key		75	0.019697
Key		45	0.022039
Key		30	0.023264
Key		20	0.023635
Chromosome	37	37	0.000925
Chromosome		23	0.001798
Chromosome		13	0.002435
Chromosome		7	0.002819

Figures. 3 and 4 show approximated polygons which represent the shape at the original shape that has been estimated in the other figures. Figure 4 depicts on how the approximation actually took place in order to reduce the number of points and refurbished the original image without compromising the presence of essential vertices.

Tables 1 and 2 show the execution time taken for both the methods, respectively. Our proposed method approximates the curve with N number of points into m (given number of line segments) with lesser time than given in Table 1.

4 Conclusion

In this paper, a polygonal approximation technique performance is demonstrated both in the presence and absence of preprocessing step. This paper demonstrates the chain code assignment steps to define the initial set of vertices. Then, it uses turn angle and the length of the consecutive line segments as a measure to define the further sets of qualified points. The analysis results reveal the importance of preprocessing step for producing the output polygon generation.

References

1. Teh, C.H., Chin, R.T.: On the detection of dominant points on digital curves. IEEE Trans. Pattern Anal. Mach. Intell. 11(8), 859–872 (1989)
2. Yin, P.Y.: Algorithms for straight line fitting using k-means. Pattern Recogn. Lett. 19(1), 31–41 (1998)
3. Kolesnikov, A., Fränti, P.: Polygonal approximation of closed discrete curves. Pattern Recogn. 40(4), 1282–1293 (2007)
4. Kumar, M.P., Goyal, S., Jawahar, C.V., Narayanan, P.J.: Polygonal approximation of closed curves across multiple views. In: ICVGIP (2002)
5. Semyonov, P.A.: Optimized unjoined linear approximation and its application for Eog-biosignal processing. In: Engineering in Medicine and Biology Society, 1990., Proceedings of the Twelfth Annual International Conference of the IEEE (pp. 779–780). IEEE (1990, November)
6. Zygmunt, M.: Circular arc approximation using polygons. J. Comput. Appl. Math. 322, 81–85 (2017)
7. Aguilera-Aguilera, E.J., Carmona-Poyato, A., Madrid-Cuevas, F.J., Medina-Carnicer, R.: The computation of polygonal approximations for 2D contours based on a concavity tree. J. Vis. Commun. Image Represent. 25(8), 1905–1917 (2014)
8. Freeman, H.: On the encoding of arbitrary geometric configurations. IEEE Trans. Electron. Comput. 10(2), 264–268 (1961)
9. Carmona-Poyato, A., Madrid-Cuevas, F.J., Medina-Carnicer, R., Muñoz-Salinas, R.: Polygonal approximation of digital planar curves through break point suppression. Pattern Recogn. 43(1), 14–25 (2010)
10. Madrid-Cuevas, F.J., Aguilera-Aguilera, E.J., Carmona-Poyato, A., Muñoz-Salinas, R., Medina-Carnicer, R., Fernández-García, N.L.: An efficient unsupervised method for obtaining polygonal approximations of closed digital planar curves. J. Vis. Commun. Image Represent. 39, 152–163 (2016)

Nonlinear System Modelling Using Programmable Hardware for Soft Computing Applications

M. Vanitha, R. Sakthivel, R. Mangayarkarasi and Suvarcha Sharma

Abstract The performance of a system depends on the performance of its components, and the component's performance depends on the effectiveness of the function performed by it. The logarithmic function is one of the most commonly used functions in the communication world. Lately, the implementation of logarithmic function is done using BJTs and MOSFETs in their weak inversion layer, but still, there is a scope for the improvement in the parameters like power, cost, speed, area and complexity. In this paper, the design for implementation of logarithmic function is proposed, and this is done by the use of three basic mathematical operations: (i) differentiation (ii) division and (iii) integration. Here, all the circuits performing such basic operations are being designed using OTA so that it can handle different arbitrary signals in a systematic manner in order to reduce the complexity of the system. Simulations of all the circuits have been done using SPECTRE.

Keywords OTA · Logarithmic function · Differentiation · Division · Integration

1 Introduction

Communication is a vast field. The technology used in communication is changing continuously. There is always a need to make the technology more and more effective in order to run parallel to the advancements. And to do so, different options are being explored. Logarithmic function is one of the most basic mathematical functions which is commonly used in the field of communication. Hence, the aim here is to design a logarithmic function generator which is more accurate and effective in terms of power consumptions. The field of operational transconductor amplifier is used to implement the logarithmic function [1]. The operational transconductance amplifier (OTA) is a voltage control current source. The OTA design is quite similar to the design of OPAMP. But some basic differences which make OTA better option for our use here in this design over OPAMP are given in Table 1.

M. Vanitha (✉) · R. Sakthivel · R. Mangayarkarasi · S. Sharma
Vellore Institute of Technology, Vellore, Tamil Nadu, India
e-mail: mvanitha@vit.ac.in

© Springer Nature Singapore Pte Ltd. 2020
K. N. Das et al. (eds.), *Soft Computing for Problem Solving*,
Advances in Intelligent Systems and Computing 1057,
https://doi.org/10.1007/978-981-15-0184-5_26

Table 1 Comparison between OTA and OPAMP

S. No.	OPAMP	OTA
1	OPAMP has output as voltage, which is proportional to difference between its two inputs	OTA output is current that is proportional to the difference in voltage between its two inputs
2	It has two inputs	It has three inputs
3	It does not usually have the biasing current as input	It has two additional biasing inputs, namely Iabc and Ibias

Fig. 1 OTA symbol

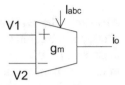

Hence, we can say that OTA is similar to OPAMP, but the mode is different, i.e., OTA is current mode device whereas OPAMP is voltage mode device, i.e., OTA takes voltage as its input and gives a proportional amount of current as an output, rather than a proportional voltage [2].

The circuit for the OTA is shown in Fig. 1.

An ideal transconductance amplifier is expected to have an infinite bandwidth, infinite input and output impedance. One of the several drawback of this device is that it has relative lower output impedance.

The ideal transfer characteristic of OTA is provided in Eqs. 1 and 2.

$$I_{\text{out}} = g_m(V1 - V2) \tag{1}$$

The pre-computed value of the voltage difference is fed as an input to the device,

$$I_{\text{out}} = g_m \times V_{\text{in}} \tag{2}$$

As the value of transconductance is constant ideally, hence it has the proportionality factor between the two parameters such as input voltage and output current. Since transconductance (g_m) itself is a function of input differential voltage, moreover, it also depends on temperature variations. To summarize, an ideal OTA structure provides a very huge input impedance for the two input voltage terminals. If this OTA structure is operated in common-mode input range, then too it provides an infinite impedance. These two differential input voltages have a control on the ideal current source which is being delivered as output. And also, the output current and input differential voltage are proportional to each other by the constant which is called as the transconductance [3].

OTA has controllable gain and controllable critical frequencies. All linear applications demand the usage of OTA in the 'open loop' mode without applying negative feedback mechanism. The optimal selection of the output resistance plays a vital role in controlling the output voltage [4, 5]. This selection of resistance value should ensure that the device does not enter into saturation mode in spite of providing a high differential input voltage.

Here in this paper, we are designing only the logarithmic circuit using OTA, but it could be used in order to design various other functions like exponential function, log with base 10, hyperbolic function, etc. Hence, further study can be continued using this concept.

2 Previous Works

Many researchers have been giving various methods and techniques for implementation and better performance of binary logarithm approximation. Mitchell suggest the usage of piecewise linear interpolation technique and lookup-table for the calculation of binary logarithm. This work develops a FPGA hardware architecture for log and antilog function. The logarithmic I–V relation of a p–n junction diode and exponential relation of Taylor's series are realized by this technique [6].

We can see from previous works that most of the logarithmic circuit design implementation is being done using a standard signal as an input. For example, sine wave, triangular wave, square wave, etc. Bhanja et al. [7] developed a function which is a ration of two other functions which involve the numerator being the differentiation of the denominator, and then the output of the integration comes as a log function. This could be applied for a parabolic input signal too using single OTA and a grounded capacitance [8].

3 Proposed Work

The work proposed in this paper is implementation of the device which can perform logarithmic operation. The circuit's basic block diagram is shown in Fig. 2.

Here, we will see that how these three individual circuits join together to perform the function ln, i.e., the logarithmic function [9].

The basic idea used here can be represented through Eq. (3).

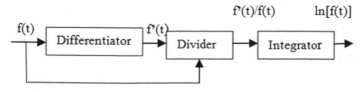

Fig. 2 Block diagram of logarithmic function

$$\int \frac{1}{f(t)} \frac{d}{dt} f(t) dt = \int \frac{f'(t)}{f(t)} dt = \ln(f(t)) \tag{3}$$

From this Eq. (3), we can understand that considering the numerator as the differentiation of the denominator, and then applying integration to this function, the logarithmic function can be obtained. Using this concept, we have implemented logarithmic function generator. Each part of the design can be discussed separately [10].

3.1 Differentiator

OTA is used for constructing a differentiator and the same is shown in Fig. 3. The output is given by the expression (4).

$$v_{\text{diff}}(t) = k_{\text{dif}} \cdot (d/dt[v_{\text{in}}(t)]) \tag{4}$$

where the differentiator coefficient $k_{\text{dif}} = g_{m1} C_1 / g_{m2} g_{m3}$ and g_{mi} denotes the transconductance of an OTA. The basic circuit of the differentiator using OTA is given in Fig. 3.

Here, in this paper, the basic function is being designed using OTA.

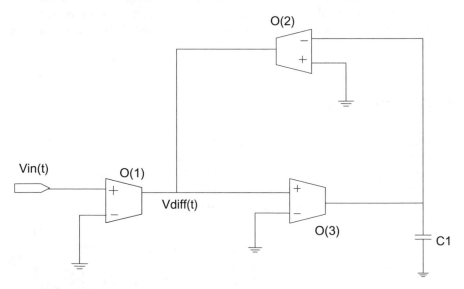

Fig. 3 OTA-based differentiator

3.2 Divider

The equation for divider is given in Eq. (5),

$$v_{\text{div}}(t) = \left[\frac{V_{\text{in3}}(t)}{V_{\text{in4}}(t)}\right] \cdot \frac{g_{m4}}{K_{\text{mul}} \cdot g_{m5} \cdot R_1} = \left[\frac{V_{\text{in3}}(t)}{V_{\text{in4}}(t)}\right] \cdot k_{\text{div}} \tag{5}$$

and the OTA-used circuit of divider is also given in Fig. 4.

We can see that multiplier is also used in the circuit of divider, and we have also implemented the circuit of multiplier using OTA, which can be given in Fig. 5.

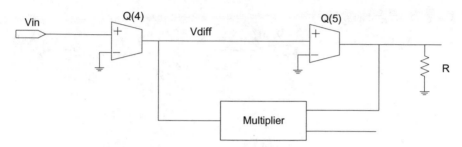

Fig. 4 OTA-based divider

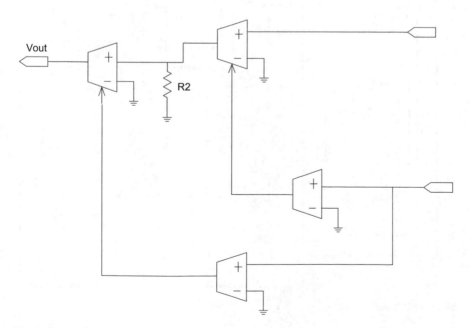

Fig. 5 OTA-based multiplier

3.3 Integrator

The integrator is driven by the equation,

$$V_o(t)\frac{g_{m10}}{C_2}\int v_i(t)\mathrm{d}t = K_{int}\int v_i(t)\mathrm{d}t \qquad (6)$$

Circuit of integrator using OTA is shown in Fig. 6.

On combining all the circuit provided above, we can obtain a main circuit of logarithmic function, which is given in Fig. 7.

Fig. 6 OTA-based integrator

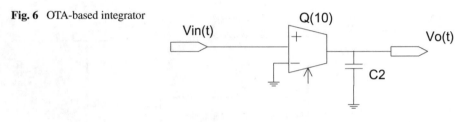

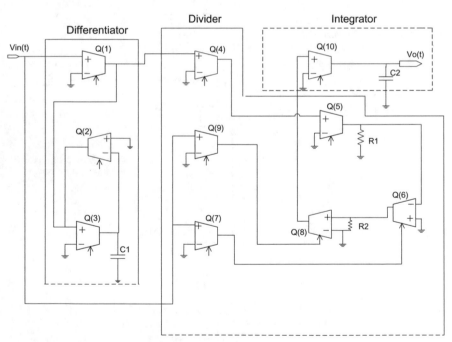

Fig. 7 OTA-based ln $[f(t)]$

4 Simulations

The proposed circuits are analysed and designed with proper sizing using design equations. The circuits were constructed using cadence virtuoso. Simulations of differentiator, divider, multiplier, and integrator and, finally, the combined circuit which perform the logarithmic function were done. The simulations were performed using 90 nm tech files using spectre in Cadence Virtuoso tool.

4.1 Simulation of OTA

We have used here two-stage OTA, the circuit of which given in Fig. 8.

For OTA circuit, we can calculate the trans-conductance

$$g_m = I_{out}/(V_{in1} - V_{in2}) \qquad (7)$$

Here, we get g_m as 13.18 uA/V and $V = V_{in1} - V_{in2}, V = 16.3$ mV

From the above graphs shown in Figs. 9, 10 and 11, we get the basic characteristics of OTA designed.

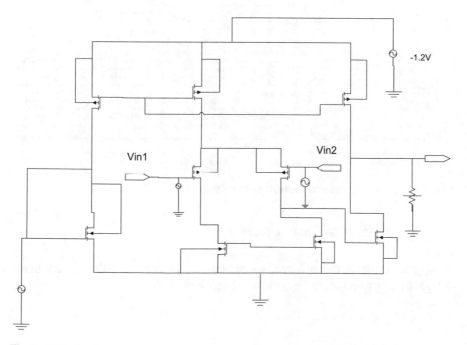

Fig. 8 OTA circuit

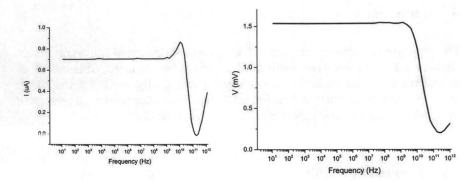

Fig. 9 g_m and V_{in} V_s frequency

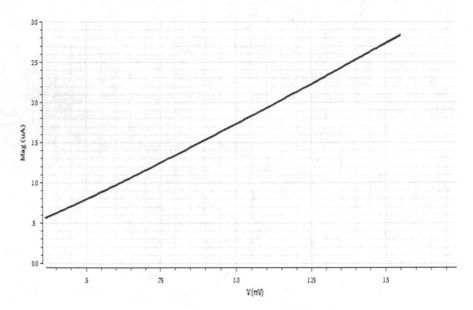

Fig. 10 Output current versus differential input voltage

4.2 Simulations of Basic Circuits

To implement Eq. 3, the differentiator, integrator, divider and multiplier are being designed using the OTA as the basic building block.

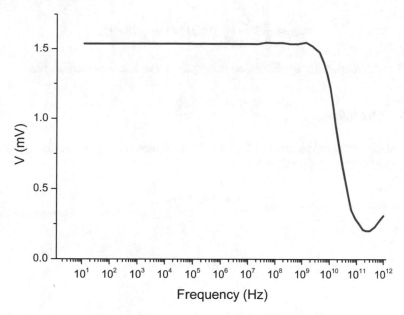

Fig. 11 Magnitude versus frequency

4.2.1 Differentiator

The OTA-based differentiator is represented in Fig. 12. The equation for the differentiator is given as

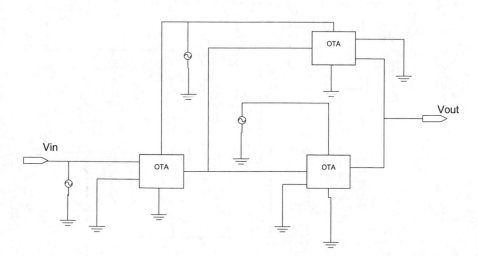

Fig. 12 Differentiator circuit

$$v_{\text{diff}} = \frac{g_{m1}C_1}{g_{m2}g_{m3}}\frac{d}{dt}[v(t)] = k_{\text{dif}}\frac{d}{dt}[v_{\text{in}}(t)] \tag{8}$$

The differentiator is implemented using passive and active components.

4.2.2 Multiplier

The multiplier is represented in Fig. 13. It multiplies the input signals $v_{\text{in}1}(t)$ and $v_{\text{in}2}(t)$ and reflects the result at Iout as,

$$I_{\text{out}} = v_o * g_m \tag{9}$$

where

$$v_o = V_{\text{in}1} * g_m * R \tag{10}$$

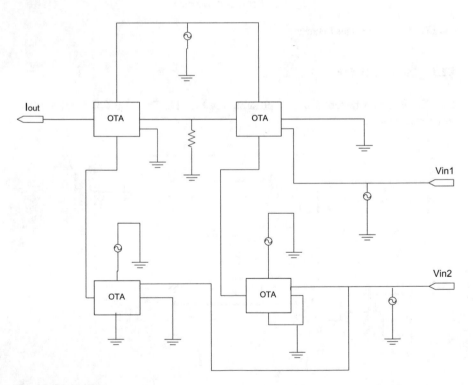

Fig. 13 Multiplier

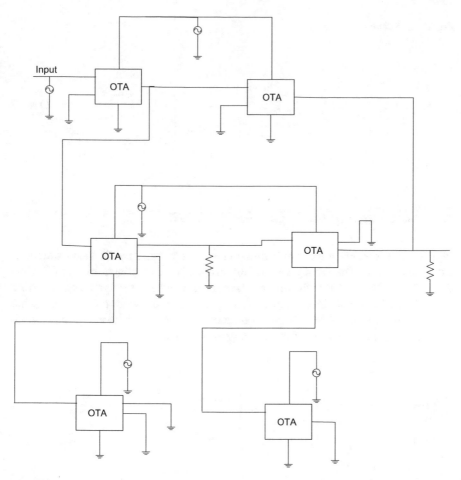

Fig. 14 Divider

4.2.3 Divider

Figure 14 represents divider circuit. The output of a basic divider is $y = x1/x2$. And, the output of the divider is given as the V_{in3}/V_{in4} * (constant).

4.2.4 Integrator

Figure 15 represents the circuit for the integrator. The working of integrator is explained in Sect. 2.

Fig. 15 Integrator

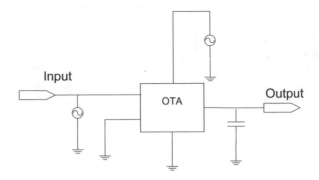

4.3 Simulation of the Logarithmic Circuit

Figure 16 represents logarithmic function circuit formed by its basic functions explained before. The input signal of the circuit is differentiated by the differentiator circuit. The output of the differentiator is provided further to the non-inverting port of the OTA in the divider circuit. The output of the divider circuit goes to the input of integrator and the final output obtained at the output of the integrating circuit is the logarithmic function of the input signal provided.

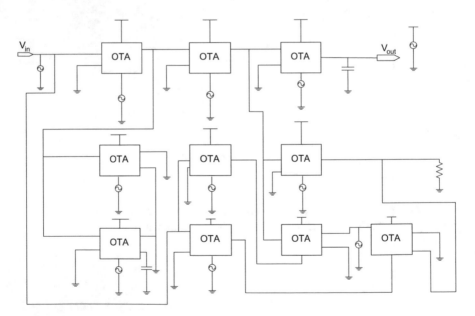

Fig. 16 $\ln f(t)$ circuit

5   Result

The circuit represented in Fig. 16 gives the design of logarithmic function using OTA. In Fig. 17, the output $\ln(f(t))$ function is provided with $f(t) =$ sinusoidal function. In Fig. 18, $f(t) =$ triangular wave. From Fig. 18, we can say that the OTA-based circuit works as a logarithmic function generator.

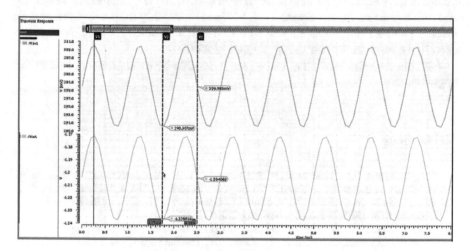

Fig. 17   Output of the logarithmic function circuit when input is sinusoidal function

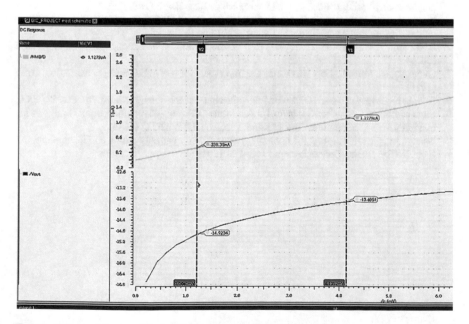

Fig. 18   Output of the logarithmic function circuit with input as ramp wave

6 Conclusion

This work had focused on logarithmic circuit and verified the output for sinusoidal and triangular input signal. The output we receive is very close to the calculated value of the input signals. Hence, we can conclude that logarithmic and exponential and other nonlinear functions can also be implemented using the OTA. The basic building blocks of a system building were realized with the help of OTA and their functionality and characteristics were analysed. OTA has more suitability for the implementation since it provides better tunability as well as programmability. Devices like Opamp can also be used for the realization of these blocks.

Nonlinear system modelling using programmable hardware for soft computing applications

References

1. Saravanakumar, O.M., Kaleeswari, N., Rajendran, K.: Design and analysis of two-stage operational transconductance amplifier (OTA) using cadence tool. 5462 NCICT Special Issue (2011)
2. Bhanja, M.: Student Member, "OTA-based logarithmic circuit for arbitrary input signal and its application. In: IEEE, and Baidya Nath Ray IEEE (2012)
3. Odame, K.M., Hasler, P.: Theory and design of OTA-C oscillators with native amplitude limiting. In: IEEE Transactions on Circuits and Systems I: Regular Papers, (vol. 56, no. 1, pp. 40–50) (2009)
4. Geiger, R.L., Sanchez-Sinencio, E.: Active filter design using operational transconductance amplifiers: a tutorial. IEEE Circuits Devices Mag. 1(2), 20–32 (1985)
5. Huang, C., Chakrabartty, S.: Current-input current-output CMOS logarithmic amplifier based on translinear Ohm's law. Electron. Lett. 47 (2011)
6. Al-Tamini, K.M., Al-Absi, M.A.: A 6.13 μW and 96 dB CMOS exponential generator. IEEE (2014)
7. Gutierrez, R., Valls, J.: Low cost hardware implementation of logarithm approximation. IEEE (2011)
8. Roh, J.: High-gain class-AB OTA with low quiescent current. Circuits Signal Process. 47 (2006)
9. Silva-Martinez, J., Sanchez-Sinencio, E.: Analogue OTA multiplier without input voltage swing restrictions, and temperature-compensated. Electron. Lett. 22(11) (1986)
10. Bhanja, M., Ghosh, K., Ray, B.: Implementation of nth order polynomial and its applications. IEEE ICIEV, Dhaka, Bangladesh, pp. 166–171 (2012)

A Study on Intrusion Detection System of Mobile Ad-hoc Networks

S. Sindhuja and R. Vadivel

Abstract Nowadays, MANET is one of the most important research topics, as the mobile users are increased every day. MANET is infrastructureless service, so it does not require any fixed infrastructure. MANET can be used for more application such as disaster management, tactical operations, and environmental monitoring, and it can provide communications in the network. In MANET, every single node acts as transmitter and also the receiver. Due to this reason, each and every node will communicate directly. The individual-configuring capability of nodes is one of the reasons to make it popular for recovery during emergency. Wireless application made it possible because of the mobility and scalability in many applications. But this flexibility has the challenged risk of security in this environment. Mobile ad hoc networks have given the low level of physical security in general. So, only the prevention mechanism is not enough to secure, thus another detection mechanism should be added. The initial line of defense for security in MANET is intrusion detection system (EDSI). It presents the challenge of designing the intrusion detection system in MANET, analyzes the existing intrusion detection techniques, and proposes further directions.

Keywords Ad hoc networks · Enhanced detection system for intrusion (EDSI) · MANET · Physical security

1 Introduction

In MANET, there exist numerous mobile nodes where each communicates with another using radio links and the nodes without any central base station [1]. In multi-hop ad hoc networks, nodes can move and communicate with one another. MANET is implemented with transmitter and receiver which is a wireless one that

S. Sindhuja (✉) · R. Vadivel
Department of Information Technology, Bharathiar University, Coimbatore 641046, India
e-mail: sindhusekar@yahoo.com

R. Vadivel
e-mail: vlr_vadivel@yahoo.co.in

© Springer Nature Singapore Pte Ltd. 2020
K. N. Das et al. (eds.), *Soft Computing for Problem Solving*,
Advances in Intelligent Systems and Computing 1057,
https://doi.org/10.1007/978-981-15-0184-5_27

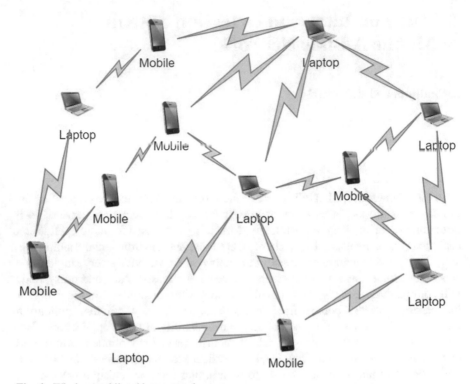

Fig. 1 Wireless mobile ad hoc network

can communicate with one another via bifacial wireless link, directly or indirectly [2]. Figure 1 shows the basic structure of wireless mobile ad hoc network.

Wireless network advantages include many, where its mobility and scalability are considered as an important advantage which helps to maintain the network success. But, this is possible only when the communication is limited to the range of transmitter. If the node is the specific diameter range, then it denotes that the communication is possible, else it is not [2].

In order to solve this problem, MANET allows intermediator to relay the data transmissions. MANET can be categorized into two types, which are (i) networks by single-hop and (ii) networks by multi-hop. Nodes can communicate in a direct manner in the single-hop network when they are in the same radio frequency. But in multi-hop networks, nodes are dependent on the intermediate node for sending and receiving communication. Shortly, the intermediate nodes are used/communicated when the destination is out of the range of radio frequency [2].

In MANET, each and every node is free to move anywhere because it is infrastructureless service. There may be a transmission problem like path loss, blockage, and interference. When comparing with other types of network, MANET works in the open medium, where it has to face many security threats. Unsecured boundaries in any node can attach, try to gather information for security attacks, and go out of the network. In MANET, the nodes tend to misbehave to perform security attacks. Thus, it will be considered as tragedy to identify the misbehaving or malicious node. Then, it is necessary to develop a secured EDSI to protect various attacks in MANET [3].

2 Types of Attacks in Manet

There are many types of attacks that affect the behavior and performance of the MANET. There are two types of attacks, namely outsider attacks and insider attacks. The insider attacks are carried out by the compromised node, which is the part of the network. The nodes which do not belong to the current network perform the attack called "outsider attack." Rather than outsider attack, insider attack is more severe because insiders know the secret information in the network and have privileged access rights [2, 3]. The attacks can be classified as active attacks and passive attacks. In passive attacks, the attacker does not disturb the process of the routing protocol but attempts to find important information with the help of analysis of traffic in the network [4]. Passive attacks will obtain exchange of data without disturbing the process, and active attacks involves in interrupting the information, modification, and by disturbing the MANETs normal functions. Some of the attacks make use of the stealth to hide their action from the individual which is monitoring the system from the intrusion detection system [2].

3 Enhanced Detection System for Intrusion (EDSI) in Manet

EDSI is defined as an application or a device that checks the traffic of the network, and it alerts the network administrator or system if any suspicious activity is found [5, 6]. It is a technique to detect attacks by analyzing and continuously monitoring network functions [7, 8]. Thus, the intrusion detection system will safeguard the system from attacks and compromises [9]. EDSI tends to (i) monitor the activity of the network, (ii) audit the network, (iii) configure the system for its vulnerability, and (iv) analyze the data for its integrity [9]. Outstanding growth and usage of Internet raise concerns about how to communicate and protect the digital information safely [10].

In general, EDSI involves three modules, which are (i) monitoring, (ii) analyzing, and (iii) responding. Controlling the collection of data is made by the monitoring module. And in the analyses module, it will decide whether the collected data is intrusion or not. And finally in the response module, it is the responsibility to manage and obey the actions for intrusion [5]. EDSI running in every mobile node checks the traffic level and detects the intrusions [11]. There are three types of EDSI as follows:

- Active and Passive EDSI.
- Network-Based and Host-Based EDSI.
- Knowledge-Based and Behavior-Based EDSI.

3.1 Active and Passive EDSI

In this, the system is designed to jam the suspected attacks in progress automatically without any involvement of the operator. And in passive EDSI, it is designed only for monitoring the traffic of the network and analysis. Then, it alerts the operator about the attacks and vulnerabilities [3].

3.2 Network-Based and Host-Based EDSI

In this, the system suspects packets from the network traffic. The operating system or application in its analysis is used for host-based EDSI [12]. The host-based EDSI caught local network traffic to the particular host. Thus, it is better to detect attack from inside [13].

3.3 Knowledge-Based and Behavior-Based EDSI

In knowledge-based EDSI, the evidences of attacks and behavior of malicious nodes are stored in the database and referred when attack or similar kind of condition arises since each and every intrusion leaves their footprint behind. To avoid the same type of attacks in the future, the footprints are stored. These footprints are called signatures. Based on these signatures, EDSI attempts to identify the intrusion [3].

In behavior-based EDSI, behavior of the system is noticed, and it assumes if there is any variation from normal or expected behavior of system. Generation of alarm is activated when the device occurs. Anything that was not related to the previously observed behavior is considered as intrusive. False high alarm rate is the disadvantage of this system because the entire scope of the system is not included in the phase of learning [3].

3.4 Architecture of EDSI

The existing EDSI architecture consists of three categories. They are (a) independent (b) cooperative (c) hierarchical

- Independent
  In independent architecture, the node acts locally and does not mingle and respond with other nodes locally. In this architecture, it has an issue of attacks on networks.
- Cooperative
  In cooperative architecture, each and every node in MANET has local EDSI system. Nodes make a decision in a distributed manner. After the determination of intrusion detection, information is shared by the nodes, also takes necessary action either in passive or active manner and eliminates the intrusion.
- Hierarchical
  The hierarchical architecture specifies the multilayered approach. In this architecture, it splits the nodes into clusters. Some nodes are selected specifically, and each node will assign responsibilities in intrusion.

3.5 Engine of EDSI

IDS engine is used to detect the local intrusions. This was made using classification algorithm. It will classify whether it is normal or abnormal intrusion [11].

4 Literature Survey

4.1 Misbehavior Routing

In MANET, routing protocols are limited to utilize all the available routing. Always the nodes in the MANET think that node combines with one another to rely on the packets [2]. This is one of the advantages to the attackers to obtain the similar impact of two compromised nodes on the network. Watchdog is fully in charge for suspecting the nodes behavior in the network to avoid attacks. Counter intended for failure is increased when a source node or intermediate node fails to send the data packet to the destination [2].

Watchdog
In watchdog mechanism, if the node sends a packet, it monitors the next node to confirm whether that node also forwards the packet correctly. In a buffer, it keeps the packets that are sent. It is confirmed that if the next nodes are forwarded to the packets, then they will be removed from the buffer [14]. If a MANET detects a failure of nodes based on threshold value, then immediately watchdog reports it as a misbehaving node [2], but it may also fail to find some malicious misbehaviors with the activity of (i) collisions with ambiguous and receiver, (ii) limited number of transmissions, (iii) reporting false misbehavior, (iv) collision, and (v) dropping partial number of data packets.

Pathrater

Pathrater acts as a response system. The pathrater cooperates with protocol, namely "watchdog," that is when it does not cooperate to report the node's status for future transmission. Many research contemplates have demonstrated that watchdog conspire is productive [5].

4.2 Acknowledgment-Based Routing Misbehavior Detection MANET

Many approaches are said by many researchers, to refer the six weaknesses in watchdog scheme. TWOACK is one of the most significant approaches among them. It is neither based on watchdog scheme nor an enhancement. TWOACK aims to defeat issues like collision and low level of transmission power [5].

Even though it solves some problems in watchdog, it has network overhead, due to the acknowledgment that is processed for the packet [14]. These methods aim to detect nodes which act as selfish that will engage in discovering the routes and maintenance process, but disagree to move the packet. In this scheme, it will acknowledge each and every packet, that is transmitted by sender to receiver, over three consecutive nodes and detect the misbehaving link [5].

By receiving the data packet, it is always necessary to acknowledge all the three consecutive nodes that are by hopping away from the route. Retrieval of TWOACK packet indicates that if it reaches within a time frame, the transmission is successful otherwise it determines that the destination and also the intermediate are malicious [5].

4.3 Enhancement of Multimedia Transmission

The intrusion detection system is also called as Adaptive Acknowledgment (AACK). It is an end-to-end scheme fully based on acknowledgment. It is a blend of TACK (identical to 2-ACK) and end-to-end acknowledgment scheme (ACK). Whenever the destination accepts the data packet successfully, it sends the acknowledgment packet to the sender [14]. To avoid the malicious node in another transmission, it detects the misbehavior of malicious nodes. When compared to TWOACK, it has more throughput of network and minimizes the network overhead [5, 14]. Thus, the detection of misbehavior in both the schemes fully depends upon acknowledgment packet from receiver. Misbehaving nodes will not have normal behavior, and it has some abnormal behaviors that can disturb the operation of the network and reply to the acknowledgment packet to the source node [5].

4.4 EAACK

MANET has vulnerabilities to attackers by default due to its nature. It is very tough to design and develop a mechanism for intrusion detection to save MANET from security attacks. This paper focuses to propose EDSI which is fully based on the nature of MANET to detect the intruders and attackers. EDSI is also based on the acknowledgment scheme, and to prevent from the attacker, it uses digital signature [2, 5].

EDSI partially supports cryptography, where it denotes the learning of mathematical technique related to (i) confidentiality, (ii) data integrity, and (iii) entity authentication [2]. To detect and prevent the attacker from developing its current state, an acknowledgment packet, digital signature is used. So, all the acknowledgment packet should be signed digitally. This EAACK scheme makes use of digital signature. It is necessary for all acknowledgment packets to be signed securely in a digital manner [5].

The three parts of EAACK are

1. ACK
2. S-ACK
3. MRA (Misbehavior Report Authentication).

ACK:

It is totally an end-to-end acknowledgment-based method [1]. Whenever the source node sends packet to the destination, the whole intermediate node accepts to forward it to the destination. If the packets reach destination, then acknowledgment packets are needed to be sent back and wait to receive the ACK packet. Within a specified duration of time frame, sender node receives the acknowledgment, and then it is considered that transmission of packet is successful. Otherwise, the source node needs to change to S-ACK mode that is by sending a S-ACK packet [2].

Secure-ACK:

Three sequential nodes work in S-ACK; it gathers information to identify getting out of misbehaving nodes. Information is gathered for every three nodes. With predefined time, if any node neglects to send affirmation, then it is noted as a set part as malicious node. EAACK needs the source node to change to MRA mode and affirms this report [5, 14].

MRA:

MRA is well designed to remove the disadvantages faced by watchdog in finding the misbehaving node. Likely, the reports may come with false information that even true nodes may act like a malicious node. When activating the mode of MRA, the source node aims to search local knowledge and also search for alternate destination [1]. In MANET, it is possible to find the alternate route in between the nodes. When MRA packet is received by the destination, it looks for the knowledge related to local base and check whether the reported packet is received. If that packet is malicious, it should be received already, then that is false misbehavior report [5].

5   Existing IDS

In this, we study and compare the existing Intrusion detection.

Home Agent: Home agent is present in every system, and it collects the information from the system in application layer.

Current Node: Home agent monitors its own system continuously. It asks classifier construction to identify the attacks when an attacker sends or collects some data.

Neighboring Node: Broadcasting is done, when the system sends information from one system to another. Before the message was transferred, the message was sent to mobile agent to neighboring node to collect the information. After that it comes back to the system, and it implements classifier rule to identify the attacks.

Collection of Data: It is used for anomaly detection and sub-system to collect the values of features. A general profile is created with the help of data collected during normal scenario. The data attacked is collected during attack scenario.

Preprocessing Data: The data is collected and is used for anomaly detection. This data preprocess is a technique that can be processed with information by testing the data.

Action Agent: Each and every node has this action agent. When each node uploads a host-based monitoring agent, it can able to identify if there are any abnormal activities. If any anomaly is detected, it will respond by blocking or terminating the user from the network.

Decision Agent: It can run only where the network monitoring agents are running. So, in this scenario, the nodes will collect the packets within radio range and determine whether it is under attack. If it identifies that node is malicious, then the network will be divided into clusters. It has a cluster head for each cluster [13].

6   Challenges in Intrusion Detection

There are some of the challenges that are faced while deploying an intrusion detection system. These are discussed below.

Guaranteeing the EDSI Implementation
To make sure in obtaining a high level of visibility of threat, it is necessary for the organizations to confirm whether the EDSI is implemented by default and optimized correctly. A lot of plan is required in the design and also in the implementation phase.

Managing Large Volume of Alerts
Both types of detection system use a combination of digital signature and anomaly-based detection techniques. When there is a known attack pattern and sensor that either detects the activity with that, the alarm is generated. Many alerts are created by the EDSI which can disturb the multiple nodes in MANET.

Understanding and Investigating the Alerts

Examining EDSI cautions may be exceptionally time and resource concentrated, needing advantageous data from different frameworks to support and decide if an alert is high. Authority skills are fundamental to understand the security framework, and numerous associations come up to the security specialists for performing this crucial function [15, 16].

7   Conclusion

In ad hoc network, identifying the intrusion is a thrust area, where this paper focuses. It usually searches on the intrusion detection by analyzing the problem in the system and also prevents various attacks. In this paper, various types of detection system for intrusion are presented. Providing the security against dropping of packet, transmission modification is discussed. Thus, future research can be made on enhanced adaptive acknowledgment EAACK to overcome some of the flaws in the existing intrusion detection system.

References

1. Spurthi, K., Narayan Shankar, T.: A survey of intrusion detection system in manets using security algorithms. Int. J. Appl. Eng. Res. **12**(24), 14408–14414. ISSN 0973-4562 (2017)
2. Rubi, A., Vairachilai, S.: A survey on intrusion detection system in mobile Adhoc networks. IJCSMC, **2**(12), 389–393 (2013)
3. Kulkarni, A., Rewagad, P., Agarwal, M.: Literature survey on EDSI of MANET. Int. J. Sci. Res. Manag. (IJSRM) **3**(9), 3549–3552 (2015)
4. Nadeem, A., Howarth, M.: A survey of MANET intrusion detection and prevention approaches for network layer attacks. Artic. IEEE Commun. Surv. Tutor. (2013)
5. Bhosale, R. J. Ambekar, R.K.: ASurvey on intrusion detection system for mobile Ad-hoc networks. Int. J. Comput. Sci. Inf. Technol. (IJCSIT), **5**(6), 7330–7333 (2014)
6. Mandala, S., Ngadi, Md.A., Hanan Abdullah, A.: A survey on MANET intrusion detection. Int. J. Comput. Secur. **2**(1), 1–11
7. Meenatchi, I., Palanivel, K.: Intrusion detection system in MANETS: a survey. Int. J. Res. Dev. Eng. Technol. www.ijrdet.vom (ISSNb2347-6433) **3**(4) (2014)
8. Meenatchi, I., Palanivel, K.: Cross layer intrusion detection In MANET's: a survey. Int. J. Adv. Res. Commun. Eng. Technol. (IJARCET) **4**(3), 622–626 (2015)
9. Nithya Karthika, M., Raj Kumar: Survey on network based intrusion detection system in MANET. Int. J. Comput. Sci. Mob. Comput. **3**(4), 660–663 (2014)
10. https://www.researchgate.net/publication/316599266_INTRUSION_DETECTION_SYSTEM
11. Amiri, E., Keshavarz, H., Heidari, H., Mohamadi, E., Moradzadeh, H.: Intrusion detection systems in MANET: a review. ICIMTR 2013. www.sciencedirect.com
12. Anand babu, G.L., Sekhar Reddy, G., Agarwal, S.: Intrusion detection techniques in mobile Ad-hoc networks. Int. J. Comput. Sci. Inf. Technol. **3**(3):3867–3870 (2012)
13. Treesa Nice, P.A.: A survey on EDSI in mobile Adhoc network. Int. J. Adv. Res. Comput. Commun. Eng. **2**(5), 2287–2291 (2013)

14. Muraleedharan, M.: A case study on intrusion detection technique in MANET. Int. Hournal Recent. Trends Eng. Res. (IJRTER) **03**(12), (2017)
15. Mahajan, K., Singh, H.: MANET, its types, challenges, goals and approaches: a review. Int. J. Sci. Res. (IJSR) **5**(5), 1591–1594 (2016)
16. Intrusion Detection: Challenges and Myths by Marcus J. Ranum

Invasive Weed Optimization Algorithm for Prediction of Compression Index of Lime-Treated Expansive Clays

T. Vamsi Nagaraju, Ch. Durga Prasad and N. G. K. Murthy

Abstract With the recent emphasis on large-scale civil engineering constructions, artificial intelligence in the construction activities has received importance. Compressibility behavior is an important property in fine soils to find out the settlements in foundation designs. However, compression index (C_c) from one-dimensional swell-consolidation test is time consuming and laborious. Many traditional prediction-stimulated models rely on simplified assumptions, leading to inaccurate C_c estimations. This paper explores, by comparison, the application of invasive weed optimization (IWO) algorithm and particle swarm optimization (PSO) to predict C_c via multiple linear regression models. The predicted model equations have been developed, uses four input parameters namely plasticity index, free swell index, rate of heave and swell potential in both methods. The results confirm that the developed models using IWO provides accurate prediction than standard particle swarm optimization (PSO) algorithm.

Keywords Compression index · Expansive clays · IWO · PSO · Swelling

1 Introduction

Expansive soils are well-covered deposits in coastal areas of Andhra Pradesh. The problems identified in expansive clays are well-documented in worldwide, which posses alternate swelling and shrinkage with variation of moisture content [1]. The distresses occurring in the sub-soils allow settlements, which further effects on the light weight structures and pavements. In this view, there is a need of empirical correlations or predicted models to predict settlements and that will be the task of the geotechnical engineer to cope with them.

T. Vamsi Nagaraju (✉)
Department of Civil Engineering, S. R. K. R. Engineering College, Bhimavaram 534204, India
e-mail: tvnraju@srkrec.edu.in

Ch. Durga Prasad · N. G. K. Murthy
Department of Electrical and Electronics Engineering, S. R. K. R. Engineering College,
Bhimavaram, India

© Springer Nature Singapore Pte Ltd. 2020
K. N. Das et al. (eds.), *Soft Computing for Problem Solving*,
Advances in Intelligent Systems and Computing 1057,
https://doi.org/10.1007/978-981-15-0184-5_28

To calculate compressibility behavior of expansive clays, laboratory oedometer tests which depict one-dimensional swell-compressibility behavior needed to be performed on clay specimens [2]. However, laboratory testing using consolidation comprises unwieldiness and time consuming [3]. To counteract time-consuming testing, laborious tasks and complicated graphical procedures some empirical correlations were developed and vogue in practice [2, 4, 5]. The drawbacks of these correlations are the reliance of C_c on either one or two parameters only. In this regard, some research has been adopted new computational prediction models to predict the compression index, and a brief account of the results is presented below:

Farazin and Afshin [6] adapted artificial neural networks (ANN) for predicting compression index from the index properties such as liquid limit and plasticity index. Onyejekwe et al. [7] evaluated the empirical equations for the index of fine-grained soils; various indices were used in the assessment including the root mean square error, ranking distance and ranking index. Puri et al. [8] adopted machine-learning techniques for the prediction of shear strength parameters such as cohesion and angle of internal friction by using SPT N-value data. The developed corellations can be usefull in the site fillings for particularly in the state of Haryana.

Recent years, in civil engineering, from foundation soil analysis to super structure materials analysis, from load conditions of building to bearing capacities of soils, from actual designs to safety factors, from design assumptions to limitations for any structural element, artificial intelligence have found an emerging tool for civil engineers to make the work more fast and economical. Especially in geotechnical engineering, for prediction of soil parameters in various aspects of designs, soft computing tools like artificial neural networks, genetic algorithm, fuzzy logic and particle swarm optimization [9–12].

This paper investigates, by comparison, prediction of compression index (C_c) using particle swarm optimization (PSO) and invasive weed optimization algorithm (IWO). The predicted model equations has been developed, uses four input parameters namely plasticity index, free swell index, rate of heave and swell potential in both methods.

2   Data Base Compilation

Following the previous trend of studies carried out on lime expansive clay blends [3], in the present study for the estimation of compression index of the expansive clays, the plasticity and swell properties such as plasticity index (PI), free swell index (FSI), rate of heave (RH) and swell potential (S) were considered. The plasticity and one-dimensional swell-consolidation data were produced by the lime expansive clay blends with different percentages of lime content by dry weight of soil (Appendix). The samples were all collected using a standard procedure and tests were carried out using odeometer.

3 Multivariable Linear Equation Models for Estimation of Compressive Index

The data information of plasticity index (PI), free swell index (FSI), rate of heave (RH), swell potential in percentage (S) presented in Sect. 2 are utilized to estimate compression index value (C_c). From literature survey, linear regression models are utilized for estimating unknown parameters from the known parameters. In this paper, multivariable linear equations are designed for compressive index estimation instead of taking direct regression models. Each parameter (PI, FSI, RH and S) is multiplied by a coefficient and the total sum is taken as C_c. Hence, the multivariable linear equation is formulated which involving all available four variables as shown in Eq. (1) and taken as model A.

$$C_{c(est)} = a_1.PI + a_2.FSI + a_3.RH + a_4.S \tag{1}$$

where a_1, a_2, a_3 and a_4 are coefficients need to be estimated properly to calculate final output variable C_c. On the other hand, an additional constant is included in the Eq. (1) to convert it into multiple linear regression models and the modified equation is shown in (2) and which is taken as model B.

$$C_{c(est)} = a_0 + a_1.PI + a_2.FSI + a_3.RH + a_4.S \tag{2}$$

In Eq. (2), a_0 is the constant added for model A. Identification of such coefficients by regular regression tools is not satisfactory for typical studies. For estimation of such coefficients, artificial neural networks (ANN) type intelligent models were used extensively. However, ANN required large data to train and test the results. Hence, an attempt is made in this paper to use meta-heuristic optimization techniques for estimation of C_c with the help of minimum data availability and obtained better results. For the entire data samples, C_c is estimated and an error is evaluated from actual data of C_c. For evaluation of error, the following formula is used.

$$E(k) = \sum_{i=1}^{N} \left(C_{c(act)}(i, k) - C_{c(est)}(i, k) \right)^2 \tag{3}$$

In Eq. (3), k represents iteration number. N is total number of data samples, $C_{c(act)}$ is actual value of compression index and $C_{c(est)}$ estimated value of compression index. The error calculated from both estimated and actual values of C_c for an iteration 'k' is represented by E. From calculated error values, the objective function (J) is formed as shown in Eq. (4)

$$J = \min(E) \tag{4}$$

The above objective function depends on prediction of coefficients estimated by optimization techniques. Limits of these variables depend on the user requirement and/or data set availability. For this problem, the limits are fixed in between -100 and 100 obviously large solution regions.

4 Invasive Weed Optimization Algorithm (IWO)

In this paper, particle swarm optimization (PSO) and invasive weed optimization (IWO) are applied for estimating the compressive index in terms of other soil parameters. IWO is also nature-inspired MHOA, inspired by spreading strategy of weeds proposed in 2006 by Alireza Mehrabian and Caro Lucas [13]. This algorithm is implemented based on capturing the adaptive properties of weeds for environmental changes. The basic idea of implementation of IWO algorithm is weed ecology mainly involves reproduction, struggle for existence with competitors and avoidance of predators. Similar to PSO, finite number of random seeds (instead of particles) are initialized in the solution space and corresponding fitness functions are evaluated. The plant produced from each seed reproduces other seeds (reproduction) based on the function fitness value (objective function). Later, the generated seeds are randomly distributed in the solution space with statistical parameters of seeds, i.e. zero mean value and variable variance value. However, standard deviation (SD), σ, of the random function will be reduced from a previously defined initial value ($\sigma_{initial}$), to a final value (σ_{final}), in every iteration/generation as shown in Eq. (5).

$$\sigma_{iter} = \frac{(iter_{max} - iter)^n}{iter_{max}^n} . (\sigma_{initial} - \sigma_{final}) + \sigma_{final} \qquad (5)$$

In Eq. (5), n stands for non linear modulation index. 'iter' is the general notation used for iteration number. This process continue until maximum iterations is reached and hopefully the plant with best fitness it the closest to the optimal solution.

5 Results and Discussion

Data set considered for estimation of C_c includes plasticity and swelling characteristics such as PI, FSI, rate of heave and swell potential for the lime-blended expansive clays. For predicting C_c estimation, IWO and PSO methods were chosen. Equation (6) is the final model equation to predict C_c using IWO where as the Eq. (7) is using PSO. The relative comparison of both PSO and IWO predictions are presented in Fig. 1. Even IWO-based equation provides close prediction results over the entire samples than PSO, the relative deviation is considerable.

$$C_{c(est)} = 0.0178PI + 0.000673.FSI + 3.958.RH - 0.8194.S \qquad (6)$$

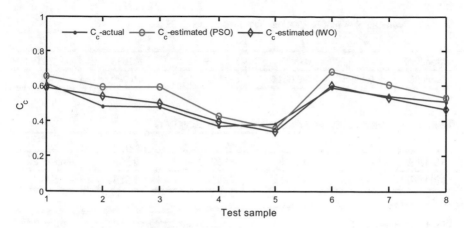

Fig. 1 Comparison of C_c for PSO and IWO estimated values with actual C_c for model A

$$C_{c(est)} = 0.0223\,\text{PI} + 0.000559.\text{FSI} + 8.7636.\text{RH} - 1.791.\text{S} \qquad (7)$$

As discussed in Sect. 3, the above models do not yield close prediction hence, model B is selected and the coefficients are obtained by using PSO and IWO. Equation (8) is the final model equation to predict C_c using IWO where as the Eq. (9) is using PSO. Using model B, IWO gives accurate prediction of C_c which is evident from Fig. 2.

$$C_{c(est)} = 0.5343 - 0.0033.\text{PI} + 0.0023.\text{FSI} + 7.9294.\text{RH} - 1.599.\text{S} \qquad (8)$$

$$C_{c(est)} = 0.8295 - 0.0114\text{PI} + 0.0026.\text{FSI} + 16.79.\text{RH} - 3.3717.\text{S} \qquad (9)$$

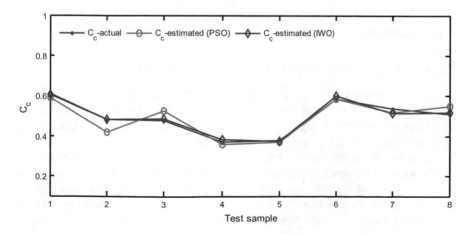

Fig. 2 Comparison of C_c for PSO and IWO estimated values with actual C_c for model B

Table 1 Coefficients of model equations identified by using PSO

	a_0	a_1	a_2	a_3	a_4
Model A	–	0.0223	5.591×10^{-4}	8.7636	−1.7910
Model B	0.8295	−0.0114	0.0026	16.7900	−3.3717

Table 2 Coefficients of model equations identified by using IWO

	a_0	a_1	a_2	a_3	a_4
Model A	–	0.0178	6.757×10^{-4}	3.958	−0.8194
Model B	0.5343	−0.0033	0.0023	7.9294	−1.599

Fig. 3 Comparison of overall errors in PSO and IWO

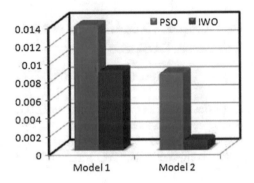

Together, coefficients for models A and B are obtained for both PSO and IWO are listed in Tables 1 and 2, respectively. For both models, the errors between actual data and predicted data are presented in Fig. 3.

6 Conclusions

The following conclusions can be drawn from the foregoing study:

1. The compression index (C_c) is one of the vital properties of expansive clays that are essential for geotechnical designs. One aim will be to give the engineer an integrated tool helping him in designing an efficient way without any time consuming, cumbersome and unwieldiness testing.
2. Regarding C_c prediction, four important parameters (PI, FSI, rate of heave and swell potential) were identified considering that the linear equations resulted from swarm-assisted multiple linear regression and invasive weed linear regression.
3. Two approaches (PSO and IWO) are used to help the geotechnical engineers to estimate the compression index (C_c) of blended expansive clays. The developed

equations are function of the PI, FSI, rate of heave and swell potential. It is evident that estimated C_c by using IWO obtained better convergent results compared to the estimated C_c by using PSO.

Appendix: Data of Plasticity and Swelling Characteristics of Lime-Blended Expansive Clays

Soil location	Properties	Lime content				
		0%	1%	2%	3%	4%
Amalapuram (AM)	Liquid limit (%)	83	77	75	73	71
	Plastic limit (%)	28	29	30.5	32	33
	Plasticity index (%)	55	48	44.5	41	38
	Free swell index (%)	189	128	88.4	72	63.6
	Rate of heave (mm)	4.01	2.88	2.81	2.76	2.75
	Swell potential	20	14.4	14	13.8	13.75
	Compression index	0.608	0.482	0.48	0.37	0.37
Bhimavaram (BM)	Liquid limit (%)	72	70	67.5	65	64.3
	Plastic limit (%)	29	31	32.3	34	35
	Plasticity index (%)	43	39	35.2	31	29.3
	Free swell index (%)	145	95	88.5	56	41
	Rate of heave (mm)	1.88	1.65	1.58	1.54	1.57
	Swell potential	9.4	8.25	7.9	7.7	7.85
	Compression index	0.587	0.54	0.51	0.48	0.33

References

1. Chen, F.H.: Foundations on Expansive Soils. Elsevier Scientific Publishing Co., Amsterdam (1988)
2. Terzaghi, K., Peck, R.B., Mesri, G.: Soil Mechanics in Engineering Practice. Wiley, New York (1967)
3. Phanikumar, B.R., Nagaraju, T.V.: Swell compressibility characteristics of expansive clay lumps and powders blended with GGBS—a comparison. Indian Geotech. J. 1–9, ISSN 2277-3347 (2018)
4. Azzouz, A.S., Krizek, R.J., Corotis, R.B.: Regression analysis of soil compressibility. Soils and Foundations. Jpn. Soc. Soil Mech. Found. Eng., 16(2), 19–29 (1976)
5. Bowles, J.E.: Physical and Geotechnical Properties of Soils. McGraw-Hill Companies, New York (1989)

6. Kalantary, F., Kordnaeij, A.: Prediction of compression index using artificial neural network. Sci. Res. Essays **7**(31), 2835–2848 (2012)
7. Onyejekwe, S., Kang, X., Ge, L.: Assessment of empirical equations for the compression index of fine grained soils in Missouri. Bull. Eng. Geol. Environ. (2014). https://doi.org/10.1007/s10064-014-0659-8
8. Puri, N., Prasad, H.D., Jain, A.: Prediction of geotechnical parameters using meachine learning techniques. Procedia Comput. Sci. **125**, 509–517 (2018)
9. Jyoti, S.T., Sandeep, N., Chakradhar, I.: Optimum utilization of fly ash for stabilization of sub grade soil using Genetic algorithm. Procedia Eng. **51**, 250–258 (2013)
10. Taskiran, T.: Prediction of California bearing ratio (CBR) of fine grained soils by AI methods. Adv. Eng. Softw. **41**, 886–892 (2010)
11. Yilmaz, I., Kaynar, O.: Multiple regression, ANN (RBF, MLP) and ANFIS models for prediction of swell potential of clayey soils. Expert Syst. Appl. **38**(5), 5958–5966 (2011)
12. Zhu, A.X., Qi, F., Moore, A., Burt, J.E.: Prediction of soil properties using fuzzy membership values. Geoderma **158**(3–4), 199–206 (2010)
13. Mehrabian, A.R., Lucas, C.: A novel numerical optimization algorithm inspired from weed colonization. Ecol. Inform. **1**(4), 355–366 (2006)

Collaborative Filtering for Book Recommendation System

Gautam Ramakrishnan, V. Saicharan, K. Chandrasekaran, M. V. Rathnamma and V. Venkata Ramana

Abstract Collaborative filtering is one of the most important techniques in the market nowadays. It is prevalent in almost every aspect of the internet, in e-commerce, music, books, social media, advertising, etc., as it greatly grasps the needs of the user and provides a comfortable platform for the user to find what they like without searching. This method has a few drawbacks; one of them being, it is based only on the explicit feedback given by the user in the form of a rating. The real needs of a user are also demonstrated by various implicit indicators such as views, read later lists, etc. This paper proposes and compares various techniques to include implicit feedback into the recommendation system. The paper attempts to assign explicit ratings to users depending on the implicit feedback given by users to specific books using various algorithms and thus, increasing the number of entries available in the table.

Keywords Recommender systems · Book recommendation · Collaborative filtering · Implicit feedback · Explicit ratings

G. Ramakrishnan (✉) · V. Saicharan · K. Chandrasekaran
National Institute of Technology, Surathkal, Karnataka, India
e-mail: 16co219.gautam@nitk.edu.in

V. Saicharan
e-mail: vsaicharan1998@gmail.com

K. Chandrasekaran
e-mail: kchnitk@ieee.org

M. V. Rathnamma
Kandula Srinivasa Reddy Memorial College of Engineering, Kadapa,
Andhra Pradesh, India
e-mail: rathnamma@ksrmce.ac.in

V. V. Ramana
Chaitanya Bharathi Institute of Technology, Proddatur, Andhra Pradesh, India
e-mail: ramanajntusvu@gmail.com

© Springer Nature Singapore Pte Ltd. 2020   325
K. N. Das et al. (eds.), *Soft Computing for Problem Solving*,
Advances in Intelligent Systems and Computing 1057,
https://doi.org/10.1007/978-981-15-0184-5_29

1 Introduction

Recommender systems are type of intelligent systems which can analyze past user behavior on items or service to make personalized recommendations on items to users. The two most widely used methods in recommendation systems are content-based algorithms and collaborative filtering memory-based methods. These are a set of methods which use the entire database (user-item database) to make a recommendation to a user. These methods can also be classified into two types: Item-based collaborative filtering: In this particular method, two items are compared and given a similarity score between the two using some sort of metrics like the cosine distance, Pearsons correlation, or Jaccard distance. User-based collaborative filtering: In this method, two users are compared for similarity and assigned a similarity score between them. The same metrics as mentioned above are used here. Using the similarity scores between the users, active recommendations are made for a user. Content-based filtering methods work by constructing a model to map a user to a set of items using either explicit features like ratings or implicit features like search history or number of page views. They compare user preferences to item description to make their predictions. Recommendation systems are very crucial for a company to be successful. Netflix claims their recommender system saves them $1bn every year [1]. In this paper, we discuss one particular problem of recommender systems in general. The problem discussed is the lack of inclusion of implicit data in recommendation systems. Elaborating, there exist two types of data on which similarity can be calculated and recommendations can be provided. The first type is explicit data, which consists of data which user explicitly provides, in the form of ratings, or small surveys. This data is generally well understood as it provides a concrete understanding of user interests. The other type of data is implicit ratings. This is the type of data which the user implicitly provides, through their actions on the platform, including but not restricted to page views, adding to wish lists, commenting on an item, etc. This type of data is generally vague and less understood, as the user does not provide a very clear indication of their preferences for that particular item. One of the significant issues faced by recommender systems is the inclusion of implicit ratings into the recommendation. Implicit feedback is a very dominant metric in determining recommendations as the innate requirements of the user are expressed very freely in their online behavior. However, due to the very vague and misunderstood nature of implicit data, it becomes a very tedious task to implement recommendation using implicit data. Also, the likelihood of the user rating an object is minimal compared to the data received from studying browsing patterns of the users.

Also, the recent trend is that many major organizations have integrated implicit ratings into their recommendation systems. For example, YouTube recommends videos by checking which videos are generally viewed by users in the same viewing session [2]. This is classified as implicit feedback as watching a video does not involve a user giving a rating. Also, most music streaming applications recommend music based on what music the user frequently listens to, rather than expecting a user to assign a rating for a song. In this paper, we discuss content-based collaborative filtering for a book recommendation system and propose solutions to improve collaborative, item–item based filtering by using implicit data. The implicit data in this case is the list of books each user has marked for future reading, but has not rated explicitly. The same method can be extended to other implicit data.

This paper is divided into five main sections. Section 2 deals with the past work that has been carried out in the domain of recommender systems and implicit feedback. Section 3 discusses the implementation of the book recommender system and various approaches for including implicit feedback. Section 4 elaborates on all the algorithms implemented. Section 5 discusses the outcome of each method and compares the efficiency of various methods stated. Section 6 concludes the paper, followed by future work.

2   Related Works

The related works described here are the two collaborative filtering methods, memory, and model-based. Similarity metrics, evaluation of the system and implicit feedback are also discussed. The two filtering methods form the core part of recommender systems and are described below, along with their submethods.

2.1   Collaborative Filtering

Collaborative-based filtering methods predict items for a particular user using data of other similar users or items. These methods are very popular on the web today and are used by E-commerce giants like Amazon, eBay, Flipkart, and others. They are also used in the domain of social media to recommend other users who are similar and also used to recommend news articles. Spotify, the music application uses this particular technique to recommend personalized music tracks and albums to its customers [3]. It has enjoyed great popularity and success mainly due to its simplicity in implementation and effectiveness. Collaborative filtering can be performed in two ways mainly, model-based and memory-based.

2.2   Memory Based

In memory-based collaborative filtering, the whole dataset is used to make a recommendation.

2.2.1 User Based

In user-based filtering, similarity between users is found based on how similar their ratings are for the books they have rated. It suffers from serious drawbacks of scalability and sparsity [4]. With a large number of users ranging in the order of millions, it is difficult to account for new users to join. Also, even the most active user would have rated less than 1% of the items in general, which makes it difficult for the nearest neighbors algorithm to make a decision.

2.2.2 Item Based

In item-based filtering, the similarity is found between every pair of items, by matching how many common users have rated the same pair of books. It was initially introduced by Amazon in 1998, due to its distinct advantages over user-based collaborative filtering as the matrix size for item similarity is generally smaller and the items are more stable compared to users [5].

2.3 Model Based

This method does not use the complete dataset for generating recommendations. Instead, this technique works by using a user-item database as training data to learn a model, which might be simple or complex depending on the learning algorithm. The model then makes recommendations on the test data. The model can be evaluated by testing its accuracy on real-world data. There exist several algorithms; for such methods, some of them include singular value decomposition, clustering, and Bayesian networks [6].

2.4 Similarity Metrics

Similarity metrics are various methods used to assign similarity to a pair of users or items. The most commonly used traditional metrics are cosine, Pearson's coefficient, Spearman's correlation, Jaccard index, Euclidean, mean squared difference, and adjusted cosine. Appropriate choice of similarity metrics is crucial as the quality of recommendation is directly related to the similarity values generated [4].

$$cosine\_sim(u, v) = \frac{\sum_{i \in I_{uv}} r_{ui} \cdot r_{vi}}{\sqrt{\sum_{i \in I_{uv}} r_{ui}^2} \cdot \sqrt{\sum_{i \in I_{uv}} r_{vi}^2}} \tag{1}$$

$$pearson\_sim(u, v) = \frac{\sum_{i \in I_{uv}} (r_{ui} - \mu_u) \cdot (r_{vi} - \mu_v)}{\sqrt{\sum_{i \in I_{uv}} (r_{ui} - \mu_u)^2} \cdot \sqrt{\sum_{i \in I_{uv}} (r_{vi} - \mu_v)^2}} \qquad (2)$$

Equations 1 and 2 illustrate the formula to calculate cosine similarity and Pearson's similarity, respectively. Here, I_{uv} refers to all books rated by both users u and v, r_{ui} refers to rating assigned to book i by user u, r_{vi} refers to rating assigned to book i by user v, μ_u and μ_v are the average ratings assigned by users' u and v, respectively on all books.

2.5 Evaluation of Recommender Systems

It is highly essential that the quality of recommendation is constantly tested and made sure that the recommendations are as accurate as possible. There are various metrics that measure the quality/accuracy of prediction. Some of them are root mean square error, mean absolute error, Z-score, coverage, and normalized mean average error [7].

$$Mean\,Average\,Error = \frac{1}{\left|\hat{R}\right|} \sum_{r_{ui} \in \hat{R}} \left| r_{ui} - \hat{r_{ui}} \right| \qquad (3)$$

$$Root\,Mean\,Square\,Error = \sqrt{\frac{1}{\left|\hat{R}\right|} \sum_{r_{ui} \in \hat{R}} (r_{ui} - \hat{r_{ui}})^2} \qquad (4)$$

Equations 3 and 4 illustrate the calculation of mean average error (MAE) and root mean square error (RMSE). Here, $\hat{R}$ refers to all the predictions in ratings and $\hat{r_{ui}}$ refers to prediction in rating by the recommender system on the rating given by user u to book i.

2.6 Implicit Feedback

Implicit feedback is the process of collecting data passively, instead of letting users give quantitative feedback in the form of a rating, score, etc. Implicit feedback may involve page views, user clicks, bookmarks, etc. A substantial amount of work has been carried in this domain. Hu et al. [8] explore the idea of combining implicit data along with explicit ratings to make better recommender systems. This paper proposes to use implicit feedback to associate it with positive or negative value with varying confidence levels. The paper proposes a model similar to the matrix factorization method to include implicit feedback.

Liu et al. [9] combine explicit feedback model and the implicit feedback model using a factorized neighborhood model. This is an extension of the traditional nearest item-based model in which the item–item similarity matrix is approximated via low-rank factorization.

Our idea is to use implicit feedback in an item–item based collaborative filtering model. We have implemented, analyzed, compared various methods and evaluated the performance of our method.

3   Implicit Rating in Book Recommendation System

According to our literature survey, the availability of recommender systems for books was minimal. In this paper, we have attempted to build a recommendation system for books using item-based collaborative filtering methods, using K-nearest neighbors algorithm to pick the recommendation. We also aim to improve the accuracy of these generic methods through our novel approaches. For this purpose, we have used the GoodBooks-10k dataset. It has data for 10,000 books and 53,424 users. It has ratings numbered 1-5, 1 implying not very interested, with 5 implying great interest in the book. It also has a separate set of data for users who have added books to their wish list. As stated before, just like the lack of book recommenders, there is a shortage of good dataset for books, but the GoodBooks-10k dataset seemed to be extremely apt.

3.1   Cleaning the Dataset

The GoodBooks-10k dataset was obtained from Kaggle. The dataset was for most of the part without any noise. There were a total of 980,000 ratings. The first step was to remove duplicate ratings. A pair of book ID and user ID with the same value was considered to be a duplicate. 4487 duplicate ratings were found. The second step involved removing all users who have rated lesser than four books. This was done because users with very few explicit ratings contribute very less to evaluation and add to the unnecessary size of data. 8364 such users were found and were removed from the dataset. The dataset was further analyzed to observe certain trends.

The distribution of ratings was plotted in Fig. 1. Most of the ratings ranged from 3 to 5, and very few ratings were found to be in the range 1-2. The number of ratings per user was also plotted in Fig. 2. The observation obtained from the dataset was that very few users rated more than ten books. This chart gives a good idea of users who are many raters and who are not, which might be useful while considering feedback.

The number of ratings per book was plotted in Fig. 3. From this data, we can conclude that most of the books in the dataset have around 18-20 ratings(rated by users) on average. Very few books have less than 1 or more than 30 ratings.

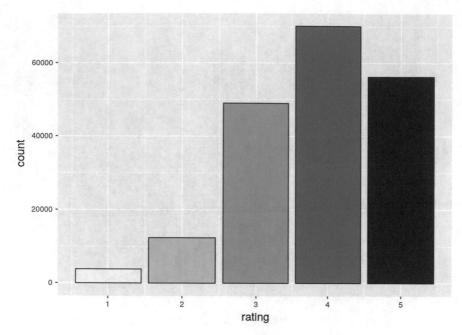

Fig. 1 Distribution of ratings

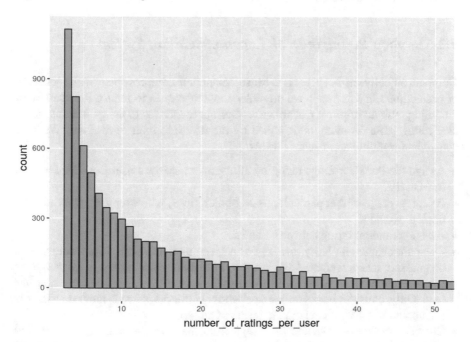

Fig. 2 Number of ratings per user

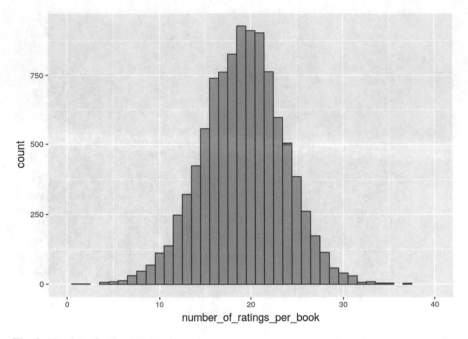

Fig. 3 Number of ratings per book

3.2 Implicit Rating in Book Recommendation System

Our main objective in this paper is to cause a meaningful improvement in the accuracy of predictions. In this paper, we present a few techniques to reduce the sparsity of the rating table to improve performance. Our approach was to assign a rating in the user rating table for every book added by the user into their to-read list. We have considered multiple approaches for this.

- Assign the book's average rating by all users' wherever a user has added a book to wish list.
- Assign average of that particular users book ratings, wherever a user has added a book to wish list.
- Assign a random rating between 1 and 5.
- Compare cosine similarity matrix. For a given user and book whose rating is to be guessed, take a weighted average of all the users book ratings, with weight as the cosine similarity of the book to be guessed and the other book the user has read. Only, consider books on average whose similarity value is more than 0.7 (as it amounts to a 45° of deviation in terms of angle).
- Consider the users deviation in ratings from the average trend. If the standard deviation is less than a threshold, assign the average rating of that book as the rating.

We constructed a book recommender using collaborative item–item filtering. For this purpose, we used an open-source recommendation system package for Python known as surprise [10]. It is currently in development by Nicolas Hug. Surprise has the standard collaborative filtering algorithms inbuilt. We took up the process of adding our novel approach to reduce the RMSE value of the pure ITEM-based filtering. The other frameworks we used were Pandas; for handling large datasets, SciKit learns for running the K-nearest neighbors algorithm and NumPy for its data structure capabilities. Our choice of programming language was Python. We chose these as it provided a convenient way to run and test our work effortlessly.

Our approach was inspired by the fact that the extrapolation of results would highly depend on the average ratings a user has given to a book and average book rating in general. We wished to combine these two factors along with the item–item similarities. The description of the method is as follows:

For a given user U and a book B, the user U has added to their wish list the following.

Find the weighted average of the ratings of the other books U has rated, considering only books in the average whose similarity with book B is greater than or equal to $1/\sqrt{2}$. We considered that value as $1/\sqrt{2}$ is equal to arcos(45) A 45 deviation of similarity seemed ideal to consider, anything about that is closer to being dissimilar than similar.

$$\hat{r_{ui}} = \frac{\sum_{v \in N_i^k(u)} sim_{u,v} \cdot (r_{vi} - \mu_v)}{\sum_{v \in N_i^k(u)} sim(u, v)} \tag{5}$$

where $\hat{r_{ui}}$ is the predicted rating that user U would give item I and sim(u, v) is the similarity of user u with user v and $r_{u,i}$ is the rating assigned to item i by user U.

But then, this method alone, even though takes into account a lot of factors for making the prediction, does not fill in a lot of values because if the user added a book out of his interest zone to the wish list, it will not assign a rating. Hence, we added a second part to this method. If there was no book with the similarity of more than, average book rating could have been used. But using average book rating directly may not be very accurate. So, to increase the strictness while picking, we calculated the standard deviation of the user book ratings with the average book rating. If the standard deviation was less than 1, we came to the conclusion that the users ratings are generally on par with overall average ratings. In that case, we found it appropriate to assign the average book rating directly.

This constitutes our method to increase the density of rating matrix, which at the same time, takes into account various factors and trends in user ratings and book similarities while making a meaningful and personalized choice to improve recommendation accuracy.

4   Experiment Conducted

Normal Item-based collaborative filtering: We ran the standard item-based collaborative filtering algorithm along with the k-nearest neighbors algorithm to give recommendations. We tabulated the RMSE value obtained. The similarity metric used in this case was cosine distance, and for prediction, we used the K-nearest neighbors approach.

4.1   Assigning Random Ratings

We assigned a random rating to the books which users have added to their to-read list. We then conducted the standard item-based collaborative filtering algorithm on the new dataset which includes the books implicitly rated by the user and the random rating assigned to them.

This method could significantly increase the density of the rating matrix. Hence, we tried this method. The disadvantage of this method is that it assigns values with absolutely no correlation to any of either user or book similarities. However, it is a good way to test how the density of the rating matrix affects recommendation and also checks the extent to which alteration in the RMSE value occurs, which will provide a baseline for comparing RMSE values.

4.2   Assigning Average Book Rating

For every book users have added to their read list, we added the books average rating in the rating table. This is a better indicator than a random rating, but then it really does not take into account the real interest of the user. This could be thought of like a popularity-based metric where a user is assigned a rating which is globally calculated considering all other users. It still picks a value based on a well-known parameter, which represents the general interest of the users. The RMSE value was tabulated.

4.3   Assigning Average of User's Book Rating

For every book users have added to their to-read list, we added the average of the users book ratings to the rating table. For example, if a user rated three books and gave them a rating of 3, 4, and 5, we assigned a value $(3 + 4 + 5)/3 = 4$. This method considers the users general ratings, but does not take into account the affinity of the given book and the user. It represents the users interests better, but fails if the user selects a book, not in their interest zone. The RMSE value was tabulated.

4.4 Assigning Values Based on Our Approach

Finally, we used the approach we formulated to assign ratings. This method takes into account all intricacies of user interest, trends in user ratings, and general opinions. The drawback this method faces is that it does not assign a rating for every book in the wish list, as it cares for the multiple other parameters and does not take a wild guess at a rating. The RMSE value was tabulated.

Algorithm 1 Assign implicit rating to book

---

procedure ASSIGNRATING(U , B) ▷ User U and book B
    for book A in user U's rated books **do**
      if similarity(A , B) > 0.7 **then**
          $score$ += similarity(A , B)*Rating(A)
          $count1$ += similarity(A , B)
      end if
      $distance = (AverageRating(A) - UserRating(A))^2$
      $count2+ = 1$
    end for
    if $count1 > 0$ **then**
        $Rating(B) = score/count1$
    end if
      else
    if $distance/count2 <= 1.0$ **then**
        $Rating(B) = AverageRating(B)$
    end if
end procedure

---

5 Observations and Analysis

We plot the RMSE values and we got with each of the results.

Each of the methods discussed above was tested three times to make sure a good value of RMSE was obtained. This was also done to eliminate any noise or randomness present in data (Fig. 4).

Even though the number of predictions made in our approach was just 247,038 entries, it still managed to fare very well. There was a great decrease in RMSE values while using book average and the users' book average. However, the number of predictions made in these methods was 912,705. Our understanding from this experiment was that the density of ratings in the matrix might have been a more dominant factor in reducing the RMSE values, over the reliability of the method. The random assignment fared very poorly, making it an undesirable action, by drastically increasing the RMSE values. It does show that increasing the density of the matrix may lead to such a large change in RMSE values, but in this case, it is for the worse.

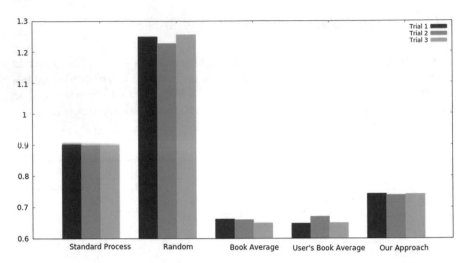

Fig. 4 Various algorithms and their RMSE values

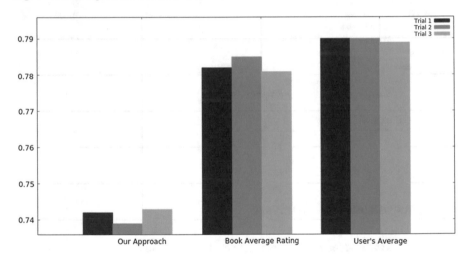

Fig. 5 Algorithms and their RMSE values

Our algorithm fares well compared given the fact that it assigned ratings only for about 30% of the total dataset.

For a better understanding and a more fair comparison of our algorithm with the book average and users rating average, we conducted the same method for a subset of 300,000 new entries in the case of book average and users rating average.

We chose to compare with 300,000 new entries as it is close to but still higher than 247,038 entries. The results are plotted in Fig. 5.

To compare the efficiency of our algorithm, we made the average user rating and average book rating method make predictions for only 300,000 entries and compared it with the results of our prediction algorithm. Our method fared much better in this case.

Thus, it implies that our assumption that the density of the matrix was more dominant compared to the efficiency of the assignment of values was correct.

6   Conclusion and Future Works

In this paper, we demonstrated a novel and innovative approach to make use of implicit feedback for a book recommendation system. In conclusion, the inclusion of implicit ratings proved to be a good improvement over the normal method of considering only explicit ratings. We also plan to improve upon the same method in future works by including other kinds of implicit data which will give a complete profile of the user. As the usage of recommendation systems becomes increasingly prevalent, it is critical to building a recommendation engine to improve the user experience. At the same time, there is also a thin line between collecting implicit feedback and respecting the privacy of the user. Implicit feedback collected must not exploit the information provided by the user in any manner.

In the future, we aim to improve upon the work discussed in this paper. Some possible extensions could be to use a user–user similarity metric. In our method, we only used item–item similarity. We could also try the same method on a different dataset, which gives a wider variety of implicit feedback. The adjustment will have to be made as multiple implicit feedback is present.

References

1. Gomez-Uribe, C.A., Hunt, N.: The Netflix recommender system: algorithms, business value, and innovation. ACM Trans. Manage. Inf. Syst. **6**(4), Article 13, 19 p. (2015). https://doi.org/10.1145/2843948
2. Davidson, J., Livingston, B., Sampath, D., Liebald, B., Liu, J., Nandy, P., Vleet, T.V., Gargi, U., Gupta, S., He, Y., Lambert, M.: The YouTube video recommendation system, Proceedings of the fourth ACM conference on Recommender systems - RecSys 10 (2010)
3. Jacobson, K., Murali, V., Newett, E., Whitman, B., Yon, R.: Music personalization at Spotify. In: Proceedings of the 10th ACM Conference on Recommender Systems (RecSys '16). ACM, New York, NY, USA, pp. 373–373 (2016). https://doi.org/10.1145/2959100.2959120
4. Sarwar, B., Karypis, G., Konstan, J., Reidl, J.: Item-based collaborative filtering recommendation algorithms, Proceedings of the tenth international conference on World Wide Web - WWW 01 (2001)
5. Linden, G., Smith, B., York, J.: Amazon.com recommendations: item-to-item collaborative filtering. IEEE Internet Comput. **7**(1), 7680 (2003)
6. Bobadilla, J., Ortega, F., Hernando, A., Gutirrez, A.: Recommender systems survey. Knowl.-Based Syst. **46**, 109132 (2013)
7. Karypis, G.: Evaluation of Item-Based Top-N Recommendation Algorithms. In Proceedings of the tenth international conference on Information and knowledge management (CIKM '01). In: Paques, H., Liu, L., Grossman, D.: (eds.) ACM, New York, NY, USA, pp. 247–254 (2001)

8. Hu, Y., Koren, Y., Volinsky, C.: Collaborative Filtering for Implicit Feedback Datasets. In: 2008 Eighth IEEE International Conference on Data Mining, Pisa, pp. 263–272 (2008). https://doi.org/10.1109/ICDM.2008.22

9. Liu, N.N., Xiang, E.W., Zhao, M., Yang, Q.: Unifying explicit and implicit feedback for collaborative filtering. In: Proceedings of the 19th ACM International Conference on Information and Knowledge Management (CIKM '10). ACM, New York, NY, USA, pp. 1445–1448 (2010). https://doi.org/10.1145/1871437.1871643

10. https://github.com/NicolasHug/Surprise

11. Choi, K., Yoo, D., Kim, G., Suh, Y.: A hybrid online-product recommendation system: combining implicit rating-based collaborative filtering and sequential pattern analysis. Electron. Commer. Res. Appl. 11(4), 309317 (2012)

12. Jain, A., Vishwakarma, S.K.: Collaborative filtering for movie recommendation using Rapid-Miner. Int. J. Comput. Appl., 169(6), 2933 (2017)

13. Lee, S.K., Cho, Y.H., Kim, S.H.: Collaborative filtering with ordinal scale-based implicit ratings for mobile music recommendations. Inf. Sci. 180(11), 21422155 (2010)

14. Gomez-Uribe, C.A., Hunt, N.: The Netflix recommender system: algorithms, business value, and innovation. ACM Trans. Manage. Inf. Syst. 6(4), Article 13, 19 p. (2015). https://doi.org/10.1145/2843948

15. Casino, F., Patsakis, C., Puig, D., Solanas, A.: On privacy preserving collaborative filtering: current trends, open problems, and new issues. In: 2013 IEEE 10th International Conference on e-Business Engineering, Coventry, pp. 244–249 (2013). https://doi.org/10.1109/ICEBE.2013.37

16. Kim, Heung-Nam, Ji, Ae-Ttie, Ha, Inay, Jo, Geun-Sik: Collaborative filtering based on collaborative tagging for enhancing the quality of recommendation. Electron. Commer. Res. Appl. 9(1), 73–83 (2010)

17. Herlocker, J.L., Konstan, J.A.: Borchers, A., Riedl, J.: An algorithmic framework for performing collaborative filtering. In: Proceedings of the 22nd Annual International ACM SIGIR Conference on Research and Development in Information (1999)

18. Casino, F., Patsakis, C., Puig, D., Solanas, A.: On privacy preserving collaborative filtering: current trends, open problems, and new issues. In: 2013 IEEE 10th International Conference on e-Business Engineering, Coventry, pp. 244–249 (2013)

19. Balabanovi, M., Shoham, Y.: Fab: content-based, collaborative recommendation. Commun. ACM 40(3), 66–72. (1997). https://doi.org/10.1145/245108.245124

Design of an Above Knee Low-Cost Powered Prosthetic Leg Using Electromyography and Machine Learning

Cyril Joe Baby, Ketan Jitendra Das and P. Venugopal

Abstract Electronic knee prosthesis provides a wide range of mobility when compared to mechanical prosthesis. Powered prosthetics are quite expensive and hence are not widely used. In this paper, we have proposed a low-cost above knee powered prosthetic leg which is reliable and the complexity of the model is less. Theprosthesis works by taking inputs from sensor placed in the model and an EMG sensor that records muscle activity of the thigh of the amputee. In order to calculate joint motion and knee angle a variety of methods can be used which include goniometer, inertial measurement units and magnetic encoders. Based on these sensor values, the actuation of the joint is determined. In this paper, we are discussing about an approach that uses sensor data along with muscle activity for actuation. The model classifies the current phase of walking based on the EMG sensor and angle values obtained from the leg. Once the gait phase is determined, the next gait phase is initiated. This enables the prosthesis to be reliable and efficient at the same time being cost effective.

Keywords Prosthesis · EMG · IMUs · Piezo · Random forest classifier · SVM

1 Introduction

According to a report released in March, 1983 by National Sample Survey Organization (NSSO), 1.8% of the population were physically disabled. Locomotive disability accounted for the majority (40%), out of those 8% was because of amputees, which is approximately 424,000(0.62 per 1000).

However, according to a recent survey by NSSO released on 2002, the number of disabled people in India has increased from 1.8 to 2%. Locomotor disability still remains to be the major cause of disability in the country, accounting to 52% in rural areas and 55% in urban India. A traditional artificial limb made in India costs around

C. J. Baby · K. J. Das · P. Venugopal (✉)
School of Electronics and Communications Engineering, VIT University, Vellore 632014, India
e-mail: venugopal.p@vit.ac.in

© Springer Nature Singapore Pte Ltd. 2020
K. N. Das et al. (eds.), *Soft Computing for Problem Solving*,
Advances in Intelligent Systems and Computing 1057,
https://doi.org/10.1007/978-981-15-0184-5_30

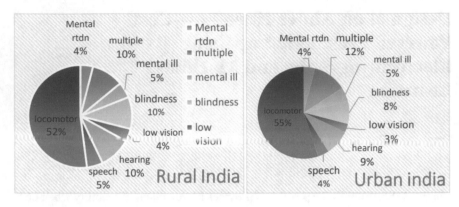

Fig. 1 Survey of urban and rural disabilities

Rs. 6000–8000, however, an imported device may cost up to 1 Lakh Rs. The low-cost devices provide less mobility, the costlier devices provide accurate gait, better ability to adapt to walking speeds, terrain surface. This paper is aimed at making those features available at an affordable price (Fig. 1).

For achieving wide range of mobility, feedback sensors are used in electronic knees, which send real-time data to a microcontroller [1]. In order to calculate joint motion and knee angle, a variety of methods can be used which include goniometer, inertial measurement units and magnetic encoders. The primary objective of this paper is to develop an adaptive control system to facilitate better and smooth movement. In the adaptive control system, a potentiometer is used to measure the knee angle, a piezo sensor, which is the feedback sensor, is placed on bottom of the feet to get information about contact with the ground. This adaptive feedback can be used instead of currently in-use passive feedback systems. While most of the prosthetic knee will use knee angle as feedback, EMG sensor along with the angle sensor will give more accurate values for the feedback [4].

Results show that the overall cost of modern electronic knee can be reduced by use of low-cost sensing methods and better algorithms.

2   Human Gait

A single stride of human gait cycle can be seen in Fig. 2. The basic model of walking comprises of two legs moving in median plane (anteroposterior) or sagittal plane alone. Gait cycle can be divided into two phases: stance and swing phase. In "stance" phase both the lower extremities are in contact with ground and in "swing" phase only one of the extremities is in contact with the ground.

Stance phase contributes to about 60% of the entire gait phase and is also referred to as "period of double support." The center of mass moves in an arc during the gait cycle and is at its highest at "mid stance" and lowest at "heel strike" and "toe-off."

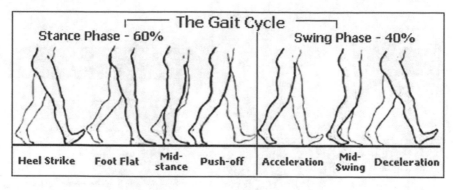

Fig. 2 Human gait

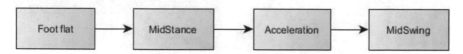

Fig. 3 Gait phases of walking pertaining to a knee-actuated prosthesis

The midstance consists of heal strike, foot flat, midstance and toe-off. The mid-swing phase consists of acceleration and mid-swing. Hence, a walking cycle can be classified into heel strike, foot flat, midstance, toe-off, acceleration and mid-swing. For a prosthesis with only actuation at the knee, the difference of motion between heel strike and toe-off is not considered as the ankle joint is torsional or mechanical. For a knee-actuated prosthesis, the phases are foot flat, midstance, acceleration and mid swing (Fig. 3).

3 Proposed Methodology

Proposed framework is shown in Fig. 4. In our framework, we use a microcontroller that takes input from all the sensors and determines the current gait phase and initiates actuation for the next gait phase. After actuation, a feedback loop is done to correct any error.

The working of electronic knee can be broadly classified into hardware, software and the microcontroller. Hardware components include all the sensing units and a linear actuator. The software part uses machine -learning algorithm, and various other algorithms for filtering the EMG. All the software processing is done in real time by the brain of the system, the microcontroller. The key component in any feedback systems are sensors. Sensors placed on knee, hip and feet send feedback signals which helps in achieving a smoother gait. The number of sensors and the algorithm determine the complexity of the system.

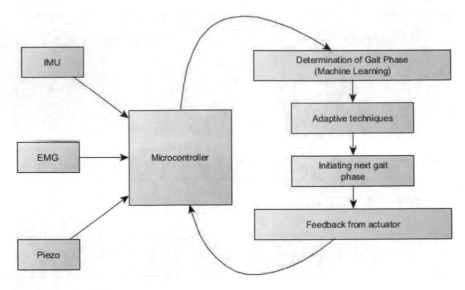

Fig. 4 Proposed framework

3.1 Angle Sensor (IMU)

The knee angle, i.e., the angle between two links can be determined by using an IMU placed on the pivot of joint [6, 7]. Indirectly, angle can be measured using accelerometer, which measures the tilt of the link and determines the angular movement of the joint.

3.2 Pressure Sensor

Pressure sensor or contact sensor is placed on the feet to determine any contact during the gait cycle. Whenever there is a contact, the sensor sends an interrupt to the microcontroller. Based on the time interval between the interrupts, the movement speed can be determined.

3.3 Actuator

Knee movement can be carried out using actuators of various kinds like linear actuator, hydraulic, pneumatic, etc. The cost of the prosthetic leg varies greatly according to the sensors and actuators used. Hydraulic actuators are widely preferred because of the safety and stability related advantages. The better the actuator, more natural is the gait. Mobility and the life time of the leg depend on the life of actuator. Research

Fig. 5 Actuator design

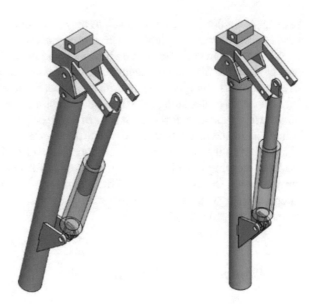

is being carried out to increase the reliability of actuators. The actuator output unit is controlled by the microcontroller, which constantly sends signals based on the current gait phase and walking speed.

3.4 Design

The design of the prosthesis is a simple crank-slider joint where the movement of the stroke of the actuator causes knee rotation. The IMU is placed on the knee for knee angle. The linear motion of the actuator is converted to rotational motion at the knee joint (Fig. 5).

3.5 EMG Sensor

Every time a muscle contracts, an electrical tension is generated, which is used as information in a technique called Electromyography. Electrodes applied on the skin to control the movements of the prosthesis, capture the tension from voluntary contracted muscles. EMG sensors are kept on the two main pairs or muscles on the thigh. The values from both the channels of the EMG along with the angle values are used to determine the current gait phase and hence initiate the next gait phase.

Every hardware requires a backbone to support its functioning, the software. Commercially available prosthetic legs use various kinds of algorithms to determine the knee angle, which acts as a feedback to the microcontroller. The angle is measured based on the change in potentiometer values, corresponding to the movement of the knee. To get more accuracy, more sensors are required, however, to make prosthetic legs more economic, a single senor and a simpler algorithm is used. But with the use of advanced machine-learning algorithms, higher feedback accuracy can be obtained [8]. The EMG values along with the angle values are used to classify the current gait phase using a random forest classifier. In this technique, ideal walking data is collected and stored in a database. Ideal data is collected by recording the sensor values obtained from the walking pattern of a normal person. Using this ideal database, a machine-learning model is trained to determine the next angle values based on the current value. Due to the ability of the algorithm to predict the next movement, the microcontroller can pass on instructions to the actuator in advance, enabling a much smoother and efficient gait.

3.6 Random Forest Classifier

Random forests also known as random decision forest is an arrangement of decision-tree predictors that ensemble a supervised machine-learning algorithm best suited for classification problems. By principle, more the number of trees in the forest, more robust and accurate the prediction results [11].

A decision tree consists of a root node subdivided into decision nodes, each branching out further and eventually ending at leaf nodes. Each leaf node corresponds to a class label or one among the possible outputs to be predicted. The branching out or division into further nodes is decided by considering the attributes to be measured and by selecting the right ones in the first place. The attributes selection is done by one of the following two measures: Information gain and Gini index [12].

The Gini index a prudential measurement parameter. That is, it measures how often a randomly chosen element would be incorrectly predicted or identified [13]. Information gain is done by obtaining the estimate of the information held by each attribute. This is done by calculating the entropy measure of each attribute. After finding the entropy measures the decision tree is drawn. An attribute with entropy higher value than other should be positioned as "root node" and the branch with entropy 0 should be considered as "leaf node." Positive entropy branches will have further branching. The distinction offered by the random forest classifier will not overfit the data, especially when there are more numbers of decision trees.

Random forest algorithm:

- Randomly identify "k" attributes for decision making
- Calculate the node "d" using the best split point, with respect to the k attributes
- Further divide the node into daughter nodes using the best split procedure
- Repeat the procedure till required number of nodes has been reached
- Repeat steps 1–4 "n" times to generate n trees.

This assimilation of random trees comprises the random forest.

Random forest prediction algorithm:

- Step 1. The test features are passed through the each of the random decision trees, with respect to the rules or decision parameters of each of the respective trees, and the predicted output is stored as a vote.
- Step 2. Tabulate the votes of each prediction
- Step 3. The prediction with highest votes is treated as the final prediction and is the output of the random forest classifier.

We used the random forest algorithm from another work of one of the authors [16]. Similar algorithm from [16] was used in a different application to classify walking pattern. The EMG and angle values obtained by collecting ideal walking data with different subjects are used to train the model. The model once trained is used to predict the gait phase of real-time input. The angle and EMG values are fed to the random forest classifier and the model predicts the current gait phase and hence the next gait phase can be initiated.

4 Computational Experimental Results

The proposed model was tested using real-time data. Real-time data was collected and the proposed algorithm was tested. The EMG data was classified using a random forest classifier. The train set and test set were 70% and 30%, respectively. The confusion matrix of the random forest classifier is shown in Fig. 6.

To test the proposed algorithm for actuation of the knee data was collected real time. The knee angle and EMG sensor data acquired real time are shown in Fig. 7. The actuator motion pattern is shown in Fig. 8.

```
[[114     0     0     0     0     0     0]
 [  0    71     0     0     0     0     0]
 [  0     0   172     0     0     0     0]
 [  0     0     0   107     0     0     0]
 [  0     0     0     0   104     0     0]
 [  0     0     0     0     0    87     0]
 [  0     0     0     0     0     0   105]]]
```

Fig. 6 Confusion matrix for random forest classifier

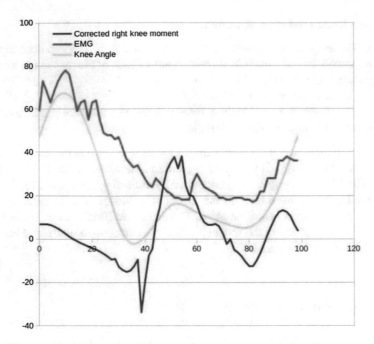

Fig. 7 Knee angle and EMG sensor data acquired in real time

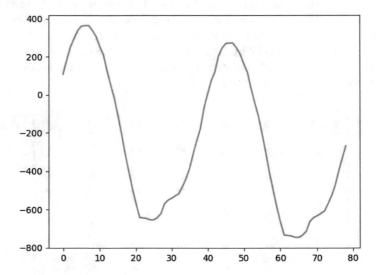

Fig. 8 Actuator motion pattern in real time

5 Concluding Remarks

The proposed model for a prosthetic leg was successfully implemented and tested. EMG data was used to classify gait along with angle values. The random forest algorithm used to classify gait phases shows promising results. This work can be implemented in current day prosthetic to make it more reliable and adaptive. The proposed model was tested using real-time data and the model was able to analyze and execute the walking pattern observed from the sensor inputs. The sensor inputs from the IMU and EMG was used to classify gate and the next gait phase was initiated. The design of the prosthesis was quite simple. The algorithm of the model was not complex and hence the proposed model for a prosthesis is really effective.

Future work can be done by using different classification and statistical learning techniques to classify gait phases and analyze walking pattern. Deep-learning techniques can be used. We believe, with this paper, we can encourage the future work by shedding some light on this topic.

References

1. Sup, F., Bohara, A., Goldfarb, M.: Design and control of a powered transfemoral prosthesis. Int. J. Robot. Res. **27**(2), 263–273 (2008)
2. Hargrove, L.J., Young, A.J., Simon, A.M., Fey, N.P., Lipschutz, R.D., Finucane, S.B., Kuiken, T.A.: Intuitive control of a powered prosthetic leg during ambulation: a randomized clinical trial. JAMA **313**(22), 2244–2252 (2015)
3. Scheme, E.P., Englehart, K.: Electromyogram pattern recognition for control of powered upper-limb prostheses: State of the art and challenges for clinical use. J. Rehabil. Res. Dev. **48**(6), 643 (2011)
4. Chan, A.D., Englehart, K.B.: Continuous myoelectric control for powered prostheses using hidden Markov models. IEEE Trans. Biomed. Eng. **52**(1), 121–124 (2005)
5. Englehart, K., Hudgins, B.: A robust, real-time control scheme for multifunction myoelectric control. IEEE Trans. Biomed. Eng. **50**(7), 848–854 (2003)
6. Seel, T., Raisch, J., Schauer, T.: IMU-based joint angle measurement for gait analysis. Sensors **14**(4), 6891–6909 (2014)
7. El-Sheimy, N., Nassar, S., Noureldin, A.: Wavelet de-noising for IMU alignment. IEEE Aerosp. Electron. Syst. Mag. **19**(10), 32–39 (2004)
8. Saridis, G.N., Gootee, T.P.: EMG pattern analysis and classification for a prosthetic arm. IEEE Trans. Biomed. Eng. **6**, 403–412 (1982)
9. Ajiboye, A.B., Weir, R.F.: A heuristic fuzzy logic approach to EMG pattern recognition for multifunctional prosthesis control. IEEE Trans. Neural Syst. Rehabil. Eng. **13**(3), 280–291 (2005)
10. Parker, P., Englehart, K., Hudgins, B.: Myoelectric signal processing for control of powered limb prostheses. J. Electromyogr. Kinesiol. **16**(6), 541–548 (2006)
11. Breiman, L., Cutler, A.: Random forests manual (version 4.0). Technical Report of the University of California, Berkeley, Department of Statistics (2003)
12. Han, H., Guo, X., Yu, H.: Variable selection using mean decrease accuracy and mean decrease Gini based on random forest. In: 2016 7th IEEE International Conference on Software Engineering and Service Science (ICSESS), pp. 219–224 (2016)

13. Kumar, A., Shahnawazuddin, S., Pradhan, G.: Exploring different acoustic modeling techniques for the detection of vowels in speech signal. In: 2016 Twenty Second National Conference on Communication (NCC), pp. 1–5 (2016)

14. Ljumović, M., Klar, M.: Estimating expected error rates of random forest classifiers: a comparison of cross-validation and bootstrap. In: 2015 4th Mediterranean Conference on Embedded Computing (MECO), pp. 212–215 (2015)

15. Ehlers, A., Rosenhahn, B., Liu, W., Baumann, F.: Sequential boosting for learning a random forest classifier. In: 2015 IEEE Winter Conference on Applications of Computer Vision, pp. 442–447 (2015)

16. Itagi, A., Baby, C.J., Rout, S., Bharath, K.P., Karthik, R., Kumar, M R : Lisp detection and correction based on feature extraction and random forest classifier. In: Microelectronics, Electromagnetics and Telecommunications (pp. 55–64). Springer, Singapore (2019)

Analysing Environmental Factors for Corporate Social Responsibility in Mining Industry Using ISM Methodology

M. Ramaganesh and S. Bathrinath

Abstract The aim of the article is to analyze about the environmental problems faced in mining industry for corporate social responsibility. Fifteen crucial factors are considered from reputed literature journals and expert's opinion from the industries. Interpretive Structural Modeling (ISM) is used to evaluate the factors identified and the results are verified by using covariance based structural equation modeling (CBSEM). Unregulated and illegal mines, Health of the local population, Acid rock drainage, Formation of sinkholes, Usage of Heavy metals and CO_2 emitted by vehicles are found as the most influencing factors and these factors are taken into consideration and mitigation technique is applied and the factors causing environmental challenge is reduced.

Keywords Environmental Factors in Mining Industry · Interpretive Structural modeling · Corporate Social Responsibility · Structural relationship · Covariance based structural equation modeling

1 Introduction

A country's economy cannot be a healthy one without industries producing valuable products. The valuable products can be manufactured with the help of minerals and ores. These minerals and ores are gathered by the process of Mining. Mining industry is one of the valuable industries for a developing country like India. It makes our economy to get more strength by giving ores to the manufacturing industries. But while doing mining there are some environmental problems also occurs. Mining industries is involved with the extraction process of precious minerals and other geological materials. The extracted materials are transformed into a mineralized from that serves an economic benefit to the prospector or miner. Mining processes generally make a harmful environment effect at the time of mining activity as well as

M. Ramaganesh · S. Bathrinath (✉)
Department of Mechanical Engineering, Kalasalingam Academy of Research and Education, Krishnankoil, Tamil Nadu, India
e-mail: bathri@gmail.com

© Springer Nature Singapore Pte Ltd. 2020
K. N. Das et al. (eds.), *Soft Computing for Problem Solving*,
Advances in Intelligent Systems and Computing 1057,
https://doi.org/10.1007/978-981-15-0184-5_31

mining activity has completed. Various environmental problems such as soil erosion, contamination of land, formation of sinkholes, loss of bio diversity, dirtiness in groundwater and surface water by chemicals from mining processes, etc. are occurred. Heavy machines are used in mining to explore, to remove, to break and remove various rocks. For digging the land Bulldozers, trucks and drills are all necessary for the mining process. It reduces the strength of soil and sometimes it will end up with earthquake. Hence many countries have made rules to decrease the mining activity. The safety level in the working area in mining industry is also an important thing to consider. Latest practices have considerably enhanced safety in mines. By considering the protection of environment from mining activities, this thesis work was taken with the objective of identification of most influencing factors and the possible solutions to elimination them.

2 Relevant Literature

Owen et al. [1] examined the social and environmental challenges faced by the mining industry while going for highly populated area and in case of high cost ore. Hilson et al. [2] examined sustainable development in the mining industry illuminating the corporate perception. They suggested that to make industry as more sustainable one, industry has to ensure the environmental clearing and taking, social remedies. Jenkins et al. [3] discussed the corporate social responsibility in the mining industry. They endorsed enclosing social and environmental information in the company's procedure while doing mining can justify their product with their description. Jenkins et al. [4] analyzed the conflicts and constructs in corporate social responsibility for mining industry. They briefed about the corporate responsibility about the environmental, sustainability factors and also these kind of taking over led to the change in the strategy of mining in how to deal it and bring out new outcomes in facing such problems. O'Regan et al. [5] developed the interaction model be-tween environmental and economic factors in the mining industry using system dynamics. This brings out decision for better under-standing of the system structure. Wang et al. [6] explained about the origin level and remediation for mercury pollution in aquatic environment. Contamination of the water bodies with mercury leads to drastic changes in the water area and the organisms living in and around. Choi et al. [7] reviewed the utilization of photovoltaic and wind power systems in mining industry. Photovoltaic and wind power systems employed in the mines to solve the energy supply problems and to have benefit from the exhausted mines. Zhengfu et al. [8] explained about the environmental problems that occur during coal mining and the reusability of waste received. It is common in some countries in using the waste products for their raw materials. After completion of mining process, the land lying vacant has been used for agricultural and some other activities. Gu [9] discussed mining, pollution and site remediation which includes environmental issues like pollution occurs during mining containing top-soils, deposits, aquifer and groundwater, and brooks. Appropriate maintenance activity in mining site should be done. Muduli K et al. [10] analyzed the obstacles to green supply chain management in Indian mining industries using graph theory. Green supply chain management system will help to reduce the energy and utilization of materials to improve the efficiency level while doing mining. Muduli et al. [11] explained the behavioral issues in green supply chain management execution in Indian mining industries. GSCM will vary according to the human behavioral changes. Barve et al. [12] established a model for the issues of green supply chain management processes in Indian mining industries. They analyzed the barriers like lack of eco-friendly consciousness, meagre regulation and insufficient burden from society. Govindan et al. [13] identified the barriers for green supply chain management execution in Indian trades using analytic hierarchy process and confirming that the green supply chain management is proficient of clearing the barriers and can manage a green enabled environment during mining. Singh et al. [14] identified and cleared the relationship among the barriers in knowledge management using ISM technology. Luthra et al. [15] analyzed the obstacles in the adoption of green supply chain management in automobile business using ISM. Pfohl et al. [16] developed an interpretive structural modeling for supply chain risks. They analyzed the relationship among the potential supply chain risks and their de-pendency among each other is observed and derived.

3 Problem Description

This study analyzes the environmental factors in mining industry. Table 1 shows the identified environmental factors which were developed from various literature surveys and experts opinions from Industry. Then, Interpretive Structural Modeling (ISM) approach was applied to evaluate these issues. Figure 1 represents the framework of the work to be carried out.

4 Methodology

Interpretive Structural Modeling (ISM) is a computer based technique to develop the graphical representations of complex systems [17]. It is a technique in which indirectly related factors are structured into one form. First step is to analyze the factors and finds the relation between them. Next, Structural self-interaction matrix is developed for solving the issues. Then, the matrix obtained is converted into a transitivity matrix and the partitioning of elements into different levels which results in the formation of ISM. Based on the results formulated Digraph is developed.

Table 1 Environmental factors in mining industry

S. No	Environmental factors	Definitions
1	Erosion of soil (EROS)	Losing the strength of soil
2	Contamination of Air (CONTOA)	Polluting the air in different way like burning
3	Ground water contamination (GRWCONT)	Improper drainage
4	Acid rock drainage (ACRD)	Chemical water mixing in the river due to improper drainage
5	Formation of sinkholes (FSINK)	Sinkholes formed in the overlaying ground due to improper drainage system of chemical water
6	Health of the local population (HEALP)	Health condition of people gets affected due to pollutants like air, water and land
7	Dumped toxic waste (DUMTW)	Unwanted materials in all forms which are dumped in the ground. It can various pollutions like air, land, water, etc.
8	Loss of Biodiversity (LBIOD)	Affects the living organisms on the land.
9	Improper backfilling mines sites (IMBFMS)	It will able to make Landslides.
10	Carbon Emissions (CAREMI)	Greenhouse gases can be emitted from working area by the combustion of fossil fuels
11	Usage of heavy metals (USHM)	Dissolution and transport of metals by run-off ground water
12	CO_2 emitted by vehicles (COEMV)	Carbon dioxide gas was emitted by the usage of heavy loaded vehicles. It will cause air pollution
13	Destruction and poison linger (DEAPL)	Irregular mining processes can make firing in coal storage, which can hurt for long time
14	Deforestation (DEFOR)	Destroying the forest area
15	Unregulated and illegal mines (URILM)	Disobeying the government rules

4.1 Identification of Factors

First step it to collect the factors which are related to the environmental problems.

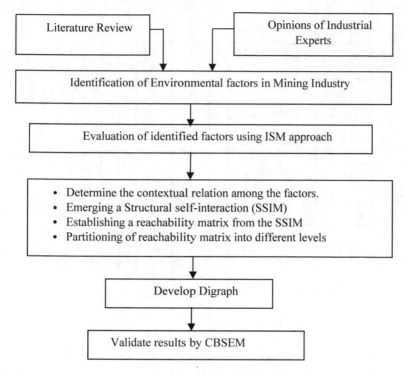

Fig. 1 Flow chart for interpretive structural modeling (ISM)

4.2 Developing a Structural Self-interaction Matrix (SSIM) of Factors

ISM technique uses methods like sharing ideas among the group members in making the relation among the variables. It is used to identify the nature of the elements. It is done in the way that all the influencing factors of them. Some of the elements there can be some factors which influences the other. The relation between the factors is represented by the four characters with sub characters (i, j) under consideration. Based on the contextual relationships, the SSIM has been made which is shown in Table 2.

4.3 Developing a Reachability Matrix from the SSIM

For developing a reachability matrix from the SSIM, the following conditions to be followed.

1. If the (i, j) score in the SSIM is V, then the (i, j) score in the reachability matrix becomes 1 and the (j, i) score becomes 0.

Table 2 Structural self interaction matrix (SSIM)

Factors nO	Factor	15	14	13	12	11	10	9	8	7	6	5	4	3	2	1
1	EROS	A	X	O	V	X	X	A	V	A	O	V	O	X	O	–
2	CONTOA	A	X	A	A	A	A	O	V	O	V	O	O	O		
3	GRWCONT	A	X	O	O	A	O	A	X	A	V	A	A			
4	ACRD	A	V	V	O	A	O	V	X	A	V	X				
5	FSNK	A	V	V	V	X	O	A	X	A	V					
6	HEALP	O	A	A	A	A	A	A	A	A						
7	DUMTW	A	X	V	O	A	V	V	V							
8	LBIOD	A	X	V	A	A	A	A								
9	IMBFMS	A	X	V	O	A	O									
10	CAREMI	A	X	V	X	X										
11	USHM	A	V	V	O											
12	COEMV	A	V	V												
13	DEAPL	O	X													
14	DEFOR	A														
15	URILM	–														

V—i helps to attain j

A—j helps to attain i

X—i and j are related to each other

O—i and j are not related to each other

2. If the (i, j) score in the SSIM is A, then the (i, j) score in the reachability matrix becomes 0 and the (j, i) score becomes 1.
3. If the (i, j) score in the SSIM is X, then the (i, j) score in the reachability matrix becomes 1 and the (j, i) score also becomes 1.
4. If the (i, j) score in the SSIM is 0, then the (i, j) score in the reachability matrix becomes 0 and the (j, i) score also becomes 0.

Table 3 represents the final reachability matrix.

4.4 Developing a Reachability Matrix from the SSIM

The reachability and antecedent sets are obtained for each factors from the final reachability matrix. The reachability set contains the factors itself and the factors which it influences. Intersection of sets and their levels are determined. If the reachability and intersection are same, then they are assigned as the top level in the ISM hierarchy. Once it reached the top layer of the hierarchy, it is discarded from the reflection. Then the iterative procedure is continued until the level of each factor is found. These levels help in creating the diagraph and ISM model. Table 4 shows the level of portioning of reachability matrix.

5 Result and Discussions

Figure 2 shows the graph between drive power and dependence matrix for the environmental factors for CSR in mining industry. In our study, no factors lies in Autonomous cluster (weak drive power and dependence). Four factors namely Acid Rain, Formation of sink holes, Usage of heavy metals and Co_2 emitted by vehicles (4, 5, 11 and 12) are lies in Dependent category (Weak driving power and strong dependence power). Nine factors such as Erosion of soil, Contamination of Air, Ground water contamination, dumped toxic waste, Loss of Biodiversity, Improper backfilling mines sites, Car-bon Emissions, Destruction and poison linger, Deforestation (1, 2, 3, 7, 8, 9, 10, 13 and 14) are lied in third cluster i.e. Linkage (Strong drive power as well as dependence power). Unregulated and illegal mines factor is lies in independent variable category (Strong drive power and weak dependence power). Figure 3 shows the di-graph which is generated from the final reachability matrix (Table 3). It is found that Health of the local population is top level barrier and Unregulated and illegal mines factor is bottom level barrier.

Table 3 Final reachability matrix

Factors	Final reachability matrix															Drive power	Rank
1	1	1	1	1	1	1	1	1	1	1	1	1	1	1	0	14	7
2	1	1	1	1	1	1	1	1	0	0	1	1	1	1	0	12	13
3	1	1	1	1	1	1	1	1	1	1	1	1	1	1	0	14	7
4	1	1	1	1	1	1	1	1	1	1	1	1	1	1	0	14	7
5	1	1	1	1	1	1	1	1	1	1	1	1	1	1	0	14	7
6	0	0	0	0	1	0	0	0	0	0	0	0	0	0	0	1	15
7	1	1	1	1	1	1	1	1	1	1	1	1	1	1	0	14	7
8	1	1	1	1	1	1	1	1	1	1	1	1	1	1	0	14	7
9	1	1	1	1	1	1	1	1	1	1	1	1	1	1	0	14	7
10	1	1	1	1	1	1	1	1	1	1	1	1	1	1	0	14	7
11	1	1	1	1	1	1	1	1	1	1	1	1	1	1	0	14	7
12	1	1	1	1	1	1	1	1	1	1	1	1	1	1	0	14	7
13	1	1	0	0	1	1	1	1	0	0	1	1	1	1	0	10	14
14	1	1	1	1	1	1	1	1	1	1	1	1	1	1	0	14	7
15	1	1	1	1	1	1	1	1	1	1	1	1	1	1	1	15	1
Dependance power	14	14	13	13	15	14	14	14	12	12	14	14	14	14	1		

Table 4 Level of partitioning

Factors	Reachability set	Antecedent set	Intersection set	Level
6	6	1 2 3 4 5 6 7 8 9 10 11 12 13 14 15	6	1
1	1 2 3 4 5 7 8 9 10 11 12 13 14	1 2 3 4 5 7 8 9 10 11 12 13 14 15	1 2 3 4 5 7 8 9 10 11 12 13 14	II
2	1 2 3 4 5 7 8 9 10 13 14	1 2 3 4 5 7 8 9 10 11 12 13 14 15	1 2 3 4 5 7 8 9 10 13 14	II
3	1 2 3 4 5 7 8 9 10 11 12 13 14	1 2 3 4 5 7 8 9 10 11 12 13 14 15	1 2 3 4 5 7 8 9 10 11 12 13 14	II
7	1 2 3 4 5 7 8 9 10 11 12 13 14	1 2 3 4 5 7 8 9 10 11 12 13 14 15	1 2 3 4 5 7 8 9 10 11 12 13 14	II
8	1 2 3 4 5 7 8 9 10 11 12 13 14	1 2 3 4 5 7 8 9 10 11 12 13 14 15	1 2 3 4 5 7 8 9 10 11 12 13 14	II
9	1 2 3 4 5 7 8 9 10 11 12 13 14	1 2 3 4 5 7 8 9 10 11 12 13 14 15	1 2 3 4 5 7 8 9 10 11 12 13 14	II
10	1 2 3 4 5 7 8 9 10 11 12 13 14	1 2 3 4 5 7 8 9 10 11 12 13 14 15	1 2 3 4 5 7 8 9 10 11 12 13 14	II
13	1 2 3 7 8 9 10 13 14	1 2 3 4 5 7 8 9 10 11 12 13 14 15	1 2 3 7 8 9 10 13 14	II
14	1 2 3 4 5 7 8 9 10 11 12 13 14	1 2 3 4 5 7 8 9 10 11 12 13 14 15	1 2 3 4 5 7 8 9 10 11 12 13 14	II
4	4 5 11 12	4 5 11 12 15	4 5 11 12	III
5	4 5 11 12	4 5 11 12 15	4 5 11 12	III
11	4 5 11 12	4 5 11 12 15	4 5 11 12	III
12	4 5 11 12	4 5 11 12 15	4 5 11 12	III
15	15	15	15	IV

6 Statistical Validation

Covariance based structural equation modeling (CBSEM) is used for verifying the ISM results. CBSEM is used to assess the fit goodness which concentrating on diminish the variations between both observed and estimated covariance matrix and also it is termed as confirmatory factor analysis. Questionnaire was created and given to the decision makers in the mining industry who are having vast experience in the mining field. Their rating and comments were obtained based on the (0-very low importance to 4-very high importance scale). Then, the collected data are solved by using covariance based structural equation modeling. CBSEM results revealed that factors 6 and 15 in the top & bottom level and it is shown in Table 5. Figure 4 represents the comparison of ISM and CBSEM results and found similar one.

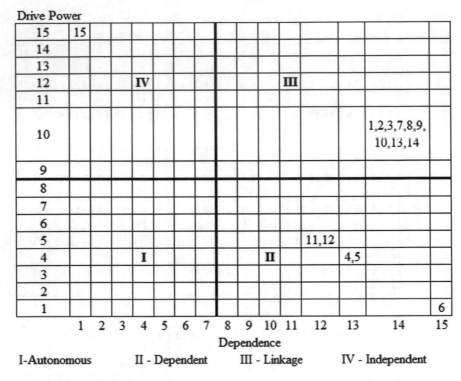

Fig. 2 Drive power- dependence matrix

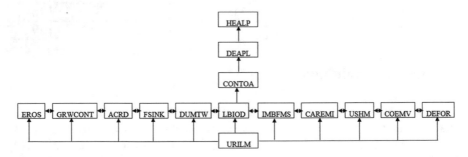

Fig. 3 ISM diagraph for environmental factors in mining

7 Conclusion

Environmental problems caused by Mining Industry is a critical one and needs necessary action to reduce the impact on the environment. Fifteen important factors which are affecting the environmental condition in mining industry is considered by reputed literature survey and experts opinion from the industry. Interpretive structural modeling (ISM) is implemented to find the relationships among the factors in mining

Table 5 CBSEM results

Factors	Mean	S.D	Level	Factors	Mean	S.D	Level
1	1.295	1.166	II	9	1.295	1.166	II
2	1.295	1.166	II	10	1.295	1.166	II
3	1.295	1.166	II	11	1.709	1.139	III
4	1.709	1.139	III	12	1.709	1.139	III
5	1.709	1.139	III	13	1.295	1.166	II
6	3.846	2.562	I	14	1.295	1.166	II
7	1.295	1.166	II	15	1.203	0.801	IV
8	1.295	1.166	II				

Fig. 4 ISM and CBSEM results comparison

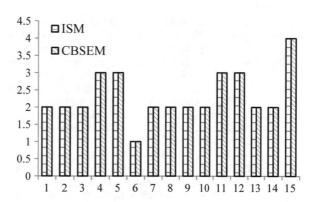

industry. From the results arrived, Health of the local population (HEALP) has been identified as top level factor and Unregulated and illegal mines (URILM) is identified as bottom level factor. Elimination of these factors will assist in the reduction of the environmental problems occurred. Health of the people is affected mainly because of the impact of air like particulate matter, gas emissions, and pollution in the drinking water, noise and vibration. The organization has to take necessary remedial actions to prevent this problems. To prevent the unregulated and illegal mines, Government has to make strict rules for mining policy. If the model is developed for other industry or country, some factors may be removed and re-placed with other factors. We have made ISM model for environ-mental issues in Indian mining industry. It can be extended to other issues related to social or economic.

References

1. Owen, J.R., Kemp, D.: Social management capability, human migration and the global mining industry. Re-sources Policy **53**, 259–266 (2017)
2. Hilson, G., Murck, B.: Sustainable development in the mining industry: clarifying the corporate perspective. Re-Sources Policy **26**(4), 227–238 (2000)
3. Jenkins, H., Yakovleva, N.: corporate social responsibility in the mining industry: exploring trends in social and environmental disclosure. J. Clean. Prod. **14**(3–4), 271–284 (2006)
4. Jenkins, H.: Corporate social responsibility and the mining industry: conflicts and constructs. Corp. Soc. Responsib. Environ. Manag. **11**(1), 23–34 (2004)
5. O'Regan, B., Moles, R.: Using system dynamics to model the interaction between environmental and economic factors in the mining industry. J. Clean. Prod. **14**(8), 689–707 (2006)
6. Wang, Q., Kim, D., Dionysiou, D.D., Sorial, G.A., Timberlake, D.: Sources and remediation for mercury contamination in aquatic systems—a literature review. Environ. Pollut. **131**(2), 323–336 (2004)
7. Choi, Y., Song, J.: Review of photovoltaic and wind power systems utilized in the mining industry. Renew. Sustain. Energy Rev. **75**, 1386–1391 (2017)
8. Zhengfu, B.I.A.N., Inyang, H.I., Daniels, J.L., Frank, O.T.T.O., Struthers, S.: Environmental issues from coal mining and their solutions. Min. Sci. Technol. (China) **20**(2), 215–223 (2010)
9. Gu, J.D.: Mining, pollution and site remediation (2018)
10. Muduli, K., Govindan, K., Barve, A., Geng, Y.: Barriers to green supply chain management in Indian mining in-dustries: a graph theoretic approach. J. Clean. Prod.-Tion **47**, 335–344 (2013)
11. Muduli, K., Govindan, K., Barve, A., Kannan, D., Geng, Y.: Role of behavioral factors in green supply chain management implementation in Indian mining industries. Resour. Conserv. Recycl. **76**, 50–60 (2013)
12. Barve, A., Muduli, K.: Modelling the challenges of green supply chain management practices in Indian mining industries. J. Manuf. Technol. Manag. **24**(8), 1102–1122 (2013)
13. Govindan, K., Kaliyan, M., Kannan, D., Haq, A.N.: Barriers analysis for green supply chain management implementation in Indian industries using analytic hierarchy process. Int. J. Prod. Econ. **147**, 555–568 (2014)
14. Singh, S.K.: Role of leadership in knowledge management: a study. J. Knowl. Manag. **12**(4), 3–15 (2008)
15. Luthra, S., Kumar, V., Kumar, S., Haleem, A.: Barriers to implement green supply chain management in auto-mobile industry using interpretive structural modeling technique-an Indian perspective. J. Ind. Eng. Manag. **4**(2), 231–257 (2011)
16. Pfohl, H.C., Gallus, P., Thomas, D.: Interpretive structural modeling of supply chain risks. Int. J. Phys. Distrib. Logist. Manag. **41**(9), 839–859 (2011)
17. Li, Y., Sankaranarayanan, B., Kumar, D.T., Diabat, A.: Risks assessment in thermal power plants using ISM methodology. Ann. Oper. Res. 1–25 (2019)

Driver Assistance through Geo-fencing, Sign Board Detection and Reporting Using Android Smartphone

S. Veni, R. Anand and D. Vivek

Abstract This work focuses on analysis, need, design and implementation of an easy to work driver-assistance system in both normal and especially in a fleet management environment. The work provides assistance to the drivers and also to the fleet managers. Considering situations such as bumpers on the road, a side road that joins the highway, etc., if alerted to driver it can save a lot of accidents. The work is implementation of the same through global positioning system (GPS) and through sign board recognition mechanism. It alerts the driver also the fleet manager about speed limit zone, honking restriction zone through geo-fencing implementation. The work proposes a special hardware that could interface with any android phone through an android application. This hardware will be used to collect necessary information like sign board information, vehicle environmental information and communicate to the android phone either through On The Go (OTG) cables or Bluetooth interface. The phone in turn uses GPS position and embeds the position data with the above vehicle data, alerts and stores them for further analysis. The android application works with the incoming information and the GPS positioned and gives its feedback as an assistance to the driver. This work includes need for driver-assistance combined with fleet management and sign board recognition concept.

Keywords Android · Geo-fencing · Sign board detection · UART module · Android for automobile

S. Veni
Amrita School of Engineering, Amrita Vishwa Vidyapeetham,
Amrita University, Coimbatore 641112, India
e-mail: s_veni@cb.amrita.edu

R. Anand (✉)
Department of Electronics and Communication Engineering, Sona Signal and Image
Processing Research Center, Sona College of Technology, Salem, India
e-mail: anand.r@sonatech.ac.in

D. Vivek
Automotive Systems, Department of Mechanical Engineering,
Robert Bosch Business and Engineering Solutions Private Ltd., Amrita
Vishwa Vidyapeetham, Amrita Nagar, Coimbatore 641112, Tamil Nadu, India

© Springer Nature Singapore Pte Ltd. 2020
K. N. Das et al. (eds.), *Soft Computing for Problem Solving*,
Advances in Intelligent Systems and Computing 1057,
https://doi.org/10.1007/978-981-15-0184-5_32

361

1 Introduction

The geo-fencing is a combination of the actual GPS coordinates, the latitude and the longitude and the radius around it. This is usually used to mark an area or a zone in the map. This can be used for a device to be aware of a particular zone. This can be a pre-existing boundary or a programmed perimeter for specific application. It is referred to as custom digitised geo-fencing.

Similar to geo-tagging in photographs, any data that we monitor in the vehicle if applied with geo-tagging it is going to be certain use later. For example, vehicle load sensing, carrier cabin door opening or closing, etc., if attached with a geo-location it will be of great use in some of the critical load carrying applications. So in this work, this kind of application will be discussed, implemented and tested. Reclus and Drouard [1] present various applications used in road transport, the control and monitoring techniques based on geo-fencing like route adherence, schedule adherence, etc. The paper concludes that geo-fencing has various advantages like flexibility and independent of vehicle types, no need of monitoring cameras, applications for safety and use of satellite positioning systems. Other geo-fencing-based researches include identification of objects under a particular geo-fence.

The work proposed by Sanjay Pramanathan and Santhosh Kumar [2] elaborates a system for location-sensitive speed adaption for vehicles. This paper provides the overview for the location sensing of the vehicle, and the speed limit of that zone from the available or preloaded maps. Based on the inputs, the driver is alerted based on the over speeding and also the flow of the fuel to the injectors is also controlled to limit the speed. This paper also deals with the details about the number of accidents due to the over speeding in the slow speed zones and the importance of the speed control in the vehicles. The above paper does not clearly explain the driver what may happen in these zones. In our system, we indicate the driver clearly stating the zones and we apply monitoring in this part. The system will monitor the speed and log the data and violations.

Further, fleet management involves managing a set of vehicles under a group across various geographical locations delivering goods at appropriate places. Ensuring timing, accurate delivery location and dynamic routing are the recent trends in this field. These are common for any fleet unit carrying any kind of goods. Fleet management systems have their fleets as their assets. Like in a production unit, where cycle time reduction improves production and in turn its profit, on time delivery and optimum utilisation of the fleets improve productivity and profit. So these become a necessary implementation for any fleet management company. Also, the stated application through this research work could be like a plug and play or a brought-in device which is easy to implement in any existing fleet management company. Going ahead, the ZigBee protocol method of sign board method is proposed. It operates on the IEEE 802.15.4 standard at 2.4 GHz, 900 and 868 MHz. The ZigBee is studied to have power saving due to short working period, low-power consumption. Low cost of the modules, short time delay like 30ms for device searching and 15ms for standby to activation and 15ms for channel access. ZigBee provides complete data integrity check and authentication function. With these characteristics, the ZigBee protocol is reliable and safe.

The research paper by Fleyeh and Dougherty [3] proposes identification of sign boards through image processing where we could find some of the disadvantages that, poor quality of images due to rain, fog, too many sign boards at the same place, etc. This will be overcome by our proposal. Also with high-range wireless transmission, it can be implemented in highways like a side road detection, a dangerous curve, etc. Our work proposes the detection via wireless transmission of data and it overcomes the above said disadvantages.

Ubidots is an open platform for implementing and demonstrating IoT – Internet of things. It works with various platforms like Android, Arduino, Raspberry pi, and many other boards through LAN (local area network), Wi-Fi (wireless fidelity) and GPRS (general packet radio service). In our work, since this is a mobile application, GPRS can be proved to be properly interfaced with an IP (Internet Protocol) address. The sensor data transferred to this portal can be monitored from any part of the world connected to Internet and having a login for the server, which is very useful for us in demonstrating the reporting part of our idea. The proposed work demonstrates how the above techniques are integrated and provides a new application that becomes useful for vehicle drivers and fleet owners which is easily implemented. Section 2 deals with design techniques and the methodology is explained in Sect. 3. Section 4 deals with the results followed by conclusion in Sect. 5.

2 Design Techniques

According to our proposal, there are three major sections of providing driver and fleet owner assistance. Firstly, sensors placed in the vehicle are many and are used internally with control units in the car itself. Sensors are placed for various functionalities of the car such as critical and essential application like engine management, sophistication like cabin temperature for AC (air conditioner) control and lot of other applications inside the automobile. Figure 1 explains the proposed with three blocks of sensors, sensor interface for them to communicate with the smartphone.

In cases of the goods that are environment sensitive, for example changes behaviour or exhibits different reaction at different temperature and pressure, these sensors at the storage space and monitoring of them both locally and remotely could be of much use during situations like carriers taking flammable goods, weapons, food items, etc.

Secondly, working on the geo-fencing part with an android with the help of google database is performed in this work. Basic idea explained in this work is based on building a database of locations of interest in our mobile phone app in that particular city. Two ways of implementation of the geo-fencing part using distance to the point of interest and using geo-fencing module in Android Table 1 are considered.

But the actual implementation is the second part but in a discrete way. The straight distance between the coordinates of location of interest and the current location

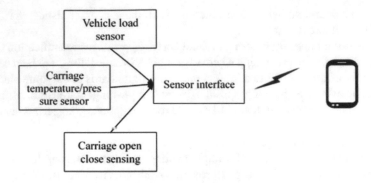

Fig. 1 Sensor interface for a geo-fence application

Table 1 Special methods associated with Android geo-fencing [4]

Method	Description
GEOFENCE_TRANSITION_DWELL	Indicates that the user enters a geo-fences for a period off-time
GEOFENCE_TRANSITION_ENTER	Indicates that the user enters the geo-fence
GEOFENCE_TRANSITION_EXIT	Indicates that the user exits the geo-fence
NEVER_EXPIRE	A value that is used to indicate a geo-fence should never expire

coordinates is calculated by a discrete method (Haversine formula as given below). In case of prevention of unwanted honking and over speeding, it creates a perfect geo-fence or a circular fence around the area of interest.

Haversine Formula [5]:

$$\begin{cases} \text{dlon} = \text{lon2} - \text{lon1} \\ \text{dlat} = \text{lat2} - \text{lat1} \\ a = \sin^2(\frac{\text{dlat}}{2}) + \cos(\text{lat1}) * \cos(\text{lat2}) * \sin^2(\frac{\text{dlon}}{2}) \\ c = 2 * \arcsin(\min(1, \sqrt{a})) \\ d = R * c \end{cases} \tag{1}$$

Whereas, 'dlon, dlat' are differences in latitude and longitude coordinates of two positions. Lat1 and lon1 are coordinates of position 1 and lat2 and lon2 are coordinates of position 2. 'd' will give mathematically and computationally exact results.

Thirdly, the sign board recognition is done using a wireless interface hardware. Implementation of sign board recognition is done using a ZigBee transmitter at the sign board and a ZigBee receiver at the automobile. Most of the current sign board recognition technology is based on image processing. It requires high processing capability and programming. Also, the time consumed by this work is more. Thus, when the vehicle is in motion, recognition sometimes does not help because we may

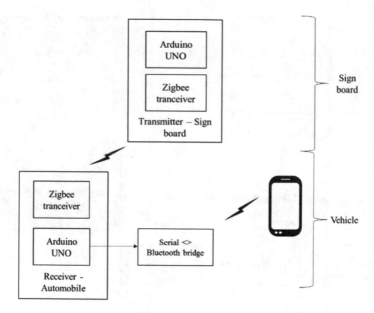

Fig. 2 Design of sign board detection method

actually cross the region of interest before something could be recognised. Also, there are other difficulties in the common implementation of the sign board recognition. The design of the current work begins with a ZigBee wireless interface between vehicle and the sign board. With unique code for each of the sign board is considered for demonstration purpose.

The block diagram of the proposed method for sign board detection is described in Fig. 2. It has two main blocks for communication device, ZigBee transceiver, and its controller, Arduino Uno. One each at the sign board and the vehicle. In the vehicle also has a bridge for converting serial data from the controller to Bluetooth to be able to send to a smartphone.

3 Proposed Method

The different design parts for implementing geo-fencing, vehicle interior sensing and the sign board recognition are described here. The main device which integrates is the smartphone controlled by an Android application. Figure 3 explains the overall block diagram of the proposal after integration of all the three proposals.

Information transfer considered between the board and automobile are (i) Code word for recognition of the sign board. This is to infer which restriction is ahead like '01 ' for a side road alert sign board, '23' for a speed breaker alert sign board. (ii) Distance to the interested zone. For example, if this is for a side road indication,

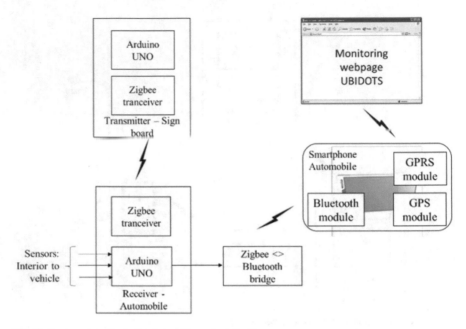

Fig. 3 Proposed method after integration of whole idea

Table 2 Code word transmitted from sign boards

Code word	Indication
0/1	Side roader
2/3	Speed breaking bump
4/5	Dangerous curve
6/7	Accident zone

how distant it is from the sign board. The reception of this is only for reference, as the reception can happen at any distance from the sign board. But the distance value transferred can only be from sign board to the actual spot. (iii) Speed limit expected. This will transfer the speed limit at the destined spot or zone. Based on the severity of the zone, for example, if the dangerous curve sign board wants to specify a speed based on the severity of the curve, it can do so in this part of message transfer.

For the code word, the data (Table 2) was formulated for demonstration purpose. The separate block diagram for the transmitter that is going to be located in the sign board is explained in Fig. 4. The battery should be a powerful one and a continuously recharging circuit similar to a solar battery charger to be present to fulfil our aim. A simple control circuit that could communicate with the transceiver through a universal asynchronous receiver/transmitter (UART) is present. The ZigBee transceiver is a basic board containing CC2500 chip and basic communication interface like UART as proposed.

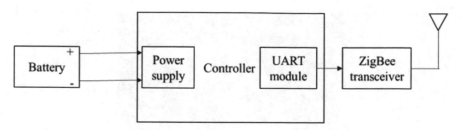

Fig. 4 Proposed transmitter for actual application

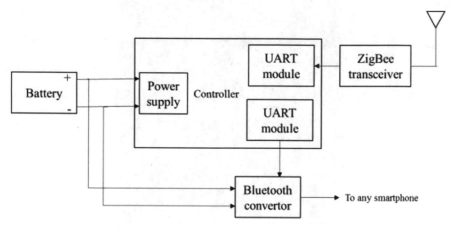

Fig. 5 Proposed receiver for actual application

The block diagram for the receiver that is going to be located inside the vehicle is explained in Fig. 5. The battery shall be the automotive battery. A simple control circuit that could communicate with the transceiver through a UART is present. The ZigBee transceiver is similar to transmitter. The controller also will communicate with the Bluetooth bridge for converting plain serial data to Bluetooth signal. In the city, wherever the restrictions are to be followed to be entered as a database in the smartphone application. In this experiment, a few hospitals and schools have been loaded to check the behaviour of the application. Drive was arranged around the city with the application running and whenever the region of interest comes within 300m (for example). It will start displaying in the warning area and when we travelled past the region this warning automatically goes off. Currently, in this work, only a picture warning is placed in the application interface, whereas it can be modified to a sound warning or even a customisation warning (Fig. 6).

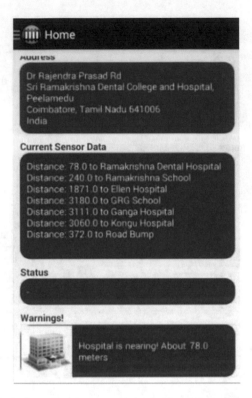

Fig. 6 Screenshot displaying a Geo-fence warning

4 Results and Discussion

4.1 Experiment 1

4.2 Experiment 2

In this part of the experiment as discussed already, we have a hardware (HW) setup to demonstrate the idea and theory (Figs. 7 and 8). The HW is assembled mainly for demonstration purpose and it is not a complete HW setup for actual application.

As described already, there are a few switches assembled for transferring of data at the same time illustrating one for each sign board. With the transmitter box and receiver box set up in line of sight, buttons are pressed randomly and this is finally read in the smartphone application. And corresponding warnings are sent to the

Fig. 7 Hardware setup for the transmission side

Fig. 8 Hardware setup for the receiving side

screen (Fig. 9). The layout is the same for both warnings rising out of a geo-fencing restriction and a sign board restriction. The application integrates both the monitoring and displays necessary warnings (Fig. 10).

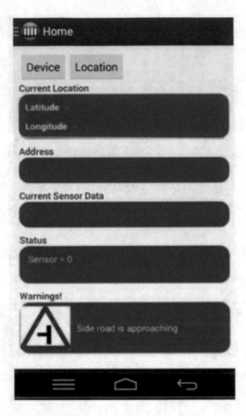

Fig. 9 Screenshot for sensing and display of a side road warning

5 Conclusion

The analysis needs design and implementation of an easy to work driver-assistance system in both normal and fleet management was performed in this work. The proposed idea as split into three major parts and can be concluded in the following way. Firstly, the geo-fencing and restriction based on that. This could be done by all available resources like the GPS in the smartphones and information from the Google maps but it is an easily implementable add on to the vehicle/fleet owners.

Secondly, the sign board detection proposal using radio frequency (RF) transmission and reception. The proposed method overcomes disadvantages like high processing time, disturbances like fog, dust, a visual block of the sign board, etc.,

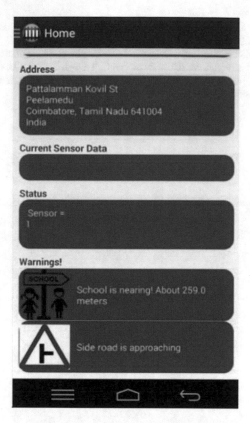

Fig. 10 Screenshots for sensing and display of sign board warnings

and establishes a stable method. Eventually, this generally would come under the 'brought in' category of devices in the automobile segment instead of the 'built in' category and as a smartphone application, which is very common.

References

1. Reclus, F., Drouard. K.: Geofencing for fleet & freight management. In: 2009 9th International Conference on Intelligent Transport Systems Telecommunications (ITST). IEEE (2009)
2. Beckert, R.D. et al.: Vehicle computer system with high speed data buffer and serial interconnect. U.S. Patent No. 6,009,363. 28 Dec 1999
3. Gatty, V., Kadam, M., Dighe, B.: Enhancement of a road sign board using wireless sensors. Int. J. Recent Innov. Trends Comput. Commun. **3**(6), 3938–3941 (2015)
4. Website: https://developers.google.com/android/reference
5. Sinnott, R.W.: Virtues of the Haversine. Sky Telesc. **68**, 159 (1984) (actual application)
6. Jeon, B., Kim, R.Y.C.: A system for detecting the stray of objects within user-defined region using location-based services. Int. J. Softw. Eng. Its Appl. **7**, 355–362 (2013)

7. Lemelson, J.H., Pedersen. R.D.: GPS vehicle collision avoidance warning and control system and method. U.S. Patent No. 5,983,161. 9 Nov 1999
8. Wang, F.-Y.: Parallel control and management for intelligent transportation systems: Concepts, architectures, and applications. IEEE Trans. Intell. Transp. Syst. **11**(3), 630–638 (2010)
9. Monk, S.: Making Android Accessories with IOIO. O'Reilly Media, Inc. (2012)
10. Website: http://ubidots.com/docs/devices/android.html

A Surrogate Forward Model Using Artificial Neural Networks in Conjunction with Bayesian Computations for 3D Conduction-Convection Heat Transfer Problem

M. K. Harsha Kumar, P. S. Vishweshwara and N. Gnanasekaran

Abstract The present work describes the determination of heat flux at the boundary for a conjugate heat transfer problem based on a coupled three-dimensional conduction-convection fin numerical model, also referred to as complete model. The model is developed using commercially available software and solved along with Navier–Stokes equation in order to acquire the required temperature distribution. An inverse analysis is proposed by treating the boundary heat flux as unknown while the temperatures of the fin are known. The inverse analysis is greatly accomplished with the help of Bayesian framework that combines the solution of the forward model and the simulated measurements. Markov chain Monte Carlo (MCMC) is applied to explore the sample space that drives samples to proper convergence and the selection or acceptance of the new samples is performed using Metropolis–Hastings algorithm. Thus, the novelty of the present work is the use of artificial neural network (ANN) as surrogate model, that not only retains the full nature of the complete model but also acts as a fast forward model in the inverse analysis, within the Bayesian framework that quantifies the uncertainty of heat flux. The results of the present work emphasize that even for noise-added temperature data the final estimates are very close to the actual values and the uncertainty of the unknown heat flux is reported in terms of standard deviation accompanied by mean and maximum a posteriori (MAP).

Keywords ANN · Bayesian · Conjugate · Fin · Inverse

M. K. Harsha Kumar
Department of Mechanical Engineering,
PESIT South Campus, Bangalore, Karnataka, India
e-mail: Harsha84.nitk@gmail.com

P. S. Vishweshwara · N. Gnanasekaran (✉)
Department of Mechanical Engineering,
NITK Surathkal, Mangalore, Karnataka 575025, India
e-mail: gnanasekaran@nitk.edu.in

P. S. Vishweshwara
e-mail: vishweshwara.ps@gmail.com

© Springer Nature Singapore Pte Ltd. 2020   373
K. N. Das et al. (eds.), *Soft Computing for Problem Solving*,
Advances in Intelligent Systems and Computing 1057,
https://doi.org/10.1007/978-981-15-0184-5_33

1 Introduction

Most of the engineering problems deal with transfer of heat into and out of the system. The concept of heat transfer is involved in almost every branch of engineering. When the information about the source is available and the effect is measured then such problem is termed as direct problem. But, when the information about the source is not available and only the measured data is present, estimating the unknown source by using the information from the measured data is called inverse problem. Inverse problem is solved using deterministic and stochastic methods, the former one deals with the gradient-based method and the latter one is probability-based method, usually data driven. Beck et al. [1] provides fundamental understanding of the inverse heat transfer problems. Ozisik and Orlande [2] provide a comprehensive solution methodology for inverse heat transfer problems. The highlights of commonly used deterministic inverse approach in several heat transfer problems can be viewed in [3–6].

Many researchers used neural network to find the solution of the unknown parameters in the inverse heat transfer approach. Deng and Hwang [7] solved the forward model using neural network with back-propagation algorithm for obtaining temperature distribution and the inverse problem is solved to find out the boundary conditions. Ghadimi et al. [8] implemented neural networks for the estimation of heat flux in a locomotive brake disc. For an inverse heat conduction problem (IHCP), the heat transfer coefficient between a solid and fluid assembly has been estimated using the knowledge of temperature based on neural networks [9]. Jambunathan et al. varied several combinations of network training and successfully estimated the convective heat transfer coefficients for experimental data [10]. A volumetric heat generation of a Teflon cylinder was estimated using the combination of asymptotic computational fluid dynamics (ACFD) as the forward model and ANN as the inverse model [11]. Principal thermal conductivities were concurrently estimated using genetic algorithm (GA) with ANN approach for a honey comb structure material [12]. Recent times, Bayesian estimation of unknown parameters is becoming more popular in solving inverse problems. The Bayesian inference approach is mostly combined with MCMC method for solving the inverse problems [13]. The technique provided satisfactory results for many examples of inverse heat transfer problems. The unknown thermal parameters of the fin are estimated using Bayeisan-MCMC approach based on inexpensive experiments on natural convection [14]. ANN is combined with Bayesian-MCMC approach to solve the IHCP [15]. A 2D conduction with heat generation was considered to estimate the convective heat transfer coefficient and the thermal conductivity of the material. Based on the aforementioned literature, it has been found that the problem of conjugate heat transfer in the inverse problem has not adequately been dealt in literature. It is also identified that many parameter estimation problems in heat transfer have been attempted using a simplified or modified conduction model. The estimation of unknown parameters using Bayesian framework is becoming popular these days due to an extensive sample space which can provide the uncertainty of the final estimates. Therefore, in this work, a conjugate heat transfer model is considered based on the information reported in literature and a surrogate model is introduced to minimize the computational cost and to retain the physics of the conjugate model during the estimation of heat flux.

2 Methodology

Figure 1 depicts the overview of the present methodology adopted in this work. Initially, the numerical model is created with the known input parameters. The heat fluxes and temperatures obtained using CFD simulations are fed in to the neural network model where training is done to develop a surrogate model. On the other hand, one can obtain measured temperature data or simulated measurements. The network temperatures and experimental/simulated measurements are given as input for the calculation of posterior probability density function (PPDF). When the required numbers of samples are obtained, calculate mean, MAP and the standard deviation else the Markov chain process explores the sample space till the condition is satisfied.

3 Forward Model

The computations are executed on a 3D numerical fin model and the dimensions are obtained from Kumar et al. [16]. The numerical model is shown in Fig. 2a, consists of an aluminum base of size $(150 \times 250 \times 8)$ mm^3 and the longer edge of the aluminum base is kept normal to the ground. On the front surface of the base plate, at the center, mild steel fin of dimensions $(150 \times 250 \times 6)$ mm^3 is placed such that its longer edge is parallel to the longer edge of the base. Navier–Stokes and energy equations are

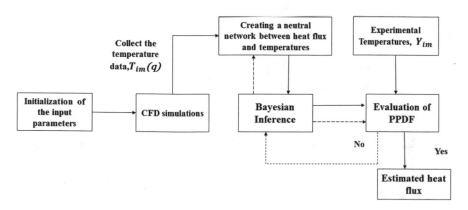

Fig. 1 Overview of the present work

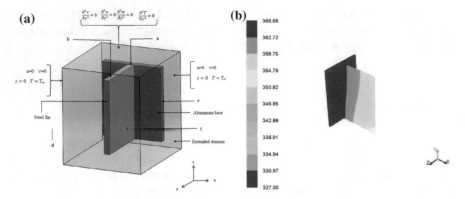

Fig. 2 **a** Schematic representation of the problem domain. **b** temperature contour along the height of fin obtained for heat flux of 1600 W/m^2

solved in conjunction with the conduction equation for the aluminum base and the mild steel fin; as a result, the present forward model becomes conjugate heat transfer problem. Air is considered to be fluid medium; hence, an extended domain is framed close to the fin to obtain the information about flow parameters. It is pertinent to mention here that the Boussinesq approximation is incorporated to treat density as the varying parameter with respect to temperature.

The mathematical representation of the problem is determined with the help of the governing differential equations expressed in 3D. All the faces of the extended domain are treated as pressure outlet for the numerical analysis. To obtain the temperature distribution, the heat flux is presumed while the boundary conditions are already available.

The governing equations used in numerical solutions are given as follows:

Continuity

$$\frac{\partial u}{\partial x} + \frac{\partial v}{\partial y} + \frac{\partial w}{\partial z} = 0 \tag{1}$$

Momentum equations

$$\rho \left(u\frac{\partial u}{\partial x} + v\frac{\partial u}{\partial y} + w\frac{\partial u}{\partial z} \right) = -\frac{\partial P}{\partial x} + \mu \left(\frac{\partial^2 u}{\partial x^2} + \frac{\partial^2 u}{\partial y^2} + \frac{\partial^2 u}{\partial z^2} \right) \tag{2}$$

$$\rho \left(u\frac{\partial v}{\partial x} + v\frac{\partial v}{\partial y} + w\frac{\partial v}{\partial z} \right) = -\frac{\partial P}{\partial y} + \mu \left(\frac{\partial^2 v}{\partial x^2} + \frac{\partial^2 v}{\partial y^2} + \frac{\partial^2 v}{\partial z^2} \right) + \rho g \beta (T - T_\infty) \tag{3}$$

$$\rho \left(u\frac{\partial w}{\partial x} + v\frac{\partial w}{\partial y} + w\frac{\partial w}{\partial z} \right) = -\frac{\partial P}{\partial z} + \mu \left(\frac{\partial^2 w}{\partial x^2} + \frac{\partial^2 w}{\partial y^2} + \frac{\partial^2 w}{\partial z^2} \right) \tag{4}$$

Energy equation for fluid

$$\rho c_p \left(u \frac{\partial T}{\partial x} + v \frac{\partial T}{\partial y} + w \frac{\partial T}{\partial z} \right) = k_f \left(\frac{\partial^2 T}{\partial x^2} + \frac{\partial^2 T}{\partial y^2} + \frac{\partial^2 T}{\partial z^2} \right) \tag{5}$$

where ρ is the density of fluid (kg/m^3), k_f is the thermal conductivity of the fluid (W/mK), C_p is the specific heat (J/kg K), g is the gravity constant (m/s^2), $\beta = 1/T$ (K^{-1}) and μ are the dynamic viscosity (Ns/m^2).

The solid walls are assigned no-slip condition; therefore, the velocity of the fluid on these walls becomes zero. A constant heat flux is specified as the thermal boundary condition for the base plate. The following are the details regarding the boundary conditions of the present numerical model [17–19].

Inlet velocity and temperature boundary conditions are

$$u = 0, v = 0, w = 0, T = T_\infty \tag{6}$$

Boundary conditions at outlet are

$$\frac{\partial^2 u}{\partial y^2} = 0, \frac{\partial^2 v}{\partial y^2} = 0, \frac{\partial^2 w}{\partial z^2} = 0, \frac{\partial^2 T}{\partial y^2} = 0 \tag{7}$$

Boundary condition is specified to the left side of the extended domain

$$\frac{\partial T}{\partial x} = 0 \tag{8}$$

Boundary condition at the right side of the computational extended domain

$$\frac{\partial u}{\partial x} = 0, \frac{\partial T}{\partial x} = 0, u = 0 \tag{9}$$

Aluminum base and mild steel fin:

$$k_{al} \left(\frac{\partial^2 T}{\partial x^2} + \frac{\partial^2 T}{\partial y^2} + \frac{\partial^2 T}{\partial z^2} \right) = 0 \tag{10}$$

At the base of the aluminum

$$- k_{al} \frac{\partial T}{\partial z} = q_0 \tag{11}$$

At region "a"

$$k_{al} \frac{\partial T}{\partial z} = k_{ms} \frac{\partial T}{\partial z} \tag{12}$$

where k_{al} is the thermal conductivity of the aluminum plate, k_{ms} is the thermal conductivity of the mild steel fin.

At region b, c, d

$$k_{ms}\frac{\partial T}{\partial x} = k_f \frac{\partial T}{\partial n} \quad \text{and } T_{ms} = T_{\text{fluid}} \tag{13}$$

At region e

$$\frac{\partial T}{\partial x} = 0 \tag{14}$$

The input to the numerical model is heat flux and the output is the temperature distribution. The temperature distribution for one such heat flux is shown in Fig. 2b. A study on grid independence is carried out to fix the number of nodes required for the present numerical analysis. The temperature values at five locations along the height of the fin are considered. Based on the study, nodes of 463116 have been chosen for further numerical simulations.

4 Fast Forward Model Using Artificial Neural Network

It is pertinent to mention that the forward model is executed to obtain the temperature distribution of the fin for the assumed value of heat flux given in Eq. (11) and known boundary conditions. Whereas, in the inverse approach, the heat flux, mentioned in Eq. (11) is treated as unknown and is estimated for the known temperature distribution which is obtained from measurements. CFD simulations for various values of heat flux are performed and corresponding temperatures are collected and a neural network is established that represents as fast forward model and used for inverse analysis. The neuron independence study was carried out and six neurons were sufficient for establishing the network.

5 Bayesian-Based Inverse Approach

The inverse model used for the present problem is a probabilistic-based method which follows the Bayes theorem, shown in Eq. (15) [20].

$$P(\mathbf{x}/\mathbf{Y}) = \frac{P(\mathbf{Y}/\mathbf{x})P(\mathbf{x})}{P(\mathbf{Y})} \tag{15}$$

where the posterior probability density function (PPDF) is given as $P(\mathbf{x}/\mathbf{Y})$; the likelihood function is $P(\mathbf{Y}/\mathbf{x})$; the prior density function is $P(\mathbf{x})$; $P(\mathbf{Y})$ is normalizing constant. The objective function of the Bayesian inference is to minimize the error between the measured temperature and the calculated temperature and assign maximum probability to the heat flux which results in zero error. Incorporating Bayesian algorithm to the present problem, the term $\mathbf{x}$ will be replaced by the heat flux, 'q'

and the observation $\mathbf{Y}$ will be replaced by temperature, $\mathbf{T}$. And the corresponding mathematical representation of the Bayesian model is shown in Eq. (16).

$$P(\mathbf{T}/q) = (2\pi\sigma^2)^{-n/2} \exp\left\{\frac{1}{2\sigma^2}[\mathbf{T}_{\text{meas}} - \mathbf{T}(q)_{\text{sim}}]^T[\mathbf{T}_{\text{meas}} - \mathbf{T}(q)_{\text{sim}}]\right\} \quad (16)$$

where n is the number of measurements

$$P(q) = (2\pi\sigma^2)^{-0.5} \exp\left\{\frac{1}{2\sigma^2}(q - \mu_p)^T(q - \mu_p)\right\} \quad (17)$$

$$P(\mathbf{x}/\mathbf{Y}) = (2\pi\sigma^2)^{-n/2} \exp\left\{\frac{1}{2\sigma^2}[\mathbf{T}_{\text{meas}} - \mathbf{T}(q)_{\text{sim}}]^T[\mathbf{T}_{\text{meas}} - \mathbf{T}(q)_{\text{sim}}]\right\}$$
$$(2\pi\sigma^2)^{-0.5} \exp\left\{\frac{1}{2\sigma^2}(q - \mu_p)^T(q - \mu_p)\right\} \quad (18)$$

Equation (16) is the likelihood function where the comparison between the simulated temperature obtained from the forward model and the experimental/simulated measurements is done. Equation (17) defines the prior model based on the a-priori information available before performing experiments. Equation (18) is termed as PPDF and as observed it is the product of likelihood density function and the prior probability density function. More importantly, solving Eq. (15) is not easy and one has to rely on MCMC samples to obtain the solution.

5.1 Metropolis–Hastings Algorithm

When one uses Markov chain Monte Carlo method to collect the samples of the unknown parameter, the parameter is accepted or rejected based on Metropolis–Hastings (MH) algorithm [21, 22]. The procedure for the algorithm is given as

1. Draw a sample u $\sim$ U(0,1), i.e., from a uniform distribution between 0 and 1.
2. Draw a sample $x^* \sim q(x^*/x^i)$
3. If u $< A(x^*, x^i)$, $x^{i+1} = x^*$, where $A(x^*, x^i) = \min\left(1, \frac{P(x^*).q(x^i/x^*)}{P(x^i).q(x^*/x^i)}\right)$
4. Else go to step 2 with $x^{i+1} = x^i$.

6 Results and Discussion

6.1 Fast Forward Model

The estimation of unknown heat flux under the inverse approach is time consuming due to multiple evaluation of forward model in the Markov chain process. Hence, it

Table 1 Comparison of ANN temperature with CFD temperature

Sl. No.	Heat flux (W/m^2)	ANN	CFD	Error (%)
		Temperature in K	Temperature in K	
1	1300	352.74	353.03	0.082
2	1700	366.89	366.90	0.002
3	2500	393.83	393.66	0.043
4	3300	419.64	419.49	0.035

is inevitable to bring in the surrogate model but in principle the information about the complete model should be retained in the surrogate model. As a result, ANN is used as a fast forward model so as to minimize the computation time of the forward model. The output from the neural network is compared with the simulation temperature which buttresses the use of ANN as the surrogate model. This comparison is shown in Table 1 and a maximum error of 0.082% has been observed.

6.2 Estimation of Heat Flux from the Simulated Measurement

The unknown heat flux is estimated for the simulated measurements. The simulated measurements are the temperature distribution obtained from forward model and later added with noise to mimic the real-time experiments. Treating the temperature distribution as known and the heat flux that caused the temperature distribution as unknown, an inverse problem is framed in estimating the unknown heat flux using the proposed Bayesian framework. Table 2 shows the retrieved values of the heat flux using Markov Chain Monte Carlo and Metropolis–Hastings algorithm. The heat flux is expressed in terms of mean, SD and MAP. The computation time required to run one CFD simulation is 20 min. From Table 2, it is observed that the estimated values of the heat flux closely match the values of actual heat flux.

Table 2 Estimated values of heat flux

Sl. No.	Heat flux (W/m^2)	Mean	SD	MAP
1	1100	1101.8	2.0789	1100.1
2	1700	1705.2	1.9336	1700
3	2150	2147.2	1.9254	2150
4	3500	3475.6	1.6911	3499.6

6.3 Retrieval of Heat Flux for the Noisy Data

To prove the soundness and efficacy of the proposed inverse solution, the unknown heat flux is retrieved for the simulated measurements added with noise. The temperature obtained from neural network is added with the Gaussian noise, expressed in Eq. (19), where T_i is the temperature obtained from neural network at $i = 1$ to n points. The retrieved heat flux value for the noisy data is represented in Table 3. The PPDF plot for the heat flux value of 3500 W/m^2 after the noise addition is shown in Fig. 3a. The maximum error as seen from Table 3 is 3.45% which signifies the fact that even with the presence of measurement error, Bayesian inverse algorithm is able to estimate the unknown values in close proximity.

$$Y_i = T_i + \varepsilon \qquad (19)$$

With the proper selection of the *a priori* information, the error in the estimation can be further reduced. When information about the prior model is available then such prior model is called as 'informative prior'. An attempt has been made to incorporate the informative prior model termed as normal prior and retrieval of the unknown heat flux is performed for the noisy temperature data. Table 4 shows the results of such an exercise. From Table 4, it observed that with the use of the normal

Table 3 Retrieved values of heat flux

Sl. No.	Actual heat flux (W/m^2)	Actual temperature (K)	Noise-added temperature data (K)	Retrieved heat flux (W/m^2)	Error (%)
1	1100	346.0001	347.5001	1138.7	3.45
2	1700	366.9029	368.4029	1742.9	2.52
3	2150	382.3875	383.8875	2199.3	2.29
4	3500	425.4793	426.9793	3545	1.28

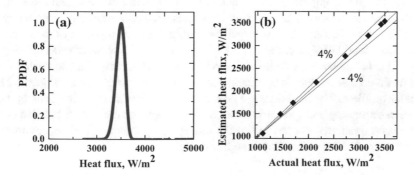

Fig. 3 **a** PPDF of 3500 W/m^2 with noise addition. **b** parity plot W/m^2

Table 4 Retrieved heat flux for the *a-priori* information

Sl. No.	Actual heat flux (W/m²)	Retrieved heat flux (W/m²)	Error (%)
1	1100	1086.1	1.26
2	1700	1728.9	1.70
3	2150	2142.4	0.35
4	3500	3496.6	0.09

prior in the Bayesian retrieval algorithm the error in estimation is still reduced as in comparison with Table 3 in which uniform prior was used. With the proper selection of a prior information the error between the actual and estimated heat flux comes down drastically in turn reduces the standard deviation of the retrieved heat flux. This is the greatest advantage of using Bayesian framework in the inverse problems over other methods. A parity plot that shows a good agreement between the retrieved and actual value of the heat flux for the noisy data is shown in Fig. 3b.

7 Conclusion

A hybrid methodology has been successfully implemented by combining ANN with Bayesian framework. A 3D numerical fin model was chosen to be the conjugate heat transfer forward model. The inverse problem was framed in such a way that the temperature distribution of the conjugate fin model is known and the heat flux that caused the temperature distribution was unknown; hence, the sole idea was to estimate the unknown heat flux and to report the uncertainty associated with the estimate in order to quantify the heat loss so that one can efficiently design a thermal system. Upon establishing the neural network, the efficiency of the network was tested and the maximum error was found to be 0.082%. Using the hybrid technique, the heat flux for the 3D conjugate fin heat transfer problem was estimated and reported in terms of mean, SD and MAP. The simulated measured data was then corrupted with the addition of the Gaussian error and the maximum error reported in the retrieval of heat flux was 3.45%. Hence, the task of assimilating ANN, which was developed by CFD simulations, with Bayesian-based MCMC algorithm evince to be an effective methodology for the determination of unknown heat flux for a conjugate heat transfer problem. The methodology presented in this work not only opens up new vistas in estimating multiple parameters with the sophisticated model but also helps reduce computational uncertainty in the emerging approach called approximation error model [23, 24] within Bayesian framework.

References

1. Beck, J.V., Blackwell, B., Clair, C.R., Jr.: Inverse Heat Conduction: Ill-Posed Problems. Wiley (1985)
2. Ozisik, M.N., Orlande, H.R.B.: Inverse Heat Transfer: Fundamentals and Applications. CRC Press (2000)
3. Alifanov, O.M., Kerov, N.V.: Determination of external thermal load parameters by solving the two-dimensional inverse heat-conduction problem. J. Eng. Phys. Thermophys. **41**(4), 1049–1053 (1981)
4. Huang, C.H., Jan, L.C., Li, R., Shih, A.J.: A three-dimensional inverse problem in estimating the applied heat flux of a titanium drilling-theoretical and experimental studies. Int. J. Heat Mass Transfer. **50**(17–18), 3265–3277 (2007)
5. Huang, C.H., Chang, W.L.: An inverse problem in estimating the volumetric heat generation for a three-dimensional encapsulated chip. J. Electron. Pack. **132**(1), 011004 (2010)
6. Pereyra, S., Lombera, G.A., Frontini, G., Urquiza, S.A.: Sensitivity analysis and parameter estimation of heat transfer and material flow models in friction stir welding. Mater. Res. **17**(2), 397–404 (2014)
7. Deng, S., Hwang, Y.: Applying neural networks to the solution of forward and inverse heat conduction problems. Int. J. Heat Mass Transfer. **49**(25), 4732–4750 (2006)
8. Ghadimi, B., Kowsary, F., Khorami, M.: Heat flux on-line estimation in a locomotive brake disc using artificial neural networks. Int. J. Ther. Sci. **90**, 203–213 (2015)
9. Sablani, S.S., Kacimov, A., Perret, J., Mujumdar, A.S., Campo, A.: Non-iterative estimation of heat transfer coefficients using artificial neural network models. Int. J. Heat Mass Transfer. **48**(3), 665–679 (2005)
10. Jambunathan, K., Hartle, S.L., Ashforth-Frost, S., Fontama, V.N.: Evaluating convective heat transfer coefficients using neural networks. Int. J. Heat Mass Transfer. **39**(11), 2329–2332 (1996)
11. Gnanasekaran, N., Kumar, S., Kumar, H.: A neural network based method for estimation of heat generation from a Teflon cylinder. Front. Heat Mass Transfer **7**(1) (2016)
12. Chanda, S., Balaji, C., Venkateshan, S.P., Yenni, G.R.: Estimation of principal thermal conductivities of layered honeycomb composites using ANN-GA based inverse technique. Int. J. Ther. Sci. **111**, 423–436 (2017)
13. Wang, J., Zabaras, N.: A bayesian inference approach to the inverse heat conduction problem. Int. J. Heat Mass Transfer. **47**(17–18), 3927–3941 (2004)
14. Gnanasekaran, N., Balaji, C.: A bayesian approach for the simultaneous estimation of surface heat transfer coefficient and thermal conductivity from steady state experiments on fins. Int. J. Heat Mass Transfer. **54**(13–14), 3060–3068 (2011)
15. Balaji, C., Padhi, T.: A new ann driven mcmc method for multi-parameter estimation in two-dimensional conduction with heat generation. Int. J. Heat Mass Transfer. **53**(23–24), 5440–5455 (2010)
16. Mk, H.K., Vishweshwara, P.S., Gnanasekaran, N., Balaji, C.: A combined ANN-GA and experimental based technique for the estimation of the unknown heat flux for a conjugate heat transfer problem. Heat. Mass. Transfer. **31**(11), 3185–3197 (2018)
17. Premachandran, B., Balaji, C.: Conjugate mixed convection with surface radiation from a vertical channel with protruding heat sources. Numer. Heat. Tr. A-Appl. **60**(2), 171–196 (2011)
18. Dorfman, A.S.: Conjugate Problems in Convective Heat Transfer. CRC Press (2009)
19. Ahamad, S.I., Balaji, C.: A simple thermal model for mixed convection from protruding heat sources. Heat. Transfer. Eng. **36**(4), 396–407 (2015)
20. Mota, C.A., Orlande, H.R., De Carvalho, M.O.M., Kolehmainen, V., Kaipio, J.P.: Bayesian estimation of temperature-dependent thermophysical properties and transient boundary heat flux. Heat. Transfer. Eng. **31**(7), 570–580 (2010)
21. Chib, S., Greenberg, E.: Understanding the metropolis-hastings algorithm. Am. Stat. **49**(4), 327–335 (1995)

22. Kumar, H., Kumar, S., Gnanasekaran, N., Balaji, C.: A markov chain monte Carlo-Metropolis hastings approach for the simultaneous estimation of heat generation and heat transfer coefficient from a teflon cylinder. Heat. Transfer. Eng. **39**(4), 339–352 (2018)
23. Berger, J., Orlande, H.R., Mendes, N.: Proper generalized decomposition model reduction in the Bayesian framework for solving inverse heat transfer problems. Inverse. Prob. Sci. Eng. **25**(2), 260–278 (2017)
24. Lamien, B., Orlande, H.R.B., Eliabe, G.E.: Particle filter and approximation error model for state estimation in hyperthermia. J. Heat. Transfer. **139**(1), 012001–012012 (2017)

Analysis of Spatial Domain Image Steganography Based on Pixel-Value Differencing Method

C. D. Nisha and Thomas Monoth

Abstract Image steganography is a technique of embedding secret information inside a cover image and transmits through a public channel without revealing the presence of a message. So that no one except the intended recipient can recognize secret message within the carrier image. Image steganography based on pixel value differencing (PVD) is one of the most important steganographic methods in spatial domain. To embed secret information, PVD method utilizes the difference value of each pixel pair on cover image. In this paper, we study the basic concepts of image steganography based on PVD method and latest research developments in image steganography using PVD. We also presented a detailed analysis of these techniques on the basis of some performance parameters such as payload capacity, imperceptibility, and robustness.

Keywords Steganography · Image steganography · Spatial domain image steganography · Pixel-value differencing

1 Introduction

Emergence of Internet made a drastic change in our communication system. In this present scenario, most of the information travel over Internet. Secure information transfer over Internet requires a great concern in today's world. The main challenging issue in Internet communication is the privacy of data. In order to provide proper security for data in Internet, a lot of security mechanisms are evolved in recent

C. D. Nisha (✉)
Department of Information Technology, Kannur University, Kannur 670567, Kerala, India
e-mail: raisacarmel@gmail.com

T. Monoth
Department of Computer Science, Mary Matha Arts & Science College, Kannur University, Mananthavady, Wayanad 670645, Kerala, India
e-mail: tmonoth@yahoo.com

© Springer Nature Singapore Pte Ltd. 2020
K. N. Das et al. (eds.), *Soft Computing for Problem Solving*,
Advances in Intelligent Systems and Computing 1057,
https://doi.org/10.1007/978-981-15-0184-5_34

years. Steganography is one of the most important strategies among them. Nowadays, steganography plays a major role in secret communication. In the business world, steganography can be used to hide some secret methodology or plans for a new invention. Steganography can also be used for sending out trade secrets of a company without knowing anyone except the intended one [1]. Terrorists can also use steganography to keep their communications secret and to coordinate attacks. In medical field, it was also used to hide the result of medical test in digital medical image [2].

Steganography is an ancient art of information hiding. The first steganographic technique was developed in primeval Greece around 440 BC [3, 4]. It was documented by the Greek historian Herodotus in his chronicles 'Histories.' In that record, Herodotus mentioned two events associated with the implementation of steganography in ancient Greece. He tells the story of Demaratus, the son of Ariston who at the time was exiled in Persia. While he was there, he learned of the imminent attack on Greece by the Persians. As the story goes, Demaratus decided to inform the king of Spartans about the attack by writing the warning, on a wooden tablet and covered it with wax and sent it to him. By doing this, he saves lives of people in Greece from a harmful attack. Steganography continued over time to develop into new levels. During the times of war, steganography is used extensively. During the period of American Revolutionary War both American and British army used invisible inks for secret communication. Some commonly used invisible inks on those days included milk, vinegar, fruit juice, and urine. To decipher these hidden messages required light or heat [4]. The classic representation of digital steganography was first stated by Simmons as *Prisoner's Problem* [5]. Here the problem was explained on the basis of the story of two individuals: Alice and Bob who were sentenced for few years to imprisonment due to the crime executed and were thrown into two different cells in jail. In order to get away from the jail, they were trying to develop an escape plan. But they were not allowed to communicate with each other by using any traditional methods because they were under the control of the warden named Wendy. To overcome this situation, they had to build up a secret communication strategy by embedding messages within an innocuous cover object and pass this cover object through a public channel. Here the messages were hidden inside a carrying object and its existence will never be noticed by warden Wendy.

2 Digital Image Steganography

Steganography is of different types. Depending upon the carrier media, we can categorize it into text steganography, image steganography, audio steganography, video steganography, and network steganography [6, 7, 10]. Due to the high redundancy in intensity values of pixels, most of the steganography methods select images as their carrier object. In image, steganography images are to be considered as cover object

[7, 13]. In literature, we can see that there exist various image steganographic methods. Generally, image steganography techniques can be classified into two classes based on their embedding domain, namely transform domain image steganography and spatial domain image steganography [6]. To embed a message in transform domain, firstly the cover image must be transformed into its equivalent frequency domain coefficients. Embedding in frequency coefficients provides better information hiding capacity and security [8, 12]. In spatial domain, embedding is applied directly on the gray levels of each pixel [9]. Some spatial domain methods include simple LSB steganography [10], LSB matching revisited (LSBMR) [11], edge-based embedding (EBE) [12], pixel value differencing (PVD) [13]. Spatial domain methods have larger embedding capacity, but they are less robust against various image manipulation operation. Among the various spatial domain methods, PVD method exhibits better performance in terms of payload capacity and imperceptibility. The succeeding section gives a short picture of the fundamentals of PVD-based image steganography scheme.

3 Image Steganography Based on PVD

PVD-based steganography was proposed by Wu and Tsai in 2003 [13]. The basic idea behind PVD method depends upon the characteristic features of human visual system. In human visual system, human eyes are not sensitive to the changes in high contrast pixels of an image, whereas they are sensitive to low contrast pixels. In PVD method, edge regions are selected to embed large amount of data because changes made in edge region are unnoticeable to human eyes. The difference value of pixels determines whether a pixel belongs to smooth area or in edge area. In this way, PVD method offers a better steganography strategy by ensuring high embedding capacity, imperceptibility, and security [14, 15]. In PVD scheme, cover image is divided into non-overlapping blocks of two successive pixels $c_i, c_i + 1$ in raster scan order. The intensity values of those pixels are represented by $v_i, v_i + 1$. The difference value $|d_i|$ of each block can be calculated by subtracting v_i from $v_i + 1$. For all pixels in grayscale image, the difference values should be in the range -255 to 255 and their absolute value is from 0 to 255. A minute value of d_i represents the existence of a smooth area, whereas a bigger value specifies the existence of the sharp region. The difference values can be classified into several regions. The lower and upper boundary of each R_i is denoted by $[l_i, u_i]$. The number of secret bits n embedded in two consecutive pixels depends on the range table, and it was calculated by using formula $n = \log_2(u_i - l_i + 1)$. Then the value of n is converted into its decimal equivalent n_d. After that new difference value d_i' is computed by $d_i' = n_d + l_i$. To embed n bits of secret data into the PVD of c_i and $c_i + 1$ we apply the function f $((v_i, v_i + 1), s)$ where $s = |d_i' - d_i|$. Then we get the modified pixel values $v_i', v_i + 1'$ [13].

PVD-based steganography is an efficient steganographic method for hiding secret message into cover image without generating perceptible changes in the cover image. Also, this method allows us to extract secret message from the stego-image without referencing original carrier image. This method not only provides a better way for hiding large amounts of data into cover images without any visual artifacts, but also offers an easy way to accomplish secrecy. The embedding and extraction procedures are outlined below [13].

Algorithm 1. Embedding procedure

1. Determine the difference of intensity values v_i and v_i+1 of two successive pixels c_i and c_i+1 in cover image. It is given by $d_i = |v_i+1 - v_i|$.
2. Calculate the width of range in which difference value d_i belongs by using the equation $w_i = u_i - l_i + 1$.
3. Compute the total number of bits to be hidden in a pixel block which can be defined as $n = (log_2 w_i)$.
4. Read n bits from binary secret data and convert it into its corresponding decimal equivalent n_d.
5. Calculate $d_i' = n_d + l_i$.
6. Modify the pixel intensity values by using following equation
   $f((v_i, v_i+1), s) = (v_i', v_i+1')$
   $= (v_i - \text{ceiling}_s, v_i+1+\text{floor}_s)$; if d is an odd figure ($v_i - \text{floor}_s, v_i+1 + \text{ceiling}_s$); if d is an even figure, where $s = |d_i' - d_i|$.
7. Repeat steps 1–6 until all secret bits are embedded into the cover image.

Algorithm 2. Extraction procedure

1. Divide the stego-image into non-overlapping blocks of two consecutive pixels.
2. Calculate new difference value $d_i' = |g_i+1' - g_i'|$
3. Find the range w_i of d_i' and calculate $n_d = |d_i' - l_i|$
4. Convert n_d into its corresponding binary value of 'n' bits, $n = (log_2 w_i)$, where 'n' bits are the hidden data obtained from the pixel block(v_i', v_i+1').

4 Recent Developments in PVD Method

This section tries to give an overview of the recent research developments in PVD-based image steganography technique during the period of 2003 to 2018. It also evaluates the merits and challenges of each method by using various performance measures.

Wu and Tsai [13] introduced an innovative steganographic method for embedding covert message into grayscale images. During this process, cover image is divided into non-overlapping blocks of two consecutive pixels. Each block is classified according to the pixel-value difference of pixels within a block. Smooth area of a cover image holds pixels with small intensity values, and edge area contains large difference values. The number of bits embedded in cover image is determined by the width of the range in each classification. Compared to simple LSB, this method offers better payload capacity and imperceptibility. Moreover, we can retrieve a hidden message from stego-image without referencing original cover.

Zhang and Wang [16] research studies show that the histogram of stego-image formed by the original PVD method consists of unusual steps. In order to prevent this vulnerability, they suggested a refined version of PVD method in which range table is designed by a randomly generated secret key. Instead of using fixed ranges, this method supports variable ranges for each pixel pairs. The investigational results illustrate that the proposed method effectively removes the chance of detecting hidden information from stego-pixels. It also preserves payload capacity and imperceptibility of original PVD.

To improve the hiding capacity and image quality of cover image, Chang et al. [17] proposed a novel steganographic scheme using tri-way pixel-value differencing method in 2008. Here the carrier image is partitioned into 2×2 blocks of four pixels. Then calculate the pixel difference of each pixel pair in all directions, i.e., one horizontal, one vertical, and two diagonals. Select three pixel pairs from each block and embed secret bits on that pixels by applying conventional PVD [13] method. The main advantage of this scheme is that it hides more data in three-directional edges than traditional PVD. And it also ensures more imperceptibility compared to other methods.

In order to overcome the shortcomings such as irregular increases and fluctuations in the PVD histogram, Joo et al. [18] devised an embedding method based on turnover policy. The proposed steganographic algorithm consists of four states: 1. pixel grouping state, 2. embedding state, 3. adjusting state, and 4. overcoming state. In the pixel grouping state, cover image is divided into non-overlapping blocks of two adjacent pixels. In embedding state, difference value of two pixels is simply adjusted by a modulus function to map with message value. The modifying value m is used to minimize the distortion of stego-image. The adjusting state is employed to destroy fluctuations shown in the PVD histogram of stego-image. The overcoming state prevents abnormal increases in the histogram by implementing turnover policy.

In 2011, Pan et al. [19] presented a PVD-based steganography method which considers pixel pairs in both horizontal and vertical direction for data embedding. This method divides the cover image into 2×2 non-overlapping blocks of pixels pairs, two in horizontal direction, and others in vertical direction. The algorithm utilizes modulus function method to embed data in horizontal pixel pairs. To embed data in vertical direction, one pair of pixel is selected among two and traditional PVD method is applied for hiding purpose. Experiments show that suggested scheme contributes a good result in embedding capacity, visual quality, and security.

In 2011, Luo et al. [20] implemented an adaptive technique in PVD scheme. According to adaptive pixel value differencing (APVD) method, firstly cover image is partitioned into squares of size AB × *AB,* where *AB* is a multiple of 3 and then rotating each square of image by a random degree of 0, 90, 180 or 270. After that, rotated image is divided into non-overlapping blocks of three consecutive pixels. Middle pixel element of each block is selected for data embedding by considering difference value of three pixels and predefined threshold value T. The number of bits embedded within the middle pixel is determined by the range value of that pixel. Investigational results show that the anticipated approach achieves better embedding capacity and security compared with the previous PVD-based methods.

In 2012, Mandal and Das [21] devised a new PVD steganographic method for color images. To execute this scheme, color components of each pixel is separated and create three M*N matrices, to represent each component. PVD method is applied in each matrix separately, in a sequencing manner. In this scheme, number of bits hidden in each channel is different from each other. This improves overall security and visual quality.

Nagaraj et al. [22] proposed a pixel value modification method by using modulo three functions. This scheme is succeeded by separating image into three color components, and each one is represented by a matrix. By applying PVD method, secret bits are embedded in three color planes of selected pixels. Implementation of pixel-value modification in PVD method provides better visual quality for stego-images. Falling-off boundary problem existing in the PVD method has been removed using this method.

In order to improve the embedding capacity and visual quality of stego-image, Tseng and Leng [23] proposed a new idea for designing quantization range table. In Wu and Tsai's [13] scheme, the quantization range table is designed by considering width of range table as power of two. The proposed method designs a new quantization range table based on a perfect square number. For each pixel value, select nearest perfect square number n^2. The experimental result proves that the proposed method ensures greater capacity and higher PSNR value than Wu and Tsai's method.

Swain [24] proposed two adaptive steganographic techniques to improve the capacity of existing PVD methods. In the first method, he suggested the partitioning of carrier image into 2 × 2 blocks of non-overlapping pixels by scanning image in raster scan manner. To embed secret data, both left upper and right bottom pixels are selected according to their association with neighboring two pixels. This method gives better hiding capacity. In second method, carrier image is partitioned into 3 × 3 non-overlapping pixel blocks. In each block, central pixel is selected for embedding and other pixels are considered as neighbor pixels. Depending on four different values in each block and predefined threshold value T, secret bits are embedded in middle pixel of each block. The second method ensures better visual quality with high PSNR value.

Hameed et al. [25] presented an adaptive directional pixel value differencing (ADPVD) method for color images by considering embedding direction and pixel-value difference of adjacent pixels in three color channels. The proposed scheme hides more secret bits in a vertical and diagonal direction because its difference values were greater than the pixel-value difference in horizontal direction. This increases hiding capacity of ADPVD method. Moreover, color channels and number of bits embedded in each channel were varying in each block. This will provide better security and hiding efficiency.

Prasad and Pal [26] proposed a new color steganography in 2017 based on overlapping block-based pixel-value method. Initially, pixels in a cover image are decomposed into three color components and represent these pixels as a combination of two successive pixels in dissimilar color planes like (R, G), (G, B), respectively. Then PVD method is executed in each block independently to hide secret information. To achieve proper color value to stego-image, intermediate color components are readjusted. The experimental observation proves that proposed method is capable to provide better visual quality to stego-images with acceptable PSNR value and high embedding rate.

Swain [27] proposed two PVD-based steganographic methods—one for uniform embedding and other for adaptive embedding scheme. In both schemes, cover image is partitioned into 1×2 overlapped pixel blocks of adjacent pixels. First method executes arithmetic addition/subtraction mechanism to embed and extract secret message by using a fixed quantization range table and second method uses adaptive quantization range table and modular arithmetic for the same purpose. The experimental results show that proposed APVD technique possesses higher embedding capacity and PSNR as compared to existing APVD techniques and proposed non-APVD technique acquires higher embedding capacity as compared to existing non-adaptive PVD techniques.

5   Analysis of Research Advances in Image Steganography Based on PVD Method

In this session, we presented a detailed analysis of recent research trends explored in the area of PVD image steganography along with its merits and demerits. The outcome of our analysis is shown in Table 1.

Table 1 Analysis of major PVD-based image steganography schemes

S. No.	Author and year	Method used	Image type	Merits	Demerits
1	Wu and Tsai [13]	PVD with fixed range width	Grayscale image	1. More imperceptible than simple LSB 2. Easy to extract the secret message without referring the original cover image	1. Vulnerable to histogram-based steganaylsis 2. Not applicable to color images 3. Embedding capacity of cover based on range width 4. Some blocks are abandoned because of falling-off boundary problem
2	Zhang and Wang [16]	PVD with variable range	Grayscale images	1. Good imperceptibility 2. Overcomes the histogram-based steganalysis	1. Only for grayscale images 2. Less embedding capacity compared to original PVD 3. Falling-off boundary problem
3	Chang et al. [17]	Tri-way pixel-value differencing	Grayscale images	1. High embedding capacity 2. Protection from dual statistical stego-analysis 3. Reduce Distortion	2. Increases the computational complexity

(continued)

Table 1 (continued)

S. No.	Author and year	Method used	Image type	Merits	Demerits
4	Joo et al. [18]	PVD with histogram preserving modulus function, Implements Turnover policy & pixel adjusting process	Grayscale images	1. Prevents the abnormal increases in the histogram values 2. Removes the fluctuations in the PVD histogram 3. Solve falling-off boundary problem	1. Complexity increases 2. Less image quality compared to Zhang's [16] method
5	Pan et al. 2011 [19]	PVD and modulus function, embedding data in vertical and horizontal pixel pairs	Grayscale images	1. Large embedding capacity 2. Easy to implement	1. Average image quality compared
6	Luo et al. [20]	Adaptive PVD	Grayscale images	1. Higher security compared to other PVD method 2. Preserve local statistical features	1. Greater computational complexity 2. Restriction for data embedding

(continued)

Table 1 (continued)

S. No.	Author and year	Method used	Image type	Merits	Demerits
7	Mandal and Das [21]	PVD in color images	24-bit color images	1. Improves the hiding capacity 2. Easy to retrieve secret message without using the original cover image 3. Support color images	1. Requires additional computation
8	Nagaraj et al. [22]	PVD with modulus function	24-bit color images	1. Less distortion 2. Provide more security 3. Better visual quality 4. Solve falling-off boundary problem	1. Difficult to implement
9	Tseng and Leng [23]	PVD with Perfect square number	Grayscale images	1. Large capacity 2. High imperceptibility	1. Requires additional computation for defining range table
10	Swain [24]	Adaptive PVD embedding horizontal and vertical edge	Grayscale images	1. Technique 1: high embedding rate 2. Technique 2: high PSNR	1. Technique1: less PSNR 2. Technique2: low embedding rate

(continued)

Table 1 (continued)

S. No.	Author and year	Method used	Image type	Merits	Demerits
11	Hameed et al. [25]	Adaptive directional pixel-value differencing (ADPVD)	Color images	1. High embedding capacity in a vertical and horizontal direction 2. Better visual quality	1. PSNR value becomes low in horizontal and vertical direction
12	Prasad and Pal [26]	Overlapping block-based pixel-value differencing	Color Images	1. High embedding rate 2. Better visual quality compared to other methods	1. Computational complexity
13	Swain [27]	Adaptive and non-adaptive PVD using overlapped pixel blocks	Color images	1. High embedding capacity 2. High PSNR	1. Increases the complexity of embedding and extracting algorithms

6 Conclusion

The foremost objective of this article is to discuss the role of PVD methods in spatial domain image steganography. By investigating the research papers, it is realized that several PVD methods are explored to increase embedding capacity and visual quality of stego-images. The hiding capacity of PVD methods depends on the difference value of pixels and width of range table in which the difference belongs. Most of the studies are based on fixed range table, and this will help the steganalysts to detect secret data from stego-image. Hence, future studies must be focused on the range table with variable size. The survey shows that the implementation of PVD steganography in color images was very few compared to grayscale images. By considering all factors discussed above, we suggest a hybrid steganography approach that can offer better payload capacity, security, and imperceptibility than existing PVD methods.

References

1. Doshi, Ronak, Jain, Pratik, Gupta, Lalit: Steganography and its applications in security. Int. J. Mod. Eng. Res. (IJMER) 2(6), 4634–4638 (2012)
2. Johnson, N.F., Jajodia, S.: Exploring steganography: seeing the unseen. IEEE Trans. Comput. 2, 26–34 (1998) (IEEE)
3. Siper, A., Farley, R., Lombardo, C.: The rise of steganography. In: Proceedings of Student/Faculty Research Day, CSIS, Pace University (2005)
4. Yunus, Y.A., Ab Rahman, S., Ibrahim, J.: Steganography: a review of information security research and development in muslim world. Am. J. Eng. Res. (AJER), 2(11), 122–128 (2013)
5. Simmons, G.J., The prisoners' problem and the subliminal channel. In: Advances in Cryptology, Proceedings of CRYPTO '83, in: Lecture Notes in Computer Science, Plenum, New York, pp. 51–67 (1983)
6. Sumathi1, C.P.., Santanam, T., Umamaheswari, G.: A study of various steganographic techniques used for information hiding. Int. J. Comput. Sci. Eng. Surv. (IJCSES) 4(6), 9–25 (2013)
7. Shih, F.Y.: Digital Watermarking and Steganography: Fundamentals and Techniques, p. 137. CRC Press, Taylor & Francis Group, HRD (2007)
8. Shelke, S.G.., Jagtap, S.K.: Analysis of spatial domain image steganography techniques. In: International Conference on Computing Communication Control and Automation, pp. 665–667. IEEE Computer Society (2015). https://doi.org/10.1109/iccubea.2015.136
9. Muhammad, K., Ahmad, J., Rehman, N.U., Jan, Z., Sajjad, M.: CISSKA-LSB: color image steganography using stegokey-directed adaptive LSB substitution method. Multimedia Tools Appl. . (2016). https://doi.org/10.1007/s11042-016-3383-5 (Springer)
10. Chan, C.K., Cheng, L.M.: Hiding data in images by simple LSB substitution. Pattern Recogn. 37(3), 469–474 (2004) (Elsevier)
11. Mielikainen, J.: LSB matching revisited. Sig. Process. Lett. 13(5), 285–287 (2006) (IEEE)
12. Luo, W., Huang, F., Huang, J.: Edge adaptive image steganography based on LSB matching revisited. IEEE Trans. Inf. Forens. Secur. 5(2), 201–214 (2010) (IEEE)
13. Wu, D.-C., Tsai, W.-H.: A steganographic method for images by pixel value differencing. Pattern Recogn. Lett. 24, 1613–1626 (2003) (Elsevier)
14. Yang, C.H., Weng, C.Y., A steganographic method for digital images by multipixel Differencing. In: Proceedings of International Computer Symposium, Taipei, Taiwan, ROC, pp. 831–836 (2006)

15. Luo, W., Huang, F., Huang, J.: A more secure steganography based on adaptive pixel-value differencing scheme. Multimedia Tools Appl. 407–430 (2010). https://doi.org/10.1007/s11042-009-0440-3
16. Zhang, X., Wang, S.: Vulnerability of pixel-value differencing steganography to histogram analysis and modification for enhanced security. Pattern Recogn. Lett. **25**, 331–339 (2004)
17. Chang, K.-C., Chang, C.-P., Huangb, P.S., Tua, T.-M.: A novel image steganographic method using tri-way pixel-value differencing. J. Multimedia **3**(2) (2008) (Elsevier)
18. Joo, J.C., Lee, H.Y., Lee, H.K.: Improved steganographic method preserving pixel-value differencing histogram with modulus function. EURASIP J. Adv. Signal Process. 1–13 (2010). https://doi.org/10.1155/2010/249826
19. Pan, F., Li, J., Yang, X.: Image steganography method based on PVD and modulus function. In: International Conference on Electronics, Communications and Control (ICECC), pp. 282–284 (2011). https://doi.org/10.1109/icecc.2011.6067590
20. Luo, W., Huang, F., Huang, J.: A more secure steganography based on adaptive pixel-value differencing scheme. Multimedia Tools Appl. **52**, 407–430 (2011) https://doi.org/10.1007/s11042-009-0440-3
21. Mandal, J.K., Das, D.: Color image steganography based on pixel value differencing in spatial domain. Int. J. Inf. Sci. Tech. **2**(4), 83–93 (2012)
22. Nagaraj, V., Vijayalakshmi, V., Zayaraz, G.: Color image steganography based on pixel value modification method using modulus function. In: International Conference on Electronic Engineering and Computer Science, pp. 17–24 (2013)
23. Tseng, H.-W., Leng, H.-S.: A steganographic method based on pixel-value differencing and the perfect square number. J. Appl. Math. Res. **Article ID 189706** 1–8 (2013). http://dx.doi.org/10.1155/2013/189706 (Hindawi Publishing Corporation)
24. Swain, G.: Adaptive pixel value differencing steganography using both vertical and horizontal edges. Multimedia Tools Appl. **75**, 13541–13556 (2015). https://doi.org/10.1007/s11042-015-2937-2 (Springer)
25. Hameed, M.A., Aly, S., Hassaballah, M.: An efficient data hiding method based on adaptive directional pixel value differencing (ADPVD). Multimedia Tools Appl. (2017). https://doi.org/10.1007/s11042-017-5056-4 (Springer)
26. Prasad, S., Pal, A.K., An RGB color image steganography scheme using overlapping block-based pixel-value differencing. Dryad Digital Repository (2017). http://dx.doi.org/10.5061/dryad.21tm5)
27. Swain, G.: Adaptive and non-adaptive PVD steganography using overlapped pixel blocks. Arab. J. Sci. Eng. (2018). https://doi.org/10.1007/s13369-018-3163-9

An ANN-Based Text Mining Approach Over Hash Tag and Blogging Text Data

Archana Tamrakar, Pradeep Mewada, Purva Gubrele, Ritu Prasad
and Praneet Saurabh

Abstract In the daily life, everybody keeps on using Internet after the waking up that includes the communication among different people of this world. On one side, Internet has made everyone's life very convenient and provides many facilities that can be used on social networking sites such as chat, messaging, comment, and blogging. This way everyone keeps on sharing the personal data at different places like Web sites, chats, social media. The text mining can be defined as the process of finding useful information from the given text. There exist various methods that remain helpful in analyzing texts and extracting the information but these often suffer from various complexities. Also, theoretically, it is quite difficult to analyze and extract the information from these raw data. This paper presents an Effective feed forward artificial neural network (FP-ANN) for text mining that generates different textual patterns from several resources and provides results very precisely with lesser computation time, lower cost, and overhead. FP-ANN approach calculates the hash label and discovers importance between inputs for text mining.

Keywords Classification · Information retrieval · Text mining · Artificial neural network

A. Tamrakar (✉) · P. Mewada · P. Gubrele · R. Prasad (✉)
Technocrats Institute of Technology (Advance), Bhopal, MP 462021, India
e-mail: archana.tamrakar91@gmail.com

R. Prasad
e-mail: rit7ndm@gmail.com

P. Mewada
e-mail: pradeepmewada07@gmail.com

P. Gubrele
e-mail: purvagubrele99@gmail.com

P. Saurabh (✉)
Technocrats Institute of Technology, Bhopal, MP 462021, India
e-mail: praneetsaurabh@gmail.com

© Springer Nature Singapore Pte Ltd. 2020
K. N. Das et al. (eds.), *Soft Computing for Problem Solving*,
Advances in Intelligent Systems and Computing 1057,
https://doi.org/10.1007/978-981-15-0184-5_35

1   Introduction

The word "text mining" itself indicates its meaning, which is to handle texts or data, so accordingly it works on the applications of data mining techniques in which it extracts out the high-quality useful information from the texts and then uses it for the future reference [1]. In the same way, "content mining" is utilized to portray the use of information mining procedures to the computerized revelation of helpful or fascinating learning from unstructured content [2]. A few systems have been introduced for content mining including applied structure, affiliation lead mining, and scene mining that involved choice trees and run different enlistment strategies. What is more, information retrieval methods have broadly utilized the "pack of words" show for errands, for example, archive coordinating, positioning, and grouping. Figure 1 illustrates the various text mining fields.

The present proliferation of computing devices led to innovation in various fundamentals as tremendous amounts of content archives become accessible. In the present scenario, there exists a considerable measure of business information which exists as substance. Content mining helps to deal with the clashing unrefined information, unstructured information that allows dealing with deficient information, irregularity, and defenselessness [3, 4].

Current state of art highlights the difficulty to analyze and extract the information from these datasets [5]. Some methods were introduced but they fall short in analyzing texts and extracting the information [6]. This paper presents a Feed forward artificial neural network (FP-ANN) that generates different textual patterns from several resources and provides results very precisely with lesser computation time, lower cost, and overhead. This paper is organized in the following manner, Sect. 2 presents the related work of the domain while Sect. 3 introduces the proposed work, Sect. 4 discusses the experimentation and result evaluation, and Sect. 5 concludes the paper.

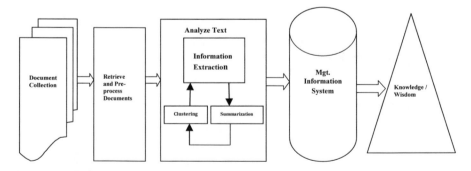

Fig. 1   Text mining fields

2 Related Work

In recent years, there is tremendous growth in web contents due to popularity and access of computing services. Minimizing capital as well as operating costs of the systems is the major objective of all approaches. Pazzani and Billsus [7] presented a system that examined content-based proposal systems, i.e., structures that endorsed anything to a customer in light of a portrayal of the thing and a profile of the customer's focal points. Subsequently, Ferrer et al. [8] portrayed weblogs as one of crucial segments in web experience issues to find relevant online journals. In their work, they introduced a weblog recommender framework that connected between weblogs as some sort of rating and thereafter evaluated in a genuine dataset. Later on, Ferragina et al. [9] demonstrated the way recommender frameworks help E-business to improve business by breaking it down in six market-driving components. These components make scientific classification of recommender frameworks, including the sources of information needed from the buyers, the extra learning required from the database, the ways the proposals are exposed to customers, the advances utilized to avail the opinion, and the degree of personalization of the recommendations. Thereafter, Bart et al. [10] presented a survey of major choices in assessing synergistic separating recommender frameworks while Bahdanau et al. [11] introduced multiple criteria analysis for recommender system. In this work, they illustrated the execution and capacity of treatment of existing recommender systems. They also measured ROC contrasted with a multiple rating collaborative filtering approach. Afterward, Bandyopadhyay et al. [12] introduced emotionally supportive network designers in quick prototyping recommender structures that utilized different case-based reasoning methodologies. CBR structures gigantically benefit by the reusing CBR frameworks. In another work, Teppan [13] puts forward the influence of constraints that can turn existing recommender frameworks into safe frameworks. This work also demonstrated the data stored because of the confinements for data through theoretic concepts, capacities, and entropies. All these methods presented, firmly illustrated, that there are various methods that remain helpful in analyzing texts and extracting the information but often suffer from various complexities. Also, theoretically, it is quite difficult to analyze and extract the information from these datasets. Some explores involving bio-inspired techniques [14, 15] can be looked upon to find more realistic solution in this domain [16, 17]. Next section introduces the proposed Feed forward artificial neural network (FP-ANN) which calculates the hash label and discovers importance between inputs. Table 1 put forwards the summary of various advances in this in the domain.

Table 1 Comparison among work with their limitations

Authors	Presented techniques	Highlights	Limitations
Pazzani and Billsus [7]	Substance-based recommendation system	Provided an interface that allowed users to construct a representation of their own interests	Required effort from the user, difficult to get large number of users to make this effort
Ferrer and Kruse [8]	Scan-based calculations for test arrangement age in practical testing	Presented a weblog recommender framework in light of connection structure of weblog diagram	Web experience issues in finding applicable online journals
Ferragina et al. [9]	Internet business recommendation applications	Helps E-business locales, increments deals and breaks down the recommender frameworks	Receives increment with the recommendations business that sometimes does not work
Bart et al. [10]	Assesses cooperative sifting recommender frameworks	Used the courses in which expectation quality is measured	Sometimes did not yield proposed results
Bahdanau et al. [11]	A recommender system in light of multiple criteria analysis	Used a multiple rating collaborative filtering (MRCF) approach for better rating	Do not always result in good rating
Bandyopadhyay et al. [12]	Prototyping recommender systems	Quick prototyping recommender structures utilized case-based reasoning methodologies	Difficult to implement and understand
Teppan [13]	The influence limiter: provably manipulation-resistant recommender systems	Can turn existing recommender frameworks into manipulation-safe frameworks	Not safe

3 Proposed Methodology

To improve the performance of text mining, this section presents Feed forward artificial neural network (FP-ANN) which calculates the hash label values and discovers importance between inputs. The proposed FP-ANN enhances FP growth calculation with neural networks to sustain the feed forward approach. Key aspects of the proposed FP-ANN are given below.

(i) Use of ANN in FP-ANN bolsters forward calculation for the hash label age and discovers the importance between the words input.

(ii)   Algorithm used additionally utilizes the utilization of regular example devel-
       opment approach for finding the word significance and their recurrence. As per
       their utilization and development score, the calculation is executed.
(iii) FP-ANN works with the substantial database and calculation works with the
       beforehand given information approach.
(iv)   Calculation creates a hash tag with high rate of exactness and in addition
       preferable outcome execution.
(v)   Implemented to an information dataset from a micro blogging site is taken and
       processed by different sub process library.

FP-ANN, feed forward layer is based on ANN, and it is utilized for calculations
to remain productive while finding the information pertinence and getting hash label
age over the substantial dataset. Figure 2 below gives an idea how the proposed
FP-ANN works.

The proposed Feed forward artificial neural network (FP-ANN) is profoundly
safe and devours less calculation time and along these lines computational rate over

Fig. 2 Working of FP-ANN

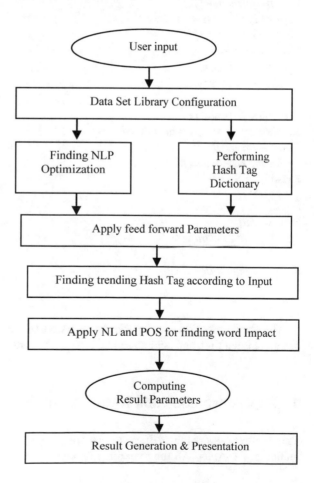

the organized accessible dataset. ANN in the FP-ANN enables it to perform quick calculations. It shows its usefulness and calculates similarity measure. These calculations likewise check for appropriate repetition, usage of more secure and dependable parameters. Pseudo code of FP-ANN is depicted underneath.

Pseudo code for FP-ANN.

Input: Input Tweets, Input data post, FP parameters.
Output: Communication process, feed forward outcome,,
Steps:
For each{
Tweet file listing {t1,t2....tN};
Input post request ();
Loading vocab Data ();
Process FP over Input & Data dictionary;
NLP performance ();
POS performance ();
Performing Frequency word ();
Finding relevancy ();
Finding feed forward outcome ();
Hashtag= TagGen ();
If(Hashtag==Accepted)
{
Substitute Hashtag ();
Set status = Active; generate statistics ();
  Generate result ();
  Plotting outcomes ();
} else

{
Status=exit;
creating data for inquire;
}
Return Computation time;
}
End.

The above pseudo code describes the proposed FP-ANN that will exhibit low computational time and computational cost with more security.

4   Experimentation and Result Analysis

This section presents the different experiments performed to underline the performance of FP-ANN. All the experiments were performed using an i3-4005U CPU

@ 1.70 GHz along with 4 Gb RAM on Windows 10. The proposed FP-ANN was implemented in Java using Eclipse IDE with feature selection algorithms. For the experiments, a real-time Twitter data fetching API is used to fetch data. Further, NLP library configuration is performed using MySQL to structure the Twitter dataset.

4.1 Computation Time Comparison

Experiments are done to evaluate the proposed FP-ANN, and then it is compared to the current state of the art that actually quantifies the performance of FP-ANN among its peers. Experimentations are carried on FP-ANN and current state of the art with Twitter dataset for bringing, marking, and plotting of results.

Underneath tables and figures give a review of the performance of both proposed FP-ANN and existing current state of the art. Experiments are performed for both proposed FP-ANN and current state of art with different dataset size to evaluate computation time (in ms) from 1 K tweets to 50 K tweets.

Table 2 and Fig. 3 represent computation time for both current state of art and the proposed FP-ANN with variation in number of tweets from 1 K to 50 K. These results very clearly state that the proposed FP-ANN outperforms the current state of art and reports lower computation time for different number of tweets. For all the values of

Table 2 Computation time comparison

Number of tweets (1000 = 1 K) (K)	Current state of art (computation time) (ms)	FP-ANN (computation time) (ms)
1	14.36	12.2
2	17.98	16.0
5	23.12	20.9
10	43.5	40.89
50	143.8	132.6

Fig. 3 Computation time comparison

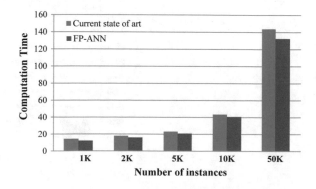

number of tweets, FP-ANN gives better and lower computation time which is very significant in determining the performance of any method. These results also show that the new integrations of feed forward and artificial neural networks are working in tandem to lower computation time.

4.2 Computation Cost Comparison

Computation cost is an important parameter to determine the performance of any algorithm. This experiment is performed to calculate the computation cost for processing different sizes of datasets ranging from 1 K to 50 K for both the current state of art and the proposed FP-ANN.

Table 3 and Fig. 4 illustrated very clearly put forward that the proposed FP-ANN performs better as compared to the current state of art and reports lower cost for different number of tweets.

For all the values of number of tweets, FP-ANN gives better and lower computation cost which is very significant in determining the performance of any method when number of tweets vary from 1 K to 50 K. Experimental results reflect that

Table 3 Computation cost comparison

Number of tweets (1000 = 1 K) (K)	Current state of art (Computation cost)	FP-ANN (Computation cost)
1	1121	910
2	2339	1289
5	2881	2201
10	3443	3046
50	8776	6989

Fig. 4 Computation cost comparison

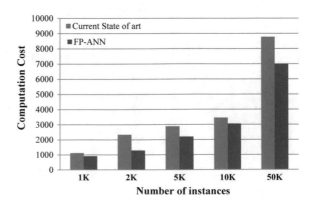

Table 4 Overhead comparison

Number of tweets (1000 = 1 K) (K)	Current state of art (overhead)	FP-ANN (overhead)
1	5454	4340
2	7845	4908
5	8081	7668
10	12,897	9880
50	19,912	16,569

Fig. 5 Overhead comparison

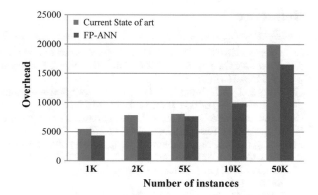

mechanism of feed forward with artificial neural network gelled well and lowered computation cost for all the variation in dataset from 1 K to 50 K.

4.3 Overhead Comparison

This experiment is performed with the intent of calculating overhead in the proposed FP-ANN and current state of the art as overhead plays detrimental effect in performance.

Experimental results presented in Table 4 and Fig. 5 indicate that proposed FP-ANN reports lower overhead which is better as compared to the existing state of the arts. Trend of result maintains its character and remains same when number of samples in the test set increases from 1 K to 5 K.

5 Conclusion

Internet has made everyone's life very convenient but has generated humungous amount of data which makes text mining complex and challenging. Various methods have evolved and proved to be helpful in analyzing texts and extracting the information but these suffered from various complexities. This paper presents a Feed forward artificial neural network (FP-ANN) that integrated concepts of feed forward and artificial neural networks and generated different textual patterns from several resources and used hash tag with high rate of exactness. Experimental results, very precisely, illustrated that FP-ANN outperformed its peers and reported lesser computation time, lower cost, and overhead.

References

1. Fong, S., Gao, E., Wong, R.: Optimized swarm search-based feature selection for text mining in sentiment analysis. In: IEEE 15th International Conference on Data Mining Workshop, pp. 1153–1162 (2015)
2. Fong, S., Deb, S., Yang, X.S., Li, J.: Feature selection in life science classification, metaheuristic swarm search. IEEE IT Prof. **16**(4), 24–29 (2014)
3. Sagayam, R.: A survey of text mining: retrieval, extraction and indexing techniques. Int. J. Comput. Eng. Res. **2**(5), 1443–1446 (2012)
4. Padhy, N., Mishra, D., Panigrahi, R.: The survey of data mining applications and feature scope. Int. J. Comput. Sci. Eng. Inf. Technol. (IJCSEIT) **2**(3), 43–58 (2012)
5. Fan, W., Wallace, L., Rich, S., Zhang, Z.: Tapping the power of text mining. Commun. ACM **49**(9), 76–82 (2006)
6. Weiss, S.M., Indurkhya, N.T., Zhang, T., Damerau, F.: Text Mining: Predictive Methods for Analyzing Unstructured Information, pp. 157–195. Springer (2010)
7. Pazzani, M.J., Billsus, D.: Content-Based Recommendation Systems, pp. 325–341. Springer (2007)
8. Ferrer, J., Kruse, P.M., Chicano, F.E., Alba, E.: Search based algorithms for test sequence generation in functional testing. Inf. Softw. Technol. **58**, 419–432 (2015)
9. Ferragina, P., Piccinno, F., Santoro, R.: On analyzing hashtags in twitter. In: Proceedings of the Ninth International AAAI Conference on Web and Social Media, pp. 110–119 (2015)
10. Bart, P., Knijnenburg, M.C., Willemsen, K.A.: A pragmatic procedure to support the user-centric evaluation of recommender systems. In: RecSys'11, pp. 321–324. ACM (2011)
11. Bahdanau, D., Cho, K., Bengio Y.: Neural machine translation by jointly learning to align and translate. In: International Conference on Learning Representations, pp. 1–15 (2015)
12. Bandyopadhyay, A., Ghosh, K., Majumder, P., Mitra, M.: Query expansion for microblog retrieval. Int. J. Web Sci. **1**(4), 368–380 (2012)
13. Teppan, E.C.: Implications of psychological phenomenons for recommender systems. In: RecSys'08, pp. 323–326. ACM (2008)
14. Saurabh, P., Verma, B.: An efficient proactive artificial immune system based anomaly detection and prevention system. Expert Syst. Appl. **60**, 311–320 (2016). Elsevier
15. Saurabh, P., Verma, B,: Immunity inspired cooperative agent based security system. Int. Arab. J. Inf. Technol. **15**(2), 289–295 (2018)
16. Saurabh, P., Verma, B., Sharma, S.: An immunity inspired anomaly detection system: a general framework a general framework. In: 7th International conference on bio-inspired computing: theories and applications (BIC-TA 2012), vol 202, AISC, pp. 417–428. Springer (2012)
17. Saurabh, P., Verma, B, Sharma, S.: Biologically Inspired Computer Security System: The Way Ahead, Recent Trends in Computer Networks and Distributed Systems Security, CCIS, vol. 335, pp. 474-484. Springer (2011)

Design and Analysis of 4-Bit Squarer Circuit Using Minority and Majority Logic in MagCAD

Saurabh Kumar, R. Marimuthu and S. Balamurugan

Abstract One of the recent advances for the beyond complementary metal-oxide semiconductor (CMOS) area is nano-magnetic logic (NML). Many researchers are developing a keen interest in the areas of circuit designing, area optimization and its application in storing the data. Because of its inherent magnetic nature, it can be used to design any magnetic circuit. When compared to the CMOS technology NML logic has potential advantages in terms of non-volatile, low power and radiation in hard. In this paper, we have proposed and implemented a 4-bit squarer circuit using the perpendicular NML (pNML). Squarer design is implemented in MagCAD tool, which embeds design rules, physical models and technological parameters. A register-transfer-level of (VHDL) of the circuit is automatically extracted by Mag-CAD. The extracted model can be simulated with fast HDL-simulators; this makes it possible to verify the behavior and extract the performance of the designed circuit. This nano-magnetic logic technology is alternative to *CMOS* technology, which is non-volatile in nature with reduced power consumption.

Keywords Nano-magnetic logic · Emerging technologies · Squarer · Majority-based logic · MagCAD

1 Introduction

We have witnessed an exponential growth in the power of the chips and an equally exponential decrease in the size of the chip in the past four to five decades. If we go by International Technology Roadmap [1], then the technological limits of CMOS

S. Kumar · R. Marimuthu · S. Balamurugan (✉)
School of Electrical Engineering, Vellore Institute of Technology, Vellore, Tamil Nadu, India
e-mail: sbalamurugan@vit.ac.in

S. Kumar
e-mail: saurabhkumar.2015@vit.ac.in

R. Marimuthu
e-mail: rmarimuthu@vit.ac.in

© Springer Nature Singapore Pte Ltd. 2020   409
K. N. Das et al. (eds.), *Soft Computing for Problem Solving*,
Advances in Intelligent Systems and Computing 1057,
https://doi.org/10.1007/978-981-15-0184-5_36

technology are not very far. To overcome this problem several alternative technologies are studied in the literature [2]. Among various emerging technologies NML seems to be a promising candidate in research field [3]. Most work on with NML has focused on the technology that couple in-plane (iNML) [4]. Recent work has also employed devices with out-of-plane, perpendicular nano-magnetic logic (pNML) [5–7]. The main advantage if using this technology is low consumption of power [8], easy designing of circuits [9] added with the ability to retain information even when it is switched off.

A binary squarer is a special case of binary multiplier used in many of the DSP and image processing algorithms [10, 11]. In this paper, we proposed a 4-bit squarer circuit using pNML technology. In CMOS architectures, NANDs and NORs are basic logic gates in synthesize tool, however, in NML logic basic elements are minority/majority logic gates [12–16]. We have implemented our design using MagCAD, a powerful tool which can be used to implement pNML logic as well as iNML logic. Currently, MagCAD supports only the technologies related to NML, however, by the use of plugins it is expected that it can be used in the other beyond CMOS technologies too [9].

The rest of the paper is organized as follows. Section 2 focuses on the background of NML technology. Section 3 and 4 deal with modeling of the circuit and results, respectively. System parameters are mentioned in Sect. 5 with the conclusion.

2 Background

The NML technology is efficient when it comes to energy which depends on field coupling of neighboring nano-magnets to propagate and process binary operation [12]. Nano-magnetic logic (NML) is an exciting modern field where computation of data and transmission are accomplished using magnetic fields instead of electrical current. NML can be divided into two categories namely pNML and in-plane nano-magnetic logic (iNML). If the magnetization vector and the magnetic plane are aligned perpendicular to each other then it is pNML technology and if the magnetization vector and the magnetic plane are parallel to each other then it is iNML technology. This work deals with pNML because presently it is considered the most suitable technological implementation of NML. One of the basic advantages of pNML technology is that it can be organized in different layers so that the circuit becomes more compact, the interconnections are minimized [9], and the fabrication of the 3D circuits is easier in the pNML technology. We have used basic blocks like magnet, inverter magnet, nucleation center, via magnet, etc.

We have used MagCAD tool to design the proposed squarer circuit by means of drag and drop of basic elements available inside a graphic framework. After we completed the design part, we exported the VHDL description of the circuit. We have used the VHDL ISim simulator-Xilinx. The test bench is automatically generated. The behavior of the circuit is verified by feeding the input to the test bench.

3 Modeling of Circuit

3.1 Half Adder

One of the basic building blocks of the 4-bit squarer circuit is the half adder. This circuit is used to add two one-bit binary numbers and it produces the output as sum and carry. The circuit diagram for the half adder and its implementation in pNML layout in MagCAD is given in Fig. 1. In MagCAD, we have implemented the circuit using majority logic [12]. Table 1 summarizes the circuit performance of half adder considering all the inputs.

Fig. 1 Half-adder circuit and its pNML layout in MagCAD

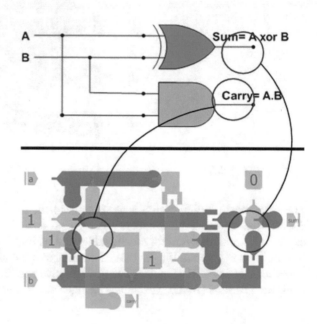

Table 1 Performance analysis of half-adder

Half adder				
Critical path = 1.953E−6				
Bounding box area = 8.19 μm²				
Input pattern		Output		Latency (s)
A	B	Sum	Carry	
0	0	0	0	1.144E−6
0	1	1	0	0.942E−6
1	0	1	0	1.751E−6
1	1	0	1	1.953E−6

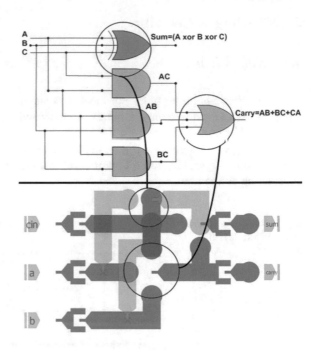

Fig. 2 Full-adder circuit and its pNML layout in MagCAD

3.2 Full Adder

The other basic building circuit used in the 4-bit squarer is the full adder. This circuit uses three one-bit numbers and then produce two output namely sum and carry. The circuit diagram for the full adder and its implementation in pNML layout in MagCAD is given in Fig. 2. Table 2 summarizes the circuit performance of full-adder considering all the inputs. Due to the effective coupling on the minority artificial nucleation centers (ANC), the latency changes according to the input pattern.

3.3 4-Bit Squarer Circuit

A binary squarer [17, 18] is a combinational circuit used in many of the DSP algorithms to perform the squaring of two-binary numbers. The circuit diagram for 4-Bit squarer and its implementation in pNML layout in MagCAD is given in Fig. 3. Table 3 summarizes the circuit performance of 4-bit squarer considering all the inputs. In this paper, we considered the parameters reported in Table 4.

Table 2 Performance analysis of full adder

Half adder					
Critical path = 1.715E−6					
Bounding box area = 4.86 μm²					
Input pattern			Output		Latency (s)
A	B	C in	Sum	Carry	
0	0	0	0	0	1.549E−6
0	0	1	1	0	0.942E−6
0	1	0	1	0	1.751E−6
0	1	1	0	1	1.549E−6
1	0	0	1	0	1.751E−6
1	0	1	0	1	1.549E−6
1	10	0	0	1	1.144E−6
1	1	1	1	1	1.751E−6

4 Simulation Results

In this section, we present the result of the majority logic-based proposed squarer design using pNML layout in MagCAD v2.2.6. All the simulation parameters are shown in Tables 1, 2 and 3 are obtained by using the default values shown in Table 4. The simulation results of the proposed 4-bit squarer design are shown in Fig. 4 for the input combination ($A3\ A2\ A1\ A0 = 1111$). The squarer result for the specified input combination is 11100001. We need to take care of the fact that both the size of the nanowire and the width of the nanowire should be comparable.

5 Conclusion

In this study, we have designed and analysis of majority and minority logic-based 4-bit squarer circuit in pNML technology using MagCAD. In this paper, we have proved the feasibility of multi-layer squarer design in pNML technology. The result has been verified and tested in Xilinx. Multilevel circuits like multipliers, filters and many complex designs can be modeled in MagCAD. The set of values for the latency we have got depends on the design parameters. Modern technology is focusing more on approximate computing which allows a permissible amount of error into consideration so that the circuit size can be reduced and speed can be increased.

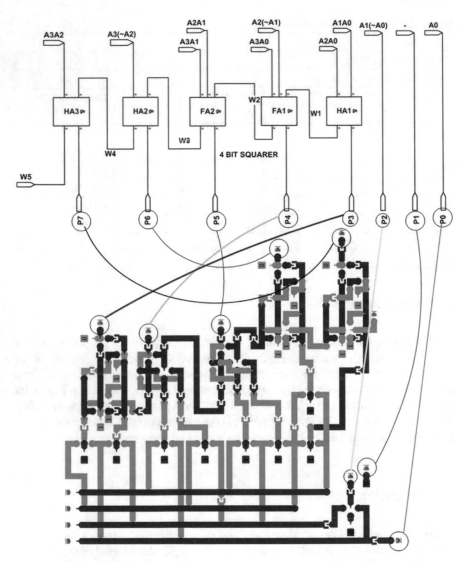

Fig. 3 4-bit squarer design and its pNML layout in MagCAD

Table 3 Performance analysis of 4-bit squarer

4-bit square

Critical path = 6.411E−6

Bounding Box area = 140.22 μm²

Input pattern				Output O								Latency (s)
A1	A2	A3	A4									
0	0	0	0	0	0	0	0	0	0	0	0	3.983E−6
0	0	0	1	0	0	0	0	0	0	0	1	3.983E−6
0	0	1	0	0	0	0	0	0	1	0	0	3.983E−6
0	0	1	1	0	0	0	0	1	0	0	1	3.983E−6
0	1	0	0	0	0	0	1	0	0	0	0	4.388E−6
0	1	0	1	0	0	0	1	1	0	0	1	4.388E−6
0	1	1	0	0	0	1	0	0	1	0	0	4.792E−6
0	1	1	1	0	0	1	1	0	0	0	1	4.792E−6
1	0	0	0	0	1	0	0	0	0	0	0	4.388E−6
1	0	0	1	0	1	0	1	0	0	0	1	4.388E−6
1	0	1	0	0	1	1	0	0	1	0	0	6.006E−6
1	0	1	1	0	1	1	1	1	0	0	1	6.411E−6
1	1	0	0	1	0	0	1	0	0	0	0	4.995E−6
1	1	0	1	1	0	1	0	1	0	0	1	4.995E−6
1	1	1	0	1	1	0	0	0	1	0	0	4.995E−6
1	1	1	1	1	1	1	0	0	0	0	1	4.995E−6

Table 4 Geometrical and physical parameters used in MagCAD

Parameters	Values
Filed pulse amplitude (Oe)	560.0
Clocking field amplitude (Oe)	560.0
Intrinsic pinning field (Oe)	190.0
Grid size (m)	3.0e−7 m
Temperature (K)	293.0
Stack thickness (m)	6.2e−6
ANC volume (m^3)	1.68e−23
Activation volume (m^3)	1.26e−23
Effective anisotropy (J/m^3)	2.0e + 5
Saturation magnetization (A/m)	1.4e + 6
Depinning field	0.98
Nucleation probability	0.50
Width of domain wall [m]	2.2e−7
Notch width [m]	5.4e−8
Coupling field for the via (Oe)	75.0
Attempt frequency (Hz)	2.0e−9

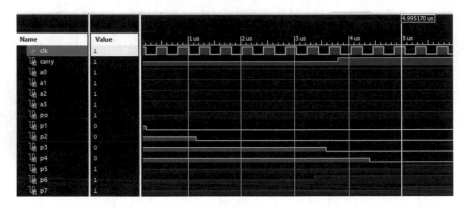

Fig. 4 Output waveform of a 4-bit squarer for the input of "1111"

References

1. International technology roadmap of semiconductors 2.0. Beyond CMOS. http://public.itrs.net (2015)
2. Conte, T.M., DeBenedictis, E.P., Gargini, P.A., Track, E.: Rebooting computing: the road ahead. Computer **50**(1), 20–29 (2017)
3. Chumak, A.V., Vasyuchka, V.I., Serga, A.A., Hillebrands, B.: Magnon spintronics. Nat. Phys. **11**(6), 453 (2015)
4. Labrado, C., Thapliyal, H., Lombardi, F.: Design of majority logic based approximate arithmetic circuits. In: IEEE International Symposium on Circuits and Systems (ISCAS), pp. 1–4 (2017)
5. Ferrara, A., Garlando, U., Gnoli, L., Santoro, G., Zamboni, M.: 3D design of a pNML random access memory. In: 13th Conference on Ph.D. Research in Microelectronics and Electronics (PRIME), pp. 5–8. IEEE (2017)
6. Turvani, G., Riente, F., Plozner, E., Vacca, M., Graziano, M., Gamm, S.B.V.: A pNML compact model enabling the exploration of three-dimensional architectures. IEEE Trans. Nanotechnol. **16**(3), 431–438 (2017)
7. Agrawal, S., Harish, G., Balamurugan, S., Marimuthu, R.: Design of high speed 5:2 and 7:2 compressor using nanomagnetic logic. In: VDAT 2018 (Accepted for Publication in Springer)
8. Riente, F., Garlando, U., Turvani, G., Vacca, M., Roch, M.R., Graziano, M.: MagCAD: tool for the design of 3-D magnetic circuits. IEEE J. Explor. Solid-State Comput. Devices Circ. **3**, 65–73 (2017)
9. Garlando, U., Riente, F., Zamboni, M., Graziano, M.: Topolinano & MagCAD: A design and simulation framework for the exploration of emerging technologies
10. Cho, K.J., Chung, J.G.: Parallel squarer design using pre-calculated sums of partial products. Electron. Lett. **43**(25), 1414–1416 (2007)
11. Kolagotla, R.K., Griesbach, W.R., Srinivas, D.H.: VLSI implementation of 350 MHz 0.35/spl mu/m 8 bit merged squarer. Electron. Lett. **34**(1), 47–48 (1998)
12. Gu, Z., Nowakowski, M.E., Carlton, D.B., Storz, R., Hong, J., Chao, W., Lambson, B., Bennett, P., Alam, M.T., Marcus, M.A., Doran, A.: Speed and reliability of nanomagnetic logic technology (2014). arXiv:1403.6490
13. Riente, F., Ziemys, G., Turvani, G., Schmitt-Landsiedel, D., Gamm, S.B.V., Graziano, M.: Towards logic-in-memory circuits using 3d-integrated nanomagnetic logic. In: IEEE International Conference on Rebooting Computing (ICRC), pp. 1–8. IEEE (2016)
14. Imre, A.: Experimental study of nanomagnets for Quantum-dot cellular automata(MQCA) logic applications, Ph.D. thesis, University of Notre Dame, Notre Dame, Indiana, Dec 2005
15. Labrado, C.: Exploration of majority logic based designs for arithmetic circuits. Master's thesis, University of Kentucky (2017)
16. Breitkreutz, S., Kiermaier, J., Eichwald, I., Hildbrand, C., Csaba, G., Schmitt-Landsiedel, D., Becherer, M.: Experimental demonstration of a 1-bit full adder in perpendicular nanomagnetic logic. IEEE Trans. Magn. **49**(7), 4464–4467 (2013)
17. Muller, J.-M.: Elementary Functions. Springer (2006)
18. Meher, P.K., Stouraitis, T.: Arithmetic Circuits for DSP Applications. Wiley (2017)

Takagi Sugeno Fuzzy-Tuned RTDA controller for pH Neutralization process Subject to Addictive Load Changes

Geetha Mani and G. Manochitra

Abstract For the control of pH value in neutralization process, three closed-loop control schemes are designed in this work, namely Proportional–Integral Derivative (PID), model predictive control (MPC) and Robustness Tracking Disturbance Overall Aggressiveness (RTDA) controller. As the title of the paper implies, the neutralization process undergoes addictive changes in its process parameters that clearly indicate the necessity of introducing this advanced control technique for this process. RTDA controller is a next generation regulatory controller which is an alternative to the popular PID control scheme. It combines the simplicity of the PID controller with the versatility of MPC. The pH neutralization is a highly non-linear and time-varying process, which has different operating regimes. The control objective in neutralization process is to sustain pH value at the prescribed level by controlling the flow rate of both acid and base. Takagi Sugeno (TS) Fuzzy-Tuned RTDA controller is employed for this process to vary the controller parameters for each operating point so that the set-point can be tracked effectively in all the operating regimes. An additive load disturbance is applied in the flow rate of acid and base to obtain the regulatory response. Thus, the paper focuses on effective disturbance rejection in each operating region and robustness in tracking the desired output. The simulation results are compared using time domain specifications, computational time and performance index like integral square error (ISE).

Keywords Fuzzy scheduling · MPC · Next generation controller · Nonlinear process · pH neutralization · RTDA

G. Mani (✉)
School of Electrical Engineering, Vellore Institute of Technology, Vellore 632014, Tamil Nadu, India
e-mail: geethamr@gmail.com

G. Manochitra
Department of Electrical & Electronics Engineering, Sri Krishna College of Engineering & Technology, Coimbatore, Tamil Nadu, India
e-mail: manochitrag@skcet.ac.in

© Springer Nature Singapore Pte Ltd. 2020
K. N. Das et al. (eds.), *Soft Computing for Problem Solving*,
Advances in Intelligent Systems and Computing 1057,
https://doi.org/10.1007/978-981-15-0184-5_37

1 Introduction

Neutralization is a chemical reaction between acid and base. The challenging part in such a process is to make the plant to regulate itself to different pH values in spite of the disturbances that arises while calculating the flow rate of titrating stream [1]. The controlling of pH neutralization process gets tedious because of its nonlinearity property inferred from its titration curve, particularly when strong acids and bases are involved [2]. The pH process control has been widely used in chemical, light industry, wastewater treatment, biotechnological industries and environmental protection. For instance, the pH of effluent streams from wastewater treatment plants must be maintained within stringent environmental limits [3]. Constricted control of pH is tough in pharmaceutical industries. Though proper controller is implemented, it is not easy to obtain better performance because of its time-varying properties due to unmeasured changes in the buffering capacity [4].

The major problems that contribute to unacceptable and inadequate control performance can be summarized as follows:

- Increases in plant complexity and strict constraints in terms of environmental and other performance requirements present a significant challenge in the applications.
- The inherent and severe nonlinearity of a pH neutralization process is a major source of difficulty in terms of robust and stable control system design.

Due to the influence of serious nonlinear, time delay and strong interference in pH neutralization process, the control of the pH neutralization process has been one of the most difficult problems in relative fields [5]. Thus, the research of the control and identification in pH neutralization process is very important.

The authors [6] proposed set range intelligent controllers to maintain the pH range in industrial waste water. The authors [7] have proposed predictive control to a pH neutralization process model obtained from Wiener model identification. A non-linear MPC context based on the sequential quadratic programming algorithm is implemented to control the pH value [8]. A comparative approach is done with linear and non-linear adaptive control, even soft computing thus neural network [9] is also implemented for this process.

Though advanced control techniques keep emerging every day, its demand in industrial application is on track only during the recent years. As the processes are getting complicated and still quality products are expected, the process industries are excited to incorporate the modern control techniques for more reliable, flexible, precise, efficient and robust control of parameters in process plant along with the consideration of the environmental factors. So it increases the necessity of continuous research to obtain different forms of control to meet those requirements. In addition, it requires advancement in control system design, a process model to represent the process plant better.

RTDA is a next generation controller which is the combined feature of classical PID controller and MPC. It utilizes digital technology to implement a simplified model prediction with transparent tuning parameters. The three tuning parameters

σ_R, σ_T and σ_D of RTDA controller are directly related to robustness, set-point tracking and disturbance rejection characteristics of the plant. The fourth tuning parameter σ_A is associated with the overall controller aggressiveness. All these four tuning parameter values are normalized between 0 to 1, which makes it much more easier to tune [10–12]. The RTDA configuration proposed in this paper is capable of handling the uncertainties and satisfying the desired references while keeping at its nominal working point.

The progress in control technology has significant impact on industrial process control. The advancements in control technology lead to a smart control method categorized as intelligent control strategy. It is a control strategy in which the solution or control is obtained by imitating human characteristics especially in a way how a human thinks and makes a decision during different situations. The intelligent control techniques are mainly categorized fuzzy logic, artificial neural networks and artificial intelligence. A new method on the selection of Q and R for optimal control is proposed in [13]. The need for optimal controllers is ever lasting. Also, optimal controllers have been used for the control of air supply in fuel cell systems [14–16].

The Fuzzy-Tuned RTDA controller contains a cluster of local controller parameter for every operating point with a scheduler. Through simple scheduling scheme it is tough to correlate the controller parameter for each operating region. In order to overcome this issue, Takagi Sugeno fuzzy inference system is implemented where operating regions where taken as membership function. This techniques provide smooth transfer of values for each region.

The three optimal control schemes used in this paper: Proportional Integral Derivative (PID), Linear Model-based Predictive Control (LMPC), minimize a multi-objective function and next generation regulatory—RTDA controller. A number of advantages make the RTDA control scheme suitable for these application : (1) less computational complexity, (2) handles constraints, (3) meets the time domain specifications and (4) allows multi-objective optimization. Furthermore, the RTDA controller can introduce robustness naturally.

The objective of proposed work is to present and compare an advanced optimization-based control schemes to maintain the pH value in the neutralization process. The three control schemes Fuzzy-Tuned RTDA, MPC and PID were implemented and compared based on time domain specifications and performance index like ISE and total computational time.

The paper is prepared and presented as follows: Section 2 gives the basic mathematical equations describing pH neutralization process. Section 3 addresses the closed-loop configurations and the performance analysis of all control schemes presented in this paper. Section 4 includes the performance of every controller and the comparative assessment of controllers is done in two different sections: (1) Servo response and (2) Regulatory response under additive load disturbances. Finally, the conclusions of the work and the scope of further work are consolidated in Section 5.

2   pH Neutralization Process Description and Modeling

The neutralization shown in Fig. 1 is a process of reducing the acidity or alkalinity by mixing acids or bases to produce neutral solution. Acidic or basic wastewater must be neutralized prior to discharge for minimum impact on environment. Neutralization is considered as a preparatory step in the wastewater treatments because many of the subsequent wastewater treatment are pH dependent. The regulations on the quality of industrial waste have become increasingly stringent in recent years. Industrial waste must be neutralized before it is discharged as effluent from manufacturing plant.

A process effluent composed of variety of components has a varying titration curve that defeats effort to attain smooth neutralization on a sound economic basis unless the pH controller can adapt to feedback control conditions ranging from oscillatory to over-damped. In the system of wastewater management, controlling the pH of an effluent stream is challenging because it is very unstable at neutral condition.

In wastewater treatment, the pH control is very difficult because the composition of flowing stream changes at every instant. The exact information about the flow, pH and alkalinity or acidity of the wastewater and about how much and how quick these parameters change are required for proper control system design. A typical pH control system consists of one or more reactors, mixer, measuring elements, controllers and reagent delivery systems.

The gain of the system at the equivalence point for the strong acid or strong base system is very high and it occurs at a pH of 7, which is neutral pH. Control of this system near pH 7 would place very high demands both on the accuracy of the control system and on the range ability of the reagent delivery system. Because of the lower gain near neutrality, the weak acid or weak base system is easier to control.

The design of the neutralization process depends mostly on factors like reaction tank size, which can affect the neutralization performance (e.g. A large vessel is required when low solubility reagents such as calcium lime are present.) for which the retention time should be minimized, mixing and agitation for complete elimination of the area of unreacted reagent which if not done properly may lead to insufficient mixing and excessive cycling and poor pH control, the relative location of the vessel

Fig. 1 Schematic diagram
of pH neutralization process

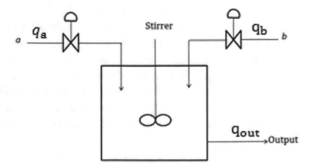

inlet, outlet and measurement probe location for maximum speed of response and the reagent delivery system and reagent addition point location for close pH control.

pH is only a measure of the concentration of dissociated hydrogen ions present in a solution. When a solution contains a weak acid, most of its hydrogen ions are undissociated. This reasoning can be applied to a weak alkali by considering hydroxide ions (OH−) instead of hydrogen ions (H+). The different operating region is selected based on static characteristics of pH neutralization performance. The pH is related to the concentration of the ions [H+] through the following logarithmic function as stated in [17].

$$pH = -\log_{10}\left[H^+\right] \tag{1}$$

The neutralization of a strong acid effluent (HCl) in a CSTR by a strong base (NaOH) is considered here [8]. First-order dynamics model is used with titration curve as the nonlinearity. The reactions that occur are

$$\begin{cases} HCl + NaOH \rightarrow H_2O + NaCl \\ \quad H^+ + OH^- \rightarrow H_2O \end{cases} \tag{2}$$

The ionic concentrations of [Cl$^-$] and [Na$^+$] in the CSTR can be related to the flows of acid q_a, and of base q_b and to the input concentrations of $\left[Cl_{in}^-\right]$ and $\left[Na_{in}^+\right]$, according to the following equations, provided the mixture is perfect and instantaneous.

$$V\frac{d}{dt}\left[Cl_{in}^-\right].q_a - \left[Cl_{in}^-\right].q_{out} \tag{3}$$

$$V\frac{d}{dt}\left[Na^+\right] = \left[Na_{in}^+\right].q_b - \left[Na^+\right].q_{out} \tag{4}$$

where, V is the volume of liquid in the CSTR. The electro-neutrality equation must be satisfied by the concentrations.

$$\left[Na^+\right] + \left[H^+\right] = \left[Cl^-\right] + \left[OH^-\right] \tag{5}$$

$$\left[H^+\right].\left[OH^-\right] = K_W = 10^{-14} \tag{6}$$

which relates these concentrations to [H$^+$] and therefore to pH.

The difference of the ionic concentrations X is

$$X \equiv \left[OH^-\right] - \left[H^+\right] \tag{7}$$

that combined with Eq. (5) results in:

$$X \equiv \left[Na^+\right] - \left[Cl^-\right] \tag{8}$$

Collecting these equations results in

$$\begin{cases} \left[H^+\right] = \frac{X}{2}\cdot\left(\sqrt{1 + \frac{4.K_W}{X^2}} - 1\right) & \text{IF } X > 0 \\ \left[H^+\right] = -\frac{X}{2}\cdot\left(\sqrt{1 + \frac{4.K_W}{X^2}} + 1\right) & \text{IF } X < 0 \\ \left[H^+\right] = \sqrt{K_W} & \text{IF } X = 0 \end{cases} \tag{9}$$

The equation describing the process dynamics is obtained by subtracting (3) from (4) and using (8), resulting in:

$$V\frac{dX}{dt} = \left[Na_{in}^+\right].q_b - \left[Cl_{in}^-\right].q_a - X.q_{out} \tag{10}$$

The time constant τ is given by,

$$\tau = \frac{v}{q_{out}} \tag{11}$$

It is dependent on the residence time.

The open-loop response of the system is shown in Fig. 2. The model of the system taken is represented in transfer function as:

$$G(s) = \frac{344.9910}{108.25s + 1} \tag{12}$$

The discrete-time state-space model of the system is given by,

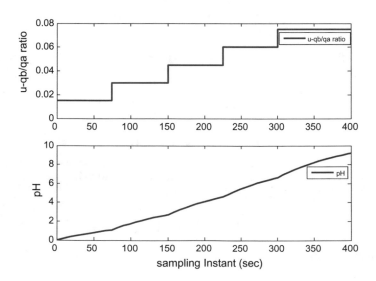

Fig. 2 Open-loop response of pH neutralization process

$$\dot{x} = 0.9991x + 0.1u \tag{13}$$

$$y = 3.1870x \tag{14}$$

2.1 Closed-Loop Control Schemes

For the purpose of controlling pH neutralization process, the closed loop control schemes discussed are PID, MPC, RTDA and Fuzzy Tuned RTDA controller design. The design of each control scheme is presented in detail.

2.2 PID Controller Design

In PID controller, the control signal $u(t)$ is produced based on the combination of proportional, integral and derivative action. The proportional term (P) is directly related to the error signal $e(t)$, integral term (I) corresponds to integral of the error and derivative term (D) is based on derivative action on the error. The resulting signals weighted and summed to obtain the control signal $u(t)$, which is applied to the plant model. The schematic block diagram of PID control scheme is shown in Fig. 3.

A mathematical representation of the PID controller is:

$$u(t) = K_p e(t) + K_i \int_0^t e(\tau)d\tau + K_d \frac{d}{dt}e(t) \tag{15}$$

where $u(t)$ is control input to the plant model and $e(t)$ is error signal, K_p is proportional gain, K_i is integral gain and K_d is derivative gain. These tuning parameters are significant for the better response of the system. Among the different tuning

Fig. 3 Block diagram of PID control scheme

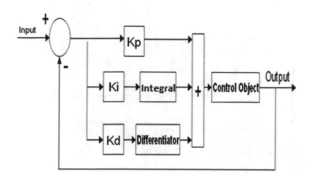

approaches available for the determination of these gain values, Ziegler–Nichols technique is utilized to tune the system [18].

2.3 Model Predictive Control

The model predictive control is an advance optimal control strategy [19] that is based on the explicit use of state-space model of the system to predict the controlled process variables over a certain time horizon and the prediction horizon [20]. The structure of a model predictive controller is shown in Fig. 4. The dynamic matrix control algorithm is discussed in this section.

Based on a step response model, it has the form

$$\hat{y}_k = \sum_{i=1}^{N-1} S_i \Delta u_{k-i} + S_N u_{k-N} \tag{16}$$

The additive disturbance is the difference between the measured output and model prediction. The corrected prediction acquired from additive difference is

$$\hat{y}_k^c = \hat{y}_k + d_k \tag{17}$$

The corrected predicted output for jth step in the future is:

$$\hat{y}_{k+j}^c = \hat{y}_{k+j} + \hat{d}_{k+j} \tag{18}$$

$$\hat{y}_{k+j}^c = \sum_{i=1}^{j} S_i \Delta u_{k-i+j} + \sum_{i=j+1}^{N-1} S_i \Delta u_{k-i+j} + S_N u_{k-N+j} + \hat{d}_{k+j} \tag{19}$$

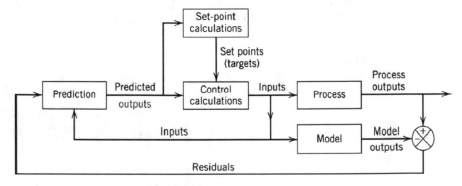

Fig. 4 Structure of a model predictive controller

The effect of the past and the future control moves can be separated as

$$\hat{y}_{k+j}^c = S_1 \Delta u_{k+j-1} + S_2 \Delta u_{k+j-2} + \cdots + S_j \Delta u_k + S_N \Delta u_{k+j-1}$$
$$+ S_{j-1} \Delta u_{k-1} + S_{j+2} \Delta u_{k-2} + \cdots + S_{N-1} \Delta u_{k-N+j+1} + \hat{d}_{k+j} \quad (20)$$

The correction term is assumed as constant in the future

$$\hat{d}_{k+j} = \hat{d}_{k+j-1} = \cdots = d_k = y_k - \hat{y}_k \quad (21)$$

A prediction horizon of m steps and a control horizon of n steps yields

$$\hat{Y}^c = S_f \Delta u_f + S_{\text{past}} \Delta u_{\text{past}} + S_N u_p + \hat{d} \quad (22)$$

The difference between set-point trajectory and the future prediction is

$$r - \hat{Y}^c = r - \left[S_{\text{past}} \Delta u_{\text{past}} + S_N u_p + \hat{d} \right] - S_f \Delta u_f \quad (23)$$

that yields

$$E^c = E - S_f \Delta u_f \quad (24)$$

The objective function is

$$\emptyset = (E^c)^{\mathrm{T}} E^c + (\Delta u_f)^{\mathrm{T}} W \Delta u_f \quad (25)$$

Minimization of this objective function is

$$\Delta u_f = \left[\left(S_f^{\mathrm{T}} W_e S_f + W_u \right)^{-1} S_f^{\mathrm{T}} W_e \right] E \quad (26)$$

The current and the future control move vector is proportional to the unforced error vector

$$\Delta u_f = K_1 * E \quad (27)$$

where

$$K_1 = \left(S_f^{\mathrm{T}} W_e S_f + W_u \right)^{-1} S_f^{\mathrm{T}} W_e \quad (28)$$

The deviation in u values is determined by minimizing an objective function including the predicted future errors. Here m represents prediction horizon and n represents control horizon, r represents the future set-point trajectory and $\hat{y}$ represents the vector of controlled outputs.

2.4 RTDA Controller

RTDA controller utilizes the parameters of a first-order process model. The model error is the difference between the model outputs from the actual process. It involves estimation of present disturbance and prediction of the future disturbance. Then the stipulated error is estimated to calculate the control input, which is given to system to track the output. The RTDA control scheme flow is illustrated in Fig. 5.

2.4.1 Reformation of Model

The actual dynamics of the process is usually approximated to first-order model, is used as it gives better approximation for the actual dynamics of the process. The transfer function model of the system is given by

$$y(s) = \frac{K}{\tau s + 1} u(s) \tag{29}$$

Here a discretized model is used for model prediction

$$\hat{y}(l + 1) = \alpha \hat{y}(l) + \beta u(l) \tag{30}$$

The control action $u(l)$ is restricted to remain the same for the whole prediction horizon. Hence, only the 1st control move $u(l)$ is allowed to be maintained for entire P-step horizon,

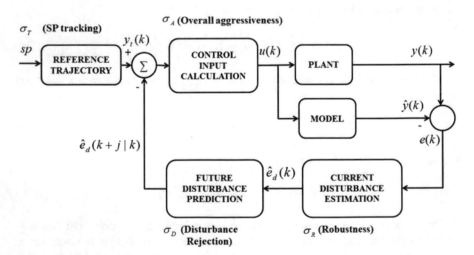

Fig. 5 RTDA control scheme flow

$$u(l + i) = u(l) \tag{31}$$

The predicted process output for each next P-step horizon is obtained as,

$$\hat{y}(l + i) = \alpha^i \hat{y}(l) + \beta \eta_i u(l) \quad \text{for } 1 \leq i \leq P \tag{32}$$

where

$$\eta_i = \frac{1 - a^i}{1 - a} \tag{33}$$

2.4.2 Error Updation

As the first-order model is an approximation of a process, the model predicted output has deviation from the actual output of the process hence the prediction needs to be updated. The model mismatch is given by,

$$e(l) = y(l) + \hat{y}(l) \tag{34}$$

$$e(l) = e_m(l) + e_d(l) \tag{35}$$

Using Bayesian estimation principle, $e_d(l)$ is estimated as

$$\hat{e}_d(l) = \sigma_R \hat{e}_d(l - 1) + (1 - \sigma_R)e(l) \tag{36}$$

The future error is estimated as

$$\hat{e}_d\left(l + \frac{j}{l}\right) = \hat{e}_d(l) + \frac{\alpha}{1 - \alpha}\left[1 - \alpha^i\right]\nabla e_d(l) \quad \text{for } 1 \leq j \leq P \tag{37}$$

where $\nabla e_d(l)$ is difference in error between two consecutive error values, is given by:

$$\nabla e_d(l) = e_d(l) - e_d(l - 1) \tag{38}$$

The parameter α is now substituted with $(1 - \sigma_d)$,

$$\hat{e}_d(l + j|l) = \hat{e}_d(l) + \frac{1 - \sigma_D}{\sigma_D}\left[1 - (1 - \sigma_D)^j\right]\nabla \hat{e}_d(l) \tag{39}$$

Here σ_d have the control response corresponds to disturbances and it is scaled to lie between 0 and 1. The corrected prediction output for P-step prediction horizon is

$$\tilde{y}(l+i) = \hat{y}(l+i) + \hat{e}_d(l+i|l) \quad \text{for } 1 \le i \le P \tag{40}$$

2.4.3 Reference Trajectory

Reference trajectory is to be defined in which sp(l) represents the set-point to be tracked to attain the desired process output. Let the desired trajectory $y_t(l)$, be given by

$$y_t(k) = \sigma_T y_t(l-1) + (1-\sigma_T) * \text{sp}(l) \tag{41}$$

When alternation in set-point values is not known in advance then the immediate past set-point values are taken for current set-point values, and then it is obtained as

$$y_t(l+j) = \sigma_T y_t(l) + \left(1 - \sigma_T^j\right) * \text{sp}(l) \quad \text{for } 1 \le j \le \infty \tag{42}$$

where σ_T is the trajectory tracking tuning parameter.

2.4.4 Control Input Calculations

The control action $u(l)$ is the minimization of the deviation of the model predicted output from the reference trajectory for P-step horizon which is needed for the model predicted output. The objective function of the controller is given as

$$\min_{u(l)} \sum_{i=1}^{N} (y_t(l+i) - \tilde{y}(l+i))^2 \tag{43}$$

Consider,

$$r_i(l) = y_t(l+i) - \tilde{y}(l+i) \tag{44}$$

From the corrected prediction output Eq. (40)

$$r_i(k) = \Psi_i(l) - \beta \eta_i u(l) \tag{45}$$

where

$$\Psi_i = y_t(l+i) - a^i \hat{y}(l) - \hat{e}_d(l+i|l) \tag{46}$$

where $\Psi_i(l)$ represents the stipulated error.

Through the objective function, the control input $u(l)$ is obtained as

$$u(l) = \frac{1}{b} \frac{\sum_{i-1}^{N} \eta_i \Psi_i(l)}{\sum_{i=1}^{N} \eta_i} \tag{47}$$

The prediction horizon length P depends on the total aggressiveness parameter σ_A is given by,

$$P = 1 - \frac{\tau}{t_s} \ln(1 - \sigma_A) \tag{48}$$

where t_s is the sampling time.

2.4.5 Fuzzy-Tuned RTDA Controller

The non-linear process is distributed into 'M' operating regions and σ_R is computed for each operating region based on the minimum performance error criteria.

Rule $l = 1$: M If $Z_1(K)$ is $M_{I.1}...Z_g(K)$ is $M_{I.g}$

Then individual σ_R value for each operating region is obtained and combined using Takagi Sugeno fuzzy inference method and the global σ_R value is calculated.

$$\sigma_R = \sum_{i=1}^{M} h_i(Z(K)) * (\sigma_R, I) \tag{49}$$

$$h_i(z(K)) = W_i(K)/\sum_{i=1}^{M} W_i(K) \tag{50}$$

The control flow of Fuzzy-scheduled RTDA is shown in Fig. 6.

3 Simulation Results and Analysis

The pH neutralization process the sampling time is taken as 5 s. The main objective of the process is to maintain the pH value.

3.1 PID Controller

In this study, PID controller parameters are optimized by the Ziegler–Nichols tuning algorithm as used in [17]. The PID-tuned values are tabulated in Table 1.

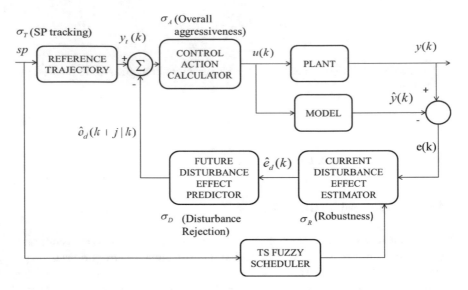

Fig. 6 Fuzzy-scheduled RTDA control scheme

Table 1 PID controller parameters for pH neutralization process

K_p	K_i	K_d
2.35	8.92	0.0187

3.1.1 Set-Point Tracking

The simulated set-point tracking response is illustrated in Fig. 9. The PID configuration provides the highest oscillations while tracking the reference value. It is the slowest response with respect to settling time (T_{ST}) and also it does not handle any constraints. However, it shows the faster in rise time (T_{RT}) because of its simplicity in design. The steady-state error for PID control scheme is less than 1% and may be considered negligible. The complete performance analysis of PID control scheme such as time domain specifications, performance indicator and total computational time are tabulated in Table 5.

3.1.2 Disturbance Rejection

The simulated disturbance rejection response along with the load disturbance profile is illustrated in Fig. 10. To obtain the regulatory response, the load disturbance is given to the process. It takes about 6 s to track the set-point after the occurrence of load disturbance. However, the PID configuration does not provide smoothen response under the load profile. For each disturbance occurred in the process, the controller involves undershoot.

3.2 MPC Controller

The simulated MPC controller settings are tabulated in Table 2. The weighted matrices W_e and W_u are tuned using the formulae $W_u = B^T B$ and $W_e = C^T C$, respecively. It is also noted that the weighted matrices W_e and W_u should be less to get the reliable output and minimum steady-state error. $0 \leq u \geq 5$ is the constraints imposed on the manipulated variable (q_b/q_a ratio).

3.2.1 Set-point Tracking

The simulated set-point tracking response of MPC is illustrated in Fig. 8. The controller provides a smooth response without any oscillations while tracking. This control scheme provides zero steady-state error. Based on settling time (T_{ST}) and rise time (T_{RT}) the repose is faster comparatively. But, it shows the poor performance in computational complexity because of its complex in design. When compared to classical PID scheme, MPC gives minimum ISE value. The complete performance analysis of MPC control scheme such as time domain specifications, performance indicator and total computational time is tabulated in Table 5.

3.2.2 Disturbance Rejection

The simulated disturbance rejection response along with the load disturbance profile of MPC is illustrated in Fig. 10. To obtain the regulatory response, the load disturbance is given to the process. It takes about very less time in seconds to track the set-point after the occurrence of load disturbance. Also, the MPC configuration ensures no steady-state error and smoothen response under the load profile. But, it shows the poor performance in computational complexity for regulatory response.

Table 2 MPC controller settings for pH neutralization process

Parameters	Values
Prediction horizon (m)	10
Control horizon (n)	1
Weighting factor (W_e)	12
Weighting factor (W_u)	7
Sample time	5
Final simulation time	400

4 Fuzzy-Tuned RTDA Controller

The process has different operating regions hence one tuning parameter cannot satisfy all the operating conditions. Hence, this technique is implemented to gain better performance in all the operating points of the system. The fuzzy technique utilized here is Takagi Sugeno [15], as the scheme requires absolute tuning value for each operating region whereas Mamdani fuzzy inference can only provide range of values. Here, the controller parameters of tracking, disturbance and aggressiveness are manually tuned according to each operating regions and these parameters are optimized by observation method. Here, only the robustness parameter is tuned using this controller.

The different operating regions and its corresponding robustness parameter σ_R value are tabulated in Table 3, which is to be feed into the fuzzy system. As it is a pH process, the operating region of the plant is limited from 0 to 10.2. The input to the controller is given in terms of membership function and the optimal robustness tuning parameter σ_R value is obtained as the result. The controller includes five membership functions which represents each operating region of the process.

The membership functions utilized for transition of controller parameters according to the regions are shown in Fig. 7. The type of membership function used for each operating region is trapezoidal. Each operating region in the input is related with corresponding controller parameter through the rule base, the rule editor serves

Table 3 Optimal σ_R values for various operating regions of pH process

Region	Range	Optimal σ_R
Region 1	2.9–3.5	0.1
Region 2	4–6	0.75
Region 3	6.4–8	0.5
Region 4	8.2–9.2	0.5
Region 5	10.2–10.7	0.1

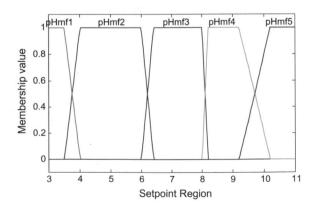

Fig. 7 Membership function for pH neutralization process

this option. The Defuzzification technique involved is weighted average method, the entire operating region is weighted and averaged. The robustness parameter σ_R value obtained as output from the controller is feed into the RTDA controller to obtain the desired performance of the system.

The variation in σ_R value according to the set-point operating region is shown in Fig. 8. Based on the set-point changes in the process, the respective operating region and its corresponding controller parameter are identified. The optimal control value for the tuning parameter is obtained by analyzing normalized integral square error of the process.

The other tuning parameters corresponding to tracking σ_T, disturbance rejection σ_D and aggressiveness σ_A are manually tuned to obtain the desired results. As each parameter is tuned according the requirements. If effective tracking performance is needed, then σ_T should be chosen nearer to 1 else nearer to zero. Similarly, σ_D values are chosen nearer to 1 if the process involves disturbance. The aggressiveness parameter σ_A corresponds to both servo and regulatory responses. The simulated RTDA controller settings these three tuning parameters are tabulated in Table 4.

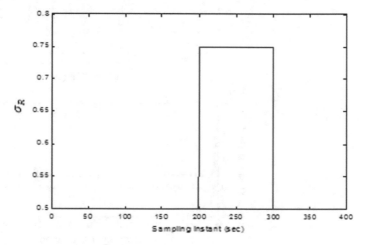

Fig. 8 Variation in σ_R for the given set-point changes

Table 4 RTDA controller settings for pH neutralization process	Region	σ_T	σ_D	σ_A
	Region 1	0.85	0.31	0.72
	Region 2	0.79	0.25	0.6
	Region 3	0.81	0.3	0.75
	Region 4	0.92	0.237	0.81
	Region 5	0.86	0.35	0.77

4.1 Set-point Tracking

The simulated set-point tracking response is illustrated in Fig. 9. The Fuzzy-Tuned RTDA configuration provides no oscillations while tracking the reference with operational input and output constraints. The controller shows fastest in the rise time (T_{RT}) because of its robustness and overall aggressiveness in design. The steady-state error of RTDA control scheme is zero by tuning σ_R. The Fuzzy-Tuned RTDA provides smooth change in tuning values for each different operating region. This controller meets requirements such as time domain specifications and minimum ISE value. The complete performance analysis of RTDA control scheme is tabulated in Table 5.

Fig. 9 Comparative servo response of all control schemes in pH neutralization process. **a** Process output, **b** manipulated variable

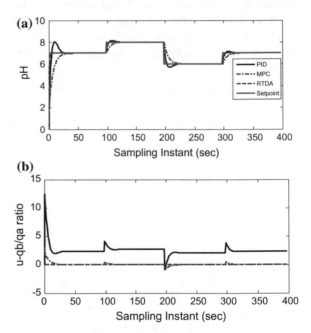

Table 5 Set-point tracking—comprehensive performance analysis of all closed-loop configurations for pH neutralization process

Control schemes	Time domain Specifications			Performance indicator	Computational complexity
	Over shoot (%)	Rise time (s)	Settling time (s)	Integral square error (ISE)	Computational time (s)
PID	9.3460	2.0588	18.2907	2.347	0.0063
MPC	1.1768e–12	4.1898	8.287	1.935	0.0243
Fuzzy RTDA	0	2.9297	5.1052	1.732	0.089

4.2 Disturbance Rejection

The simulated disturbance rejection response along with the load disturbance profile of RTDA Controller is illustrated in Fig. 10. To obtain the regulatory response, the load disturbance is given to the process. It takes about very less time in seconds to track the set-point after the occurrence of load disturbance. Also, the RTDA configuration ensures no steady-state error and smoothens response under load profile. As disturbance is given in only one operating region, single set of tuning values is enough to attain the result.

To emphasize the Fuzzy-Tuned RTDA controller in terms of computational time, its output response is compared with other two controllers. Model predictive controller and PID controllers are taken into account as the next generation regulatory control is developed by combining these two controllers. From Fig. 9, it is observed that PID controller exhibits huge oscillations and takes much time to settle, though MPC controller ensures no steady-state error and zero overshoot, it takes much time to track the set-point whereas Fuzzy-Tuned RTDA encounters all the above-mentioned issues and provides the better result.

Fig. 10 Comparative disturbance rejection response of all control schemes in pH neutralization process. **a** Process output, **b** manipulated variable

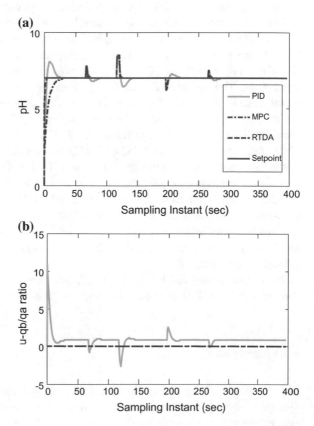

The PID controller takes much less time for computation as it does not include any complex computations. In consideration to performance measures, it exhibits good rise time but becomes poor in both settling time and overshoot. It can not handle constraint and obtains high integral square error. It is evident from Fig. 9 that PID controller exhibits more oscillations in case of disturbance and it encounters both overshoot and undershoot.

Though the controller is simple, its regulatory response is found to be poor. As the pH process needs high accuracy even in presence of disturbance, the controller could not satisfy the requirements. This indicates the lagging of PID controller which can overcome in other controllers.

From Fig. 10, it is observed that MPC provides nonoscillatory and smooth response. MPC controller exhibits better settling time and overshoot than the PID controller but high in rise time. It is reasonably good in integral square error and it can handle constraints.

Even in the presence of disturbance, the controller performs well and it can be seen that it has zero steady-state error. It is evident from Fig. 10 that the encountered disturbance is compensated as soon as possible. But in case of computational time, MPC lags as it involves complex computations.

In case of Fuzzy-Tuned RTDA controller, the response is faster with no oscillations and error-free steady state. In some controllers, overshoot is unavoidable but RTDA guarantees no overshoot. Also it provides better settling time and rise time. In integral square error point of view also RTDA stands better than MPC and PID controller. The control signal required from this controller is also minimum compared to other controllers. The amount of control signal to be provided to the system to avoid disturbance is evaluated effectively. In case of regulatory control, the time taken for disturbance rejection is found to be very minimum. The amount of control signal required for MPC is lower than PID and for RTDA is lower than MPC. The magnified view of controller output of MPC and RTDA controller is shown in Fig. 11 as it is minimum compared to controller output of PID. It indicates that RTDA controller gives better performance with the lower value of control signal.

The computational time of the Fuzzy-Tuned RTDA is found to be less than MPC controller and higher than PID controller which is due to the fact that RTDA controller too involves prediction technique. Unlike MPC controller, the prediction horizon need not to be very high to obtain desired response, prediction length less than % is enough to get better results.

From the results, the Fuzzy-Tuned RTDA is recommended for the non-linear process involving different operating regions (Table 6).

5   Conclusion

In pH neutralization process, the aim is to keep pH value in prescribed limit. The dynamics of pH process is required to manage nonlinearities and uncertainties. The challenge encountered here is to handle the addictive changes that happen because of

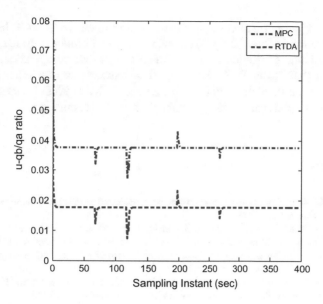

Fig. 11 Controller output of MPC and RTDA in disturbance rejection

Table 6 Disturbance rejection—comprehensive performance analysis of all closed-loop configurations for pH neutralization process

Control schemes	Time domain specifications			Performance indicator	Computational complexity
	Steady-state error (%)	Rise time (s)	Time taken to anticipate disturbances (s)	Integral square error (ISE)	Computational time (s)
PID	0.76	3.6414	8	2.7570	0.0072
MPC	0	6.1898	5	2.0821	0.0440
Fuzzy RTDA	0	3.2202	4	1.9379	0.0237

nature of the process by providing a proper controller. One such controller is RTDA whose functionality is based on a combination of PID and MPC. RTDA control scheme gives the fast set-point tracking under load without increasing the computational effort. As pH neutralization is a non-linear process it has different operating regions, so the controller parameters need to be tuned for each operating region. Thus, instead of a simple RTDA controller Fuzzy-Tuned RTDA controller is implemented, which provides satisfactory performance in all the operating regions. Hence, it is ensured that the process remains in desired value in all the different operating regions and the controller itself capable of handling the disturbance by being less sensitive to the parameter variations. The comparative performance analysis of the controller is done with other control schemes like PID and MPC.

An appropriate optimal control scheme is developed for the acid-base ratio of pH neutralization is done. The performance of each controller is evaluated in terms of time domain specifications, integral square error and computational time and it was found that Fuzzy-Tuned RTDA performs superiorly over other controllers. Fuzzy scheduling in addition to the inherent capability of RTDA controller makes the controller to perform better even in the presence of disturbances.

References

1. Kim, D.K.: Control of pH neutralization process using simulation based dynamic programming. Korean J. Chem. Eng. (2004)
2. Asuero, A.G., Michalowski, T.: Comprehensive formulation of titration curves for complex acid-base systems and its analytical implications. Crit. Rev. Anal. Chem. **41**, 151 (2011)
3. Henson, M.A., Seborg, D.E.: Adaptive nonlinear control of a pH neutralization process. IEEE Trans. Control Syst. Technol. **2**, 169 (1994)
4. Ibrahim, R.: Practical modelling and control implementation studies on a pH neutralization process pilot plant. Doctorate thesis, University of Glasgow, March 2008
5. Chen, X., Chen, J., Lei, B.: Identification of pH neutralization process based on the T-S fuzzy model. Adv. Compu. Sci. Environ. Ecoinform. Educ. **2**, 579 (2011)
6. Ahmed, D.F.: On-line control of the neutralization process based on fuzzy logic. Ph.D. thesis, University of Baghdad (2003)
7. Gomez, J.C., Baeyens, E.: Wiener model identification and predictive control of a pH neutralization process. IEEE Control Theory Appl. **151**, 329 (2004)
8. Sanaz Mahmoodia, Javad Poshtana, Mohammad Reza Jahed-Motlaghb, Allahyar Montazeria, "Nonlinear model predictive control of a pH neutralization process based on Wiener–Laguerre model," Chemical Engineering Journal, 146(2009), 328
9. Abd Al Kareem, D.I.: Implementation of neural control for neutralization process. Master's thesis, University of Technology (2009)
10. Srinivasan, K., Anbarasan, K.: Fuzzy scheduled RTDA controller design. ISA Trans. **52**, 252 (2013)
11. Ogunnaike, B.A., Mukati, K.: An alternative structure for next generation regulatory controllers Part I: basic theory for design, development and implementation. J. Process Control **16**, 499 (2006)
12. Kapil, M., Ogunnaike, B.: An alternative structure for next generation regulatory controllers. Part II: stability analysis and tuning rules. J. Process Control **19** 272 (2009)
13. Oral, O., Çetin, L., Uyar, E.: A novel method on selection of Q and R matrices in the theory of optimal control. Int. J. Syst. Control **1**, 84 (2010)
14. Hasikos, J., Sarimveis, H., Zervas, P.L., Markatos, N.C.: Operational optimization and real-time control of fuel-cell systems. J. Power Sour. **193**, 258 (2009)
15. Mani, G., Pinagapani, A.K.: Design and implementation of a preemptive disturbance rejection controller for PEM fuel cell air-feed system subject to load changes. J. Electr. Eng. Technol. **11**, 1449 (2016)
16. Rodatz, P., Paganelli, G., Guzella, L.: Optimization air supply control of a PEM fuel cell system. Am. Control Conf. 2043 (2003)
17. Jacobs, O.L.R., Hewkin, M.A., While, C.: Online computer control of pH in an industrial process. IEE Proc. **127**, 161 (1980)

18. Yua, Z., Wanga, J., Huangb, B., Bi, Z.: Performance assessment of PID control loops subject to setpoint changes. J. Process Control **21**, 1164 (2011)
19. Prakash, J., Senthil, R.: Design of observer based nonlinear model predictive controller for a continuous stirred tank reactor. J. Process Control **18**, 504 (2008)
20. Prakash, J., Srinivasan, K.: Design of nonlinear PID controller and nonlinear model predictive controller for a continuous stirred tank reactor. ISA Trans. **48**, 273 (2009)

Passive Bandwidth Estimation Techniques for QoS Routing in Wireless LANs

Rajeev Kumar

Abstract Recent research in the wireless network has shown that contending traffic and interference may degrade the performance of multimedia application. Performance degradation includes low-quality image, frame loss, and multiple rebuffering. The performance of multimedia application is directly affected by availability of the bandwidth. Estimation of available bandwidth has shown much interest in the real-time multimedia application that needs some QoS guarantees. Estimation of availability of the bandwidth is a part admission control to provide quality of service routing in wireless LANs. Accurate estimation of available bandwidth is essential for resource management. Bandwidth estimation can be categorized into three categories: active probing technique, passive technique, and model-based technique. This paper provides a detail survey of "passive bandwidth estimation techniques" presented and a new bandwidth estimation technique (BWE-AODV) is implemented. The throughput of proposed and best-effort AODV is compared, and result shows that BWE-AODV performs better than AODV.

Keywords Bandwidth estimation · Active technique · Passive technique and admission control

1 Introduction

Wireless devices are becoming popular due to their flexible support for mobile communication. Many applications, like voice and real-time multimedia, need low packet loss and delay. If minimum bandwidth required for multimedia application is not available than transmission of data packet waste bandwidth and effects on other traffic. In wireless network, due to shared medium, attenuation, and fading, accessing wireless medium is more complex than wired network [1]. Bandwidth estimation techniques have generated several contributions to wireless networks. Active (intrusive) approaches: Active technique uses probe to measure the bandwidth available on

R. Kumar (✉)
CSE, VTU, Belagavi, India
e-mail: raj83it@gmail.com

© Springer Nature Singapore Pte Ltd. 2020
K. N. Das et al. (eds.), *Soft Computing for Problem Solving*,
Advances in Intelligent Systems and Computing 1057,
https://doi.org/10.1007/978-981-15-0184-5_38

a route. Active technique estimates the available bandwidth from one end to another end by propagating equal size packet and determines the one-way delay of probing packets. In fact, transmission delay of probe packet is more than the theoretically calculated maximum delay than wireless medium suffers due to congestion. Passive (non-intrusive) approaches: Passive technique uses only neighbor information and the utilization of the bandwidth. Node monitors the usage of wireless medium by sensing the medium. This mechanism can switch information through from node to its neighbor node. But this switching of information can be done by appending this information in the hello packet, which is used by routing protocol to get information about local topology [2]. If this switching of information does not consume much bandwidth, it is termed as passive technique [3].

2 Passive (Non-intrusive) Bandwidth Estimation Techniques

The radio medium is monitored over a time interval. To estimate available bandwidth, channel usage ratio is used without affecting the existing flows [4, 5]. In this section, we have focused on different passive techniques for wireless networks such as cPEAB, DBE, APBE, DABE, and ABEWN.

2.1 Cognitive Passive Estimation of Available Bandwidth (CPEAB)

CPEAB estimates the bandwidth available on a link by measuring the amount of DIFS and back-off delay, probability of collision, "acknowledgement delay and idle period" of channel in the measurement interval [6]. CPEAB is developed on multiband Athros software (driver) for Wi-Fi (MadWi-Fi).

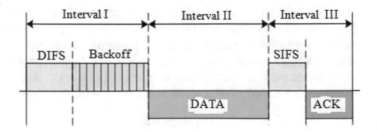

Fig. 1 IEEE 802.11 frame exchange process

If medium is busy in the "interval I," (Fig. 1) then transmission process is interrupted for DIFS and goes to back-off process and repeats if medium is idle again. Therefore, available bandwidth can be estimated as:

$$ABW_{cPEAB} = (1 - K) \times T_i/T \times C \tag{1}$$

where T_i is the channel idle time, T is the measurement period, C is the channel capacity, and K is the amount of consumed bandwidth due to waiting and back-off process.

$$\overline{Backoff} = \sum_{K=0}^{R} P(X = K) \times \frac{\min(CW_{min}; 2^k CW_{min}) - 1}{2} \tag{2}$$

where

CW_{min} indicates minimum value of contention window.
$CW_{max} = 2^k CW_{min}$
R indicates highest number of retransmission.
X indicates retransmission attempt of a given frame.

The value of interval II can be obtained from counter register of Wi-Fi chipset. Proportion of the consumed bandwidth in "interval II" can be calculated as:

$$K = ACK_{timeout} + SIFS/T \tag{3}$$

where $ACK_{timeout}$ is the value of lost acknowledgement.
Therefore, Eq. (1) can be written as:

$$ABW_{cPEAB} = (1 - ACK) \times (1 - K) \times T_i/T \times C \tag{4}$$

Impact of hidden/exposed nodes:
Packet collision rate depends on the size of frame.
If P_C is the probability of collision than Eq. (4) can be written as:

$$ABW_{cPEAB} = (1 - ACK) \times (1 - K) \times (1 - P_C) \times T_i/T \times C \tag{5}$$

cPEAB provides more accurate result in "IEEE 802.11 Wi-Fi WLANs."

2.2 Accurate Passive Bandwidth Estimation (APBE)

APBE estimates the "available bandwidth" by considering the DIFS and back-off, ACK delay, probability of packet collision, and idle time of channel in the measurement period and also includes RTS/CTS mechanism [7]. The overload of RTS/CTS can be measured as:

$$R/C = (\text{RTS} + \text{CTS}) + \text{SIFS}/T \quad \text{if RTS/CTS is used otherwise.}$$
$$R/C = 0 \tag{6}$$

Available bandwidth can be estimated as:

$$\text{AB}_{\text{APBE}} = (1 - K) \times (1 - R/C) \times (1 - \text{ACK}) \times (1 - P_C) \times T_i/T \times C \tag{7}$$

where K indicates the bandwidth used by the "waiting and back-off".

P_C indicates probability of packet collision of RTS message.
T_i indicates idle time of the wireless channel.
T is the measured period.
C is the capacity of the medium.

The performance of APBE is compared with cPEAB, and the result shows that APBE performs better in terms of accuracy.

2.3 Dual Bandwidth Estimation (DBE)

Method of estimation: the proper way is to measure utilization of resource and deduct this from the capacity of the channel. The basic methods to measure the utilization of resources are busy time of channel, delay, congestion window, and collision [8]. The busy time metric of wireless medium is a passive measurement. The available bandwidth can be calculated as:

$$B_{\text{avl}} = (1 - U) \times B_{\text{max}} \tag{8}$$

Where B_{avl} is the available bandwidth.
U is the resource utilization.
B_{max} indicates maximum bandwidth of the medium.

Due to the nature of IEEE 802.11, total capacity of the medium cannot be used. The overhead caused by DIFS and SIFS and back-off mechanism considered for calculating the bandwidth. The demerit of busy time method is that host cannot release the consumed bandwidth, when route breaks [3]. Therefore, another method is used in AODV routing protocol called "hello message" that updates their neighbor caches. To include two new fields, hello message is modified. The first field contains node address, timestamp, and utilized bandwidth, and the second one includes neighbor address, timestamp, and utilized bandwidth. This method utilizes both busy time and "hello message" according to condition of the channel [9, 10]. Result shows that use of hello messages increases with respect to increase in mobility and utilization of listen method decreases.

2.4 Bandwidth Estimation Based on Retransmission (BER)

BER is a method for link available bandwidth estimation. It is based on frame retransmission prediction. Node broadcasts hello message, which includes information about utilization of channel [11]. In the same period, node evaluates probability of collision and number of retransmission based on the "average size of the contention window." Then, bandwidth available on a link can be calculated with this information and related back-off information can be recorded by nodes. Node cannot send frame in the interval SIFS and DIFS. Therefore, node that listens to RTS frame can estimate the busy time of channel using (9).

$$T_B = T_{RTS} + NAV_{RTS} + T_{DIFS} + T_{backoff} \tag{9}$$

where T_{RTS}, NAV_{RTS}, and T_{ACK} are constant. $T_{backoff}$ changes according to topology of the network.

Evaluation of frame collision: If the collision occurs between frames because of hidden terminal or any other reason, nodes double the contention window size. Thus, size of contention window may give information about collision. Frame transmission procedure can be expressed as:

$$CW_{max} = 2^N \times CW_{min} \tag{10}$$

Back-off time can be calculated as:

$$Backoff_{Time} = Random\ () \times aSlot_{Time} \tag{11}$$

where Randon () is a random integer chosen between $[0, CW]$

$aSlot_{Time}$ is a slot duration determined by PHY characteristics.

The statistical average method to predict back-off duration based on previous records can be estimated as:

$$T_{Backoff} = T_M \times \frac{1}{N} \times \sum_{i=1}^{N} T_i \tag{12}$$

where T_M is the period of monitor.

T_i is the duration of back-off in each second.

$T_{Backoff}$ is the time of back-off in the observation period.

Link available bandwidth is calculated as:

$$BW_L = \frac{T_M - \sum_{i=1}^{n}(NAV_i + T_{DIFS} + T_{RTS}) - T_{Backoff}}{T_M} \times C \tag{13}$$

where C indicates capacity of the channel.

NAV$_i$ is the total busy time of channel calculated from NAV, and it includes NAV period of other nodes in hello message and NAV period locally.

BER accurately estimates the available bandwidth and the frame collision status under different network load.

2.5 Available Bandwidth Estimation in IEEE 802.11-Standard Wireless Networks (ABEWN)

ABEW estimates the effect of contention, interference, and channel fading. ABEWN estimates the available bandwidth of a host to each of its neighbors. Throughput for transmission of a packet is estimated as:

$$TP = P_z/t_r - t_s \tag{14}$$

where P_z indicates size of the packet.

t_r is the time for receiving ACK.

t_s is the time at which frame is ready at MAC layer.

$(t_r - t_s)$ is time for busy period of channel and contention period.

If contention is high, throughput decreases and $(t_r - t_s)$ increases. This technique can captures the phenomenon of interference and fading error since these errors may effect on RTS and data packet. In this case, they need to be retransmitted. Result shows that at a minimum cost, each data flow receives its minimum requested bandwidth.

2.6 Distributed Available Bandwidth Estimation in Mobile Ad Hoc Network (DABE)

The bandwidth available on a link is determined using multiple factors such as utilization of channel, probability of collision, back-off period, and frame retransmission time. Channel Monitoring: IEEE 802.11 MAC protocol uses "virtual carrier sensing" to monitor channel. Nodes send RTS frame that contains time to complete this data frame and the time needed to transmit the remaining data, one ACK frame and one "CTS frame" [4]. The destination host replies with a small "CTS frame" in which the time duration is equal to the value from received "RTS minus the time needed to send CTS and SIFS frame." The node that hears "RTS or CTS" frame sent by neighbor node and keep NAV to the value obtained from time-period field [12]. All the neighbor nodes of communicating node are excluded from transmission by NAV setting. When NAV reaches to zero, nodes are allowed to access the channel. In this four-way handshake, the collision from hidden terminal is eliminated.

Collision Estimation

If collision occurs because of hidden nodes or any other reason or any other reason, contention window size is doubled until the window size reaches the maximum threshold.

Link Available Bandwidth Estimation

To estimate bandwidth availability, node obtains the NAV information and back-off duration of their neighbors and computes the busy period of the channel within the observation period [13]. To estimate amount of bandwidth available between two nodes, the common "idle time of two nodes" must be obtained. The available bandwidth can be obtained as:

$$B = C - \frac{C}{T_m} \times \sum_{i=1}^{N} \left(S_{\text{Busy}}[i] + D[i] \right) \tag{15}$$

where C is the capacity of the channel.

$S_{\text{Busy}}[i]$ *and* $D[i]$ are the busy duration of host S and host D.

Result shows that the error ratio of the "estimated and the measured" value of chain, regular and random topology are "3.474, 4.575 and 6.895" and when "node density" is below 8%, the error ratio is "3.74, 3.04, 4.973." Thus, this technique performs better in case of topology that has low node density.

3 Comparative Evaluation

The objective of this section is to evaluate and compare the performance of best-effort AODV protocol with BWE-AODV (bandwidth estimation AODV). This approach use collaboration between MAC, physical and network layer [5]. Available bandwidth is the maximum data rate that can be supported without degrading the performance of ongoing flow in the network. Most of the bandwidth estimation techniques in the "IEEE 802.11-based ad hoc network" use passive bandwidth estimation. BWE uses admission control to admit new flow and avoid degradation of ongoing flow in the network. To estimate the available bandwidth on link (s, r), each node s and r monitors the channel at the MAC layer. Every Δ time (set to 1 s) the ratio of free and busy time of two nodes are computed and the available bandwidth of the link is estimated as:

$$b_{(s,r)} = (t_s.t_r).C \tag{16}$$

where $b_{s,r}$ is the available bandwidth of the link (s, r), t_s and t_r are the ratio of idle time on MAC layer of node S and R. C is the capacity of the medium. BWE considers the probability of non-existence of collision $(1 - C_P)$, where C_P is the probability of collision. IEEE 802.11 provides the back-off mechanism to minimize collision phenomenon BWE considers the proportion of bandwidth

(K_p) consumed by the back-off mechanism.

$$K_p = \frac{\text{DIFS} + \overline{\text{backoff}}}{T_m}$$

where T_m is the time separating the emission of two consecutive frames.

The final available bandwidth on link (s, r) is

$$\text{BWE}_{(s,r)} = (1 - k).(1 - C_P).b_{(s,r)} \tag{17}$$

BWE is added to AODV, the throughput needed by the application is added in the RREQ packet, when a node broadcasts RREQ, admission control is performed at each node which receives this message using Eq. (17). This technique uses an exchange of hello standard message periodically (for BWE in Δ seconds) for connectivity management. The link break is detected as the technique used in AODV. BWE also exchange "hello" packet to gain the bandwidth information between neighbor nodes that is the node r should have the value of b_s to run Eq. (17) (Table 1).

In Fig. 2, each flow is composed of 512 bytes packets with a data rate of 600 kbps using a data rate capacity of 2 Mbps. Flow 1 start at time 1.4 s, flow 2 at 10 s, flow 3 at 20 s, and flow 4 at 30 s. First, we observe that all four flows have been accepted. Flow 1 achieved a throughput more than 500 kbps; when flow 2 is admitted, it also achieved a throughput more than 500 kbps. When flow 3 admitted, it also achieved a throughput of 500 kbps, when fourth flow is admitted, it has achieved a throughput of 350 kbps, and all the flows throughput decreased and reached to a lower throughput between 200 and 350 kbps.

In Fig. 3, first flow has achieved a throughput of more than 800 kbps, flow 2 has achieved 600 kbps, flow 3 has achieved more than 500 kbps, and flow 4 has achieved more than 300 kbps. When all the flows admitted, all flows achieved a minimum throughput between 250 kbps and 400 kbps.

Table 1 Simulation Parameters

Parameters	Value
Number of nodes	30
Simulation time	80 s
Simulation area	$500 \times 500 \ m^2$
Traffic source	CBR
Agent	UDP
No of flows	4
Routing protocol	AODV, BWE-AODV
Packet size	512 Bytes
Basic rate	1 Mbps
Data rate	2 Mbps

Fig. 2 Throughput of each
flow using AODV

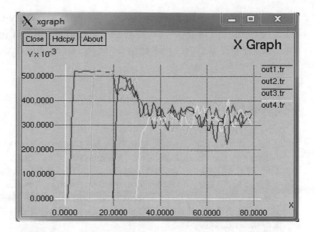

Fig. 3 Throughput of each
flow using BWE-AODV

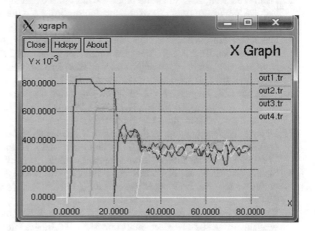

4 Conclusion and Future Works

Bandwidth estimation is an issue in wireless network because every node does not
have complete information of the network status. This paper has provided various
"passive bandwidth estimation techniques" in wireless networks. Each technique
has its own merits and limitations. In this paper, a BWE technique has been implemented and incorporated into AODV. As we can observe that the throughput using
BWE-AODV is much better than best-effort AODV, AODV does not incorporate any
bandwidth estimation technique and admission control. In future, we plan to incorporate bandwidth estimation techniques into NS-2.35 simulator. In this embedded
NS-2.35, we will be able to see the performance of routing protocols in different
network topology. The aim of forming bandwidth estimation method is to improve

resource utilization in a proper way in mobile ad hoc network. The method of bandwidth estimation will be combined with admission control and resource reservation method. In this way, we can improve QoS in highly dynamic network.

References

1. Kumar, R.: A comprehensive analysis of MAC protocols for Manet. In: IEEE International Conference on Electrical, Electronics, Communication, Computer and Optimization Techniques (ICEECCOT), December 2017, pp. 56–58. GSSSIETW, Mysure
2. Kumar, R., Kumar, R.: Reactive unicast and multicast routing protocols for Manet and issues a comparative analysis. In: International Conference on Internet of Things, Journal of Engineering and Technology (IJERT), vol. 4, Issue 29, pp. 135–137, August 2016. APS College of Engineering, Bangalore
3. Park, H.J., Roh, B.J.: Accurate passive bandwidth estimation (APBE) in IEEE 802.11 wireless LANs. In: Proceedings of the 5th International Conference on Ubiquitous Information Technologies and Applications, pp. 1–4 (2010)
4. Chen, L., Heinzelman, W.B.: QoS-aware routing based on bandwidth estimation for mobile ad hoc networks. IEEE J. Sel. Areas Commun. 23(3), 561–572 (2005)
5. Sivakumar, R., Sinha, P., Bharghavan, V.: CEDAR: a core-extraction distributed ad hoc routing algorithm. IEEE J. Select. Areas Commun. 17, 1454–1465 (1999)
6. Tursunova, S., Inoyatov, K., Kim, Y.-T.: Cognitive passive estimation of available bandwidth (cPEAB) in overlapped IEEE 802.11 WiFi WLANs. In: IEEE Network Operations and Management Symposium, pp. 448–454 (2010)
7. Dapeng, W., Zhen, Y., Bing, S., Chunxiu, X., Muqing, W.: Improving accuracy of bandwidth estimation based on retransmission predicting in MANET. In: 4th International Conference on Wireless Communications, Networking and Mobile Computing, pp. 1–4 (2008)
8. Shah, S.H., Chen, K., Nahrstedt, K., Available bandwidth estimation in IEEE 802.11-based wireless networks, In: Proceedings of 1st ISMA/CAIDA Workshop on Bandwidth Estimation (2003)
9. Yan, Z., Dapeng, W., Bin, W., Muqing, W., Chunxiu, X.: A novel call admission control routing mechanism for 802.11e based multi-hop MANET. In: 4th International Conference on Wireless Communications, Networking and Mobile Computing, pp. 1–4 (2008)
10. Latif, L.A., Alliand, A., Ooi, C.C.: Location based geo-casting and forwarding (LGF) routing protocol in mobile ad hoc network. In: Proceedings on the Advanced Industrial Conference on Telecommunications/Service Assurance with Partial Intermittent Resources, pp. 536–541 (2005)
11. Peng, Y., Yan, Z.: available bandwidth estimating method in IEEE802.11e based mobile ad hoc network. In: 9th International Conference on Fuzzy Systems and Knowledge Discovery, pp. 2138–2142 (2012)
12. de. Renesse, R., Friderikos, V., Aghvami, A.H.: Cross-layer cooperation for accurate admission control decisions in mobile ad hoc networks. IET Commun. 1(4), 577–586 (2007)
13. Zhao, H., Garcia-Palacios, E., Wei, J., Xi, Y.: Accurate available bandwidth estimation in IEEE 802.11-based ad hoc networks. Comput. Commun. 32, 1050–1057 (2009)

Dynamic Monitoring of Health Using Smart Health Band

Viraj Puntambekar, Shreyas Agarwal and P. Mahalakshmi

Abstract Smart health band would be a smart assistive innovation-focused solution for the detection. It is also a great tool for analysing the health statistics of the person who is considered to be in grave danger. It is also a means for a network-synced ambulant and also considered as a health-monitoring bracelet which is capable enough of reading the user's human vitals such as (pulse rate and body temperature). In this bustling world, where the activities of human are restricted to sedentary lifestyle and working on their desktops for the all day long, would be evident to create a burden of problems including obesity, heart problems, no exercise hence respiratory problems, and lot more.

Keywords Health analysis · Human vitals · Smart health

1 Introduction

Regulating the healthy and salubrious motives in daily life, would result in far more beneficial aspects. But understanding the study of the unhealthy routine is a necessity to work on appropriate solution.

Examining the past work, various parametric associations are taken into consideration: time complexity, accurate measurement, cost efficiency, operating range, sensitivity range and high portability. Prior examination revealed that studies are based on the static analysis with low-precision instruments. With the development in the microprocessing ability, today, we can parallel process more than three sensors at the same time as opposed to the previous technology. Considering the conventional practice of data analysis, more focus is made in data simplification and presentation by parallel analysis.

V. Puntambekar · S. Agarwal
Vellore Institute of Technology, Vellore, Tamil Nadu 632014, India

P. Mahalakshmi (✉)
School of Electrical Engineering, Vellore Institute of Technology, Vellore, Tamil Nadu 632014, India
e-mail: pmahalakshmi@vit.ac.in

© Springer Nature Singapore Pte Ltd. 2020   453
K. N. Das et al. (eds.), *Soft Computing for Problem Solving*,
Advances in Intelligent Systems and Computing 1057,
https://doi.org/10.1007/978-981-15-0184-5_39

Sometimes the recorded mental imploding needs are supposed to be controlled. Most of the times it has resulted into resulted in unpredictable suffering. Sporadic mental pressure detectability helps in reduction and prevention of the stress related health issues. The objective here is to design an IoT base wearable, which would be endorsed with cost-effective featurability and less power workable smart band which helps to measure for health care and the necessary benefits that detect physical tension based on skin reactance. This device helps in the real-time application to monitor the wearer's physical strain incessantly and transmit the stress related data wirelessly to user's smartphone device. It just not only helps the patient for a clearer understanding of their stress configuration but also provides the physicians and the doctors with reliable data assisting in much better treatment.

2 Hardware Requirements

2.1 Pulse Rate Monitorization Using PPG (Photoplethymogram)

Pulse rate monitoring would be considered to be the main feature of the project. With each recorded cardiac cycle, it has been observed that the heart would push the blood to the periphery. This initiates a pressure on the periphery walls. Even though this pressure pulse is assumed to be damped, but until time it would reach the outer periphery, it would be considered sufficient to distend the arteries and also the arterioles in the subcutaneous tissue. A pressure pulse can also be observed from the venous plexus, if the pulse oximeter is attached without compressing the epidermis as a small secondary peak.

A photoplethysmogram (PPG) would be considered to be optically derived from plethysmogram, which would be considered to be a volumetric quantification of an organ. A PPG is acquired by using a pulse oximeter which would brighten up the skin and would quantify the changes in absorbed light.

2.2 Drowsiness Alarm System

If the device is on car mode, it will analyse the photoplethysmogram **(PPG)** signals and inform whether the user is fit to drive or not with the help of the alarm.

2.3 Real-Time Data Transfer Using WI-FI Module ESP8266

The data analysed would be stored in the cloud via the use of Wi-Fi module ESP8266. The data will help the doctor to analyse the patient's condition on daily or weekly basis (based on the time span required by the patient).

Application intended to use is "ThingSpeak," because "ThingSpeak" is an open-source platform for IoT application and APIs to store and retrieve data from the process in motion by using HTTP protocol on the Internet or via a local area network (LAN). "ThingSpeak" helps with the aspect of enabling the creation of sensor logging applications and location-tracking applications which bind as a social network of things with status updates.

2.4 Atmospheric Temperature and Pressure Measurement

The sensor used for this purpose is BMP180 which helps user to know the temperature and pressure.

2.5 Body Temperature

The body temperature is found with the help of LM35. The LM35 is precision integrated circuit temperature measurement sensor, whose output voltage varies linearly in proportionality to the Celsius temperature.

3 Methodology

All the data which is received by the microcontroller unit, i.e. Arduino Nano, is sent by the sensors like BMP 180, DM11, LM35, IR, etc. and is processed by it. This information is further sent to the cloud through Wi-Fi module ESP8266. This information is then further accessed by Internet and acted upon.

Inputs to the smart assistive band are the various signals received from disparate sensors. By insightfully deducing the relationship in-between these signals using machine learning algorithm, this band predicts that whether the patient of the user would be suffering from stress or not. With the significant progress with the working and assembly of the project, and have brought out, significant results, which are classified as data, for tracking human health.

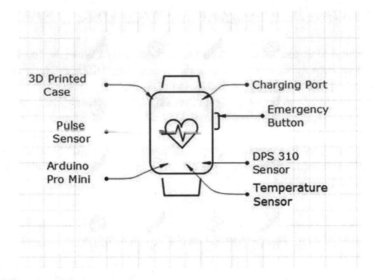

Fig. 1 Representation of the final blueprint of the dynamic smart health band

3.1 Photoplethysmogram

The modification in volume is perceived by the pressure pulse and is detected by lightening up the skin with the light from a light-emitting diode (LED) and then it would be measuring the intensity of light which can be considered to be either transmitted or reflected to a photodiode. Each cardiac cycle appears as a peak. This signal will be conditioned with the help of the conditioning circuits (Fig. 1).

3.2 Internet of Things

The Internet of Things involving as the **key** technology is said to include the following parametric axioms: electronic product code (EPC), the information identification system (ID system), the EPC middleware (implementation information filtering and sampling), discovery service (information discovery service), and the EPC information service (EPCIS) [1].

With the promising technological as well as economic, and social prospects, the IoT revolution is redesigning the modern health care.

Wearable technical advancements have displayed a successful growth upliftment in recent years, with almost millions and millions of devices which are sold to customers and as well steady development which are being made in the technological capabilities. Weighing on the Shannon and Thorpe's 1961 experiment, the current form and ability of the contemporary wearables have said to be changed. When integrating the futuristic technologies which are subjected to be torn away of the same design issues have to be taken into account. Although wearables have profited from the advances in portable technologies, functionality still remains restricted compared to smartphones accessibility. Additionally, the smartphones are not considered to be well-furnished devices due to its structuring and the limiting criterial measurement, and they have been more suited to aesthetic pleasure requirement. To empower consumers with this self-knowledge, wearables present a market-wide opportunity for developing an incessant line of data about our physiology and kinesiology (Fig. 2).

Wearables can offer insights that smartphones cannot on Human health and fitness. This would be considered to be inevitable from the tremendous popularity of fitness trackers (e.g., the Fitbit Blaze, Jawbone UP, and Nike + Fuel-Band) and also smartwatches (e.g., the Apple Watch and Samsung Gear) being used by customers for self-monitoring physical activity. Health conditions such as hypertension and stress can be self-monitored and prevented using wearable.

Wearables are considerably and steadily becoming one of the most prevalent personal devices, by offering the users, all the ability to interact with other sensors as well as physical conditions and environment around them [2]. Worked on creation a conventionally connected world—common ways via which the Internet and connected devices may shift to more active activity of the content and environment, specifically in the health care is possible once the IoT becomes more widely adopted.

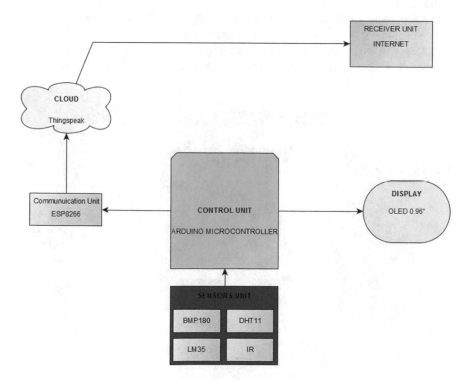

Fig. 2 Diagrammatic representation of the cloud-based platform

Now having a look at a few other case studies that showcase how the wearables and other advancements integrate to perform an IoT solution with different domains in nature.

There is almost limitless application for the implementation of the IoT in the health industry, and these steps will help in reducing the human intervention and ultimately the human error.

4 Implementation

Our initial aim happened to be on sensing the heartbeat. For this, we used infrared sensor which was helpful in sensing the blood flow very precisely. For more precision, we calibrated the IR using the potentiometer for the better distance analysis and focus in reading the actual blood flow. In contrast to the remainder of the equipment available in the market, this watch not just focuses for the heartbeat through the wrist strain; here, the focus would be also on the dorsal vertebrae for precise and accurate heartbeat analysis [3].

The prototype for the smart health is still in motion for the complete output PCB fabrication (Fig. 3).

The data collected from the sensor are self-analysed with the assigned parametric limitations. The device would also progress the exposure of the external environmental factors with the human limitations. With respect to the atmospheric pressure and temperature fluctuations, the watch could be also able to predict the limitations of

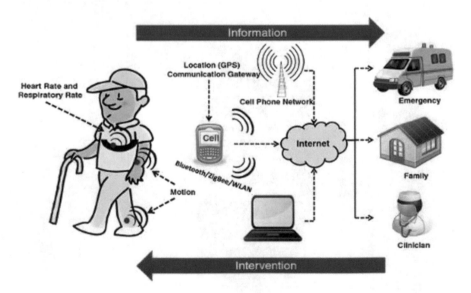

Fig. 3 Diagrammatic representation of the communication system for health monitoring

the users to within the sudden change in the environment. The watch would take into consideration the change in the reading accuracy due to the excess hand movement. The error predicted would cover up with 10 bpm reading affecting the accuracy not less than 91% disturbance [4, 5].

Considering the excess environmental factors, as the one due to the excess hand movement and sudden temperature change factorization, the watch precision would be characterized with respect to the surrounding atmosphere it is exposed to. A person exerting in the gym and a person undergoing a hike both experience elevated heartbeat, but the rate of elevation is far higher in a gym due to incessant workout regime rather than an inconsistent stressful environment during a hike. Hence, the precision and data updating would be a fast rate in a gym workout than a sporadic time interval compared to a less stressful environment, such as a gym [6].

With the regular health sensing parameters, progress is also made to link the input digital data for the machine learning.

4.1 Machine Learning

Machine learning is considered to be more perceivable and more immediately useful than the rest of the productive solutions. It establishes its attention on its ability of AI to learn and adapt, which is teaching computers to make predictable analysis based on examples or training supplementary data. Apart from the programming assembly, this type of learning helps in proper organization of the static and dynamic data processing that not just grows but trains and learns with the system architecture for appropriate functionality of the system's nature.

The input data are classified using the regression modes. This would help in understanding how much the data types are varied from the healthy data set.

Table 1 gives in the actual measured reading from four subjects. Making comparison with the most reliable and the most used fitness readings, machine learning

Table 1 Observational reading of the data analysed in the first phase

Sr. No.	Observational readings			
	Subjects volunteered	Pulse (per/min)	Body temperature (°C)	Atmospheric temperature (°C)
1.	Subject 1 (20 years male)	78	35	28
2.	Subject 2 (21 years male)	83	34	29
3.	Subject 3 (19 years male)	80	36	30
4.	Subject 4 (20 years male)	77	36	28

algorithms can be utilized to provide proper suggestion and data analysis. This would help in the development of the health benefits without human intervention and would significantly reduce the human errors.

This research utilizes the usage of support vector machines in order to classify the ECG signal generated [7]. A colossal amount of data is analysed with respect to the average mean to merit the assessment of the boundary threshold. Then, the performance of the feature extraction is analyzing the output of the ECG graph. With these, we can aim for a singularity spectrum which helps to concentrate the figurative output into a categorized analysis.

5 Results

Based on the functionality, the linkage of the processed data with the "Thingspeak" application helps with the communication aspect. The applicational use of the ESP 8266 module makes use of the TCP/IP protocol. The ESP8266 Wi-Fi module is considered to be a self-contained SOC with integrated TCP/IP protocol stack which can access to the given Wi-Fi network. The ESP module would allow the Arduino board to connect with the router and then access Internet network. The Arduino is programmed to communicate with the cloud platform, i.e. ThingSpeak over TCP/IP protocol. The Arduino can implement TCP/IP protocol by passing AT commands serially to the ESP8266 module [8].

The parameters analysed were atmospheric pressure, pulse rate, body temperature, and atmospheric temperature.

The resultant graphical representation for the same would be: -

This graphical analysis helps with the estimation of the real-time dynamic changes in the human health. The data as presented above also can be calibrated with respect to the time period specified, which can be expanded over the user specific requirements (Fig. 4).

The observed data are within 4% accuracy rate according to the highly sensitive and precise devices. With the advent in the technological advances, the gap can be

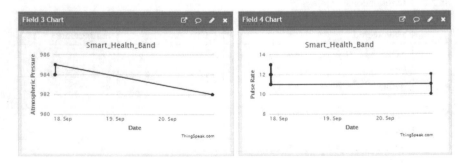

Fig. 4 Graphical representation of atmospheric pressure and pulse rate

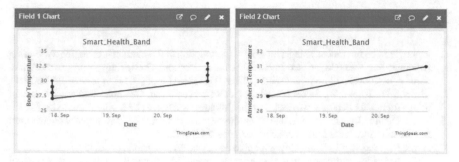

Fig. 5 Graphical representation of body temperature and atmospheric temperature

reduced and the precise reading can be obtained. The body temperature was measured with respect to the thermometer. The temperature measurement of the surrounding was observed with respect to the sensor such as RTD and thermistors whose precision level were well established (Fig. 5).

6 Conclusion

Since the development in the biomedical fields, embedded systems relating to health band associations have impacted significantly in the engineering research field. Recently, focus on such applicational study helps to maintain the advancement in the monitoring health and wellness. Since 1980s, the advancement the focus was shifted from remote monitoring to portable health assessment, which assisted to the convenience of the health analysis.

Consequently, the target of remote monitoring at any location could be achieved, it has witnessed that a heavy deal of work towards the assembly of wearable technologies and communication and data analysis technologies [9, 10]. Besides, when monitoring task is executed in the home, researchers and clinicians have already developed ambient sensors in the remote monitoring systems. It has witnessed quite a growing interest for the emerging need for establishing a part of telepresence in the home setting so as to replace implement clinical interventions [11]. The goal of establishing a telepresence in the home environment could be achieved with the home robots which will soon be integrating into home automation for facilitating healthy monitoring system.

Acknowledgements I would like to thank my faculties who supported me in my research, my colleagues and my seniors for supporting and being there throughout the whole time, whenever I needed.

All the experimentation has been conducted ethically in the safe environment. All the subjects involved were humans and no humans were harmed in any way [12]. Consent was taken from the human subjects before conducting the experiments.

References

1. Evans, D.: The Internet of Things: how the next evolution of the Internet is changing everything [Internet] San Jose (CA): Cisco Internet Business Solutions Group. http://www.cisco.com/c/dam/en_us/about/ac79/docs/innov/IoT_IBSG_0411FINAL.pdf (2011). Accessed 25 Jan 2011
2. Metcalf, D., Khron, R., Salber, P. (eds.): Health-e everything: wearables and the internet of things for health: Part one: wearables for healthcare. Orlando, FL: Moving Knowledge (2016)
3. World Health Organization factsheets: cardiovascular diseases (CVDs). http://www.who.int/mediacentre/factsheets/fs317//en/ (2015). Accessed Apr 2015
4. Apple Inc.: Apple watch. https://www.apple.com/watch/. Accessed Apr 2015
5. FitBit Inc.: Flex: wireless activity + sleep wristband. https://www.fitbit.com/flex. Accessed Apr 2015
6. Hu, F., Xie, D., Shen, S.: On the application of the internet of things in the field of medical and health care. In: IEEE International Conference on and IEEE Cyber, Physical and Social Computing Green Computing and Communications (GreenCom), (iThings/CPSCom), Aug 2013, pp. 2053–2058
7. AF classification from a short single lead ECG recording: the PhysioNet/computing in cardiology challenge. https://physionet.org/challenge/2017/ (2017)
8. http://fab.cba.mit.edu/classes/865.15/people/dan.chen/esp8266/
9. Olorode, O., Nourani, M., Reducing leakage power in wearable medical devices using memory nap controller. In: Circuits and Systema Conference (DCAS), pp. 1–4. IEEE Dallas, Oct 2014
10. http://ieeexplore.ieee.org/document/7113786/?part=1
11. https://www.ncbi.nlm.nih.gov/pmc/articles/PMC5334130/
12. https://drive.google.com/file/d/1OiV8ZgJNpIX1st2Ro_gTCpctFnPqOOIw/view?usp=sharing

Agent-Based Modeling of the Adaptive Immune System Using Netlogo Simulation Tool

Snehal B. Shinde and Manish P. Kurhekar

Abstract The biological immune system is the progressive complex adaptive systems (CASs) that consist of inhomogeneous and adaptive agents. It is an important defense mechanism in the human beings that generates an complex cellular response against the foreign disturbances. It also exhibits prominent properties such as emergence and self-organization. There is an immediate need of immune system modeling to understand its complex inbuilt mechanisms and to see how it keep up the homeostasis. Various mathematical and computational simulation approaches have been proposed for the modeling of the intricate dynamics of the complex biological immune system. There are two kinds of modeling techniques that are used to simulate the immune system: equation-based modeling and agent-based modeling. Due to some drawbacks of the equation-based simulation, agent-based modeling technique is used. In this paper, we propose an agent-based model of the adaptive immune system and it is developed with the help of Netlogo simulation tool. This model helps researchers to understand the structure of the immune response and verify hypotheses. This model can be easily employed as an educational tool in academics and a research tool in scientific disciplines for developing medicines that can keep a disease under control.

Keywords Adaptive immune response · Agent-based modeling · Complex adaptive systems · Computational modeling · Immune system · Netlogo tool

S. B. Shinde (✉) · M. P. Kurhekar
Department of Computer Science and Engineering, Visvesvaraya National Institute
of Technology, Nagpur, Maharashtra, India
e-mail: snehalbankatrao@students.vnit.ac.in

M. P. Kurhekar
e-mail: manishkurhekar@cse.vnit.ac.in

© Springer Nature Singapore Pte Ltd. 2020
K. N. Das et al. (eds.), *Soft Computing for Problem Solving*,
Advances in Intelligent Systems and Computing 1057,
https://doi.org/10.1007/978-981-15-0184-5_40

463

1 Introduction

Complex systems consist of various components with interactive and adaptive nature that cause the emergence and self-organization in the system. These outcomes are often impossible to predict by looking for the interactions at the individual or population level. The complex systems that change their behavior and configuration over time as they evolve are referred to as complex adaptive systems (CAS) [2, 3]. The immune system is also considered as an advanced CAS [3] that consists of cells and molecules. It is an inbuilt defense system that protects the body against the foreign disturbances [9] by providing a response against them. Due to the interactive and adaptive nature of the immune cells, the immune system exhibits various predominant properties, namely [5]:

- An adaptation in CAS refers to the change in the state of the entities because of the perturbations to get the stable state of the system.
- The emergence refers to the appearance of various features and variations in the CAS when the system is considered as a whole.
- Self-organization is a process that organizes the system by itself without external effects.

Systems biology is the field that studies interrelatedness of different biological components and helps to understand the complex interactions among these components such as molecules, cells, tissues, organisms, and the entire species [4, 14]. The aim of this approach is to study the emergent behaviors that are observed in the CASs. Mathematical and computational modeling approaches are part of the systems biology approach to study biological phenomena. There are two kinds of simulation techniques that are proved to be helpful to model the biological dynamics in systems biology [3, 6, 10, 11]. The first top-down simulation technique called as the equation-based modeling (EBM) that mainly comprises ordinary differential equation and partial differential equation-based models. The second bottom-up simulation technique is the agent-based modeling (ABM) that is comprised of different agents and rules. EBM has certain limitations that are as follows:

- EBM assumes sufficiently large population sizes. It considers rates of creation, death, binding, and diffusion for the entire population.
- Spatiality and topology that depend on individual interactions get ignored.
- The modeling of the immune system involves non-linearities, due to which it is very difficult to get solutions for differential equations.

Although both EBM and ABM modeling techniques have some pros and cons, the ABM is mainly used for simulating various complex biological systems. EBMs are more suitable for homogeneously distributed systems and they are less used where spatiality is one of the main concerns. ABM has an inbuilt spatial component that helps to depict the local interactions and environmental heterogeneity [8]. Agent-based modeling is based on the synthesis of complex interactions from the activities at the microscopic level. There are different ABM tools that are helpful during the simulation of the immune system dynamics such as Repast, Mason, NetLogo, and Swarm [14]. As compared to other tools, it is easier to understand and to simulate the complex dynamics with the help of Netlogo. It is one of the best tools for novice programmers. It is thus widely used simulation tool for immune system modeling [6].

The main focus of the paper is on the modeling of the complex cellular immune dynamics that are involved during the immune response to foreign invaders. This paper is organized as follows: Sect. 2 provides particulars of the immune system and agent-based modeling. Section 3 presents our proposed idea of the adaptive immune response simulation through the Netlogo simulation tool. Section 4 provides the discussion of our model and the results. The conclusions drawn and future research directions are provided in Sect. 5.

2 Immune System and Agent-Based Modeling

This section provides preliminary knowledge about the immune system and agent-based modeling.

2.1 Immune System

The immune system protects the body from invading harmful foreign pathogens by providing complex responses to them [9, 12]. This immune response is of two kinds [3, 9]:

1. *Innate response:*
   This first line of response is of surface defense, and it also includes the chemical systems (i.e., complement system and interferons) and provides protection against any foreign molecules without knowing them. Thus, this innate kind of response by the immune system is called as the non-specific cellular attack. Surface defense is comprised of the skin and mucous membrane. Cellular response consists of the attack of killer cells, such as macrophages, natural killer cells (NK), and neutrophils.
2. *Adaptive immune response:*
   Lymphocytes (i.e., B-cells and T-cells) provide the adaptive immune response that is also referred to as the foreign molecule-specific immune response. During this response, antigen-presenting cells (APCs) interact with the foreign invaders and break them down into the antigens to showcase these antigens to lymphocytes. T-helper cells (i.e., a type of T-cells) get activated after their interactions with the APCs. These activated T-helper cells are referred as effector cells that interact with naive B-cells and cytotoxic T-cells (TC) (i.e., another type of T-cells) to activate them and provide humoral and cell-mediated attack, respectively [8, 9].

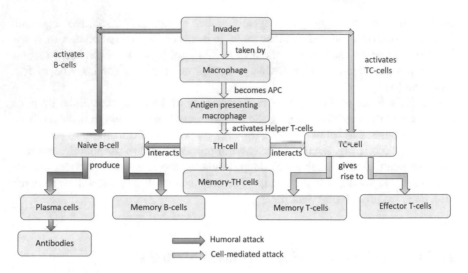

Fig. 1 Adaptive immune response

After the activation of naive B-cells, the plasma cells (antibody-secreting B-cells) are produced. These plasma cells further produce a large number of antibodies that act as labels on the pathogens so that they get eliminated from the body. This type of immune attack by B-cells through the production of plasma cells and antibodies is called as humoral attack. Tc-cells kill infected cells. The attack by Tc-cells is called as cell-mediated attack. During these attacks memory cells are generated that remain in the body for a longer time. These attacks are shown in Fig. 1.

2.2 Agent-Based Modeling

Agent-based modeling simulates the dynamics of the immune system and observes the interactions among systems' entities that lead to a global property called as the emergence of patterns [6, 7]. As ABMs are implicit and universal in structure, they are proved to be useful during the immune system modeling [14]. The models in this technique are mainly modeled on a grid. The ABM consists of three elements:

- Agents with attributes and behavior;
- Rules for interactions among the agents;
- Environment where the agents interact with each other and share spatial information.

Figure 2 depicts the agents and interactions with regard to the target system and ABMs.

Fig. 2 Entities and their interactions for target and ABM system

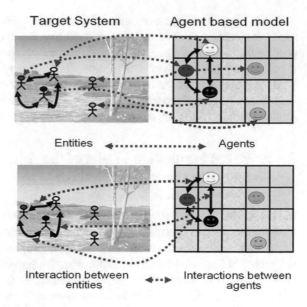

It has been observed that the immune system is simulated through the ABM due to its following features:

1. The immune cells are considered as agents and the cell–cell biological interactions are mimicked (i.e., direct cell–cell contacts with other cells) through the rules defined during the modeling;
2. Complexity in the immune system arises from the number of immune cells along with their interactions;
3. Global properties during the immune response emerge from the local interactions between the immune cells as agents are efficiently observed through agent-based modeling.

Table 1 Analysis of some agent-based tools

Repast	Mason	NetLogo	Swarm
Written in Java, C++, python, and others	Written in Java	Written in Logo language.	Written in Objective-C or Java
It is easy to extend for complicated models	Good choice for analyzing both evolution learning with the help of network causality	It is difficult to extend and it is a good choice for novice programmers	It supports complex models
Supports 2D and 3D modeling	Supports 2D and 3D modeling	Supports 2D and 3D modeling	Supports 2D and 3D modeling

Table 1 provides the information of the existing simulation tools that are mainly used for modeling the immune system. NetLogo is a computational and multi-agent programmable modeling environment that uses Logo programming language. It was developed by Uri Wilensky at the Center for Connected Learning (CCL) at Northwestern University of the and is available as open-source downloadable software [15]. In this proposed model, we have used Netlogo simulation tool for modeling the immune system dynamics during its response against the foreign molecules.

3 Proposed Simulation Model of the Adaptive Immune System with the Help of Netlogo

In [14], we have provided the review of agent-based modeling approach that models the immune system dynamics. In this work, we postulated that this approach has an inherent representation of the immune system dynamics. We have also provided the survey of immune system simulators that are proved to be helpful to understand homeostasis. It is observed that during the simulation, ABM considers the immune cells as agents and interactions between these cells as the biological rules that are applied to the agents. This rule-based technique exhibits built-in emergent behavior that helps to simulate the emergence of the immune system [1, 7].

Our proposed model provides the simulation of the cell-mediated and humoral immune response though the Netlogo simulation tool. This model uses "Moore neighborhood" (i.e., eight neighborhood) model, a type of on-lattice square model. For example, for 2×2 grid, if an immune cell agent is present on site $[i, j]$, then the corresponding neighbors of this agent are shown in Fig. 3 and they are as follows:

$$\text{cell}[i-1][j-1], \text{cell}[i-1][j], \text{cell}[i-1][j+1], \text{cell}[i][j-1],$$
$$\text{cell}[i][j+1], \text{cell}[i+1][j-1], \text{cell}[i+1][j], \text{cell}[i+1][j+1].$$

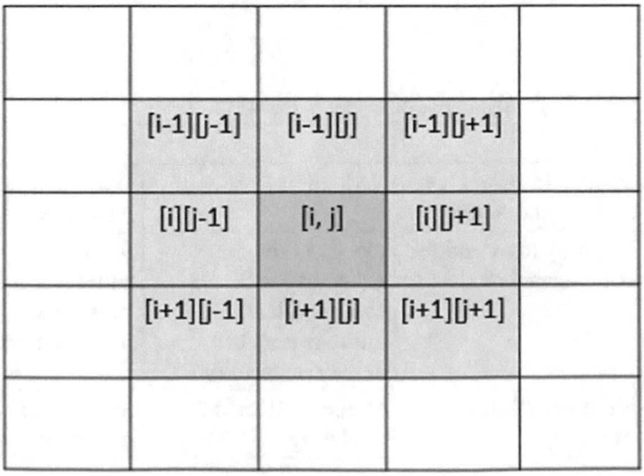

Fig. 3 Neighbors of the immune cell agent on site $[i, j]$ are shown in gray [13]

	[i-1][j-1]	[i-1][j]	[i-1][j+1]	
	[i][j-1]	[i, j]	[i][j+1]	
	[i+1][j-1]	[i+1][j]	[i+1][j+1]	

This model consists of immune cells that are referred to as the agents during simulation that are as follows:

1. **Macrophage**, denoted by MF. It contributes greatly during the initialization of adaptive immune response. MF interacts with the neighborhoods to get activated through foreign agents like invaders. After activation, it changes its state to AMF (i.e., antigen-presenting MF).
2. **T-helper cell**, denoted by TH. This cell type interacts with AMF, i.e., it changes its state to effector Th (eTH) if one of its neighborhoods is AMF. These ETH agents further help to generate cell-mediated and humoral immune attacks.
3. **T-killer cell**, denoted by TC-cell. This type of cell is activated by AMF and it changes its state to effector TC-cell (ETC) that further generates the cell-mediated attack.
4. **Naive B-cell**, denoted by NB-cell. This cell type gets activated if one of its neighborhoods is found to be AMF. After getting activated by AMF, this cell type generates the humoral attack.
5. **Plasmablast cell**, denoted by PB. This cell type is created by activated NB-cell agents.
6. **Antibody**, denoted by AB. This agent type is generated from the PB-cell agents.
7. **Memory B-cell**, denoted by MBC. This cell type is created after the activation of NB cell agents by ETH.
8. **Memory T-cell**, denoted by MTC. This cell agent is added to the grid after the activation of TC agents.
9. **Healthy cell**, denoted by HC. If this agent encounters with invaders, then it changes its state type to the infected cell (IC) state.

Following are the rules applied on these agents:

1. **MF**:

$$\text{grid[i][j].type} = \begin{cases} \text{AMF,} & \text{if \{grid[i][j].type = MF} \\ & \text{\& NH(grid}[i][j]) = \text{ID\}} \\ \text{MF,} & \text{otherwise.} \end{cases} \qquad (1)$$

2. **TH**:

$$\text{grid[i][j].type} = \begin{cases} \text{ETH,} & \text{if \{grid[i][j].type = TH} \\ & \text{\& NH(grid}[i][j]) = \text{AMF\}} \\ \text{TH,} & \text{otherwise.} \end{cases} \qquad (2)$$

3. **TC**:

$$\text{grid[i][j].type} = \begin{cases} \text{ETC,} & \text{if \{grid[i][j].type = TC} \\ & \text{\& NH(grid}[i][j] = \text{AMF\}} \\ \text{TC,} & \text{otherwise.} \end{cases} \qquad (3)$$

4. **HC**:

$$grid[i][j].type = \begin{cases} IC, & \text{if } \{grid[i][j].type = HC \\ & \& \ NH(grid[i][j] = ID\} \\ HC, & \text{otherwise.} \end{cases} \tag{4}$$

5. **NB**:

$$\begin{array}{ll} CreatePB() & \text{if } \{grid[i][j].type = NB\} \\ CreateMB() & \text{it } \{grid[i][j].type = ND\} \end{array} \tag{5}$$

Here, CreatePB() and CreateMBC() add more agents into the grid. These functions consider a variable size of the grid for adding PBs and MBCs at the neighboring empty positions of current NB. For example, if NB is at location $[i, j]$ then the corresponding grid of the neighborhoods is considered as of size 5×5 to add PB and MBC. PBs and MBCs count are randomly generated with the help of the slider in the Netlogo. Here, PB count is generated as a random number between the maximum and minimum value that are provided in the slider.

1. **MTC**:

$$CreateMTC() \qquad \text{if } \{grid[i][j].type = ETC\} \tag{6}$$

2. **AB**:

$$CreateAB() \qquad \text{if } \{grid[i][j].type = PB\} \tag{7}$$

CreateMTC() and CreateAB() functions work same as the functions createPB() as mentioned in Eq. 6. Here, variable grid size is considered to add more cells into the grid.

This model uses the same number of agents and rules that are previously used in our model developed using C-programming [13]. Our proposed model that is implemented using Logo language in Netlogo tool provides the quality visualizations of the adaptive immune response. This type of model-based methodology that provides emergence visualization has not been provided in [13] with the help of C-programming.

4 Results and Discussion

In this paper, we have proposed an agent-based model of intricate cellular dynamics during the process of adaptive immune response. The results are observed for timesteps in the simulation. These cellular level rules lead to the generation of the adaptive immune response at global level using agent-based modeling technique with the help of Netlogo.

Figure 4 provides the initial window of the immune response simulation. Separate sliders for MF, NB, TH, TC, HC, ID, MTC, PB, and AB indicate the minimum and

Fig. 4 Input during the model

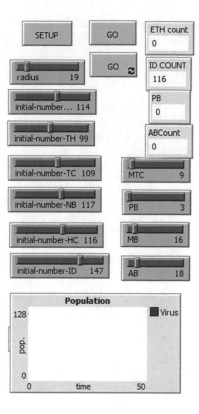

maximum number of the agents that are added during the simulation. The radius slider set the length for neighboring locations, i.e., if it is 1 then 8 neighboring locations are considered. The monitors for ETH and others keep the track of their numbers that are in the environment. Population plot checks the ID population during the simulation.

Figure 5 presents the final result of the simulation. In this figure, the monitor is added for checking the invader count during the simulation. This monitor shows how infection gets increased due to the replication of invaders and due to the addition of MTCs and ABs, system goes to initial state, i.e., invader free state. The model easily demonstrates the procedure of immune response and has the following properties:

1. Initially, all agents are randomly put into the grid based on the numbers that are set on the sliders of the agents. All rules mentioned in Eqs. 1–7 are applied from timestep 0. If the IDs are introduced into the grid, then the immune response is generated by allowing the type change and the rapid increase in the number of the cells.
2. Decision for changing the type of cell depends on the local environment of the cell, i.e., neighborhoods of the cell. For example, suppose a cell is inserted into the environment, its neighborhoods are calculated based on which the type of the cell is changed. This type change is presented by the rules in Eqs. 1–4.

Fig. 5 Final result of the
model

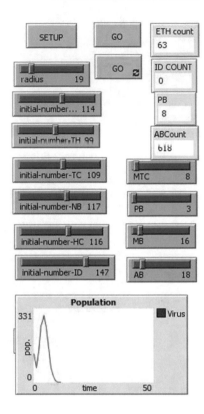

3. At timestep = 0, at the very beginning, MF state is changed to AMF, i.e., by the rule of MF to AMF conversion as defined in rule 1. This rule is executed when the IDs are present into the grid.
4. If timestep = 1, then the rule of TH to ETH conversion is executed only if AMF agents are present in the neighborhood locations. It is shown in rule 2.
5. ETH generates the humoral and cell-mediated attack. It is thus noted that at timestep = 2, TC agents is activated as shown in equation for rule 3. Similarly, NB agents produce plasma cells and memory B-cells as shown in rule 5.
6. Healthy cells convert to infected cells by following the rule in Eq. 4. Invader (ID) replicates inside the infected cell. After certain timestep, some of the ICs burst and new IDs will be created in the grid.

tot_IC ⇒ the total number of infected cells at a timestep.
MAX_ID ⇒ Maximum number of IDs
MIN_ID ⇒ Minimum number of IDs
New_IDs ⇒ Count of new IDs that are generated through the burst of particular IC.
New_IDs ⇐ rand()%(MAX_ID − MIN_ID) + MIN_ID
When "(timestep%3==0)", then ICs bursts

{
// Randomly select the ICs (i.e., tot_IC/2) to burst
//Put the newly generated IDs in the neighborhood grid of variable size for particular IC.
}
7. At timestep = 3, ETC cell agents will create MTC agents as shown in rule 6.
8. Following these rules same type of model is obtained for multiple runs of the model.
9. The movements of B-cells and T-cells in our model are independent of the number of cells in the particular location, and they particularly follow the rules based on the "type" to evolve over the time.
10. Our model is flexible and it allows perturbation in the size as we can change the value of parameters during the simulation.

Our model is able to mimic various biological properties of the immune response during the infection. It reproduces or imitates the biological facts of complex cellular immune system dynamics [13].

5 Conclusions

In this paper, we discussed the immune system which is one of the CASs that generates a complex immune response to the foreign molecules. Due to the intricate immune cellular dynamics during the response, there is a requirement of understanding the system through modeling and simulation. The modeling helps to analyze the intricate immune cellular dynamics and to study the immune system homeostasis. In this paper, we discussed a systems biology approach of the immune system modeling, which is referred to as an agent-based modeling. ABM requires less time and cost and helps in facilitating experiments and measurements that are supposed to be infeasible in a laboratory setting. It is also used to simulate the complex immune cellular behaviors after the invader entry into the body. The proposed model of the immune system is implemented using Logo language with the help of Netlogo modeling tool. This model provides the quality visualizations of the cell-mediated and humoral immune response by incorporating a large number of immune cells along with their interactions. The proposed model is seen as an analytical, iterative, and model-based methodology through which an emergence of the cells and their behaviors can be effectively observed. At the same time, this model will also help to create new visions and hypotheses for further research and development.

An agent-based model that incorporates cell movements based on the space for both innate and adaptive immune response are being developed. We can also think of the agent-based modeling of the effects of physical and psychological stress on the immune system; the immune system cells dynamics during the stroke; the complex interactions between the immune cells and microbiota during the inflammation.

References

1. Abar, S., Theodoropoulos, G.K., Lemarinier, P., OHare, G.M.: Agent based modelling and simulation tools: a review of the state-of-art software. Comput. Sci. Rev. (2017)
2. Anderson, P.: Perspective: complexity theory and organization science. Organ. Sci. **10**(3), 216–232 (1999)
3. Bianca, C., Pennisi, M.: Immune system modelling by top-down and bottom-up approaches. Int. Math. Forum **7**, 109–128 (2012)
4. Broderick, G., Craddock, T.J.A.: Systems biology of complex symptom profiles: capturing interactivity across behavior, brain and immune regulation. Brain Behav. Immun. **29**, 1–8 (2013)
5. Chavali, A.K., Gianchandani, E.P., Tung, K.S., Lawrence, M.B., Peirce, S.M., Papin, J.A.: Characterizing emergent properties of immunological systems with multi-cellular rule-based computational modeling. Trends Immunol. **29**(12), 589–599 (2008)
6. Chiacchio, F., Pennisi, M., Russo, G., Motta, S., Pappalardo, F.: Agent-based modeling of the immune system: Netlogo, a promising framework. BioMed Res. Int. (2014)
7. Christley, S., An, G.: Agent-based modeling in translational systems biology. In: Complex Systems and Computational Biology Approaches to Acute Inflammation, pp. 29–49. Springer (2013)
8. Jacob, C., Sarpe, V., Gingras, C., Feyt, R.P.: Swarm-based simulations for immunobiology. In: Information Processing and Biological Systems, pp. 29–64. Springer (2011)
9. Janeway, C.A., Travers, P., Walport, M., Shlomchik, M.J.: Immunobiology: The Immune System in Health and Disease, vol. 1. Current Biology (1997)
10. Li, X.-H., Wang, Z.-X., Lu, T.-Y., Che, X.-J.: Modelling immune system: principles, models, analysis and perspectives. J. Bionic Eng. **6**(1), 77–85 (2009)
11. Mei, Y., Hontecillas, R., Zhang, X., Bisset, K., Eubank, S., Hoops, S., Marathe, M., Bassaganya-Riera, J.: ENISI Visual, an agent-based simulator for modeling gut immunity. In: 2012 IEEE International Conference on Bioinformatics and Biomedicine (BIBM), pp. 1–5. IEEE
12. Pappalardo, F., Palladini, A., Pennisi, M., Castiglione, F., Motta, S.: Mathematical and computational models in tumor immunology. Math. Model. Nat. Phenom. **7**(3), 186–203 (2012)
13. Shinde, S.B., Kurhekar, M.P.: Complex biological immune system through the eyes of dual-phase evolution. J. Biol. Syst. **26**(03), 473–493 (2018)
14. Snehal, S., Manish, K.: Review of the systems biology of the immune system using agent-based models. IET Syst. Biol. (2018)
15. Wilensky, U.: NetLogo NetLogo 6.0.3 User Manual (2016) [Online]. Available at: http://ccl.northwestern.edu/netlogo/docs/

Cardiac Arrhythmia Detection Using Ensemble of Machine Learning Algorithms

R. Nandhini Abirami and P. M. Durai Raj Vincent

Abstract An ECG signal is a bioelectrical signal which records the electrical activity of the heart. ECG signals are used as the parameter for detecting various heart diseases. Cardiac arrhythmia can be detected using ECG signals. Arrhythmia is a condition in which the rhythm of the heart is irregular, too slow or too fast. The data for this work is obtained from the University of California, Irvine machine learning repository. The data obtained from the repository is preprocessed. Feature selection is made, and machine learning models are applied to the preprocessed data. Finally, data is classified into two classes, namely normal and arrhythmia. Feature selections were made to optimize the performance of machine learning algorithms. Features with more number of missing values and which showed no variation for all the instances have been deleted. Accuracy achieved using ensemble of machine learning algorithms is 85%. The objective of this research is to design a robust machine learning algorithm to predict cardiac arrhythmia. The prediction of cardiac arrhythmia is performed using ensemble of machine learning algorithms. This is to boost the accuracy achieved by individual machine learning algorithms. The technique of combining two or more machine learning models to improve the accuracy of the results is called ensemble prediction. More accurate results can be achieved using ensemble methods than the results achieved using single machine learning model.

Keywords Electrocardiogram · Machine learning · Cardiac arrhythmia and classification

1 Introduction

'The heart is a muscular organ in humans and other animals, which pumps blood through the blood vessels of the circulatory system [1].' The heart is in the size of the fist and is located in the middle of the chest. Heart is surrounded by a membrane called pericardium which holds the heart in its position. Pericardial fluid provides

R. Nandhini Abirami (✉) · P. M. Durai Raj Vincent
Vellore Institute of Technology, Vellore, India
e-mail: nandhini.raj25@gmail.com

© Springer Nature Singapore Pte Ltd. 2020
K. N. Das et al. (eds.), *Soft Computing for Problem Solving*,
Advances in Intelligent Systems and Computing 1057,
https://doi.org/10.1007/978-981-15-0184-5_41

lubrication to the heart. The function of heart is to pump the blood around the body through the blood vessels. Blood is essential to provide nutrients to the body and also essential to remove the metabolic wastes from the body. Heart pumps the blood through arteries and veins, and this network is called the cardiovascular system. The left side of the heart contains oxygenated blood from the lung. The heart pumps the oxygenated blood around the body, in order to provide each cell with oxygen. Some of the oxygen is used by the cells for respiration. The usage of oxygen by the cells in the body causes the blood to become deoxygenated. The deoxygenated blood then travels to the right side of the heart and then from there to the lungs to be oxygenated.

The ECG is a bioelectric signal which measures electrical activity of the heart at rest. Heart diseases, namely myocardial infarction which indicates previous heart attacks and hypertension which is the enlargement of heart due to blood pressure can be detected from the analysis of ECG signal. ECG recorded at rest is different from ECG recorded at stress [2].

An ECG test can be performed by placing electrodes from electrocardiograph to the patient's skin on arms, legs and chest. The signals from the heart are read by the electrodes and sent to the electrocardiograph. The information received from the electrodes is then displayed in the electrocardiograph monitor. Resting ECG and Exercise ECG are some of the methods of measuring ECG. Resting ECG is the standard ECG to measure the electrical activity of the patient when the patient is lying still. Exercise ECG which is also called as the stress test is measured while the patient pedals a stationary bicycle or walks on a treadmill. This test is conducted to test the heart during stress or exercise.

Figure 1 shows two periods of normal ECG signal. ECG is a measure of electrical activities occurring in each cardiac cycle of heart. It has five visible waveforms, namely P wave, Q wave, R wave, S wave and T wave. QRS complex is the high peak in the signal, and it is a combination of Q wave, R wave and S wave. The P wave in the

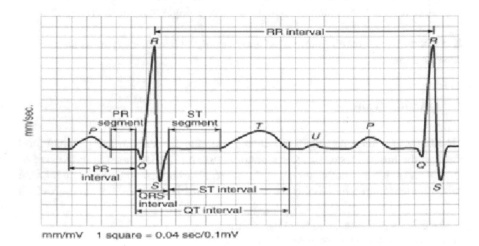

Fig. 1 Detection of different waves in ECG signal [3]

ECG signal is the first component of a normal ECG signal. It is an upward wave which represents right and left atrial depolarization. The Q wave is the downward deflection occurring after P wave which represents ventricular depolarization. The R wave is the upward deflection occurring after Q wave. It is normally easy to identify in an ECG signal. S wave is the downward deflection occurring after R wave representing late ventricular depolarization. T wave is upright and rounded representing ventricular repolarization. U wave is upright and rounded which follows T wave and is not present in all rhythm strips [1].

The activity of the heart occurs during various intervals in ECG, namely PR, ST, QT interval and QRS complex. The RR interval represents the duration between each heartbeat. Irregular beat phase of heart is called arrhythmia, some of which may be dangerous to the patient. QT interval represents the time taken for ventricular depolarization and repolarization cycle. The end of ventricular depolarization is represented by the ST segment. QRS complex represents the ventricular depolarization, and it follows the P wave.

Figure 2 shows normal sinus rhythm. A normal ECG means all the important ECG intervals, namely PR interval, QRS complex, QT interval and ST interval lie in normal range. The normal ranges of ECG intervals are shown in Table 1.

Figure 3 represents ECG with abnormality. There are several abnormalities which can be detected with ECG some of which are atrial fibrillation, bundle branch block,

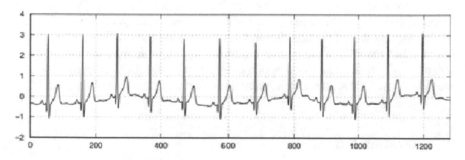

Fig. 2 Normal ECG [4]

Table 1 Normal range of ECG intervals and events during the intervals

S. No.	ECG intervals	Normal duration		Events during ECG intervals
		Average	Range	
1	PR interval	0.18	0.12–0.20	Atrial depolarization
2	QRS complex	0.08	0.05–0.10	Ventricular depolarization and atrial repolarization
3	QT interval	0.40	0.4–0.43	Ventricular depolarization and ventricular repolarization
4	ST interval	0.32	0.30–0.35	Ventricular repolarization

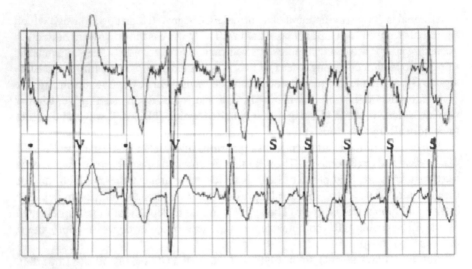

Fig. 3 ECG with abnormality [5]

pacemaker failure to pace, pacemaker failure to capture, premature atrial capture, sinus arrest, sinus arrhythmia and sinus tachycardia.

The two basic states of arrhythmia are bradycardia and tachycardia. The states detected are based on the heart rate. When the heart rate is between 70 and 100 beats per minute, the state of the heart is normal; when the heart rate is below 60 beats per minute, then the state of the heart is bradycardia, and when the heart rate is above 100 beats per minute, then the state of the heart is tachycardia. Bradycardia is due to the slow beat of the heart while tachycardia is due to the fast beat of the heart [6].

2 Literature Survey

Sarkaleh et al. [7] classified ECG arrhythmias using neural networks. Discrete wavelet transform (DWT) is used for processing the acquired ECG signal and to extract feature. Classification is performed by multilayer perceptron neural network. ECG signals obtained from MIT-BIH database have been used for the neural network-based classification. Testing and training have been performed using ten files which included two arrhythmias. The classifier is capable of detecting two types of arrhythmias and achieved an accuracy of 96.5%.

Jaiswal et al. [8] classified ECG signals using artificial neural network. They extracted various features of the ECG signals and analyzed the duration of ECG intervals, namely PR interval, ST segment, QRS complex and ST interval. Heart rate is used as a parameter for detecting the illness of heart. The data has been obtained from MIT-BIH database.

Sadr et al. [9] compared the performance of different machine learning algorithms in predicting ECG signal. They employed a database with 50 signals obtained from 50 patients. Wavelet transform has been used to remove noise from the signal. After normalization, the entire dataset is divided into three segments for training, testing and validation. The analysis showed that RBF outperformed MLP with an accuracy of 94%.

Tang et al. [10] performed the classification of ECG signal using RS and quantum neural networks (QNN). Initially, the signals were normalized and feature extraction is performed using wavelet transform. Redundant attributes are deleted using RS reduction technique. Classification modeling based on QNN is performed after attribute reduction. The study revealed that classification using RS–QNN is better than conventional methods.

Belgacem et al. [11] performed classification of ECG signal by employing two types of neural network classifiers, namely multilayer perceptron (MLP) classifier and learning vector quantization (LVQ) neural network. The two different classifiers were tested with ECG signals, and the result shows that MLP classifier outperformed LVQ neural network.

Kshirsagar et al. [12] classified ECG signal using artificial neural network (ANN). Data obtained from MIT-BIH arrhythmia database is used for the analysis. Wavelet transform is used for preprocessing the ECG signal. Various abnormalities in an ECG signal are predicted by detecting QRS complex of ECG signal. GUI was developed to detect the abnormality in the ECG signal.

Sao et al. [13] performed ECG signal analysis using artificial neural network. Parameters of the ECG signal, namely Poincare plot, Lyapunov exponent and spectral entropy are used for the analysis. An artificial neural network is used for detecting the abnormalities in the ECG signal.

Mitra and Samanta [14] performed a cardiac arrhythmia classification using neural networks. Correlation-based feature selection is used and classification is done using incremental back propagation. Data is obtained from UCI database. 87.71% accuracy is achieved using 100 simulations.

Gayathri and Sumathi [15] detected ECG arrhythmia and classification using relevance vector machine. Relevance vector machine is used to perform automatic classification of ECG beats. The experiment was conducted using the data from MIT-BIH database. Feature extraction is performed using signal processing methods. Simulation results for various abnormalities, namely atrial flutter, premature atrial contraction, premature ventricular contraction and right bundle branch blocks were shown.

Acharya et al. [16] classified heartbeats by developing a nine-layer deep convolutional neural network. CNN automatically identifies different categories of heart beat in an ECG signal. After noise removal from a publicly available dataset, an accuracy of 94.03% has been achieved.

Acharya et al. [17] presented a new deep learning approach to efficiently classify cardiac arrhythmia. The research has been performed using data MIT-BIH database from 45 patients. Deep 1D-CNN algorithm has been used which produced an accuracy of 91.33%.

Pławiak et al. [18] presented a methodology for efficient classification of 17 different classes of cardiac disorders based on ECG signals using evolutionary neural system. The research has been performed using data MIT-BIH database from 45 patients. The research is performed by developing evolutionary neural system based on SVM classifier which gave an accuracy of 98.85%.

Pławiak et al. [19] presented a methodology where ensemble of classifiers was used for efficient classification of 17 different classes of cardiac disorders based on ECG signals using evolutionary neural system. The research has been performed using data MIT-BIH database from 29 patients.

3 Materials and Methods

The cardiac arrhythmia dataset is obtained from UCI machine learning repository. The dataset has 452 patient records and 279 attributes for each record. Each of the 452 patients is classified into one of the 16 classes and each class representing a condition. Class 01 represents normal ECG while classes 02 to 15 represent different classes of arrhythmia, and unclassified patients are represented by class 16. The data is preprocessed where the missing values are replaced with null values. The attributes which had large number of missing values were removed. Attributes J angle and P angle had significant number of missing values so they were deleted from the dataset. Also, since the class 16 represents unclassified patients, the records corresponding to class 16 were removed to improve accuracy of classification. Classes with value 1 represents normal patients and so classified as 'normal,' and classes 02 to 15 were merged and represented as 'arrhythmia.'

Figure 4 shows the methodology to predict arrhythmia. Data is acquired from UCI machine learning repository and preprocessed. Feature selection is made using Boruta library, and machine learning models are applied to the preprocessed data. The entire dataset is split into training and testing data in the ratio of 75:25. Cross validation has been used on the training data to arrive on the best model. Finally, data is classified into two classes, namely normal and arrhythmia.

Feature extraction and reduction play a major role in classification since the performance of the classifier will be highly affected if the features were not chosen well. The reduction in dimension speeds up the process of inference in case of large database. Few columns had single value for all the instances, and such attributes were deleted since they do not represent any variation between the patients and found to be irrelevant for the study. Since there were more number of attributes, considering all the attributes to train the model would affect the performance of the model. Hence, to avoid over fitting, only important features were selected, and the model was trained using these features. The features having maximum correlation with the output were used for training the model.

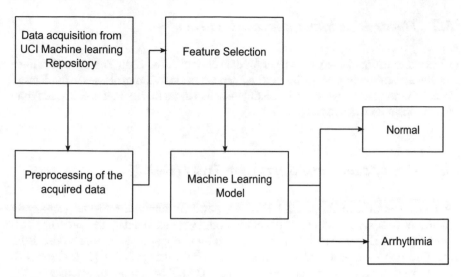

Fig. 4 Methodology to predict arrhythmia

3.1 Support Vector Machines

SVM is one of the supervised machine learning algorithms. Support vector machines can perform regression, outlier detection, linear and nonlinear classification. SVM builds a highly accurate model and overcomes the local optima. SVM will develop a model with the training dataset in a way that the data points that belong to different groups are separated by a distinct gap. The data samples which lie on the margin are called the support vectors. Figure 5 shows support vector machines.

Fig. 5 Support vector machines

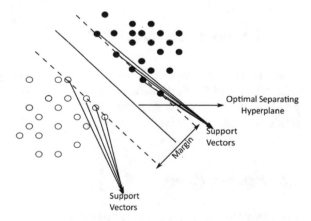

3.2 Linear Discriminant Analysis (LDA)

Linear discriminant analysis is used for data classification [20]. The algorithm finds the linear combination of attributes which separates the different classes in the dataset. When the number of classes to be distinguished is more than two, it is called multiple discriminant analysis (MDA).

3.3 Classification and Regression Trees (CART)

A Classification and Regression Tree is a powerful analytic tool that recursively partitions data space and fits a prediction model with each data partition. The Classification and Regression Tree is represented by simple graphical tree model. Both classification trees and regression trees can be built using CART [21]. In order to avoid overfitting, the trees have to be pruned. CART is capable of discovering the complex interactions between predictors.

3.4 Naïve Bayes

Naïve Bayes classifiers are probabilistic classification supervised machine learning algorithm based on Bayes theorem. The algorithm is used when the inputs have high dimensionality. The model is easy to build and is suitable for large datasets. Though this model is simple to implement, often this, it outperforms other complicated classification models [22].

3.5 K-Nearest Neighbor (KNN)

Nearest neighbor algorithm is the simplest of all existing machine learning algorithms. It is used for classification and regression and is an instance-based algorithm where a training dataset is given, based on which the new input data may be classified by simply comparing with the data point in the training dataset [23].

3.6 Random Forest

Random forest creates a bunch of decision trees from the subset of training set. The final decision is made by aggregating the votes from different decision trees. Alternatively, the model can make the final decision by applying a weight for each decision tree. Decision trees with higher number of errors are given least weight values while decision trees with the least error are given higher weight values. The basic parameter for random forest would be the number of trees to be generated and other decision tree-related parameters.

4 Results

The system for diagnosis of heart disease using machine learning algorithm is implemented using R language. In this system, the entire database is divided into training and testing data. Seventy-five percent of the total records are used for training while the remaining 25% of the total records are used for testing. Rows with missing values in the data are omitted. The performance of the algorithm is evaluated using accuracy, sensitivity, specificity and precision.

Accuracy is the ratio of accurately predicted samples to the total number of available samples.

$$\text{Accuracy} = \frac{TP + TN}{TP + FP + TN + FN}$$

Kappa, also called as Cohen's kappa, is a measure of agreement between two individuals. It is always less than or equal to 1. A value equal to 1 implies perfect agreement between two individuals.

$$\text{Cohen's Kappa} = \frac{P_o - P_e}{1 - P_e}$$

where

$$P_o = \text{observed accuracy} = \frac{TP + TN}{TP + FP + FN + TN}$$

$$P_e = \text{expected accuracy} = \frac{[(TP + FP)(TP + FN) + (FN + TN)(FP + TN)]}{[(TP + FP + FN + TN)^2]}$$

where

TP (True positive)   is the number of samples that is true and is correctly classified as true,

TN (True Negative)   is the number of samples that is false and is correctly classified as false,

FP (False Positive)   is the number of samples that is false and is wrongly classified as true, and

FN (False Negative)   is the number of samples that is true and is wrongly classified as false.

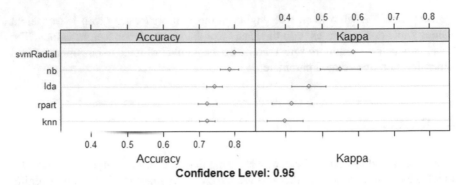

Fig. 6 Dotplot comparing the model results

Figure 6 shows dot plot comparing the model results using accuracy and kappa. Table 2 shows the performance analysis of the base learners. Various performance metrics, namely sensitivity, specificity, precision, accuracy, positive predictive value (PPV), negative predictive value (NPV) and F1 score are analyzed.

Figure 7a shows the ROC curves of base models. Figure 7b shows ROC curve of random forest algorithm as a combiner of base models. An ROC curve is a graph which is used to evaluate the performance of various classification models. The curve basically plots two parameters, namely false positive rate (FPR) and true positive rate (TPR). AUC is a measure of the area under the entire ROC curve. The higher is the value of AUC, better is the performance of the model.

The predictions of classifier are combined using random forest algorithm. The accuracy is lifted from 80.30 to 85.15% which is an impressive improvement on accuracy.

Table 3 shows the accuracy achieved using random forest as combiner for the base models. Tenfold cross validation is performed. Accuracy was used to select the optimal model using the largest value. The final value used for the model was mtry = 3.

5 Conclusion

The paper presents the detection of cardiac arrhythmia using ensemble of machine learning algorithms. Various techniques were adopted to preprocess the data to suite the requirement of analysis. Feature selections were made to optimize the performance of machine learning algorithms. Features with more number of missing values and which showed no variation for all the instances have been deleted. The objective of this research is to design a robust machine learning algorithm by combining five different machine learning algorithms using stacking of machine learning algorithms. This is to boost the accuracy achieved by individual machine learning algorithms. More accurate results can be achieved using ensemble methods than

Table 2 Base learner's performance analysis

Algorithm	Sensitivity	Specificity	Precision	Accuracy	PPV	NPV	F1
LDA	0.5932203	0.8518519	0.8139535	0.7168142	0.8139535	0.6571429	0.6862745
KNN	0.5254237	0.8888889	0.8378378	0.6991150	0.8378378	0.6315789	0.6458333
SVM	0.8135593	0.8333333	0.8421053	0.803008	0.8421053	0.8035714	0.8275862
CART	0.5423729	0.8888889	0.8421053	0.7079646	0.8421053	0.6400000	0.6597938
NB	0.6101695	0.8888889	0.8571429	0.7433628	0.8571429	0.676056	0.7128713

(a)   **(b)**

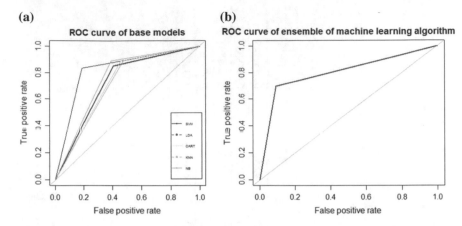

Fig. 7 **a** ROC curves of the base models. **b** ROC curve of random forest algorithm

Table 3 Accuracy achieved using random forest algorithm as combiner	Mtry	Accuracy	Kappa
	2	0.8515172	0.6970743
	3	0.8515236	0.6966339
	4	0.8439877	0.6809248

the results achieved using single machine learning model. The maximum accuracy achieved among the five different machine learning algorithms is 80.30% using SVM algorithm. The algorithms are combined using random forest algorithm as combiner which boosted the accuracy to 85.15%.

References

1. Kelwade, J.P., Salankar, S.S.: Prediction of cardiac arrhythmia using artificial neural network. Int. J. Comput. Appl. **115**(20) (2015)
2. http://www.mydr.com.au/tests-investigations/electrocardiogram-ecg
3. https://lifeinthefastlane.com/ecg-library/basics/pr-interval/
4. https://physionet.org/physiobank/database/nsrdb/
5. https://physionet.org/physiobank/database/svdb/
6. Rai, H.M., Trivedi, A.: Classification of ECG waveforms for abnormalities detection using DWT and back propagation algorithm. Int. J. Adv. Res. Comput. Eng. Technol. (IJARCET) **1**(4), 517 (2012)
7. Sarkaleh, M.K., Shahbahrami, A.: Classification of ECG arrhythmias using discrete wavelet transform and neural networks. Int. J. Comput. Sci. Eng. Appl. **2**(1), 1 (2012)
8. Jaiswal, G.K., Paul, R.: Artificial neural network for ECG classification. Recent Res. Sci. Technol. **6**(1) (2014)
9. Sadr, A., Mohsenifar, N., Okhovat, R.S.: Comparison of MLP and RBF neural networks for prediction of ECG signals. Int. J. Comput. Sci. Netw. Secur. (IJCSNS) **11**(11), 124–128 (2011)

10. Tang, X., Shu, L.: Classification of electrocardiogram signals with RS and quantum neural networks. Int. J. Multimed. Ubiquitous Eng. **9**(2), 363–372 (2014)
11. Belgacem, N., Chikh, M.A., Reguig, F.B.: Supervised Classification of ECG Using Neural Networks (2003)
12. Kshirsagar, P.R., Akojwar, S.G., Dhanoriya, R.: Classification of ECG-Signals Using Artificial Neural Networks
13. Sao, P., Hegadi, R., Karmakar, S.: ECG signal analysis using artificial neural network. Int. J. Sci. Res. National Conference on Knowledge, Innovation in Technology and Engineering, pp. 82–86 (2015)
14. Mitra, M., Samanta, R.K.: Cardiac arrhythmia classification using neural networks with selected features. Procedia Technol. **10**, 76–84 (2013)
15. Gayathri, B.M., Sumathi, C.P.: Comparative study of relevance vector machine with various machine learning techniques used for detecting breast cancer. In: 2016 IEEE International Conference on Computational Intelligence and Computing Research (ICCIC), pp. 1–5. IEEE (2016)
16. Acharya, U.R., Oh, S.L., Hagiwara, Y., Tan, J.H., Adam, M., Gertych, A., San Tan, R.: A deep convolutional neural network model to classify heartbeats. Comput. Biol. Med. **89**, 389–396 (2017)
17. Yıldırım, Ö., Pławiak, P., Tan, R.S., Acharya, U.R.: Arrhythmia detection using deep convolutional neural network with long duration ECG signals. Comput. Biol. Med. **102**, 411–420 (2018)
18. Pławiak, P.: Novel methodology of cardiac health recognition based on ECG signals and evolutionary-neural system. Expert Syst. Appl. **92**, 334–349 (2018)
19. Pławiak, P.: Novel genetic ensembles of classifiers applied to myocardium dysfunction recognition based on ECG signals. Swarm Evol. Comput. **39**, 192–208 (2018)
20. Banu, G.R.: Predicting Thyroid Disease Using Linear Discriminant Analysis (LDA) Data Mining Technique
21. Chaurasia, V.: Early Prediction of Heart Diseases Using Data Mining Techniques (2017)
22. Medhekar, D.S., Bote, M.P., Deshmukh, S.D.: Heart disease prediction system using naive Bayes. Int. J. Enhanced Res. Sci. Technol. Eng. **2**(3) (2013)
23. Deekshatulu, B.L., Chandra, P.: Classification of heart disease using k-nearest neighbor and genetic algorithm. Procedia Technol. **10**, 85–94 (2013)

Characterization of Top Hub Genes in Breast and Lung Cancer Using Functional Association

Richa K. Makhijani and Shital A. Raut

Abstract Identification of disease gene in cancer is a complex problem, not due to the lack of methods, but because of the lack of standard validation procedure for identified genes. This issue is addressed in this paper, where we first select disease markers from both microarray and RNA-seq gene expression data of breast and lung cancer, and then validate the obtained gene set by the method of functional similarity with the known cancer genes. The classification procedure using kernel support vector machine (KSVM) is applied to find functional similarity of the hub genes from the protein–protein-interaction (PPI) network of identified genes, and a set of top genes of interest is extracted. We then construct a common cancer network and observe its functional characteristics in cancer. The results highlight the enrichment of this network in important cancer pathways, and thus validate the association of the identified gene set with the disease. It is anticipated that including more cancer types in the experimentation may help validate the drug targets, particularly common to these cancer types.

1 Introduction

Uncovering the role of major genes in diseases for their functional regulation is a wide study involving experts from molecular biology, computer science, and statistics. Due to the undiscovered complexity and behavior of the disease, cancer remains to be the primary choice of research for these scientists. Gene identification in cancer has seen comprehensive evaluation of huge omics data, which seems to be challenging. Moreover, the analysis of various types of omics data definitely produces interesting results, but also bring in discrepancy due to methods used. Hence, current researchers focus on investigating different cancer types incorporating var-

R. K. Makhijani (✉) · S. A. Raut
Visvesvaraya National Institute of Technology, Nagpur, India
e-mail: richa_makhijani@yahoo.co.in

S. A. Raut
e-mail: saraut@cse.vnit.ac.in

© Springer Nature Singapore Pte Ltd. 2020   489
K. N. Das et al. (eds.), *Soft Computing for Problem Solving*,
Advances in Intelligent Systems and Computing 1057,
https://doi.org/10.1007/978-981-15-0184-5_42

ious in-silico data and methods for gene expression analysis [2, 6, 12, 23, 28]. To understand cancer, biomarkers are identified which are a result of executing a bioinformatics pipeline [3, 27]. Based on the number of deaths, the highly risked cancers are *breast* among women and *lung* among both men and women (Cancer facts and figures 2017 and WHO cancer country profiles). Identification of significant marker genes in these cancers is essential so as to develop a deeper understanding of the functional changes occurring in the diseased condition. Recent research involves analysis of gene expression data from both microarray and RNA-seq platforms so as to obtain a combined result for gene identification. The Gene Expression Omnibus (GEO) [1] and The Cancer Genome Atlas (TCGA) [24] are the public repositories where a wide variety of such data are available. The advantages of such multi-platform analysis are high reproducibility among biological replicates [16]. Most of the studies focus on the similarity between the two data platforms, whereas, a study demonstrated difference between the two and provided a comparison using RNA samples [26]. Greater benefits for RNA-seq were observed involving identification of large number of genes at higher fold change (FC) levels and a higher correlation between the genes. Also, it was found to be more accurate in identification of perturbed genes, as compared to microarray analysis [11]. Despite these facts, researchers find themselves more comfortable with microarray data, as RNA-seq is new, complex to understand and incurs more expense in storage and analysis. Similar analysis on both data platforms yields consistent and correlated results [4]. A consistent set of differential expressed genes was observed in breast, lung, and prostate cancer using a bioinformatics pipeline on microarray and RNA-seq gene expression data and heterogeneity across the cancers was observed [15]. Also, the importance of fold change in the characterization of common disease genes was evident by experimental evaluation [14]. Such analysis is extended further to network-based experiments for identifying complexities in terms of dis-regulation of pathways and processes [25]. The issue of results inconsistency arises while identifying diseased genes, which is difficult to handle for computer scientists. This is due to the lack of clinical association and expertise, as the validation of being diseased gene has to be performed using clinical procedures. Another way of validating the identified genes could be finding their association with the known cancer genes. This is achieved by finding functional similarity of genes which can be complex to find.

Hence, we present an approach to find association of observed diseased hub genes with those of the known cancer genes, and thus obtain top 20 genes in two cancer types. These genes further help in characterizing a common cancer network significantly enriched in cancer pathways.

2 Materials and Methods

2.1 Experimental Data

Experimentation was performed on two biological datasets, human PPI dataset, and gene expression dataset. The data was extracted for two most popular cancers among women and men, breast and lung. Gene expression data belonging to both microarray and RNA-seq platforms from Gene Expression Omnibus (GEO) were used. Accession numbers (number of tumor samples/number of normal samples) for microarray data are GSE45827 (149/11) [5], GSE48984 (3/9) [22], GSE26910 (6/6) [17], GSE19804 (60/60) [13] and GSE10072 (57/50) [10]. The RNA-seq data GSE62944 (1621/164) [18] contained gene expression values for 24 cancer types out of which we extracted data for the two cancers of interest. Normalization of expression is required to scale the values in microarray data to a certain range. We used robust multichip average (RMA) for this purpose due to its advantages in terms of efficient detection of expression change, variance stability, and low-false positives [7]. The PPI data used by our algorithm was extracted from the STRING database [21].

2.2 Selection of Diseased Genes

Linear modeling method, LIMMA, was used for efficient extraction of differentially expressed genes. This method uses statistics based on empirical Bayes and proves to be strong in terms of its power, time required for execution, less false positives, and is easy to align with the analysis [8, 20]. A normalization method *voom* is included in the LIMMA package for RNA-seq normalization. We applied LIMMA with *voom* to extract genes of interest from the two cancer types, breast and lung (P-value < 0.05 with BH correction). Its advantages are best explained in [19]. A comparison of LIMMA with other models demonstrates its effectiveness in multifactored experiments [8]. Along with its capability to be applied to microarray and RNA-seq, it is well suited for identification of differential genes [20].

2.3 Identification of Top Hub Genes

After successful selection of a diseased gene set from both microarray and RNA-seq gene expression data platforms, we extract their corresponding PPI data from the STRING database. These interactions are extracted for both the diseases under consideration. This PPI data is transformed into an igraph object in R environment. Then, genes with high degree and betweenness centrality were identified out of this network, which are termed as hub nodes. An overlap of the list of these hub genes was performed with those of the curated cancer gene lists from Bushman's laboratory and COSMIC. We furthered our attention on those nodes which did not fall into the intersection with the curated genes, as these need some validation of their association with the disease. This could be verified if the functions of these proteins are similar to the known diseased genes. Hence, we used PPInfer R package for classification of functionally closely related proteins networks [9]. It employs the kernel support vector machine (KSVM) based on the regularized Laplacian matrix for a graph. The kernel matrix K is used with a classification algorithm for predicting the class of vertices in the given dataset. The complete approach is shown in Algorithm 1.

Algorithm 1 Steps for characterization of top hub genes in Breast and Lung cancer

Require: *igraph, PPInfer* R package
1: **Input:** Microarray and RNA-seq gene expression datasets as mentioned in Sect. 2.1
2: **Output:** Common cancer network
3: STEP 1: Apply LIMMA to all individual microarray data for breast cancer.
4: STEP 2: Save microarray DEGs for breast cancer as a union of all the DEGs obtained from step 1.
5: STEP 3: Apply LIMMA to all individual microarray data for lung cancer.
6: STEP 4: Save microarray DEGs for lung cancer as a union of all the DEGs obtained from step 3.
7: STEP 5: Apply LIMMA to RNA-seq data for breast cancer.
8: STEP 6: Save RNA-seq DEGs for breast cancer obtained from step 5.
9: STEP 7: Apply LIMMA to RNA-seq data for lung cancer.
10: STEP 8: Save RNA-seq DEGs for lung cancer obtained from step 7.
11: STEP 9: Save common DEGs in breast cancer by performing intersection of results from step 2 and step 6.
12: STEP 10: Save common DEGs in lung cancer by performing intersection of results from Step 4 and STEP 8.
13: STEP 11: Construct PPI networks for breast and lung cancer by extracting STRING PPI data for obtained genes in step 9 and 10.
14: STEP 12: Select hub genes from both networks with high degree and betweenness centrality.
15: STEP 13: Select common hubs found in both networks by performing intersection of lists obtained in step 12.
16: STEP 14: Find the intersection of these common hubs with the known cancer gene lists.
17: STEP 15: Save the list of novel genes which are not reported in known cancer gene lists.
18: STEP 16: With reference to the novel genes obtained in step 15, extract top 20 functionally similar genes from the PPI networks of breast and lung cancer.
19: STEP 17: Plot a common cancer PPI network from STRING for the combined top genes from step 16 and verify the enrichment of this network in cancer processes and pathways.

---

3 Experimental Results

This section provides the results of experimental evaluation of our algorithm.

3.1 Diseased Genes from Gene Expression Data

LIMMA was used to extract diseased genes from individual microarray and RNA-seq dataset of the two cancer types. For microarray data, diseased gene lists obtained from individual datasets of breast cancer were merged by performing a union operation. Similar operation was performed for lung cancer microarray datasets. After obtaining genes from both data platforms, an intersection of the results is performed to select the genes evident from both platforms. This reduces the chance of any technological bias and improves the confidence over the results. The results are shown in Table 1.

3.2 Association of Novel Hub Genes with Cancer

Individual PPI networks were constructed for these genes by extracting their protein-interaction partners form the STRING database. This resulted in breast cancer PPI network with 1001 nodes-6108 edges, and lung cancer network with 512 nodes-2609 edges. From the produced networks, hub genes were identified which showed high degree and betweenness centrality values. The lists of hub genes from both the networks was intersected upon to produce a list of common hubs. Hence, 34 genes were obtained as hub genes. The intersection of these genes with the known cancer gene lists from Bushman's laboratory and COSMIC database was performed. The results are shown in Fig. 1. It could be seen from the figure that 23 genes were not identified as cancer genes in the curated list. Thus, an association of these 23 genes was evaluated by using PPInfer R package. The original PPI networks of diseased genes were first decomposed into a network of minimum 50 nodes. Hence, breast cancer network was transformed to have 980 nodes-6096 edges and lung cancer network was transformed to have 507 nodes-2607 edges. Then, taking our 23 novel genes as reference, Laplacian kernel matrix was calculated for the networks which was further used to infer the top genes of interest in the networks. We obtained 20 such genes of interest from both these networks. The list of these genes is provided in Table 2.

Seven genes were found to be common in both the top gene lists provided in Table 2. Hence, we characterized a union of these gene lists so as to ascertain the association of our 23 reference genes with these top genes of interest. Using the combined top genes in both cancer conditions, we extract a common PPI network from the STRING database. The Gene Ontology analysis for significant biological processes

Table 1 Number of DEGs obtained from microarray and RNA-seq data

	DEGs microarray	DEGs RNA-seq	Common
Breast	2629	8639	1236
Lung	1080	7234	860

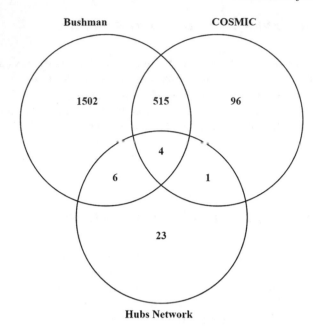

Fig. 1 Intersection of common hub genes in breast and lung cancer with curated gene lists

Table 2 List of top 20 genes obtained from breast and lung cancer network

Sr. No.	Genes in breast cancer	Genes in lung cancer
1	CNN1	SDPR
2	SDPR	ATP1A2
3	PPP1R14A	CAV2
4	MFAP4	MFAP4
5	ATP1A2	SORBS1
6	DEPDC1	CNN1
7	LMOD1	LMOD1
8	EHD2	MYH10
9	PMP22	PMP22
10	TAGLN	MYH11
11	FAM64A	DES
12	SORBS3	ITGA1
13	TRIM29	TIMP3
14	MYH11	IL1B
15	EGFR	CAV1
16	MITF	RHOB
17	DTL	MGAM
18	KRT15	F3
19	MYO16	GATA6
20	NCAPG2	CDH5

(BP), molecular functions (MF), and cellular components (CC) highlighted the role of these genes in cancer functions. Moreover, the most hit KEGG pathways were *proteoglycans in cancer, focal adhesion, and endocytosis*. The obtained common cancer network with the significant KEGG pathways is shown in Fig. 2 and Table 3.

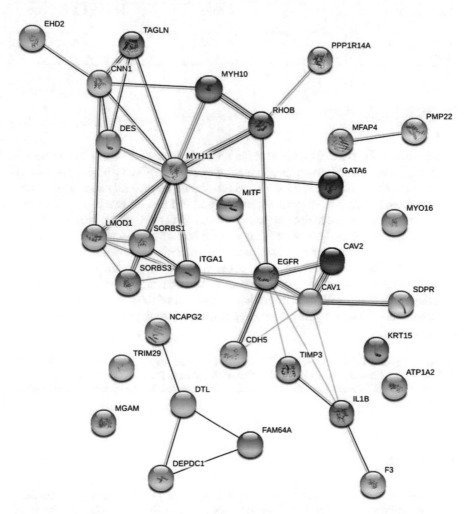

Fig. 2 Compact PPI network of common functionally similar genes in breast and lung cancer

Table 3 List of enriched KEGG pathways

Pathway	Observed gene count	False discovery rate	Matching proteins in network
Endocytosis	4	0.0385	CAV1, CAV2, EGFR, EHD2
Focal adhesion	4	0.0385	CAV1, CAV2, EGFR, ITGA1
Proteoglycans in cancer	4	0.0385	CAV1, CAV2, EGFR, TIMP3

4 Conclusion

Identification of common disease genes in multiple cancer types may lead to significant improvements in the therapeutic plan of cancer. The drug designing is heavily dependent on the drug target selection, which should be carefully addressed. Using computational evaluation of gene expression and PPI data of two cancer types, we could identify the suitable drug targets common to the two types, breast and lung cancer. Topological analysis of PPI network for individual cancer networks lead to characterization of hub genes, which require a validation for their function in cancer. The association of these novel hub genes with the already known cancer genes was validated by successful construction of a common cancer network enriched in functions associated with cancer. Further inspection of these genes by integrating lncRNA data can be performed to get deeper insight into the complexity of multiple cancer types.

Acknowledgements We acknowledge Dr. Hemant Purohit, Engineering Genomics Division, CSIR-National Environmental Engineering Research Institute [Nagpur (MS), India] for his incessant support and motivation during the course of this research. We also thank Dr Dhananjay V. Raje for valuable guidance.

References

1. Barrett, T., Wilhite, S.E., Ledoux, P., Evangelista, C., Kim, I.F., Tomashevsky, M., Marshall, K.A., Phillippy, K.H., Sherman, P.M., Holko, M., et al.: NCBI GEO: archive for functional genomics data sets update. Nucleic Acids Res. **41**(D1), D991–D995 (2012)
2. Carson, M.B., Gu, J., Yu, G., Lu, H.: Identification of cancer-related genes and motifs in the human gene regulatory network. IET Syst. Biol. **9**(4), 128–134 (2015)
3. Chen, D., Yang, H.: Integrated analysis of differentially expressed genes in breast cancer pathogenesis. Oncol. Lett. **9**(6), 2560–2566 (2015)
4. Fumagalli, D., Blanchet-Cohen, A., Brown, D., Desmedt, C., Gacquer, D., Michiels, S., Rothé, F., Majjaj, S., Salgado, R., Larsimont, D., et al.: Transfer of clinically relevant gene expression signatures in breast cancer: from affymetrix microarray to illumina RNA-sequencing technology. BMC Genomics **15**(1), 1008 (2014)

5. Gruosso, T., Mieulet, V., Cardon, M., Bourachot, B., Kieffer, Y., Devun, F., Dubois, T., Dutreix, M., Vincent-Salomon, A., Miller, K.M., et al.: Chronic oxidative stress promotes H2AX protein degradation and enhances chemosensitivity in breast cancer patients. EMBO Mol. Med. **8**(5), 527–549 (2016)
6. Huang, Y., Tao, Y., Li, X., Chang, S., Jiang, B., Li, F., Wang, Z.M.: Bioinformatics analysis of key genes and latent pathway interactions based on the anaplastic thyroid carcinoma gene expression profile. Oncol. Lett. **13**(1), 167–176 (2017)
7. Irizarry, R.A., Hobbs, B., Collin, F., Beazer-Barclay, Y.D., Antonellis, K.J., Scherf, U., Speed, T.P.: Exploration, normalization, and summaries of high density oligonucleotide array probe level data. Biostatistics **4**(2), 249–264 (2003)
8. Jeanmougin, M., De Reynies, A., Marisa, L., Paccard, C., Nuel, G., Guedj, M.: Should we abandon the t-test in the analysis of gene expression microarray data: a comparison of variance modeling strategies. PLoS ONE **5**(9), e12336 (2010)
9. Jung, D., Ge, X.: PPInfer: a bioconductor package for inferring functionally related proteins using protein interaction networks. F1000Research **6** (2018)
10. Landi, M.T., Dracheva, T., Rotunno, M., Figueroa, J.D., Liu, H., Dasgupta, A., Mann, F.E., Fukuoka, J., Hames, M., Bergen, A.W., et al.: Gene expression signature of cigarette smoking and its role in lung adenocarcinoma development and survival. PLoS ONE **3**(2), e1651 (2008)
11. Li, J., Hou, R., Niu, X., Liu, R., Wang, Q., Wang, C., Li, X., Hao, Z., Yin, G., Zhang, K.: Comparison of microarray and RNA-seq analysis of mRNA expression in dermal mesenchymal stem cells. Biotechnol. Lett. **38**(1), 33–41 (2016)
12. Li, T., Huang, H., Liao, D., Ling, H., Su, B., Cai, M.: Genetic polymorphism in HLA-G 3′ UTR 14-bp ins/del and risk of cancer: a meta-analysis of case-control study. Mol. Genet. Genomics **290**(4), 1235–1245 (2015)
13. Lu, T.P., Tsai, M.H., Lee, J.M., Hsu, C.P., Chen, P.C., Lin, C.W., Shih, J.Y., Yang, P.C., Hsiao, C.K., Lai, L.C., et al.: Identification of a novel biomarker, SEMA5A, for non-small cell lung carcinoma in nonsmoking women. Cancer Epidemiol. Prev. Biomarkers **19**(10), 2590–2597 (2010)
14. Makhijani, R.K., Raut, S.A., Purohit, H.J.: Fold change based approach for identification of significant network markers in breast, lung and prostate cancer. IET Syst. Biol. (2018)
15. Makhijani, R.K., Raut, S.A., Purohit, H.J.: Identification of common key genes in breast, lung and prostate cancer and exploration of their heterogeneous expression. Oncol. Lett. **15**(2), 1680–1690 (2018)
16. Nookaew, I., Papini, M., Pornputtapong, N., Scalcinati, G., Fagerberg, L., Uhlén, M., Nielsen, J.: A comprehensive comparison of RNA-seq-based transcriptome analysis from reads to differential gene expression and cross-comparison with microarrays: a case study in *Saccharomyces cerevisiae*. Nucleic Acids Res. **40**(20), 10084–10097 (2012)
17. Planche, A., Bacac, M., Provero, P., Fusco, C., Delorenzi, M., Stehle, J.C., Stamenkovic, I.: Identification of prognostic molecular features in the reactive stroma of human breast and prostate cancer. PLoS ONE **6**(5), e18640 (2011)
18. Rahman, M., Jackson, L.K., Johnson, W.E., Li, D.Y., Bild, A.H., Piccolo, S.R.: Alternative preprocessing of rna-sequencing data in the cancer genome atlas leads to improved analysis results. Bioinformatics **31**(22), 3666–3672 (2015)
19. Rapaport, F., Khanin, R., Liang, Y., Pirun, M., Krek, A., Zumbo, P., Mason, C.E., Socci, N.D., Betel, D.: Comprehensive evaluation of differential gene expression analysis methods for RNA-seq data. Genome Biol. **14**(9), 3158 (2013)
20. Ritchie, M.E., Phipson, B., Wu, D., Hu, Y., Law, C.W., Shi, W., Smyth, G.K.: limma powers differential expression analyses for RNA-sequencing and microarray studies. Nucleic Acids Res. **43**(7), e47–e47 (2015)
21. Szklarczyk, D., Franceschini, A., Wyder, S., Forslund, K., Heller, D., Huerta-Cepas, J., Simonovic, M., Roth, A., Santos, A., Tsafou, K.P., et al.: STRING v10: protein–protein interaction networks, integrated over the tree of life. Nucleic Acids Res. **43**(D1), D447–D452 (2014)
22. Timmerman, L.A., Holton, T., Yuneva, M., Louie, R.J., Padró, M., Daemen, A., Hu, M., Chan, D.A., Ethier, S.P., vant Veer, L.J., et al.: Glutamine sensitivity analysis identifies the xCT

antiporter as a common triple-negative breast tumor therapeutic target. Cancer Cell **24**(4), 450–465 (2013)

23. Wang, Z., Arat, S., Magid-Slav, M., Brown, J.R.: Meta-analysis of human gene expression in response to *Mycobacterium tuberculosis* infection reveals potential therapeutic targets. BMC Syst. Biol. **12**(1), 3 (2018)

24. Weinstein, J.N., Collisson, E.A., Mills, G.B., Shaw, K.R.M., Ozenberger, B.A., Ellrott, K., Shmulevich, I., Sander, C., Stuart, J.M., Network, C.G.A.R., et al.: The cancer genome atlas pan-cancer analysis project. Nat. Genet. **45**(10), 1113 (2013)

25. Wu, H., Dong, J., Wei, J.,: Network-based method for detecting dysregulated pathways in glioblastoma cancer. IET Syst. Biol. **12**(1), 39–44 (2018)

26. Zhao, S., Fung-Leung, W.P., Bittner, A., Ngo, K., Liu, X.: Comparison of RNA-seq and microarray in transcriptome profiling of activated T cells. PLoS ONE **9**(1), e78644 (2014)

27. Zhao, Y., Fu, D., Xu, C., Yang, J., Wang, Z.: Identification of genes associated with tongue cancer in patients with a history of tobacco and/or alcohol use. Oncol. Lett. **13**(2), 629–638 (2017)

28. Zheng, S., Zheng, D., Dong, C., Jiang, J., Xie, J., Sun, Y., Chen, H.: Development of a novel prognostic signature of long non-coding RNAs in lung adenocarcinoma. J. Cancer Res. Clin. Oncol. **143**(9), 1649–1657 (2017)

Selection of a Green Marketing Strategy Using MCDM Under Fuzzy Environment

Akansha Jain, Jyoti Dhingra Darbari, Arshia Kaul and P. C. Jha

Abstract The concept of green manufacturing has gained cognizance among manufacturers due to regulations imposed by the government and rising environmental consciousness of customers. Acknowledging the fact that green manufacturing can yield long-term economic and environmental gains with significant efforts channelized toward green marketing, firms are reinventing their marketing strategies. Although many researchers have discussed the importance as well as theory building of green marketing, none have analyzed the strategies under multi-criteria environment and with multi-stakeholder perspective, which is the novelty of this study. Here, a real-life case of a manufacturing firm has been considered, who wants to select an appropriate green marketing strategy for promoting its newly introduced green product, from four available strategies, namely (i) Lean Green, (ii) Defensive Green, (iii) Shaded Green, and (iv) Extreme Green. The firm's objective is to select the most appropriate strategy which is ideal for targeting green consumers. Integrated methodology of Fuzzy Analytical Hierarchy Process (Fuzzy AHP) and Fuzzy Technique for Order of Preference by Similarity to Ideal Solution (Fuzzy TOPSIS) is implemented for strategy selection, based on criteria such as green market size, price parity, and top management's commitment. It confirms accurate selection even with difference of opinion among stakeholders.

Keywords Green marketing · Strategy selection · Fuzzy AHP · Fuzzy TOPSIS

A. Jain (✉) · P. C. Jha
Department of Operational Research, University of Delhi, Delhi, India
e-mail: akansha.269@gmail.com

P. C. Jha
e-mail: pcjhadu@gmail.com

J. D. Darbari
Department of Mathematics, Lady Shri Ram College, University of Delhi, Delhi, India
e-mail: jydbr@hotmail.com

A. Kaul
Asia-Pacific Institute of Management, New Delhi, India
e-mail: arshia.kaul@gmail.com

© Springer Nature Singapore Pte Ltd. 2020
K. N. Das et al. (eds.), *Soft Computing for Problem Solving*,
Advances in Intelligent Systems and Computing 1057,
https://doi.org/10.1007/978-981-15-0184-5_43

1 Introduction

Recent years have witnessed the emergence of green manufacturing, owing mainly to the concerns of environmentally conscious consumers. To ensure that potential customers are not lost to competition, manufacturers need to adopt appropriate green marketing strategies through which effective communication is made with the target customers for achieving business profits. For manufacturing firms, green marketing strategies are persuasive marketing techniques to promote and sell their green products. The strategy selected must be such that the firm is not classified as a 'greenwasher' but as 'propagator of greenness'. 'Greenwashers' are those firms who positively communicate their environmental performance even when their performance, in reality, is poor [1]. However, any firm claiming to be a promotor of greenness must first ensure that the products manufactured are environmentally less harmful or in other words can be claimed as green [2]. Some of the common set of attributes which define products as green product(s) are: (i) they are energy efficient, (ii) do not contain ozone-depleting chemicals, (iii) may contain recycled components, and (iv) are biodegradable or can be recycled. Essentially, it is important to note that any one product is not entirely green; they possess some of the green attributes defined above [3].

Green marketing has been defined by different authors in different ways [4, 5] but essentially the idea of green marketing could be summarized by the definition given by Mishra and Sharma [2], '*Green Marketing refers to holistic marketing concept wherein the production, marketing consumption and disposal of products and services happen in a manner that is less detrimental to the environment with growing awareness about the implications of global warming, non-biodegradable solid waste, harmful impact of pollutants etc.*'

Green marketing strategies are classified by Ginsberg and Bloom [6] as *lean green, defensive green, shaded green,* and *extreme green*. The classification is based on the degree of aggressiveness of promotion. A lean green strategy on one extreme is about promotion being carried out passively, and on the other hand, extreme green strategy is about promotion which is carried out aggressively. Defensive and shaded green strategies fall in between the two extreme strategies. Out of these, the green marketing strategy selected for promotion must be in relation to the products manufactured. Some examples of the green products launched by different firms are: LED E60 and E90 series monitors launched by LG which consume 40% less energy than traditional monitors; ZSLK eco-friendly tubeless tyres' series launched by MRF; and Green crematoriums invented by ONGC which produce less smoke and consume less oxygen [7].

It is a really challenging task for the manufacturer to select appropriate green marketing strategy through which customers can be persuaded to purchase their green products. In addition to the nature of the product, the green marketing strategies also depend upon the customer segment under consideration. The customer segment for case of green products is broadly categorized into: *True blue greens, Greenback greens, Sprouts, Grousers,* and *Basic browns* [6]. True blue greens are those customers who buy green products as they are genuinely concerned about the environment. Greenback greens are those who are inclined toward buying a green product but are not aware of its availability. Sprouts are those customers who hardly buy a green product because of the notion that they are costly. They basically believe in environmental concerns only theoretically and not practically. Grousers are the customers who are unaware of the environmental issues and the positive impact of 'buying a green product' on the environment. Basic browns are not concerned with the environmental issues at all. Clearly, target customers must be identified by the firm before introducing green product in the market, and consequently, appropriate green marketing strategy must be selected.

Further, the decision of selection of a green marketing strategy is affected by the opinions of various stakeholders. As described by Freeman [8], '*A stakeholder is traditionally defined as any individual, organization, or institution that is associated with a firm, and is either affected by the firm in some way or affects the firm's actions and goals.*' There is a direct relationship that exists between the selection of a green marketing strategy and the stakeholders of the firm [9]. Therefore, the voice of each stakeholder also plays a crucial role in the decision-making process, and thus, selection of green marketing strategy is a complex group decision-making problem with conflicting opinions.

In the similar context, objective of the research has been set to select an appropriate green marketing strategy for the manufacturer of a green product under consideration based on a case study. This is a multi-criteria decision-making problem, which is affected by various stakeholders of the firm. When dealing with group decision-making problem, fetching linguistic assessment becomes easier. But in such a case, conventional methods of strategy selection do not work as they cannot handle the imprecise nature of such assessments. To curb this limitation, fuzzy multi-criteria decision-making methods have been adopted for the selection of an appropriate green marketing strategy. An integrated Fuzzy Analytical Hierarchy Process (Fuzzy AHP)–Fuzzy Technique for Order of Preference by Similarity to Ideal Solution (Fuzzy TOPSIS) approach for the selection of best suitable green marketing strategy has been utilized. The evaluation is carried out on the basis of multiple criteria such as green market size, price parity, competitors' green performance, and so on; incorporating the opinions of various stakeholders. To the best of our knowledge, there is limited research that proposes a structured mathematical approach for green marketing strategy evaluation and selection based on customer segmentation and more so, none of them have considered the role of stakeholders in the strategy selection.

The rest of the paper is organized as follows; Sect. 2 presents review of literature and research gap leading to the motivation of this study. Section 3 discusses research methodology employed. In Sect. 4, the case study is presented followed by elaboration of case implementation in Sect. 5. Section 6 concludes the paper and provides avenues for future research.

2   Literature Review

It is quite apparent from the existing literature that different nuances of green marketing have already been explored by many researchers. While some authors have deliberated upon the basic definitions of green marketing [2, 4, 5], others have discussed different aspects related to green marketing like linkages of the components of the marketing mix to green marketing, influence of stakeholders on the green marketing strategy, effect of that strategy on customers, etc. Relating marketing mix to green marketing, Ginsberg and Bloom [6] analyzed relationship between elements of green marketing mix and of the four green marketing strategies under consideration. Davari and Strutton [10] examined that out of the four elements of the marketing mix, 'green promotion' and 'green price' have higher positive effect on consumers' green satisfaction as compared to 'product' and 'place.'

Few authors have also considered the concepts of segmentation, targeting, positioning, and differentiation under the gamut of green marketing. Ginsberg and Bloom [6] and Freeman [8] segmented the market into five segments on the basis of customers' environmental awareness, namely *True Blue greens*, *Greenback greens*, *Sprouts, Grousers,* and *Basic browns*. Banyte et al. [11] proposed that consumers can also be segmented into five groups based on their willingness to contribute towards environmental improvement, namely *'loyal green consumers,' 'less devoted green consumers,' 'consumers developing towards green,' 'conservative consumers unwilling to change,'* and *'consumers completely unwilling to change.'* Modi and Patel [12] discussed the concept of segmentation in relation to consumers' green activism wherein they divided the market into two segments: *active green activists* and *passive green activists.*

From the perspective of targeting, Polonsky and Rosenberger [13] suggested that technological tools, like the Internet, could help organizations in targeting green consumers. D'Souza et al. [14] put forth positioning strategies that communicate green marketing mix elements to strengthen the reputation of the organization. Dangelico and Vocalelli [15] discussed that positioning in green marketing could be either functional or emotional. Although both are effective for green marketing, the best way forward is to integrate the two [15]. D'Souza et al. [16] highlighted that internal green practices of a firm-like green supplier selection, green product development, and so on are affected by green marketing. Some authors have also discussed the role of stakeholders while considering green marketing. Delmas and Burbano [1] pointed out that stakeholders of a firm are affected by the type of green marketing strategy it chooses and thus identified four groups of stakeholders that are affected by the green marketing strategy selected by the firm: *Market stakeholders* (competitors, customers, distributors, and suppliers), *Social pressure group* (press and media, environmental organizations, and local people), *Immediate providers* (owners, shareholders, and labor unions), and *Legal stakeholders* (international and national regulations and voluntary agreements). Xu et al. [17] proposed a grey-based rough set theory approach to identify and manage those internal activities that are linked to green marketing. They discussed that there needs to be a trade-off between achieving greenness and the cost incurred for achieving this greenness.

Although conceptual discussion regarding selection of green marketing strategy has been done in literature [6], none of the studies have proposed a structured methodology for green marketing strategy selection [15]. This gap in the extant literature has led to the motivation of our research. Thus, we propose an integrated Fuzzy AHP–Fuzzy TOPSIS approach for the selection of the best suited green marketing strategy from those available. The novelty of the proposed research is in the fact that green marketing strategy selection is carried out mathematically, considering stakeholders' perspective.

3 Research Methodology

The research methodology used in this paper involves the following stages:

1. Identification of criteria for evaluation of green marketing strategies
2. Identification of stakeholders participating in the evaluation process
3. Utilizing integrated Fuzzy AHP–Fuzzy TOPSIS methodology for selection of an appropriate green marketing strategy based on target customers.

After identification of criteria for evaluation and stakeholders involved in the decision-making process, an integrated Fuzzy AHP–Fuzzy TOPSIS methodology is adopted. The evaluation criteria are prioritized using Fuzzy AHP method proposed by Van Laarhoven et al. [18]. The pairwise comparison matrices are developed using Triangular Fuzzy Numbers (TFNs), and Chang's extent analysis approach is utilized for determining the weights of criteria [19]. Next step is to find the weights of four green marketing strategies using Fuzzy TOPSIS introduced by Chen and Hwang [20]. An extended TOPSIS approach for fuzzy data given by Chen [21] and Chen et al. [22] is used for finding out the most suitable green marketing strategy for the firm. The overview of the complete methodology is given in Fig. 1 in brief. For detailed steps of the methodology, the reader may refer to [23].

4 Case Study

The case of a manufacturing firm (name not disclosed due to reasons of confidentiality), which has ventured into manufacturing a new green product: an all-purpose cleaning solution is presented in this paper. The cleaning solution produced by the firm is a green product as: (i) it is free of harsh solvents that pollute the air, (ii) it is biodegradable, and (iii) it is packed in recyclable plastic bottles. To promote its new product, the firm wants to select an appropriate green marketing strategy, which will target the prospective customers effectively.

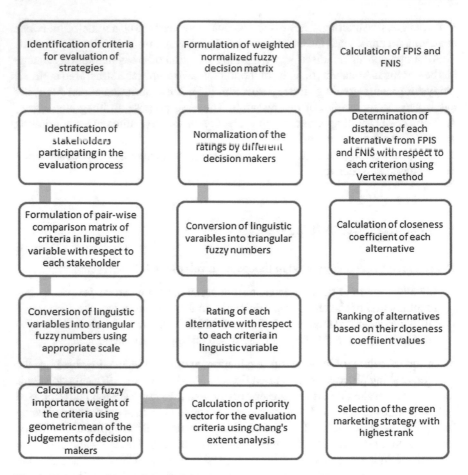

Fig. 1 Flowchart of research methodology

As discussed earlier, there are four types of green marketing strategies. The challenge in front of the firm is to decide the strategy best suited for promoting their green product. Since the decision of adoption of right strategy is a strategic decision having long-term impact, it must be deliberated upon extensively. There are many issues to ponder upon while selecting the appropriate strategy [6] such as: (i) How substantial is the green customer segment? (ii) Can the firm increase revenue by improving on perceived greenness? (iii) Will business suffer if customer's judgment changes or if the customers are indifferent? (iv) How competitors are pursuing potential target segments? (v) Can the firm be differentiated on green dimension? (vi) Does the firm have resources and internal commitment to achieve greenness? and (vii) Can competitors be beaten on this dimension?

In view of the above concerns, it is imperative for the decision makers to understand firms' financial position, resource capability, customer expectation, competitive position, and expected profits while taking such a crucial decision. Additionally, since selection of a green marketing strategy not only affects the firm itself, but also other stakeholders involved; decisions regarding marketing strategy should be taken considering their opinions.

This brings us to the purpose of the present study which aims to achieve the following: (i) Selecting the best green marketing strategy out of *Lean Green*, *Shaded Green*, *Defensive Green*, and *Extreme Green* for promoting the green product, (ii) Identification of appropriate factors that must be considered for evaluating the strategies for a conclusive decision which has a long term positive impact, and (iii) Utilizing an effective methodology for evaluation of the best green marketing strategy under multiple criteria considerations and stakeholders' participation.

5 Case Implementation

Based on the research methodology described in Sect. 4, seven criteria are determined after discussion with top management on the basis of their expectations and concerns, given in Table 1. Table 2 lists the stakeholders involved in the decision-making process.

First step in the integrated methodology is to identify weights of importance of seven evaluation criteria. Each of the decision makers is asked to fill up a questionnaire for formulating the pairwise comparison matrices of criteria by giving relative weight of each criterion based on the linguistic variable scale for Fuzzy AHP given by Tolga et al. [24] where EI represents 'Equal Importance', MI represents 'Moderate Importance', SI represents 'Strong Importance', VSI represents 'Very Strong

Table 1 Criteria for the evaluation of strategies

	Criteria	Description
C1	Green market size	Type and size of green market segment to be catered to by the firm
C2	Potential revenue	Revenue which can be earned by green promotion
C3	Price parity	The probable price differential between the green product and the traditional counterpart as perceived by the target segment
C4	Competitors' green performance	The green image of the competitor(s) as viewed by the customers
C5	Availability of resources	The availability of resources for manufacturing and promotion of the green product
C6	Top management commitment	The level of commitment of top management towards the green initiatives of the firm
C7	Government incentives	The benefits provided by the government to the firm if it incorporates green initiatives in its business

Table 2 Stakeholders involved in the evaluation of strategies

	Stakeholders	Role
S1	Supplier	Provides desired raw materials for green operations
S2	Customer	The green marketing strategy affects their purchase intentions
S3, S4, S5	Marketing Manager (2)/Senior Marketing Manager (1)	Takes decisions regarding the marketing strategies and their implementation for the firm
S6	Financial Manager	Allots capital for marketing activities of the firm
S7	Purchase Manager	Ensures that the desired raw materials are of a set standard by the firm
S8	Distributor	Provides logistical support for green operations
S9	Shareholder	Affected by the revenue earned by the firm
S10	Government/NGO	Provides financial assistance to the firm

Importance' and DI represents 'Demonstrated Importance'. For instance, the pairwise comparison matrix based on the ratings given by S1 is represented in Table 3. For calculations, these pairwise comparison matrices are mathematically represented using TFNs. Table 4 gives the mathematical representation of pairwise comparison matrix given in Table 3. To check steadiness of each decision maker's judgment, consistency of all the pairwise comparison matrices is checked using algorithm given in [25]. Next, the fuzzy importance weights of criteria are calculated by taking geometric mean of responses given by the stakeholders [26] as given in Table 5.

Priority vector for all the evaluation criteria is calculated using extent analysis method [27]. The priority weights obtained after this procedure are:

$$W' = (0.874, 0.878, 0.893, 0.876, 0.749, 1.000, 0.799)$$

Table 3 Pairwise comparison matrix using linguistic variables given by S1

Criteria	C1	C2	C3	C4	C5	C6	C7
C1	EI	MI	DI	VSI	SI	1/MI	VSI
C2	1/MI	EI	VSI	SI	MI	1/SI	SI
C3	1/DI	1/VSI	EI	1/MI	1/SI	1/DI	1/MI
C4	1/VSI	1/SI	MI	EI	1/MI	1/VSI	EI
C5	1/SI	1/MI	SI	MI	EI	1/SI	MI
C6	MI	SI	DI	VSI	SI	EI	VSI
C7	1/VSI	1/SI	MI	EI	1/MI	1/VSI	EI

Table 4 Pairwise comparison matrix using TFNs given by S1

Criteria	C1	C2	C3	C4	C5	C6	C7
C1	(1, 1, 1)	(1/2, 1, 3/2)	(2, 5/2, 3)	(3/2, 2, 5/2)	(1, 3/2, 2)	(2/3, 1, 2)	(3/2, 2, 5/2)
C2	(2/3, 1, 2)	(1, 1, 1)	(3/2, 2, 5/2)	(1, 3/2, 2)	(1/2, 1, 3/2)	(1/2, 2/3, 1)	(1, 3/2, 2)
C3	(1/3, 2/5, 1/2)	(2/5, 1/2, 2/3)	(1, 1, 1)	(2/3, 1, 2)	(1/2, 2/3, 1)	(1/3, 2/4, 1/2)	(2/3, 1, 2)
C4	(2/5, 1/2, 2/3)	(1/2, 2/3, 1)	(1/2, 1, 3/2)	(1, 1, 1)	(2/3, 1, 2)	(2/5, 1/2, 2/3)	(1, 1, 1)
C5	(1/2, 2/3, 1)	(2/3, 1, 2)	(1, 3/2, 2)	(1/2, 1, 3/2)	(1, 1, 1)	(1/2, 2/3, 1)	(1/2, 1, 3/2)
C6	(1/2, 1, 3/2)	(1, 3/2, 2)	(2, 5/2, 3)	(3/2, 2, 5/2)	(1, 3/2, 2)	(1, 1, 1)	(3/2, 2, 5/2)
C7	(2/5, 1/2, 2/3)	(1/2, 2/3, 1)	(1/2, 1, 3/2)	(1, 1, 1)	(2/3, 1, 2)	(2/5, 1/2, 2/3)	(1, 1, 1)

Table 5 Fuzzy geometric mean of pairwise comparison

	C1	C2	C3	C4	C5	C6	C7
C1	(1, 1, 1)	(0.7, 0.9, 1.2)	(0.7, 1, 1.6)	(0.8, 1.1, 1.8)	(1, 1.2, 1.4)	(0.6, 0.8, 1.3)	(0.9, 1, 1.4)
C2	(0.8, 1, 1.4)	(1, 1, 1)	(0.7, 1, 1.4)	(0.8, 1.1, 1.7)	(0.8, 12, 1.5)	(0.6, 0.9, 1.4)	(0.8, 1, 1.3)
C3	(0.6, 1, 1.5)	(0.7, 1, 1.4)	(1, 1, 1)	(0.7, 1, 1.6)	(0.9, 1.3, 1.7)	(0.6, 0.9, 1.3)	(0.7, 1.1, 1.6)
C4	(0.6, 0.9, 1.3)	(0.6, 0.9, 1.2)	(0.6, 0.9, 1.4)	(1, 1, 1)	(0.7, 1.1, 1.6)	(0.6, 0.8, 1.2)	(0.8, 1, 1.4)
C5	(0.7, 0.8, 1)	(0.7, 0.8, 1.7)	(0.6, 0.8, 1.2)	(0.6, 0.9, 1.4)	(1, 1, 1)	(0.5, 0.7, 1)	(0.7, 0.9, 1.3)
C6	(0.8, 1.2, 1.7)	(0.7, 1.1, 1.6)	(0.8, 1.1, 1.6)	(0.8, 1.3, 1.8)	(1, 1.5, 2.1)	(1, 1, 1)	(0.8, 1.3, 2)
C7	(0.7, 0.9, 1.2)	(0.8, 1, 1.3)	(0.6, 0.9, 1.4)	(0.7, 1, 1.3)	(0.8, 1.1, 1.5)	(0.5, 0.8, 1.2)	(1, 1, 1)

Table 6 Rating of green marketing strategies in linguistic variables (TFNs) given by M1

Criteria	Lean green	Defensive green	Shaded green	Extreme green
C1	P (1, 2, 3)	G (7, 8, 9)	MG (5, 6.5, 8)	F (4, 5, 6)
C2	VP (0, 0, 2)	MP (2, 3.5, 5)	MG (5, 6.5, 8)	VG (8, 10, 10)
C3	VG (8, 10, 10)	G (7, 8, 9)	P (1, 2, 3)	P (1, 2, 3)
C4	F (4, 5, 6)	MG (5, 6.5, 8)	MP (2, 3.5, 5)	G (7, 8, 9)
C5	G (7, 8, 9)	F (4, 5, 6)	MP (2, 3.5, 5)	P (1, 2, 3)
C6	G (7, 8, 9)	G (7, 8, 9)	VG (8, 10, 10)	VG (8, 10, 10)
C7	P (1, 2, 3)	F (4, 5, 6)	G (7, 8, 9)	G (7, 8, 9)

After normalization, final priority weights for all the seven criteria are determined, which is given as:

$$W = (0.144, 0.145, 0.147, 0.144, 0.123, 0.165, 0.132)$$

Next step is to prioritize the four green marketing strategies based on the seven evaluation criteria. After finding the importance weights of criteria, Fuzzy TOPSIS is used to rank the green marketing strategies. A committee of three marketing managers (two managers (M1, M2) and one senior manager (M3)) was established who, based on the identified criteria, provided their judgments for the four green marketing strategies. The linguistic variable scale defined by Shukla et al. [23] for Fuzzy TOPSIS was used to rate the alternatives with respect to each criterion where VP, P, MP, F, MG, G and VG stand for 'Very Poor', 'Poor', 'Medium Poor', 'Fair', 'Medium Good', 'Good' and 'Very Good'. Using the scale defined in [23], the linguistic ratings provided by three marketing managers for the four green marketing strategies with respect to each criterion are represented in form of TFNs. The ratings given by M1 in terms of linguistic variables and their corresponding TFN representation are shown in Table 6.

Fuzzy decision matrix is formulated using the aggregated rating for the four green marketing strategies by implementing the steps of Fuzzy TOPSIS method. Table 7 gives the fuzzy rating of strategies for each criterion. This decision matrix

Table 7 Aggregated fuzzy rating of strategies

Criteria	Lean green	Defensive green	Shaded green	Extreme green
C1	(0, 1.8, 5)	(5, 7.5, 9)	(4, 6, 8)	(2, 5, 8)
C2	(0, 0.7, 3)	(2, 3.5, 5)	(4, 6, 8)	(7, 9.3, 10)
C3	(7, 9.3, 10)	(7, 8, 9)	(1, 4, 6)	(1, 2, 3)
C4	(4, 5.5, 8)	(5, 7, 9)	(1, 3.5, 6)	(5, 8.2, 10)
C5	(4, 7, 9)	(4, 6, 9)	(2, 3.5, 5)	(1, 2, 3)
C6	(4, 6, 9)	(7, 8, 9)	(5, 7.7, 10)	(8, 10, 10)
C7	(0, 1.8, 5)	(4, 5, 6)	(5, 8.2, 10)	(7, 8, 9)

Strategy	Closeness coefficient	Rank
Lean green	0.333	4
Defensive green	0.636	1
Shaded green	0.459	2
Extreme green	0.369	3

Table 8 Closeness coefficient and rank of each alternative

is normalized according to the method given in [21]. After normalization, weighted normalized fuzzy decision matrix is formed.

The next step is to calculate Fuzzy Positive Ideal Solution (FPIS) and Fuzzy Negative Ideal Solution (FNIS). For this purpose, the chosen seven criteria are classified into benefit criteria and cost criteria. Out of the seven criteria considered; 'price parity' and 'competitor's green performance' are cost criteria and all the others are benefit criteria. Vertex method is used to calculate the distance of each alternative from FPIS and FNIS with respect to each criterion [21].

Next, closeness coefficient of each of the four alternative strategies is calculated as described in [21] and the result obtained is shown in Table 8. The closeness coefficient highlights the distance of the alternative to FPIS and FNIS.

As can be observed, the closeness coefficient of *Defensive Green* strategy is 0.636, which is the highest in this case; thus, it is the best suitable strategy for the firm. However, *Defensive Green* strategy was not the first choice based on each criterion of evaluation. This clearly justifies the use of integrated Fuzzy AHP–Fuzzy TOPSIS methodology which effectively handles the complexity in the selection procedure due to the conflict in stakeholders' opinions and presence of different nature of multiple criteria.

6 Conclusion

Growing concerns of customers towards the environment along with regulations imposed by the government has made it imperative for manufacturing firms to reorient their business strategies. Consequently, it calls for devising a marketing strategy plan by the manufacturing firms for targeting the green consumers. This requires decision making involving all the participants of the supply chain. In this research, case of a manufacturing firm which is producing a green product is considered. The research objective of the study is evaluation and selection of the appropriate green marketing strategy for promotion of the green product. The strategies considered are *Lean green*, *Shaded green*, *Defensive green*, and *Extreme green*. The novelty of the study lies in an exhaustive evaluation process considering multiple criteria such as green market size, potential revenue, price parity, and so on and incorporating multiple stakeholders' perspectives. An integrated Fuzzy AHP–Fuzzy TOPSIS approach has been used for the selection of the green marketing strategy, based on the target customers. Fuzzy AHP is used for finding the weights for the criteria considered in the study and

Fuzzy TOPSIS is used to reflect the 'voice' of firm's various stakeholders regarding their preferences in terms of final weights of the four strategies. It was found that defensive green strategy is the most suitable strategy for the firm even if it was not the first choice with respect to some criteria of evaluation. Thus, the use of Fuzzy AHP–Fuzzy TOPSIS approach is justified. A limitation of the proposed approach is that interdependencies between the criteria and various stakeholders have not been considered. In future, strategy selection can be done over a dynamic planning horizon, where market scenarios change over time and selection of strategy for each time period is different.

Compliance with Ethical Standards

This is an independent and non-funded research study, thus, there are no potential source(s) of conflict of interests. Also, informed consents were taken from all the respondents of the questionnaire.

References

1. Delmas, M.A., Burbano, V.C.: The drivers of greenwashing. Calif. Manag. Rev. **54**(1), 64–87 (2011)
2. Mishra, P., Sharma, P.: Green marketing: challenges and opportunities for business. J. Mark. Commun. **8**(1), 35–41 (2012)
3. Ottman, J.A., Stafford, E.R., Hartman, C.L.: Avoiding green marketing myopia: ways to improve consumer appeal for environmentally preferable products. Environ. Sci. Policy Sustain. Dev. **48**(5), 22–36 (2006)
4. Wymer, W., Polonsky, M.J.: The limitations and potentialities of green marketing. J. Nonprofit Public Sect. Mark. **27**(3), 239–262 (2015)
5. Peattie, K., Charter, M.: Green marketing. In: Baker, M.J. (ed.) The Marketing Book, pp. 726–755. Butterwoth-Heinemann, Oxford (2003)
6. Ginsberg, J.M., Bloom, P.N.: Choosing the right green-marketing strategy. MIT Sloan Manag. Rev. **46**(1), 79–84 (2004)
7. Choudhury, R.: Top 10 Green Companies of India. FIINOVATION (2016). https://fiinovationblogs.wordpress.com/2016/02/29/top-10-grccn-companies-of-india/
8. Freeman, R.E.: Strategic Management: A Stakeholder Approach. Pitman, Boston (1984)
9. Rivera-Camino, J.: Re-evaluating green marketing strategy: a stakeholder perspective. Eur. J. Mark. **41**(11/12), 1328–1358 (2007)
10. Davari, A., Strutton, D.: Marketing mix strategies for closing the gap between green consumers' pro-environmental beliefs and behaviors. J. Strateg. Mark. **22**(7), 563–586 (2014)
11. Banyte, J., Brazioniene, L., Gradeikiene, A.: Investigation of green consumer profile: a case of Lithuanian market of eco-friendly food products. Econ. Manag. **2**, 374–383 (2010)
12. Modi, A.G., Patel, J.D.: Classifying consumers based upon their proenvironmental behaviour: an empirical investigation. Asian Acad. Manag. J. **18**(2), 85–104 (2013)
13. Polonsky, M.J., Rosenberger III, P.J.: Reevaluating green marketing: a strategic approach. Bus. Horiz. **44**(5), 21–30 (2001)
14. D'Souza, C., Mehdi, T., Sullivan-Mort, G.: Environmentally motivated actions influencing perceptions of environmental corporate reputation. J. Strateg. Mark. **21**(6), 541–555 (2013)
15. Dangelico, R.M., Vocalelli, D.: "Green marketing": an analysis of definitions, strategy steps, and tools through a systematic review of the literature. J. Clean. Prod. **165**, 1263–1279 (2017)

16. D'Souza, C., Taghian, M., Sullivan-Mort, G., Gilmore, A.: An evaluation of the role of green marketing and a firm's internal practices for environmental sustainability. J. Strateg. Mark. **23**(7), 600–615 (2015)

17. Xu, Z.J., Liu, X., Bai, C., Hu, L.: Green marketing: a grey-based rough set theory analysis of activities. Int. J. Innov. Sci. **7**(1), 27–38 (2015)

18. Van Laarhoven, P.J.M., Pedrycz, W.: A fuzzy extension of Saaty's priority theory. Fuzzy Sets Syst. **11**, 229–241 (1983)

19. Chang, D.Y.: Application of the extent analysis method on Fuzzy AHP. Eur. J. Oper. Res. **95**, 649–655 (1996)

20. Chen, S.J., Hwang, C.L.: Fuzzy Multiple Attribute Decision Making. Methods and Applications. Springer, Berlin (1992)

21. Chen, C.T.: Extension of the TOPSIS for group decision-making under Fuzzy environment. Fuzzy Sets Syst. **114**(1), 1–9 (2000)

22. Chen, C.T., Lin, C.T., Huang, S.F.: A fuzzy approach for supplier evaluation and selection in supply chain management. Int. J. Prod. Econ. **102**, 289–301 (2006)

23. Shukla, R.K., Garg, D., Agarwal, A.: An integrated approach of Fuzzy AHP and Fuzzy TOPSIS in modeling supply chain coordination. Prod. Manuf. Res. **2**(1), 415–437 (2014)

24. Tolga, E., Demircan, M., Kahraman, C.: Operating system selection using Fuzzy replacement analysis and analytical hierarchy process. Int. J. Prod. Econ. **97**, 89–117 (2005)

25. Ramik, J.: Consistency of pair-wise comparison matrix with fuzzy elements. In: IFSA/EUSFLAT Conference, pp. 98–101 (2009)

26. Lee, A.H.: A Fuzzy supplier selection model with the considerations of benefits, opportunities, costs, and risks. Expert Syst. Appl. **36**, 2879–2893 (2009)

27. Kulak, O., Kahraman, C.: Fuzzy multi-attribute selection among transportation companies using axiomatic design and analytic hierarchy process. Inf. Sci. **170**, 191–210 (2005)

Stacked Convolutional Autoencoder for Detecting Animal Images in Cluttered Scenes with a Novel Feature Extraction Framework

S. Divya Meena and L. Agilandeeswari

Abstract Detection of animals from a cluttered scene is not a trivial task. So far, convolutional neural network (CNN) architectures have served this purpose. We introduce stacked convolutional autoencoders (SCAE) for this purpose. It is an unsupervised stratified feature extractor that could be used for high-dimensional input images. We also introduce a hybrid feature extraction technique based on Fisher Vectors (FV) and stacked autoencoders (SAE). SCAE learns significant features utilizing plain stochastic gradient descent and finds a good initialization for CNNs so as to eliminate the various unique local minima of exceptionally non-convex target functions emerging in virtually all deep learning problems. We have proposed a parallel pipeline for both detecting animals in both visible and infrared images. The framework model has achieved 97% accuracy.

Keywords Convolutional neural network · Stacked convolutional autoencoder · Animal detection · Fisher Vector · Infrared images · Fuzzy

1 Introduction

Usually, an image may consist of several objects, and in such case, each item must be distinctively identified. Along with image category techniques, we can give labels to each of the instances in the image; however, this is cumbersome and the result can be inaccurate. Alternatively, object detection can be a more appropriate and efficient one to uniquely identify each instances of the image. In addition to this, they also aid in object localization.

Object detection is one of the interesting problems in computer vision (CV). Computer vision has other interesting applications like image categorization, detection of particular items, object or face detection, pose estimation, and many more. Among all, object detection is all about uniquely identifying and tracking down each instance in the image. Object detection has seen many techniques for the purpose but

S. D. Meena (✉) · L. Agilandeeswari
School of Information Technology and Engineering, VIT University, Vellore, India
e-mail: sdivya.meena2017@vitstudent.ac.in

© Springer Nature Singapore Pte Ltd. 2020 513
K. N. Das et al. (eds.), *Soft Computing for Problem Solving*,
Advances in Intelligent Systems and Computing 1057,
https://doi.org/10.1007/978-981-15-0184-5_44

yet there is no single technique that could be claimed as the best, which means, there is certainly still so much to improve. A few of the common issues in object detection includes having more than one instance in the image, size, and modeling problem.

Despite the issues pointed out above, object detection has its own useful applications like face detection, animal detection, counting, visual search engine, airborne image analysis, and more. Animal detection is one of the most required and interesting applications of object detection. The animal detection can be applied in many life-saving scenarios, like detecting the animals on the road to avoid animal–vehicle collision, detecting in the forest for animals survey, detecting in the forest village border to avoid human–animal conflict, and many. Animal image detection has seen many different techniques in recent times and every of the technique has achieved a different performance.

Dewan et al. [1] have used edge-based approach for detection. Though the model reduces the risk of false alarm due to illumination change and camera motion, however, the overall detection rate was low. Dabarera and Rodrigo [2] used appearance-based recognition model which was not efficient in case of occluded animal images. Goswami et al. [3] used a supervised visual identification of individual variations in tusk, ear fold, and lobe shape of elephant for detection. In real time, the capture of elephant's front image is not possible. Dhande [4] used IRIS recognition system which was highly accurate but costs high to implement. Ardovini [5] used shape comparison of the nicks characterizing the elephant's ears to detect elephants. But the model was not suitable for deformed images or when the image is partially visible. Chen et al. [6] used convolutional-based deep neural network for detection but the overall accuracy of the system was only about 38.8%. Figueroa et al. [7] also used convolutional neural network for detection purpose but the model was based on manual feature extraction and was trained with only few thousands of images. Norouzzadeh et al. [8] used different deep learning architectures like ALexNet, ResNet, NiN, VGG, and GoogleNet. The best model produced an accuracy of about 92% for Snapshot Serengeti dataset. Gomez et al. [9] also used various deep learning architectures like ALexNet, ResNet (50, 101, 52), NiN, VGG, and GoogleNet and the best model achieved an accuracy of about 88.9% in Top-1 category and about 98.1% in Top-5 category.

With the above shortcomings in mind, we work toward building an efficient animal image detector model.

The following are our contributions:

1. Developing a novel feature extraction model based on Fisher Vector and stacked autoencoders.
2. Developing a hybrid autoencoder by stacking convolutional autoencoders to form stacked convolutional autoencoder.
3. Initializing the convolutional neural network with the weights obtained from stacked convolutional autoencoder.
4. Parallel pipeline to detect animals from both visible and infrared images.

The remaining of the paper is organized in the following way: Sect. 2 describes the proposed methodology. Section 3 discusses the experimental framework and performance metrics. Section 4 details the results obtained. Conclusion and future work are presented in Sect. 5.

2 Proposed Methodology

In this section, we discuss about our parallel pipeline for animal detection. Initially, the images are preprocessed and the area of interest is segmented. The features are extracted from the segmented image. The extracted features are fed into the stacked convolutional autoencoder. Each convolutional layer in SCAE is followed by the sparse Maxpooling layer, in order to reduce the dimensions. This produces a latent space representation in the bottleneck layer, which the output layer uses to detect the type of animal.

2.1 Framework of Stacked Convolutional Autoencoder

When it comes to object recognition task, animals are the hardest [10]. Efficient animal detection system with huge intra-class variance is still a challenging task. Hence, we propose a parallel model that detects animals in both visible and infrared images. Though infrared images are most useful in nighttime, it can be used for day vision too. But, visible images are not good at detecting animals in nighttime without proper lighting [11]. Figure 1 depicts the proposed animal detection system.

The original image is preprocessed and the area of interest is segmented. We follow different preprocessing techniques for both visible and infrared images. Features are then extracted from the segmented images using Fisher Vector-based stacked autoencoder. The animals are detected using stacked convolutional autoencoder.

Step 1: **Preprocessing of visible and infrared images**

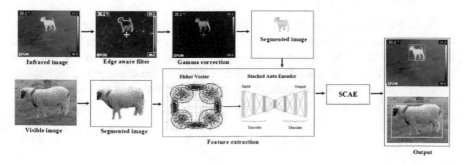

Fig. 1 Proposed animal detection system

(1) *Visible images*

Initially, the images are preprocessed by resizing its size to 224 × 224, despite its original size and its proportion. To avoid too much of computation time, we segment the area of interest for extracting the features. We use graph-cut-based segmentation [12] followed by deriving its active contour to get the area of interest.

(2) *Infrared images*

The steps involved in thermal images substantially vary from visible image. For thermal images, we first use gradient-based guided edge-aware smoothing filter to remove the noise. This is followed by image enhancement using gamma correction. We segment the area of interest using fuzzy-based edge detection. Figure 2 illustrates the pre-processing of the infrared images.

(a)   Gradient-based guided edge-aware smoothing filter

Guided filter works on the basis of a second image called guide image, which preserves the edges of the input image, thus retaining the structure of the object in spite of smoothing. The input image could itself be taken as guide image and in such case, the edges will be perfectly retained and the final output will be similar to the input image. For a given original image I, we denote the guidance image as G and the output image as O. The guided filter can then be defined as [13];

$$O_i = a_k G_k + b_k, \quad \forall i \in w_k$$

The i in the above equation denotes the pixel's index and the w denotes the square window and the k denotes the window's index. The filtered output image O is given by;

$$O_i = \overline{a}_i G_i + \overline{b}_i$$

$\overline{a}_i$ and $\overline{b}_i$ denotes the average values of a and b on w at index i.

(b)   Gamma correction

Gamma correction involves a nonlinear operation on the pixels of the original image, causing a variation in the saturation. The brightness of the image can be controlled with a gamma correction, but an incorrect gamma value can have poor contrast. The luminance of the image is not liner and they depend on the capturing or displaying device. So, we use gamma correction to correct the luminance of the image. Gamma correction I' for image I is given by [14].

$$I' = 255 \times \left(\frac{I}{255}\right)^{\gamma}$$

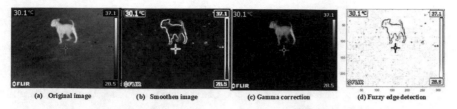

(a) Original image   (b) Smoothen image   (c) Gamma correction   (d) Fuzzy edge detection

Fig. 2 Image pre-processing techniques **a** Original; **b** noise removal through image smoothening using gradient-based guided edge-aware smoothing filter; **c** image enhancement using gamma correction; and **d** Fuzzy based edge detection

(c) Fuzzy-based edge detection

We designed a Mamdani-based fuzzy inference system (FIS) that takes the gradients of the images as input. Initially the gradients of the image is calculated along the X and Y axis and these are convolved with the input image to obtain Grad_x and Grad_y. These are fed as input to the FIS and a unit membership function is defined for both. The membership function for both the inputs is of type trimf and this was chosen solely on the basis of trial and error [15]. The parameters of the inputs were set to $[-1\ 0\ 1]$ and the range was defined to be $[-1\ 1]$. For the output, we defined two membership functions each corresponding to black and white. The parameters for white was set to $[0.1\ 1\ 1]$ and for the black $[0\ 0\ 0.6]$. Again, we used trimf function for the output. The rules for the inference system were defined as;

$$(\mathrm{Grad}_x == \mathrm{Zero})\ \&\ (\mathrm{Grad}_y == \mathrm{Zero}) => (\mathrm{Output} = \mathrm{White})$$

$$(\mathrm{Grad}_x \cong \mathrm{Zero})\ \&\ (\mathrm{Grad}_y \cong \mathrm{Zero}) => (\mathrm{Output} = \mathrm{Black})$$

Step 2: Unsupervised feature learning via Fisher Vector-based stacked autoencoder

With the advancement in deep learning, one can easily extract the activations of a pre-trained neural network model. But, when these features are used as Global features [11], and then the outcome may not be optimal. To leverage the power of CNN, we use the activations of CNN as local features instead of global. We introduce a novel feature extraction framework based on Fisher Vector (FV) and stacked autoencoders and we name it as Fisher Vector-based stacked autoencoder (FVSAE). The algorithm for FVSAE is given below:

Algorithm: Fisher Vector-based stacked autoencoder for feature extraction.
Input: Feature sample for training, labels for training, odd number of hidden layers for SAE, extracted features from SAE e_1, and features from Fisher Vector f_1.
Output: Overall features extracted f_2

1. Pre-train the hidden layers based on the sample feature x with a constraint that the bottleneck layer to contain only e_1 neurons

2. Tune the network based on the training label y with stochastic gradient descent and back propagation techniques
3. Extract the node values of e_1 from the bottleneck layer and try reconstructing them on a new dataset
4. Compute the Fisher Vector for the features in the new dataset
5. Rearrange the feature in descending order based on Fisher Vector value and select the first f_2 features.

We apply the features extracted from FVSAE to the convolutional layers

Step 3: **Image representation with convolutional layer and Maxpooling**

Convolutional layer: In our network, we use three convolutional layers with 100, 150, and 200 filters each of size 5×5, 5×5, and 3×3, respectively. We use two Maxpooling layers with a filter size of 2×2 for dimensionality reduction.

Maxpooling: In our proposed method, we use Maxpooling as a means to introduce sparsity in the latent space representation. All other values other than the maximum one (non-maximal values) are erased and this brings in sparsity in the bottleneck layer. With such a sparse representation, the detector can easily avoid non-trivial solutions. In addition to this, the sparse latent representation also decreases the mean filters that contribute to the decoder in the reconstruction phase.

Step 4: **Detection of animals via Softmax classifier**

The Softmax layer has 13 neurons, one for each type of animal. The compressed feature map from the Maxpooling layer is the input to Softmax layer. The Softmax layers outputs the closest match from the 13 types of animal as the output for the test image.

3 Experimental Framework

For the experimentation part, we worked with our own dataset which is discussed in Table 1. The experiments were carried out in single NVIDIA GeForce 940M Version 376.82 GPU-based laptop with Intel(R) Core(TM) i5-5200U CPU @ 2.20 GHz processor and 8 GB RAM. For the storage purpose, we used 1 TB Seagate hard

Table 1 Comparison with existing model

Metrics	[6]	[7]	[8]	[9]	Our approach
Accuracy	87.1	82.7	82.5	72.1	97.8
Specificity	90.1	87.9	79.5	75.1	98.7
Sensitivity	89.1	91	76.5	74.1	96.5
Precision	92.3	92.3	73.5	77.3	94.3

Fig. 3 Sample images **a** infrared images captured with FLIR thermal camera; **b** visible images from FLIR; and **c** ImageNet dataset

disk drive. The experiments were carried out on Jupyter Notebook based on various machine learning libraries and for the dimensionality reduction part, we used MATLAB R2018a.

3.1 Dataset

The dataset is manually created, so that it would be more accurate for our problem. We have captured real-time animals using FLIR e40 thermal imaging camera. The images captured with FLIR can be used as both infrared and visible images. The sample images are given in Fig. 3.

3.2 Performance Metrics

The performance of our model is assessed with few usual metrics. For assessing the performance, we use the metrics like accuracy, specificity, sensitivity, and precision. Accuracy is the proportion to which the model is correct or perfect. Specificity is the percentage to which the model is exact. Precision is the positive predictive value of the model, i.e., being accurate. Sensitivity is the ability of the model to recall.

Fig. 4 Sample results for correct detection

4 Results and Discussion

This section discusses about the results and is evaluated against the performance metrics. We have achieved an accuracy of 97% from our proposed model. To the best of our knowledge, autoencoders have not been employed in detection applications, specifically animal detection. Moreover, autoencoders are commonly used for classification problem and for dimensionality reduction. We have attempted it for detection problem. The results are quite promising and can be very well applied to other similar detection problems. The sample results for correct detection is represented in Fig. 4.

Most of the works in image classification has either taken camera trap images or online datasets. For a fairer comparison, we have considered our own dataset and implemented the methods given in [6–9].

So far, animal image detection has been approached with CNN in most of the cases. Fine-tuning the CNN for our application may not always yield better results. Our approach was based on feature extraction from proposed hybrid framework. Feeding these features into the autoencoders resulted in better performance.

5 Conclusion

We introduced stacked convolutional autoencoders (SCAE), an unsupervised stratified feature extractor that may be used for high-dimensional input images. We also introduced a hybrid feature extraction technique based on Fisher Vectors and stacked convolutional autoencoders. With sparse Maxpooling, the best filters emerged, without having used any regularization functions. Features extracted from our own hybrid feature extractor tend to outperform the pre-trained features of CNN. The SCAE learns significant features utilizing plain stochastic gradient descent and finds the good initialization for CNNs and eliminates the various unique local minima associated with exceptionally non-convex target features. We proposed a framework model for animal image detection and achieved a good accuracy of 97%. The particular proposed methodology is pretty guaranteeing to various detection problems.

References

1. Dewan, A.M., Islam, M.M., Kumamoto, T., et al.: Water Resource Manag. **21**, 1601 (2007). https://doi.org/10.1007/s11269-006-9116-1
2. Dabarera, R., Rodrigo, R.: Vision based elephant recognition for management and conservation. In: Proceedings of the 2010 5th International Conference on Information and Automation for Sustainability, ICIAfS 2010 (2010). https://doi.org/10.1109/ICIAFS.2010.5715653
3. Goswami, A.V., et al.: Enhanced J-protein interaction and compromised protein stability of mtHsp70 variants lead to mitochondrial dysfunction in Parkinson's disease. Hum. Mol. Genet. **21**(15), 3317–3332 (2012)
4. Vinod A.D., Kantilal. P.R.: Identification of Animal using IRIS Recognition. Int. J. Adv. Technol. Eng. Sci. **3**(1) (2015)
5. Ardovini, R.: Ardovini R., 2008 *Bela africana* sp.n. dal West Africa, Senegal. Malacologia Mostra Mondiale **XX**, 12–13 (2008)
6. Chen, G., Han, T.X., He, Z., Kays, R., Forrester, T.: Deep convolutional neural network based species recognition for wild animal monitoring. In: IEEE International Conference on Image Processing (ICIP), pp. 858–862 (2014). https://doi.org/10.1109/icip.2014.7025172
7. Figueroa, K., Camarena-Ibarrola, A., García, J., Villela, H.T.: Fast automatic detection of wildlife in images from trap cameras. In: Iberoamerican Congress on Pattern Recognition, pp. 940–947. Springer (2014). https://doi.org/10.1007/978-3-319-12568-8_114
8. Norouzzadeh, M.S., Nguyen, A., Kosmala, M., Swanson, A., Palmer, M.S., Packer, C., Clune, J.: Automatically identifying, counting, and describing wild animals in camera-trap images with deep learning, pp. 1–17. Available from: https://www.semanticscholar.org/paper/Automatically-identifying%2C-counting%2C-and-describing-Norouzzadeh-Nguyen/2bff54fb3f6aacb0b89323da8db49491c5e1e4a5
9. Gomez, A., Salazar, A., Vargas, F.: Towards automatic wild animal monitoring: identification of animal species in camera-trap images using very deep convolutional neural networks. Ecol. Inform. **41**, 24–32 (2017). https://doi.org/10.1016/j.ecoinf.2017.07.004
10. Masci, J., Meier, U., Cireşan, D., Schmidhuber, J.: Stacked convolutional auto-encoders for hierarchical feature extraction. In: Honkela, T., Duch, W., Girolami, M., Kaski, S. (eds.) Artificial Neural Networks and Machine Learning – ICANN 2011. ICANN 2011. Lecture Notes in Computer Science, vol. 6791. Springer, Berlin, Heidelberg (2011)
11. Erhan, D., Bengio, Y., Courville, A., Manzagol, P.A., Vincent, P.: Why does unsupervised pre-training help deep learning? J. Mach. Learn. Res. **11**, 625–660 (2010)
12. Cruz-Roa, A.A., Arevalo Ovalle, J.E., Madabhushi, A., González Osorio, F.A.: A deep learning architecture for image representation, visual interpretability and automated basal-cell carcinoma cancer detection. In: Mori, K., Sakuma, I., Sato, Y., Barillot, C., Navab, N. (eds.) Medical Image Computing and Computer-Assisted Intervention – MICCAI 2013. MICCAI 2013. Lecture Notes in Computer Science, vol. 8150. Springer, Berlin, Heidelberg (2013)
13. Lapuschkin, S., Binder, A., Montavon, G., Müller, K.-R., Samek, W.: Analyzing Classifiers: Fisher Vectors and Deep Neural Networks, pp. 2912–2920 (2016). https://doi.org/10.1109/cvpr.2016.318
14. Guo, X., Liu, X., Zhu, E., Yin, J.: Deep Clustering with Convolutional Autoencoders. ICONIP (2017)

15. Wang, P., Liu, L., Shen, C., Huang, Z., van den Hengel, A., Tao Shen, H.: Multi-attention network for one shot learning. In: 2017 IEEE Conference on Computer Vision and Pattern Recognition (CVPR), pp. 6212–6220 (2017)

Personality Identification from Social Media Using Deep Learning: A Review

S. Bhavya, Anitha S. Pillai and Giuliana Guazzaroni

Abstract **Social media** helps in sharing of ideas and information among people scattered around the world and thus helps in creating communities, groups, and virtual networks. Identification of personality is significant in many types of applications such as in detecting the mental state or character of a person, predicting job satisfaction, professional and personal relationship success, in recommendation systems. Personality is also an important factor to determine individual variation in thoughts, feelings, and conduct systems. According to the survey of Global social media research in 2018, approximately 3.196 billion social media users are in worldwide. The numbers are estimated to grow rapidly further with the use of mobile smart devices and advancement in technology. Support vector machine (SVM), Naive Bayes (NB), Multilayer perceptron neural network, and convolutional neural network (CNN) are some of the machine learning techniques used for personality identification in the literature review. This paper presents various studies conducted in identifying the personality of social media users with the help of machine learning approaches and the recent studies that targeted to predict the personality of online social media (OSM) users are reviewed.

Keywords Deep learning · Five-factor model · Conscientiousness · Openness · Extraversion · Neuroticism · Personality recognition · Online social media

S. Bhavya (✉) · A. S. Pillai
School of Computing Sciences, Hindustan Institute of Technology and Science,
Chennai, Tamil Nadu, India
e-mail: bhavyaclt@gmail.com

A. S. Pillai
e-mail: anithasp@hindustanuniv.ac.in

G. Guazzaroni
Marche Polytechnic University, Ancona, Italy
e-mail: juli.guazzaroni@gmail.com

© Springer Nature Singapore Pte Ltd. 2020
K. N. Das et al. (eds.), *Soft Computing for Problem Solving*,
Advances in Intelligent Systems and Computing 1057,
https://doi.org/10.1007/978-981-15-0184-5_45

1 Background

Studies in psychology have recommended that the conduct and desires of individuals can be described by underlying psychological constructs: The most widely accepted model of personality, Big Five or Five-Factor Model, Costa and McCrae [1] embraces five traits: openness, conscientiousness, extraversion, agreeableness, and neuroticism [1].

- **Openness (O)**: This measures the imagination, creativity, tolerance, and appreciation power of a person in predicting if he/she adapts to change and appreciate new and unfamiliar ideas.
- **Conscientiousness (C)**: This helps in determining if a person is organized, reliable, and consistent. People who have high value for this trait tend to plan their day to do activities to achieve their aims and objectives.
- **Extraversion (E)**: The ability to socialize, participate in social gatherings mingle with people are identified using this. If a person scores high on this it indicates that he/she can easily get along with people and they enjoy interacting with others, participating in social meetings, and are energetic.
- **Agreeableness (A)**: Agreeableness factor helps to maintain positive social relations and these people will be friendly, affectionate, and cooperative. Generally, people belonging to this category trust others and are open in accepting and adapting the ideas and needs of the trusted ones.
- **Neuroticism (N)**: A high score on this is an indication of mood swings and experience the negative feelings. They will be sad, anxious, tensed, depressed, and very sensitive.

1.1 Deep Learning

Deep learning and machine learning are the two subsets of artificial intelligence that has gained popularity. Deep learning is an artificial neural network that can imitate the functions of a human brain in data processing and making decisions [2]. The term 'deep' in deep learning refers to the number of hidden layers it has. As we go deeper in the layers, the accuracy improves and hence can solve complex problems. A deep learning network can have as many numbers of hidden layers and it can learn the features on its own, this is the advantage of the deep learning over machine learning. The architecture of deep neural network follows an input layer, hidden layers, and an output layer.

1.2 Why Deep Learning: The Motivation Behind

From the literature review, it is identified that the deep learning algorithms are started using in the research work of personality prediction of online social media users from the year of 2015 onwards. As the volume of data is enormous manual methods are impossible and so an automated approach is planned. Machine learning/deep learning algorithms are used by authors, as they are capable of learning 'n' high-level features from the data. This in turn eliminates the domain expertise and feature extraction.

1.3 Applications of Deep Learning

Many deep learning applications are already made an impact in our life. In the next few years, the deep learning software package will become standard parts of every program development toolkit [3]. Following are the some of the applications of deep learning:

Image recognition: Refers to identify and find the objects and persons in an image and also understand its ideas and context. Image recognition is being used in sectors like gaming, social media, etc.

Natural language processing: Deals with developing the algorithms which is diagnosed and able to generate human language automatically. Amazon's Alexa and OK Google are examples of NLP-based systems.

Healthcare: Deep learning plays a big role in the healthcare industry. Identifying and detecting the various diseases like cancer, Alzheimer's, etc., in early stage will make a high impact on human life. Image processing is used for the diagnosis.

Chatbots: Chatbots are conversational UI and the implementation of chatbots in different sectors is the current trend in the artificial intelligence field. If you are clicking on the support link on your favorite shopping Web site/restaurant Web site, there will be a Chatbot in action by asking 'how can I help you?' It is an automated program that can read your text and give reply accordingly or a simple bot can redirect you to a live representative.

Social media analysis: Deep learning algorithms are used to analyze the social network for users' personality identification which can predict their mental illness and stress factor, etc.

Autonomous driving: Teaching a computer how to drive! is another surprising approach using artificial intelligence. Here, digital sensor mechanisms are used instead of the sense of humans.

Predicting earthquakes: Scientists are started using deep learning algorithms to predict the earthquakes. Timing factor is very important in earthquake calculation and according to studies deep learning technique improved this.

Translation: Translation of one language to another. With the use of deep neural networks, translation will be more accurate and faster.

2   Related Work

Recognizing an individual's personality efficiently and with high accuracy is a worthwhile goal. The traditional approaches to predict the personality are interviews conducted by psychologist and self-report inventories or questionnaires. These are expensive and hardly possible in social media platforms since there is a necessity of automated system to predict the personality of online social media users [4].

A study by Golbeck et al. [5] used the machine learning principle for the personality prediction of Facebook users. The study was conducted to answer a question as to whether the users' account data in online social network can predict their personality. To collect the data, they have created a Facebook application interface which consisted of two functions. One was to give a set of questions to the users based on the five-personality factors; another function was to collect the all publicly available profile information from users. So, in total they have gathered 161 sample data for the analysis. The features they have collected are: 1. Structural features (to get the information of users' network, they have gathered users' friends list to determine the density of the network) 2. Personal information (username, date of birth, status, country of residence, relationship, gender, religion, etc.) 3. Activities and preferences, and 4. Language features. Authors [5] performed a linguistic analysis on the sample data and as a result, they did identify many correlations in the data. For the personality trait prediction, they have trained and designed two ML (machine learning) algorithms: M5sup' rules and Gaussian process by using the user profile as a feature dataset. As per authors, on a normalized 0-1 scale, the mean absolute error for each personality factor was roughly 11% [5]. They have performed this study only on two features: number of friends and network density. The authors suggested that by taking more features into account may reveal so many personality factors of users' profile.

Another study was on the Twitter data by Golbeck et al. The approach [6] was applied by gathering 2000 public Twitter posts. This was combined, measured, and processed through a text analysis tool in order to get a feature dataset. For the personality prediction, by using these feature sets, they have performed a regression analysis using Gaussian process and ZeroR machine learning algorithms. As per authors, on a normalized 0-1 scale, the mean absolute error for each personality factor was between 11 and 18% [6]. This analysis predicted the personality on each five factors to within between 11 and 18% of the actual values [6].

Qiu et al. [7] study was focused on identifying the relationship between the usage of microblog and the user's personality. They examined the relationship between the personality expressions in writing samples from users' Twitter profile. They have identified the valid linguistic features associated with personality features from users Twitter posts. One hundred and forty-two participants' Twitter data sample (gender, age, hometown, and ethnicity) was collected over a month period of time. They have used employees to rate and the linguistic analysis software [7] to identify the personality and linguistic features from tweets. Authors observed significant correlation between self-report and aggregated observer rating on agreeableness and neuroticism [7]. So, they have accurately predicted the degree of those two personality factors (agreeableness and neuroticism) from the tweets.

Kosinski et al. [8] did a study on psychological variables on Facebook. Here, the study was conducted on a dataset of over 58,466 volunteers. The study was based on the Facebook likes. The study included the Facebook profile data, the likes list, detailed personal information, and the scores of various psychological measurements [8]. Numeric variables such as age or intelligence were predicted using a linear regression model, whereas variables such as gender or sexual orientation were predicted using logistic regression [8]. For the personality feature, openness, prediction accuracy was close to the test-retest reliability for a standard personality test [8].

A study by Farnadi et al. [9] predicted personality traits using Facebook status update and machine learning techniques. The sample data collected from 250 Facebook users and 9917 status updates [9]. Personality metric questionnaires were given to each user and as depending on the answers, and they were categorized to one or more personality traits. The study was to predict these personality factors of each user. The analysis was based on user's status post update (used LIWC—Linguistic Inquiry and Word Count tool to extract the feature), time-related features and social network properties. As there is more than one personality trait present in each user, they have introduced a binary classifier to classify each trait.

Three-machine learning algorithms used were support vector machine with linear kernel-nearest neighbors with $k = 1$, and Naive Bayes [9]. The most interesting findings were [9]:

– The study could perform better than the majority class baseline algorithms even with a small sample training dataset
– A single kind of feature cannot give the best results for all personality traits
– ML-based methods for personality recognition generalize across domains.

By taking the advantage of generalization property of ML-based methods for personality recognition across the domains, they could combine the dataset from different social network platform to train and design a more accurate model. With the help of this model, work on the social media site which has no training data available was possible [9]. According to the authors, the classification results which used Linguistic Inquiry and Word Count (LIWC) method could be improved by concentrating more on feature selection method.

Authors, Alam et al. [10] suggested a strong correspondence between users' personality and their behavior on social media. Alam et al. [10] analyzed automatic identification of Big-5 personality traits on the social network Facebook using customers' status text. For the study, they considered the following classification methods: 1. Sequential minimal optimization for support vector machine (SMO); 2. Bayesian logistic regression (BLR); 3. Multinominal Naïve Bayes (MNB). They measured the performance of the systems using macro-averaged precision, recall and F1; weighted average accuracy (WA) and un-weighted average accuracy (UA). Moreover, their study pointed out that MNB performs better than BLR and SMO for personality traits identification using social media data. The authors followed a bag-of-words approach and different sets of tokens (e.g., internet-slang, smiles, emoticons, etc.). In fact, the tokens, presented in the corpus, carry specific information regarding personality trait recognition. They conducted many experiments to examine the performance of different classification methods for their automatic recognition of main personality traits. Their main contribution consists in the observation that MNB sparse modeling performs better than SMO and BLR [10] ones.

Ana Carolina et al. described a method, PERsonality prediction in SOcial Media datA-PERSOMA [11]. The study was on to a group of tweets, based on the five-factor personality model [11]. According to Authors, this was a unique model, since it identified the personality factor included in the group of tweets rather than of single one. In this study, individual's Twitter profile has not been taken for the analysis. One advantage of this study stated that since the study was on to group of tweets, they could able to work effectively on a very large dataset. The system followed three phases: preprocessing, transformation, and classification [11].

In preprocessing module, the metadata is extracted from a large set of group of tweets. The outcome of the preprocessing module was a multi-label problem, since each tweet may contain up to five-personality traits. In the transformation module, the multi-label problem is transformed into a set of five binary classification problems [11]. The third module is classification; here, the semi-supervised learning algorithms are used to classify the text. The algorithms used in the study are: Naive Bayes (NB), support vector machine (SVM) and multilayer perceptron (MLP) [11]. The performance of the method was measured by standard multi-label evaluation measures. The overall accuracy calculated was 83% [11]. Authors suggested that performance of this method could be improved with the use of automatic data clustering instead of a manual clustering task.

A study by Nie et al. investigated personality traits prediction by using the data from users' microblog [12]. The invited volunteers completed the five-factor personality online questionnaires. The experiment was conducted by using the microblog-users' dataset [12]. Features are extracted from the microblog-users' training data. The extracted features were: users' personal information, users' friends and followers' details, users' social planning and users' social habit such as at time they are posting a status update. Stepwise feature selection method was used to test all the five-personality traits of users'. For the personality prediction, they have implemented a local linear kernel regression approach [12]. To deal with the unlabeled data, a local linear semi-supervised regression models [12] were developed. These two models are used to solve the problem. Mean absolute error (MAE) was used to measure the performance of the method and both the models reduced the MAE by 2.5 to 7% [12]. Also, depicted relative absolute error (RAE) to measure the model and the accuracy obtained was 75%. Authors suggested that the performance could be improved if tested on a large dataset.

Authors, Peng et al. conducted a study to identify personality traits from Chinese text, but they focused on extroversion factor [13]. The sample data was gathered with text post and personality feature scores of 222 Facebook users who use their main written language as Chinese [13]. After the collection of data, a tokenizer has been used for text segmentation task. They have used a text segmentation tool for Chinese language for the segmentation task. To predict the personality factors, authors used support vector machine (SVM) algorithm. The result of their study revealed that extroverts tend to share their feelings and life with the world than introverts [13]. Accuracy obtained for classifying extraversion was 73.5% [13].

Plank and Hovy [14] studied personality features on the social media Twitter. They explored the use of social network as a source for a large-scale and an open-vocabulary personality identification. They pointed out which attributes are significant of personality traits and presented a corpus of 1.2 M tweets with personality and gender annotation. Natural language processing (NLP) was used to classify extra-linguistic features based on textual input to extract personality types. According to these authors [14] predicting personality is interesting both for commercial use and health care. In fact, there is a link between personality types, social network behavior, and psychological disorders. They claimed that specific personality traits can predict mental illness. They used vast amounts of personalized data, produced daily on social networks, to collect useful information. They combined this source with self-assessed Myers-Briggs Type Indicators (MBTI) to classify costumers as intro-vert–extrovert, intuitive–sensing, thinking–feeling, and judging–perceiving. Their results suggested that certain personality distinctions, especially introvert–extrovert and thinking–feeling, can be accurately identified early from social network data (e.g., Twitter) and the accuracy obtained was 72.5% for introvert–extrovert and 61.2% for thinking–feeling. Learning the other two dimensions, namely sensing–intuitive and judging–perceiving turns out to be difficult. Moreover, according to the research [14] the meta-information regarding gender, number of followers, statuses, or list of membership, add additional significant information. Large-scale, open-vocabulary analysis of customer attributes improves accuracy in the classification and important insights into personality profiles [14].

The study conducted by Poria et al. on sarcastic analysis based on CNN (convolutional neural network) [15] made use of 9497 tweets for sentiment analysis; 5205 sentences labeled by emotions for emotion analysis; 2400 essays labeled by one of the five-personality traits each for personality analysis. CNN and SVM (support vector machine) were used for this study. They obtained an accuracy of 97.71% for sentiment analysis, 94.00% for emotion analysis, and 93.30% for personality analysis [15].

Personality identification based on deep learning-based document modeling [16] was performed by Majumder et al. They have used the dataset named 'essays' which consists of 2468 anonymous essays tagged with authors' personality traits [16]. Their method consists of five steps: preprocessing, document-level feature extraction, filtering, word-level feature extraction, and classification for which they have used a CNN [16]. They have applied the document modeling technique. According to authors, their output outperformed current state of the art for all the five traits [16]. Authors claimed that their network got a training accuracy of 98% after 50 epochs and to get better accuracy, a large volume of data to be used [16].

A work on personality prediction from Facebook posts by da Silva and Paraboni [17]; their study was to recognize personality features in Portuguese language. At first, the authors made a collection of texts labeled with the personality features.

Thereafter, they used the trained dataset, for some supervised learning models of personality recognition. They have performed experiments by using different text representations and machine learning algorithms and so their discussion limited to six models. These six models and a majority class baseline were functioned to the prediction of five-personality traits. The models depended on text-based techniques, for example, word embedding [17]. Results suggested that word embedding models performed well than any other lexical resources [17] and the study analyzed that there was no optimal algorithm to predict all the personality factors of social network users. As per authors, future work would be to improve the models by using word embedding [17] with deep neural networks (DNNs).

Another work was on personality identification of social network profiles using deep learning principle by Xue et al. developed a combination of neural networks [18]. They have collected one set of dataset from a project and the other they have developed a Facebook application, which allows users to answer the online psychology test and send their scores and profile data for the study. The study was based on the Facebook users' profiles and the available scores of personality test. From the dataset, they have created an input array and it was fed to a deep neural network [18] to get the feature set. Results suggested that an advantage of the study was they have evaluated deep semantic features [18]. According to the study, one method of performance improvement of the personality prediction approach is that by applying these semantic features as the input of traditional regression algorithms [18]. They have compared the accuracy result with the mean absolute error (MAE). The lowest average MAEs (O-0.358; C-0.425, E-0.477, A-0.386, and N-0.487) were obtained for the proposed model when compared to other approaches [18]. As per authors, prediction accuracy can be improved by applying some special designed regression algorithms by utilizing their deep semantic features.

3 Conclusion

Personality identification/prediction is the most interesting as well as challenging research field. Authors have worked and still working on it by accessing the publicly available information on online social platforms. It can automatically differentiate an individual's personality features by their posts, shares, and uploads on social networks. These findings can lead in detecting depression and mental illness on social media users. Identifying individual's personality efficiently and with high accuracy is a worthwhile goal. Traditional approach to predict the personality are interviews conducted by a psychologist and self-report evaluation studies or questionnaires. Deep learning algorithms are found to be the best method in personality prediction. Though the authors used different ML approaches to computational personality identification, all studies focused on identifying the personality traits of individuals based on the dominant prototype in psychology which is The Five-Factor Model.

Table 1 depicts the summary of the related work mentioned in the paper with the result observed.

Table 1 Overview of related work in terms of sample data, methods used, and the results obtained from each

Related work	Sample data	Methods used	Results
Golbeck et al. (2011)—Predicting Personality with Social Media	167 Facebook users'	M5sup' rules and Gaussian process	As per authors, on a normalized 0-1 scale, the mean absolute error for each personality factor was roughly 11% [5]
Golbeck et al. (2011)—Predicting Personality from Twitter	2000 public Twitter posts	A regression analysis using Gaussian process and ZeroR ML algorithms	As per authors, on a normalized 0-1 scale, the mean absolute error for each personality factor was between 11 and 18% [6]
Golnoosh Farnadi et al. (2013)—Recognising Personality Traits Using Facebook Status Updates	Data of 250 Facebook users and 9917 status updates	Support vector machine (SVM) with linear kernel-nearest neighbor with $k = 1$ (KNN), and Naive Bayes (NB)	Worked well with the majority class baseline algorithms even with a small sample training data; A single kind of feature cannot give the best results for all personality traits ML-based methods for personality recognition generalizes across domains
Firoj Alam et al. (2013)—Personality Traits Recognition on Social Network—Facebook	Used tokens as features from the Facebook status	Sequential minimal optimization (SMO) for support vector machine), Bayesian logistic regression (BLR), multinominal Naïve Bayes (MNB).	The main contribution consists in the observation that MNB sparse modeling performs better than SMO and BLR [10] ones. The best accuracy obtained was 61.79% for MNB model
Ana Carolina E.S. et al. (2014)—A multi-label, semi-supervised classification approach applied to personality prediction in SM	41 Group of tweets—total 18,435 tweets (not on individual profile)	Naive Bayes (NB), Support vector machine (SVM) and multilayer perceptron (MLP)	The overall accuracy calculated was 83%. openness followed by conscientiousness were the most difficult traits to predict

(continued)

Table 1 (continued)

Related work	Sample data	Methods used	Results
Dong Nie et al. (2014)—Predicting Personality on Social Media with Semi-Supervised Learning	1792 microblog-users' personal information, users' friends and followers' details, users' social planning, and users' social habit	Local linear kernel regression (LLKR) approach and a local linear semi-supervised regression (LLSSR) models	Both the models reduced the mean absolute error (MAE) by 2.5 to 7%. The accuracy calculated was 75% for LLSSR
Kuei-Hsiang Peng et al. (2015)—Predicting Personality Traits of Chinese Users Based on Facebook Wall Posts	222 Facebook users' post and number of friends as extra feature	Support vector machine (SVM) as classification algorithm	The best accuracy obtained for extraversion was 73.5%
Barbara Plank and Dirk Hovy (2015)—'Personality Traits on Twitter – or – How to Get 1500 Personality Test in a Week'	1.2 M English tweets—1500 users	NLP and logistic regression classifier	They claimed that specific personality traits can predict mental illness. Their results suggested that certain personality distinctions, especially introvert–extrovert and thinking–feeling, can be accurately identified early from social network data (e.g., Twitter) and the accuracy obtained was 72.5% for introvert–Extrovert and 61.2% for thinking–feeling
Soujanya Poria et al. (2016)—A Deeper Look in to Sarcastic Tweets Using Deep Convolutional Neural Networks	9497 tweets for sentiment analysis; 5205 sentences labeled by emotions for emotion analysis; 2400 essays labeled by each five-personality traits each for personality analysis.	Convolutional neural network (CNN) and support vector machine (SVM)	They obtained an accuracy of 97.71% for sentiment analysis, 94.00% for emotion analysis and 93.30% for personality analysis

(continued)

Table 1 (continued)

Related work	Sample data	Methods used	Results
Navonil Majumder et al. (2017)—'Deep Learning-Based Document Modeling for Personality Detection from Text'	2468 anonymous essays tagged with authors' personality traits	Multiple layer perceptron (MLP), support vector machine (SVM), and CNN	Claimed that results outperformed current state of the art for all the five traits. Authors mentioned that their network received a training accuracy of 98% after 50 epochs
Barbara Barbosa Claudino da Silva and Ivandré Paraboni (2018)—'Personality Recognition from Facebook Text'	1031 participant's Facebook text post, contains 2.2 million words total (used Portuguese language)	Random forest classification with different text representations and models such as bag-of-words (BoW), majority baseline, psycholinguistics, word2vec-cbow-600 word2vec-skip-600 doc2vec LSTM-600	Their findings/observation: There is no optimal model for all the personality traits, authors observed that the model word embedding with the use of deep neural network could be outperform the current model
Di Xue et al. (2018)—Deep learning-based personality recognition from text posts of online social network	115,864 Facebook users, 11,494,862 text posts	Convolutional neural network (CNN), recurrent convolutional neural network (RNN), regression algorithms	The lowest average MAEs (O-0.358, C-0.425, E-0.477, A-0.386, and N-0.487) were obtained for the proposed model

The table summarizes that majority of the authors have chosen the two popular social media sites such as Facebook and Twitter for their study to predict users' personality. Machine/deep learning algorithms such as support vector machine (SVM), convolutional neural network (CNN), recurrent convolutional neural network (RNN), Naïve Bayes (NB), Gaussian, and ZeroR were used widely for the studies. It is also observed that the deep learning algorithms: CNN, MLP, and SVM gave the better accuracy on a large volume of dataset. It also can observe that the deep learning method started implementing in the research work on predicting personality of social media users' in 2015 onwards.

References

1. Costa, Jr., P.T., McCrae, R.R.: The SAGE Handbook of Personality Theory and Assessment, vol. 2. The Revised NEO Personality Inventory (NEO-PI-R) (2008)
2. https://www.investopedia.com/terms/d/deep-learning.asp/https://deeplearning4j.org

3. https://medium.com
4. Xu, D., Hong, Z., Guo, S., Gao, L., Wu, L., Zheng, J., Zhao, N.: Personality Recognition on Social Media with Label Distribution Learning. IEEE (2017). https://doi.org/10.1109/access. 2017.2719018
5. Golbeck, J., Robles, C., Turner, K.: Predicting personality with social media. In: CHI'11 Extended Abstracts on Human Factors in Computing Systems, Vancouver, 7–12, May 2011, pp. 253–262. ACM (2011)
6. Golbeck, J., Robles, C., Edmondson, M., Turner, K.: Predicting personality from Twitter. In: 2011 IEEE Third International Conference on Privacy, Security, Risk and Trust and 2011 IEEE Third International Conference on Social Computing, Boston, 9–11 Oct 2011, pp 149 156 IEEE (2011)
7. Qiu, L., Lin, H., Ramsay, J., Yang, F.: You are what you tweet: personality expression and perception on twitter. J. Res. Pers. (2012)
8. Kosinski, M., Stillwell, D., Graepel, T.: Private traits and attributes are predictable from digital records of human behavior. Proc. Natl. Acad. Sci. USA (2013)
9. Farnadi, G., Zoghbi, S., Moens, M.-F., De Cock, M.: Recognising Personality Traits Using Facebook Status Updates. Association for the Advancement of Artificial Intelligence (2013). www.aaai.org
10. Alam, F., Stepanov, E.A., Riccardi, G.: Personality Traits Recognition on Social Network – Facebook. AAAI Technical Report WS-13-01. Computational Personality Recognition (2013). www.aaai.org
11. Lima, A.C.E.S., de Castro, L.N.: A multi-label, semi-supervised classification approach applied to personality prediction in social media. Neural Netw. (2014)
12. Nie, D., Guan, Z., Hao, B., Bai, S., Zhu, T.: Predicting personality on social media with semi-supervised learning. In: IEEE/WIC/ACM International Joint Conferences on Web Intelligence (WI) and Intelligent Agent Technologies (IAT) (2014)
13. Peng, K.-H., Liou, L.-H., Chang, C.-S., Lee, D.-S.: Predicting Personality Traits of Chinese Users Based on Facebook Wall Posts (2015). ieee.org
14. Plank, B., Hovy, D.: Personality traits on Twitter – or – how to get 1500 personality test in a week. In: Proceedings of the 6th Workshop on Computational Approaches to Subjectivity, Sentiment and Social Media Analysis, pp. 92–98 (2015)
15. Poria, S., Cambria, E., Hazarika, D., Vij, P.: A Deeper Look in to Sarcastic Tweets Using Deep Convolutional Neural Networks (2016). http://creativecommons.org/licenses/by/4.0
16. Majumder, N., Poria, S., Gelbukh, A., Cambria, E.: Deep Learning-Based Document Modeling for Personality Detection from Text (2017)
17. da Silva, B.B.C., Paraboni, I.: Personality Recognition from Facebook Text. Springer (2018)
18. Xue, D., Wu, L., Hong, Z., Guo, S., Gao, L., Wu, Z., Zhong, X., Sun, J.: Deep Learning-Based Personality Recognition from Text Posts of Online Social Networks. Springer (2018)

An Improved Gaussian Mixture Model Based on Prior Probability Factor for MR Brain Image Segmentation

J. B. Ashly⑩, S. N. Kumar⑩, A. Lenin Fred⑩, H. Ajay Kumar⑩ and V. Suresh⑩

Abstract Computer-aided algorithms play an inevitable role in medical field for disease diagnosis and preplanning for surgery. The improved Gaussian mixture model based on prior probability factor was proposed here for the segmentation of white matter (WM), gray matter (GM), and cerebrospinal fluid (CSF) on MR brain data sets. The performance of the classical Gaussian mixture model was enhanced by the incorporation of prior probability smoothing factor and the results outperform the classical FCM and K-means algorithm. Experimental analysis was carried out by performance metrics like Jaccard index, probabilistic rand index, and percentage of classification ratio. The algorithms are developed in Matlab2015a and validated on BrainWeb public database images.

Keywords Segmentation · White matter · Gray matter · Gaussian mixture model · FCM · K-means

1 Introduction

Image segmentation role is vital in computer vision, medical field, and remote sensing for the analysis of data. The different types of segmentation algorithms with their characteristics and parameter tuning for medical images are described in

J. B. Ashly (✉) · S. N. Kumar · H. Ajay Kumar · V. Suresh
School of Electronics and Communication Engineering, Mar Ephraem College
of Engineering and Technology, Kanyakumari, India
e-mail: ashlyvilayil95@gmail.com

S. N. Kumar
e-mail: appu123kumar@gmail.com

H. Ajay Kumar
e-mail: ajayhakkumar@gmail.com

A. Lenin Fred
School of Computer Science and Engineering, Mar Ephraem College
of Engineering and Technology, Kanyakumari, India
e-mail: leninfred.a@gmail.com

© Springer Nature Singapore Pte Ltd. 2020
K. N. Das et al. (eds.), *Soft Computing for Problem Solving*,
Advances in Intelligent Systems and Computing 1057,
https://doi.org/10.1007/978-981-15-0184-5_46

535

[1, 2]. The segmentation of white matter and gray matter is really a challenging task in MR images due to the presence of Rician noise and intensity inhomogeneity effect. In the improved Gaussian mixture model with bias field correction was proposed for segmentation of WM and GM in MR images. The legendary polynomial was employed for bias field correction and the fuzzy c-means clustering algorithm was incorporated in the expectation-maximization technique to estimate the parameter of Gaussian mixture model [3]. In [4], adaptive distance-based FCM (ADFCM) and adaptive distance expectation-maximization algorithms (ADEM) were proposed for the segmentation of GM, WM, and CSF. The Euclidean distance in the conventional FCM was replaced by a dynamic and weighted distance and Mahalanobis distance was employed in the pixel classification phase of EM algorithm. Ariyo et al. proposed a hybrid segmentation algorithm comprising of Gaussian mixture model and spatial fuzzy c-means algorithm for the segmentation of MR brain images; better efficiency was observed when compared with traditional FCM, GMM, and spatial FCM algorithms. Prior to segmentation preprocessing was performed by wiener filter and Otsu's thresholding algorithm for the removal of additive noise [5].

Cui et al. proposed a fuzzy algorithm based on global energy and local energy with an adaptive weight function that considers the contextual information [6]. The accuracy of the segmentation was improved and addresses the intensity in homogeneity problem. The deep learning neural network along with Gaussian mixture model efficiently performs 3D brain segmentation for the extraction of gray matter, white matter, and cerebrospinal fluid [7]. A hybrid algorithm comprising of fuzzy clustering and Markov random field was utilized for the segmentation of MR brain images [8]. Markov random field exploits the contextual information and the results outperform fuzzy c-means (FCM) algorithm, fast generalized fuzzy c-means (FGFCM) algorithm, fuzzy local information c-means (FLICM) algorithm [9]. Rough probabilistic clustering integrates the merits of rough sets and stomped normal (SN) distribution; hidden Markov model, when coupled with rough probabilistic modeling efficiently segments the HEP-2 cell and brain MR images.

A detailed study has been performed on the different supervised, unsupervised, and semi supervised algorithms for the segmentation of brain MR images [10]. The unsupervised technique does not rely on any prior information and accuracy is low when compared with supervised methods. Supervised algorithms need expert guidance and hence consume more time. An improved expectation-maximization algorithm with the incorporation of non-sampled contourlet transform was employed for the fully automatic segmentation of MR brain images [11]. The fuzzy local Gaussian mixture model algorithm efficiently performs the segmentation of MR brain images and handles the noisy images, low contrast, and bias field effect [12]. The constrained Gaussian mixture model improves the segmentation accuracy and it does not require prior anatomical atlas [13]. A survey on Gaussian mixture model for MR brain image segmentation gives the comparative analysis of merits, demerits, and parameter tuning [14]. A hybrid algorithm comprising of fuzzy c-means and Gaussian mixture model was proposed for the segmentation of WM/GM for improving the accuracy. Efficient results were produced, when compared with the traditional FCM, GMM, and spatial FCM algorithms [15]. The GMM with reversible jump Markov Chain

Monte Carlo (RJMCMC) hybrid algorithm was proposed for WM/GM segmentation, RJMCMC decides the number of classes. The Gibbs function was used prior to segmentation for improving the noise immunity. Better results were produced, when compared with the spatially variant finite mixture model (SVFMM) [16]. The gray matter segmentation in spinal cord MR image is a challenging task and a detailed study on various automatic and semi-automatic algorithms were analyzed [17]. The WM/GM segmentation in 4D CT images was done by atlas-based model followed by the geodesic active contour model [18]. The fuzzy clustering and Markov random field (MRF) was coupled together for the WM/GM segmentation by utilizing the contextual information and efficient results were produced, when compared with the FCM, FGFCM, and FLICM algorithms [8]. The Gaussian mixture model along with deep convolution neural network was employed for 3D brain segmentation; the neural network-based classifier gives efficient results when compared with classical GMM model especially for indeterminate pixels [19]. The Markov random field model along with rough probabilistic algorithm was found to generate accurate segmentation results for Hep-2 cell and MR brain images [20]. The maximum a posterior probability and Bayesian rule algorithm was applied based on the global and local fuzzy energy for the segmentation of MR brain images. The incorporation of special neighborhood pixels minimizes the effect of noise [21].

2 Proposed Methodology

The classical segmentation algorithms like K-means and fuzzy c-means are described initially. The proposed segmentation approach improved Gaussian mixture model is based on prior probability factor and the various stages are also described.

2.1 Data Acquisition

The algorithms are validated on MR brain data sets from BrainWeb public database. It comprises of normal and lesion images. In this work, four healthy subjects are used and the input images in MINC format was converted to DICOM format for analysis of segmentation algorithms.

2.2 K-means Clustering Algorithm

K-means is an unsupervised clustering algorithm and it is termed as hard clustering algorithm, as each pixel can belong to any one of the K-classes. The number of classes is user defined and Euclidean distance metrics is used to determine the distance between each pixel and centroid of each class. Each cluster has a centroid and is initialized randomly.

The steps in classical K-Means algorithm are summarized as follows

1. Initialize the number of clusters and centroids randomly
2. Determine the Euclidean distance between the cluster centroids and each pixel $I(x, y)$ as follows:

$$D = ||I(x, y) - C_k||^2 \tag{1}$$

3. Group pixels into a cluster with minimum Euclidean distance between them
4. Cluster centers are updated

$$C_k = \frac{\sum_{x \in k} \sum_{y \in k} I(x, y)}{n \in k} \tag{2}$$

5. The steps 3 and 4 are repeated, until convergence is reached; difference in Euclidean distance is less than the threshold value.

For convergence criteria, Bayesian information criterion (BIC) or minimum description length (MDL) can also be employed for the estimation of K. The classical K-means algorithm is sensitive to noise and an appropriate preprocessing algorithm is required prior to segmentation.

2.3 Fuzzy C-means Clustering Algorithm

The FCM is a soft clustering algorithm and is widely used in many applications. The pixels are grouped into clusters based on the fuzzy membership function value. The Euclidean distance is used for the grouping of pixels into a class. Minimum the Euclidean distance, more the membership value and has high preference in a class. The steps in classical FCM are summarized as follows:

1. Initialize randomly the clusters centers and fuzzy membership function value
2. Estimate fuzzy membership of each pixel with respect to cluster center

$$\mu_{ij} = \frac{1}{\sum_{k=1}^{C} \left(\frac{||y_i - C_j||}{||y_i - C_k||} \right)^{\frac{2}{f-1}}} \tag{3}$$

where f is fuzziness index and $f \in [1, \infty]$ and μ_{ij} is the membership of ith pixel with jth cluster.

3. Update cluster centers as follows:

$$C_j = \frac{\sum_{i=1}^{n} \mu_{ij} * y_i}{\sum_{i=1}^{n} \mu_{ij}} \tag{4}$$

Objective function minimization takes place until the convergence is reached.

$$J(u, v) = \sum_{i=1}^{n} \sum_{j=1}^{C} \mu_{ij} * ||y_i - C_j||^2 \tag{5}$$

2.4 Gaussian Mixture Model Algorithm

In the case of magnetic resonance images, the gray-scale pixel values follow the Gaussian distribution. The classical Gaussian mixture model is based on probability density function that determines distribution in each class.

Let $I = \{x_i, x_2, \ldots, x_n\}$ represents an image with n pixels. The image is splitted into C classes and each class is represented by a Gaussian distribution with mean and variance values.

The probability density distribution for each is represented as follows:

$$P(x_i/C_n) = \frac{1}{\sqrt{2\pi}\sigma} * \exp\left(\frac{-(x_i - \mu_n)^2}{2\sigma_n^2}\right) \tag{6}$$

The expression for Gaussian mixture model is expressed as follows:

$$P(C_n/y_i) = \frac{P(x_i/C_n) * P(C_n)}{P(x_i)} \tag{7}$$

$$P(x_i) = \sum_n P(x_i/C_n) * P(C_n) \tag{8}$$

2.5 Improved Gaussian Mixture Model Algorithm

Many improved segmentation models based on GMM are there, however, many of them not preserve spatial information does and are sensitive to noise. The influence of noise has to be considered in medical image processing, since it will be reflected in the subsequent operations like classification and compression. The MR brain images are considered in this research work, they are corrupted by Rician noise and bias field effect is also there. The prior probability factor incorporated in the GMM model improves the segmentation accuracy and preserves the image details. Let $y_i\ i = 1, 2, \ldots, N$ represents the pixels of an image with dimension D. The neighborhood of the ith pixel is represented by f_i. The labels are denoted by $\{\varnothing_1, \varnothing_2, \ldots, \varnothing_k\}$. For the segmentation of an image into C clusters, the conditional probability has to be determined. The first step is the random initialization of mean, standard deviation, and prior probability values. The cluster is depicted by Gaussian distribution; the probability of belonging to any class is represented as follows:

$$P(C_n/y_i) = \frac{P(y_i/C_n) * P(C_n)}{P(y_i)} \tag{9}$$

where $P(y_i/C_n)$ is likelihood probability and $P(C_n)$ is the prior probability factor.

$$P(y_i/C_n) = \frac{1}{\sqrt{2\pi}\sigma} * \exp\left(\frac{-(y_i - \mu_n)^2}{2\sigma_n^2}\right) \tag{10}$$

$$P(y_i) = \sum_n P(y_i/C_n) * P(C_n) \tag{11}$$

The prior probability factor is expressed as follows:

$$f_{ik} = \frac{\exp\left[\sum_{m \in \partial_i^{s_n^*}} (Z_{mn}^{(t)} + \pi_{mn}^{(t)}) \Big/ N_i^{s_n^*}\right]}{\sum_{h=1}^{N} \exp\left[\sum_{m \in \partial_i^{s_n^*}} (Z_{mh}^{(t)} + \pi_{mh}^{(t)}) \Big/ N_i^{s_h^*}\right]} \tag{12}$$

where $\partial_i^{s_n^*}$ represents the neighborhood of pixel i at direction s_n^* for cluster n that comprises of $N_i^{s_n^*}$ pixels.

$$s_n^* = \underset{s=1}{\arg\max} \sum_{m \in \partial_i^s} \mathrm{dist}(y_i, \mu_n^{(t)}) \tag{13}$$

$$P(C_n) = Z^{-1} \exp\left\{\frac{1}{T} \sum_{i=1}^{N} \sum_{k=1}^{K} f_{ik} \log \pi_{ik}\right\} \tag{14}$$

where distance $(y_i, \mu_n^{(t)})$ is the Euclidean distance between point i and cluster center μ_n at iteration step t and ∂_i^s is the neighborhood of pixel i at direction S for cluster n. Here, the value of $S = 4$ is considered four directions horizontal, vertical, and two-diagonal directions are taken into account. Figure 1 depicts the filter masks for four directions and is based on 3×3 neighborhood connectivity. The convolution operation is performed on the image with the prior probability factor and the four filters are adaptively applied with minimum difference between the intensity values in the current directions.

Finally, the mean and standard deviation values are updated as follows:

$$\mu_n = \frac{\sum_i P(C_n/y_i) * y_i}{\sum_i P(C_n/y_i)} \tag{15}$$

Fig. 1 Spatial filters in four directions corresponding to the prior probability factor

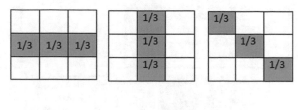

$$\sigma_n = \frac{\sum_i P(C_n/y_i) * (y_i - \mu_n)^2}{\sum P(C_n/y_i)} \tag{16}$$

$$P(C_n) = \frac{\sum_i P(C_n/y_i)}{k} \tag{17}$$

The values are updated until the convergence condition is achieved, and the change in mean value is less than the threshold value. The steps of the proposed improved GMM algorithm for image segmentation are as follows:

Step 1: Initialization

1. Initialize randomly the mean values μ_n and the standard deviation σ_n.
2. Determine the prior distributions f_{ik} based on the initialized mean values and standard deviation.
3. Initialize the number of clusters C and the neighborhood dimension, here $C = 3$ and neighborhood dimension is 3×3 (Fig. 2).

Step 2: Estimation of probability values

Determine the probability values as per the above equations and parameters are updated.

1. Update the mean values, μ_n
2. Updating the standard deviation values, σ_n.
3. Updating prior probabilities values f_{ik}.

Step 3: Termination

The stages of improved GMM model are executed until the convergence criterion is met; a threshold value is defined as convergence value for the successive change in mean and standard deviation. The convergence criterion can also be defined by maximum likelihood estimation algorithm.

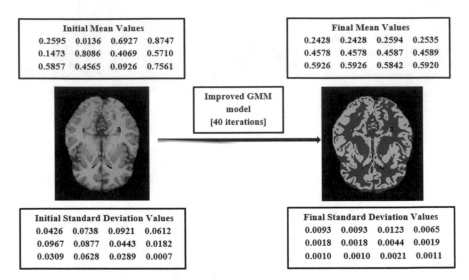

Fig. 2 Initial and final mean, standard deviation values of test image 1 in Fig. 3 and the segmentation result

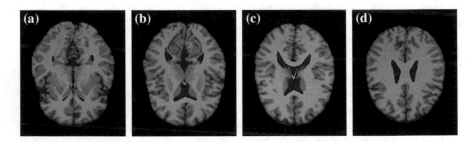

Fig. 3 Input MR brain images from BrainWeb data base

3   Results and Discussion

The algorithms are developed in Matlab2015a and tested on MR brain data sets from BrainWeb database. The algorithms are tested in system with specification of Intel Core i3 @3.30 GHz processor with 4 GB RAM (Figs. 3 and 4).

The improved GMM model generates efficient results for the extraction of ROI in MR brain images. The performance of the proposed algorithm was compared with the classical K-means and FCM algorithm. Superior results were produced by improved GMM model, when compared with the classical K-means and FCM algorithm. For performance validation, the following metrics are used. The parameter tuning is simple in improved GMM model, number of clusters is 3 for brain images; since there are only 3 ROI. The number of iterations is set to 40 and it can be increased up

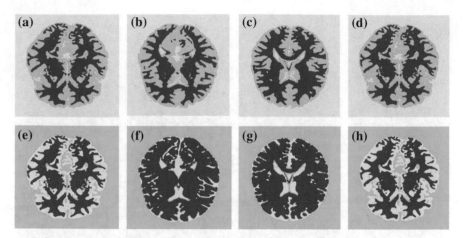

Fig. 4 First row depicts the FCM segmentation result and second row depicts the K-means segmentation result corresponding to the input images in Fig. 3

to 100; however, when the convergence limit is reached, there will not be any change in the resultant output.

The performance metrics are expressed as follows; the classification ratio (CR) is also termed as accuracy and represents the correctly classified pixels with respect to the ground truth image. Here, S represents the algorithm generated result and G represents the ground truth image.

$$CR = \frac{G \cap S}{G} \qquad (18)$$

Higher the classification rate, better the efficiency of algorithm. The improved GMM has higher classification rate, when compared with the classical FCM and K-means algorithm (Fig. 5).

The value of JC when higher than 0.9 is an indication of efficient segmentation algorithm [22].

$$JC = \frac{G \cap S}{G \cup S} \qquad (19)$$

The Probabilistic Rand Index (PRI) is expressed as follows:

$$PRI = \frac{1}{(N/2)} \sum_{\substack{i,\, j \\ i < j}} [C_{ij} P_{ij} + (1 - C_{ij})(1 - P_{ij})] \qquad (20)$$

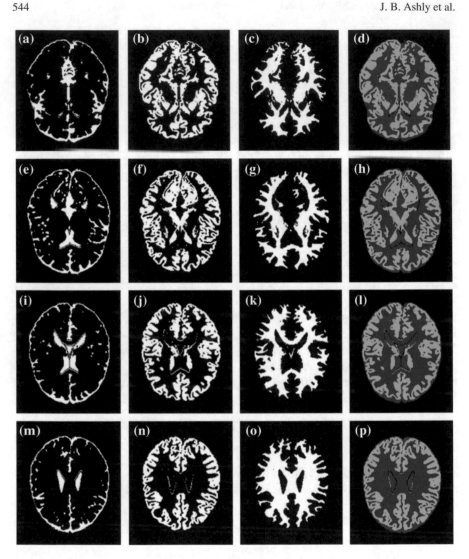

Fig. 5 CSF, WM, and GM segmentation results corresponding to the input images in Fig. 3

The P_{ij} represents the pixels of the ground truth image (G) and C_{ij} represents the pixels of the segmented image(S). The value of PRI when greater than 0.9 indicates the efficiency of good segmentation algorithm [23] (Tables 1, 2, and 3).

The clustering algorithm was found to be proficient, when compared with the traditional segmentation techniques like region growing, thresholding, and edge detection. The number of clusters is the only tunable parameter and in almost all application the cluster number lies in the range 3–6. The less user interaction and the performance metrics evaluation states that clustering algorithm is an excellent option for medical image segmentation.

Table 1 Classification ratio of segmentation algorithms

Dataset ID	Segmentation algorithms		
	GMM (%)	FCM (%)	K-means (%)
ID1	91.12	87.45	78.23
ID2	90.66	85.64	76.59
ID3	92.06	88.53	78.82
ID4	92.74	90.25	79.27

Table 2 Jaccard index of segmentation algorithms

Dataset ID	Segmentation algorithms		
	GMM	FCM	K-means
ID1	0.8354	0.7861	0.7378
ID2	0.8280	0.7594	0.6749
ID3	0.8510	0.8079	0.7648
ID4	0.8630	0.8167	0.7694

Table 3 Probabilistic rand index of segmentation algorithms

Dataset ID	Segmentation algorithms		
	GMM	FCM	K-means
ID1	0.9033	0.8681	0.8147
ID2	0.8981	0.7564	0.7189
ID3	0.9132	0.8378	0.7954
ID4	0.9200	0.8573	0.8167

4   Conclusion

This work proposes an improved GMM model for the segmentation of CSF, WM, and GM on MR brain data sets. The inclusion of the prior probability factor makes the algorithm insensitive noise. The validation of the improved GMM model was done by performance metrics like Jaccard coefficient, probabilistic rand index, and classification ratio. The results reveal that improved GMM model outperforms the classical FCM and K-means algorithm.

Acknowledgements The authors would like to acknowledge the support provided by DST under IDP scheme (No.: IDP/MED/03/2015).

References

1. Kumar, S.N., Muthukumar, S., Kumar, A., Varghese, S.: A voyage on medical image segmentation algorithms. Biomed. Res. (2018)

2. Kumar, S.N., Fred, A.L., Varghese, P.S.: An overview of segmentation algorithms for the analysis of anomalies on medical images. J. Intell. Syst. https://doi.org/10.1515/jisys-2017-0629

3. Chena, Y., Zhaoa, B., Zhanga, J., Zhengb, Y.: Brain MR image segmentation and bias correction using an improved Gaussian mixture model. In: Proceedings of International Conferences on ISA, CIA, pp. 95–102 (2014). http://dx.doi.org/10.14257/astl.2014.48.17

4. Kalti, K., Mahjoub, M.A.: Image segmentation by Gaussian mixture models and modified FCM algorithm. Int. Arab J. Inf. Technol. **11**(1), 11–18 (2014)

5   Ariyo, O., Zhi-guang, Q,, Tian, L.: Fusion of Gaussian mixture model and spatial fuzzy C-means for brain MR image segmentation. DEStech Trans. Comput. Sci. Eng. (2017) (case). https://doi.org/10.12783/dtcse/csae2017/1756

6. Cui, W., Wang, Y., Lei, T., Fan, Y., Feng, Y.: Brain MR image segmentation based on an adaptive combination of global and local fuzzy energy. Math. Probl. Eng. **2013**, Article ID 316546, 10 p. (2013). http://dx.doi.org/10.1155/2013/316546

7. Nguyen, D.M., Vu, H.T., Ung, H.Q., Nguyen, B.T.: 3D-brain segmentation using deep neural network and Gaussian mixture model. In: 2017 IEEE Winter Conference on Applications of Computer Vision (WACV), pp. 815–824. IEEE (2017). https://doi.org/10.1109/wacv.2017.96

8. Chen, M., Yan, Q., Qin, M.: A segmentation of brain MRI images utilizing intensity and contextual information by Markov random field. Comput. Assist. Surg. **22**(Suppl. 1), 200–211 (2017). https://doi.org/10.1080/24699322.2017.1389398

9. Banerjee, A., Maji, P.: Rough-probabilistic clustering and hidden Markov random field model for segmentation of HEp-2 cell and brain MR images. Appl. Soft Comput. **46**, 558–576 (2016). https://doi.org/10.1016/j.asoc.2016.03.010

10. Ahmadvand, A., Daliri, M.: Brain MR image segmentation methods and applications. OMICS J. Radiol. **3**, e130 (2014). https://doi.org/10.4172/2167-7964.1000e130

11. Prakash, R.M., Kumari, R.S.S.: Modified expectation maximization method for automatic segmentation of MR brain images. MIDAS J., 1–8 (2013). http://hdl.handle.net/10380/3445

12. Ji, Z., Xia, Y., Sun, Q., Chen, Q., Xia, D., Feng, D.D.: Fuzzy local Gaussian mixture model for brain MR image segmentation. IEEE Trans. Inf. Technol. Biomed. **16**(3), 339–347 (2012). https://doi.org/10.1109/TITB.2012.2185852

13. Greenspan, H., Ruf, A., Goldberger, J.: Constrained Gaussian mixture model framework for automatic segmentation of MR brain images. IEEE Trans. Med. Imaging **25**(9), 1233–1245 (2006). https://doi.org/10.1109/TMI.2006.880668

14. Balafar, M.A.: Gaussian mixture model based segmentation methods for brain MRI images. Artif. Intell. Rev. **41**(3), 429–439 (2014). https://doi.org/10.1007/s10462-012-9317-3

15. Ariyo, O., Zhi-guang, Q., Tian, L.: Fusion of Gaussian mixture model and spatial fuzzy C-means for brain MR image segmentation. DEStech Trans. Comput. Sci. Eng. (2017) (case). https://doi.org/10.12783/dtcse/csae2017/17560

16. Shi, X., Zhaoa, Q.H.: Gaussian mixture model and RJMCMC based RS image segmentation. Int. Arch. Photogramm. Remote Sens. Spatial Inf. Sci. **42**(2/W7) (2017). https://doi.org/10.5194/isprs-archives-xlii-2-w7-647-2017

17. Prados, F., Ashburner, J., Blaiotta, C., Brosch, T., Carballido-Gamio, J., Cardoso, M.J., Conrad, B.N., Datta, E., Dávid, G., De Leener, B., Dupont, S.M.: Spinal cord grey matter segmentation challenge. Neuroimage **15**(152), 312–329 (2017). https://doi.org/10.1016/j.neuroimage.2017.03.010

18. Manniesing, R., Oei, M.T., Oostveen, L.J., Melendez, J., Smit, E.J., Platel, B., Sánchez, C.I., Meijer, F.J., Prokop, M., van Ginneken, B.: White matter and gray matter segmentation in 4D computed tomography. Sci. Rep. **7**(1), 119 (2017). https://doi.org/10.1038/s41598-017-00239-z

19. Nguyen, D.M., Vu, H.T., Ung, H.Q., Nguyen, B.T.: 3D-brain segmentation using deep neural network and Gaussian mixture model. In: 2017 IEEE Winter Conference on Applications of Computer Vision (WACV), 24 Mar 2017, pp. 815–824. IEEE (2017). https://doi.org/10.1109/wacv.2017.96

20. Banerjee, A., Maji, P.: Rough-probabilistic clustering and hidden Markov random field model for segmentation of HEp-2 cell and brain MR images. Appl. Soft Comput. **1**(46), 558–576 (2016). https://doi.org/10.1016/j.asoc.2016.03.010
21. Cui, W., Wang, Y., Lei, T., Fan, Y., Feng, Y.: Brain MR image segmentation based on an adaptive combination of global and local fuzzy energy. Math. Probl. Eng. **2013** (2013). http://dx.doi.org/10.1155/2013/316546
22. Real, R., Vargas, J.M.: The probabilistic basis of Jaccard's index of similarity. Syst. Biol. **45**(3), 380–385 (1996)
23. Carpineto, C., Romano, G.: Consensus clustering based on a new probabilistic rand index with application to subtopic retrieval. IEEE Trans. Pattern Anal. Mach. Intell. **34**(12), 2315–2326 (2012). https://doi.org/10.1109/TPAMI.2012.80

Unimodal Medical Image Registration Based on Genetic Algorithm Optimization

J. V. Alexy John⬧, **S. N. Kumar**⬧, **A. Lenin Fred**⬧, **H. Ajay Kumar**⬧ **and W. Abisha**⬧

Abstract This research work proposes unimodal image registration based on genetic algorithm. The intensity-based image registration is employed here, and normalized cross-correlation is used as the similarity index, and for choosing the optimal values of image registration parameters, genetic algorithm was employed. The performance of the image registration was validated by the performance metrics and tested on MR brain images of BrainWeb database. The performance metrics peak-to-signal noise ratio (PSNR), mean squared error (MSE), normalized cross-correlation (NCC), and mutual information (MI) reveals the superiority of the image registration algorithm.

Keywords Registration · Genetic algorithm · Mutual information · Normalized cross-correlation

1 Introduction

In [1], a medical image registration algorithm termed as VoxelMorph based on convolution neural network was proposed for the MR brain data sets. Efficient results were produced and applicable for cardiac MR scan and lung CT images and can be

J. V. Alexy John (✉) · S. N. Kumar · H. Ajay Kumar · W. Abisha
School of Electronics and Communication Engineering, Mar Ephraem College of Engineering and Technology, Kanyakumari, India
e-mail: alexyjohn48@gmail.com

S. N. Kumar
e-mail: appu123kumar@gmail.com

H. Ajay Kumar
e-mail: ajayhakkumar@gmail.com

W. Abisha
e-mail: abishawilson01@gmail.com

A. Lenin Fred
School of Computer Science and Engineering, Mar Ephraem College of Engineering and Technology, Kanyakumari, India
e-mail: leninfred.a@gmail.com

© Springer Nature Singapore Pte Ltd. 2020   549
K. N. Das et al. (eds.), *Soft Computing for Problem Solving*,
Advances in Intelligent Systems and Computing 1057,
https://doi.org/10.1007/978-981-15-0184-5_47

used for multimodal registration also. The joint histogram of two images was estimated for multimodal image registration with Jensen–Arimoto (JA) divergence as the similarity measure for the statistical relationship between two images [2]. The 3D MR brain image registration helps to estimate the growth of tumor; sum of squared difference metric is used as similarity metric and regular step gradient optimizer is used for optimization [3]. In [4], particle swarm optimization (PSO) and artificial bee colony (ABC) were employed for intensity-based image registration; ABC generates superior results, when compared with PSO algorithm. The improved PSO generates superior results when compared with classical PSO in medical registration [5].

The PSO with modified mutual information similarity index performs robust medical image registration for CT/MR images [6]. In [7], detailed study has been performed on the vascular image registration and outcome of this work will be an aid for researchers in vascular registration algorithms. An improved CNN-based multimodal registration algorithm was proposed for the MR and ultrasound images; compared with classical registration techniques, efficient results were produced [8]. A detailed study on medical image registration and fusion for medical images was described in [9]. Mutual information was found to be an efficient similarity index for the rigid registration of medical images. In [10], the mutual information for registration was determined from the joint histogram technique; here in this work, mutual information is determined from the pixel intensity value. Genetic algorithm-based multimodal image registration comprising of MR and PET images of brain by the maximization of image registration gives efficient results [11]. The efficiency of genetic algorithm when coupled with image registration techniques was tested in 16 registration methodology in MR images [12]. In [13], mutual information-based image registration was proposed for the CT/MR and PET images. The statistical dependency between the voxels is explored, and it is based on histogram analysis. The registration procedure was automatic and applicable for multimodal images. The deep learning architecture was analyzed, and its applicability in image registration was analyzed. The advantages and challenges in the deep learning architectures are also analyzed [14]. The depth image registration based on expectation maximization algorithm was explored, and the inclusion of feedback improves the accuracy of registration [15]. The DIRBOOST, a machine learning algorithm, was employed for the deformable image registration [16]. The deep learning neural network gains importance in the multi-atlas registration for multimodal pathological images [17]. Section 2 describes the image registration procedure and role of genetic algorithm in fining the optimal value for the registration. Section 3 depicts the result and discussion, and conclusion is drawn Sect. 4.

2 Materials and Methods

2.1 Data Acquisition

BrainWeb, simulated brain database, was used in this research work for the analysis of image registration algorithm. It comprises of 20 normal anatomical models, and in this research work, input images are taken from MR brain data sets.

2.2 Medical Image Registration

Image registration has pivotal role in image processing where images captured from different sensors or different view points or at different times are matched. Image registration role is inevitable in different sub-domains like robotics, medical image processing, and satellite image processing. The objective of image registration is to align images with respect to each other. Image registration methods are widely divided into intensity-based and feature-based techniques. In image registration procedure, a reference image is used for the alignment of source image. Medical image registration refers to the technique of determining the appropriate mapping between the input and the reference images.

In image registration process, two images are considered: a reference image and the model image. The two images are represented by I_P and I_Q, respectively. The outcome of the registration methodology is in such a manner that the model image I_Q and the transformed image $f(I_p)$ are as similar as possible. The image registration is characterized by three main features: the transformation model, the similarity metric, and the optimization process. The transformation model depicts the type of transformation used to align the images. The rigid transformation consists of translation and rotation operations, and the similarity transformation uses scaling. The similarity metric determines the quality of the solution of an image registration problem.

The input image I_P is transformed in accordance with f and the measure of transformed query image $f(I_p)$, and the model image. I_Q represented by $\varnothing$ is determined so $F(I_P, I_Q, f) = \varnothing(I_Q, f(I_P))$. The metrics employed in the feature-based technique are based on the distance measures such as mean squared error (MSE). The MSE corresponds to the average square distance between features in the query and the model images. In intensity-based approaches, sum of squared differences, normalized correlation, and mutual information metrics are used. Unimodal registration computes and carries out a transformation between two or more pictures of the same subject.

2.3   Image Registration Based on Genetic Algorithm

Genetic algorithm (GA) is a bio-inspired optimization algorithm that works on the principle of survival of the fittest concept for the generation of optimized solution, and the performance and complexity of GA rely on the following features: coding, population size, fitness quality of solution, and selection criteria. The CT and PET image modalities depict the structural information, and MR images give the functional information. Let P and Q be two images, the image registration search space comprises of a linear combination of affine transformations and a similarity measurement (fitness function).

The affine transformation model is represented as 3×3 matrix represented as follows:

$$P' = T P \tag{1}$$

$$\begin{vmatrix} p' \\ q' \\ 1 \end{vmatrix} = \begin{vmatrix} s_p \cos \theta & -s_q \sin \theta & t_p \\ s_p \sin \theta & s_q \cos \theta & t_q \\ 0 & 0 & 1 \end{vmatrix} \begin{vmatrix} p \\ q \\ 1 \end{vmatrix} \tag{2}$$

where p and q represent the input images and p' and q' represent the transformed images. The relationship between pixels is given by the mapping T, which is represented as a 3×3 affine model. The affine model consists of five degrees of freedom $[t_x, t_y, \theta, s_x, s_y] t_x$ and t_y are the translation vectors, s_x and s_y are scaling parameters, and θ is the angle of rotation. The above set of parameters defines the chromosome in genetic algorithm. For chromosome representation, binary coding is used. It describes how an image is deformed to match other image. The optimization algorithm finds its role in image registration process. The flow diagram of the optimization-based medical image registration is depicted in Fig. 1.

The feature extraction is performed, and the optimization algorithm seeks for a match between the two images, and the transformation parameters are determined. The iteration process continues until the convergence occurs within a tolerance threshold of the taken similarity metric.

The GA comprises of the following stages

Coding: This is the first step and defines the random population with N-dimensional case.

Initialization: From the N-dimensional value, randomly select the initial population based on their fitness function. Fitness function is objective functions that need to get maximized value. In search space, the population are defined by all possible encoding of the solution.

$$Y = \{y_1, y_2, y_3, \ldots, y_n\} \tag{3}$$

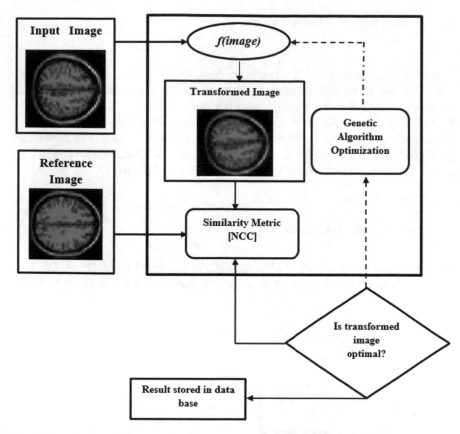

Fig. 1 Flow diagram of image registration based on genetic algorithm

Selection: The selection operation copies a single individual probabilistically selected based on fitness into the next generation of the population. The initial population are selected through the fitness-based process, where the fitness value is measured and typically selected the best value. Mainly used selection methods are roulette wheel selection and tournament selection. The flow diagram of the genetic algorithm is depicted in Fig. 2.

The pseudo-code for the genetic algorithm is depicted below:

```
Function GA ()
  {
  Initialization of population;
  Estimation of fitness value;
  While (fitness function value = termination condition)
    {
    Selection;
    Crossover;
    Mutation;
```

```
Estimation of fitness value;
 }
}
```

Crossover: Crossover combines two individuals to create new individuals for possible inclusion in the next generation. Crossover points are randomly selected, and generally, chance of crossover is between 0.6 and 1.0.

Mutation: It creates slightly random changes in the individual's population which offers genetic diversity; it allows GA to search a broader space. The above-said procedure is repeated until the termination state is reached. Condition for termination is total number of fitness evaluation, when it reaches a given limit.

3   Results and Discussion

The algorithms are developed in MATALB 2010a and tested on MR brain data sets. The algorithms are tested in system with the specification of Intel Core i3 @3.30 GHz

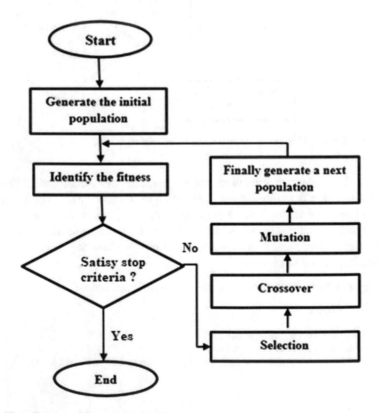

Fig. 2 Flow diagram of the genetic algorithm

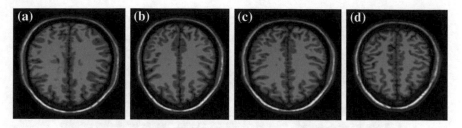

Fig. 3 Input images for registration

processor with 4 GB RAM. The input images corresponding to four MR brain data sets are depicted in Fig. 3, and the reference images for registration are taken into account for registration procedure. The proposed GA-based image registration is an intensity-based technique with normalized cross-correlation as similarity metric. The role of GA is to find an optimum value for translation, rotation, and scaling parameters with respect to normalized cross-correlation (NCC).

The unimodal registration is adopted in this work, and here, MR data sets of brain are used. Table 1 depicts the number of iterations and convergence value for NCC for each input image with respect to the reference image. The image registration result was validated by performance metrics like peak-to-signal noise ratio (PSNR), mean squared error (MSE), normalized cross-correlation (NCC), and mutual information (MI). The performance metrics are determined between input and reference images before and after registration. Higher value of PSNR, MI, NCC and lower value of MSE reveal the efficiency of GA-based image registration technique (Figs. 4, 5, 6, and 7).

$$PSNR = 20 \log \frac{R^2}{MSE} \tag{4}$$

$$MSE = \frac{1}{MN} \sum_{i=1}^{M} \sum_{j=1}^{N} [g(m, n) - r(m, n)]^2 \tag{5}$$

Table 1 Number of iterations and convergence value for GA-based image registration

Reference image	ID1		ID2		ID3		ID4	
	N	Convergence value	N	Convergence value	N	Convergence value	N	Convergence value
RD1	58	1.015	58	1.015	68	1.106	68	1.12
RD2	57	1.13	57	1.13	59	1.124	53	1.01
RD3	59	1.135	53	1.1.26	53	1.126	53	1.13
RD4	60	1.174	76	1.123	100	1.13	56	1.096

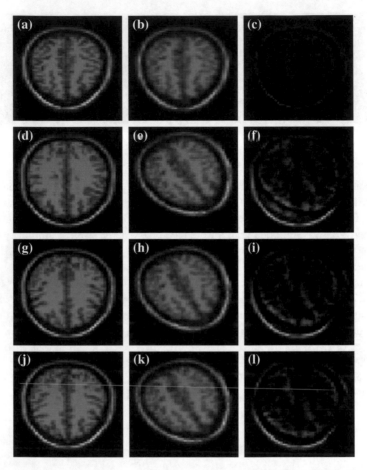

Fig. 4 First column represents the reference images; second column represents the registration result with respect to the input image (a) in Fig. 3; and third column represents the error image

$$\text{NCC} = \sum_{m=1}^{M} \sum_{n=1}^{N} \frac{g(m,n)r(m,n)}{[r(m,n)]^2} \tag{6}$$

$$\text{MI}(g,r) = \sum_{g,r} p(g,r) \frac{\log p(g,r)}{p(g)p(r)} \tag{7}$$

The performance metrics for input images with respect to reference images before and after registration are depicted in Tables 2, 3, 4, and 5.

The optimization technique role is inevitable in the search of best alignment for images. The genetic algorithm is already a proven algorithm for many complex

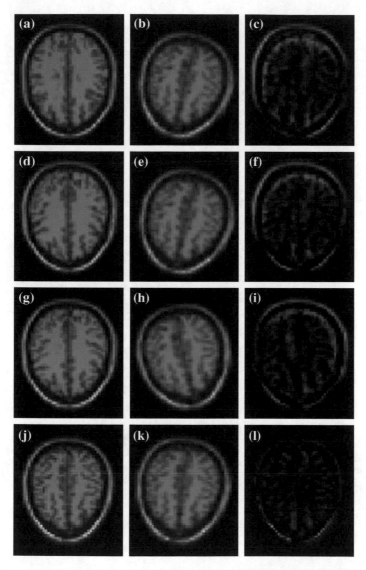

Fig. 5 First column represents the reference images; second column represents the registration result with respect to the input image (b) in Fig. 3; and third column represents the error image

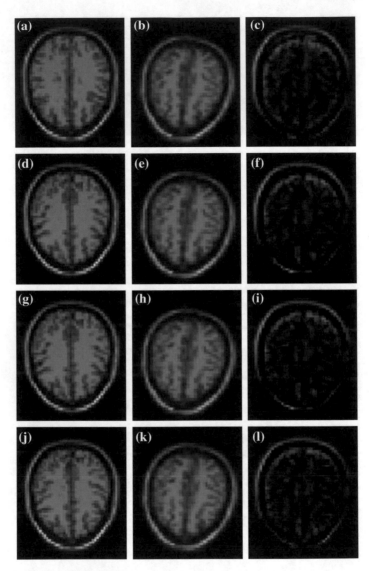

Fig. 6 First column represents the reference images; second column represents the registration result with respect to the input image (c) in Fig. 3; and third column represents the error image

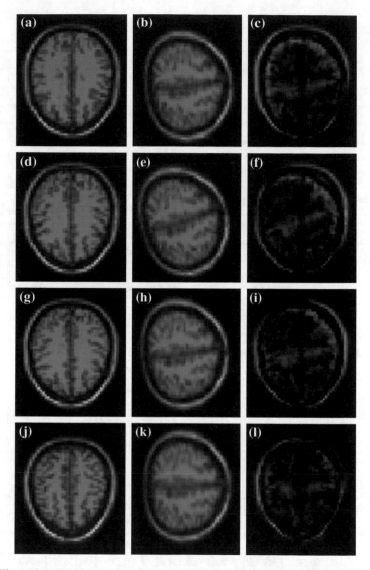

Fig. 7 First column represents the reference images; second column represents the registration result with respect to the input image (d) in Fig. 3; and third column represents the error image

Table 2 Image registration performance metrics with respect to first reference image

Image details	Before registration				After registration			
	MSE	PSNR	NCC	MI	MSE	PSNR	NCC	MI
ID1	10.51	29.65	0.78	9.94	29.40	33.45	0.99	9.89
ID2	99.64	28.15	0.77	10.16	93.96	28.40	0.88	10.46
ID3	92.01	28.49	0.78	10.10	80.49	29.07	0.90	10.11
ID1	88.31	28.67	0.78	10.13	75.32	29.36	0.89	10.22

Table 3 Image registration performance metrics with respect to second reference image

Image details	Before registration				After registration			
	MSE	PSNR	NCC	MI	MSE	PSNR	NCC	MI
ID1	28.35	95.08	0.82	10.54	28.46	92.71	0.89	10.59
ID2	28.35	95.08	0.82	10.54	28.46	92.71	0.89	10.59
ID3	28.69	88.09	0.82	10.41	29.05	80.91	0.89	10.46
ID4	28.94	83.06	0.82	10.43	29.25	77.33	0.89	10.49

Table 4 Image registration performance metrics with respect to third reference image

Image details	Before registration				After registration			
	MSE	PSNR	NCC	MI	MSE	PSNR	NCC	MI
ID1	28.30	96.13	0.80	10.43	28.85	84.82	0.78	10.32
ID2	28.39	94.11	0.88	10.51	29.07	80.64	0.88	9.94
ID3	28.63	89.15	0.79	10.30	29.69	69.77	0.76	10.14
ID4	28.89	84.05	0.89	10.37	31.79	43.23	0.97	10.20

Table 5 Image registration performance metrics with respect to fourth reference image

Image details	Before registration				After registration			
	MSE	PSNR	NCC	MI	MSE	PSNR	NCC	MI
ID1	28.23	97.76	0.76	10.09	28.53	91.28	0.85	9.67
ID2	28.68	88.16	0.81	9.96	29.01	81.66	0.89	9.99
ID3	28.86	84.48	0.84	9.98	29.11	79.72	0.89	9.70
ID4	30.30	60.64	0.89	9.79	30.37	59.73	0.91	9.38

problems in the determination of optimum solution. The intensity-based image registration based on optimization algorithm was found to be an efficient one for unimodal medical image registration and was applicable for multimodal images.

4 Conclusion

The unimodal intensity-based image registration is proposed in this research work for the MR brain images. The genetic algorithm is coupled with the intensity-based image registration algorithm for the selection of optimum registration parameters. The manual tuning of the parameters is replaced by the fine-tuning by GA, and the results were validated by performance metrics. The proposed image registration algorithm can be extended to multimodal image modality by considering the mutual information as similarity metric. Image registration based on multi-resolution helps in the disease diagnosis and preoperative planning.

Acknowledgements The authors would like to acknowledge the support provided by DST under IDP scheme (No.: IDP/MED/03/2015).

References

1. Matl, S., Brosig, R., Baust, M., Navab, N., Demirci, S.: Vascular image registration techniques: a living review. Med. Image Anal. **35**, 1–17 (2017). https://doi.org/10.1016/j.media.2016.05.005
2. Li, B., Yang, G., Liu, Z., Coatrieux, J.L., Shu, H.: Multimodal medical image registration based on an information-theory measure with histogram estimation of continuous image representation. Math. Probl. Eng. **2018** (2018)
3. Irmak, E., Turkoz, M.B.: A useful implementation of medical image registration for brain tumor growth investigation in a three dimensional manner. Int. J. Comput. Sci. Netw. Secur. **17**(6), 155–161 (2017)
4. Sarvamangala, D.R., Kulkarni, R.V.: Swarm intelligence algorithms for medical image registration: a comparative study. In: International Conference on Computational Intelligence, Communications, and Business Analytics, pp. 451–465. Springer, Singapore (2017). https://doi.org/10.1007/978-981-10-6430-2_35
5. Maddaiah, P.N., Pournami, P.N., Govindan, V.K.: Optimization of image registration for medical image analysis. Int. J. Comput. Sci. Inf. Technol. **5**(3), 3394–3398 (2014)
6. Abdel-Basset, M., Fakhry, A.E., El-Henawy, I., Qiu, T., Sangaiah, A.K.: Feature and intensity based medical image registration using particle swarm optimization. J. Med. Syst. **41**(12), 197 (2017). https://doi.org/10.1007/s10916-017-0846-9
7. Balakrishnan, G., Zhao, A., Sabuncu, M.R., Guttag, J., Dalca, A.V.: VoxelMorph: A Learning Framework for Deformable Medical Image Registration (2018). arXiv:1809.05231
8. Hu, Y., Modat, M., Gibson, E., Li, W., Ghavami, N., Bonmati, E., Ourselin, S.: Weakly-supervised convolutional neural networks for multimodal image registration. Med. Image Anal. **49**, 1–13 (2018). https://doi.org/10.1016/j.media.2018.07.002
9. El-Gamal, F.E.Z.A., Elmogy, M., Atwan, A.: Current trends in medical image registration and fusion. Egypt. Inform. J. **17**(1), 99–124 (2016)
10. Williams, C., Lalush, D.D.: Rigid-body image registration using mutual information, pp. 1–6 (2004) (Report). Accessed 10 Nov 2018
11. Panda, S., Sarangi, S.K., Sarangi, A.: Biomedical image registration using genetic algorithm. In: Intelligent Computing, Communication and Devices, pp. 289–296. Springer, New Delhi (2015)
12. Valsecchi, A., Damas, S., Santamaría, J.: An image registration approach using genetic algorithms. In: 2012 IEEE Congress on Evolutionary Computation (CEC), pp. 1–8. IEEE (2012)

13. Maes, F., Vandermeulen, D., Suetens, P.: Medical image registration using mutual information. Proc. IEEE **91**(10), 1699–1722 (2003)
14. Litjens, G., Kooi, T., Bejnordi, B.E., Setio, A.A., Ciompi, F., Ghafoorian, M., Van Der Laak, J.A., Van Ginneken, B., Sánchez, C.I.: A survey on deep learning in medical image analysis. Med. Image Anal. **1**(42), 60–88 (2017)
15. Li, X., Li, D., Peng, L., Zhou, H., Chen, D., Zhang, Y., Xie, L.: Color and depth image registration algorithm based on multi-vector-fields constraints. Multimed. Tools Appl., 1–9 (2019)
16. Muenzing, S.F., van Ginneken, B., Pluim, J.P.: DIRBoost: an algorithm for boosting deformable image registration. In: 2012 9th IEEE International Symposium on Biomedical Imaging (ISBI), 2 May 2012, pp. 1339–1342. IEEE
17. Tang, Z., Yap, P.T., Shen, D.: A new multi-atlas registration framework for multimodal pathological images using conventional monomodal normal atlases. IEEE Trans. Image Process. **28**(5), 2293–2304 (2019)

Control Strategies Applied in Solar-Powered Water Pumping System—A Review

T. Poompavai and M. Kowsalya

Abstract Agriculture is a well-versed procedure adapted in many farms of our world which helps to increase the crop yields and its diversification. Moreover, these practices include the use of conventional electric motors and generators. In advance to that solar PV-based water pumping system (SWP) came into practice which offers fabulous solutions compared to other conventional systems as it needs lesser maintenance, ease for installation, free of fuel cost, zero emanation of greenhouse gases, portable and most important reliable too. Henceforth, it greatly motivated the researchers to design a more efficient and controlled system. This manuscript takes an initiative work to represent SWP system control strategies wherein the details of input source PV, boosting PV voltage by various converters, charge controllers, and also energy management by supervisory controllers for SWP system are discussed.

Keywords Solar photovoltaic water pump · Maximum power point tracking · Power conditioning unit · Charge controller · Supervisory controller

1 Introduction

In the world of fluctuating traditional energy prices, renewable, and sustainable energies are gaining a higher value of interest because of its zero emission. Thus, researchers and industry people turning their attention highly toward the development of products which keep running by non-conventional energy sources. Water pumping system runs by solar energy becomes promising standalone systems. In India, the market for solar water pumping scheme estimated growth rate is about 18.7% compound annual growth rate (CAGR) for the period of the financial year (FY) 2017–22 [1]. The problem with the diesel pump and other gasoline pumps is that it needs more maintenance and causes noise, dust, and fume pollution to the environment [2, 3].

T. Poompavai · M. Kowsalya (✉)
School of Electrical Engineering, VIT University, Vellore 632014, India
e-mail: mkowsalya@vit.ac.in

T. Poompavai
e-mail: paavai.oct@gmail.com

© Springer Nature Singapore Pte Ltd. 2020
K. N. Das et al. (eds.), *Soft Computing for Problem Solving*,
Advances in Intelligent Systems and Computing 1057,
https://doi.org/10.1007/978-981-15-0184-5_48

563

Technology enrichment toward SWP for the benefit of society had been started by the researchers from 50 years back itself. Generally, solar water pump (SWP) consists of PV array, power conditioning unit (PCU), and a motor fixed with the pump. The PCU unit comprises of either DC–DC converter or DC–AC inverter based on the required type of electric motor. In [4], the authors analyzed several methodology procedures that include PV array peak power, water tank capacity, total head, and water consumption profiles in Algeria. The authors concluded that SWP seriously depends on the total head of the pump and the maximum power output from the PV array. Hence for the extraction of maximum power, MPPT circuitry can be merged with the system which tunes the direct current (DC) operating point at all the time, recent reviews on different MPPT techniques detailed in [5–7]. Martiré et al. [4] studied a system-based certain population needs, depth of well, and metrological profile. Thereby the PV modules supply continuous current to inverter it converts DC current into three-phase AC current and then it fed to induction motor drive coupled with a centrifugal pump. Sontake and Kalamkar [8] had reviewed various classifications of solar water pump based on the energy storage, input electric power, type of pump and different tracking mechanisms. In addition to that, the authors also evaluated the socioeconomic benefits, efficiency, performance, and longevity of different SWP systems. As far as now many researchers reviewed the status of SWP in different approaches depending on the specific region and application but still there exist a gap in research in the area of control units which needed to the system such as maximum power point tracking (MPPT) for extracting maximum power from PV array at any time (MPPT controller), battery charging and discharging control (charge controller), supervisory controllers for making efficient use of battery as well as to store excess energy acquired from PV array, motor and pump control (motor controller) for delivering water at desired flow rate.

2 Solar Photovoltaic Array Water Pumping System

The major components utilized for the SWP system are solar panels, charge regulator, pump controller, batteries, inverter, pump/motor, pipes, and storage tank. Depending upon the motor electric power input SWP is classified into DC and AC drove pumping system. SWP can also be classified according to the energy storage such as battery coupled or direct driven pumping system (obtaining energy directly from PV array). Figure 1 shows the layout of SWP system in that for DC pump motor power obtained from PV array is directly given and for AC motor pump an inverter connected to make the conversion of input DC into AC power output [9]. Another component to be addressed is charge regulator or charge controller which is used for preventing the battery from overcharging and deep discharging. It has the capability of reducing heat dissipation and the aging effect that simultaneously increases life and efficiency of battery [10, 11].

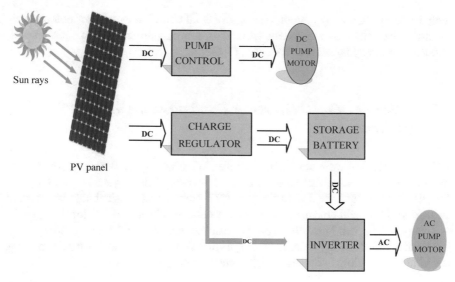

Fig. 1 Schematic layout of PV array fed water pumping system

3 Control Strategies in SWP System

For ensuring reliability and optimal operation of the water pumping system assisted by PV array, control units are very much necessary. For SWP system, power conditioning unit (PCU) comprises of required control unit stages. Depending upon the application, DC–DC converter and DC–AC inverter would be preferred for DC or AC motor pump.

3.1 MPPT Control for SWP System

In the past decade, most of the researches were carried out in MPPT to obtain a solution to consume the maximum power output from PV array at all changing environmental and partial shading conditions. In general, PV array has a point on its current/voltage characteristics which is referred to be MPP. The generated power at MPP is larger voltage and smaller current which is quite opposite to the requirement of motor pump hence to solve this mismatch problem and appears to increase efficiency by transforming the power into larger current and smaller voltage which fulfill the motor pump characteristics. The conventional MPPT techniques which are used so far to validate the voltage where PV array extracts the maximum power output are perturb and observe (P&O), ripple correlation control, incremental conductance (IC), constant voltage control, table lookup MPPT, fractional short-circuit current, and fractional open circuit voltage (FOCV) [12–16]. The modern intelligent MPPT

techniques are fuzzy logic, neuro-fuzzy, artificial neural network, ant colony optimization, artificial immune system, genetic algorithm, particle swarm optimization,
and the new firefly algorithm (FA), etc [17–22].

3.2 Different DC–DC Converter Configurations for MPPT Implementation

DC–DC converters were very much helpful in extracting maximum power output
from PV array at various insolation stages for water pumping system. The converter
families commonly used for tracking MPPT are a boost, buck, buck–boost, zeta
converter; their main concern is to raise the voltage gain as needed for the specific
application. The family of high voltage ratio conversion for PV-based application
has been detailed in [23]. Coelho et al. [24] had derived the equation for finding out
tracking system global efficiency (n_g) which can be expressed as

$$n_g = n_s n_H \tag{1}$$

From Eq. (1), n_s represent the software efficiency, additionally referred as tracking factor, and n_H is the hardware efficiency. The software part refers to the tracking
algorithms that use intelligent and conventional MPPT techniques, whereas the hardware refers to the power circuit of the converter. Hereby tracking efficiency factor is
decided by the aspects such as load which coupled to the PV module terminal and the
converter static characteristic. El Khateb et al. [25] recently found a new feature to
minimize the PV current ripple without a filtering capacitor using a constant current
converter. The ripple content in current causes a great effect on a PV curve because
when it gets maximizes, the operating point tends to reach a constant voltage section
that creates excess drop over average power simultaneously. Hence, the role of the
power converter is very much essential for getting efficient PV power. Caracas et al.
[26] had conducted an experiment on a low-cost highly efficient modified DC–DC
two-inductor boost converter (TIBC) which design relies on current fed converter
type. The model also provided with a snubber circuit with hysteresis controller to
improve converter efficiency, and it functions by permitting the part of the energy
from input to output directly by using snubber without going for the transformer. The
total system had gained an overall efficiency of about 91 and 93.64% attained by the
modified DC–DC TIBC stage. The problem with the buck and boost converters is
that it causes a big impact in PV output at lower irradiance conditions such as giving
discontinuous current, semiconductor switching losses, higher ripple current at output. To resolve those troubles, a new type of two-cell interleaved boost converter had
experimented and studied in [27]. Khadmun and Subsingha [28] also tested a high
voltage gain interleaved boost converter in the output which they got 130 V from
24 V DC voltage. Even though interleaved boost converters had the advantages such
as improved power quality, reliability, cancelation of ripple, it still requires a larger

gate driver circuit that may lead to the complex circuit. In recent years, Kumar and Singh [29–34] contributed their work related to various DC–DC converter configurations especially applied for SWP system run through permanent magnet brushless DC (BLDC) motors.

In [35], the authors had chosen boost converter in-between the PV array and VSI to track MPPT at an irradiance of 1000 W/m², and the efficiency was found to be 81.5% and for the buck–boost converter at the same irradiance, the efficiency raised to 86.6%. In [36] for an SWP system, the buck–boost converter initiates soft starting of the motor with a better control for raising the voltage and to level down. Even if the classical buck–boost converter and its family offer better conversion efficiency, they suffer from increased stresses over the power switches along with added reactive components makes the system size and cost-wise high [37, 38]. Other converters which have been analyzed under buck–boost family are Landsman, Luo, Zeta, and SEPIC. By construction, the Landsman converter is free from external filter besides that the snubber circuit attached with power switches gives lesser oscillation in the output current. It provides abundant tracking of MPPT with high efficiency and merges well with the BLDC motor water pumping system. The Luo converter ensures safe starting of BLDC motor by running with the abridged current. It produces smooth current output with ripple filter at the input stage. The efficiency of different converters is shown in Fig. 2, and it can be seen that the interleaved boost converter and TIBC provided higher efficiency of approximately 96.15 and 93.64% than the other converter types. In the case of buck–boost, boost–buck and Luo converters, it brought 86.6%. For Landsman and zeta, it delivered about 83% and finally the SEPIC and boost converter given 81%.

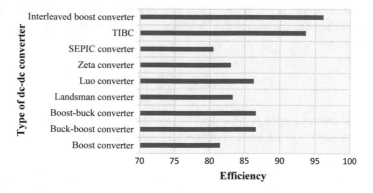

Fig. 2 Comparative analysis of various converters

4 Energy Storage for SWP System

The excess power acquired from the photovoltaic array for a water pumping system can be stored in two ways. One is by storing the pumped water in the storage tank and another one is storing the generated PV energy using batteries. The latter one has more advantage because storing water in a tank may lead to contamination and other problems but this is not in the case of using batteries. A battery has the capability to store energy during nighttime and cloudy weather conditions, and it also meets out power fluctuation issues at pumping side. The commonly used batteries in solar electric systems are lead–acid, nickel–cadmium (NICAD), nickel–iron (NIFE), Lithium–ion (Li-ion), nickel–metal hydride (NiMH). To analyze the battery behavior such as charging, discharging, and change in parameters such as voltage, current, temperature, density, resistivity, CIEMAT (Copetti) model has been chosen by Achaibou et al. [39]. The same mathematical model for the SWP system with different MPPT methods such as P&O, fuzzy control and neuro-fuzzy optimization was investigated by Rahrah et al. [40]. They analyzed the system using the model and concluded that neuro-fuzzy results to give a better dynamic performance. The equivalent circuit model for CIEMAT is given in Fig. 3. The features of source voltage E_{batt} and internal resistance R_{batt} depend upon the external temperature and battery charging state. The voltage equation for the number of cells (n_{batt}) is given by

$$V_{batt} = n_{batt} \cdot E_{batt} \pm n_{batt} \cdot R_{batt} \cdot I_{batt} \tag{2}$$

Here the V_{batt} represents the voltage of the battery, I_{batt} is battery current, E_{batt} is the electromotive force which relies on battery charging state, and R_{batt} signifies internal resistance that changes according to charging state. Here the battery tends to be complex impedance Z_{batt} which develops resistance of R_{batt} and a reactance of X_{batt} to disturbance.

$$|Z_{batt}| = \left[\frac{V_{batt}}{I_{batt}} \right] \tag{3}$$

For DC solar water pumping application vanadium redox flow battery (VRFB) storage system regulated by DC–DC converter assigned mainly to preserve the constant

Fig. 3 CIEMAT equivalent circuit model

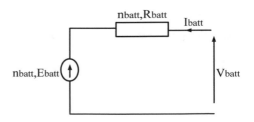

power in [41]. The peculiar VRFB battery had selected because of its numerous features such as high durability, reliability, extensibility of energy and power density, and capability of deep discharging.

Here the modeling for VRFB battery is expressed where the relation among electrical parameters are specified by

$$V_{cell} = V_{eq} + (2RT|F)\ln\{SOC|1 - SOC\}\text{Volts} \tag{4}$$

For that stack voltage is

$$V_{stack} = n * V_{cell} \tag{5}$$

The battery terminal voltage is represented by

$$V_b = V_{stack}I_{stack}(R_{reactance} + R_{resistance}) \tag{6}$$

For VRFB state of charge (SOC) estimation,

$$SOC = (P_{stack} * T_{step})/E_{capacity} \tag{7}$$

Internal losses in battery

$$I_{stack}(V_{stack} - V_{terminal}) \tag{8}$$

where

V_{cell} = Single cell voltage of VRFB (V)
V_{eq} = Cell equilibrium voltage (V)
R = Gas Constant 8.3148 (J K^{-1} Mol^{-1})
T = Battery stack temperature (°C)
F = Faraday's number
SOC = State of charge
V_{stack} = Battery stack voltage (V)
n = No. of cells
V_b = Terminal voltage of battery (V)
I_{stack} = Current through the VRFB stack (A)
$R_{reactance}$ = Equivalent resistance of reaction inside stack (Ω)
$R_{resistance}$ = Equivalent resistance of parasitic losses (Ω).

This VRFB power management mathematical model had purposely designed to estimate the state of charge and protects from repeated deep discharging conditions to ensure a longer life for the battery. Interestingly, the smoothing operation of the pump is very much determined by the concerned parameters; they are current, speed, torque, and its fluctuation problem greatly reduced when using VRFB storage system.

5 Battery Bank Charging and Discharging Control

A charge controller is very much essential for the maintaining the battery performance efficiently and to extend its lifetime. Figure 4 describes the layout of the PV system with charge controller meant for battery operation. Charge controllers terminate charging whenever the battery is full by restricting the current flow. When a battery fails to fulfill the supply to the pump side and when there is no generation of power from PV side, the controller automatically disconnects the batteries to prevent excessive discharge. The main features of a charge controller are the capability of blocking reverse current, preventing overcharge, low voltage disconnect, overload protection, etc. [42].

Dakkak and Hasan [43] had designed a charge controller based on a microcontroller for a photovoltaic system. The benefits which they identified are the possibility of correcting setpoint values for different battery type, the option of changing the algorithm, reduced power consumption, and elimination of analog-type feedback circuitry. Furthermore, the lifetime of each battery is guaranteed by some optimum charging control algorithms which keep the battery at a high state of charge that was discussed in [44]. In charging mode [45], the battery gets charged with MPPT between two predefined voltage edges and the charging get stopped and when the battery reaches a maximum voltage at level 1, then again charging initiated when it arrives at minimum voltage. In discharging mode, the same thing repeats to make certain the battery bank not to attain deep discharging.

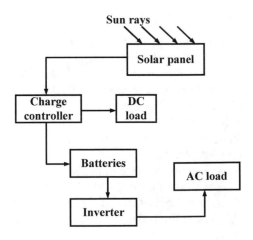

Fig. 4 Layout of a PV system with battery storage

6 Supervisory Controller

The photovoltaic pumping system in a closed-loop system permits the way to gain excessive energy that utilized for battery unit. Khiareddine [46–50] employed supervisory controllers which are a neuro-fuzzy controller, fuzzy controller, and algorithm controller for the energy management of the SWP system. The main objectives of this controller are:

- To ensure the design of MPPT.
- To guarantee the control speed required for the motor.
- To regulate the water in the tank.
- To ensure the precise operation for all conversion.

The foresaid controller main inputs are water level in the storage tank, state of charge (SOC) of battery, solar illumination G, and the temperature T and the outputs are reference speed required for the motor, duty cycle, and the controller switches. In order to achieve a better pump, flow the reference speed of the machine should go with the set point without any overshoot and error. To validate the performance of controllers, the authors computed daily maximum energy value obtained from PV array with three control strategies reference speed along with respect to maximum output power. The resulted energy gain from the neuro-fuzzy controller was 99.5%, 92.4% for the fuzzy controller and finally 82% for algorithm controller.

7 Conclusion

In this article, various control strategies which are applicable for solar water pumping system investigated by several types of research are impressively reviewed with its specific working features. It seems that solar water pumping technologies are very much reliable and economically viable alternative to electric and diesel water pumps. Countries which receive good sunshine can make avail of this technology. The main conclusions and recommendations regarding each control units are specified below

- The battery charging algorithms ensure that the operating voltage is maintained at a constant level and increases the lifetime of the battery. But the biggest issue is for small-scale pumping applications making use of battery may not affordable, it surely leads the system cost-wise high.
- The controller part in motor gives the way to improve the transient and dynamic performance of the pumping system. By controlling the speed of a motor, flow rate, and pressure of water pump regulated at a desired rate and energy get saved accordingly.
- The supervisory control enhances the system smartly and fulfills the control strategies of the entire SWP system.
- Finally, the better model sizing procedures, effective energy management, and optimal storage planning help to attain higher reliability of the system.

References

1. India Solar Water Pumping System Market (2017–2022). http://www.6wresearch.com/market-reports/india-solar-water-pumping-system-swp-market-2017-2022-market-forecast-by-power+rating-competitive+landscape-design_type-surface+submersible
2. Al-Smairan, M.: Application of photovoltaic array for pumping water as an alternative to diesel engines in Jordan Badia, Tall Hassan station: case study. Renew. Sustain. Energy Rev. **16**(7), 4500–4507 (2012)
3. Elhadidy, M.A., Shaahid, S.M.: Parametric study of hybrid (wind + solar + diesel) power generating systems. Renew. Energy **21**(2), 129–139 (2000)
4. Martiré, T., Glaize, C., Joubert, C., Rouvière, B.: A simplified but accurate prevision method for along the sun PV pumping systems. Sol. Energy **82**(11), 1009–1020 (2008)
5. Jordehi, A.R.: Maximum power point tracking in photovoltaic (PV) systems: a review of different approaches. Renew. Sustain. Energy Rev. **30**(65), 1127–1138 (2016)
6. Ramli, M.A., Twaha, S., Ishaque, K., Al-Turki, Y.A.: A review on maximum power point tracking for photovoltaic systems with and without shading conditions. Renew. Sustain. Energy Rev. **31**(67), 144–159 (2017)
7. Aliyu, M., Hassan, G., Said, S.A., Siddiqui, M.U., Alawami, A.T., Elamin, I.M.: A review of solar-powered water pumping systems. Renew. Sustain. Energy Rev. **87**, 61–76 (2018)
8. Sontake, V.C., Kalamkar, V.R.: Solar photovoltaic water pumping system—a comprehensive review. Renew. Sustain. Energy Rev. **30**(59), 1038–1067 (2016)
9. Solar Water Pumping. http://www.sunelco.com/planning_pumping.html
10. Ashiquzzaman, M., Afroze, N., Hossain, M.J., Zobayer, U., Hossain, M.M.: Cost effective solar charge controller using microcontroller. Can. J. Electr. Electron. Eng. **2**(12), 571–576 (2011)
11. Hiwale, S., Patil, M.V., Vinchurkar, H.: An efficient MPPT solar charge controller. Int. J. Adv. Res. Electr. Electron. Instrum. Eng. **3**(7) (2014)
12. Mohapatra, A., Nayak, B., Das, P., Mohanty, K.B.: A review on MPPT techniques of PV system under partial shading condition. Renew. Sustain. Energy Rev. **80**, 854–867 (2017)
13. Subudhi, B., Pradhan, R.: A comparative study on maximum power point tracking techniques for photovoltaic power systems. IEEE Trans. Sustain. Energy **4**(1), 89–98 (2013)
14. Soulatiantork, P., Cristaldi, L., Faifer, M., Laurano, C., Ottoboni, R., Toscani, S.: A tool for performance evaluation of MPPT algorithms for photovoltaic systems. Measurement **128**, 537–544 (2018)
15. Logeswaran, T., SenthilKumar, A.: A review of maximum power point tracking algorithms for photovoltaic systems under uniform and non-uniform irradiances. Energy Procedia **1**(54), 228–235 (2014)
16. Hohm, D.P., Ropp, M.E.: Comparative study of maximum power point tracking algorithms using an experimental, programmable, maximum power point tracking test bed. In: Photovoltaic Specialists Conference, 2000. Conference Record of the Twenty-Eighth IEEE 2000, pp. 1699–1702. IEEE
17. Yahyaoui, I., Nafaa, J., Charfi, S., Chaabene, M., Tadeo, F.: MPPT techniques for a photovoltaic pumping system. In: Renewable Energy Congress (IREC), 2015 6th International, 24 Mar 2015, pp. 1–6. IEEE
18. Garraoui, R., Sbita, L., Hamed, M.B.: MPPT controller for a photovoltaic power system based on fuzzy logic. In: 2013 10th International Multi-conference on Systems, Signals & Devices (SSD), 18 Mar 2013, pp. 1–6. IEEE
19. El Telbany, M.E., Youssef, A., Zekry, A.A.: Intelligent techniques for MPPT control in photovoltaic systems: a comprehensive review. In: 2014 4th International Conference on Artificial Intelligence with Applications in Engineering and Technology (ICAIET), 3 Dec 2014, pp. 17–22. IEEE
20. Zainuri, M.A., Radzi, M.A., Soh, A.C., Rahim, N.A.: Development of adaptive perturb and observe-fuzzy control maximum power point tracking for photovoltaic boost dc–dc converter. IET Renew. Power Gener. **8**(2), 183–194 (2013)

21. Dzung, P.Q., Lee, H.H., Vu, N.T.: The new MPPT algorithm using ANN-based PV. In: 2010 International Forum on Strategic Technology (IFOST), 13 Oct 2010, pp. 402–407. IEEE
22. Mellit, A., Kalogirou, S.A.: Artificial intelligence techniques for photovoltaic applications: a review. Prog. Energy Combust. Sci. **34**, 574–632 (2008)
23. Arunkumari, T., Indragandhi, V.: An overview of high voltage conversion ratio DC–DC converter configurations used in DC micro-grid architectures. Renew. Sustain. Energy Rev. **30**(77), 670–687 (2017)
24. Coelho, R.F., dos Santos, W.M., Martins, D.C.: Influence of power converters on PV maximum power point tracking efficiency. In: 2012 10th IEEE/IAS International Conference on Industry Applications (INDUSCON), 5 Nov 2012, pp. 1–8. IEEE
25. El Khateb, A.H., Rahim, N.A., Selvaraj, J., Williams, B.W.: DC-to-DC converter with low input current ripple for maximum photovoltaic power extraction. IEEE Trans. Ind. Electron. **62**(4), 2246–2256 (2015)
26. Caracas, J.V., Farias, G.D., Teixeira, L.F., Ribeiro, L.A.: Implementation of a high-efficiency, high-lifetime, and low-cost converter for an autonomous photovoltaic water pumping system. IEEE Trans. Ind. Appl. **50**(1), 631–641 (2014)
27. Kumar, R., Singh, B.: BLDC motor driven water pump fed by solar photovoltaic array using boost converter. In: India Conference (INDICON), 2015 Annual IEEE, 17 Dec 2015, pp. 1–6. IEEE
28. Khadmun, W., Subsingha, W.: High voltage gain interleaved dc boost converter application for photovoltaic generation system. Energy Procedia **1**(34), 390–398 (2013)
29. Singh, B., Bist, V.: A BL-CSC converter-fed BLDC motor drive with power factor correction. IEEE Trans. Ind. Electron. **62**(1), 172–183 (2015)
30. Parackal, R., Koshy, R.A.: PV powered zeta converter fed BLDC drive. In: 2014 Annual International Conference on Emerging Research Areas: Magnetics, Machines and Drives (AICERA/iCMMD), 24 Jul 2014, pp. 1–5. IEEE
31. Mohan, N., Undeland, T.M.: Power Electronics: Converters, Applications, and Design. Wiley (2007)
32. Singh, B., Kumar, R.: Solar photovoltaic array fed water pump driven by brushless DC motor using Landsman converter. IET Renew. Power Gener. **10**(4), 474–484 (2016)
33. Kumar, R., Singh, B.: Solar photovoltaic array fed Luo converter-based BLDC motor driven water pumping system. In: 2014 9th International Conference on Industrial and Information Systems (ICIIS), 15 Dec 2014, pp. 1–5. IEEE
34. Singh, B., Bist, V.: Solar PV array fed water pumping system using SEPIC converter based BLDC motor drive. IEEE Trans. Ind. Appl. **51**(2) (2015)
35. Khoucha, F., Benrabah, A., Herizi, O., Kheloui, A., Benbouzid, M.H.: An improved MPPT interleaved boost converter for solar electric vehicle application. In: 2013 Fourth International Conference on Power Engineering, Energy and Electrical Drives (POWERENG), 13 May 2013, pp. 1076–1081. IEEE
36. Kumar, R., Singh, B.: Buck-boost converter fed BLDC motor drive for solar PV array based water pumping. In: IEEE International Conference on Power Electronics, Drives and Energy Systems (PEDES) (2014)
37. Kumar, R., Singh, B., Chandra, A., Al-Haddad, K.: Solar PV array fed water pumping using BLDC motor drive with boost-buck converter. In: Energy Conversion Congress and Exposition (ECCE), 2015 IEEE, 20 Sep 2015, pp. 5741–5748. IEEE
38. Poompavai, T., Priya, P.V.: Comparative analysis of modified multilevel DC link inverter with conventional cascaded multilevel inverter fed induction motor drive. Energy Procedia **117**, 336–344 (2017)
39. Achaibou, N., Haddadi, M., Malek, A.: Modeling of lead acid batteries in PV systems. Energy Procedia **1**(18), 538–544 (2012)
40. Rahrah, K., Rekioua, D., Rekioua, T., Bacha, S.: Photovoltaic pumping system in Bejaia climate 786 with battery storage. Int. J. Hydrogen Energy **40**(39), 13665–13675 (2015)
41. Bhattacharjee, A., Mandal, D.K., Saha, H.: Design of an optimized battery energy storage enabled solar PV pump for rural irrigation. In: IEEE International Conference on Power Electronics, Intelligent Control and Energy Systems (ICPEICES), 4 Jul 2016, pp. 1–6. IEEE

42. Ingole, J.N., Choudhary, M.A., Kanphade, R.D.: PIC based solar charging controller for battery. Int. J. Eng. Sci. Technol. (IJEST) **4**(02), 384–390 (2012)
43. Arias, N.B., Franco, J.F., Lavorato, M., Romero, R.: Metaheuristic optimization algorithms for the optimal coordination of plug-in electric vehicle charging in distribution systems with distributed generation. Electr. Power Syst. Res. **31**(142), 351–361 (2017)
44. Dakkak, M., Hasan, A.: A charge controller based on microcontroller in stand-alone 799 photovoltaic systems. Energy Procedia **1**(19), 87–90 (2012)
45. Chang, W.Y.: The state of charge estimating methods for battery: a review. ISRN Appl. Math. **23**, 2013 (2013)
46. Khiareddine, A., Salah, C.B., Mimouni, M.F.: Power management of a photovoltaic/battery pumping system in agricultural experiment station. Sol. Energy **28**(112), 319–338 (2015)
47. Khiareddine, A., Salah, C.B., Mimouni, M.F.: Strategy of energy control in PVP/battery water pumping system. In: 2014 International Conference on Green Energy, 25 Mar 2014, pp. 49–54. IEEE
48. Khiareddine, A., Salah, C.B., Mimouni, M.F.: Determination of the target speed corresponding to the optimum functioning of a photovoltaic system pumping and regulation of the water level. In: 2013 International Conference on Electrical Engineering and Software Applications (ICEESA), 21 Mar 2013, pp. 1–5. IEEE
49. Serir, C., Rekioua, D., Mezzai, N., Bacha, S.: Supervisor control and optimization of multi-sources pumping system with battery storage. Int. J. Hydrogen Energy **41**(45), 20974–20986 (2016)
50. Ouachani, I., Rabhi, A., Yahyaoui, I., Tidhaf, B., Tadeo, T.F.: Renewable energy management algorithm for a water pumping system. Energy Procedia **31**(111), 1030–1039 (2017)

Realization of 2-D DCT Using Adder Compressor

Raunak R. Lahoti, Shantanu Agarwal, S. Balamurugan and R. Marimuthu

Abstract The increased usage of image and video compression in real-time applications has led to pressing need for efficient compression algorithms and dedicated hardware circuits. The aim of this proposed paper is to design an efficient adder compressor for approximated two-dimensional discrete cosine transform realization. This paper deals with enhancing the existing XOR-based adder compressor to a better XOR-XNOR-based architecture. The proposed compressors offer less delay, low power with the penalty of area. The proposed method provides less power consumption and offers high speed as compared to conventional design by 11.5% and 2%, respectively.

Keywords XOR-XNOR gate · Compressor · Image processing · Adder and discrete cosine transform (DCT)

1 Introduction

Image compression is the technique to minimize the size of the image without affecting its quality. This allows user to put more files in the memory. Using adder compressor is one of the most efficient ways of achieving low power consumption and less time delay in the implementation of the discrete cosine transform as mentioned in [1].

There have been numerous DCT architectures over the years. In [2], low power rounded DCT architecture for image compression is proposed, which utilizes 11 and 29 multiplications, additions, respectively. Later, butterfly-based 2-D DCT structure was proposed in [3]. Multiplier-free architecture for DCT was proposed in [4]. The proposed structure is applicable in high-efficiency video-coding (HEVC) system. Several authors have proposed approximated DCT to reduce the complexity of the computation of the circuit and time. An extensive review about various types of DCT implementations are presented in [5–7]. Even though there is a large amount

R. R. Lahoti · S. Agarwal · S. Balamurugan · R. Marimuthu (✉)
Vellore Institute of Technology, Vellore, India
e-mail: rmarimuthu@vit.ac.in

© Springer Nature Singapore Pte Ltd. 2020   575
K. N. Das et al. (eds.), *Soft Computing for Problem Solving*,
Advances in Intelligent Systems and Computing 1057,
https://doi.org/10.1007/978-981-15-0184-5_49

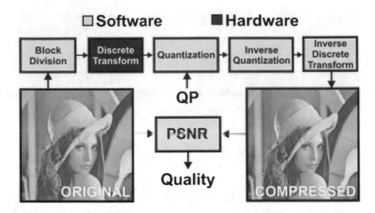

Fig. 1 DCT flow

of work that leads to the reduction of the computational effort, and/or the reduction of arithmetic operators of the DCT, there is only one paper which uses adder compressors [1]. This paper provides the right explanation for the use of adder compressor, however, uses the conventional models of 4-2 using XOR gates. Several authors proposed various compressors for the efficient multipliers and FFT [8, 9]. In [8], high-order compressors 15-4 were proposed and utilized in the image processing applications. In [9], FFT structure was implemented using various signed compressors and results are compared with existing structures and proved the efficiency of the proposed design. But, the performance of the DCT can be enhanced using XOR-XNOR-based adder compressors as in [10]. Figure 1 shows the design flow of the image compression system.

Image compression involves block division that is breaking down of images into multiple blocks of 8×8. Then, it is passed on to the only hardware block in the entire compression process, which is the DCT block. This is used to calculate the frequency components of the signal. The XOR-XNOR-based adder compressor reduces the delay, power consumption and occupies more area as compared to conventional design [1].

2 Compressor Design

DCT architecture involves a multiple adder compressor [11]. Most widely used compressor is 4-2 compressor. Several authors have proposed 4-2 compressor design. MUX-based compressor circuit is used in [1] and shown in Fig. 2. This structure is popularly used by several researchers.

Using the structure of approximation as stated in [6], 99.34% of total average energy is in top six lines of matrix. It is recommended to discard the S6 and S7 outputs. Figure 3 shows the approximate 1-D architecture.

Fig. 2 1-D DCT architecture

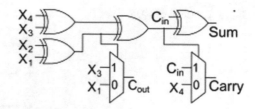

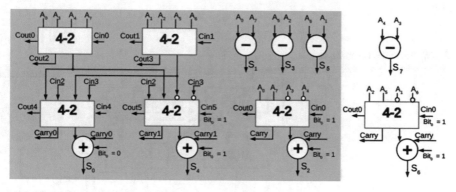

Fig. 3 1-D approximated architecture

To achieve the desired enhanced output use of XOR-XNOR, MUX-based architecture was used as proposed in [10]. The XOR-XNOR gates are used to provide two outputs. The first output is XOR gate and the second output is XNOR gate. The XNOR gate output is acted as input to the MUX for the later stages. This leads to provide low power than conventional design. Figures 4 and 5 show the transistor implementation of XOR-XNOR and MUX.

The proposed architecture reduces the number of XOR gates in a conventional 4:2 adder compressor by half, that is, conventional model required 4 XOR gates, and the proposed model requires only 2 XOR-XNOR gates for a 4:2 compressor [12]. Figure 6 shows the design of 4-2 compressor using XOR-XNOR gates.

Fig. 4 XOR-XNOR

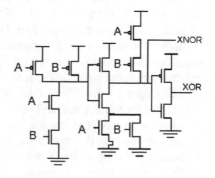

Fig. 5 Structure of MUX

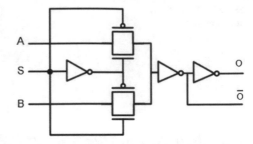

Fig. 6 4:2 adder compressor

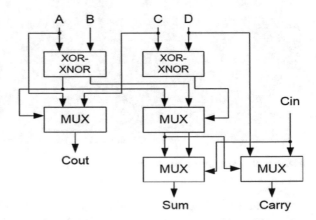

Using this new gate and MUX architecture as mentioned in [10] for a 4:2 adder compressor, the result can be achieved. In a similar process for a 8:2 adder compressor, four 4:2 adder compressors are cascaded to give a 8:2 compressor [1].

3 Results

A transposition buffer is designed based on two 8 × 8 buffers, the buffer stores the output of the first stage that is 1-D DCT from stage A, stage B serves the past result to the new 1-D transform. And thus, DCT is implemented using adder compressors.

In order to achieve the 2-D output, a transition buffer block is needed. Transposition buffer is utilized to convert the row transformed data of the first DCT to the column transform circuit, that is, the second DCT. This is a conventional method of achieving a 2-D transformation [1, 10]. The proposed architecture of the DCT is then tested by Xilinx software in order to get all the syntax corrections. The code is then run on the cadence software. Here, conventional as well as the proposed models are compared to get the result in its area, time delay, and power consumption of the architecture. Figure 7 shows the structure of 2-D DCT. The following result was achieved for the proposed model.

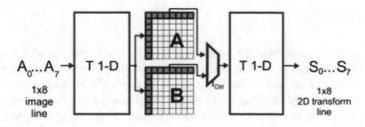

Fig. 7 Structure of 2-D DCT

Figure 8 and Table 1 show the output of the proposed DCT architecture and its comparison with the conventional model used in [1]. Hence, we can say that proposed model is faster than the conventional model and also it requires a low power consumption although the area consumed has been increased.

- The power consumption is reduced by 11.5%
- The time delay is reduced by 2%.

But the proposed adder compressor required more area than the conventional compressor.

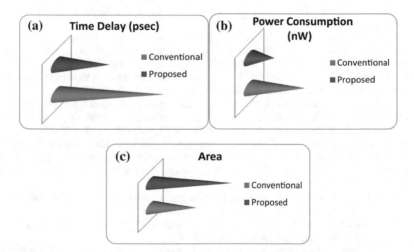

Fig. 8   a Time delay. **b** Power consumption (nW). **c** Area consumed (μm^2)

Table 1 Cadence output result comparison

DCT	Time (ps)	Power (nW)	Area (μm^2)
Conventional [1]	1520	25,308.313	183.456
Proposed	1490	22,392.167	201.802

4 Conclusion

The proposed adder compressor uses less time delay and lower power consumption with a slight increase in the area consumed. The proposed DCT architecture can be used in image, video and audio compression and as well as in biomedical image and video processing. The design was implemented and synthesized using Cadence RTL compiler and the output was verified. Several approximate compressors and adders can be designed for applications like image compression, video processing, and signal processing applications in order to get the better results in terms of circuits and errors.

References

1. Schiavon, T., Paim, G., Fonseca, M., Costa, E., Almeida, S.: Exploiting adder compressors for power-efficient 2-D approximate DCT realization. In: 7th Latin American Symposium on Circuits & Systems (LASCAS), pp. 383–386 (2016)
2. Bayer, F.M., Cintra, R.J.: Image compression via a fast DCT approximation. IEEE Latin Am. Trans. 8(6), 708–713 (2010)
3. Edirisuriya, A., Madanayake, A., Cintra, R.J., Bayer, F.M.: A multiplication-free digital architecture for 16 × 16 2-D DCT/DST transform for HEVC. In: 27th Convention of Electrical & Electronics Engineers in Israel (IEEEI), pp. 1–5 (2012)
4. Bayer, F.M., Cintra, R.J., Madanayake, A., Potluri, U.S.: Multiplier less approximate 4-point DCT VLSI architectures for transform block coding. Electron. Lett. 49(24), 1532–1534 (2013)
5. Potluri, U.S., Madanayake, A., Cintra, R.J., Bayer, F.M., Kulasekera, S., Edirisuriya, A.: Improved 8-point approximate DCT for image and video compression requiring only 14 additions. IEEE Trans. Circuits Syst. I: Regul. Pap. 61(6), 1727–1740 (2014)
6. Coutinho, V.D.A., Cintra, R.J., Bayer, F.M., Kulasekera, S., Madanayake, A.: Low-complexity pruned 8-point DCT approximations for image encoding. In: International Conference on Electronics, Communications and Computers (CONIELECOMP), pp. 1–7 (2015)
7. Madanayake, A., Cintra, R.J., Dimitrov, V., Bayer, F., Wahid, K.A., Kulasekera, S., Edirisuriya, A., Potluri, U., Madishetty, S., Rajapaksha, N.: Low-power VLSI architectures for DCT/DWT: precision vs approximation for HD video, biomedical, and smart antenna applications. IEEE Circuits Syst. Mag. 15(1), 25–47 (2015)
8. Marimuthu, R., Rezinold, Y.E., Mallick, P.S.: Design and analysis of multiplier using approximate 15-4 compressor. IEEE Access 5, 1027–1036 (2017)
9. Marimuthu, R., Mallick, P.S.: Design of efficient signed multiplier using compressors for FFT architecture. J. Eng. Sci. Technol. Rev. 10(2) (2017)
10. Tonfat, J., Reis, R.: Low power 3-2 and 4-2 adder compressors implemented using ASTRAN. In: Symposium on Third Latin American in Circuits and Systems (LASCAS), pp. 1–4 (2012)
11. Marimuthu, R., Bansal, D., Balamurugan, S., Mallick, P.S.: Design of 8-4 and 9-4 compressors for high speed multiplication. Am. J. Appl. Sci. 10(8), 893–900 (2013)
12. Agrawal, S., Harish, G., Balamurugan, S., Marimuthu, R.: Design of high speed 5:2 and 7:2 compressor using nanomagnetic logic. In: VDAT 2018 (in press)

Object Classification from Shape Detection

Pragya Nagpal and Ankush Mittal

Abstract We evaluate the problem of object detection and classification based on a single model for five diverse classes. The class detection problem is implemented by enabling a method which detects the presence or absence of every shape-based model in every instance of a class. Low-level feature extraction is also performed to facilitate object categorization on one-dimensional information of the dataset. We compute the categorization performance for both the modalities in separate as well as combined representations to produce improved experimental results. The combined obtained solutions provide a classification of the object into the five classes. We have evaluated our approach on the ETHZ dataset and found that it performs with an accuracy of 88.2% in classification based on object detection.

Keywords Canny edge detection · Image classification · MPEG-7 · Object classification · Object detection · Shape detection · MPEG-7 · Structural similarity · Template matching

1 Introduction

Object classification by object detection is a field which has a lot of applications in the real world. Analysis of objects detected can be done to gather information which can be used for various purposes.

We aim at detecting the shape present in a real image given a single hand-drawn model of the shape. This method is implemented with the help of features generated from the MPEG-7 tool and by the usage of template matching algorithm. These two approaches detect the model present in the images and classifies them into five distinct classes of objects.

The task presents various challenges. The shape of a similar type of object present in a particular class varies considerably due to difference in shapes and in angles at which they are observed. The parts of images are sometimes missing and do not

P. Nagpal (✉) · A. Mittal
Graphic Era Hill University, Dehradun, Uttarakhand, India
e-mail: nagpalpragya85@gmail.com

© Springer Nature Singapore Pte Ltd. 2020
K. N. Das et al. (eds.), *Soft Computing for Problem Solving*,
Advances in Intelligent Systems and Computing 1057,
https://doi.org/10.1007/978-981-15-0184-5_50

make up to the whole shape of the object. They are only recognized when seen from a wider perspective to detect the whole shape. The images have been extracted from natural scenes and, therefore, add extra objects which increase ambiguity and decrease visibility of the object to be found.

In this paper, we present a new approach for image classification by object detection on the basis of shape detection by combining the capabilities of a powerful feature, that is, MPEG-7 and the template matching algorithm. MPEG-7, a suite of standards for description and search of audio, visual, and multimedia content, is being used to produce the low-level visual description characteristics. MPEG-7 is actually called Multimedia Content Description Interface. It contains six MPEG-7 visual descriptors of rudimentary features such as texture, color, and shape.

Operating on the template matching algorithms and the structural similarity index brings to us the key advantage of finding the similarity score between the model and the input real image provided as input. The real images are compared with the model to find the shape present in image. Accuracy of this is further increased by extracting the portion of the image where that model is detected. The extracted fragment is again compared to all the five models, and a structural similarity index is calculated. The similarity index of all the models is compared to find the best model which is matched by template matching.

The power of the MPEG 7 features is combined with the template matching technique for further improvement in the detection of class.

This paper is structured as Sect. 2 exhibits the literature and overview of the previous work conducted in the same field. Section 3 describes the theory and algorithms which cover the MPEG-7 feature extraction tool and the template matching with structural similarity implemented in the work. Section 4 represents the experimentation conducted. Section 5 presents the results generated over the work done. Section 6 concludes and discusses the future work in the same area of research.

2 Literature and Overview

Much previous work on object class detection through shape matching has been proposed [1, 2]. They can classify objects into different classes while allowing for deformations within a class. The drawback in these techniques is that they assume the image to be a clear image, thereby not addressing the problem of localization.

Chamfer matching methods [3] have been successful in detecting shapes in cluttered images. However, as mentioned by Leibe [4], Thayananthan et al. [5], they require a large number of templates to handle shape variations. Over a thousand templates are required in [3] and might produce around 1–2 per image false positives.

Another model, [6], which proposed a point-matching-based method on inter quadratic programming, uses real images as models, keeps us unknown to the fact that how it would perform if provided with less-informative hand-drawn models. Same is the case with [7], which uses an approach based on edge patches producing unknown results on hand drawings.

In the recent years, such work has only been done in the field of deep leaning [8–10]. In variance with the previous work done, our method allows classification of cluttered images which also allows shape variability. Small objects in large cluttered images can also be detected. We work from five models [11] (a single image for each lass) to the classification of 255 images belonging to these classes.

3 Theory and Algorithms

3.1 System Overview

Figure 1 shows the series of steps that have been conducted to implement the two techniques and then combing their probabilities. The dataset is evaluated on MPEG-7 features and the template matching algorithm separately. The output generated is then combined in the third step of evaluation.

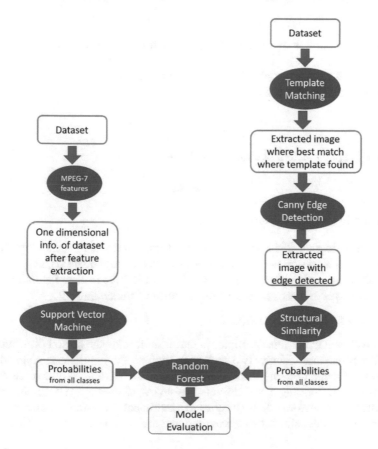

Fig. 1 System overview

3.2 MPEG-7

MPEG-7 is a low-level feature extraction tool for multimedia content. It was developed by Moving Picture Experts Group. The MPEG is a working group of International Standards Organization/International Electro-technical Committee (ISO/IEC). MPEG-7 is formally called Multimedia Content Description Interface [12]. It is used for multimedia content description. The description, associated with the content, allows reliable and quick search for the material that is useful for the user. The visual descriptors provided by MPEG-7 provide elementary features such as color, texture, and shape. The six descriptors are as follows:

(a) **Color Structure Descriptors (CSD)**

It represents the spatial layout of colored images. This is used for differentiating between the images of the basis of color distributions present in an image and the structure of the image.

An 8×8 structuring element is used where the number of samples is 64, and the sub-sampling factor and spatial extent vary with the image size. The spatial extent (s) is:

$$s = \max\{0, \text{round}(0.5 * l \log_2 \text{hw} - 8)\}$$

where w and h are width and height of the image, respectively
   k is the sub-sampling factor:

$$k = 2^s$$

$T \times T$ is spatial extent of structuring element.

$$T = 8k$$

(b) **Scalable Color Descriptor (SCD)**

It is a color histogram. This descriptor is used for image to image matching based on color feature. This descriptor contains color space, color quantization, and histogram descriptor. It is a color histogram quantized into 256 bins in Hue-Saturation-Value (HSV) color space. It uses Haar transform coefficient encoding [8].

(c) **Color Layout Descriptor (CLD)**

The main advantage of this descriptor is that it represents the spatial distribution of color. The procedure includes two steps: (a) dividing the input picture into 64 parts (8×8) blocks and deriving their average colors, each average color represents its block; (b) by performing 8×8 discrete cosine transformation (DCT), the average colors are then transformed into a series of coefficient. Low-frequency coefficients are then filtered out using zig-zag scanning and are quantized to form CLD.

CLD measures the matching between two CLD's using distance measure which is as follows.

Suppose, we have two CLD's $\{DY_j, DCr_j, DCb_j\}$ and $\{DY'_j, DCr'_j, DCb'_j\}$. Then, the distance measure D is as follows:

$$D = \sqrt{\sum_j w_{yj}\left(DY_j - DY'_j\right)^2} + \sqrt{\sum_j w_{rj}\left(DCr_j - DCr'_j\right)^2}$$
$$+ \sqrt{\sum_j w_{bj}\left(DCb_j - DCb'_j\right)^2}$$

where (DY_j, DCr_j, DCb_j) depict the jth coefficients of the respective color components.

The retrieval efficiency and the functionality of visual signal matching are provided by the compactness of the spatial distribution. It provides the features of sequence to sequence matching and image to image matching.

(d) Dominant Color Descriptor (DCD)

This feature is used for representing the dominant color in a particular region of interest. Color quantization is used to extract a small number of representing colors in each region.

(e) Homogenous Texture Descriptor (HTD)

It is used for categorizing the image and characterizing the properties of texture present in an image. Scale and orientation sensitive filters are used for computing HTD. The filtered outputs are then used to calculate the quantitative measures, mean and standard deviation.

(f) Edge Histogram Descriptor (EHD)

This descriptor primarily performs image to image matching on the basis of edges present in them. Edge is an essential part of an object. Using it in classification can give significantly improved results even in the case of natural images, where edge distribution is highly non-uniform. There are five types of edge orientations (one non-directional and four directional), as shown in Fig. 2, on which the descriptor works. It can return instances with similar semantic meaning. It is used specially for real images non-uniform edges. This descriptor combined with other descriptor significantly improves the retrieval efficiency.

All the MPEG-7 features give different features each as an output, which sum up to 512 in all. The features all combined are sent as input to SVM implemented in Python 3.6. This in turn generates the five probabilities of each image for every class.

Fig. 2 Edge masks:
a horizontal, **b** vertical,
c diagonal 135°, **d** diagonal
45°, **e** non-directional

(a)

1	1
-1	-1

(b)

1	-1
1	-1

(c)

0	$\sqrt{2}$
$-\sqrt{2}$	0

(d)

$\sqrt{2}$	0
0	$-\sqrt{2}$

(e)

2	-2
-2	2

3.3　Support Vector Machine (SVM)

Support Vector Machine is a supervised machine learning algorithm which is mostly used for solving classification problems. N-dimensional space is plotted with each data item. Here, n is the number of features we have. In our case, we have 512 features + 1 label which span five classes of data. Classification is then performed by finding the hyper planes that differentiates the five classes. This algorithm was proposed by Cortes and Vapnik [13]. Its two sub parts are:

(A)　Kernel Trick

The main purpose of the kernel is that it is used for pattern analysis. It helps us to find the similarity among classifications, correlations, clusters, etc. The kernel provides the facility of operating data in higher dimensional, implicit feature space. Hence, no computation of coordinates required. This is known as "Kernel Trick." Python provides us with four kernels: linear, poly, sigmoid, and RBF.

(B)　Support Vectors

Vectors which define the hyper plane that is used to distinguish between classes in a SVM are known as support vectors [13]. In SVM, the training involves the minimization of error function:

$$\frac{1}{2}v^t v + m \sum_{a=1}^{n} X_a$$

Subjected to:

$$y_a(v^t \varphi(x_a) + b) \geq 1 - X_a \text{ and } X_a \geq 0, \ a = 1, \ldots, n$$

Table 1 Probabilities generated for five example instances after application of MPEG-7 features and classification using Support Vector Machine

S. No.	Apple logo	Bottle	Giraffe	Mug	Swan	Label
1	0.344	0.180	0.134	0.182	0.158	0
2	0.105	0.424	0.150	0.287	0.032	1
3	0.004	0.028	0.792	0.018	0.156	2
4	0.118	0.117	0.260	0.338	0.164	3
5	0.104	0.202	0.150	0.123	0.419	4

m is capacity constant, b is a constant, v is vector of coefficients, and X_a represents parameters for handling non-separable data provided as input. The index a labels the n training cases.

x_a represents the independent variables and represents the class labels. The kernel, polynomial kernel used here, is used to transform data from input to the feature space.

Polynomial kernel function:

$$k(x, y) = \left(\gamma . x^s y + u\right)^d, \gamma > 0$$

Here, u, d, and γ are hyper parameters. γ is used to control the locality of the kernel function.

We have implemented SVM in Python with the help of Python's "sklearn library" to generate the probabilities for each image over each **class.**

We have specifically utilized probabilities of each class, as shown in Table 1, generated by SVM rather than class label. This helps us to combine with template matching in case of false alarms and false negatives.

3.4 Template Matching and Structural Similarity

Template Matching

Template matching is the method of finding the areas of an image that are similar to another image provided as template. It has varied applications like object detection, shape detection, etc.

A template is basically a small image with certain features. The goal of this technique is to find the portions in the larger image which are similar to the smaller image, that is, the template as shown in Fig. 3.

The algorithm requires two inputs:

- The source image: the image in which the template is to be found.
- The template image: the smaller image to be found in the larger image.

In template matching, the technique that is followed is that the smaller image or template is made to convolve over the larger image which is varied in size from

(a)   **(b)**   **(c)**

Fig. 3 Template matching performed on an example, **a** is source image, **b** template image, **c** matched template on image

100% of the image to 20% of its size for a best match to be found. This is done by looping over the input image at multiple scales, making it progressively smaller. The match of it is found by every time comparing the region beneath the template to the template. Python's OpenCV library is used for implementing this technique.

Every time, the model is convolved over the rescaled input image, and the value of the largest correlation coefficient is maintained and is considered the best match for the template in the input image.

We use the fast Fourier transform implementation done in the OpenCV library. In this method, the signals representing each pixel are evaluated into its equivalent waveform making the process of comparison much faster than other template matching techniques.

$$f \otimes w = \text{iFFT}[\text{FFT}[f(a, b)]^* . \text{FFT}[w(a, b)]]$$

where iFFT is the inverse and $[f(a, b)]^*$ is the complex conjugate of $[f(a, b)]$.

The procedure followed for the implementation of template matching, as shown in Fig. 4, is:

(a)   **(b)**   **(c)**

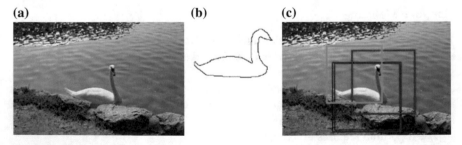

Fig. 4 Template matching of model performed on the images multiple times to find the best fit, **a** represents original image, **b** represents the model, **c** represents all template matching performed on the image

(a)   (b)

Fig. 5 Extraction of the best fit during the template matching, **a** represents all template matching, **b** represents best match

- Convolution of the model image on the images of test dataset.
- Detection of possible places where the object might be present by repeating the process of template matching.
- Identification of the region of the image which matches the most with the model provided.
- Extraction of the portion with highest matching as shown in Fig. 5.

Canny Edge Detection

Canny edge detection [14] is an edge detection algorithm that uses multi-stage technique for the detection of a wide range of edges present in that image.

- The portion extracted by the best fit generated by template matching is operated on by applying the Canny edge detection.
- This makes the extracted portion to be efficiently compared to the original test models as shown in Fig. 6.

(a)   (b)   (c)

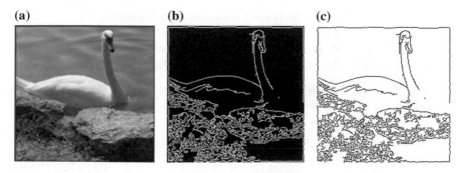

Fig. 6 Extracted image (**a**) is sent as input to Canny edge detector which detects the edges of the image (**b**). The colors of the returned image are inverted to match it to the model image (**c**)

(a)   (b)

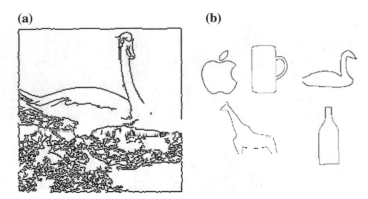

Fig. 7 Images with edges generated is compared with all the five models to generate probabilities, **a** is edge-detected image, **b** all models to be compared to the image

Structural Similarity

The SSIM was developed by Wang et al. in 2004 [15]. It is a perception-based model that finds out the perceived changes in the structure of the information. Structural information defines that the pixels possess strong inter dependency. His inter dependency is a noteworthy point especially when the pixels are spatially close.

As per the observation, the structural similarity between the model and the template matched, and the edge-detected image is the probability of having similar features.

The dataset formed after Canny edge detection offers a platform for better comparison with the model images, as shown in Fig. 7, as they are nearly in the same format. For implementation of this, we use the following techniques by the structural similarity index calculation algorithm:

- It is a method used for measuring how similar two images are.
- SSIM tries to model the perceived changes in the structural information of the image.
- It does not compare the entire images but the small sub-samples.
- The approach of not combining the entire images but the sub-samples is a more robust one that is able to account for changes in the structure of the image.
- The SSIM value varies between -1 and 1, where 1 indicates perfect similarity.

The algorithm is for multi-scale structural similarity for images to be compared [16]. The measure between images a and b of common size is as follows:

$$\text{SSIM}(a, b) = \frac{(2\mu_a\mu_a + g_1)(2\sigma_{ab} + g_2)}{(\mu_a^2 + \mu_b^2 + g_1)(\sigma_a^2 + \sigma_b^2 + g_2)}$$

where

- μ_a is the average of a

Table 2 Probabilities of classes generated for image in Fig. 4

Class	Apple logo	Bottle	Giraffe	Mug	Swan
SSIM	0.392	0.421	0.395	0.41	0.43

- μ_b is the average of b
- σ_a^2 is the variance of a
- σ_b^2 is the variance of b
- σ_{ab} is the co-variance of a and b
- $g_1 = (k_1 L)^2$ and $g_2 = (k_2 L)^2$
- L is the dynamic range of pixel values
- $k_1 = 0.01$ and $k_2 = 0.03$ by default.

Example

The highest structural similarity index obtained, shown in Table 2, is that of the model of the swan. This classifies the input image to be a part of the fifth class.

4 Experimentation

4.1 Dataset

This dataset contains images of five shape-based classes collected from Flickr and Google images by ETHZ Zurich [17]. It majorly consists of 255 real test images and features five different shape-based classes and five distinct hand-drawn models, one of each class, shown in Fig. 8.

The main challenge that this dataset poses includes:

- Clutter, shape variability, and scale changes.
- In some images, the object comprises only a small portion of the image.
- The objects are appearing at a wide range of scales.
- Most of the objects are taken from the same viewpoint.

Apple Logo   Bottle   Giraffe   Mug   Swan

Fig. 8 The model images of the five classes on which the dataset is **based**

Table 3 Total dataset of 255 images

Name of class	Number of images
Apple logos	40
Bottles	48
Giraffes	87
Mugs	48
Swans	32

Models

The test dataset, referred to in Table 3, is further split into 176 images (70% of total) for being trained over the model. The remaining 79 images (30% of total) are used for being tested on the model images.

4.2 Implementation Framework

The MPEG-7 and the template matching implementation were done on Python 3.6. Further, the two probabilities were combined using Orange 3.

5 Result and Discussion

5.1 Results Generated by MPEG-7

The confusion matrix shown in Table 4 generated by SVM applied on the MPEG-7 features shows that the class of giraffes has been predicted correctly using SVM.
    The overall accuracy obtained is as follows:

Table 4 Confusion matrix generated by SVM

	A	B	C	D	E	$\sum$
A	7	1	0	1	0	9
B	3	5	5	4	0	17
C	1	1	25	1	0	28
D	3	2	3	7	1	16
E	3	0	1	0	2	6
$\sum$	17	9	34	13	3	76

A Apple logo, *B* Bottles, *C* Giraffes, *D* Mugs, *E* Swans

Table 5 Confusion matrix by template matching

	A	B	C	D	E	$\sum$
A	8	1	0	1	0	10
B	0	8	4	1	0	13
C	0	4	23	0	1	28
D	0	4	5	6	0	15
E	0	4	1	0	5	10
$\sum$	8	21	33	8	6	76

$$accuracy = \frac{number\ instances\ classified\ correctly}{total\ number\ of\ instances} \times 100 = \frac{46}{76} \times 100$$
$$= 60.52\%$$

5.2 Results Generated by Template Matching and Structural Similarity

The confusion matrix shown in Table 5 generated by template matching shows improved results over majority classes than SVM.

The overall accuracy obtained is as follows:

$$accuracy = \frac{number\ instances\ classified\ correctly}{total\ number\ of\ instances} \times 100 = \frac{50}{76} \times 100$$
$$= 65.78\%$$

5.3 Result Generated by Combining the Two Above-Mentioned Approaches

Table 6 gives a summary of classification on test data over four classifiers: K-Nearest

Table 6 Results generated on various classifiers

Method	AUC	Accuracy	Precision	Recall
kNN	0.952	0.842	0.854	0.842
Tree	0.906	0.829	0.837	0.829
SVM	0.973	0.868	0.867	0.868
Random forest	0.952	0.882	0.881	0.882

Neighbor, Tree, SVM, and Random Forest.

It can be observed that Random Forest-based model outperforms every other model.

$$\text{accuracy} = \frac{\text{number instances classified correctly}}{\text{total number of instances}} \times 100 = \frac{67}{76} \times 100$$
$$= 88.2\%$$

5.4 ROC Curve

Confusion Matrix generated by Random Forest is given in Table 7. From the matrix, we can observe that the number of correctly classified instances has increased to a significant number after combining the two classifiers in Orange, thus generating the maximum accuracy as 88.2%. The ROC curve, shown in Fig. 9, shows that the maximum area under curve (AUC) is generated by Random Forest algorithm in case of all classes.

Table 7 Confusion Matrix generated by Random Forest

	A	B	C	D	E	$\sum$
A	12	0	1	1	0	9
B	0	14	1	2	0	17
C	0	1	22	0	0	28
D	2	1	0	9	0	12
E	0	0	0	0	10	10
$\sum$	14	16	24	12	10	76

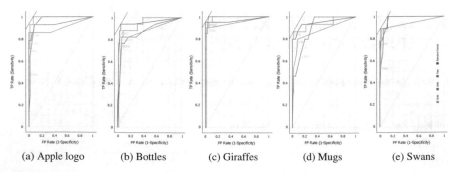

(a) Apple logo   (b) Bottles   (c) Giraffes   (d) Mugs   (e) Swans

Fig. 9 ROC curve for all the classes on the ETHZ dataset shows that best results have been obtained by Random Forest classification Algorithm. **a–e** represent the ROC on all four classification algorithms for each class

6 Conclusion and Future Work

In this paper, we proposed a generic approach for image classification on the basis of the shape present in it. We have showed the effectiveness of MPEG-7 features. We have worked on only 255 images in this classification problem and achieved a good accuracy on this dataset. We have been successful in proving that even small databases can be used for classification on the basis of shapes. On increasing the number of classes and on adding more complicated structures like doors, chairs, animals, etc., more efficient results can be obtained. This is going to serve the purpose of searching or locating entities. The primary purpose is classification of objects into different categories. The generic approach works in a way that machine learning framework will learn the pattern of images and will classify the results with a good accuracy.

References

1. Basri, R., Costa, L., Geiger, D., Jacobs, D.: Determining the similarity of deformable shapes. Vision. Res. **38**(15–16), 2365–2385 (1998)
2. Belongie, S., Malik, J., Puzicha, J.: Shape matching and object recognition using shape contexts (2002)
3. Gavrila, D.M., & Philomin, V.: Real-time object detection for "smart" vehicles. In: The Proceedings of the Seventh IEEE International Conference on Computer Vision, vol. 1, pp. 87–93. IEEE (1999)
4. Leibe, B., Seemann, E., Schiele, B.: Pedestrian detection in crowded scenes. In: Null, pp. 878–885. IEEE (2005)
5. Thayananthan, A., Stenger, B., Torr, P. H., Cipolla, R.: Shape context and chamfer matching in cluttered scenes. In: IEEE Computer Society Conference on Computer Vision and Pattern Recognition, 2003. Proceedings 2003, vol. 1, pp. I–I. IEEE (2003)
6. Berg, A.C., Berg, T.L., Malik, J.: Shape matching and object recognition using low distortion correspondences. In: IEEE Computer Society Conference on Computer Vision and Pattern Recognition, CVPR 2005, vol. 1, pp. 26–33. IEEE (2005)
7. Nelson, R.C., Selinger, A.: A cubist approach to object recognition. In: Sixth International Conference on Computer Vision, pp. 614–621. IEEE (1998)
8. Chen, L.C., Barron, J.T., Papandreou, G., Murphy, K., Yuille, A.L.: Semantic image segmentation with task-specific edge detection using CNNs and a discriminatively trained domain transform. In: Proceedings of the IEEE Conference on Computer Vision and Pattern Recognition, pp. 4545–4554 (2016)
9. Badrinarayanan, V., Kendall, A., Cipolla, R.: Segnet: a deep convolutional encoder-decoder architecture for image segmentation. IEEE Trans. Pattern Anal. Mach. Intell. **39**(12), 2481–2495 (2017)
10. Chen, L.C., Papandreou, G., Kokkinos, I., Murphy, K., Yuille, A.L.: Deeplab: Semantic image segmentation with deep convolutional nets, atrous convolution, and fully connected crfs. IEEE Trans. Pattern Anal. Mach. Intell. **40**(4), 834–848 (2018)
11. Ferrari, V., Jurie, F., Schmid, C.: From images to shape models for object detection. Int. J. Comput. Vision **87**(3), 284–303 (2010)
12. Manjunath, B.S., Salembier, P., Sikora, T.: Introduction to MPEG-7: Multimedia Content Description Interface, vol. 1. Wiley (2002)
13. Cortes, C., Vapnik, V.: Support-vector networks. Mach. Learn. **20**(3), 273–297 (1995)

14. Canny, J.: A computational approach to edge detection. IEEE Trans. Pattern Anal. Mach. Intell. **6**, 679–698 (1986)
15. Wang, Z., Bovik, A.C., Sheikh, H.R., Simoncelli, E.P.: Image quality assessment: from error visibility to structural similarity. IEEE Trans. Image Process. **13**(4), 600–612 (2004)
16. Wang, Z., Simoncelli, E., Bovik, A.: Multi-scale structural similarity for image quality assessment. In: Asilomar Conference on Signals Systems and Computers, vol. 2, pp. 1398–1402. IEEE (1998)
17. ETHZ Shape Classes. http://www.vision.ee.ethz.ch/en/datasets/

Analysis of Multichannel SSVEP for Different Stimulus Frequencies

V. Sankardoss and P. Geethanjali

Abstract Steady-state visual evoked potential (SSVEP)-based methods are gaining popularity in different fields like cognition and brain–computer interface (BCI) studies. The SSVEP is used in BCI due to less training time, more accuracy, and high information transfer rate. The properties of SSVEP depend on factors like shape, size, frequency, and color of the stimuli. In this work, the white/black square stimuli are studied from O_z, O_1, O_2, P_z, P_1, and P_2 channels to analyze the power and energy at four different frequencies. The Psychophysics Toolbox in MATLAB is used to design the stimuli. The average power and average energy from white/black square stimuli of different channels help in frequency selection of visual stimuli at different channels. Further, signals with high energy may be effective in BCI.

Keywords Electroencephalogram · Repetitive visual stimulus · Steady-state visual evoked potential

1 Introduction

Amyotrophic lateral sclerosis (ALS), brain injury, brainstem stroke, cerebral paralysis, muscular dystrophies, numerous sclerosis, and different sicknesses disable the neural pathways that disable the muscles or control muscles [1]. There are different techniques accessible to assist the physically disabled people in controlling communication devices, a computer or a wheelchair. The autonomous portability from an electric wheelchair with the hands-free interface is exceptionally vital for the neuromuscular disabled subject and achieved using a brain–computer interface (BCI) [2].

In noninvasive BCI, electroencephalography (EEG) is applied due to high temporal resolution, easy acquisition, and less cost compared with the other brain movement checking modalities. Noninvasive electrophysiological sources such as event-related

V. Sankardoss · P. Geethanjali (✉)
School of Electrical Engineering, VIT, Vellore, Tamil Nadu, India
e-mail: pganjali78@hotmail.com

© Springer Nature Singapore Pte Ltd. 2020
K. N. Das et al. (eds.), *Soft Computing for Problem Solving*,
Advances in Intelligent Systems and Computing 1057,
https://doi.org/10.1007/978-981-15-0184-5_51

synchronization/desynchronization (ERS/ERD), P300 evoked potentials, slow cortical potentials (SCPs), visual evoked potentials (VEPs), steady-state visual evoked potentials (SSVEPs), and μ and β rhythms are used in BCI control [3, 4]. The SSVEP-based BCI is preferred due to the high information transfer rate (ITR) and requires less training time for the subject [5].

The steady-state visual evoked potential in the visual cortex is generated, when the subject is given visual stimuli. The evoked potential frequency is equal to the fundamental frequency (or harmonics) of visual stimulus flickering [6, 7]. The SSVEP evoked at different frequency bands such as low (5–12) Hz, medium (12–25) Hz, and high (25–50) Hz [8]. The 64-channel electrophysiological SSVEP is related to visual-spatial attention with high-density signal for independent BCI [9]. The 1-D as well as 2-D cursor control BCI designed volitional control of multichannel EEG frequency spectrum or SSVEP using kernel partial least squares classification to attain high accuracy [10]. The biphasic stimulation method describes the phase drifts of SSVEP BCI with eight target visual stimuli [11]. The medium and high stimulus frequency effects as well as the demographics response of SSVEP-based BCI navigate the small robot [12]. The paradigm of reliable BCI, employing frequency-shift keying (FSK) method, increases visual stimulus number with different properties. Its application is with a limited number of frequencies [13]. The SSVEP-based BCI with 12 subjects and 9 targets is displayed in the black background in a random sequence with white disk stimulus flickers in a clockwise direction at a constant frequency [14].

The cathode ray tube (CRT) screen is used to evoke the SSVEP with the refresh rate of 60 Hz [15]. The SSVEP evoked at a frequency range of (1–90) Hz using light-emitting diode (LED) [16]. The SSVEP-based BCI provides higher duty cycle using six visual target LED flickers [17]. The custom-built BCI stimulator consists of 8×8 matrix of green LEDs and produces a higher SSVEP response compared with ordinary color LEDs [18]. The SSVEP based BCI is implemented with 5 repetitive visual stimuli (RVS) as well as with LEDS for varying flicker frequencies using single-channel EEG with 250 Hz sample rate, 24 bits resolution and 4 μV of noise level [19]. The LCD is backlit through LED arrays having 4×2 matrix having a distance of 0.8 cm using microcontroller [20]. The SSVEP EEG signal was recorded with software by BrainTech Ltd. and a cap of 8 water-based electrodes placed on occipital as well as parietal positions with 15 subjects.

The square stimuli flickers on LCD screen using frequency and phase mixed code in SSVEP based BCI [21]. The SSVEP-based BCI with 11 subjects, 16 channels, and 9 checkerboard visual stimulus spatial flicker frequencies is displayed on the LCD screen [22]. The 40-target BCI speller is employed in LCD monitor with frequency as well as phase values of all targets by applying joint frequency and phase modulation method [23]. The SSVEP-based BCI uses wearable in-the-ear EEG system. It has six white and black circle visual stimulus frequency ranges from 4 to 7.2 Hz on the LCD screen at the 144 Hz of refresh rate [24]. The SSVEP based BCI experiment was attempted with 13 subjects, wearing the EEG cap containing 6 electrodes and 3 targets having circle HoloLens, flickers at a frequency of 10, 12, and 15 Hz [25].

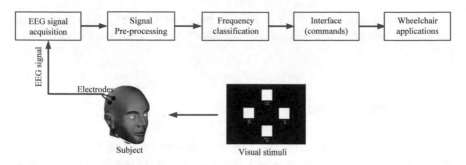

Fig. 1 Block diagram of SSVEP-based BCI

The schematic block diagram of SSVEP-based BCI is shown in Fig. 1. The user's SSVEP is evoked for his/her attention on a visual stimulus and exhibits an oscillatory response in the users' EEG.

A BCI builds on the conceptual idea of translating brain activity patterns into different classes and uses these as control commands to a computer or an external device/environment. In more detail, a complete BCI system consists of several parts: signal acquisition system, signal preprocessing, frequency classification, interface, and wheelchair applications.

The BCI control starts with the signal acquisition setup that acquires EEG, such that BCI rightly understands it. For extracting the bands of the frequency of the relevance, a bandpass filter is utilized on the EEG as a preprocessing of the signal. Usually, 8–30 Hz range bandpass filter is utilized by those in pursuit of research. In performing the pattern classifications for BCI applications, a wide choice of classifier techniques is employed. In this, the activities of the brain to be gathered, from the choice of proper visual stimulus technique and extraction of information, consider the type of algorithm utilized for BCI.

There are interfacing devices utilized for communicating from BCI classification to wheelchair actuating arrangements, such as a microcontroller, ADC, encoder, optocoupler, and motors. The wheelchair will receive the translational commands from classifier through the interface devices. Based on the essential translation commands, the wheelchair gets mobilized in forward, reverse, left, and right directions.

The visual stimulation and its wide range of properties play a crucial role in the BCI framework. The BCI performance is obtained by information transfer rate, which really measures how much information can be moved in 60 s. In order to enhance BCI performance, stimulation properties of size, color, frequencies have received consideration despite the fact that they can greatly influence the performance. The visual stimuli utilizing emotional human faces fundamentally improved the amplitude of the SSVEP signals when compared with checkerboard stimuli [26, 27]. The various types of stimulus patterns like the high duty cycle stimuli, high-frequency stimuli, and image-based stimuli have been introduced for decreasing visual fatigue while

keeping up a robust performance of SSVEP-based BCI [28–30]. This paper introduces an investigation of the impacts of various stimulus frequencies at six different
channels.

2 Stimulator Design

The white/black square visual stimuli can be generated from two different methods
of pattern reversal: (i) display of revered images—display of one image over another
image after a time delay—and (ii) repeating the screen region—pattern reversal at
frequency f Hz is accomplished by changing patterns to all T ms. In this work, the
visual stimuli are programmed using the Psychophysics Toolbox in MATLAB for
various frequencies 7.5, 11, 3, and 10 Hz in frame-based technique. Each white/black
square has a size of 3.5 cm × 3.5 cm flickers with equal space on either side. The
visual stimuli generated are shown on an LCD screen with the 60 Hz refresh rate.

The constant intensity evokes a powerful SSVEP response for a stimulus flickering
frequency in 3–12 Hz. The refresh rate plays an important role in the selection of
flicker frequencies. In a frame-based approach, the number of frames in the flickering
period is fixed for each stimulation frequency. The LCD screen with refresh rate of
60 Hz, produces black and white stimuli flickering rate at 3 Hz for the frame rate of
20 frames/period, 7.5 Hz for the 8 frames/period, 10 Hz for the 6 frames/period, but
11 Hz can be built with the averaging two frequencies, 10 and 12 Hz. Theoretically,
the maximum flicker frequency can be half the refresh rate. A frame structure of
generating visual stimuli is shown in Fig. 2. Each square flickers with a white/black
pattern of different frequencies.

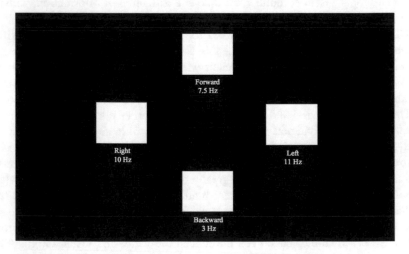

Fig. 2 White/black square flickers

3   Experimental Setup

In this work, the EEG signals are acquired for four visual stimuli. The experiment has been conducted with the approval from Institutional Human Ethics Committee with the participant consent. Eleven volunteers participated in the offline work in a dark room without electromagnetic shielding. The subjects were situated 70 cm from the LCD screen, and EEG was acquired utilizing TruScan EEG device with AgCl/Ag electrodes from the scalp. During acquisition, six electrodes were used. The sites are chosen by the 10-20 international system. The six EEG signals are acquired from O_z, O_1, O_2, P_z, P_1, and P_2. The reference electrode is placed on the mastoid bone behind the left ear and the ground electrode is placed at a nasion point on the forehead. The Nuprep Skin Prep Gel is used to prepare the skin, and Ten20 Conductive Paste is used to decrease the impedance.

The subject was seated in a comfortable chair from the LCD screen and was instructed to focus on the center of the square, with no or negligible eye flickering. The subject was instructed to see each square for 60 s. A little time took to account the subject to rest the eyes in the middle of the trials. The sampling frequency is 256 Hz. The recorded information was put away in the computer for further analysis.

The power (P) and energy (E) of the EEG signal are obtained using Eqs. (1) and (2) from the amplitude spectrum.

$$P = \frac{1}{K} \sum_{k=0}^{k-1} x^2[k] \tag{1}$$

$$E = T \sum_{k=0}^{k-1} x^2[k] \tag{2}$$

where

T Duration of the signal
$x[k]$ Discrete samples of the signal.

The six-channel EEG signals of O_z, P_1, P_z, P_2, O_1, and O_2 for a subject at the stimulation frequency of 11 Hz are shown in Fig. 3. The power spectral density of EEG signal from one subject for 11 Hz at O_z channel is shown in Fig. 4. The harmonics are not significant. The average power and average energy at different frequencies are calculated from the amplitude spectrum. Table 1 shows the power and energy computed from each subject at different frequencies.

4   Results and Discussion

The average power and average energy of six-channel EEG signal were computed at different frequencies 3, 7.5, 10, and 11 Hz.

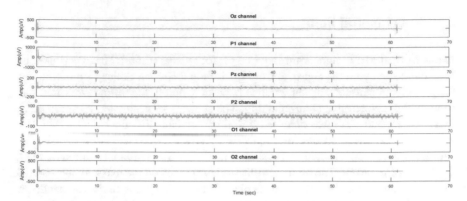

Fig. 3 Six-channel EEG signals from one subject for 11 Hz

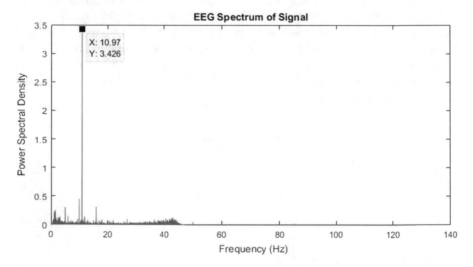

Fig. 4 Power spectral density of EEG signal from one subject for 11 Hz at O_z channel

It is clear from Table 1 the average power and energy produced from each subject are different. This may be due to varying factors like concentration and electrode positioning. However, the mean of power and energy is calculated for 11 subjects and shown in Figs. 5 and 6. From Figs. 5 and 6, it is clear that the energy and power of the O_z channel are higher for frequencies 10 and 3 Hz compared to other channels. The response of P_z channel is high for 11 and 7.5 Hz compared to O_z, but not significant. Therefore, O_z and P_z may be chosen for this BCI study for the four frequencies considered in this work.

Table 1 Average power and the average energy of SSVEP at different stimulus frequencies

Subjects	Average power (W)				Average energy (J)			
	3 Hz	7.5 Hz	10 Hz	11 Hz	3 Hz	7.5 Hz	10 Hz	11 Hz
Subject 1	7.51E−04	7.13E−04	5.24E−04	1.15E−03	2.40E−02	2.28E−02	1.68E−02	3.67E−02
Subject 2	9.54E−04	2.56E−03	7.85E−05	5.79E−04	6.10E−02	8.19E−02	2.51E−03	1.85E−02
Subject 3	0.009387	0.001	0.008322	0.001522	0.300407	0.031991	0.266349	0.048708
Subject 4	0.00405	0.002493	0.003819	0.005857	0.129618	0.079789	0.122239	0.187435
Subject 5	0.001126	0.001704	0.002925	0.000828	0.036039	0.054522	0.093624	0.026484
Subject 6	0.000331	0.000745	6.35E−05	4.15E−05	0.010579	0.047663	0.002031	0.001329
Subject 7	3.41E−05	0.000233	0.001558	0.000752	0.001093	0.007449	0.049867	0.024057
Subject 8	0.001315	0.000905	0.002593	0.000382	0.084187	0.057938	0.165955	0.02444
Subject 9	0.000293	0.000615	0.000993	0.000519	0.009383	0.019694	0.031777	0.016611
Subject 10	3.54E−05	0.000305	0.000143	2.11E−05	0.001133	0.019527	0.004592	0.000676
Subject 11	0.001293	0.011767	0.00929	0.005483	0.041393	0.376598	0.297303	0.175486
Mean	0.001779	0.002095	0.002755	0.001558	0.063531	0.072716	0.0957315	0.0509478

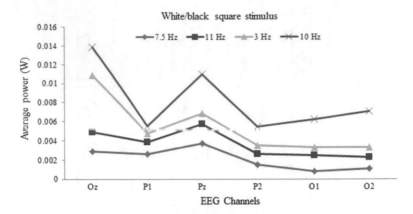

Fig. 5 Average power versus EEG channels

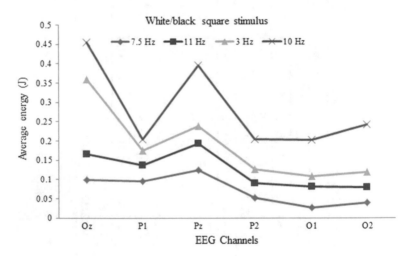

Fig. 6 Average energy versus EEG channels

5 Conclusion

In this work, the visual stimuli in BCI at different frequencies for multichannel SSVEP were analyzed briefly. The impact of frequency and location of electrodes was analyzed. This work helps to select the appropriate stimuli which may improve the performance of SSVEP-based BCI. This work may be further extended to different types and frequencies to identify the electrode location.

Acknowledgements Thanks to the Science and Engineering Research Board (SERB), Government of India, for the funding support: SB/FTP/ETA-54/2013.

References

1. Kubler, A., Kotchoubey, B., Wolpaw, J.R., Birbaumer, N.: Brain-computer communication: unlocking the locked in. Psychol. Bullet. **127**(3), 358–375 (2001)
2. Tanaka, K., Matsunaga, K., Wang, H.: Electroencephalogram-based control of an electric wheelchair. IEEE Trans. Robot. **21**(4), 762–766 (2005)
3. Bashashati, A., Fatourechi, M., Ward, R.K., Birch, G.E.: A survey of signal processing algorithms in brain-computer interfaces based on electrical brain signals. J. Neural Eng. **4**(2), R32–R57 (2007)
4. Zhu, D., Bieger, J., Molina, G.G., Aarts, R.M.: A survey of stimulation methods used in SSVEP-based BCIs. Hindawi Publ. Corp. Comput. Intell. Neurosci. **702357**(12) (2010)
5. Parini, S., Maggi, L., Turconi, A.C., Andreoni, G., A robust and self-paced BCI system based on a four class SSVEP paradigm: algorithms and protocols for a high-transfer-rate direct brain communication. Comput. Intell. Neurosci. **864564**(11) (2009)
6. Geethanjali, P., Prakash, P.R., Kothari, S.: Investigation of multiple frequency recognition from single channel SSVEP for efficient BCI application. IET Signal Process. **12**(3), 255–259 (2018)
7. Lin, Z., Zhang, C., Wu, W., Gao, X.: Frequency recognition based on canonical correlation analysis for SSVEP-based BCIs. IEEE Trans. Biomed. Eng. **54**(6), 1172–1176 (2007)
8. Regan, D.: Human brain electrophysiology: evoked potentials and evoked magnetic fields in science and medicine. Elsevier (1989)
9. Kelly, S.P., Lalor, E.C., Reilly, R.B., Foxe, J.J.: Visual spatial attention tracking using high-density SSVEP data for independent brain-computer communication. IEEE Trans. Neural. Syst. Rehabil. Eng. **13**(2), 172–178 (2005)
10. Trejo, L.J., Rosipal, R., Matthews, B.: Brain-computer interfaces for 1-D and 2-D cursor control: designs using volitional control of the EEG spectrum or steady-state visual evoked potentials. IEEE Trans. Neural. Syst. Rehabil. Eng. **14**(2), 225–229 (2006)
11. Wu, H.Y., Lee, P.L., Chang, H.C., Hsieh, J.C.: Accounting for phase drifts in SSVEP-based BCIs by means of biphasic stimulation. IEEE Trans. Biomed. Eng. **58**(5), 1394–1402 (2011)
12. Volosyak, I., Valbuena, D., Luth, T., Malechka, T., Graser, A.: BCI demographics II: How many (and what kinds of) people can use a high-frequency SSVEP BCI? IEEE Trans. Neural. Syst. Rehabil. Eng. **19**(3), 232–239 (2011)
13. Kimura, Y., Tanaka, T., Higashi, H., Morikawa, N.: SSVEP-based brain–computer interfaces using FSK-modulated visual stimuli. IEEE Trans. Biomed. Eng. **60**(10), 2831–2838 (2013)
14. Maye, A., Zhang, D., Engel, A.K.: Utilizing retinotopic mapping for a multi-target SSVEP BCI with a single flicker frequency. IEEE Trans. Neural. Syst. Rehabil. Eng. **25**(7), 1026–1036 (2017)
15. Lyskov, E., Ponomarev, V., Sandstrom, M., Mild, K.H., Medvedev, S.: Steady-state visual evoked potential to computer monitor flicker. Int. J. Psychophys. **28**, 285–290 (1998)
16. Herrmann, C.S.: Human EEG responses to 1–100 Hz flicker: resonance phenomena in visual cortex and their potential correlation to cognitive phenomena. Exp. Brain Res. **137**, 346–353 (2001)
17. Lee, P.L., Yeh, C.L., Sung Cheng, J.Y., Yang, C.Y., Lan, G.Y.: An SSVEP-based BCI using high duty-cycle visual flicker. IEEE Trans. Biomed. Eng. **58**(12), 3350–3359 (2011)
18. Waytowich, N.R., Yamani, Y., Krusienski, D.J.: Optimization of checkerboard spatial frequencies for steady-state visual evoked potential brain–computer interfaces. IEEE Trans. Neural Syst. Rehabil. Eng. **25**(6), 557–565 (2017)
19. Ajami, S., Mahnam, A., Abootalebi, V.: An adaptive SSVEP-based brain-computer interface to compensate fatigue-induced decline of performance in practical application. IEEE Trans. Neural Syst. Rehabil. Eng. **26**(11), 2200–2209 (2018)
20. Chabuda, A., Durka, P., Żygierewicz, J.: High frequency SSVEP-BCI with hardware stimuli control and phase-synchronized comb filter. IEEE Trans. Neural Syst. Rehabil. Eng. **26**(2), 344–352 (2018)
21. Jia, C., Gao, X., Hong, B., Gao, S.: Frequency and phase mixed coding in SSVEP-based brain–computer interface. IEEE Trans. Biomed. Eng. **58**(1), 200–206 (2011)

22. Wang, Y., Chen, X., Gao, X., Gao, S.: A benchmark dataset for SSVEP-based brain–computer interfaces. IEEE Trans. Neural Syst. Rehabil. Eng. **25**(10), 1746–1752 (2017)
23. Waytowich, N.R., Krusienski, D.J.: Multiclass steady-state visual evoked potential frequency evaluation using chirp-modulated stimuli. IEEE Trans. Hum. Mach. Syst. **46**(4), 593–600 (2016)
24. Ahn, J.W., Ku, Y., Kim, D.Y., Sohn, J., Kim, J.-H., Kim, H.C.: Wearable in-the-ear EEG system for SSVEP-based brain–computer interface. Electron Lett. **54**(7), 413–414 (2018)
25. Si-Mohammed, H., Petit, J., Jeunet, C., Argelaguet, F., Spindler, F., Evain, A., Roussel, N., Casiez, G., Lécuyer, A.: Towards BCI-based interfaces for augmented reality: feasibility, design and evaluation. IEEE Trans. Visual. Comp. Graph (2018)
26. Cheng, M., Gao, X., Gao, S., Xu, D.: Design and implementation of a brain–computer interface with high transfer rates. IEEE Trans. Biomed. Eng. **49**(10), 1181–1186 (2002)
27. Bakardjian, H., Tanaka, T., Cichocki, A.: Emotional faces boost up steady-state visual responses for brain–computer interface. NeuroReport **22**(3), 121–125 (2011)
28. Gao, S., Wang, Y., Gao, X., Hong, B.: Visual and auditory brain–computer interfaces. IEEE Trans. Biomed. Eng. **61**(5), 1436–1447 (2014)
29. Hwang, H.J., Kim, D.H., Han, C.H., Im, C.H.: A new dual frequency stimulation method to increase the number of visual stimuli for multi-class SSVEP-based brain–computer interface (BCI). Brain Res. **1515**, 66–77 (2013)
30. Wang, Y., Wang, R., Gao, X., Hong, B., Gao, S.: A practical VEP based brain-computer interface. IEEE Trans. Neural Sys. Rehabi. Eng. **14**(2), 234–240 (2006)

Design and Implementation of a Blood Vessel Identification Algorithm in the Diagnosis of Retinography

Kumarkeshav Singh, Kalugotla Raviteja, Viraj Puntambekar and P. Mahalakshmi

Abstract In this work, diverse condition of craftsmanship strategies for retinal vein division was executed and investigated. Right off the bat, an administered strategy in light of dark level and minute invariant highlights with neural system was investigated. Alternate counts considered were an unsupervised strategy in view of dark level co-event framework with nearby entropy and a coordinated separating technique in light of first request subordinate of Gaussian. Amid the work, openly accessible picture database DRIVE was used for assessing the execution of the calculations which incorporates affectability, specificity, exactness, positive prescient esteem and negative prescient esteem. The accuracy for the blood vessel segmentation was very near to which was given in the literature which was referred. Now because the sensitivity of all the methods used by us was lower, it leads to lower number of correctly classed vessels from the images taken. The results achieved have tremendous potential for application in real life, and in practical use, only a little bit of modification is required to get better segmentation between vessels and the corresponding background.

Keywords Matched filtering · Performance measure · Retinal image · Retinal vessel segmentation · Supervised method · Unsupervised method

1 Introduction

The only place where all the blood vessels related to the eyesight can be analyzed directly in situ is in the retina. As in the last 20 years, the technology has advanced to great heights, and with it, the digital imaging methods have also caused a revolution in this retinal imaging process. The resolution of conventional imaging has not been reached yet, and the latest technology imaging systems gives us some of the most high-resolution images that are more than sufficient for most of the medical scenarios.

K. Singh · K. Raviteja · V. Puntambekar · P. Mahalakshmi (✉)
School of Electrical Engineering, Vellore Institute of Technology, Vellore,
Tamil Nadu 632014, India
e-mail: pmahalakshmi@vit.ac.in

© Springer Nature Singapore Pte Ltd. 2020
K. N. Das et al. (eds.), *Soft Computing for Problem Solving*,
Advances in Intelligent Systems and Computing 1057,
https://doi.org/10.1007/978-981-15-0184-5_52

The images of the retina have been extensively used for detection in vascular and non-vascular pathology in the field of medicine [1]. The images captured gives us knowledge on the change in the pattern of retinal blood vessels which actually shows us the type of disease like diabetes, occlusion, glaucoma, hypertension, cardiovascular disease and stroke [2, 3]. These images send us information about the change in pattern of blood vessels and the contrast and the reflectivity. This helps in the early detection, and these early detections help us to cure it early.

The segregation or segmentation of the blood vessels of the retina from its images will help us in quick and accurate medical studies and researches. For this same reason, this method has to be precise and reliable as far as possible. This segregation is to first declassify the given image to ungroup the components and then only select the data needed to us.

1.1 Background

1.1.1 Classification of Methods

A couple of methodologies for the division of retinal picture have been represented in composing. In perspective of machine learning systems, retinal vein division can be disconnected into two social occasions: oversaw procedures [1, 4–7] and unsupervised strategies [8–11]. Coordinated procedures rely upon the prior checking information which orders whether a pixel has a place with a vessel or non-vessel class. However, unsupervised methodologies do not utilize prior stamping information and have ability to learn and mastermind information in solitude to find the illustrations or packs that take after the veins. Isolation or partition based strategies [12–16] use a Gaussian-shaped curve to show the cross-zone of a vessel and turn the organized channels to recognize veins with different presentations. Particular shaped Gaussian channels, for instance, clears Gaussian model [12–15] and backup of Gaussian limit [16], have been used for vessel area. Another procedure in light of logical morphology [17, 18] misuses known vessel features and points of confinement and addresses them in numerical sets. By then, using morphological chairmen, the vessels are removed from the establishment.

1.1.2 Motivation

Manual division of the retinal veins is troublesome and dull, and affecting a point by direct division, can be worrying if the diverse nature of the vascular framework is too high [19]. Along these lines, robotized division is vital, as it lessens the time and effort required, and in the best circumstance, a motorized computation can give as awesome or better division comes to fruition as an authority by manual naming [20]. For convenient applications, it is more intelligent to have counts that do not essentially depend after outlining various parameters, so, moreover, non-experts may utilize this advancement effortlessly [28]. Automated vein division has defied challenges related to low separation in pictures, broad classification of vessel diameters and extensive range of features in retinal pictures, for instance, retinal picture limits, optic circle and retinal bruises caused by afflictions [29]. In spite of the way that unmistakable systems are open for retinal division, there is scope for betterment.

2 Methodology

In this, we will first capture and get the image and process the final image to get a processed image which will be helpful to find out the defect using different techniques, which are thoroughly discussed. The later step will give us the image which is the required image for us, which will be helpful in the last part, which is the analysis part and it requires us to compare the output with the training data set [16].

The given steps with all the essential steps have been ascertained below. Figure 1 shows the transition of the image from retinal image to the output showing the defect.

2.1 Converting Image to Grayscale

All the digital images can be manipulated mathematically in simple and complex form due to their makeup structure. For any grayscale image, according to the intensity, any given locus of the pixel has a range from 0 (black) to 255 (white). The image we can say therefore is composed of pixels of different intensities stored in an array form, with the intensity indicating the grayness of any particular point.

Fig. 1 Flow diagram of the steps followed

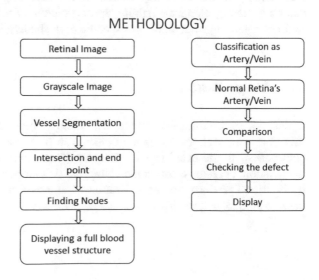

Various kinds of segmentation for the blood vessels were studied during the making of this report including three methods which are considered to be very good. During the study and research in the following work, they are referred.

This conversion is done using the function rgb2gray (). This allows us to do processing of the grayscale image [16].

2.2 Image Processing

It causes change of dim estimations of the pixels by utilizing three essential systems. Initially, the dark estimations of the pixels are changed without causing any change or procedure of the encompassing pixel esteems. Second, by neighborhood preparing the estimation of the given pixel is consolidated in a little neighborhood of the pixel. Third, a total change of the picture is required because of the unpredictable kind of change, yet what might as well be called the picture continues as before, however, it is extraordinary. This takes into account more prominent adaptability and intense handling of the picture.

2.3 Image Enhancement

Amid image capture, one of the challenges confronted is the picture quality, which is influenced by defocus, nearness of ancient rarities or different components. Hence, this progression includes advancement or change of the picture with the goal that its further utilization is significantly more dependable, appropriate and similar. It implies it is more reasonable for survey, examination or handling. Enhancing differentiation can be a strategy alongside lighting up of a picture. The table put away gives us a thought regarding what number of specific dark shading happens.

2.4 Image Restoration

Evacuation of meddling examples or deblurring of pictures fall in this classification. The outer aggravations can prompt change in pixel esteems, in this way creating commotion. In the event that the imaging gear experiences electronic rehashing aggravation, occasional commotion happens. At that point by changing this picture by Fourier Transform and by utilization of commotion channels before changing back to unique picture, this clamor can be decreased.

2.5 *Image Segmentation*

Division includes partitioning pictures into subsections that are exceptionally compelling, for example, characterizing zones of a picture that are fitting to be in this manner examined, or discovering circles, lines or different states of intrigue. Division can stop when such protests of intrigue have been segregated. Division calculations for monochrome pictures are for the most part in light of brokenness of picture forces, for example, edges in a picture, or on similitudes judged by predefined criteria.

There are many ways of image segmentation and to analyze and predict the defect using the images of the retina. Many writers have different methods to solve this problem. Some of them are listed below:

(1) Pattern recognition techniques
(2) Matched filtering
(3) Numerical morphology
(4) Multiscale approach
(5) Vessel following
(6) Model related approach
(7) Parallel/hardware related approach.

Figure 2 shows the types of algorithms in a tabulated form.

We would utilize and examining the example acknowledgment methods and different calculations in this classification.

Example of this, the techniques manages characterization of retinal veins and non-vessels together with foundation, in light of key highlights. This approach has two techniques: supervised and unsupervised. In the event that an earlier data is utilized

Fig. 2 Various types of algorithms

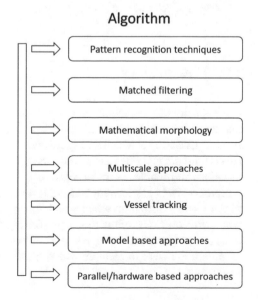

Algorithm

Pattern recognition techniques

Matched filtering

Mathematical morphology

Multiscale approaches

Vessel tracking

Model based approaches

Parallel/hardware based approaches

to decide a pixel as a vessel or not, at that point that technique is managed, else it is unsupervised strategy. The following is the exchange of the directed techniques utilized for handling.

2.6 Pattern Recognition Methods

It is the procedure by which the information is isolated into various classes or questions by looking at the example the information has or the property, that is, has. In light of the given strategies or system, estimation of the protest is done to group highlights and contrast it with known examples with deciding the class to which it has a place.

Some of the procedures are listed below:

i.   Classification classes' definition: According to our aim and features of the data obtained from image, the class for each pixel is determined, for which it would be assigned.
ii. Selection of particular feature: The components like texture, gray band value, etc. are chosen for categorization.
iii. Characterization of class: One data set for training of classifier and the other for testing are usually defined with known class memberships.
iv. Defining the constraints for classifier: The training data obtained is used by the classification algorithm to formulate decision rules.
v.   Perform classification: The data generated after testing is classified to the classes using the trained classifier.
vi. Result evaluation: The test data classification results help us to judge the accuracy and reliability of the classifier.

2.7 Thresholding

This capacity of thresholding enables a picture to be classified into various parts by changing over it into a parallel picture. This is finished by setting an edge estimation of the power, relating to which the higher and lower force pixels are named white or dark. This is fundamental since it evacuates the subtle elements which are superfluous or are variations and enables us to concentrate just on points of interest of our advantage. A comprehensively acknowledged limit level can likewise be chosen naturally or based on the picture histogram we can pick our own particular incentive for smooth and exact partition. Some of the time a perplexing force criterion can likewise be picked. Once in a while numerous edge levels can be connected for a solitary picture utilizing neighborhood edge in pictures which have diverse levels of foundation power and brightening [16].

2.8 Filters

Because of neighborhood handling, the productivity and value of the prepared calculation get upgraded by contemplating the estimations of contiguous pixels in figures.
A framework characterized by the client known as veil is characterized to cover the
pixel being referred to as well as its neighboring pixels. Every pixel that is concealed
is connected to a relating capacity. This combination of veil and capacity is called
channel. Along these lines because of veil the capacity of a pixel mulls over its
neighboring pixel esteems too [12–15].

2.9 Linear Discriminant Analysis (LDA)

LDA is somewhat related to ANOVA and regression analysis. This basically tries to
show that one dependent variable is linearly dependent on a combination of features.
    LDA is a summed up adaptation of Fisher's straight discriminant—this strategy is
utilized as a part of different fields to discover direct mix that really gives us a thought
whether the given classes have any likeness in their qualities or not. The outcome
consequently got can be utilized for dimensionality diminishment. Calculated relapse
and restrict relapse are more like LDA than ANOVA is. They both endeavor to clarify
utilizing the ceaseless factors which are free a variable known as downright factor.
    LDA is more in resemblance to PCA. They both are on lookout for combinations
that ought to be linear. PCA does not take into consideration any difference of class.
LDA likes to formulate a table of the similarity that is not to be, of classes of data.
    There are numerous procedures for arrangement of information which can be
utilized. Chief component analysis (PCA) and linear discriminant analysis (LDA)
are two normally utilized strategies for information arrangement. LDA can without
much of a stretch handle the case for which the class frequencies are not equivalent
and their attributes have been analyzed on created information.

2.10 Storing the Standard Data

- The code has to undergo a training process for which a subsidiary MATLAB file
  main_training has been created.
- During the training of the images, each image is selected subsequently, and the
  same steps are applied for each image.
- The histogram feature stores the whole data in a gradient form in the X- and Y-axis
  which can be used.
- This data set is stored in the variable tables created, i.e., svm_trained.mat,
  type_id.mat, and after the whole process is over, these compared images are used
  to find whether the image is a normal one or not.

2.11 Comparison

The AVR ratio is finally compared, and with it, the pattern of veins and artery is also compared with the test cases. The two methods enlisted above are used to fully confirm whether the image is of good eye or bad eye. Finally, the output of comparison is displayed.

3 Results and Outputs

The results are according to the process specified, and each step or process gives us an output image which is used as an input for the next process. In this way, the whole process is completed.

3.1 Background Normalized Image

The first part of the process combining conversion to grayscale and background normalization has been done, and the test results are as follows (Fig. 3) [9].

Fig. 3 Normalized image taken from the grayscaled image of retina due to the fact that green color gives us a better contrast in vein to background ratio, green color image is taken out from the red/blue/green channel and it is used for the segmentation purposes

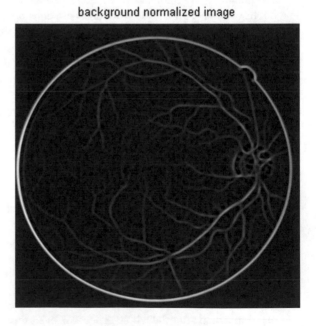

background normalized image

Fig. 4 Vessel segmentation image

vessel segmentation image

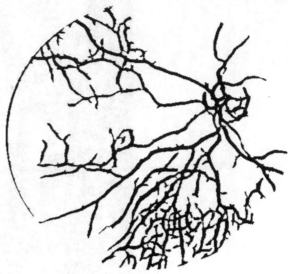

3.2 Vessel Segmentation Image

The vessel image is segmented by giving particular threshold brightness for blood vessel. Thus, rest of the part turns white and only vessel remains.

This segmented image shown in Fig. 4 acts as a raw image for the processing steps.

3.3 Vessel Centerline Extraction

Using a thinning algorithm in MATLAB—using Testthin () function—the vessel segmentation image is converted into a centerline image (Fig. 5) for purpose of objectivity and precision.

3.4 Pattern Recognition

Using a function findend junctions (), we find out the end points and intersection points in the particular image which would be used to plot the straight line image of the blood vessel. According to the algorithm, if the number of transitions for a particular point from 0 to 1 is greater than or equal to 6, then it is intersection. And if transition = 2, then it is end point.

Fig. 5 Vessel centerline
extraction

Using this pattern recognition algorithm, we get Fig. 6 with the prominent blood
vessels displayed.

Fig. 6 Image after joining
intersection points

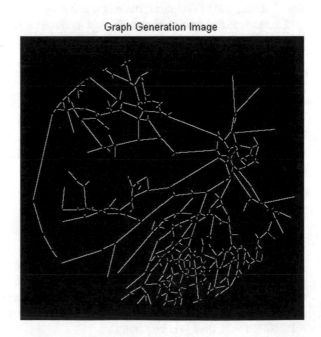

Fig. 7 Final comparison image

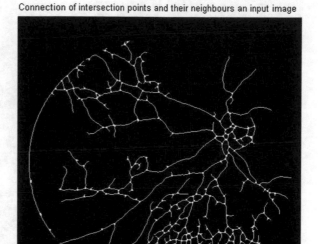

Connection of intersection points and their neighbours an input image

3.5 Comparison Image

The intersection and end points are found out. They are displayed separately, and then, their degree is calculated. Distance of each node has to be found.

Each node is covered, and all the vessel segments are displayed one by one, eventually resulting in a single image with all the segments.

This image is processed to find the AVR and the pattern of artery and vein is judged based on this image and the final output is given out as shown in Fig. 7.

3.6 AVR and AV Images

The final images based on AVR and dimensions of artery and vein are generated. They are compared with the test cases which are initially run. The final results are based on the comparison of these images. Figure 8 shows veins and artery after categorization.

Fig. 8 Vessels classified as
vein and artery

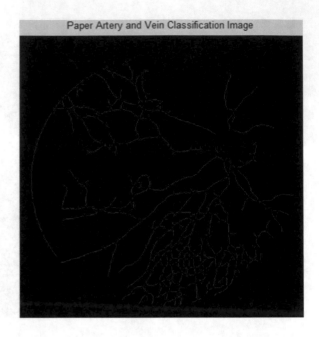

Paper Artery and Vein Classification Image

4 Conclusion

In this work, the study materials related to various methods for retinal blood vessel
segmentation were examined, and three separate high accuracy algorithms were
selected [4, 8].

The calculations can distinguish vessels and foundation from fundus pictures just
to a specific degree. As we probably are aware, the manual division of veins is hard
and tedious, and it is smarter to utilize a quick, computerized framework which
could recognize higher measure of veins. Will it spare time as well as decline the
quantity of specialists general required. It can likewise portion colossal quantities
of fundus pictures in a brief span. And at the present, utilized calculations have
generally low arranged vessel rates, and they are not so much pertinent for use in
automatized frameworks. In spite of the fact that the exactness rate of the calculation
is equivalent, huge extent of change in the vessel and foundation discovery should
be possible. We can state that vessel identification framework is not entirely reliable
as the outcomes bring up.

5 Future Direction

In this proposal, the literary works with respect to various strategies for retinal vein
division were considered and three diverse best in class techniques were actualized
[4, 8, 16]. In view of the investigation of writing and determination of three distinct

strategies, the issues tended to in this proposal are: (1) How precise are the chosen vein division techniques in separating veins from fundus pictures, and (2) Are comes about because of computerized vein extraction frameworks reliable contrasted with manual divisions done by specialists.

Retinal advanced examination of picture enables us to use easily with which retinal flow of vessels can be seen, captured and investigated without even a moment's pause itself. The most usually performed quantitative estimation by utilizing the advanced retinal picture examination must be the AVR. In spite of the fact that this has ended up being an intense and valuable research strategy to gauge narrowing of supply routes, extensive measure of study has been required to set up relationship of this element with foundational factors, measurably. It is likewise indistinct from current examinations whether the identification of retinal microvascular changes has extra prescient incentive above current institutionalized strategies (Wong 2004). As of late, 'reconsidered' formulae for the AVR (Knudtson et al. 2003) may hold more noteworthy guarantee for future investigations to discover weaker relationship with more prominent measurable power. While the AVR has been generally utilized, different kinds of retinal vascular geography as not utilized that much.

With an undeniably matured populace and expanded strain on restorative assets, the utilization of procedures, for example, telemedicine and across the board screening of people in danger of specific sicknesses will increment. Retinal vascular advanced picture investigation will assume an ever more noteworthy part in clinical ophthalmology.

References

1. Soares, J.V., Leandro, J.J., Cesar, R.M., Jelinek, H.F., Cree, M.J.: Retinal vessel segmentation using the 2-D Gabor wavelet and supervised classification. IEEE Trans. Med. Imaging **25**, 1214–1222 (2006)
2. Fathi, A., Naghsh-Nilchi, A.R.: Automatic wavelet-based retinal blood vessels segmentation and vessel diameter estimation. Biomed. Signal Process. Control **8**(1), 71–80 (2012)
3. Fang, B., Hsu, W., Lee, M.U.: On the Detection of Retinal Vessels in Fundus Images. http://hdl.handle.net/1721.1/3675. (04.05.2016)
4. Ricci, E., Perfetti, R.: Retinal blood vessel segmentation using line operators and support vector classification. IEEE Trans. Med. Imaging **26**, 1357–1365 (2007)
5. Staal, J., Abramoff, M.D., Niemeijer, M., Viergever, M.A., Ginneken, B.: Ridge-based vessel segmentation in color images of the retina. IEEE Trans. Med. Imaging **23**, 501–509 (2004)
6. Sinthanayothin, C., Boyce, J., Williamson, C.T.: Automated localisation of the optic disk, fovea, and retinal blood vessels from digital colour fundus images. Br. J. Ophthalmol. **83**(8), 902–910 (1999)
7. Villalobos-Castaldi, F.M., Felipe-Riveron, E.M., Sanchez-Fernandez, L.P.: A fast, efficient and automated method to extract vessels from fundus images. J. Vis. **13**, 263–270 (2010)
8. Kande, G.B., Subbaiah, P.V., Savithri, T.S.: Unsupervised fuzzy based vessel segmentation in pathological digital fundus images. J. Med. Syst. **34**, 849–858 (2009)
9. Rahebi, J., Hardalac, F.: Retinal blood vessel segmentation with neural network by using gray-level co-occurrence matrix-based features. J. Med. Syst. **38**(8), 85–97 (2014)
10. Tolias, Y., Panas, S.: A fuzzy vessel tracking algorithm for retinal images based on fuzzy clustering. IEEE Trans. Med. Imaging **17**(2), 263–273 (1998)

11. Al-Rawi, M., Qutaishat, M., Arrar, M.: An improved matched filter for blood vessel detection of digital retinal images. Comput. Biol. Med. **37**, 262–267 (2006)

12. Chaudhuri, S., Chatterjee, S., Katz, N., Nelson, M., Goldbaum, M.: Detection of blood vessels in retinal images using two-dimensional matched filters. IEEE Trans. Med. Imaging **8**, 263–269 (1989)

13. Zolfagharnasab, H., Naghsh-Nilchi, A.R.: Cauchy based matched filter for retinal vessels detection. J. Med. Signals Sens. **4**(1), 1–9 (2014)

14. Hoover, A., Kouznetsova, V., Goldbaum, M.: Locating blood vessels in retinal images by piecewise threshold probing of a matched filter response. IEEE Trans. Med. Imaging **19**(3), 203–210 (2000)

15. Pattona, N., Aslam, T.M., MacGillivray, T., Deary, I.J.: Retinal Image Analysis: Concepts, Applications and Potential (2006)

16. Staal, J.J., Abrmoff, M.D., Niemeijer, M., Viergever, M.A., van Ginneken, B.: Ridge based vessel segmentation in color images of the retina. IEEE Trans. Med. Imaging **23**(4), 501–509 (2004)

17. Mendonca, A.M., Campilho, A.: Segmentation of retinal blood vessels by combining the detection of centerlines and morphological reconstruction. IEEE Trans. Med. Imaging **25**(9), 1200–1213 (2006)

18. Chauduri, S., Chatterjee, S., Katz, N., Nelson, M., Goldbaum, M.: Detection of blood vessels in retinal images using two-dimensional matched filters. IEEE Trans. Med. Imaging **8**(3), 263–269 (1989)

19. You, X., Peng, Q., Yuan, Y., Cheung, Y., Lei, J.: Segmentation of retinal blood vessels using the radial projection and semi-supervised approach. Pattern Recogn. **44**, 2314–2324 (2011)

20. Nguyen, U.T., Bhuiyan, A., Park, L.A., Ramamohanarao, K.: An effective retinal blood vessel segmentation method using multi-scale line detection. Pattern Recogn. **46**, 703–715 (2013)

A Survey on Load/Power Flow Methods and DG Allocation Using Grasshopper Optimization Algorithm in Distribution Networks

Kola Sampangi Sambaiah and T. Jayabarathi

Abstract Emerging smart grid technologies in electric distribution network (EDN) planning, operation, and automation require fast and efficient power flow methods. This imposes the continual search for speedy and effective power flow algorithms for EDNs. The present paper illustrates a survey and summary of recent developments in power flow methods. Several technical approaches have been presented for optimal power flow methods for power loss identification and distributed generation (DG) allocation in EDNs. The main contribution of the present survey paper is calculation of network power loss using power flow methods in radial distribution networks (RDNs). Here two major power flow methods used are forward and backward sweep (FBS) and direct approach (DA). The methods are compared on the basis of loss reduction and minimum voltage bus identification capability. The present survey provides the researchers a clear idea of power flow methods utilized in EDN performance enhancement. The power flow analysis is carried out on different EDNs, i.e., small, medium and large scale. Later DG allocation is carried out by using a novel meta-heuristic algorithm called grasshopper optimization algorithm (GOA). The results obtained are compared with existing techniques.

Keywords Power flow methods · Radial distribution networks · Distributed generation · Power loss

1 Introduction

Smart grid era started in several developed countries moving towards clean and sustainable energy using sophisticated technologies. Smart grid implementation for electric distribution networks (EDNs) includes distribution network automation, distributed generation (DG) and advanced metering infrastructure (AMI) integrating

K. S. Sambaiah · T. Jayabarathi (✉)
School of Electrical Engineering, Vellore Institute of Technology, Vellore, India
e-mail: tjayabarathi@vit.ac.in

K. S. Sambaiah
e-mail: sambaiahks@gmail.com

© Springer Nature Singapore Pte Ltd. 2020
K. N. Das et al. (eds.), *Soft Computing for Problem Solving*,
Advances in Intelligent Systems and Computing 1057,
https://doi.org/10.1007/978-981-15-0184-5_53

functions. Distribution system management (DSM) is the basic tool for control and managing of smart grid. In general, EDNs are radial structure. Hence power losses are more compared to meshed/looped networks. Appropriate power flow or load flow analysis has to be opted for evaluation of loss and voltage profile of buses. Power flow analysis is a method of obtaining steady-state voltages and phase angles of EDNs at fundamental frequency [1].

A power flow solution obtained from any algorithm must have fast convergence character, numerically robust solution, and better computational efficiency (minimum time) for all the solution scenarios. Power flow analysis in transmission networks are evaluated by using Gauss–Seidel (GS), Newton–Raphson (NR) and fast decouple methods. The ill-conditioned nature of EDNs are due to the following features:

- Structure is radial or weakly meshed
- High R/X ratios
- Buses and branches are extremely in large number
- Multiple phases and unbalanced process
- Unbalanced distributed load

Hence traditional power flow methods are failed to meet aforementioned characteristics. The standard fast-decoupled version of NR method requires certain assumptions for network simplification are often not valid for EDNs. Therefore, it is suggested to have novel power flow algorithms for EDNs. It is desired to quantify all the aforementioned characteristics before the power flow algorithm utilization. Several researchers have proposed various power flow methods for EDNs in [2–8]. In the present paper significance of power flow methods on EDNs is well-illustrated. In addition to this, a novel grasshopper optimization algorithm (GOA) is used to allocate DG in distribution networks.

Rest of the paper is structured as follows: Section 2 explains power flow methods, Sect. 3 explains and compares the power flow methods solution, Sect. 4 presents the DG allocation in RDN using GOA Sect. 5 presents the conclusion of the paper.

2 Power Flow Methods

Several power flow methods have been implemented for obtaining optimal power flow of an EDN. In [2], the author proposed a current-based power flow method for distribution networks. A new method is proposed for solving power flow analysis in weakly meshed transmission and distribution networks using a multi-port compensation technique in [3]. A rigid approach for large scale distribution networks operational analysis have been presented and well-illustrated using a conductor model in [4, 5]. In [6], author proposed a novel power flow method for a network with simultaneous phase changing loads. A direct extension method of compensation-based power flow algorithm for weakly meshed EDNs is presented in [7]. Zimmerman et al. in [8] proposed a new power flow method considering comprehensive model which includes switches, lines, transformers, co-generators, shunt capacitor banks and various loads. In [9], author proposed a simple and efficient power flow method for small and large scale RDNs.

In [10], the author proposed two novel methods which are extension of existing bus admittance summation method for analyzing the radial and weakly meshed networks. A simple and efficient power flow method has been proposed using receiving end voltages expression in [11].

Conventional dist-flow branch equations are used for power flow analysis after the distribution network is realized to an equivalent single-source network in [12]. A new ratio flow method has been proposed for solving complex distribution networks using voltage ratio in [13]. In [14], author proposed a forward and backward voltage updating method based on classical forward-backward ladder technique. In recent years several power flow methods have been proposed the summary of power/load flow methods are presented in Table 1.

3 Results and Discussion

3.1 Test Systems

In 12—bus RDN has 12 bus and 11 distribution lines. The system total active and reactive loads are 435 kW and 405 kVA, respectively. The base values are 11 kV and 10 MVA. Figure 1 shows the single line diagram (SLD) of 12—bus RDN. In 15—bus RDN has 15 bus and 14 distribution lines. The system total active and reactive loads are 1226 kW and 1251 kVA, respectively. Figure 2 shows the SLD of 15—bus RDN. The 33—bus system is an RDN with the real and reactive power demand of 3715 kW and 2300 kVAr, respectively. Figure 3 shows the SLD of 33—bus RDN. The 69 bus system is also a RDN similar to 33 bus, with the real and reactive power demand of 3802 kW and 2694 kVAr, respectively. Figure 4 shows the SLD of 69—bus RDN. The bus and line data for all the test systems are taken from [15]. In 85—bus EDN has 85 bus and 84 distribution lines. The system total active and reactive loads are 2570.28 kW and 2621.936 kVAr, respectively. The base values are 11 kV and 100 MVA. Figure 5 shows the SLD of 85—bus RDN. The bus and line data of present RDNs are taken from [16, 17].

Power loss and a minimum voltage obtained by DA and FWS power flow methods are presented in Table 2.

4 DG Allocation Using GOA

Grasshopper optimization algorithm (GOA) was proposed by Saremi et al. in 2017 [20]. The proposed algorithm is used for solving several challenging optimization

Table 1 Summary of power/load flow methods

Author(s) [Ref]	Power/load flow method	Phases	Network type	Software	Year
Teng et al. [1]	Bus injection to branch current (BIBC) and branch current to bus voltage (BCBV)	Unbalanced three phases	Raial and weakly meshed	Borland C++ language	2003
Teng et al. [2]	Current-based power flow method	Single-phase and unbalanced three phases	Radial and weakly meshed	NA	1994
Shirmohammadi et al. [3]	Multi-port compensation technique	Single-phase and unbalanced three phases	Radial and weakly meshed transmission system	NA	1988
Chen et al. [4]	Generalized distribution analysis systems (GDAS)	Single-phase and unbalanced three phases	Radial and weakly meshed	NA	1991
Zimmerman et al. [8]	Fast decoupled power flow	Single-phase and unbalanced three phases	Radial	MATLAB	1995
Ghosh et al. [11]	Basic ladder	Three-phase	Radial	NA	1999
Haque et al. [12]	Branch current injection method	Three-phase	Radial	NA	2000
Liu et al. [13]	Ratio-flow method (forward-backward ladder method)	NA	Radial	MATLAB	2002
Eminoglu et al. [14]	Improved forward-backward ladder method	Single-phase and balanced three phases	Radial	MATLAB	2005
Teng et al. [21]	BIBC and BCBV	Three-phase	Radial and weakly meshed	Borland C++ language	2008
Schneider et al. [22]	Forward-backward method	Single-phase and balanced three phases	Radial	GridLAB-D	2009
Chen et al. [23]	Loop frame reference	Three-phase	Radial	MATLAB	2010

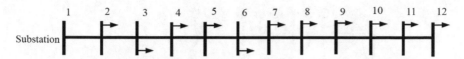

Fig. 1 Single line diagram of 12—bus RDN

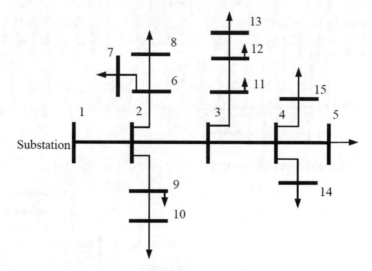

Fig. 2 Single line diagram of 15—bus RDN

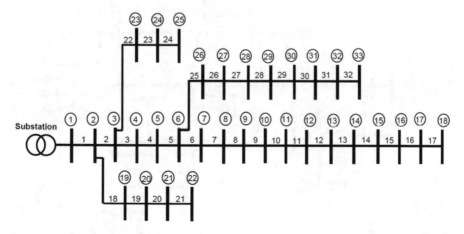

Fig. 3 Single line diagram of 33—bus RDN

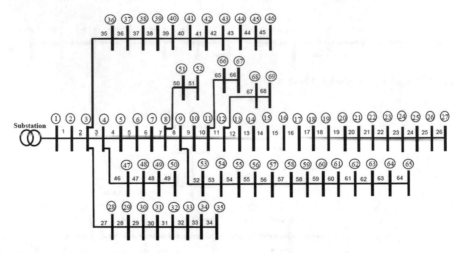

Fig. 4 Single line diagram of 69—bus RDN

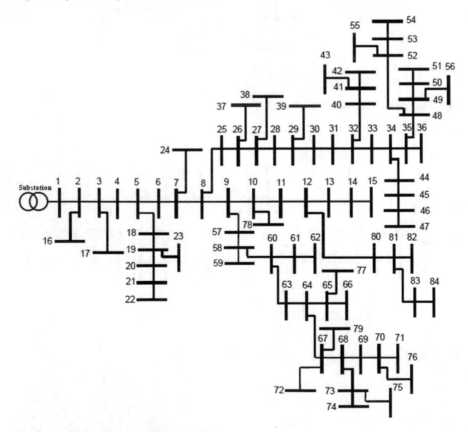

Fig. 5 Single line diagram of 85—bus RDN

Table 2 A comparative study between DA and FBS methods

Test system	Direct approach			Forward-backward sweep		
	Power loss (kW)	V_{min}@ bus	Time (in sec)	Power loss (kW)	V_{min}@ bus	Time (in sec)
12 RDN	20.7135	0.9434 @ 12	0.010317	20.6891	0.9444 @ 12	0.03299
15 RDN	95.4948	0.9313 @ 13	0.001120	93.5666	0.9314 @ 13	0.04861
33 RDN	202.6666	0.9131 @ 18	0.009257	202.6188	0.9131 @ 18	0.06744
69 RDN	224.9583	0.9092 @ 65	0.010553	225.3939	0.9099 @ 65	0.07664
85 RDN	316.0971	0.8713 @ 54	0.009723	314.5195	0.8743 @ 54	0.07592

@ refers to bus location

problems. In the present paper, GOA is used for optimal DG allocation in RDNs. The GOA is applied to 85—bus RDN and the obtained results are compared existing techniques. The simulation results are tabulated in Table 3. The power loss evaluated before DG allocation using DA is 316.09 kW with minimum bus voltage of 0.8713 at bus 53. After DG allocation the power loss is reduced to 149.93 kW with minimum voltage of 0.9527 at bus 53. Figure 6 shows the power loss obtained by various techniques. The convergence curve of power loss of GOA for 85—bus RDN is shown in Fig. 7.

Table 3 Simulation results obtained by GOA for 85—bus RDN

	Base case	ALO [17]	GABC [18]	HGWO [19]	GOA
Power loss (kW)	316.09	224.05	191.73	165.55	**149.93**
DG size (kW) @ bus	–	946.35 @ 55	1838 @ 36	1876 @ 9 416 @ 48 198 @ 70	**1064 @ 8** **781 @ 32** **435 @ 67**
Total installed capacity of DG (kW)	–	946.35	1838	2490	**2280**
% real power loss reduction	–	29.118	39.34	47.62	**52.56**
V_{min}(p.u.) @ bus	0.8713 @53	0.9109*	0.9293*	0.9513@54	**0.9527 @ 53**

Fig. 6 Comparison of
percentage power loss
obtained by GOA with other
techniques

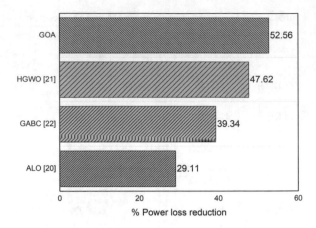

Fig. 7 Convergence curve
of power loss of GOA for
85—bus RDN

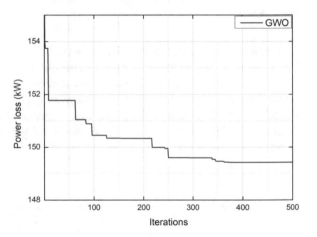

4.1 Objective Function

The main objective of present work is power loss minimization [19].

4.1.1 Power Loss

The power loss in a network with N number of lines is given by

$$f(x) = P_N^{Tloss} = \sum_{b=0}^{N-1} \left(\frac{P_{b,b+1}^2 + Q_{b,b+1}^2}{|V_b|^2} \right) * R_{b,b+1} \tag{1}$$

where $P_{b,b+1}$ and $Q_{b,b+1}$ are the active and reactive power flow between the buses b and $b+1$, respectively, in kW and kVAr; V_b is the voltage at the bus-b; $R_{b,b+1}$ is the line resistance between the buses b and $b+1$.

5 Conclusion

The present survey has tried to present the various load/power flow methods for evaluation of EDNs and allocation of various electrical components used to enhance EDNs performance. In addition, role of power flow methods in different network structures also illustrated. The two major power flow methods namely BIBC-BCBV and FBS have been tested for standard distribution networks for power loss, voltage magnitudes and computational time. The suitability of power flow methods for different frameworks and network structures also discussed. In addition to this, a novel GOA is used to allocate DG in optimal location with size and the obtained results are compared with the other existing techniques.

Acknowledgements The second author thanks VIT for providing 'VIT SEED GRANT' for carrying out this research work.

References

1. Teng, Jen-Hao: A direct approach for distribution system load flow solutions. IEEE Trans. Power Deliv. **18**(3), 882–887 (2003)
2. Teng, J.J.H., Lin, W.M.: Current-based power flow solutions for distribution systems. In: Proc. IEEE Int. Conf. Power Syst. Technol., pp. 414–418. Beijing, China (1994)
3. Shirmohammadi, D., Hong, H.W., Semlyen, A., Luo, G.X.: A compensation-based power flow method for weakly meshed distribution and transmission networks. IEEE Trans. Power Syst. **3**, 753–762 (1988)
4. Chen, T.-H., Chen, M.-S., Hwang, K.-J., Kotas, P., Chebli, E.A.: Distribution system power flow analysis—A rigid approach. IEEE Trans. Power Deliv. **6**, 1146–1152 (1991)
5. Chen, T.S., Chen, M.S., Inoue, T., Chebli, E.A.: Three-phase cogenerator and transformer models for distribution system analysis. IEEE Trans. Power Deliv. **6**, 1671–1681, 2 October 1991
6. Chen, T.H., Chang, J.D.: Open wye-open delta and open delta-open delta transformer models for rigorous distribution system analysis. In: Proceedings of the Institution of Electrical Engineers, vol. 139, pp. 227–234 (1992)
7. Cheng, C.S., Shirmohammadi, D.: A three-phase power flow method for real-time distribution system analysis. IEEE Trans. Power Syst. **10**, 671–679 (1995)
8. Zimmerman, R.D., Chiang, H.D.: Fast decoupled power flow for unbalanced radial distribution systems. IEEE Trans. Power Syst. **10**, 2045–2052 (1995)
9. Das, D., Kothari, D.P., Kalam, A.: Simple and efficient method for load flow solution of radial distribution networks. Int. J. Electr. Power Energy Syst. **17**(5), 335–346 (1995)
10. Rajičić, D., Taleski, R.: Two novel methods for radial and weakly meshed network analysis. Electr. Power Syst. Res. **48**(2), 79–87 (1998)

11. Ghosh, S., Das, D.: Method for load-flow solution of radial distribution networks. IEE Proc.-Gener. Transm. Distrib. **146**(6):641–648 (1999)
12. Haque, M.H.: A general load flow method for distribution systems. Electr. Power Syst. Res. **54**(1), 47–54 (2000)
13. Liu, J., Salama, M.M.A., Mansour, R.R.: An efficient power flow algorithm for distribution systems with polynomial load. Int. J. Electric. Eng. Educ. **39**(4), 371–386 (2002)
14. Eminoglu, U., Hocaoglu, M.H.: A new power flow method for radial distribution systems including voltage dependent load models. Electric Power Syst. Res. **76**(1–3), 106–114 (2005)
15. Kayal, P., Chanda, C.K.: Placement of wind and solar based DGs in distribution system for power loss minimization and voltage stability improvement. Int. J. Electr. Power Energy Syst. **53**, 795–809 (2013)
16. Bansal, A., Kumar, A., Kumar, N.: Loss optimization of IEEE 12 bus radial distribution system integration with wind weibull distribution function using PSO technique. In: 2016 IEEE 7th Power India International Conference (PIICON), pp. 1–6 (2016)
17. Reddy, P.D.P., Reddy, V.V., Manohar, T.G.: Ant Lion optimization algorithm for optimal sizing of renewable. In: Electrical Power & Energy Systems, vol. 28, pp. 669–678 (2017)
18. Dixit, M., Kundu, P., Jariwala, H.R.: Incorporation of distributed generation and shunt capacitor in radial distribution system for techno-economic benefits. Eng. Sci. Technol. Int. J. **20**(2), 482–493 (2017)
19. Sanjay, R., Jayabarathi, T., Raghunathan, T., Ramesh, V., Mithulananthan, N.: Optimal allocation of distributed generation using hybrid grey wolf optimizer. IEEE Access. **5**, 14807–14818 (2017)
20. Saremi, S., Mirjalili, S., Lewis, A.: Grasshopper optimisation algorithm: theory and application. Adv. Eng. Softw. **105**, 30–47 (2017)
21. Teng, J.-H.: Modelling distributed generations in three-phase distribution load flow. IET Gener. Transm. Distrib. **2**(3), 330–340 (2008)
22. Schneider, K.P., Chassin, D., Chen, Y., Fuller, J.C.: Distribution power flow for smart grid technologies. In: Power Systems Conference and Exposition, PSCE'09. IEEE/PES, pp. 1–7 (2009)
23. Chen, Tsai-Hsiang, Yang, Nien-Che: Loop frame of reference based three-phase power flow for unbalanced radial distribution systems. Electr. Power Syst. Res. **80**(7), 799–806 (2010)

Performance Analysis of Convolutional Neural Network When Augmented with New Classes in Classification

K. Teja Sreenivas, K. Venkata Raju, M. Bhavya Spandana,
D. Sri Harshavardhan Reddy and V. Bhavani

Abstract Classification of images is one of the important goals of artificial neural networks. Due to increasing efficiency and accuracy of neural networks today, neural networks have been doing more than image classification, and they are used for image captioning, text detection, and recognition. Deep learning models such as convolutional neural network, recurrent neural network, autoencoders, restricted Boltzman machines, modular neural network, and deep belief networks are widely used across the various domain of problems. The convolutional neural network has been proved successful in computer vision tasks such as object recognition and classification. This paper analyzes how the accuracy and performance of the convolutional neural network are affected while increasing number of classification classes, by augmenting with a new dataset. Through the analysis, it is propounded that the augmentation resulted in an increase in the accuracy and performance of convolutional neural network. In our experimental study, MNIST is used as the primary dataset and Fashion-MNIST is used as the augmented dataset. In our analysis, we observed five times faster convergence time for the MNIST dataset.

Keywords Convolutional neural network · MNIST · Fashion-MNIST · Data augmentation

1 Introduction

Convolutional neural networks (ConvNet) have been widely used in the field of image recognition and classification. Convolutional networks are simply neural networks that use convolution in place of general matrix multiplication in at least one of their layers [5]. The ConvNet have been inspired by the working of the mammalian brain [7]. The work of K. Fukushima on Neocognitron [2] is considered as the predeces-

K. Teja Sreenivas (✉) · K. Venkata Raju · M. Bhavya Spandana · D. Sri Harshavardhan Reddy · V. Bhavani
Koneru Lakshmaiah Education Foundation, Vaddeshwaram, Guntur, Andhra Pradesh, India
e-mail: teja.sreenivas.05@gmail.com

© Springer Nature Singapore Pte Ltd. 2020
K. N. Das et al. (eds.), *Soft Computing for Problem Solving*,
Advances in Intelligent Systems and Computing 1057,
https://doi.org/10.1007/978-981-15-0184-5_54

sor of the ConvNet. In 1990, LeCun [9] pioneered the work on convolutional neural network through the work on LeNet (Fig. 2) but due to lack of computation, the development of ConvNet fell into abeyance until the emergence of AlexNet architecture, introduced by Krizhevsky et al. [8] which is often credited for popularizing the convolutional neural networks(CNN) in computer vision. This ConvNet architecture won the ImageNet Large Scale Visual Recognition Challenge (ILSVRC) of 2012. Later, the introduction of deconvolutional layer by Zeiler and Fergus [21] helped them to tweak the AlexNet to create ZF Net which won ILSVRC-2013. The later architecture was outperformed by GoogLeNet and VGG which are a deeper convolutional neural networks with more than fifteen convolutional layers; VGG [15] has 16 layers and GoogLeNet [18] has 22 layers convolution layers.

From the mentioned architectures, it is evident that using deep convolutional neural network will produce better performance, but it is hard to determine the maximum accuracy that can be achieved by any given architecture. There are other methods proposed for increasing the performance of a neural network. This paper introduces one such possible method in training convolutional neural network. In this paper, we analyze the performance of a simple convolutional neural network using our proposed method. This method has provided a better accuracy when compared to the traditional approach used in training convolutional neural network.

2 Related Work

Many methods are proposed to increase accuracy in classification problems using the convolutional neural network. Some of them include increasing number of layers, data augmentation, regularization (dropout), ensemble methods, pseudolabeling, and transfer learning.

Increasing the number of layers to increase accuracy is the general approach that is being adopted, since a greater number of parameters help to classify the data better. The AlexNet by Krizhevsky [8], in 2012, made a significant breakthrough in machine learning by using a wider and deeper CNN which produced state-of-the-art accuracy results. Similar to AlexNet, VGG is able to successfully implement a convolutional neural network with 16 layers which was the runner up in 2014 ILSVRC. The winner of 2014 ILSVRC GoogLeNet was able to produce state-of-the-art accuracy with error percent of 6.7, which is second to later proposed ResNet. The ResNet [6] which won 2015 ILSVRC, is an ultradeep neural network with 49 convolutional layers and one fully connected layer, uses residual architecture. This holds the current state-of-the-art accuracy with least accuracy of 3.57% error. Figure 1 better illustrates the increase of performances of deep neural networks over years.

Another method which is often used is data augmentation where on existing train data, we perform simple techniques, such as cropping, rotating, and fliping input images. Using this process, we can create double or twice as many data from existing data. These techniques are future illustrated in the paper, The Effectiveness of Data Augumentation in Image Classification using Deep Learning [14].

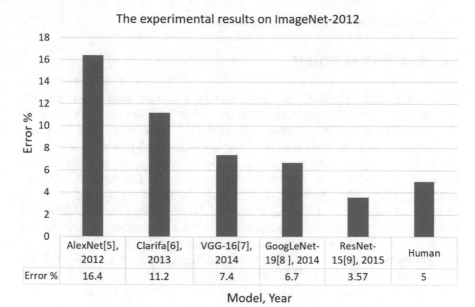

Fig. 1 Accuracy of different models in ImageNet Challenge [1]

The experimental results on ImageNet-2012

	AlexNet[5], 2012	Clarifa[6], 2013	VGG-16[7], 2014	GoogLeNet-19[8], 2014	ResNet-15[9], 2015	Human
Error %	16.4	11.2	7.4	6.7	3.57	5

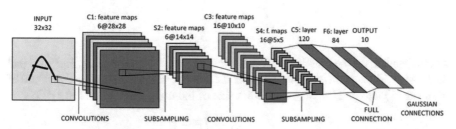

Fig. 2 The Architecture of Lenet-5

Another way to improve neural network accuracy is by employing semi-supervised learning techniques such as pseudolabeling [10]. In the pseudolabeling method, the neural network is trained simultaneously using both labeled and unlabeled data. During training, for an unlabeled data, its highest probability class is treated as its true label and weights are updated.

Transfer Learning [20] is another topic of ongoing interest in the machine learning community. This involves knowledge learned from one or more tasks is fine tuned and applied to the target task. One of the most effective and successful ways to train deep neural network is using unsupervised learning to initialize the weights. This phase of training is called unsupervised pretraining.

3 Convolutional Layer Components

3.1 Convolutional Layer

The convolutional layer is an important part of the convolutional neural network. A convolutional neural network should consist of at least one convolutional layer which usually has the first layer as a convolutional layer. It is suited for image classification due to its properties of sparse interaction and parameter sharing. In this layer, we define a number of filters. Unlike in image processing, these filters or kernels are randomly initialized or can be pretrained. The network will learn to detect features using these filters. The values of filters are updated using the backpropagation algorithm. The initial layer of the network learns to detect simple features in the images such as edges. As the number of convolutional layers increases, at higher layers, the filters learn to detect complex features.

The convolutional neural networks provide additional options such as strides(S) and padding(P) [13]. The stride is used to control the overlap during convolutional operation using filters. During convolutional step, we lose the information in the border. When using a $F \times F$, filter on $N \times N$, image the output size(O) is given by the Eq. (1), where F is filter size and N is size(height or length) of the image.

$$O = 1 + \frac{N - F}{S} \tag{1}$$

To compensate for the drawback of convolutional step, we use padding. Here, we pad the image with zeros before applying convolution operation; thus, using the values of the stride(S) and padding(P), the output size of the image is determined by Eq. (2).

$$O = 1 + \frac{N + 2P - F}{S} \tag{2}$$

3.2 Activation Function

The activation function is essential to impart nonlinearity in neural networks. The time for training increases significantly as the depth of the network increases. The AlexNet takes 90 epochs in five or six days to train on two GTX 580 GPUs [8]. RELU [11] (rectified linear activation unit), given by Eq. (3), is being widely used for its apparent advantages. Firstly, RELU is simple and takes less time when using nonlinear activation functions such as tanh or sigmoid. Secondly, it is popular due to its reduced likelihood of vanishing gradient [3].

$$Relu(x) = max(x, 0) \tag{3}$$

In the final layer, we use sigmoid activation function, given by Eq.(4). A sigmoid activation function produces probability distribution between the values zero and one which is ideal for classification. There are other activation functions such as tanh, given by the Eq.(5), which is a stepped up version of sigmoid, having a range of -1 to $+1$.

$$sigmoid(x) = \frac{1}{1 + e^{-x}} \tag{4}$$

$$tanh(x) = \frac{2}{1 + e^{-2x}} - 1 \tag{5}$$

3.3 Pooling Layer

In a typical CNN, we observe convolutional and pooling layer alternating. The primary function of pooling layer is to reduce the spatial dimensions of activation maps and thus by reducing the number of parameters and overall computational complexity. While there are various pooling operations such as Max pooling, stochastic pooling, average pooling and spectral pyramid pooling [12], multiscale orderless pooling [4], Max pooling is predominently used in image classification. It is noted by Dosovitskiy [16] that Max pooling can be replaced by convolutional layers with a stride of two.

3.4 Fully Connected Layer(FC)

In a fully connected layer, all the neurons in the previous layers are connected to every neuron in the current layer. The extracted features of the filters are inferred in this layer for classification. Since, neurons in this layer lose spatial information, by arranging linearly, we do not use any convolutional layer after this layer. The fully connected layers are used to draw an inference from the features extracted from the convolutional layers.

3.5 Regularization

We use regularization techniques to prevent overfitting in the deep neural network. Dropout [17] and DropConnect [19] are widely used regularization techniques. In dropout, we randomly set output of neurons as zero during backward and forward propagation. On the other hand, DropConnect randomly sets weights linking the neurons to zero.

4   Implemented Convolutional Architecture

4.1   Convolutional Layer:

In this architecture, we have used two convolutional layers. For simplicity, we have used 64 filters with kernel size of 5×5 for each convolutional layer. The kernels are initialized randomly with a mean of zero and a standard deviation of 0.1. After every convolutional layer, we use Max pooling layer, having a window size of 2×2. Figures 3 and 4 represents the convolutional layer 1 and layer 2.

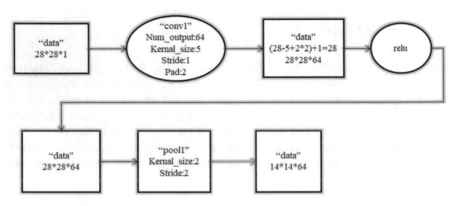

Fig. 3   Data flow diagram of convolutional layer 1

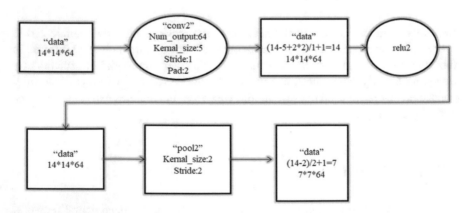

Fig. 4   Data flow diagram of convolutional layer 2

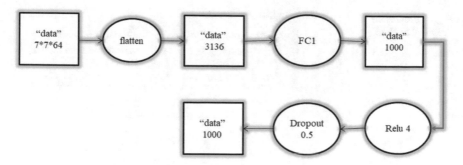

Fig. 5 Data flow diagram of Flattening and Dense layer

Fig. 6 The data flow diagram of output layer(individual)

Fig. 7 The data flow diagram of output layer(combined)

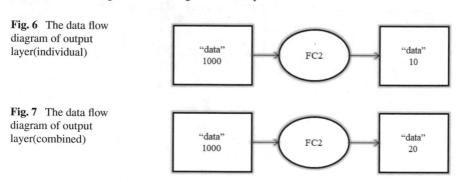

4.2 Fully Connected:

After the convolutional layer, the obtained feature maps from the final convolutional layer are flattened as shown in Fig. 5. We then add a fully connected layer (hidden layer) followed by another fully connected layer with nodes, equal to number of classes in the classification task. When training the CNN on a single dataset, we have total of ten classes as shown in Fig. 6 and when using combined dataset, we have twenty classes as shown in Fig. 7. The hidden layer consists of thousand neurons. We have used Adam Optimizer with learning rate of 0.01 while training the neural network. The proposed method to train the above described CNN architecture is illustrated in Fig. 8.

5 Implementation

Neural network architectures can be implemented using different frameworks such as TensorFlow, Pytorch, Keras, Caffe, MATLAB, and MXNet. Among these, TensorFlow is being adopted by large companies due to its flexible system architecture and scalability. It effectively handles the distribution of computation on multiple

Fig. 8 Flow chart of
Proposed method

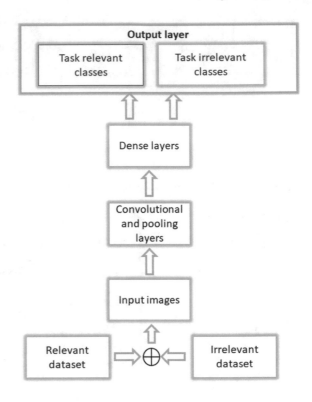

GPU and CPU. TensorFlow offers a suit of visualization tools through its Tensor-
Board. To implement the above architecture we have used TensorFlow framework.
The hardware used is a Tesla K80, 12 GB Memory GPU.

6   Results

Using the above described architecture, we have trained each dataset individually,
during which we trained it over multiple epochs until the model overfits the train
data. For the MNIST dataset, we have trained the model for 1, 10, 15, 30, and 50
epochs. Figure 9 illustrates train and test accuracy at various epochs for MNIST
dataset. Figure 10 shows the variation of validation accuracy over 50 epochs for
MNIST. Similarly, we have train and test accuracy at various epochs for fashion
dataset and the corresponding results are shown i Fig. 11. We observe that validation
accuracy starts to overfit around 30 epochs for the MNIST dataset using the given
architecture. Similarly, we observe the occurrence of overfitting for Fashion-MNIST
dataset after 30 epochs. Figures 12 and 13 represent confusion matrix for MNIST
and Fashion-MNIST datasets, respectively, when trained separately (Fig. 14).

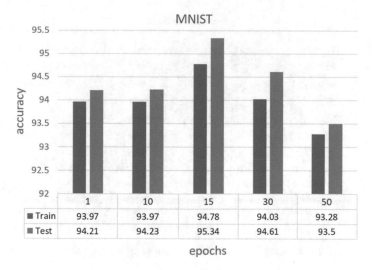

Fig. 9 Train and Test accuracy for MNIST dataset across different epochs

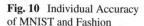

Fig. 10 Individual Accuracy of MNIST and Fashion

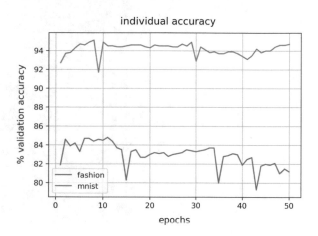

The obtained results observed in Figs. 9 and 11 are obtained when training the model separately on each dataset, each having ten classes to be classified by the model. Clearly, from the accuracy, we can posit that the classification of Fashion-MNIST images is more complex when compared to MNIST dataset. Thus, classifying both the datasets simultaneously on the same model will result in a decrease in the overall performance due to parameter constraints.

We have observed that when combining the two datasets into a single dataset, cumulatively having twenty classes, and training on the proposed architecture, the model produces better accuracy on one of the datasets within a smaller number of epochs. The obtained results justify this observation, shown in Fig. 15. It can be further noted that the number of parameter in the previously used architecture(where

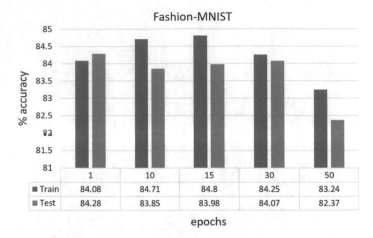

Fig. 11 Train and Test accuracy for Fashion dataset across different epochs

Fig. 12 Confusion Matrix
for MNIST test data

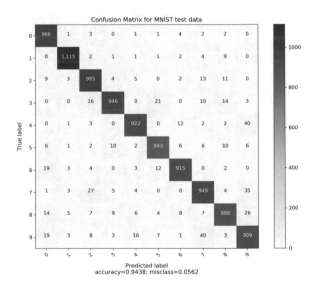

total classes are ten) and the current architecture(where total classes are twenty) is not strictly same since adding ten neurons, for additionally classes, in the classification(output) layer leads to increase in number of parameter by sum of 10,000 (10 × 1000, where 10 is additional classes and 1000 is number of neuron in hidden layer). Adding three neurons in the hidden layer of previous CNN architecture could produce a nearly same increase of parameters without any significant improvement in performance. Thus, it is reasonable to compare the performance of the network even if they differ in a number of parameters by a small number.

In Fig. 15, we observe that the rate of convergence and accuracy of MNIST classes increased significantly when compared with the initial performance (when trained in-

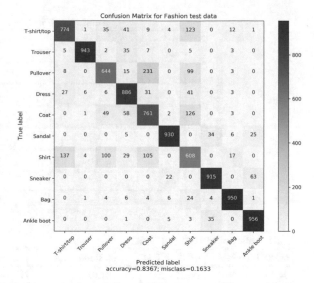

Fig. 13 Confusion Matrix for Fashion test data

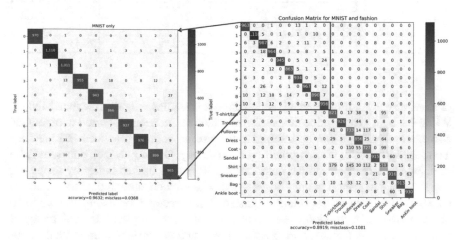

Fig. 14 Comparision Of MNIST and Fashion Confusion Matrix

dividually), since we have obtained highest accuracy within three epochs for MNIST data. While we find a significant drop in performance of fashion-MNIST classes, which is a consequence of bottleneck caused due to parameter constraint. The validation accuracy trends of individual datasets can be inferred from Fig. 15. From Table 1, we can infer that as a number of epochs increases, the increase in accuracy vacillates between the two sets of classes (MNIST and fashion-MNIST). The confusion matrix for the combined datasets is shown in Fig. 14. Further, there are less than 45 images that were misclassified across datasets, i.e, the CNN has misclassified MNIST images as Fashion and vice versa. Since, the number of such misclassifica-

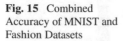

Fig. 15 Combined
Accuracy of MNIST and
Fashion Datasets

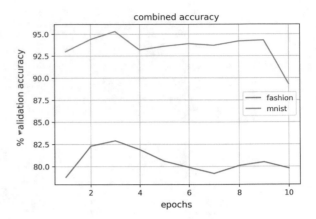

Table 1 Train and Test accuracy of MNIST and Fashion

Epochs	MNIST Train	MNIST Test	fashion Train	fashion Test
1	94.12	94.51	81.91	81.09
2	95.59	96.18	81.67	80.93
3	94.38	95.00	82.74	82.27
5	94.44	95.22	81.50	81.06
10	93.35	93.35	79.79	79.41

tion are very few it implies that Fashion dataset is significantly diverse than MNIST. While it is hard to determine the precise number of parameters required to obtain the necessary accuracy it can be inferred that by using various practices and methods mentioned in Sect. 2, we can obtain good accuracy. Owning to their inherent issue of being stuck at local optima, neural networks can be trained better by supplying large varied data. But it is not always feasible to provide enough data to the neural network. In such cases, especially in case of image recognition using convolutional neural networks, by providing non-domain specific, yet complex data, which require to extract complex features during training of neural network, we can force the neural network to better generalize. In other terms, we can force the neural network out of its local optima. Hence, our model obtained test accuracy of 96.18% within three epochs when compared to traditional approach where the highest accuracy of 95.34% was obtained around 15 epochs. Thus, we have obtained increased accuracy with five times faster convergence time.

7 Conclusion

Conclusively, we can propose from our results, that supplying complex image data, that can be relevant or irrelevant to the target task, apart from task relevant data could boost the performance because complex data enable the kernels in the convolutional layer to extract better feature maps. This architecture is illustrated using Fig. 8. Here, the task relevant data (primary dataset) is the MNIST dataset and task irrelevant data (augmented dataset) is fashion-MNIST. These two datasets are merged to form a single dataset which used to train a convolutional neural network which has a series of convolutional layers and Max-pooling layers followed by dense layers. The final output layer has 'n' classes from task relevant dataset and 'm' classes from the task irrelevant dataset, cumulatively the output layer has $m+n$ classes (neurons). The main criteria for choosing the augmented dataset are that they need to be more complex and disparate than the target task.

8   Future Work

We would further like to investigate how to increase the accuracy of a model for a given dataset using a minimum number of irrelevant classes because using more irrelevant classes might cause bottlenecking in the performance due to unused classes. Since irrelevant classes do not contribute to target tasks, using more number of them is inefficient.

References

1. Alom, M.Z., Taha, T.M., Yakopcic, C., Westberg, S., Hasan, M., Esesn, B.C.V., Awwal, A.A.S., Asari, V.K.: The history began from alexnet: a comprehensive survey on deep learning approaches. CoRR. abs/1803.01164 (2018)
2. Fukushima, K.: Neocognitron: a self-organizing neural network model for a mechanism of pattern recognition unaffected by shift in position. Biol. Cybern. **36**, 193–202 (1980)
3. Glorot, X., Bordes, A., Bengio, Y.: Deep sparse rectifier neural networks. In: AISTATS (2011)
4. Gong, Y., Wang, L., Guo, R., Lazebnik, S.: Multi-scale orderless pooling of deep convolutional activation features. In: ECCV (2014)
5. Goodfellow, I., Bengio, Y., Courville, A.: Deep Learning. MIT Press (2016). http://www.deeplearningbook.org
6. He, K., Zhang, X., Ren, S., Sun, J.: Deep residual learning for image recognition. In: 2016 IEEE Conference on Computer Vision and Pattern Recognition (CVPR), pp. 770–778 (2016)
7. Hubel, D.H., Wiesel, T.N.: Receptive fields and functional architecture of monkey striate cortex. J. Physiol. **195**(1), 215–43 (1968)
8. Krizhevsky, A., Sutskever, I., Hinton, G.E.: Imagenet classification with deep convolutional neural networks. In: NIPS (2012)
9. LeCun, Y., Boser, B.E., Denker, J.S., Henderson, D., Howard, R.E., Hubbard, W.E., Jackel, L.D.: Handwritten digit recognition with a back-propagation network. In: NIPS (1989)
10. Lee, D.H.: Pseudo-Label : The Simple and Efficient Semi-supervised Learning Method for Deep Neural Networks (2013)
11. Nair, V., Hinton, G.E.: Rectified linear units improve restricted boltzmann machines. In: ICML (2010)
12. Nguyen, A.M., Yosinski, J., Clune, J.: Deep neural networks are easily fooled: high confidence predictions for unrecognizable images. In: 2015 IEEE Conference on Computer Vision and Pattern Recognition (CVPR), pp. 427–436 (2015)

13. O'Shea, K., Nash, R.: An introduction to convolutional neural networks. CoRR. abs/1511.08458 (2015)
14. Perez, L., Wang, J.: The effectiveness of data augmentation in image classification using deep learning. CoRR. abs/1712.04621 (2017)
15. Simonyan, K., Zisserman, A.: Very deep convolutional networks for large-scale image recognition. CoRR. abs/1409.1556 (2014)
16. Springenberg, J.T., Dosovitskiy, A., Brox, T., Riedmiller, M.A.: Striving for simplicity: The all convolutional net. CoRR. abs/1412.6806 (2014)
17. Srivastava, N., Hinton, G.E., Krizhevsky, A., Sutskever, I., Salakhutdinov, R.: Dropout: a simple way to prevent neural networks from overfitting. J. Mach. Learn. Res. 15, 1929–1958 (2014)
18. Szegedy, C., Liu, W., Jia, Y., Sermanet, P., Reed, S.E., Anguelov, D., Erhan, D., Vanhoucke, V., Rabinovich, A.: Going deeper with convolutions. In: 2015 IEEE Conference on Computer Vision and Pattern Recognition (CVPR), pp. 1–9 (2015)
19. Wan, L., Zeiler, M.D., Zhang, S., LeCun, Y., Fergus, R.: Regularization of neural networks using dropconnect. In: ICML (2013)
20. Zamir, A.R., Sax, A., Shen, W.B., Guibas, L.J., Malik, J., Savarese, S.: Taskonomy: disentangling task transfer learning. CoRR. abs/1804.08328 (2018)
21. Zeiler, M.D., Fergus, R.: Visualizing and understanding convolutional networks. In: ECCV (2014)

A Numerical Representation Method for a DNA Sequence Using Gray Code Method

M. Raman Kumar and Vaegae Naveen Kumar

Abstract The exceptional speed in increase of genomic data at public databases requires advanced computational tools to perform quick gene analysis. The tools can be devised with the aid of genomic signal processing. The pivotal task in genomic signal processing is numerical mapping. In numerical mapping, the string of nucleotides is transformed into discrete numerical sequence by assigning optimum mathematical descriptor to a nucleotide. The descriptor must be compatible with the further stages of genomic application in order to achieve high efficiency. In this work, a simple numerical mapping method is proposed in which the optimum descriptor value is obtained by applying Gray code concept. The proposed method is evaluated on benchmark databases HRM195 and ASP67 for an identification of protein coding region application. The proposed method exhibits improved exon prediction efficiency in terms of performance accuracy and equal error rate when compared with similar methods.

Keywords Exon identification · Genomic signal processing · Gene encoding · Gray code · Numerical mapping · Three-base periodicity

1 Introduction

The success of the human genome project resulted in rapid growth of sequenced data at public databases and it is in strong demand of automated tools for quick and easy analysis of genome. The genome is made of biomolecule called Deoxyribonucleic acid (DNA). The DNA molecule encodes the genetic information of any living organism and it is made up of string of nucleotides called adenine (A), guanine (G), cytosine (C), and thymine (T). The gene sequence consists of two main regions called exonic region and intronic region. The exonic region is the prime source for

M. Raman Kumar · V. Naveen Kumar (✉)
School of Electronics Engineering, Vellore Institute of Technology, Vellore, India
e-mail: vegenaveen@vit.ac.in

M. Raman Kumar
e-mail: ramankumarm@gmail.com

© Springer Nature Singapore Pte Ltd. 2020
K. N. Das et al. (eds.), *Soft Computing for Problem Solving*,
Advances in Intelligent Systems and Computing 1057,
https://doi.org/10.1007/978-981-15-0184-5_55

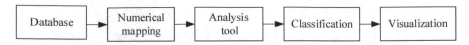

Fig. 1 General block diagram for gene analysis

protein synthesis, whereas the intronic region is known as junk DNA. The exons are randomly intermixed with introns and thus identification of exons is a challenging task in computational biology [1]. The exonic regions exhibit 3-base periodicity due to codon bias. This is the key property for engineering domains to contribute for genomic domain. The digital signal processing tools and algorithms are gaining large attention in the field of genomics to explore the hidden biological information [2]. The signal processing tools till used in genomic signal processing (GSP) like Fourier transform [3], short-time Fourier transform [4], wavelet transform [5] aim to explore hidden 3-base periodicity regions [6]. However, the signal processing tool works for numerical data rather than symbolic data. Therefore, the conversion of symbolic data into a numerical data is essential.

Figure 1 illustrates the typical stages involved in exploration of specific genomic information using signal processing algorithm. The main stages are (a) extraction of DNA sequence from a database, (b) numerical mapping or gene encoding which converts DNA string into compatible form, (c) application of signal processing tool for gene analysis, (d) classification of gene information into desired and undesired parts, and (e) projection or elucidation of result for visualization and analysis.

In the five stages, the numerical mapping stage plays a pivotal role since it has effect on overall performance of the system. The numerical mapping transforms the DNA string into numerical format by assigning descriptor value. The determination of proper descriptor value is must to achieve high prediction efficiency. The architect of mapping method has to consider two main devising factors, mainly (i) reflectance of biological behavior by descriptor (ii) compatibility between mathematical descriptor and analysis tool used in subsequent stages of the model. Therefore, numerous mapping methods discussed in literature [7] concentrate on design of optimum numerical descriptor.

The numerical mapping methods are categorized into two types based upon the nature of descriptor. They are static mapping and variable mapping.

In static mapping method, a fixed numerical value will be assigned to a nucleotide whenever the nucleotide alphabet appears along the sequence based on a predefined criterion [8], whereas in dynamic mapping, the numerical value of nucleotide will be varied according to a criterion along the sequence [9, 10]. The criterion is framed either upon structural or functional behavior of nucleotides. One may attribute binary value or integer value or real value. The numerous mapping methods were discussed in literature; however, in this work, the mapping methods which are similar to proposed method are presented as follows. They are (a) binary method, (b) integer method, (c) real method, and (d) Galois field method.

1.1 Binary Method

The widely used binary numerical mapping method [11] for conversion of DNA sequence. It decomposes the whole DNA sequence into four subsequences $\{X_A, X_G, X_T, X_C\}$ of original length. The presence of corresponding nucleotide at a position "k" is represented by binary "1" and absence by binary "0" as given in Eq. (1). The quadruple dimensionality of obtained binary subsequences for a DNA sequence will increase the computational overhead.

$$X_x(k) = \begin{cases} 1 \text{ if nucleotide } x \text{ exist at } k^{th} \text{ position} \\ 0 \text{ otherwise;} \quad \text{where } x \in \{A, G, C, T\} \end{cases} \tag{1}$$

1.2 Integer Method

The integer method encodes the DNA sequence with integer or real values. The resulting signal is a discrete value signal. The integer method proposed in [12] attributes integer values $\{1, 2, 0, 3\}$ to nucleotides C, A, T, and G, respectively. This sort of representation demonstrates that A is greater than T and G is greater than C. Another integer method proposed in [13] assigns $C = 3, A = 1, T = 4$, and $G = 2$ values for biological bar boding application.

Later, a biologically inspired gradient source localization algorithm-based integer method is proposed in [14] to map nucleotides $\{C, A, T, G\}$ as $\{1, 0, 2, 3\}$, respectively.

1.3 Real Method

The real method replaces alphabets of DNA sequence with real values. The real method proposed in [15] mapped pulse amplitude modulation (PAM) values to nucleotides such as $C = 0.5, A = 1.5, T = -1.5, G = -0.5$, and projected onto constellation diagram for graphical visualization.

1.4 Galois Field Method

The nucleotides of a DNA are mapped using a Galois field (GF)-based encoding scheme to uncover the redundant information of a linear DNA sequence. The sequence is converted into (n, k) orthogonal code words. The Galois field concept is exploited to label base pairs to a finite field of four GF(4). The polynomial of GF(2) is given in Eq. (2) as

$$\chi^2 + \chi + 1 = 0 \tag{2}$$

The polynomial of order 2 is modified under GF properties, i.e., commutative property for addition and associative property for both addition and multiplication. The definitions given in Eq. (3) illustrate the assignment of GF(4) to four nucleotides A, T, G, and C in terms of polynomial representation and corresponding numerical assignment is also defined.

$$0 = 0 \Leftrightarrow 0 \Leftrightarrow A$$
$$\chi^0 = 1 \Leftrightarrow 1 \Leftrightarrow C$$
$$\chi^1 = \chi \Leftrightarrow 2 \Leftrightarrow T$$
$$\chi^2 = \chi + 1 \Leftrightarrow 3 = G \tag{3}$$

The mapping method is applied to investigate the redundant information of DNA sequence and also applied for mutant analysis [16]. Furthermore, similar GF concept is developed by other researchers such as modifying the weights of each nucleotide of codon like multiplying them with 2^0, 2^1, and 2^2, respectively [17].

The discussed mapping methods have not achieved the complete efficiency from an application perspective. This motivates to design a novel mapping method for numerical conversion of a DNA sequence. The proposed mapping method is developed by combining GF method and Gray code concept.

1.5 Gray Code

The alternative name of Gray code is reflected binary code and it was introduced to solve mathematical puzzles in the beginning. Later, it was developed to resolve engineering issues. The Gray code is a binary symbol attributed to each family member of a set of integer values. The code should not be same or repeated in a set and any two adjacent codes have to be separated by one bit, i.e., the Hamming distance should be one. Hence, these codes are also called single-distance codes [18].

The motivation for Gray code consideration in our work is the reflectance of complementary nature by adjacent codes similar to the DNA strands shows in double helix structure. Table 1 elucidates this behavior for a set of four real numbers. The

Table 1 Demonstration of Gray code conversion for real and binary numbers

Real number	Binary equivalent	Gray code equivalent
0	00	00
1	01	01
2	10	11
3	11	10

Gray code is calculated by performing Ex-OR operation between present bit with neighboring bit.

The polynomial of order 2 is modified under GF properties, i.e., commutative for addition and associative for both addition and multiplication. The definitions given in Eq. (3) illustrate the assignment of GF(4) to four nucleotides in terms of polynomial representation and corresponding numerical assignment.

In this work, the Gray code for each GF value of nucleotide is determined and assigned to nucleotide base pairs for a DNA sequence in order to enhance the prediction efficiency of protein coding regions.

This paper is organized as follows. Section 2 presents the methodology for numerical representation, Sect. 3 is about experimental results of implementation of proposed method for the identification of protein coding region, and Sect. 4 has concluding remarks with future scope.

2 Proposed Method

In this work, the numerical mapping is carried in two stages in order to achieve better efficiency. The steps are (a) apply GF method onto a DNA sequence and (b) apply Gray code concept onto the numerical value of nucleotide resulted from GF method. The flowchart of proposed method is illustrated in Fig. 2. First, a DNA sequence is extracted from a public database and given as an input sequence to a GF-based numerical mapping method. The GF method converts the input DNA string into numerical form by assigning an integer value to each nucleotide as defined in Eq. (3). In the further step, the determined integer value of each nucleotide is given

Fig. 2 Flowchart of proposed method

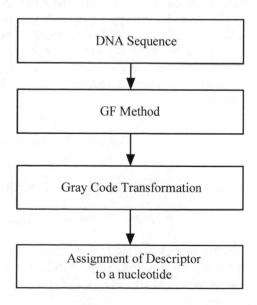

Table 2 Numerical equivalents of nucleotides

Method	Numerical equivalent of nucleotides			
	A	T	G	C
GF	00	10	11	01
Proposed	00	11	10	01

to Gray code method as explained in Sect. 1.5. Here, the Gray code value for each nucleotide is calculated as illustrated in Table 2.

After the determination of Gray code for a complete sequence, it is given to analysis tool for processing. The processed sequence is further fed to remaining stages based upon the application requirement, as explained in Fig. 1. The amount of compatibility and optimum nature of devised descriptor is evaluated in terms of performance measures.

3 Experiments

The proposed mapping method is tested for the identification of exonic regions in genes of eukaryotes. The experiment is performed on MATLABR2017b environment. First, the DNA sequence is collected from a popular National Center for Biotechnology for Information (NCBI) database and given to GF method for numerical conversion. The mapped sequence is again transformed into Gray code sequence using proposed method. The mapped Gray code sequence is fed to DFT tool for spectrum visualization and further classified into exons and introns. The accuracy of exon prediction is done by performance measures.

The numerically mapped Gray code sequence $f(n)$ of finite length N is windowed and further given to DFT tool for spectrum determination. The windowing stage is repeated till the end of sequence. The peak present in spectrum represents the presence of exonic region. The peak resolution can be improved by eliminating background noise with optimum preprocessing method. The DFT for a given sequence $f(n)$ is defined as

$$F(n, m) = \sum_{p=0}^{L-1} f(n + p)\omega(p)e^{-j\,2\pi mp/L} \tag{4}$$

where $\omega(p)$ is a Hamming window of length L and defined as

$$\omega(p) = 0.54 - 0.46\cos(2\pi p(L - 1)), 0 \le p \le L - 1 \tag{5}$$

The exonic region of eukaryotes exhibits period-3 behavior due to unequal involvement of codons in amino acid formation, whereas introns do not have. Therefore, the magnitude of DFT at $m = L/3$ is higher in exonic region compared to intronic region. The DFT of at $m = L/3$ is given as in Eq. (6).

$$F\left(n, \frac{L}{3}\right) = \sum_{k=0}^{L-1} f(n + k)\omega(k)\,e^{-j2\pi k/3} \tag{6}$$

3.1 Performance Measures

Digital signal processing methods are used to convert the obtained raw data into useful information. The performance of the proposed method is verified for identification of exonic region in eukaryotes. For performance measure, accuracy and equal error rate are used and defined in terms of true positive rate, and true negative rate as defined in Eqs. (7) and (8).

Performance accuracy

$$PA = \frac{TP + TN}{TP + TN + FP + FN} \tag{7}$$

Equal error rate

$$ERR = \frac{FP + FN}{TP + TN + FP + FN} \tag{8}$$

where TP denotes the true positive that defines the number of exons correctly identified as exons, TN denotes the true negative that defines the number of introns identified as introns, FP denotes false positive that defines the number of introns which are misidentified as exons, and FN denotes false negative that defines the number of exons which are misidentified as introns.

3.2 Results and Discussions

The experimental study is carried out on benchmark sequence Caenorhabditis Elegans chromosome III gene whose accession number is F56F11.4 and extracted from standard public database called NCBI. The gene sequence has five exonic regions ranging from 7021 to 15,022 bps. Figure 3 illustrates identification of protein coding region, in that fourth peak at 5464 is missed by GF method and it is clearly detected by proposed method and this is because of optimum assignment of descriptor to nucleotide. The extra peaks that are appearing around first and second peaks are due to background noise (e.g., after 2527 location) and this yields into false identification of exons.

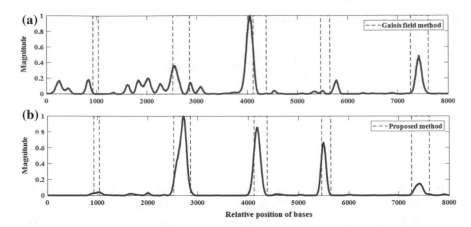

Fig. 3 Spectrum for identification of exonic region for sequence AF56F11.4 using (**a**) GF method (**b**) proposed method for comparison. The dotted lines denote actual exons location. For both the plots, the abscissa denotes relative position of bases and ordinate denotes normalized squared magnitude

This effect of false identification can be alleviated by devising compatible descriptor and by employing filtering mechanism. Table 3 presents actual location of exons and detected location of exons for a gene AF56F11.4 using proposed method.

The proposed method has been tested on two more popular databases which are widely used in literature for identification of exons in gene of eukaryotes. They are HRM195 and ASP67 databases [19]. The HRM195 database consists of single gene sequences of human, rat, and mouse in 103:82:10 ratio, respectively, and ASP67 database consists of 67 multiple gene sequences of fungi. The evaluated metrics are presented in Table 4; the PA for HRM195 is more compared to ASP67 database since ASP67 has lengthy introns.

Similarly, the PA is evaluated for other widely used numerical mapping methods for a sequence AF56F11.4 and depicted in Fig. 4 for comparison. From the figure, it is observed the increase in prediction efficiency than GF method.

Table 3 Prediction of exons for a gene (AF56F11.4)	Actual exon location (length)	Proposed method	% Exons predicted
	7949–8059 (112 bps)	7985–8059 (112 bps)	100
	9548–9877 (330 bps)	9548–9852 (304 bps)	92.1
	11,134–11,397 (264 bps)	11,134–11,299 (165 bps)	62.5
	12,485–12,664 (180 bps)	12,485–12,604 (119 bps)	66.1
	14,275–14,625 (351 bps)	14,242–14,483 (308 bps)	87.7

Table 4 A measure of performance metrics for different databases by GF and proposed method

Database	GF method		Proposed method	
	PA (%)	ERR (%)	PA (%)	ERR (%)
HRM195	62.5	37.5	80.6	19.4
ASP67	57.16	42.84	65.5	34.5

Fig. 4 Comparison of numerous mapping methods for identification of exonic region for sequence AF56F11.4, where x-axis denotes name of the mapping method and y-axis denotes PA

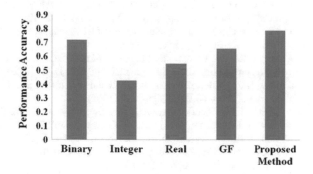

4 Conclusion

A novel static type numerical mapping method is proposed based on Gray code concept. The method is developed by modifying the GF field mapping method to enhance an optimum numerical representation. The representation is done by calculating Gray code for an each nucleotide of a DNA sequence. The proposed method is tested for the identification of exonic region for a given sequence. The exonic region shows period-3 property, whereas it is absent in intronic region. This property is exploited using DFT tool to test the performance of proposed mapping method. The GF method is unable to identify short exons, whereas our method successfully predicted it. The PA and ERR for gene sequence AF56F11.4 using proposed method and GF method is 0.82%, 0.18% and 0.66%, 0.34%, respectively. Furthermore, our method can be used for other applications of computational biology like hot spot identification, mutant analysis, and detection of structure and function of proteins.

References

1. Vaidyanathan, P.P., Yoon, B.J.: The role of signal-processing concepts in genomics and proteomics. J. Franklin Inst. **341**(1–2), 111–135 (2004)
2. Anastassiou, D.: Genomic signal processing. IEEE Signal Process. Mag. **18**, 8–20 (2001)
3. Akhtar, M., Epps, J., Ambikairajah, E.: On DNA numerical representations for period-3 based exon prediction. In: GENSIPS'07—5th IEEE International Workshop on Genomic Signal Processing and Statistics (2007)
4. Ahmad, M., Jung, L.T., Bhuiyan, A.A.: A biological inspired fuzzy adaptive window median filter (FAWMF) for enhancing DNA signal processing. Comput. Methods Programs Biomed.

149, 11–17 (2017)

5. Marhon, S.A., Kremer, S.C.: Prediction of protein coding regions using a wide-range wavelet window method. IEEE/ACM Trans. Comput. Biol. Bioinform. **13**(4), 742–753 (2016)

6. Rao, K.D., Swamy, M.N.S.: Analysis of genomics and proteomics using DSP techniques. IEEE Trans. Circuits Syst. I Regul. Pap. **55**(1), 370–378 (2008)

7. Yu, N., Li, Z., Yu, Z.: Survey on encoding schemes for genomic data representation and feature learning—from signal processing to machine learning. Big Data Min. Anal. **1**(3), 191–210 (2018)

8. Das, B., Turkoglu, I.: A novel numerical mapping method based on entropy for digitizing DNA sequences. Neural Comput. Appl. **29**(8), 207–215 (2018)

9. Mo, Z., et al.: One novel representation of DNA sequence based on the global and local position information. Sci. Rep. **8**(1), 1–7 (2018)

10. Singha Roy, S., Barman, S.: Polyphase filtering with variable mapping rule in protein coding region prediction. Microsyst. Technol. **23**(9), 4111–4121 (2017)

11. Voss, R.F.: Evolution of long-range fractal correlations and 1/f noise in DNA base sequences. Phys. Rev. Lett. **68**(25), 3805–3808 (1992)

12. Cristea, P.D.: Genetic signal representation and analysis. In: Proc. SPIE Conference on International Symposium on Biomedical Optics (BIOS'02), vol. 4623, pp. 77–84 (2002)

13. Hebert, P.D.N., Cywinska, A., Ball, S.L., DeWaard, J.R.: Biological identifications through DNA barcodes. In: Proceedings of the Royal Society of London. Series B: Biological Sciences, vol. 270, no. 1512, pp. 313–321 (2003)

14. Rosen, G.L.: Biologically-inspired gradient source localization and DNA sequence analysis. Georg. Inst. Technol., August, 2006

15. Chakravarthy, N., Spanias, A., Iasemidis, L.D., Tsakalis, K.: Autoregressive modeling and feature analysis of DNA sequences. EURASIP J. Appl. Signal Process. **1**, 13–28 (2004)

16. Rosen, G.L., Moore, J.D.: Investigation of coding structure in DNA. In: IEEE International Conference on Acoustics, Speech, and Signal Processing (ICASSP'03), 6 April 2003

17. Cristea, P.D.: Conversion of nucleotides sequences into genomic signals. J. Cell. Mol. Med. **6**(2), 279–303 (2002)

18. Lucal, H.M.: Arithmetic operations for digital computers using a modified reflected binary code. IRE Trans. Electron. Comput. **EC-8**(4), 449–458 (1959)

19. HRM195 and ASP67dataset. http://www.vision.ime.usp.br/jmena/MGWT/datasets/2010

Adaptive Way of Particle Swarm Algorithm Employing the Fuzzy Logic

Rajesh Eswarawaka, C Subash Chandra, Vadali Srinivas and Kanumuri Viswas

Abstract The Image Swarm Intelligence calculations, in numerous enhancement issues, have always filled a need of worldwide hunt strategy. One of the issues went up against amid advancement is bunching issue. Contribution for a bunching procedure is an arrangement of information, which are then composed into various sub-gatherings. Current investigations have suggested that divided or isolated bunching calculations are being more fitted for grouping of wide and colossal data objects or datasets. A standout among the most and best regular partitional bunching calculations is k-means. K-implies calculation demonstrates a more quick union than PSO, however, then against nearby ideal territory is for the most part caught relying upon the arbitrary estimations of introductory centroids. A proficient crossbreed technique is displayed in this paper, specifically molecule swarm improvement with fluffy rationale or versatile molecule swarm enhancement (APSO) to determine information grouping issue. The PSO calculation finds a decent or close ideal arrangement in sensible time, however, its introduction was upgraded by seeding the underlying swarm with fuzzifier work. The versatile fluffy molecule swarm enhancement calculation (APSO) is contrasted and *k*-implies utilizing all-out execution time and bunching bunch blunder. It is found that the aggregate execution time for APSO technique outflanks the *k*-implies and had higher arrangement quality as far as bunching bunch blunder.

Keywords Hybridization · Fuzzy logic · Particle swarm optimization · K-means

R. Eswarawaka
Jawaharlal Nehru Technological University (JNTUK), Kakinada, India

C. Subash Chandra
KIET, Kakinada, India

V. Srinivas (✉)
JNTUK, Kakinada, India
e-mail: vadalisrinivas16@gmail.com

K. Viswas
IIIT Hyderabad, Hyderabad, India

© Springer Nature Singapore Pte Ltd. 2020
K. N. Das et al. (eds.), *Soft Computing for Problem Solving*,
Advances in Intelligent Systems and Computing 1057,
https://doi.org/10.1007/978-981-15-0184-5_56

655

1 Introduction

Grouping [1] is a broadly examined learning strategy, discovers are used for breaking down information. Bunching calculations can be utilized for bioinformatics applications, design acknowledgment [2], and report classification [3]. The main aim out of sight of bunching challenges is to produce dissimilar to bunches, in spite of having need in information with respect to the case marks. A grouping calculation helps in discovering the comparable information, sharing a high level of likeness, which has basic traits. With the help of comparability lattices information occurrences parts into various bunches. It has been discovered that to work the plans which are smooth and are executed different occasions gives ideal outcomes when contrasted with the plans which are mind boggling and called for run once in a while.

The least complex worldview went for bunching is k-implies [4] calculation. To get great bunching result Euclidean separation [5], it is also required to be diminished. The major issue of k-implies calculation is that it stalls out and gets caught at neighborhood ideal zone and do not give ideal arrangement.

Swarm intelligence (SI) [6] is similarly a novel rule, in which, that includes the learning of collective exercises of the people in populace. That is being connected in look into setting to hoist the control and administration of extraordinary no. of working together substances, for example, PC and sensor systems, correspondence, and different others. Since the swarm knowledge calculations refuse the downsides of k-implies calculation by untimely union and additionally from beginning no. of centroids and give extremely proficient and great bunching results.

The center thought behind the swarm knowledge calculation is to find the comprehensive clarification for discover improvement issue.

A prominent streamlining strategy, PSO is characterized by Eberhart et al. PSO utilizes an idea by mimicking a couple of thoughts got from angle tutoring, winged animal running, and other social gatherings to locate the ideal arrangement. Every molecule in the swarm uses information held by it and the actualities common between the gatherings to accomplish its goal. PSO takes a shot at the standards of social conduct. The PSO calculation gives close ideal arrangement in sensible time yet its execution was upgraded by identifying the underlying particle swarm with fuzzifier work. The PSO calculation in the proposed work is hybridized utilizing fluffy rationale. What's more, the proposed calculation is confirmed over genuine datasets with not at all like structures to approve the bunching highlight. The outcomes are looked at and discovered that APSO outflanks from k-means calculation

Particle swarm knowledge is another approach to manage and handle streamlining issues. It is animated by the social practices of animals or bugs. The difference between whole based stochastic upgrade systems, for instance, formative estimations, PSO has comparable or even overwhelming interest execution with higher and all the more relentless association rates for some hard change issues. Also, PSO has memory; heretofore, visited best positions are reviewed, which is not the same as headway identifying that do not keep the basic information as the masses change. PSO has been used as a piece of different present-day regions, for instance, control structures, parameters learning of neural frameworks control, arranging, and speak issues showing et cetera. Regardless, observations reveal that PSO regularly joins distinctly to start with time of the looking for system, by then inundates or even closures in the later stage. It carries on like the regular close-by looking methods that trap in neighborhood optima.

Starting late, exceptional hybrid PSO systems have been proposed to overcome the weakness of getting in neighborhood optima. The blend PSO was first proposed, in which a standard decision segment was composed with PSO. Another hybrid tendency dive PSO (HGPSO), which misused the slant information to achieve speedier association without getting captured in neighborhood minima, was proposed. However, the computational demand of HGPSO is extended by the strategy of edge fall. Additionally, it is poor to manage multimodal issues that contain various neighborhood minima. A creamer PSO estimation named HGAPSO was proposed, which combined the innate count (GA) formative exercises of crossover, change, and age. A blend PSO named HPSOM was proposed, in which a steady changing space was used as a piece of change. In both HGAPSO and HPSOM, the course of action space can be researched by performing change exercises on particles along the chase, and troublesome joining will presumably be evaded. In any case, the changing space is kept unaltered all through the interest, and the space for the phase of particles in PSO is furthermore settled. It tends to be improved by contrasting the changing space along the interest. Starting late, a blend PSO with wavelet change errand (HPSOWM) was proposed in, of which the changing space vacillated in view of the wavelet theory. The course of action quality and plan reliability is gained ground. Fuzzy-1 versatile controller has been proposed for nonlinear MIMO frameworks. Fluffy controller has been intended to tune the parameters of the criticism channel gain of 1 versatile controller. Multiobjective particle swarm streamlining calculation has been utilized to discover ideal factors for information and yield participation capacities in view of best bargain arrangement between two clashing destinations. Criticism channel parameters of the 1 versatile controller were tuned by FLC with a specific end goal to enhance the vigor edges. Very nonlinear MIMO framework was utilized to demonstrate the adequacy of the proposed approach. Results approve the viability and power of the proposed approach on nonlinear framework with time-differing vulnerabilities. The smooth tuning of the criticism channel upgrades the vigor edge and decreases the control flag go. Also, quick shut circle elements have been accomplished with better heartiness execution [7]. A parallel versatile speed limit PSO neural system-based four-piece expansion utilizing five cell neuron equipment engineering to acknowledge how natural neural system capacities is built, which amplifies the execution, exactness, and offers a high level of parameterization with quicker assembly rate. In execution of 4-bit full snake utilizing neural systems by FPGAs the weights, data sources, and edge work got from versatile speed edge PSO is extremely imperative and thus arrange. Henceforth, a neural system demonstrate comprising of neuron cell joining the enhancement property of versatile speed edge PSO is executed which plays out an inquiry to locate the worldwide ideal from a gathering of nearby minima. Every neuron unit is fabricated utilizing Dertouzos' technique joined with versatile speed edge molecule swarm enhancement; hence, making a harmony between the investigation and misuse approach of versatile speed limit PSO and slope seek approach of neural systems. Five neural cell arrange performing NN calculation is proposed. Moreover, a versatile speed edge PSO equipment center is abused to acquire limit esteems so as to prepare arrange weights [8].

A two-variables and multiorders versatile fluffy time arrangement guaging model with PSO and fluffy coherent connections trees, in which the strength of lower orders fluffy intelligent connections and the exactness of higher requests fluffy legitimate connections are used at the same time to enhance the determining precision. We contrast the proposed model and some current anticipating models including univariate and multivariate models. Examination results demonstrate that the proposed display has preferred execution over many existing fluffy time arrangement determining models [9]. The exhibitions of the proposed molecule swarm enhancement-based bother and watch (PSO-P&O) are researched when PV cluster performs under somewhat shaded conditions. In this work, PV exhibition is demonstrated in view of four arrangement associated PV modules. The created PSO-P&O calculation is tried under three unique cases and its exhibitions in improving the yield control are contrasted with the regular P&O calculation. From the recreation results, PSO-P&O calculation can advance the age of PV framework by following the GMPP quicker when the surrounding conditions (sunlight-based irradiance and temperature) are changed. At the point when the yield control is moving toward GMPP, PSO-P&O calculation will choose littler voltage irritation size to limit the vacillation of the yield control in the consistent state. What's more, the proposed calculation can control the PV framework to perform at a more exact working voltage. In light of the reproduction discoveries [10], the PSO-P&O calculation has figured out how to track GMPP and gives better consistent state reaction over the traditional P&O calculation, especially amid the PSC. In the future, the execution and vigor of the proposed calculation will be tried in reasonable.

2 Particle Swarm Optimization

PSO an extremely well known unforeseen streamlining technique [11, 12] was found by Eberhart and Kennedy in 1995. Many analysts have considered that PSO is a proficient pursuit and enhancement strategy which is affected by social conduct of feathered creature run, flying together to discover the fines, the disturbance by modifying their courses of action, and separation for enhanced inquiry. PSO just deploy crude scientific administrators [13] to locate the ideal arrangements. Every one of the components is run turbulently to locate the best arrangement by modifying their ages. PSO takes a shot at double entries, the best is known as intellectual; the next one is public. Subjective term considers as the "thought" of the molecule and causes the swarm to guarantee a vigorous inquiry limit with keeping away from a nearby least esteem. The parts and swarm distributed among the entities and p swarm give a decent arrangement is emulated by the public term. The particles which are arbitrarily introduced before all else have its own particular speed and position.

Among every one of the particles, each molecule attempts to save its nearby best area (Pbest) and worldwide best area (Gbest) [14].

$$Vi(t + 1) = Vi(t) + C1 * r1(\text{Pbest} - ni(b)) + C2 * r2(\text{Gbest} - Xi(t)) \quad (1)$$

$$Xi(t + 1) = Xi(t) + Vi(t + 1) \quad (2)$$

where speed and position of the molecule are spoken to be Vi and Xi. The location of the molecule is signified by Pbest, and Gbest signifies gloal exact location for PSO, $r1$ and $r2$ are self-assertive ranges [0, 1], $c1$ and $c2$ are learning factors.

3 The Basic Pseudocode for PSO Algorithm

The Basic Pseudocode for PSO Algorithm

Step1: Start
Step2: For i in the particles with random numbers with in range:
        END for.
Step3: While
        Termination criteria are satisfied
        Do:
        For I in all particles in the population
        Do:
        Calculation fitness
        END for
        Choose the finest particles as Gbest
        For I in all particles in the population
        Do:
        Update velocity and location using Eqs. 1 and 2
        END for'
        END While
Step4: Stop.

4 Fuzzy Logic

Fluffy framework is intended to hold the hypothesis of incomplete fact and it was development of Boolean rationale [15]. Boolean rationale has double standards or steps that are expressed as a genuine or false, on or off, and likewise begin or stop.

Be that as it may, in practical world under couple of conditions in which the outcome is neither dark nor white yet is a few shades of dim. Fluffy rationale is among the unending sort of rationale that backings characterizing the shades of dim.

The idea of fluffy rationale was found by Zadeh [16]. In fluffy rationale, preparing of the information is finished by allowing incomplete set enrollment rather than fresh set participation or non-enrollment. It offers an unobtrusive and powerful strategy to achieve a specific conclusion based on erroneous, inconclusive, loose, blasting, or lost info data. Fluffy rationale offers a surmising technique, which encourages assessed human thinking know-how's to be put on to information-based framework. It can manage the issue of equivocalness and phrasal dubiousness. The primary highlights of fluffy rationale are as per the following:

(a) It grants goals with surveyed information under the nearness of deficient or inconsistent data.
(b) It is not confined to few criticisms at most two and inputs control yields.
(c) It grants arrangements inside precise, uproarious information sources. So it can said to be normally solid.

5 Proposed Algorithm

The expectation of the proposed work is to get more noteworthy or better perfect outcomes for bunching. In this manner for grouping, we are here using hybridization of particle swarm optimization with cushy reason, which depends on the hypothesis of incomplete truth and is a development of Boolean rationale. This hybridization of particle swarm optimization with fluffy rationale can defeat the meeting speed deficiency and brings about expanding the proficiency of PSO calculation. Grouping process having the pre-information of introductory bunch focuses and a number of groups can continue with great beginning begin and the bunching arrangement winds up exact and productive.

In the common PSO, at first round of the calculation, totally arbitrary components were utilized to instate the particles. This approach is considered as an all-inclusive technique for introducing the clarifications in transformative calculation. In any case, arbitrariness is not generally bearing the cost of the great begin and not gives the best answer for developmental-based advancement systems. Arbitrary variables, which are utilized totally, self-decide the obstruction to be comprehended which heads to moderate and early union. The merging lays on the relative wellness esteem f/fmax where most extreme wellness esteem in the present swarm are indicated by $f$max. The procedure will keep making the most extreme number of age or accomplishing the ideal arrangement.

In our proposed technique, we instate the particles with the assistance of fuzzifier and swarm improvement rate. The two primary explanations behind these two capacities introduction as take after:

These capacities assorted more than any arbitrary capacity in the arrangement space. Thus, our global best molecule would assist different particles with initiating betterly and our look generally advantageous and ideal arrangement should be possible more precisely and speedier.

6 Experimentation

This segment speaks to the standard investigation to look at the capability of the APSO calculation in information grouping. For dependable examination, greatest wellness esteem in the present swarm, the outcomes got have been represented utilizing the APSO calculation on eminent datasets. We assess their execution by exhibiting the examination of the APSO with k-implies which have been utilized in the writing.

A. Condition:

The examination work been completed bygone MATLAB with the system configuration like RAM, operating system, and core to dual processor.

B. Used trained set:

To test the attributes of the relating gathering approaches, four exploratory datasets, Fisher iris, k-means, wine and seed are used. These information articles or datasets are got from the UCI ML store [17] addressing occurrences of data with little, immediate, and broad point. Table 1 studies the features of picked up information objects or datasets. The delineation of real dataset is taken and they are given underneath.

Informational collection (1): Fisher's iris data object or dataset ($d = 5, n = 151, k = 4$), that involves three differing sorts of iris blooms specifically iris virginica, iris setosa, and iris versicolor. A sum of fifty tests was ordered from the four angles, for every specie, that incorporates sepal measure, sepal diameter, petal size, and petal diameter.

Informational indcx (2): K-means dataset ($n = 560, d = 4, k = 4$) which is an inbuilt dataset in MATLAB. It comprises of 560 items conveyed in four highlights.

Table 1 The characteristics of dataset

Dataset name	Number of characteristics	Number of classes	Size of the data object or data set
Iris	4	3	160
K-means	4	4	570
Seed	7	3	210
Wine	13	3	178

Informational collection (3): The wine data object or dataset ($n = 179$, $k = 4$, $d = 14$). This dataset includes 178 articles disseminated in the 13 parameters including substance of the liquor, malic corrosive and slag, alkalinity of powder, magnesium fixation, add up to phenols, flavanoids, non-flavanoid, phenols, and additionally proanthocyanius and shading sum, sort and OD280/OD315 of weakened wines and pralines. These attributes are framed in a similar zone in Italy, however, are subsidiary of three divergent cultivars and procured by looking at wines artificially. The three arrangements of the informational indexes include the measures of items as: class 1(59 articles), class 2(71 articles), class 3(48 articles).

Informational collection (4): The seed data object or dataset ($n = 210$, $d = 7$, $k = 3$) includes of pieces which has a place with three varieties of wheat to be specific Kama and Canadian. All the three classes that contain 70 components. To build information, seven estimated parameters of wheat parts are considered. The parameters are: territory, edge meant by A and P individually, compactness characterized as $c = 4 * pi * A/P^2$, width and length of bit, asymmetry coefficient, and the length of piece groove.

7 Results

The initial

A. Test for measurable hugeness:

To show abilities of information grouping, results of the experimentation were compared with k-implies calculation. The accompanying foundation is utilized to quantify the nature of individual grouping:

(1) Time complexity: In our work count, the general time multifaceted nature depends generally on the level of time that required to locate the surmised set for every particle. Besides, when all is said in done time also depends in the dimensionality of the dataset used close by the particle estimation and most noteworthy numeral accentuations.

Blunder rate: The actual number of focuses partitions the quantities of lost focuses is separated by as spoken to by the condition.

Blunder = Here n speaks to the general numeral of information focuses.

B. Experimental Results:

Table 2 tells us the survey of execution time that is achieved and mistake rates produced from the two calculations speaking to the 4 datasets. The best outcomes are featured with intense textual styles (Fig. 1).

Table 2 Juxtaposition of gassing time and glitch scale

Data object or dataset	Criteria	Algorithm	
		K-means	APSO
IRIS	Glitch (%)	18	**10**
	Gassing time (in seconds/minute)	5.17	**1.59**
WINE	Glitch (%)	21.8	29.8
	Execution time (in seconds/minute)	1.59	**1.32**
SEED	Glitch (%)	15.2	**11.4**
	Gassing time (in seconds/minute)	3.61	**2.31**
K-MEANS	Glitch (%)	NOT A	NOT A
	Gassing time (in seconds/minute)	6.83	**2.63**

Fig. 1 Correlation between APSO and k-means

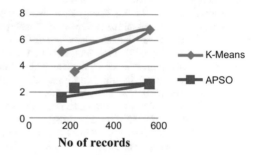

No of records

8 Conclusion

Natural to sidestep nearby pinnacles and acquire worldwide ideal outcome we have utilized mix of PSO calculation with fluffy rationale. With the assistance of worldwide inquiry limit, our proposed calculation conquers k-implies deficiencies. The displayed system beats the moderate consolidating pace of particle swarm algorithm and k-infers packing breaking free from adjacent perfect. Four open typical UCI information articles or information documents are used to investigate the execution of this work. The unproven outcomes uncover that APSO bunching calculation accomplishes an ostensible blunder rate and are fixated of the foolhardy joining and most extreme constancies of results.

References

1. Tan, P., Steinbach, M., Kumar, V.: Introduction to Data Mining, 1st edn. Addison Wesley, May 2005
2. Baraldi, A., Blonda, P.: A survey of fuzzy clustering algorithms for pattern recognition. IEEE Trans. Syst. Man Cybern. Part B: Cybern. **29**(6), 778–785, Dec 1999

3. Silic, A., Moens, M.-F., Zmak, L., Basic, B.: Comparing Document Classification Schemes Using k-means Clustering, vol. 5177, pp. 615–624 (2008)
4. MacQueen, J.: Some methods for classification and analysis of multivariate observations. In: Proceedings of the Fifth Berkeley Symposium on Mathematical Statistics and Probability, vol. 1, pp. 281–297. Univ. of Calif. Press (1967)
5. Jain, A.K., Murty, M.N., Flynn, P.J.: Data clustering: a review. ACM Comput. Surv. 31(3), September 1999
6. Fleischer, M.: Foundations of Swarm Intelligence: From Principles to Practice. Swarming: Network Enabled C4ISR 2003 By Mark Fleischer
7. Hashim, H.A., El-Ferik, S., Ayinde, B.O., Abido, M.A.. Optimal Tuning of Fuzzy Feedback Filter for L1 Adaptive Controller using Multi-Objective Particle Swarm Optimization for Uncertain Nonlinear MIMO Systems. arXiv preprint arXiv …,2017 - arxiv.org
8. Singh, D., Prasad, S., Srivastava, S.: Implementation of artificial intelligence cognitive neuroscience neuron cell using adaptive velocity threshold particle swarm optimization (AVT-PSO) on FPGA. In: 2017 6th International Conference on Reliability, Infocom Technologies and Optimization (Trends and Future Directions) (ICRITO), Noida, pp. 548–552 (2017). https://doi.org/10.1109/icrito.2017.8342488
9. Hao, X., Liu, Y., Li, X., Zhang, Y.: A two-factors and multi-orders self-adaptive fuzzy time series model based on fuzzy logical relationships trees and particle swarm optimization. In: 2017 3rd IEEE International Conference on Computer and Communications (ICCC), Chengdu, pp. 2537–2543 (2017). https://doi.org/10.1109/compcomm.2017.8322993
10. Teo, K.T.K., Lim, P.Y., Chua, B., Goh, H.H., Tan, M.K.: Particle swarm optimization based maximum power point tracking for partially shaded photovoltaic arrays. Int. J. Simul. Syst. Sci. Technol. 17(34), 20.1–20.7, 7p (2016)
11. Cui, X., Potok, T.E., Palathingal, P.: Document Clustering using Particle Swarm Optimization. 0-7803-8916- 6/05/$20.00 ©2005IEEE
12. Kennedy, J., Eberhart, R.C., Particle Swarm Optimization. In: Proceedings of the IEEE International Conference on Neural Networks, pp. 1942–1948 (1995)
13. Hyma, J., Jhansi, Y., Anuradha, S.: A new hybridized approach of PSO & GA for document clustering. Int. J. Eng. Sci. Technol. 2(5), 1221–1226 (2010). Anderberg, M.R.: Cluster Analysis for Applications. Academic Press, Inc., New York, NY (1973)
14. Alam, S., Dobbie, G., Koh, Y.S., Riddle, P., Rehman, S.U.: Research on particle swarm optimization based clustering. Swarm Evol. Comput. 17, 1–13 (2014). Journal homepage: www.elsevier.com/locate/swevo
15. Asai, K., Sugeno, M., Terano, T.: Applied Fuzzy Systems. Academic Press, New York (1994)
16. Zadeh, L.: Fuzzy sets. Inf. Control, pp. 338–353 (1965)
17. http://archive.ics.uci.edu/ml/

Low-Dimensional Spectral Feature Fusion Model for Iris Image Validation

Manjusha N. Chavan and Prashant Patavardhan

Abstract Iris images are a primal source of biometric security system. Iris images are used as input details in providing access to secure information and operating systems. In the development of biometric security through iris detection, algorithms were developed for extracting multiple features to improve accuracy. Methods were also developed toward validation of iris image in the detection of live or fake samples. Wherein focus tends more toward accuracy, less effort is observed for delay performance and overhead. Feature overhead impacts the system performance. In this paper a feature fusion representing spectral energy and Harlick features is proposed. An approach for dimensional reduction using band spectral correlation with iris image is proposed. The accuracy of the developed approach is validated with the conventional approaches.

Keywords Feature fusion · Spectral feature · Iris validation · Dimension reduction

1 Introduction

Security has become a major concern in the rapid development of new applications. As the technology has moved to digital data processing, the advantage of storage of large volume of data and accessing has become an easy task. This gave the advantage of data exchange over large distributed network globally. The advantage of data exchange has, however, given a new threat to new security concerns. Data thefting, data alteration, attacks, vulnerability, etc., have limited its usage for critical applications. Security is developed by different means in digital data domain. Secure key mechanism, voice detection, face detection, fingerprint detection, and iris detection

M. N. Chavan (✉)
ADCET, Ashta, India
e-mail: manju3205@gmail.com

P. Patavardhan
GIT, Belgaum, India
e-mail: pppatavardhan@git.edu

© Springer Nature Singapore Pte Ltd. 2020   665
K. N. Das et al. (eds.), *Soft Computing for Problem Solving*,
Advances in Intelligent Systems and Computing 1057,
https://doi.org/10.1007/978-981-15-0184-5_57

are some of these applications. Biometric security has an advantage of liveness inter-action with security processing system. Among various biometric applications, in this paper, an iris-based security system is focused.

2 Literature Survey

The concern on biometric security is the detection of live data over forged or syn-thetic data. Toward improving the accuracy of live detection of iris image, a realistic model of iris image on attack is presented in [1]. Realistic evaluation for iris liveness detection at spoofing attack is presented for testing the binary models. Single-class classifier for live only samples is used in training while complete unknown spoof sam-ples used for testing. For identification, authenticity of the eye in iris liveness detec-tion using liveness symptoms and cooperation from user is implemented. Change in pupil size is registered under visible light stimuli over static properties of the eye or its tissue. Challenge [2] in biometric is to deal with the fake reconstructed sample or self-manufactured synthetic samples that are presented. Many biometric systems do not have that much capability to detect such artificially created input as a fake input. To rectify such problem for detection of fake biometric, a liveness assessment method using various image quality assessment measures which play important role to detect fake samples and terminate is presented. In [3], threats and direct attacks on biometric verification system are outlined. Efforts are put in for developing "live-ness detection" module to reduce risk of spoof attacks, using anatomical properties to distinguish between real and fake irises. Integrating static and dynamic liveness detection modules for protection of iris biometric system is attempted. In [4], the framework for detection of multiple types of fake iris images based on texture analy-sis is proposed. Hierarchical Visual Codebook (HVC) for iris pattern representation is proposed where distinct and robust texture primitives of genuine and fake iris images are encoded by creating vocabulary tree and use of linear coding of local features. The issue liveness detection problem in biometrics is focused in [5]. Fake iris detection using co-occurrence matrix based on selected features, measurement of iris edge sharpness, and visual primitives of iris texture are carried out for counterfeit color contact lens iris with textures printed onto them. A novel iris texture represented using statistical features with bag of words as a classifier model. For encoding this texture, Hierarchical Visual Codebook (HVC) is proposed [6]. Statistical gray level pixel relations of both local and global eye region are extracted to develop a new anti-spoofing approach [7]. In [8], for creating new database, iris capture device which registers change in pupil size under visible light is outlined. Natural reaction and spontaneous oscillations are simultaneously recorded to detect the liveness of the iris and classify with the help of linear and nonlinear support vector machines which is investigated. In [9], a real biometric acceptable feature against fake manu-factured artificial or reconstructed trial is outlined. Fake iris and fingerprint detection recognition method to detect different types of fake access attempts are presented. A natural motion detection algorithm for the detection of natural eye movement

(Blinking of eyes and left–right movements) and reflection detection algorithm to detect reflection from retina which not occurred in fake scanned images of iris is outlined in [10]. This enhanced significantly the performance of the system in terms of security and reliability. In [11], iris image of person with cosmetic contact lens is captured with dual-band spectral imaging system. Natural iris textural patterns from cosmetic contact lens are distinguished by using the independent component analysis and legitimate iris patterns are restored. A tool is developed using MATLAB to determine the recognition performance of the system for databases of grayscale eye images [12]. Emphasis is on performing recognition, and not on capturing of an eye image using hardware. A rapid application development (RAD) approach is employed to produce results quickly. Different framework in the area of iris fake detection is presented in [13, 14]. Diverse approaches of iris fake detection based on attack modeling and applicator relevance are outlined. In [15, 16], iris localization and presentation attack detection (PAD) are simultaneously performed by multitask convolution neural network. By localizing iris with bounding box, parameters computation and presentation attack probability from the input ocular image are proposed. In [17], a novel algorithm using a combination of deep learning-based features and handcrafted features for detecting iris presentation attacks is presented. Harlick texture features which form local iris and global eye image in multi-level redundant discrete wavelet transform domain are combined. VGG features along with Harlick features are encoded describing textural dissimilarity between real and spoofed iris images. Feature-based fusion model is outlined in [15], where, a new spoof attack resilience method is presented, which explores the mapping of iris localization and textural feature using spatial gray level-dependent matrix and the gray level run length matrix in defining the attack of iris image in fake image detection. In [18], the issue of unimodal approach in iris authenticity is explored. Unimodal provides limited accuracy for the recognition of a person due to some limitations like error rate is high, noise contained in sensed data. Multimodal biometric systems overcome limitations of unimodal which uses additional trait of the user for authentication. A multimodal approach in iris liveness detection is presented. In [19], NIR camera image for an iris recognition system is outlined. Circular edge detection is used to extract iris from input image. Image features using deep learning-based and handcrafted-based methods are extracted from localized iris. The input iris images are classified into real and attack class using support vector machines (SVM). An iris-based authentication system, enhancing the security to prevent the iris spoofing is outlined in [20]. It uses two databases that contain the iris code and user code. Iris code is the length between the pupil and the sclera. User code is the QR code which acts as the user's password generated by using MD5 algorithm [20]. Although the approach of iris detection for different attacks and validation of the originality were developed, the system lacks the constraint of precision, speed of decision, and optimal feature overhead. In decision making of fake modeling, it is required to have a faster and robust decision system, wherein features are critical in any decision system; representation of iris feature for decision making is to be optimal with number of feature counts and its validity. A new approach of iris image feature representation and its validity in classifying live and fake sample is presented. To present the work,

this paper is outlined in six sections. Section 2 presents the conventional iris validation approach. Section 3 presents the proposed feature representation and classifying
model. The result of the developed approach is outlined in Sect. 4. Section 6 presents
the conclusion for the developed work.

3   Fusion Feature Model

In the validation of iris image authenticity, various approaches were developed in past.
Images were processed and normalized; feature extraction for classification process
is performed to make a decision. Toward the fake iris detection, in recent approach,
a redundant DWT-based fusion approach is presented in [17]. This approach defines
a new fusion approach of direct input image with segmented iris region to develop
a classification model. A multi-level Harlick feature and VGG model (MHVF) are
presented. Thirteen Harlick features over a gray level co-occurrence matrix (GLCM)
are extracted [21]. It computes second-order texture features from image subregions
for different distances and angular spatial relationship of pixels with specific gray
level. Co-occurrence matrix formed represents the features in four directions (e.g.,
horizontal (0°), vertical (90°), and diagonal (45° and 135°). Most significant image
textural characteristics can be extracted from various Harlick features such as energy,
angular second moment, correlation, inverse difference moment (IDM), sum average, sum variance, sum entropy, entropy, difference variance, difference entropy,
difference average, and information measure of correlation 1 and 2.

From the co-occurrence matrix, most dominant following Harlick features are
computed.

1. Contrast (CTR):

$$\text{CTR} = \sum_{n=0}^{N-1} n^2 \left\{ \sum_{i=1}^{N} \sum_{j=1}^{N} P(i,j); |i-j| = n \right. \tag{1}$$

2. Energy (E): Energy for a given image is calculated as

$$E = \sum_{i,j=0}^{N-1} P(i,j)^2 \tag{2}$$

3. Local Homogeneity (LH): It provides information for same types of pixel values
    in a given image calculated as

$$\text{LH} = \sum_{i,j=0}^{N-1} \frac{P_{i,j}}{1 - \left((i-j)^2 \right)} \tag{3}$$

4. Custer Shade (CS): This measure gives information about smoothness of image, calculated for the image P at the coordinates (i, j) as

$$CS = \sum_{i,j=0}^{n-1} (i - \mu i) + (j - \mu j)^3 P(i, j) \tag{4}$$

5. Cluster Prominence (CP): It finds group or clusters of pixel appearing repetitively in given image calculated as

$$CP = \sum_{i,j=0}^{n-1} (i - \mu i) + (j - \mu j)^4 P(i, j) \tag{5}$$

6. Entropy (H): The information measure of an image is called as entropy which is calculated as

$$H = \sum_{i,j=0}^{n-1} P(i - j) \log P(i - j) \tag{6}$$

The fusion model of iris image and its segmented iris region gives the feature fusion of global and local feature variant which is then classified by support vector machine (SVM). The observation for the developed approach illustrates an improvement of tolerance of attack resilience. However, the feature overhead is large, and no effort is made on decision delay. The overhead in such system is large and decreases the redundancy effort under semantic spectral details.

4 Proposed Spectral Fusion Model in Iris Validation

With motivation from fusion feature model, to overcome feature overhead and to improve the classification accuracy using soft computing techniques, modified spectral feature model (SFM) is proposed. This approach develops a new energy interpolation of transformed energy information for segmented iris region and map with the original image feature to develop a new feature representation which is lower in count and result in more accurate classification. In proposed spectral fusion model as follows:

(i) Input iris image is resized to 256×256 and for representation of the energy interpolation; a standard deviation of live and fake iris sample histogram is used. The feature F_{sample} is computed as

$$F_{\text{sample}} = \text{std}(\text{hist}(I_{\text{test}})) \tag{7}$$

(ii) Resized iris image is then segmented by coarse segmentation using region grow-
      ing algorithm for faster computation. In the segmentation, a region homogeneity
      property of the pupil, iris, and sclera region is utilized. The approach defines
      a marking of the gray color mapping where a reference value is captured from
      the center of each of the detected image region. An image region of human eye
      is shown in Fig. 1.

   The histogram feature of a test sample is shown in Fig. 2 and their corresponding
histogram plot is shown in Table 1. The planar histogram of each plane and the
standard deviation (STD) of the histogram distribution are taken as a representative
feature for color image representation.

(iii) The transformation is performed using a set of pyramidal decomposition of iris
       image sample to RDWT domain by high-pass and low-pass filter banks to two
       decomposition levels using db4 wavelet as shown in Fig. 3 (Fig. 4).

   Each of these detail bands reflects a spectral information and spectral variation
that can be noticed in each decomposed band. Spectral band density variation for
given sample is illustrated in Fig. 5. Not all bands are dominantly informative and
can be eliminated to minimize the overhead. This can be achieved by a spectral band
selection.

(iv) Spectral band selection, where all the decomposed bands are computed for
      spectral density and an optimal band with the highest value of concentrated
      energy is selected. This minimizes the overhead and makes the estimation
      faster. Selection is performed as

Fig. 1 Regions of a human
eye

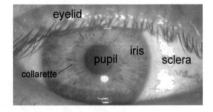

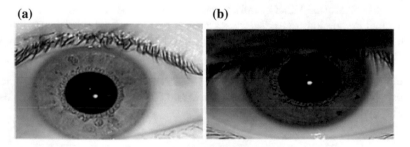

Fig. 2 Test sample (**a**) Live iris image, (**b**) Fake sample

Table 1 Corresponding STD histogram feature for the test sample

Sample	Image	Histogram (H)	STD (H)
Live		histogram-R-plane data	44.8587
Fake		histogram-G-plane data	34.7821

Fig. 3 Pyramidal decomposition of using wavelet transform

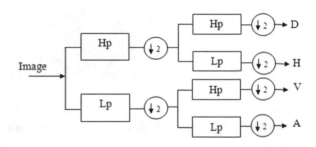

For k = {LL, LH, HL, HH}//types of bands
For b = 1: L//number of band for each type
For i = 1: l
For j = 1: m

$$cor_b^k(i, j) = correlation\left(B_b^k(i, j), B_{b\pm1}^k(i, j)\right)$$

End
End
End
End

$[mc\ b] = \min(cor)$

Fig. 4 Decomposition of
sample iris image using db4
wavelet transform

Decomposed images

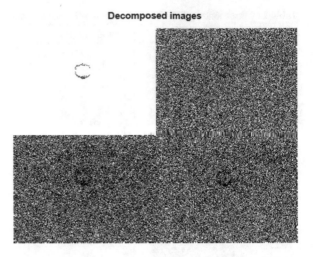

Fig. 5 Spectral density of a
decomposed detail
coefficient

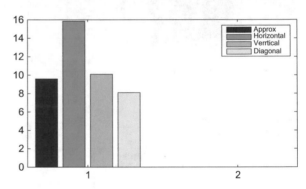

Spectral bands with the highest energy and minimum correlation among them are
selected.

(v)   Harlick features are extracted from dominant sub bands.
(vi) Feature vector is generated by concatenating energy histogram for each plane
      of resized whole eye image with texture features computed over selected bands
      for resized and segmented iris images. The feature set is defined as a set of
      {energy histogram, Harlick feature}.
(vii) Binary class SVM classifier model is trained using live and fake iris image
      from training set, and database is created. If features extracted from test image
      are more than 80%, then they are classified and assigned to live or fake class.

5   Simulation Result

To validate the proposed system, a test validation on a set of iris dataset is performed. A combined spoofing database [22] is used. Dataset includes typical printed iris images, textured contact lenses, and synthetic iris images. The samples of the data base are shown below. (Fig. 6).

MATLAB tool is used with Intel core i5 as processor operating at 2.5 GHZ for simulating the results of proposed system. The system is randomly trained for live and fake images. Each of the images from a live and fake set is chosen for feature extraction, where the extracted features are then passed to the database for classification. The SVM network model developed is a radial feedback network with three-layer interface. Time elapsed between presenting new sample and validating result is 0.9357 s. The classification result for the developed system is illustrated below. Comparative analysis over existing algorithms was made in [17] taking VGG [23], MHVF [17] algorithms. The attack detection performance is evaluated using the metrics of

- Total Error: Defined as the total misclassified iris sample in the test
- Fake Classification Error Rate (APCER): Error rate of misclassified sample of fake iris class.
- Real Classification Error Rate (BPCER): Error rate of misclassified class of live iris images (Tables 2 and 3).

Eight most discriminative Harlick features that are extracted from two sub bands at the second level of decomposition in four orientations are 64 along with Harlick features in spatial domain. Features related to energy histogram are concatenated. Therefore, feature vector is of the size 75. With reference to MHVF [17], where features are extracted from all bands are 117 for VGG and for fusion model, it

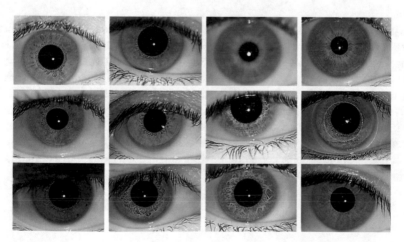

Fig. 6 Database samples

Table 2 Extracted feature of the iris samples

Sample	Degrees (D)	Contrast (CTR)	Energy (E)	Local Homo-geneity (LH)	Maximum Proba-bility (MP)	Entropy (EN)	Cluster Shade (CS)	Cluster Promi-nence (CP)
Live	0°	1260	458.49	132.042	35,655	2610	11.0745	331.0513
	45°	1490	762.04	422.175	23,455	1742	44.0314	255
	90°	1564	127.30	127.035	23,454	5516	31.02354	264.054
	135°	1367	884.51	276.48	20,545	4597	10.244	422.2354
Fake	0°	3254	332.021	221.332	2134.02	1475	59.235	632.1
	45°	3124	365.0214	250.36	2547.78	1245	94.326	687.21
	90°	3007	254.0214	232.065	2698	1269	95.36	598.3214
	135°	3481	432.021	129.0325	2475.12	1987	75.32	635.254

Table 3 Analysis of different attack resilience model with proposed approach

Algorithm	Total Error	APCER	BPCER
VGG [23]	1.33	2.11	0.75
MHVF [17]	1.23	1.82	0.78
Proposed SFM	0.76	1.53	0.03

counts up to 182. Our proposed modified spectral fusion model reduces overhead by 40–60%.

6 Conclusion

This paper presented a new approach of feature fusion using spectral correlation and energy of iris image validation. In the proposed approach, the system overhead is the compatibility lowered due to lower feature count performed over the selected bands. The Harlick features were computed over the selected bands to reduce the overhead. The accuracy of estimation is, however, improved due to optimal selection of informative bands. The histogram energy feature gives an addition spectral deviation for a live and fake image in training process. This knowledge reflects the iris image content with respect to density of spectral information in iris image. The overall performance of the proposed system illustrates a higher accuracy in estimation and minimizes the system overhead.

References

1. Sequeira, A.F., Thavalengal, S., Ferryman, J., Corcoran, P., Cardoso, J.S.: A realistic evaluation of iris presentation attack detection. In: 39th International Conference on Telecommunications and Signal Processing (TSP), IEEE (2016)
2. Ambadkar, A.S., Mante, R.V., Chatur, P.N.: Detection of fake biometric using liveness assessment method. Int. J. Adv. Res. Comput. Sci. 6(1) (2015)
3. Wei, Z., Qiu, X., Sun, Z., Tan, T.: Counterfeit iris detection based on texture analysis. In: IEEE, 19th International Conference on Pattern Recognition (2008)
4. Joy, J., Rekha, K.S.: Learning liveness detection and classification of iris images: a detailed survey. Int. J. Comput. Sci. Inf. Technol. 6(5) (2015)
5. Tan, C.-W., Kumar, A.: Integrating Ocular and Iris Descriptors for Fake Iris Image Detection, IEEE (2014)
6. Czajka, A.: Pupil dynamics for iris liveness detection. IEEE Trans. Inf. Forensics Secur. 10(4) (2015)
7. Shende, P.M., Sarode, M.V.: Fake biometric detection using liveness detection system applications: iris and fingerprint recognition system. Int. J. Res. Advent Technol. Special Issue, 1st International Conference on Advent Trends in Engineering, Science and Technology (2015)
8. Hsieh, S.-H., Li, Y.-H., Wang, W., Tien, C.-H.: A novel anti-spoofing solution for iris recognition toward cosmetic contact lens attack using spectral ICA analysis. mdpi Sens. J. 18 (2018)
9. Hombalimath, A., Manjula, H.T., Khanam, A., Girish, K.: Image quality assessment for iris recognition. Int. J. Sci. Res. Publ. 8(6) (2018)
10. Czajka, A., Bowyer, K.W.: Presentation attack detection for iris recognition: an assessment of the state of the art. ACM J. Comput. Surv. (CSUR) 51(4) (2018)
11. Chen, C., Ross, A.: A multi-task convolutional neural network for joint iris detection and presentation attack detection. In: 1st Workshop on Cross-Domain Biometric Recognition (CDBR), IEEE Winter Conference on Applications of Computer Vision (2018)
12. Hoffman, S., Sharma, R., Ross, A.: Convolutional neural networks for iris presentation attack detection: toward cross-dataset and cross-sensor generalization. In: IEEE Conference on Computer Vision and Pattern Recognition Workshops (CVPRW) (2018)
13. Kaur, B., Singh, S., Kumar, J.: A study on fake iris detection under spoofing attacks. J. Eng. Appl. Sci. 13(8) (2018)
14. Rani, M., Kant, C.: A hybrid approach for raising biometric system security by fusing fingerprint and iris traits. IOSR J. Comput. Eng. 20(3) (2018)
15. Yadav, D., Kohli, N., Agarwal, A., Vatsa, M., Singh, R., Noore, A.: Fusion of handcrafted and deep learning features for large-scale multiple iris presentation attack detection. In: IEEE Conference on Computer Vision and Pattern Recognition (CVPR) Workshops (2018)
16. Suvarchala, P.V.L., Srinivas Kumar, S.: Feature set fusion for spoof iris detection. Eng. Technol. Appl. Sci. Res. 8(2) (2018)
17. Nguyen, D.T., Baek, N.R., Pham, T.D., Park, K.R.: Presentation attack detection for iris recognition system using NIR camera sensor. mdpi Sens. J. 18 (2018)
18. Ahmad, H.M., Abdulkareem, B.J.: Integrate liveness detection with iris verification to construct support biometric system. J. Comput. Commun. 4(1) (2016)
19. Zhang, H., Sun, Z., Tan, T., Wang, J.: Learning Hierarchical Visual Codebook for Iris Liveness Detection, IEEE (2011)
20. Ajirtha, S., Anitha, S.M., Anusha, R.A., Jeni, P., Karthik, S.: An iris-based authentication system to prevent iris spoofing. J. Netw. Commun. Emerg. Technol. (JNCET) 8(4) (2018)
21. Haralick, R.M., Shanmugam, K., Dijnstein, I.: Textural features for image classification. IEEE Trans. Syst. Man Cybern. (1973)
22. Kohli, N., Yadav, D., Vatsa, M., Singh, R., Noore, A.: Detecting medley of iris spoofing attacks using DESIST. In: IEEE International Conference on Biometrics: Theory, Applications and Systems (2016)
23. Simonyan, K., Zisserman, A.: Very Deep Convolutional Networks for Large-Scale Image Recognition (2014). arXiv preprint arXiv:1409.1556

Models for Predictions of Mechanical Properties of Low-Density Self-compacting Concrete Prepared from Mineral Admixtures and Pumice Stone

B. Arun Kumar, G. Sangeetha, A. Srinivas, P. O. Awoyera, R. Gobinath and V. Venkata Ramana

Abstract This study applies the principle of artificial neural networks for modelling the mechanical characteristics of a lightweight self-compacting concrete containing pumice and mineral admixtures. Models for predicting compressive strength, split tensile strength and flexural strengths were developed based on several measures of the materials as obtained from the experimental stage. The input parameters for the model were contents of cement, ground granulated blast furnace slag (GGBS), rice husk ash (RHA), fine aggregates, coarse aggregates, pumice stone, water, super-plasticizers and micro-silica. Three output parameters, including compressive strength, tensile strength and flexural strength were considered. The data were trained, tested and validated using the feedforward backpropagation algorithm. The study established the best model for the tested concrete, based on the minimal error criteria, as 9 (input), 12 (hidden layer) and 3 (output layer). This model is expected to serve as a useful tool for concrete designers and constructors.

Keywords Compressive · Split tensile · Flexural strengths · Artificial neural network · Mineral admixtures · Pumice stone

1 General Introduction

The mechanical properties of concrete depend largely on the mix design and proportioning of constituent materials. With the yearly increase in the consumption of natural materials used for concrete production, and global demand for Portland

B. Arun Kumar · G. Sangeetha · A. Srinivas · R. Gobinath (✉) · V. Venkata Ramana
S R Engineering College, Warangal, Telangana, India
e-mail: r.gobinath2013@vit.ac.in

B. Arun Kumar · G. Sangeetha · V. Venkata Ramana
Center for Artificial Intelligence and Deep Learning, S R Engineering College, Warangal, Telangana, India

P. O. Awoyera
Department of Civil Engineering, Covenant University, Ota, Nigeria

© Springer Nature Singapore Pte Ltd. 2020
K. N. Das et al. (eds.), *Soft Computing for Problem Solving*,
Advances in Intelligent Systems and Computing 1057,
https://doi.org/10.1007/978-981-15-0184-5_58

677

cement concrete currently estimated at 10 tons yearly [1, 2], other potential sources of materials are constantly being explored by researchers. This involves using supplementary materials such as mineral and chemical admixtures to increase the strength and durability in the fresh or hardened concrete [2–5]. Subsequently, it is considered necessary to develop tools to ascertain the optimized mix proportions of concrete [6–10].

Studies that focus on proportioning the concrete materials aim to find a cost-effective material which can nearly meet the concrete properties [10–12]. As a result, different approaches such as regression analysis and simulations have been adopted for developing predictive models for concrete.

The engineering characteristics of cement-based materials mainly depend on several parameters like non-homogeneous nature, several elemental properties and also the effects of some materials [5, 7, 8]. The compressive, split tensile and flexural strengths are the vital mechanical properties in the design of concrete structures. It is expected that, after 28 days of curing, concrete should attain about 75% of its final strength. Thus, the 28-day strength is used as benchmark for computing the strength at any other given age.

Adequate prediction of concrete strengths is very necessary as it offers an essential modification to the mix proportion and also helps to avoid possibility of concrete not attaining its design strength [3, 8, 13, 14]. Also, it aids reduction in construction failures and ensures economic use of raw materials, reducing the construction cost in the long run. Numerous methods of modelling, like mechanical modelling, statistical method and artificial intelligence for developing prediction models of concrete incorporating several ingredients are being explored by researchers [13, 15–18]. The artificial neural network (ANN) technique is a fast-grown area in artificial intelligence having many engineering applications [19–24]. ANN is utilized in several civil engineering applications such as concrete mix proportion, strength prediction, detection of damage in the structure, groundwater monitoring and settlement of foundation prediction [25–27]. However, the present study aims to develop predictive models, based on the artificial neural network technique, for predicting mechanical properties of a lightweight self-compacting concrete.

2 Materials and Methods

2.1 Materials

The materials and respective levels, as utilized at the experimental phase of this study are listed in Table 1. A total of 16 different mix proportions were considered, by varying both the contents of mineral admixtures and pumice stone. The aggregates were treated by air-drying in the laboratory for seven days before use. This can facilitate to dry off the in situ moisture content from the raw materials before they are utilized. Concrete samples were prepared and cut as per the dimensions shown

Table 1 Materials and substitution levels used during preparation of concrete

S. No	Input data	Minimum	Maximum
1	Cement (kg/m^3)	424.2	466.7
2	GGBS (kg/m^3)	0.0	196.9
3	Rice Husk Ash (kg/m^3)	0	196.85
4	Fine Aggregates (kg/m^3)	899.24	999.15
5	Coarse Aggregates (kg/m^3)	429.24	643.86
6	Pumice Stone (kg/m^3)	25.04	100.16
7	Water (Litre)	173.2	190.49
8	Super-plasticizers (kg/m^3)	4.22	4.64
9	Micro-silica (kg/m^3)	0	63.6

in Fig. 1 after attaining maturity, for determination of concrete compressive, split tensile and flexural strengths. In order to develop predictive models for strength characteristics, the experimental data obtained from various tests like compression, split tensile and flexural tests of concrete made from various mix proportions were utilized.

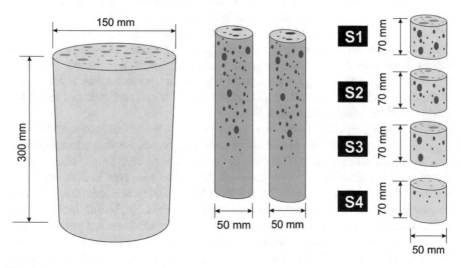

Fig. 1 Development of samples and segregations

2.2 Model Development Through Artificial Neural Networks

Generally, the ingredients of concrete determine the strength, workability and durability. The performance of manufactured concrete depends on the constituent materials like cement, ground granulated blast furnace slag (GGBS), rice husk ash, fine aggregates, coarse aggregates, pumice stone, water, super-plasticizers and micro-silica. So, to develop a model to capture the inherited characteristics of this system, a data-driven model is needed [15, 16, 19].

Artificial neural networks are a family of massively parallel architecture, which can capture the inherited characteristics through highly interconnected networks called artificial neurons. The network learns by comparing its output for each input pattern with a target output and then calculates the error as well as propagation of error through a backward function. To train the network, the weights of connections are modified according to the valuable information provided to the network. Once the model is identified, the performance of the model has to be tested. To test the accuracy of the trained network, the model predictions are statistically evaluated by calculating the coefficient of determination (R2). The ANN principles have diverse application in solving scientific and engineering problems [20, 23–29]. For instance, it uses cut across fields such as structural damage assessment, structural system identification, modelling of material behaviour and structural optimization [25, 26]. In this study, ANN modelling was performed using MATLAB software. The input data for the modelling, as obtained from the experimental stage, is presented in Table 2. The materials used in the experiment, which were also included in the model development were: cement, ground granulated blast furnace slag (GGBS), rice husk ash (RHA), fine aggregates, coarse aggregates, pumice stone, water, super-plasticizers and micro-silica. The data were trained, tested and validated using the feedforward backpropagation algorithm.

Various mix proportions were developed, by partial replacement of pumice stone by 10%, 20%, 30% and 40% in amount of coarse aggregate, varying mineral admixture and micro-silica content. The different compositions were termed according to the mineral admixture and binder material. M1 refers mix type 1 with GGBS as mineral admixture, M2 refers mix type 2 with RHA as mineral admixture, PS refers pumice stone and MS refers micro-silica. As shown in the table, 16 different mixtures are prepared with varying both the contents of mineral admixtures and pumice stone.

While the output parameters were compressive strength, split tensile strength test and flexural strength, other inputs such as curing regimes (7, 14 and 28 days) were included as represented in Table 3.

Table 2 Composition of manufactured concrete for all the identified aggregates

Mixtures	Cement	GGBS	RHA	FA	CA	Pumice Stone	Water	S.P	Micro-silica
M1	466.7	196.9	0.0	899.2	500.5	75.1	190.5	4.6	0.0
M1 + 10%PS	424.2	178.9	0.0	999.2	500.5	75.1	173.2	4.2	0.0
M1 + 20%PS	424.2	178.9	0.0	999.2	572.3	50.1	173.2	4.2	0.0
M1 + 30%PS	424.2	178.9	0.0	999.2	429.2	100.2	173.2	4.2	0.0
M1 + 40%PS	424.2	178.9	0.0	999.2	643.9	25.0	173.2	4.2	0.0
M2	466.7	0.0	196.9	899.2	500.5	75.1	190.5	4.6	0.0
M2 + 10%PS	424.2	0.0	178.9	999.2	500.5	75.1	173.2	4.2	0.0
M2 + 20%PS	424.2	0.0	178.9	999.2	572.3	50.1	173.2	4.2	0.0
M2 + 30%PS	424.2	0.0	178.9	999.2	429.2	100.2	173.2	4.2	0.0
M2 + 40%PS	424.2	0.0	178.9	999.2	643.9	25.0	173.2	4.2	0.0
M1 + 30%PS + 5%MS	424.2	178.9	0.0	999.2	429.2	100.2	173.2	4.2	21.2
M1 + 30%PS + 10%MS	424.2	178.9	0.0	999.2	429.2	100.2	173.2	4.2	42.4
M1 + 30%PS + 15%MS	424.2	178.9	0.0	999.2	429.2	100.2	173.2	4.2	63.6
M2 + 30%PS + 5%MS	424.2	0.0	178.9	999.2	429.2	100.2	173.2	4.2	21.2
M2 + 30%PS + 10%MS	424.2	0.0	178.9	999.2	429.2	100.2	173.2	4.2	42.4
M2 + 30%PS + 15%MS	424.2	0.0	178.9	999.2	429.2	100.2	173.2	4.2	63.6

Table 3 Compressive strength split tensile strength test and flexural strength tests of concrete for different mix proportions against curing time of 7, 18 and 28 days

Mixture	Compressive strength test			Split tensile strength test			Flexural strength test		
	7 Days	14 Days	28 Days	7 Days	14 Days	28 Days	7 Days	14 Days	28 Days
M1	16.8	31.2	48.2	1.3	2.42	3.8	5	6.5	8.5
M1 + 10%PS	14.38	26.81	41.24	0.95	2.04	3.14	4	6	7.5
M1 + 20%PS	14.1	25.9	40.61	0.91	1.99	3.04	3.5	6	7
M1 + 30%PS	13.5	23.8	39.2	0.88	1.96	2.95	4	5	6.5
M1 + 40%PS	12.1	22.6	37.7	0.83	1.96	2.89	3	5	5.5
M2	16.6	28.2	46.1	1.21	2.45	3.67	5.5	6.5	8.5
M2 + 10%PS	15.56	25.9	42.35	0.98	2.09	3.23	3.5	6	7
M2 + 20%PS	14.6	25.2	41.4	0.96	2.06	3.2	3.5	5.5	7
M2 + 30%PS	14	24.1	40.1	0.91	2.01	3.15	4	5	6.5
M2 + 40%PS	12.6	23.2	38.6	0.9	1.99	3.11	3	4	5
M1 + 30%PS + 5%MS	13.8	23.6	39.6	0.89	1.96	2.96	3.5	5.5	6.5
M1 + 30%PS + 10%MS	13.9	24	39.4	0.89	1.95	2.94	3	6	7
M1 + 30%PS + 15%MS	14.2	23.8	40.3	0.9	1.96	2.95	4	5.5	6
M2 + 30%PS + 5%MS	14.1	24.2	40.1	0.93	2.02	3.14	4	6	6.5
M2 + 30%PS + 10%MS	14.4	24.1	40.3	0.93	2.03	3.18	3.5	6	7
M2 + 30%PS + 15%MS	14.3	24.8	39.8	0.95	2.03	3.18	3.5	6.5	8.5

3 Results and Discussion

From the experimental results, it was observed that as the percentage of pumice stone was increased for mix with GGBS as mineral admixture, the compressive strength decreased overall all the curing days. Similarly, with RHA as mineral admixture, as the pumice was increased, compressive strength decreased subsequently. Whereas, with GGBS as mineral admixture and 30% PS used, and increase in micro-silica (MS) from 5% to 15%, there was no significant change in the compressive strength. Similar trend was also observed when RHA was used as mineral admixture, and similar composition of 30% PS and 5–15% MS. The trend observed for split tensile strength test and flexural strength test was similar to that observed with compressive strength.

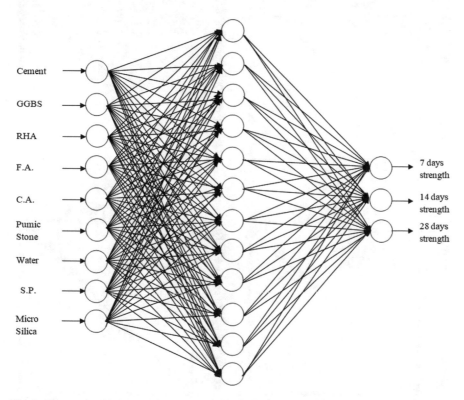

Fig. 2 Proposed artificial neural network architecture

3.1 ANN Architecture for Model Prediction

Since the performance of the ANN is dictated by the number of neurons in the hidden layers. So, this model was verified for different numbers of hidden layers, which result in higher R2 and lower root mean square error (RMSE). Among different numbers of hidden layers, the 9-12-3 topology has resulted in $R^2 = 0.9872$ and RMSE = 0.0232. Therefore, the best possible ANN architecture found was (Nine input neurons, a single hidden layer having twelve neurons and three output neurons) The optimal topology (see Fig. 2) is used for training, testing and validation.

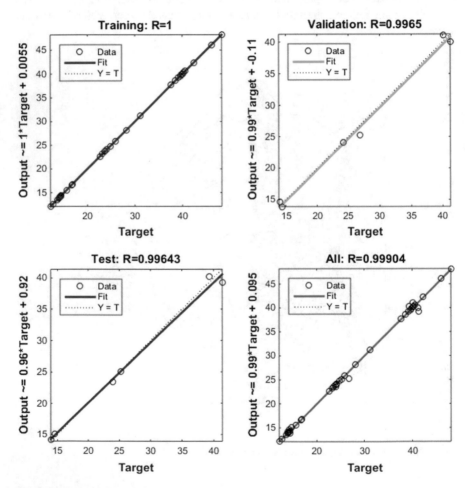

Fig. 3 Performance of selected ANN for training, testing, validation and all data set, using compressive strength data

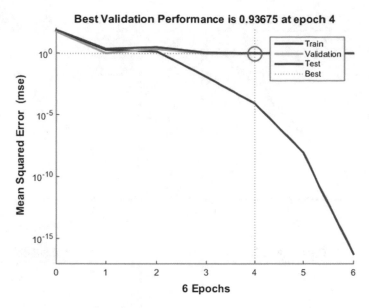

Fig. 4 Mean square error variation per epochs using compressive strength data

3.2 Compressive Strength

Based on the available experimental data, the data distribution and utilization for modelling were as follows: 70%, 15% and 15%, for training, testing and validation, respectively. From the analysis, an optimum model having 9 input layer, 12 hidden layer and 3 output layer topology was selected based on its significant average regression value $R^2 = 0.9996$, for training, testing and validation dataset. This high value itself suggests that the model predictions are exactly matched with the experimental values. Figure 3 shows the performance of the selected model. Also, Fig. 4 shows the mean square error (MSE) per different epochs for training, testing and validation data sets. As shown, the best validation performance was achieved at 0.9365 (epoch 4).

3.3 Tensile Strength

Based on the available experimental data, the data distribution and utilization for modelling were as follows: 70%, 15% and 15%, for training, testing and validation, respectively. From the analysis, an optimum model having 9 input layer, 12 hidden layer and 3 output layer topology was selected based on its significant average regression value $R^2 = 0.9998$ for training, testing and validation dataset. This high value itself suggests that the model predictions are exactly matched with the experimental

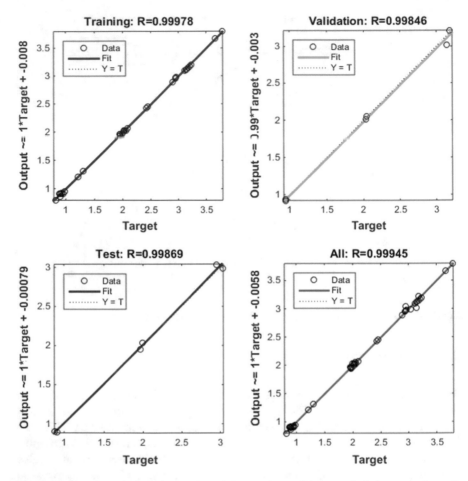

Fig. 5 Performance of selected ANN for training, testing, validation and all data set, using split tensile strength data

values. Figure 5 shows the performance of the selected model. Also, Fig. 6 shows the mean square error (MSE) per different epochs for training, testing and validation data sets. As shown, the best validation performance was achieved at 0.0031 (epoch 3).

3.4 Flexural Strength

The flexural strength or modulus of rupture of prisms fabricated from the mixes was determined, following the experimental design. Based on the available experimental data, the data distribution and utilization for modelling were as follows: 70%,

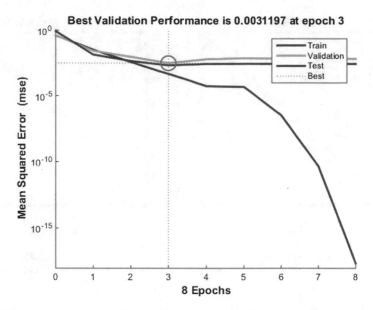

Fig. 6 Mean square error variation per epochs using split tensile strength data

15% and 15%, for training, testing and validation, respectively. From the analysis, an optimum model having 9 input layer, 12 hidden layer and 3 output layer topology was selected based on its significant average regression value $R^2 = 0.9532$ for training, testing and validation dataset. This high value itself suggests that the model predictions are exactly matched with the experimental values. Figure 7 shows the performance of the selected model. Also, Fig. 8 shows the mean square error (MSE) per different epochs for training, testing and validation data sets. As shown, the best validation performance was achieved at 0.1578 (epoch 4).

4   Conclusion

In order to predict the compressive, split tensile and flexural strengths of the concrete containing supplementary components like mineral admixtures and pumice stone, an artificial neural network model is developed. The model was trained with the experimental input and output data. Using only input parameters data, the respective strengths for 7, 14 and 28 days were predicted and they were very close to the experimental results. For the selected neural network, the average regression values for compressive, tensile and flexural strengths are 0.9996, 0.9998 and 0.952, respectively. The best validation performance was observed at epoch 4 for compressive and flexural strengths and at epoch 3 for tensile strength. It was observed that the predicted strength data was having a strong degree of coherency with experimental

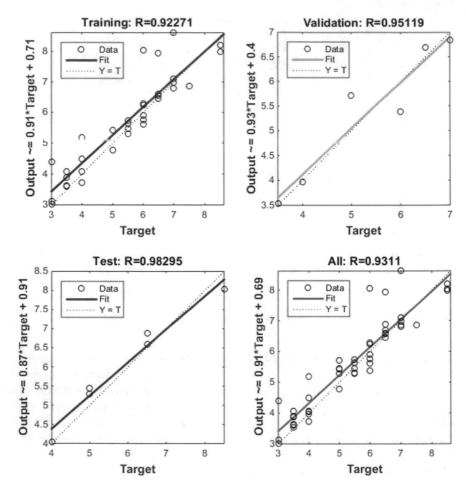

Fig. 7 Performance of selected ANN for training, testing, validation and all data set, using flexural strength data

data obtained for compressive and split tensile strength tests. So, it can be concluded that ANN approach is capable of predicting strengths of concrete with high accuracy.

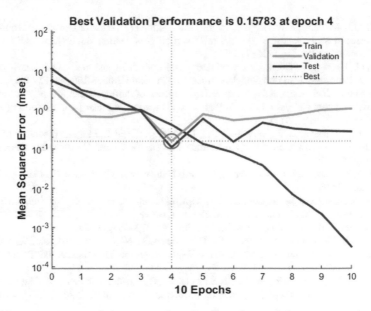

Fig. 8 Mean square error variation per epochs using flexural strength data

References

1. Imbabi, M.S., Carrigan, C., McKenna, S.: Trends and developments in green cement and concrete technology. Int. J. Sustain. Built Environ. **1**(2), 194–216 (2012)
2. Schneider, M., Romer, M., Tschudin, M., Bolio, H.: Sustainable cement production—present and future. Cem. Concr. Res. **41**(7), 642–650 (2011)
3. Rujanu, M., Diaconu, L.I., Babor, D., Plian, D., Diaconu, A.C.: Study on the optimization of some cement based mixing binders' characteristics. Proc. Manuf. **22**, 114–120 (2018)
4. Murthi, P., Awoyera, P., Selvaraj, P., Dharsana, D., Gobinath, R.: Using silica mineral waste as aggregate in a green high strength concrete: workability, strength, failure mode, and morphology assessment. Aust. J. Civ. Eng., 1–7 (2018)
5. Karthika, V., Awoyera, P.O., Akinwumi, I.I., Gobinath, R., Gunasekaran, R., Lokesh, N.: Structural properties of lightweight self-compacting concrete made with pumice stone and mineral admixtures. Revista Romana de Materiale/Rom. J. Mater. **48**(2), 208–213 (2018)
6. Chen, J.J., Ng, P.L., Kwan, A.K.H., Li, L.G.: Lowering cement content in mortar by adding superfine zeolite as cement replacement and optimizing mixture proportions. J. Clean. Prod. **210**, 66–76 (2019)
7. Karri, R.R.: Evaluating and estimating the complex dynamic phenomena in nonlinear chemical systems. Int. J. Chem. Reactor Eng. **9** (2011)
8. Busahmin, B., Maini, B., Karri, R.R., Sabet, M.: Studies on the stability of the foamy oil in developing heavy oil reservoirs. Defect Diffus. Forum **371**, 111–116 (2016)
9. Maini, B.B., Busahmin, B.: Foamy Oil Flow and Its Role in Heavy Oil Production, pp. 103–108 (2010)
10. Anandaraj, S., Rooby, J., Awoyera, P.O., Gobinath, R.: Structural distress in glass fibre-reinforced concrete under loading and exposure to aggressive environments. Constr. Build. Mater. (2018)
11. Muthukanf, C., Suji, D., Mariappan, M., Gobinath, R.: Studies on recycling of sludge from bleaching and dyeing industries in cement industries. Pollut. Res. **34**(1), 209–214 (2015)

12. Gobinath, R., Ganapathy, G.P., Akinwumi, I.I.: Evaluating the use of lemon grass roots for the reinforcement of a landslide affected soil from Nilgiris district, Tamil Nadu, India. J. Mater. Environ. Sci. 6(10), 2681–2687 (2015)
13. Liu, G., Cheng, W., Chen, L.: Investigating and optimizing the mix proportion of pumping wet-mix shotcrete with polypropylene fiber. Constr. Build. Mater. 150, 14–23 (2017)
14. Abusahmin, B.S., Karri, R.R., Maini, B.B.: Influence of fluid and operating parameters on the recovery factors and gas oil ratio in high viscous reservoirs under foamy solution gas drive. Fuel 197, 497–517 (2017)
15. Rao, K.R., Srinivasan, T., Venkateswarlu, C.: Mathematical and kinetic modeling of biofilm reactor based on ant colony optimization. Process Biochem. (Amsterdam, Neth.) 45(6), 961–972 (2010)
16. Rao, K.R., Rao, D.P., Venkateswarlu, C.: Soft Sensor Based Nonlinear Control of a Chaotic Reactor (2009)
17. Karri, R.R., Sahu, J.N., Jayakumar, N.S.: Optimal isotherm parameters for phenol adsorption from aqueous solutions onto coconut shell based activated carbon: error analysis of linear and non-linear methods. J. Taiwan Inst. Chem. Eng. 80, 472–487 (2017)
18. Lanka, S., Madhavim, R., Abusahmin, B.S., Puvvada, N., Lakshminarayana, V.: Predictive data mining techniques for management of high dimensional big-data. J. Ind. Pollut. Control 33, 1430–1436 (2017)
19. Madhavi, R., Karri, R.R., Sankar, D.S., Nagesh, P., Lakshminarayana, V.: Nature inspired techniques to solve complex engineering problems. J. Ind. Pollut. Control 33(1), 1304–1311 (2017)
20. Lingamdinne, L.P., Singh, J., Choi, J.S., Chang, Y.Y., Yang, J.K., Karri, R.R., Koduru, J.R.: Multivariate modeling via artificial neural network applied to enhance methylene blue sorption using graphene-like carbon material prepared from edible sugar. J. Mol. Liq. 265, 416–427 (2018)
21. Lingamdinne, L.P., Koduru, J.R., Chang, Y.Y., Karri, R.R.: Process optimization and adsorption modeling of Pb(II) on nickel ferrite-reduced graphene oxide nano-composite. J. Mol. Liq. 250, 202–211 (2018)
22. Karri, R.R., Tanzifi, M., Tavakkoli Yaraki, M., Sahu, J.N.: Optimization and modeling of methyl orange adsorption onto polyaniline nano-adsorbent through response surface methodology and differential evolution embedded neural network. J. Environ. Manage. 223, 517–529 (2018)
23. Karri, R.R., Sahu, J.N.: Modeling and optimization by particle swarm embedded neural network for adsorption of zinc (II) by palm kernel shell based activated carbon from aqueous environment. J. Envi. Manage. 206, 178–191 (2018)
24. Karri, R.R., Sahu, J.N.: Process optimization and adsorption modeling using activated carbon derived from palm oil kernel shell for Zn (II) disposal from the aqueous environment using differential evolution embedded neural network. J. Mol. Liq. 265, 592–602 (2018)
25. Eskandari-Naddaf, H., Kazemi, R.: ANN prediction of cement mortar compressive strength, influence of cement strength class. Constr. Build. Mater. 138, 1–11 (2017)
26. Eskandari, H., Tayyebinia, M.: Effect of 32.5 and 42.5 cement grades on ANN prediction of fibrocement compressive strength. Proc. Eng. 150, 2193–2201 (2016)
27. Azimi-Pour, M., Eskandari-Naddaf, H.: ANN and GEP prediction for simultaneous effect of nano and micro silica on the compressive and flexural strength of cement mortar. Constr. Build. Mater. 189, 978–992 (2018)
28. Awoyera, P.O.: Mechanical and Microstructural Characterization of Ceramic-Laterized Concrete Composite. Ph.D. Thesis, Covenant University, Ota, Nigeria (2018)
29. Awoyera, P.O.: Predictive models for determination of compressive and split-tensile strengths of steel slag aggregate concrete. Mater. Res. Innov. 22, 287–293 (2018)

Text Feature Space Optimization Using Artificial Bee Colony

Pallavi Grover⊙ and Sonal Chawla

Abstract A text classification system's learning is substantially dependent on the input features and their process of extraction and selection. The solitary drive encouraging feature selection practice is to lessen the dimensionality of the problem at hand; thus, facilitating the process of classification. Among several problem areas, text categorization is one area where feature selection plays a vital role. It is well-known that text categorization suffers from the curse of dimensionality. This results in the creation of feature space which may have redundant or irrelevant features leading to the creation of a poor classifier. Therefore, to build an intelligent classifier feature, selection is an important process. This paper has a fourfold objective: Firstly, it aims to create a word to vector space using a widely used score. Secondly, it intends to optimize text feature space using a nature-inspired algorithm. Thirdly, it aims at comparing classification performance of three prominently used classifiers, SVM, Naïve Bayes, and k-Nearest Neighbors in the area of text classification. Lastly, it targets to compare metrics. Besides accuracy, to understand the consequence of optimizing feature space using nature-inspired algorithm. Standard text classification dataset, Reuters-21578, was used, and the classification accuracies reached 95.07%, 92.23%, 87.37% for SVM, Naïve Bayes, and k-Nearest Neighbors, respectively. Besides accuracy, precision, recall, and F-measure were the performance metrics. Considering the encouraging results achieved using the ABC algorithm, this method seems promising for other applications of text classification.

Keywords Artificial Bee Colony · Feature optimization · Text mining

P. Grover (✉) · S. Chawla
Department of Computer Science & Application, Panjab University, Chandigarh, India
e-mail: grover.plv@gmail.com

S. Chawla
e-mail: sonal_chawla@yahoo.com

© Springer Nature Singapore Pte Ltd. 2020
K. N. Das et al. (eds.), *Soft Computing for Problem Solving*,
Advances in Intelligent Systems and Computing 1057,
https://doi.org/10.1007/978-981-15-0184-5_59

1   Introduction

Data mining and analysis in general and text mining in particular aim to recognize peculiarities within data by extracting and simulating information content. Features can be thought of as representative of vital characteristic from original content. Feature extraction from textual information in large amounts is imperative and necessary. This is because text data suffers from the curse of dimensionality. Every single element of text, i.e., a word contributes to the creation of feature space. These features are arranged into vector space with their size going up to n-dimension. Values of this n-dimensional vector range from a few hundred to thousands, in the case of text data. The idea is to extract a feature subset from the original set to reduce irrelevance, noise, or redundancies in features. This makes the process less compute-intensive and improves classification accuracy.

This paper proposes a methodology for feature selection via Artificial Bee Colony (ABC). To the best of our familiarity and knowledge, this work is the foremost to present use of ABC-based technique to optimize text feature space. Artificial Bee Colony has been used in a range of optimization problems and has produced excellent results; however, it finds limited application in the area feature selection. In our work, the entire feature set is input to the ABC algorithm which helps in weight optimization of every feature. This result in the generation of a reduced feature set based on their maximum contribution in feature space. This feature space was used to train three classifiers. Based on the experiments conducted, it was verified that classification accuracy achieved using reduced feature dataset was better than that achieved with the entire feature set.

This paper is arranged as follows: The paper commences with an introduction to some pertinent concepts followed by a review of the state of the art. It then details out the proposed methodology. A description of performance measures and empirical results obtained using this methodology follows. The last section concludes with remarks and directions for future work.

2   Feature Selection

A vital step, done with the intent of defining most relevant, consistent and significant feature space, is feature selection. Several techniques have been used such as—traversal of complete search space to discover the optimal set of features is an elementary approach; however, this approach cannot be practiced for a large number of features. Also, standard AI approaches fail to provide a solution in deterministic time. Therefore, a method using a stochastic search is a viable option in such a case. It considers features as input, for evaluation, generating a random set of pertinent features [22]. An archetypal feature selection process goes through four stages, i.e., feature subset generation, feature subset evaluation, stopping criterion, and result validation [27].

According to the literature, feature selection follows three major paradigms: *filter* approach [8, 10, 20, 30, 31], *wrapper* approach [4, 5, 9, 18, 19], and *embedded* approach. In the filter approach, an evaluation function is applied to every feature and subset selection is done based on the score achieved [25]. Measures such as information gain, term strength, document frequency, odds ratio, $\chi 2$ statistic, and mutual information have been used [3, 6, 25, 26, 28], and features with higher values form the subset indicative of the first subset. In *Wrapper approach*, feature subset is generated by adding or removing the feature and employing accuracy measure to evaluate them. It has been observed that results achieved using the wrapper method have been better than filter methods. In *Embedded* approach, feature subset creation is generally done specifically to given learning machine as a part of the learning process [27]. The classifier is constructed keeping the search criteria for an optimal subset in mind with features added to it iteratively. In some variants, the classifier is built using all features in the first go and then removes the least contributing features iteratively [7].

Based on initialization and behavior, feature selection search can be classified using three different approaches [21]:

Forward: An empty set is taken initially, and features are added to it during the process of selection;
Backward: The subset is taken as a set with all features, in the beginning, omitting one feature at a time to create a final feature set;
Bidirectional: Allows features to be inserted or omitted during the feature selection process.

3 Artificial Bee Colony

Proposed by Karaboga, Artificial Bee Colony (ABC) algorithm is a population-based stochastic optimization technique [1, 11, 14] that finds its inspiration in intelligent foraging behavior of honey bee swarms. This algorithm finds its application in solving optimization problems in several research areas [13, 14, 17, 23]. Following pseudocode provides the essence of the basic version of Algorithm 1.

Algorithm1: Pseudocode for ABC

1. Initial Phase
2. Repeat

   EBee Phase
   OBee Phase
   SBee Phase

   Remember the best solution achieved so far

3. Until C = Max Number of Cycles

The process begins when a group of bees (EBees/employed bees) leave the honeycomb to look for a food source (nectar). On discovering it, the bees take the nectar from the flowers and fly back to their hives carrying nectar in their stomach. They then share information like nectar quality, distance, and direction of nectar source from the hive with other bees (OBees/Onlooker Bee), performing waggle dance.

Further, they deploy new bees to discover the most abundant food sources [12]. For a bee swarm to exhibit collective intelligence, it requires three components:

Food Sources: The most likely solutions to the given problem;
Employee Bees: The group of bees that look for food sources and store its information in terms of quality. Further, it is their responsibility to share information with bees waiting to get deployed;
Unemployed Bees: They are categorized into two kinds—*Onlooker Bees*: They are the bees that wait in honeycomb for the Employee Bees to return and transfer the information. They are then deployed to extract good quality food sources from the neighborhood based on the information conveyed [16] and *Scout Bees*: Once a food source neighborhood has been explored to its maximum limit, the employed bees become Scout Bees [2].

The algorithm has three artificial groups of bees, and hence, the phases form the ABC algorithm, i.e., EBee (Employee Bees), OBee (Onlooker Bees), and SBee (Scout Bees) [15]. As specified by the creator, the size of the employed bee colony equals the size of the Onlooker Bees' colony. Furthermore, the number of EBee or OBee equals the number of solutions (NFS) in the population. Also, Onlooker Bee waits back in the hive (in the dance area) to decide on the selection of food source. An OBee is referred to as an EBee when it gets deployed in EBee's role. Further, an EBee that has consumed a food source completely turns into SBee. SBee is then responsible for discovering new resources using random search.

3.1 Initialization Phase

Food sources are generated randomly at the initial stage. Assuming $w^s = (w_1^s, w_2^s, \ldots, w_D^s)$ as the sth food source, this process can be defined as follows:

$$w_p^s = w_{p\_min} + \text{rand}(0, 1)\left(w_{p\_max} - w_{p\_min}\right) \tag{1}$$

where $s = 1, \ldots$, NFS, $p = 1, \ldots, D$, such that NFS is the number of food sources and D is the number of parameters to be optimized.

3.2 Employee Bee Phase

Each EBee (Employed Bee) assesses food source quality by evaluating objective function value allocated to it. It is also for a better food source around. To determine a new food source $w^{s'} = (w_1^{s'}, w_2^{s'}, \ldots, w_D^{s'})$ around the old one $w^s = (w_1^s, w_2^s, \ldots, w_D^s)$, the following expression is used:

$$w_p^{s'} = w_p^s + \phi_p^s\left(w_p^s - w_p^t\right) \tag{2}$$

where $t \in \{1, 2, \ldots, \text{NFS}\}$ and $p \in \{1, 2, \ldots, D\}$ are indexes chosen randomly; u is unequal and different from s $(u \neq s)$; and ϕ is a random number falling in the range of $-1, 1$. At this point in the algorithm, there will always be a choice between two solutions. The greedy approach is followed to select the solution. If the newly produced solution holds a better value, it replaces the old one, else old one is retained. For every iteration, an Employee Bee is permitted to search for a better food source only once with the direction of search selected randomly. This concludes the Employee Bee phase.

3.3 Onlooker Bee Phase

Once Employee Bees have explored the search area, they come back to the beehive to share details about the suitability (fitness) and distance of food source from honeycomb with the OBees (Onlooker Bees), doing waggle dance. Based on this information, food sources are supplied with Onlooker Bees, in count proportionate to the probability associated with its fitness, that is

$$P(s) = \frac{\text{fitness}(s)}{\sum_{s=1}^{\text{NFS}} \text{fitness}(s)} \tag{3}$$

where fitness(s) is the fitness value of sth food source. Using Eq. (2), food sources with better fitness values are searched around the food sources allotted to Onlooker Bees. Food source with better fitness value is memorized. If the newly produced solution holds a better value, it replaces the old one, else old one is retained.

3.4 Scout Bee Phase

After both Employee and Onlooker Bees have explored search space thoroughly, the algorithm checks for any exhausted food source that should be discarded. The choice to discard a food source is made referring to values of two parameters, i.e., LIMIT-control parameter for the algorithm and a counter-variable to keep track of changes/updates made during the previous two phases. If the value of this variable is higher than the set value of LIMIT value, then the source associated with the counter is assumed to be exhausted hence abandoned.

Scout Bee then determines a new food source and replaces the abandoned one. For every iteration, only one *scout bee* activates. The new randomly created food source is associated with SBee. It may happen so that the newly created food source may violate the boundary conditions of the search space. Following checks help in overcoming this violation:

$$w_p^{s\prime} = w_{p\_min}, \quad \text{if } w_p^{s\prime} < w_{m\_min}$$
$$w_p^{s\prime} = w_{m\_max} \quad \text{if } w_p^{s\prime} > w_{m\_max}. \tag{4}$$

4 ABC for Feature Selection

Classification of text (or any) dataset is affected negatively with the presence of less distinctive features in search space. Such data decreases the speed and overall efficiency of the system significantly. A text categorization system characteristically comprises of some essential fragments, including preprocessing, feature extraction selection. The process begins with preprocessing or data cleaning of text documents where the text is punctuations, stop words are removed, the text is converted into a single case, and tokenization is done; subsequently, converted into a feature vector. Finally, ABC is used to reduce the dimensionality.

To explore all subsets from the given superset, ABC is used. Each subset is subjected to evaluation against the performance metrics. The feature subset set with the highest score on evaluation metrics is considered as an optimized subset for classification and hence recommended for classification.

5 Experimental Work

To demonstrate the utility of the proposed feature selection algorithm, several sequences of experiments were conducted. A machine with 2.60 GHz CPU and 64 GB of RAM was used for all experiments. Entire framework was developed in MATLAB R2016a.

The proposed feature selection algorithm has the following core stages:

1. Load training samples
2. Initialize ABC parameters
3. Initial population generation, $w^s = 1, \ldots, NFS$, using (1)
4. Fitness evaluation of the population and use accuracy as fitness
5. $C = 1$
6. Repeat
7. Construct solutions by EBee
8. Determine neighbors using (2)

   a. Send feature subset to the classifier
   b. Assess fitness
   c. Calculate probability $P(s)$ using (3)

9. Construct solutions by Onlooker Bee

   d. Select feature according to the probability
   e. Compute $w^{s'}$
   f. Assess fitness
   g. Apply greedy selection

10. Determine Scout Bee by comparing to LIMIT value
11. Memorize optimal feature set
12. $C = C + 1$
13. Until $C = $ max number of cycles
14. Use same searching procedure to create the feature subset configurations.

5.1 Datasets

Several standard datasets are available that can be used in text classification as test collections. In general, Reuters-21578 is a dataset that has been used most extensively to text classification. It consists of news stories from Reuter's new agency, categorized under different classes but generally relate to economics. This dataset is chosen because of its popularity in the field of text classification. It is available at UCI machine learning repository [29]. Both subcollections have skewed distribution of data. The class with maximum training samples had 2840 documents, i.e., 43.6% of the training set.

Table 1 Document count for most frequent training and testing classes R52 dataset

Class name	# Training documents	# Testing documents
Acquisition	1596	1392
Crude	253	121
Earn	2840	1083
Interest	190	81
Money-fx	206	87
Ship	108	36
Sugar	194	25
Trade	251	75

Similarly, the class with a minimum number of documents has only one document; occupying 0.01% of the training set. This skewness in data would have led to more error. Therefore, this work adopted the top eight classes for R52 subcollections. Therefore, there were 5638 training set and 2900 testing set for R52. Table 1 shows the most frequent classes for R52, along with its training and testing document count in each.

5.2 Feature Extraction

A classifier cannot directly construe a text document; therefore, they have to be uniformly mapped into a representation for their processing. Several indexing procedures have been suggested in the literature. In this work, the normalized *tfidf* function [24] has been used. A training document d_i can be represented as a vector of terms weights of words (or features) contained in it as given in Eq. (5)

$$d_i = \left\{ w_{1i}, w_{2i}, \ldots, w_{|T|i} \right\} \tag{5}$$

where T is a set of terms occurring in a document at least once. There are a number of documents belonging to a class. Therefore, for every class, a term cross-document matrix containing how much term t_j semantically contributes to the document d_i is generated. This representation is called a bag of words, with each word weight contribution is calculated using Eq. (6)

$$W_{ij} = \frac{tfidf(t_j, d_i)}{\sqrt{\sum_{k=1}^{|T|} (tfidf(t_k, d_i))^2}} \tag{6}$$

where

$$tfidf(t_j, d_i) = \#(t_j, d_i) \cdot \log \frac{|\text{Tr}|}{\#_{\text{Tr}}(t_j)} \tag{7}$$

Table 2 Contingency matrix

Class C_Z	Predicted categories		
	Yes	No	
Actual categories	Yes	TP_Z	FN_Z
	No	FP_Z	TN_Z

where $\#(t_j, d_i)$ is occurrence count of term t_j in document d_i. Tr denotes the training set and |Tr| number or length of documents in training set. $\#_{Tr}(t_j)$ refers to count of documents in training set Tr containing term t_j.

6 Performance Measure

Any text classification problem is evaluated by calculating values of precision (π), recall (ρ), and F-measure ($F1$). These values can be computed using the contingency matrix as given in Table 2. For a given class C, there can be two types of values, the actual categories and the categories given by the classifier. TP_Z is the number of test documents that belonged to class Cz and has been classified with that label. FP_Z is a number of test documents that did not belong to class Cz, but the classifier falsely labeled them as Cz. FN_Z is a number of test documents that belong to Cz, but classifier failed to categorize them as Cz. TN_Z is the number of documents that neither belong to Cz nor the classifier classified them as Cz.

Values of the stated parameters are computed for every class using the following equations:

$$\Pi_z = \frac{TP_Z}{TP_Z + FP_Z} \tag{8}$$

$$\rho_z = \frac{TP_Z}{TP_Z + FN_Z} \tag{9}$$

$$F1 = \frac{Two * \pi * \rho}{\pi + \rho} \tag{10}$$

7 Classification Setup

According to the literature, text classification performance is also governed by the classifier used. The accuracy of the process was evaluated with tenfold cross-validation was used with Support Vector Machine, Naïve Bayes, and k-Nearest Neighbors as classifiers. In tenfold cross-validation, the entire dataset is randomly partitioned into like-sized ten folds. One part of that dataset is used for testing the accuracy of the classifier, and the remaining nine parts are used to train the classifier. The process is reiterated as many numbers of times as the value of k, keeping a new section of data, for testing, each time. Accuracy is measured as the sum of correctly classified documents by a total number of documents as in Eq. 11. Average of these scores provides estimated accuracy value.

$$\text{Accuracy} = \frac{\text{TP}_Z + \text{TN}_Z}{\text{TP}_Z + \text{TN}_Z + \text{FP}_7 + \text{FN}_7} \tag{11}$$

8 Results

The proposed work was tested for different values of control parameters. Highest predictive accuracy was achieved for parameter values shown in Table 3. These values were found after conducting several experiments in no way should be taken as final values for control parameters. Determining the control parameters or hyperparameters for the algorithm is another topic of research.

Table 4 shows the results obtained from different classifiers for other evaluation measures using the proposed methodology. Precision is the probability that the

Table 3 Control parameters of ABC

Control parameters	Values
Population size	2 * No. of features
Limit	10
Food sources	30
No of iterations	100

Table 4 Precision and recall values for SVM, Naïve Bayes, and nearest neighbor

	Support vector machine		Naïve Bayes		Nearest neighbor	
	Precision (Π)	Recall (ρ)	Precision (Π)	Recall (ρ)	Precision (Π)	Recall (ρ)
Acquisition	91.27	93.16	89.38	90.66	85.92	89.46
Crude	90.91	7837	92.33	87.57	87.29	82.22
Earn	87.03	90.13	89.13	85.61	82.48	79.77
Interest	95.67	92.44	87.32	84.96	88.70	84.83
Money-fx	70.51	68.51	72.21	70.01	59.69	58.29
Ship	64.25	72.05	62.38	65.25	60.66	59.61
Sugar	88.62	90.90	86.47	88.75	82.03	84.34
Trade	89.75	84.37	85.11	82.94	84.58	79.48
Average	84.75	83.74	83.04	81.96	78.91	77.25

Table 5 Averaged F-measure and accuracy values for SVM, Naïve Bayes, and nearest neighbor

	Support vector machine	Naïve Bayes	Nearest neighbor
Average F-Measure	96.08	82.48	78.04
Accuracy	95.07	92.23	87.37

sample is positive, given a positive test result. So primarily, it is calculated over predicted condition positive. Mathematically, it is described in Eq. 8. The overall precision value achieved using SVM was comparable to Naïve Bayes but higher than the values given by nearest neighbor classifier. This could probably because the nearest neighbor classifier does not perform well on soft margin boundaries.

Recall (Eq. 9), also known as true positive rate or sensitivity is the probability of a positive result for a given positive example. Primarily, it is calculated on actual positive conditions. Referring to Table 4, it can be observed that SVM's performance was better than the other two classifiers.

F-Measure (Eq. 10) is the measure between exactness (precision) and completeness (recall). Accuracy (Eq. 11) is a ratio of correctly predicted classifiers to total observations. Table 5 gives averaged values of F-measure and accuracy for three classifiers. From the table, it is observed that SVM outperforms the other two classifiers. Since data is unevenly distributed, F-measure values are more reliable than accuracy.

9 Conclusion

This research demonstrates the use of metaheuristic algorithm, Artificial Bee Colony, in feature selection. Compared to work in literature, the results show, an overall all better performance concerning other metaheuristic algorithms applied to the text.

In the future, this work can be combined with filter approaches. In addition to that, this research also intends to combine deep learning techniques with metaheuristic algorithms.

References

1. Akay, B.: A modified artificial bee colony algorithm for real-parameter optimization. Inf. Sci. **192**, 120–142 (2012)
2. Bao, L., Zeng, J.: Comparison and analysis of the selection mechanism in the artificial bee colony algorithm. In: Ninth International Conference on Hybrid Intelligent Systems, HIS'09, pp. 411–416 (2009)
3. Brank, J., Grobelnik, M., et al.: Interaction of feature selection methods and linear classification models. In: Workshop on Text Learning Held at ICML (2002)

4. Caruana, R.: Greedy attribute selection. Mach. Learn. Proc. **1994**, 28–36 (1994)
5. Dy, J.G.: Feature subset selection and order identification for unsupervised learning. In: ICML, pp. 247–254 (2000)
6. Forman, G.: An experimental study of feature selection metrics for text categorization. J. Mach. Learn. Res. **3**(1), 1289–1305 (2003)
7. Guyon, I.W.: Gene selection for cancer classification using support vector machines. Mach. Learn. **46**(1–3), 389–422 (2002)
8. Hall, M.A.: Correlation-based feature selection of discrete and numeric class machine learning. In: 17th International Conference Machine Learning, pp. 359–366 (2000)
9. Hruschka, E.R.: Feature selection for cluster analysis: an approach based on the simplified Silhouette criterion. In: Computational Intelligence for Modelling, Control and Automation, 2005 and International Conference on Intelligent Agents, Web Tech Internet Commerce, International Conference on IEEE, pp. 32–38 (2005)
10. Jiang, Y., Ren, J.: Eigenvector sensitive feature selection for spectral clustering. In: Joint European Conference on Machine Learning and Knowledge Discovery in Databases, pp. 114–129 (2011)
11. Karaboga, D.: A comparative study of artificial bee colony algorithm. Appl. Math. Comput. **214**(1), 108–132 (2009)
12. Karaboga, D., Akay, B.: A survey: algorithms simulating bee swarm intelligence. Artif. Intell. Rev. **31**(1–4) (2009)
13. Karaboga, D.: Neural networks training by artificial bee colony algorithm on pattern classification. Neural Netw. World **19**(3), 279 (2009)
14. Karaboga, D.: A novel clustering approach: artificial bee colony (ABC) algorithm. Appl. Softw. Comput. **11**(1), 652–657 (2011)
15. Karaboga, D.G.: A comprehensive survey: artificial bee colony (ABC) algorithm and applications. Artif. Intell. Rev. **42**(1), 21–57 (2014)
16. Karaboga, D.O.: Artificial bee colony programming for symbolic regression. Inf. Sci. **209**, 1–15 (2012)
17. Karaboga, D.O.: Cluster-based wireless sensor network routing using an artificial bee colony algorithm. Wireless Netw. **18**(7), 847–860 (2012)
18. Kim, Y.S.: Evolutionary model selection in unsupervised learning. Intell. Data Anal., 531–556 (2002)
19. Kohavi, R.: Wrappers for feature subset selection. Artif. Intell. **97**(1–2), 273–324 (1997)
20. Liu, H.: A probabilistic approach to feature selection-a filter solution. ICML **96**, 319–327 (1996)
21. Liu, H.: Toward integrating feature selection algorithms for classification and clustering. IEEE Trans. Knowl. Data Eng. **17**(4), 491–502 (2005)
22. Nakamura, R.Y.: BBA: a binary bat algorithm for feature selection. In 2012 25th SIBGRAPI Conference on Graphics, Patterns, and Images, pp. 291–297 (2012)
23. Ozturk, C.K.: Probabilistic dynamic deployment of wireless sensor networks by an artificial bee colony algorithm. Sensors **11**(6), 6056–6065 (2011)
24. Salton, G., Buckley, C.: Term Weighting Approaches in Automatic Text Retrieval. Cornell University (1987)
25. Soucy, P., Mineau, G.W.: Feature selection strategies for text categorization. In: Conference of the Canadian Society for Computational Studies of Intelligence, pp. 505–509 (2003)
26. Sousa, P.A., Pimentão, J.P., Santos, B.R.D., Moura-Pires, F.: Feature selection algorithms to improve documents' classification performance. In: International Atlantic Web Intelligence Conference, pp. 288–296 (2003)
27. Subanya, B. &. (2014). Feature selection using Artificial Bee Colony for cardiovascular disease classification. 1–6
28. Torkkola, K.: Discriminative features for text document classification. Formal Pattern Anal. Appl. **6**(4), 301–308 (2004)
29. UCI—Machine Learning Repository. (2013). Retrieved from http://archive.ics.uci.edu/ml/. http://archive.ics.uci.edu/ml/

30. Yu, L., Liu, H.: Feature selection for high-dimensional data: a fast correlation-based filter solution. In Proceedings of the 20th International Conference on Machine Learning (ICML-03), pp. 856–863 (2003)
31. Zhang, C., Hu, H.: Ant colony optimization combining with mutual information for feature selection in support vector machines. In: Australasian Joint Conference on Artificial Intelligence, pp. 918–921 (2005)

Secure Multipath Routing for Efficient Load Balancing and Jamming Attack Protection

Diksha Singhal, Ritu Prasad and Praneet Saurabh

Abstract Mobile ad hoc network (MANET) brings simplicity and flexibility but also introduces constraints due to mobility, dynamic configuration, limited resources, and security. Security is among the most important limitations in MANET. Because of its distinctive features, MANET creates a variety of serious challenges. The various protection schemes aligned with routing makes efforts in lowering threat perception. In this paper, a novel security mechanism named secure multipath routing (SMR) for MANET is proposed for jamming attacks that also takes care of routing and load balancing. The proposed technique will first identify the attacked node, and then, it will not allow other attacker node/s. Thereafter, as a corrective measure it blocks the activity of offender and drops it from the routing. SMR is then judged against conventional AODV and AOMDV protocols in NS-2.35. Experimental results show that the proposed SMR demonstrates better results in terms of throughput, average end-to-end delay, packet delivery ratio, and routing overhead.

Keywords AODV · AOMDV · Routing · Multipath routing · MANET

1 Introduction

Mobile ad hoc network (MANET) nodes are able to wander without any limitation and can get connected to dissimilar nodes in order to send data from source to destination [1]. Multipath routing is the routing via various dissimilar paths throughout

D. Singhal (✉) · R. Prasad
Technocrats Institute of Technology Advance, Bhopal, MP 462021, India
e-mail: diksha.seenu13@gmail.com

R. Prasad
e-mail: rit7ndm@gmail.com

P. Saurabh (✉)
Technocrats Institute of Technology, Bhopal, MP 462021, India
e-mail: praneetsaurabh@gmail.com

© Springer Nature Singapore Pte Ltd. 2020
K. N. Das et al. (eds.), *Soft Computing for Problem Solving*,
Advances in Intelligent Systems and Computing 1057,
https://doi.org/10.1007/978-981-15-0184-5_60

705

a system that can withstand fault tolerance, bandwidth constraints, and threat perception [2]. These paths may overlap and can be disjointed. Concurrent multipath routing (CMR) is the mechanism that takes mean and focuses on the utilization of multiple available paths. In this technique, one stream is allocated a single non-overlapping stream for transmission. Overall CMR strives to provide better resource utilization. Routing algorithms try to remain competent for load balancing and power competence. Incompetent load balancing results in higher routing overhead that consequently lowers packet delivery rate affecting Quality of Service (QoS). Due to its flexibility and dynamic character, MANET also introduces various challenges such as load balancing, jamming, and security. Load balancing is a significant limitation in MANET as nodes at the center need to cater large amount of data as compared to other nodes. Jamming in MANET is another major limitation, as it denies legitimate users from proper utilization of resources at a given time, thus lowering QoS. Since all the nodes in MANET survive and live on their own and execute multiple tasks. So to ensure non-proliferation of malicious nodes in MANET, it employs various techniques, but somehow these fall short in providing adequate security to curb threat perception. This paper introduces a novel security mechanism named secure multipath routing (SMR) for MANET for jamming attacks that also takes care of routing and load balancing.

This paper is organized in the following manner: Sect. 2 presents the related work of the domain while Sect. 3 introduces the proposed work; Sect. 4 discusses the experimentation and result evaluation; and Sect. 5 concludes the paper.

2   Related Work

Many explores have been introduced in recent years as there is tremendous growth in mobile devices due to popularity and access of computing services. Reliability and security remain a major objective in this domain. Salem and Yadav [3] discussed wireless links of mobile nodes for communication and put forward different categories of routing protocols for MANETs. They also talked about complexity and consequences of load balancing in MANET. Thereafter, Thakur and Sankaralingam [4] built a system to avoid and overcome jamming and jamming attacks in MANET. Jamming is an interference attack in wireless networks that tries to manufacture denial of service or resources. In another work, Kotzanikolaou et al. [5] proposed secure multipath routing for MANET that made substantial efforts to overcome his challenge. They recognized various attacks that leave routing protocol vulnerable and escalate the impact of malicious nodes. They have integrated their work in on-demand protocols like AODV and DSR. Nasser and Chen [6] introduced secure and energy-efficient multipath routing protocol (SEER) that used multipath as an alternative for communication between source and destination to extend network lifetime. SEER demonstrated its resistive behavior and outperformed its peers. Stavrou and Pitsillides [7] investigated various routing protocols in wireless sensor networks. In this work, routing protocols for wireless sensor network are classified into categories with respect to objectives and threat model. Major works have identified security as a prime challenge as it is not attended properly. This paper also suggested multipath routing to overcome security challenges in wireless sensor network. Vasudevan and Sanyal [8] put forward a new algorithm that embarked security in transmission for MANET without encryption. The proposed algorithm employed multipath routing with the concepts of polynomials. Li et al. [9] critically evaluated wireless sensor network and identified reasons for its failure that ranges from harsh environmental conditions, hardware failures, and limitations of software due to new and evolving challenges. Khan et al. [10] presented a weight-based secure approach for identifying selfish behavior of node in MANET. In this work, the proposed method makes the discovery of secure routing path and detects selfish nodes in MANET. In recent times, some researchers explored this domain through various bio-inspired techniques [11] that have successfully attained different objectives in this domain [12–14]. All these limitations actually point out the need of a secure multipath routing for jamming attacks in MANET. This paper introduces a secure multipath routing (SMR) for jamming attacks in MANET that also takes care of routing and load balancing.

3 Proposed Work

This section presents a novel security mechanism named secure multipath routing (SMR) for MANET to overcome the limitation of jamming attacks that also takes care of routing and load balancing. The proposed technique first identifies the attacked node and then prevents it from trying to intrude into the network next time. Proposed SMR deploys mobile nodes and initiates routing of multipath for efficient routing that tries to minimize congestion in network. SMR then tries to minimize the excessive traffic in the network. The proposed SMR is designed to defend network against jamming attacks and provide security to it. The moment SMR identifies attack, it blocks that particular node and initiates a new route from source to destination for securing the delivery of packet. Algorithm for the proposed SMR is given in Fig. 1.

Fig. 1 Flowchart of
proposed method

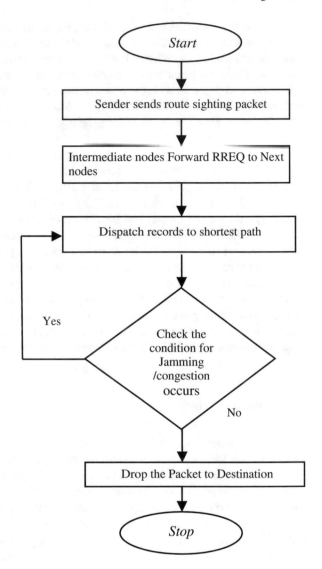

Algorithm for Secure Multipath Routing

Step 1: Senders Ss send route sighting packet intended for seek out destinations R_s.

Step 2: I_i Intermediate nodes Forward RREQ to Next nodes.

Step 3: if (Ss found through Ii established further than single pathway)

    {

               Dispatch records to shortest path

                  If (Ii Node like a Jammer Attacker)

                     {

                         Produce not needed packets

                         Enhanced unwanted traffic

                         Consume Maximum amount of bandwidth

                         Drop data packets

                     }

Step 4: Calculate packets capacity of nodes Ii for Attacker free routing

        (

               (Data receiving = = 0) && (PDR •70)

)

         {

               Jamming Attacker found

               Preventer Node (PN) node transmit note near jammer for jammed manage

               If (Ii = = jamming Attacker)

                  {

                     Broadcast their Id (Node Number)

                     Block their routing participation

                     PN mail note near sender/s node Ss on behalf of mail records throughout previous (I_i - IJ_A) nodes

                  }

SMR is introduced in ad hoc on-demand multipath distance vector (AOMDV) routing, and it utilizes the delay between delivery of packets and initiates data packets based on average bandwidth for particular paths. It also checks percentage of data delivery rate so that lightweight routes can be repaired and new secure path can be established. SMR continuously monitors the network channels for accessibility and subsequently handling jamming attacks in the network.

4 Experimentation and Result Analysis

This section put forwards the results of different experimentations performed on proposed SMR and current state of art like AODV and AODV when introduced to jamming attacks. Experiments are carried out to calculate packet delivery, overhead, packet drop, throughput, average end-to-end delay for SMR and AODV and AODV when introduced to jamming attacks, and then, comparative analysis has been done. All the experiments are done on NS 2.35 network simulator.

4.1   Packet Delivery Ratio Comparison

Packet delivery represents the percentage of packets successfully delivered from source to destination.

Packet delivery ratio of SMR, AODV, and AODV when introduced to jamming attacks are calculated with variation in number of nodes in the MANET while keeping scenarios like pause, queue, rate, source, speed, terrain, transmission range constant. Experimental results from Table 1 and Fig. 2 very clearly state that SMR reports higher packet delivery ratio as compared to traditional AODV and AODV when introduced with jamming attacks when the number of nodes in the network is increased from 5 to 70. All the results demonstrate that SMR is better equipped to take on the challenges posed by jamming attacks in the network. Results also stated that the new integrations in SMR are pivotal in efficiently delivering higher number of data packets from source to destination.

Table 1 Packet delivery ratio comparison

Number of nodes	AODV	AODV with jamming	SMR
5	0.9	0.35	0.96
10	0.9	0.64	0.99
20	0.78	0.76	0.95
30	0.77	0.57	0.8
40	0.79	0.75	0.9
50	0.81	0.75	0.84
60	0.78	0.64	0.83
70	0.6	0.44	0.8

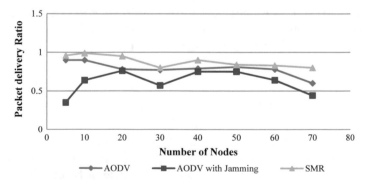

Fig. 2   Packet delivery ratio comparison

4.2 Overhead Comparison

This experiment is carried to observe the overhead required in sending packets from source to destination in proposed SMR and AODV and AODV when introduced to jamming attacks when the number of nodes increased from 5 to 70 in MANET.

Experimental results presented in Table 2 and Fig. 3 illustrate lower overhead obtained by proposed SMR as compared to traditional AODV and AODV when introduced with jamming attacks when the number of nodes in the network is increased from 5 to 70. Under all the test conditions, SMR remained stable while AODV when introduced to jamming attack shows instability. SMR for all test conditions reported lower overhead; this facilitated more network traffic in the network and also avoided chances of congestion.

Table 2 Overhead comparisons

Number of nodes	AODV	AODV with jamming	SMR
5	0.76	0	0.58
10	0.76	1	0.4
20	0.5	0.9	0.41
30	0.5	0.97	0.5
40	0.37	0.61	0.3
50	0.43	0.82	0.38
60	0.45	0.85	0.39
70	0.59	0.79	0.43

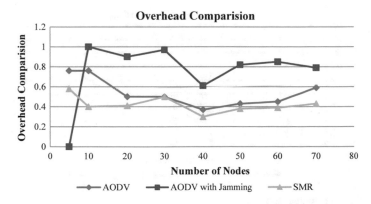

Fig. 3 Overhead comparison

4.3 Packet Drop Comparison

Packet drop represents number of packets dropped in its delivery from source to destination. Experiments are carried to calculate packets dropped for SMR, AODV, and AODV when introduced to jamming attacks with variation in number of nodes in the MANET while keeping scenarios like pause, queue, rate, source, speed, terrain, and transmission range constant.

Experimental results from Table 3 and Fig. 4 demonstrated that SMR reported lower number packets dropped when the number of nodes in the network is increased from 5 to 70, as compared to traditional AODV and AODV when introduced with jamming attacks. SMR in all experimental results overwhelmingly overcomes the challenge of jamming attacks and accounts low percentage of packets dropped.

Table 3 Packet drop comparison

Number of nodes	AODV	AODV with jamming	SMR
5	0.10	0.69	0.04
10	0.10	0.31	0.01
20	0.23	0.29	0.10
30	0.24	0.43	0.2
40	0.21	0.30	0.10
50	0.19	0.28	0.15
60	0.22	0.36	0.16
70	0.40	0.54	0.21

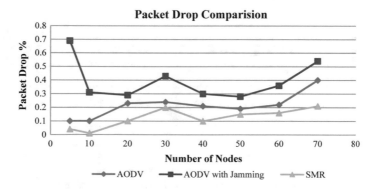

Fig. 4 Packet drop comparison

4.4 *Average End-to-End Delay Comparison*

It is the average time taken by all the packets of any message to get delivered from source to destination. Average end-to-end delay accounts for time period during which a packet is hold up while searching for route detection. This experiment is carried to find average end-to-end delay in proposed SMR and AODV and AODV when introduced to jamming attacks when the number of nodes is increased from 5 to 70 in MANET.

Table 4 and Fig. 5 present experimental results of average end-to-end delay comparison for proposed SMR as compared to traditional AODV and AODV when introduced with jamming attacks. Under all the test conditions, SMR remained reported lower average end-to-end delay and stayed stable as compared to other methods.

Table 4 Average end-to-end delay comparison

Number of nodes	AODV	AODV with jamming	SMR
5	0.16	0.20	0.16
10	0.04	0.22	0.04
20	0.48	0.25	0.27
30	0.25	0.34	0.23
40	0.24	0.14	0.20
50	0.65	1.43	0.40
60	0.26	0.13	0.14
70	0.25	0.11	0.11

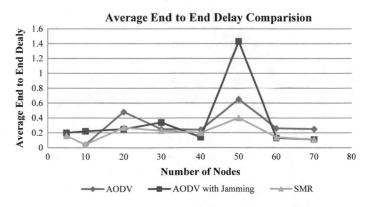

Fig. 5 Average end-to-end delay comparison

Table 5 Throughput comparison

Number of nodes	AODV	AODV with jamming	SMR
5	0.15	0.00	0.45
10	0.16	0.02	0.54
20	0.75	0.58	1.3
30	0.73	0.47	0.73
40	0.80	0.60	0.95
50	0.63	0.53	0.65
60	0.65	0.54	1.2
70	0.49	0.45	0.96

4.5   Throughput Comparison

Throughput is defined as the total number of packets successfully delivered between source and destination in given simulation time period. Higher throughput means efficient network and is always desired. In this experiment, throughput is calculated and subsequently compared between SMR, AODV, and AODV when introduced to jamming attacks with variation in number of nodes.

Table 5 and Fig. 6 illustrate throughput obtained for SMR traditional AODV and AODV when introduced with jamming attacks when the number of nodes in the network is increased from 5 to 70. Results show that SMR reported higher throughput as compared to other methods in all experiments which are desired. Higher throughput enhances the performance of network.

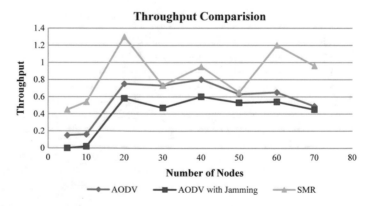

Fig. 6 Throughput comparison

5 Conclusion

MANET brings simplicity and suppleness but also introduces challenges due to its distinctive features. This paper introduces a novel security mechanism named secure multipath routing (SMR) for MANET in the presence of jamming attacks. Experiments are performed to calculate packet delivery, overhead, packet drop, throughput, and average end-to-end delay for SMR and AODV and AODV when introduced to jamming attacks, and then, comparative analysis has been done. SMR showed overwhelming results and reported high throughput, low average end-to-end delay, high packet delivery, and low routing overhead as compared to current state of art and remained stable in all scenarios.

References

1. Cao, Q., Abdelzaher, T., Stankovic, J., Whitehouse, K., Luo, L.: Declarative tracepoints: a programmable and application independent debugging system for wireless sensor networks. In: Proceedings of the ACM SenSys, Raleigh, NC, USA, pp. 85–98 (2008)
2. Shu, T., Krunz, M., Liu, S.: Secure data collection in wireless sensor networks using randomized dispersive routes. IEEE Trans. Mobile Comput. 9(7), 941–954 (2010)
3. Salem, A.M., Yadav, R.: Efficient load balancing routing technique for mobile Ad Hoc networks. (IJACSA) Int. J. Adv. Comput. Sci. Appl. 7(5), 249–254 (2016)
4. Thakur, N., Sankaralingam, A.: Introduction to Jamming attacks and prevention techniques using Honey pots in wireless networks. IRACST Int. J. Comput. Sci. Inf. Technol. Secur. (IJCSITS) 3(2), 202–207 (2013)
5. Kotzanikolaou, P., Mavropodi, R., Douligeris, C.: Secure multipath routing for mobile Ad Hoc networks. In: Wireless On-demand Network Systems and Services, WOWMOM, pp. 581–587 (2005)
6. Nasser, N., Chen, Y.: Secure multipath routing protocol for wireless sensor networks. In: 27th International Conference on Distributed Computing Systems Workshops (ICDCSW'07) (2007)
7. Stavrou, E., Pitsillides, A.: A survey on secure multipath routing protocols in WSNs. Comput. Netw. Int. J. Comput. Telecommun. Netw. 54(13), 2215–2238 (2010)
8. Vasudevan, A.R., Sanyal, S.: A novel multipath approach to security in mobile Ad Hoc networks (MANETs). In: Proceedings of International Conference on Computers and Devices for Communication (CODEC'04), pp. 1–4 (2004)
9. Li, X., Ma, Q., Cao, Z., Liu, K., Liu, Y.: Enhancing visibility of network performance in large-scale sensor networks. In: Proceedings of the IEEE ICDCS, Madrid, Spain, pp. 409–418 (2014)
10. Khan, S., Prasad, R., Saurabh, P., Verma, B.: Weight-based secure approach for identifying selfishness behavior of node in MANET, information and decision sciences. Adv. Intell. Syst. Comput. 701, 387–397 (2018) (Springer, Heidelberg)
11. Saurabh, P., Verma, B.: An efficient proactive artificial immune system based anomaly detection and prevention system. Expert Syst. Appl. 60, 311–320 (2016). (Elsevier)
12. Saurabh, P., Verma, B.: Immunity inspired cooperative agent based security system. Int. Arab J. Inf. Technol. 15(2), 289–295 (2018)

13. Saurabh, P., Verma, B., Sharma, S.: An immunity inspired anomaly detection system: a general framework. In: Proceedings of Seventh International Conference on Bio-Inspired Computing: Theories and Applications (BIC-TA 2012), vol. 202 of the series Advances in Intelligent Systems and Computing, pp 417–428. Springer, Heidelberg (2012)
14. Saurabh, P., Verma, B., Sharma, S.: Biologically inspired computer security system: the way ahead. Recent Trends Comput. Netw. Distrib. Syst. Secur. **335**, 474–484 (2011). (CCIS, Springer)

An IOT-Based Vehicle Tracking System with User Validation and Accident Alert Protocol

S. Balaji, Abhinav Gumber, R. Santhakumar, M. Rajesh Kumar,
Agastya Tiwari and Himayan Debnath

Abstract The world around us is becoming more technologically sound and advancing now than ever before. We owe this change to our engineers and the innovation they bring to our daily lives, not only in big industries. One of them is the advent of smart cars and the features that make it a smart product overall. There have been many debates about the usefulness of such cars. Recent car accident reports and other related mishaps confirm that we are not close to a world of smart cars alone. The main idea is to determine the need for such cars first. There are two major car accident issues. First is the driver's negligence, whether driving drunk or driving beyond the speed limit. Second is the time taken to provide assistance or respond to such fatal, minor and major incidents by the emergency service teams. The paper here deals with the above cases including theft prevention and by real-time tracking and surveillance, in the form of an intelligent, first of its kind product, puts forward a solution to all these problems.

Keywords Vehicle · Theft prevention · Real time · Tracking · Surveillance · IOT · Geo location

1 Introduction

The paper presents an IOT-based vehicle tracking system with user authentication and accident alert protocol [1]. The work deals with identity verification, over-speed, tracking and alert system for accidents. First, it checks whether the fingerprint of the driver is stored in the database of the fingerprint sensor, if it is, the authentication will be successful and a positive ID message will be displayed, which in turn will rotate the servomotor. The servomotor will have in its off position a metal (cardboard in our case) guarding key access to the car engine [2]. This provides a two-step user verification system.

S. Balaji (✉) · A. Gumber · R. Santhakumar · M. Rajesh Kumar · A. Tiwari · H. Debnath
School of Electrical Engineering, Vellore Institute of Technology, Vellore 632014, India
e-mail: sbalaji@vit.ac.in

© Springer Nature Singapore Pte Ltd. 2020   717
K. N. Das et al. (eds.), *Soft Computing for Problem Solving*,
Advances in Intelligent Systems and Computing 1057,
https://doi.org/10.1007/978-981-15-0184-5_61

The next part is the tracking of the vehicle. The real-time tracking [3] will be performed using the ThingSpeak platform in case of over-speeding. You can see the car's location in each instance using the number plate ID as shown on the screen. To get the location, the device uses geo-satellite imagery [4] and simple math coding. It also allows the driver to share in more than one person/places the real-time location [5] at fault. Finally, the vibration sensor alerts the authorities in shake/jolt scenarios exceeding a certain limit, and the exact location and follow-up are similar to the over speed case.

2 System Model and Overview

The main objective is to create an IOT-Arduino-based vehicle tracking system with user authentication and accident alert protocol. The work is majorly divided into the following parts, which are the authentication or verification of identity, over-speeding, real-time tracking and accident alert system. The paper provides a two-step verification process for preventing vehicle theft or stopping un-authorized people from driving the car, such as a person who does not have a driving license.

The first is the authentication part of our work. When the person who will drive enters the car and sits in the driver's seat, he/she has to verify his/her fingerprint. If the fingerprint is stored in the database, the authentication will be successful and a positive ID message will be displayed which will, in turn, make the servomotor rotate. The servomotor will have a metal (cardboard in our case) in its off position guarding the key access to the car engine. This provides a double layer of protection. One is the fingerprint sensor authentication, and the second is the fact that the driver needs to have the car key in his possession to start the car. If someone steals the key, since his/her fingerprint will not be on that car's fingerprint sensor, and the car key access will be denied [6].

The next part is the vehicle tracking. The real-time tracking will be done using the ThinkSpeak platform. In case of over-speeding, the car can be tracked using the number plate. The main station from where we will monitor will have a screen where the ID of a defaulter will appear as soon as he/she over-speeds. The ID can then be used by the personnel at the station to track the real-time location in a different ThinkSpeak window. We have used Hypertext Markup Language to make shareable document over any electronic device such as laptop, computer, mobiles and tabs which will be very useful in sharing intelligence of the whereabouts of the offender [7].

Using the.html document, one does not need to be in constant connection to the main center while going after the person at fault. The only criterion is that the device the second level of structure will use will have to have a Hypertext Markup Language reader to work properly. This will save precious time and keep everyone in the police monitoring force on their feet.

Fig. 1 Fingerprint car
security system [8]

Fingerprint Car System

The final part is the vibration sensor incorporation. The sensor alerts the authorities in scenarios of shake/jolt exceeding a certain limit. We can set a minimum limit on which the sensor will send the panic signal.

Overall, this is a product development work which will help us keep our streets safer, help us catch miscreants on roads and serve the people in need quicker than ever. It is a stand-alone system which needs to be integrated with the car to make it fully functional. Also, it is an open-ended work with possibilities galore and what more features can be added in it is restricted by our imagination alone. The block diagram of IoT-based vehicle tracking systems is shown in Fig. 1.

3 Design of the Scheme

The design approach to the model is given in the flowchart which is given in Fig. 2. We begin by taking in the fingerprint sensor input. The data of the print is sent to the Arduino Uno where the processing takes place. If the fingerprint matches, it means the authentication is successful, the flow moves down to the NodeMCU block else it goes back to the fingerprint sensor block which waits for another input to be verified. If the authentication is successful, it also activates the vibration sensor at the same time.

The vibration sensor input is also fed to the Arduino Uno which processes the data [9]. It matches the shake or jolt amplitude against the pre-defined user-fed value for providing alert. If the jolt is above the specified limit, then the signal or alert is send to NodeMCU, which receives the signal and sends it over the Internet. Final process which takes place on successful verification of the fingerprint is the opening of the car key access as the servomotor is turned on, and it makes a 180° turn.

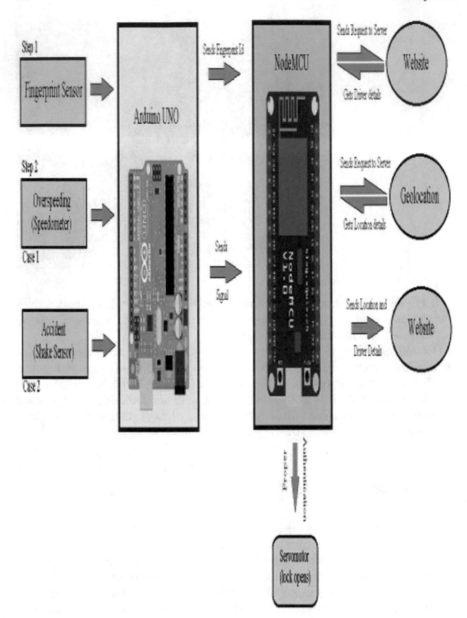

Fig. 2 IoT-based vehicle tracking system block diagram

We see hereby that three processes are set into action on authentication result of fingerprint sensor being positive. NodeMCU then gets the location details from the Internet, and in the next step it is checked whether an accident alert signal is received. If it is received, the MCU sends the driver ID, car number and the location details in the form of latitude and longitude coordinates with the alert to the corresponding Web site. The NodeMCU at the same time continues monitoring the speed of the vehicle too [10]. If the speed goes above a certain limit, the MCU again sends the details of the coordinates and drives and car ID to the Web site. This is done continuously, and hence, is a reliable system, and we can keep track of the car after it has over-sped at all times [11].

Also since GPS is not used when the car is inside a tunnel or under a shed or in a building, tracking in terms of over-speeding and accident is not an issue. The use of geo-satellite enables us to keep track in all conditions and under any circumstance.

4 Results and Discussion

The first step is the enrollment of fingerprint in the fingerprint sensor by the user of the vehicle. The user has to keep his/her finger on the sensor three times to enroll. This is done to increase sensing efficiency and enable it to predict results accurately.

Once the enrolling is successful, a message is displayed. The ID number is given on the first line and is accompanied by a hashtag. The exact display is shown in Fig. 3. In the next step after enrolling, we check the fingerprint against the ones stored in the sensor. If it is a match, a message is displayed saying: "Verified ID x." The servomotor rotates and lock opens allowing us to put in our car key and start the engine.

Till now, we have seen the first part of the work which deals with verification of identity. After authentication, the next step is to calculate speed. As we see in Fig. 4 that the speed is calculated continuously and if it goes above a certain limit, the message: "You over-sped!!!" is displayed [12]. Along with it, the ID is also shown, and the data is sent to the Web site via NodeMCU. For demonstration purposes, we have used the speed limit as 0.85 km/h in this work.

As seen from Fig. 4 after over-speeding, the location details are continuously sent to Web site. The ID and car number of the person who over-sped are displayed on the screen can be easily monitored. The latitude and longitude data of the over-sped car is sent continuously which allows real-time tracking. The location details in ThinkSpeak Web site are shown in Fig. 5 from where we can monitor using the car number from the earlier window.

The location details in ThinkSpeak Web site and Google maps are shown in Figs. 5 and 6. Therefore, whenever a car over-speeds, we can get the ID of the driver along with the car number; thereafter to locate the car, we use the car number to find out the location which is continuously being posted by our device in the aforementioned channel. So, the device provides complete security so that any over-speeding driver

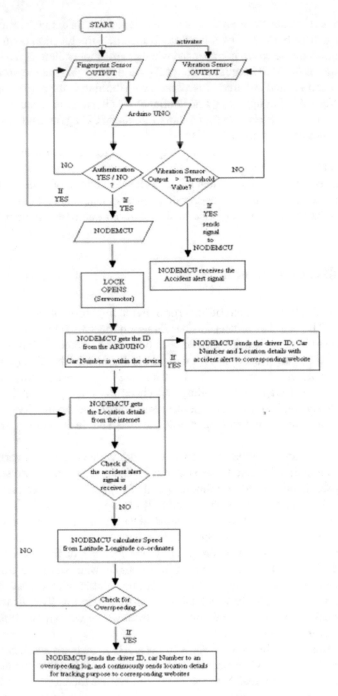

Fig. 3 Design approach for IoT-based vehicle tracking

Fig. 4 Monitoring

cannot go past the law and is immediately brought to justice by means of real-time tracking.

Even if the police fail to catch the defaulter readily, we have the ID of the driver which can be cross-referenced with the database of all personnel having driving license. Therefore, the defaulter can be tracked down, and the tracking system makes sure that he is always being followed and monitored by the authorities.

4.1 Results in Case of an Accident

The other part of the paper is providing immediate support to the driver who has faced an accident. For this purpose, we are using a vibration sensor which will continuously send the device an amplitude of vibration, for demo purposes we have used a threshold value of 18,000, and when that value is crossed the device will know an accident has occurred. In real life, for an accident to occur, the vibration output needs to be way bigger, but for demonstration purposes we are using a smaller threshold value. Given below are the values from vibration sensor in a continuous manner. The ID is being scanned firstly. Without the driver inside the vehicle, the vibration sensor will not send any alert. After the ID is scanned, the device knows that there is a driver inside and then only the alert system is turned on.

```
'□ □-□□□.
WiFi connected

Verified ID: 1
Lock opens!
Latitude - 0.000000
Longitude - 0.000000
Latitude - 12.973258
Longitude - 79.159118
Latitude = 12.973250          Latitude = 12.973243
Longitude = 79.159088         Longitude = 79.159068
Distance=3.400294             Latitude = 12.973256
Time=22148.00                 Longitude = 79.159119
Speed=0.552694                Distance=5.683426
Latitude = 12.973245          Time=23170.00
Longitude = 79.159137         Speed=0.883053
Latitude = 12.973259
Longitude = 79.159133
Distance=1.610098             You oversped!!!
Time=21809.00                 Your ID: 1
Speed=0.265778
```

Fig. 5 Output in serial monitor

An accident is detected as the car exceeds the threshold value; here, we took it as 18,000 units. Initially, the values were less than the threshold value, so it was just getting location coordinates and calculating the speed, but when the threshold value is crossed, the "Accident!" alert is initiated. The vibration sensor output is shown in Fig. 7.

It then sends the ID to the Accident log of the ThinkSpeak platform and keeps sending the location data to the corresponding car number. After accident, location details are continuously sent to the Web site, and this allows the authorities to take quick action and send help at the place of mishap.

No speed is being calculated here, as the accident protocol is activated, and it is directly sending the location details to the corresponding channel associated with the car number as shown in Fig. 8. Location details in Google maps are shown in Fig. 9. The demonstration is being performed in D-Annex block of VIT Men's hostels, and the location coordinates are being sent to the ThinkSpeak channel directly from the device. To make that fact clear, the aforementioned readings can be revisited. From the serial monitor window, we can clearly see two readings: (12.973254, 79.129121). In the results of Fig. 8 shown from the ThingSpeak channel, we can see the same reading as well as we can see that as the marker location in the Google Maps window.

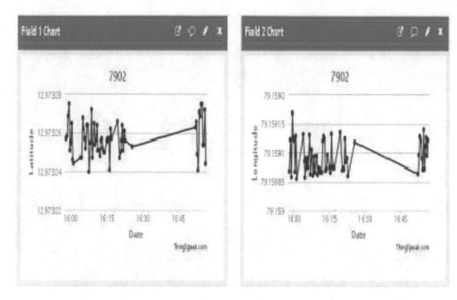

Fig. 6 Location details in ThingSpeak log

5 Constraints and Trade-off

In any research work or product which we make, schedule, cost and scope are the main constraints which we have to deal with. The quality of the output is affected by these factors, and hence, it is important to find significant balance in the desired results and trade-offs for these. In our paper, we have two main soft constraints which can be always removed with technological and hardware advancement.

5.1 Functional Constraint

There is a delay in response from the fingerprint sensor and the servomotor running and opening the lock. This delay can be attributed to the fact that Arduino Uno is verifying the fingerprint sensor identification stored in the memory and sending it to NodeMCU. The authentication and feedback of data take time accordingly to the type of sensor used and its response time to sense and give actionable output.

5.2 Sensory Design and Integration Constraint

This basically deals with the sensor response to physical sensing and the entire integration of the product with the car. The fingerprint sensor may give false readings

Fig. 7 Output in Google Maps

of negative ID when the finger of the user is smudged or too oily for it to detect the prints. The car space has to be enough around the steering wheel ideally for us to install the product and make it workable. The trade-offs are mentioned below.

While going for authentication if the fingerprint is not registered, then the access to the car will be denied which can be an issue in case of an emergency situation where a person who knows to drive but has not got a license has to drive.

The vibration sensor gives an alert when the input is above a certain limit of amplitude. The limit can be set by us during programming but there is where the issue arises. There is a chance that the car goes over a speed breaker at a good speed, and it shakes the car in the same way it might have shaken it during an actual accident or collision. Any other such incident can happen on the road, and it has to be worked upon.

```
Verified ID:1
Verified ID:1
Vibration detected - 17487
Vibration detected = 15979
Vibration detected = 101
Vibration detected - 3379
Vibration detected - 5353
Vibration detected = 2652
Vibration detected - 4144
Vibration detected - 4305
Vibration detected = 1086      Verified ID: 1
Vibration detected - 8693
Vibration detected - 1198       Lock opens!
Vibration detected = 1736       Latitude = 12.973247
Vibration detected - 1730
Vibration detected - 7395       Longitude = 79.159082
Vibration detected = 18017      Accident!
Vibration detected = 8192
Vibration detected - 1558       Latitude = 12.973263
Vibration detected = 2132       Longitude = 79.159064
```

Fig. 8 Vibration sensor output in serial monitor

6 Future Scope and Conclusion

The future scopes for the model are given below.

(A) We are not using GPS for getting speed details of a moving vehicle because the major disadvantage of using such a system is that the satellite signal will not be able to breach through tunnels, and therefore, it will not be effective in such situations [13, 14]. Hence, we are using the latitude–longitude coordinates to get speed details. After taking in the coordinates and finding the distance moved by it over a period of time, we are using simple mathematics of speed equals distance covered in the interval divided by the time period of the interval. This gives near precise reading but the calculations can be improved upon by studying geographic coordinates and deriving even more precise formulas.

(B) Instead of using Esp8266 alone and accompanying it with other Internet modules for communication, we have used NodeMCU which has not only reduced the cost of the research but has also made it less bulky and space efficient for installing in a car. More integration of technologies is a possibility in the future.

(C) The systems already in the market come with either accident alert and notification or validation of identity alone, but here we have brought both of these things together. As the need arises more, such methodologies can be incorporated in the times to come.

Fig. 9 Accident location in Google Maps

The advent of IoT and communication technologies translates the livelihood of the people and developing a promising system with these technologies expected to help the people to lead a safe life. The developed model aids to rescue people at a faster rate thus reducing the road fatality rate [15, 16].

References

1. Nasr, E., Kfoury, E., Khoury, D.: An IoT approach to vehicle accident detection, reporting and navigation. In: Multidisciplinary Conference on Engineering Technology, IEEE International (2016)
2. Raghav, A., Shanshanka, D., Sumukha Chandra, P.S., Tejas, D.C., Panda, S.K.: The safest key-smart key. In: Indian Educators' Conference (TIIEC), Texas Instruments (2013)
3. Rehman, M.M., Mou, J.R., Tara, K., Sarkar, M.I.: Real time google map and arduino based vehicle tracking system. In: Electrical, Computer and Telecommunication Engineering (ICECTE), International Conference, December 2016

4. Barco, A.A., Barahona, B.B., Arcentales, A.M., Vargas, W.V.: Geolocation and security perimeter for bicycle care within university area. IEEE Latin Am. Trans. **15**(6), 1137–1143 (2017)
5. Chetan, H.D., Hedge, P.V.: Real time traffic speed controlling, collision detection and vehicle tracking system. In: Smart Technologies For Smart Nation (SmartTechCon), International
   Conference (2017)
6. Sikri, V.: The Smart Number Plate. In: International Conference Communication and Signal
   Processing (ICCSP) (2016)
7. Xu, C., Chen, F.: Research on map-positioning technology based on W3C Geolocation API. In:
   2nd International Conference Consumer Electronics, Communication and Networks (CECNet)
   (2012)
8. http://teachrange.blogspot.com/2011/11/fingerprint-car-security-system.html
9. Al-Kuwari, M., Ramadan, A., Ismael, Y., Al-Sughair, L., Gastli, A., Benammar, M.: Smart-
   home automation using IOT-based sensing and monitoring platform. In: IEEE 12th International
   Conference on Compatibility, Power Electronics and Power Engineering (CPE-POWERING)
   (2018)
10. Sehwani, N.S., Sangle, S.R., Vadhavkar, Y.N.: Real time automobile tracking system with
    an automated security algorithm. In: International Conference on Communication and Signal
    Processing (ICCSP) (2017)
11. Fiaz, A.B., Imtiaj, A., Chowdhury, M.: Smart vehicle accident detection and alarming system
    using a smartphone. In: 2015 International Conference Computer and Information Engineering
    (ICCIE) (2015)
12. Kumar, V.P., Rajesh, K., Ganesh, M., Kumar, I.R.P., Dubey, S.: Overspeeding and rash driving
    vehicle detection system. In: Texas Instruments India Educators' Conference (TIIEC) (2014)
13. Khan, Z.: Wireless speed monitoring system using GNSS technology. In: Fifth International
    Conference on Aerospace Science & Engineering (ICASE) (2017)
14. Amin, M.S., Bhuiyan, M.A.S., Reaz, M.B.I., Nasir, S.S.: GPS and Map matching based vehi-
    cle accident detection system. In: IEEE Student Conference on Research and Development
    (SCOReD), Putrajaya, pp. 520–523 (2013)
15. Government of India. (2014, April 30) (Online). Available: https://data.gov.in/catalog/total-
    number-persons-killed-road-accidents-india
16. Balaji, S., Nathani, K., Santhakumar, R.: IoT technology, applications and challenges: a con-
    temporary survey. In: Wireless personal communications, 1–26 (2019)

Continuous Monitoring of Electricity Energy Meter Using IoT

R. Narmadha, Immanuel Rajkumar, R. Sumithra and R. Steffi

Abstract Nowadays, electrical energy stealing becomes customary which leads to many theft influences, causes abnormality including voluntary tariff globally. Impact of stealing electricity results distribution losses and it is required to accuse extra charges to customers. To overcome this problem, data acquisition of consumed power from the residential places in terms of voltage and current is retrieved and stored in the handheld device while loading the data in cloud. Internet of things (IoT) with energy meter interpretation system has been premeditated to record the energy meter reading continuously. Apart from the automation of electric bill charging for stipulated time period, it also enhances charging accountability, if there is any manipulation error. If there is any controversies were found in the energy meter reading, mismatch in the electric bill tariff for the consumed power, this information can be fed back to the consumer. In addition, notification of power tapping is sent to the cloud, reverted back to the mail, and intimated in the mobile alert message as well.

Keywords Energy meter · Power consumption · IoT · Arduino · Raspberry Pi

1 Introduction

This paper describes the significance of introducing smart electricity meters in developing countries [10–13]. An electricity meter is a maneuver castoff for gauging the amount of utilized electrical energy for a profitable small-scale industries or a residential building [4]. Owing to the cumulative cost of electricity, safety and interfering in electric meters take a main concern to the government agencies over the world in the related papers [1–3]. Stealing of power from the prohibited areas like military

R. Narmadha · I. Rajkumar · R. Sumithra (✉) · R. Steffi
Sathyabama Institute of Science and Technology, Chennai, India
e-mail: sumithraer98@gmail.com

R. Narmadha
e-mail: narmadhar2014om@gmail.com

I. Rajkumar
e-mail: imman047@gmail.com

© Springer Nature Singapore Pte Ltd. 2020
K. N. Das et al. (eds.), *Soft Computing for Problem Solving*,
Advances in Intelligent Systems and Computing 1057,
https://doi.org/10.1007/978-981-15-0184-5_62

armed force station, navy power station, air force power station comes under electricity stealing in the related papers [5–7]. Theft of electricity is the criminal practice of stealing electrical power. It is a crime and is punishable by fines and/or incarceration. It belongs to the losses. The prevailing wireless communication system of energy meter has been implemented using Zigbee, relay control, and GPRS in this paper [8–16]. The flow arrangement of electrical vitality charging is time consuming. Mistakes presented at each stage are reflected in electro-mechanical meters, human blunders while taking note of down the meter perusing It lessens the sending of labor for taking meter readings. To understand the difference between Arduino and Raspberry pi, Arduino is a simple microcontroller [9], and it cannot be supported with the operating system of a simple PC, rather Raspberry Pi, a fully functional mini-computer, equally cheap solutions for attaching the Internet of Things.

1.1 System Description

In this proposed project, IoT perception with electricity meter interruption has been used to eradicate hominid connection in energy conservation. The major components used in the system are Raspberry Pi, Arduino, ZMPT voltage sensor, ACS712 current sensor, and energy meters inclusive of submeter, transmit driver circuit along with display. The real-time appraisals from the signal pulses are composed through the microcontroller. This could be pragmatic over a demonstration which remains associated to the controller. From this demonstration, stealing status of the interpreted meter can be avoided. Throughout usual process, interpretation of the core energy pulse of the signal must be identical to the amount of interpretations of deputize energy meters.

Meanwhile, by means of only one representation of energy meter, the interpretation of both core and deputize energy meters should be identical. To interpret this system, in practical, four modules are provided. Ambiguity in the interpretations of both the meter reading and control room is identical, and the communication "NO THEFT" message will be shown in the LCD display. After energy theft happens, assembly remains commissioned from the spread line preceding to deputize energy meter again. Determination of discrepancies among the core and deputize energy meter reading "THEFT OCCURRED" exhibited in the display. Before the interconnection of circuit activates, load can be detached from the main supply.

The projected scheme eradicates the hominoid contribution in energy maintenance. Similarly, it eliminates imperfect and unproductive charges. IoT helps in associating short message services and nursing of core meters at inferior cost. Energy meter maintenance were checked through web site gateway.

The main aim of this work is to analyse complete monitoring and consumption of power using smart electricity meter. The following flowchart explains the overview of the system (Fig. 1).

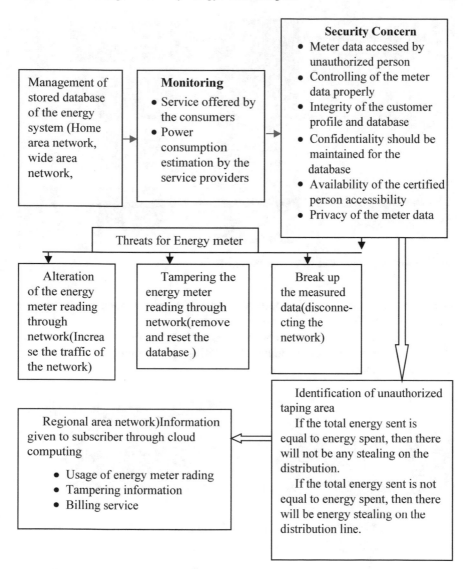

Fig. 1 Overview of system description

2 Theory Behind Energy Meter

Progressive metering infrastructure uses one-way communication capable of controlling load data remotely in a captured device (Fig. 2).

In the proposed system, current and voltage sensors reading were interfaced with Arduino. Arduino and Raspberry Pi both remain connected through an USB cable

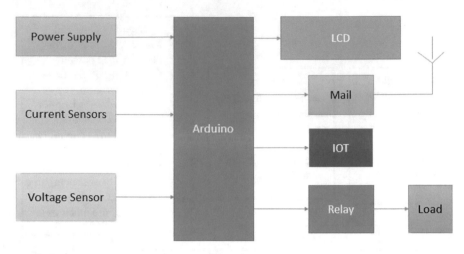

Fig. 2 Proposed system—core energy system with IoT

directly. The communication between Arduino UNO and computer is a serial communication (Fig. 3).

From the main supply, the connection holder was connected parallelly to the socket connection. From the Arduino, the readings are recorded and uploaded for the application purpose.

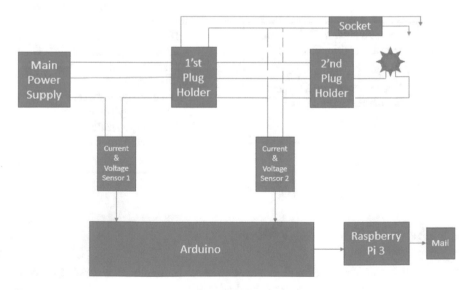

Fig. 3 Connections and block diagram of proposed system

2.1 Framework for Smart Metering

Arduino is not an entirely useful computer; It needs to be connected to a computer first to program it. Nevertheless, at half size of the Raspberry Pi, it is a minor and inconspicuous sensor. To control the power distributed to a load, pulse width modulation (PWM) signals triggered through a servomotor speed control.

The physical link is ended through the Raspberry connector. In addition, Raspberry Pi is the GPIO pin array. The reset (RST) to GPIO serves to reset the Arduino.

Arduino could reboot itself without push the reset button or watchdog to reset the board or using the watchdog timer. The Arduino programming can yield via the IDE for personal computer, adapted to switch an Arduino associated to the network by pretending a remote port.

The global smart electricity meter is expected to provide economic and environmental benefits for all users. The security features were added and implemented. And also, it includes quality of service, high efficiency, and installation cost has reduced.

Meter deployment includes installation of SIM configuration, collection of data, recovery of data, management techniques, and the optimization of the network used. Theft detection consists of integration of data, end to end security, and management process.

3 Prototype Model for Theft Detection System

As the consequence of the proposed framework, the smart energy meter with theft detection system fundamentally works for distinguishing energy sent and energy consumed. If there is conflict between these values happens, within the short period of time, message is displayed on the LCD screen and parallelly informed to the consumer. The framework works for energy stealing and furthermore enables client to utilize smart vitality meter usefulness. Utilizing this usefulness, client will be ready to see all the stealing messages with photos of that place on their mail IDs given to the fundamental station, and they can utilize the definite measure of energy consumption. The framework begins with IoT module connection, first designs the client mail Id, and offers specialist to that Id. The framework is comprised of Arduino, Raspberry Pi, current sensors, voltage sensor, LCD show, hand-off, and stacks (Fig. 4).

In this prototype model, the normal unit power which consumes the normal power supply given by the energy meter through the LCD display. The displayed value of the power, voltage and current readings should be equal for input as well as output. Arduino, Raspberry Pi, current sensor, and voltage sensor were connected to the main supply. The power consumption reading were displayed using LCD screen (Fig. 5).

Proposed system shows Fig. 6, IP power value, and OP power value in amperes, and it is a real-time value.

Table 1 shows theft detection truth table and status is by the unit of ON and OFF. Also, the status is indicated by REGULAR or THEFT NOTIFICATION (Fig. 7).

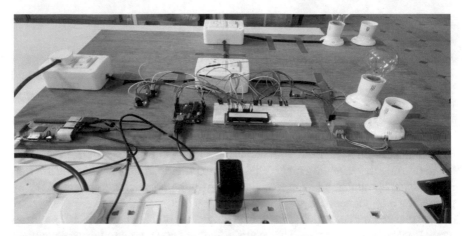

Fig. 4 Bench set up to the power supply of main system

Fig. 5 Line 1 connected to the main supply

Fig. 6 Proposed system model with LCD display

Table 1 Theft detection truth table

Normal unit	Theft detection Unit	Status
OFF	OFF	REGULAR
ON	OFF	REGULAR
OFF	ON	THEFT NOTICED
ON	ON	THEFT NOTICED

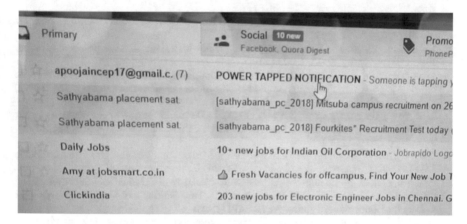

Fig. 7 Mail notification

Mail notification is given by the alert message of POWER TAPPED NOTIFICA-TION. The IP and OP power value are also messaged in the mail notification.

4 Conclusion

Stealing of electricity is a difficult and complicated issue in power grid. The improvement of automatic and smart meter increases more complex situation, and it leads to stealing of energy. Many new technologies were developed with GSM, IoT, etc. Energy theft computation is reliant on the calculation of pulses and utilizing Arduino, Raspberry Pi, current sensor, and voltage sensor in framework space. In future, using power line communication, controlling and management of energy consumption were designed, and it is a very fruitful and challenging area.

References

1. Department for Business Energy and Industrial Strategy. "Smart Meters, Quarterly Report to end December 2016, Great Britain," Tech. Rep. (2017)
2. Mohassel, R.R., Fung, A., Mohammadi, F., Raahemifar, K.: A survey on advanced metering infrastructure. Int. J. Electr. Power Energy Syst. **63**, 473–484 (2014)
3. Yang, J., Zhao, J., Luo, F., Wen, F., Dong, Z.Y.: Decision-making for electricity retailers: brief survey. IEEE Trans. Smart Grid (99), 1–1 (2017)
4. Liu, X., Heller, A., Nielsen, P.S.: CITIES data: a smart city data management framework. Knowl. Inf. Syst. **53**(3), 699–722 (2017)
5. Smith, T.B.: Electricity Theft: A Comparative Analysis Energy Policy, vol. 32, Issue 18, pp. 2067–2076 Publisher: science direct (2004)
6. Chim, T.W., Yiu, S.-M., Hui, L.C., Li, V.O.: Privacy preserving advance power reservation. IEEE Commun. Mag. **50**(8), 18–23 (2012)
7. Luan, S.W., Teng, J.H., Chan, S.Y., Hwang, L.C.: Development of a smart power meter for AMI based on ZigBee communication. In: Power Electronics Drive Systems, PEDS 2009, Taiwan, pp. 661–665 (2010)
8. Luan, S.W., Teng, J.H., Chan, S.Y., Hwang, L.C.: Development of a smart power meter for AMI based on ZigBee communication. In: Power Electronics and Drive Systems, PEDS 2009, Taiwan, pp. 661–665 (2009)
9. Tao, W., Zhang, Q., Cui, B.: The Design of Energy Management Terminal Unit based on double MSP430 MCU, Electricity Distribution CICED 2008, China, pp. 1–4 (2008)
10. Skarman, S.E., Georgiopoulos, M., Gonzalez, A.J.: Short-term electrical load forecasting using a fuzzy art map neural network. In: Aerospace/Defense Sensing and Controls. International Society for Optics and Photonics, pp. 181–191 (1998)
11. Rottondi, C., Verticale, G., Capone, A.: Privacy preserving smart metering with multiple data consumers. Comput. Netw. **57**(7), 1699–1713 (2013)
12. Molina-Markham, A., Shenoy, P., Fu, K., Cecchet, E., Irwin, D.: Private memoirs of a smart meter. In: Proceedings of the 2nd ACM Workshop on Embedded Sensing Systems for Energy-Efficiency in Building, pp. 61–66. ACM (2010)
13. Pyasi, A., Verma, V.: Improvement in electricity distribution efficiency to mitigate pollution IEEE ISEE. In: Proceedings of the IEEE International Symposium on Electronics and the Environment, San Francisco, California, pp. 1–1 (2008)
14. Depuru, S., Wang, L., Devabhaktuni, V.: Support vector machine based data classification for detection of electricity theft. In: IEEE/PES Power Systems Conference and Exposition, pp. 1–8 (2011)
15. Depuru, S.S.S.R., Wang, L., Devabhaktuni, V., Gudi, N.: Measures and setbacks for controlling electricity theft. In: IEEE North American Power Symposium, pp. 1–8 (2010)
16. Berthier, R., Sanders, W.H.: Specification-based intrusion detection for advanced metering infrastructures. In: IEEE Pacific Rim International Symposium on Dependable Computing, pp. 184–193 (2011)

Frequent Item Set Mining of Large Datasets Using CUDA Computing

Peddi Karthik and J. Saira Banu

Abstract Frequent item set mining is a very popular method in data mining and is used extensively to find out the most recurring items in mainly market basket analysis. It is commonly used for association rule learning. The market basket analysis is seen to be used in many fields like Web mining and intrusion detection. There are many algorithms like Apriori algorithm, etc., to find out the frequent items in given data but these are more of a sequential approach and take huge amounts of time for large sets of data. Existing Soft computing-based approaches for solving frequent item set mining like genetic algorithm and fuzzy logic systems are proven to reduce execution time but not in a scale in which a massively parallel system does. So the main objective of this paper is to accelerate frequent item set mining process using GPU's CUDA architecture. We have also performed a comparative study with parallel version of frequent item set mining using openMP. Our results show speedup of 2.2 for CUDA over serial implemented using genetic algorithm and 1.8 for CUDA over OpenMP.

Keywords GPU Aprori · CUDA Ariori · Open MP · Bitmap representation · Genetic algorithm · Item set mining · Performance comparison

1 Introduction

Frequent item set mining is the process of filtering out patterns from the collected history of user purchases also called as transactions. This is done based on the support threshold and the confidence threshold. Some of the most popular algorithms used in this case are Apriori algorithm where the item sets which are repeated in the transactions greater than the given support are taken and are called as frequent item sets. In the Apriori algorithm, once the frequent item sets are found out, they are taken

P. Karthik · J. Saira Banu (✉)
Vellore Institute of Technology, Vellore, India
e-mail: jsairabanu@vit.ac.in

P. Karthik
e-mail: peddi.karthik2016@vitstudent.ac.in

© Springer Nature Singapore Pte Ltd. 2020   739
K. N. Das et al. (eds.), *Soft Computing for Problem Solving*,
Advances in Intelligent Systems and Computing 1057,
https://doi.org/10.1007/978-981-15-0184-5_63

and based on the given confidence threshold, they are used to find out association rules which are used to describe which items are more likely to be picked together in a single transaction. So this pattern discovery can be used to link two or more similar products together, to increase sales.

The Apriori algorithm is highly parallelizable due to its simple method and representation of data which can be seen in the algorithm in Sect. 4. This high parallelizability can be exploited by using a popular GPU-based platform called CUDA which was brought into picture by NVIDIA. Many serial soft computing-based approaches like genetic algorithm for frequent item set mining are successful in reducing execution time, but a more efficient approach to increase the performance would be to parallelize this process using CUDA computing [1]. CUDA in its core is fundamentally an interface or platform that allows users to hide the GPU hardware from view. CUDA essentially is an extension of the C, C++, and OpenCL libraries to support parallel computing by the use of GPU hardware. The kernel is the main part of the program that runs on the GPU and is parallelized by using threads and blocks in the GPU. The kernel function runs in parallel in the CUDA cores of the GPU. Since the number of CUDA cores in GPUs is very high, we can achieve very high levels of parallelism.

In this paper, we have implemented the Apriori algorithm to utilize the CUDA cores and we have also performed a comparative study of the speedup achieved by implementing the Apriori algorithm in CUDA over OpenMP version and serial version of Apriori using genetic algorithm.

The ordering of the sections of the paper is as follows:

Section 2 of the paper consists of the Literature review, then the proposed methodology is explained in Sect. 3 followed by the results, evaluation in Sect. 4 and conclusion in Sect. 5.

2 Literature Review

Zhang et al. in [2] have implemented a GPU-based frequent item set mining and market basket analysis on an NVidia Tesla T10 GPU. They have used a static biset memory structure for the transactional database. They have used complete intersection to load the matrix into the GPU's memory beforehand which when compared to equivalent class clustering has improved the performance due to the decrease in the latency to access memory. CUDA computation is done on different threads to perform tree traversals to search item sets. Like so they have found out the support count and based on the previously set threshold were able to come up with the frequent item sets.

Chon et al. in [3] and Djenouri et al. in [4] have implemented a fast GPU-based frequent item set mining method for finding patterns in large-scale data. They also try to find a solution to the workload skewness problem faced in maximum of the algorithms. Their results show multiple times performance increase. In [4], two methods namely GA-Apriori and PSO-Apriori were proposed. They found out that using vertical bitmaps is more efficient on GPUs than on CPUs. They presented a new algorithm namely MGPUCPM which supports multiple GPU architectures and also provides a memory-aware solution.

Li et al. in [5] have found out frequent item sets by using a vertical data structure that can contribute to the aspect of saving storage by using a concept called as a multi-layer index. Their algorithm is shown to have using much less time than other algorithms. Their algorithm MVCG is based on directed item set graph.

Li in [6] considers the problem of mining frequent item sets in an incremental manner. Their method employs CUDA architecture to increase the mining speed exponentially by representing frequent item sets using bitmap representation. They also used an inverse tree in order to prune more effectively and efficiently. Their experimental results show that the algorithm they proposed has achieved better performance when running times were compared.

The paper by Rathi [7] concentrates mainly on the methods for the preprocessing of the transactional database. Deserialization phase and the sorting phase were also focused in order to make the dataset easier to work with. To achieve scalability and handle large datasets, they have used graphics processor in collaboration with CPU. Their experimental observations have shown a huge amount of performance increase when dealing with large amounts of data.

Zois et al. in [8] perform the support counting process on a GPU with the help of a highly parallel and work-efficient algorithm .They have developed a less memory space-consuming data layout scheme that makes it possible for high chip-memory bandwidth usage. They were seen to have achieved maximum throughput for the implementation of their parallel two-phase algorithm, while is faster than that of a non-work effective methods on a multi-core CPUs and a GPUs by almost 40 times.

The paper [9] presents extension of one GSP algorithm using GPUs, minor hardware and software tweaks were made to implement the algorithm in GPU. Due to these minor changes made the authors have evaluated the implementation on a "CUDA parallel-computing system" and also on a Tesla card. Solution proposed in the paper uses GSP algorithm which is a part of the Apriori family of algorithms, it does a number of scans on the dataset to calculate the value of support for the item sets, making it well suited for finding the solutions to real-time "GPGPU" problems for very large datasets.

In the paper [10], a parallel FIM algorithm named APFMS, which stands for "Accelerating Parallel Frequent Item set Mining on Graphic Processors with Sorting" (APFMS), has been discussed. The authors of the paper implemented the algorithm on the newest generation of GPUs available at the time of publication for better results. In the paper, more than one GPUs were together used to speed up the frequent item method of verification on the OpenCL platform. The results that were obtained from the implementation clearly demonstrated that the proposed algorithm had drastically reduced computation time when compared with previous methods.

It can be noticed through the literature review that the three most common frequent item set mining algorithms are Apriori, FP-Growth, and Eclat. Each of them has their own advantages and disadvantages. FP-Growth uses a complex trie-based structure which makes it harder to reduce the amount of memory used by the data structure.

Since the Apriori algorithm consumes less memory space to store the transactional database, we have chosen Apriori algorithm. To store the transactional database in less memory space we are using int4 representation format. From the literature review, we can see that comparative study of CUDA vs OpenMP speedup was not done anywhere so we focus on implementing the algorithm in Apriori and then comparing the speedup of those two implementations.

3 Proposed Work

The algorithm used in this paper is divided into two parts namely candidate generation and item set frequency calculation. In this paper, we compared serial, OpenMP, and GPU computation times for calculating frequent item sets (of size 1, 2) using Apriori algorithm.

3.1 Candidate Generation Phase

This part of the system is done in the CPU itself due to its extreme non parallelizability. The code takes the frequent 1 item sets found out by the CUDA kernel and then finds the item sets set C2 by taking the combinations of item sets in L1 (frequent 1 item set) found out by CUDA kernel. This phase is called candidate generation.

3.2 Transactional Data Representation

The transactional data is represented using a bitmap representation where the rows are the items and the columns represent the transactions in which the item exists. We have used a representation where the amount of memory required for storing the transactional database is decreased by 4 times, i.e.., instead of using an integer array which stores the transactional database as shown in Fig. 1 , we used a format called int4 where each item row is represented using int4 numbers. The int4 consists of 4 bytes or 32 bits that means each chunk can store transaction data of 32 transactions, if the number of transactions is N then the number of int4 format required to store the transactional database is ceil(N/32).Then the last bits are padded with zeroes. For example, a transactional database can be represented in bitmap representation as shown in Fig. 1.

Fig. 1 Transactional database representation

TID	ITEMS
T100	{I1, I3, I4}
T200	{I2, I5}
T300	{I3, I5}
T400	{I2, I3, I5}
T500	{I1, I4}

$$\begin{pmatrix} 1 & 0 & 1 & 1 & 0 \\ 0 & 1 & 0 & 0 & 1 \\ 0 & 0 & 1 & 0 & 1 \\ 0 & 1 & 1 & 0 & 1 \\ 1 & 0 & 0 & 1 & 0 \end{pmatrix}$$

3.3 Parallel Frequency Calculation

This part of the system is run in the GPU CUDA cores where each thread takes an item and calculates its frequency by using the bitmap representation and adding the items of column for 1-item set count and multiplying two item set columns and adding for 2-item set count.

3.4 Generating Frequent Item Sets

This is done in the CPU where the C1 and C2 are taken and based on the given support count threshold the sets L1 and L2 are formed. By checking the condition if count(C1[i])>=support.

3.5 Datasets

The following datasets are considered for evaluation of the results of the proposed implementation of Apriori in CUDA environment: Chess, Mushrooms, Connect, T40I10D100K, T10I4D100K, and Kosarak. They have been taken from [11].

3.6 Algorithm

The Apriori algorithm is given below:

Step 1: Assign the set of item sets with one element to F1:
Step 2: Set the value of $F_2 = \emptyset$
Step 3: For every item I in F_{j-1}
Step 4: For every L in F_1
Step 5: $A = I \cup L$
Step 6: If the support value Support(A) is greater than given threshold
Step 7: Add the item set A to F_j
Step 8: if F_j is empty break
Step 9: return $\cup_j F_k$

3.7   Areas of Parallelism

The following areas are parallelized:

- The process of finding whether an item (initially during frequent item set mining of item sets of size 1) is frequent or not is done by different threads in the different CUDA cores of the GPU.
- The kernel in the CUDA programming counts the frequency of the item sets by summing the bitmap representation. The kernel that is used to calculate the frequency of 2-sized item sets takes to frequent 1 item sets bit map and multiplies them and then follows the same process of counting of frequent item set mining of size one.
- The process of checking whether the count is greater than the support is done by sending the array of counts to a kernel which is also given the support value and it checks if the corresponding element in the thread Id position of the passed array is greater than or equal to that of the value of support.
- In the OpenMP implementation of the algorithm, the parts of the code like for loops which calculate the count of the item sets and multiplication of the vectors of two item sets to find the equivalent item set vector is done in parallel using OpenMP constructs.

4   Results

Figure 2 shows the properties of the datasets taken for evaluation of the three implementations namely CUDA, OpenMP, and serial.

Dataset	Density	No. of items	No of total entries	No. of transactions	Characteristics	Data Size
Chess	49%	75	118252	7,196	Dense/	~340 KB
Mushrooms	19%	119	186852	8,124	Sparse/ Real	~570 KB
Connect	33%	129	2904951	67,557	Sparse/ Real	~9.5 MB
T40I10D100K	4%	1000	3960507	1,00,000	Highly sparse/ Synthetic	~15 MB
T10I4D100K	1%	1000	1010228	1,00,000	Highly sparse/ Synthetic	~4 MB
Kosarak	1%	41270	8019015	9,90,002	Highly Sparse/ Synthetic	~32 MB

Fig. 2 Properties of datasets used

Figure 3 depicts the execution times of the serial, OpenMP, and CUDA versions of the code in a scatter plot. It can be clearly seen that as the density of the database increases the time of execution in the GPU also increases.

It is observed that the time of execution in serial program which implements genetic algorithm version of the Apriori is approximately 2.2 times that of the parallel version that is executed in the GPU.

Figure 4 depicts the speedup that was observed when comparing the results for different datasets.

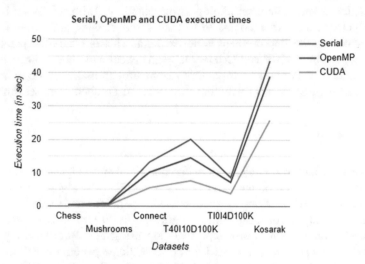

Fig. 3 Execution time comparison of serial versus OpenMP versus CUDA

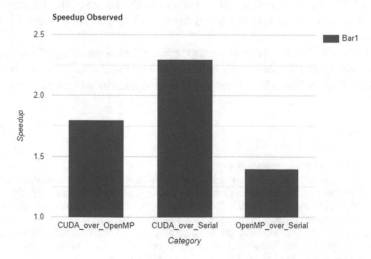

Fig. 4 Speedup observed

It is also observed that if the number of items in a database is more and comparative of the size of the number of transactions the time taken is more since the GPU parallelism is done on the vertical representation this can be observed in the case of Kosarak dataset.

It can also be observed that in the Kosarak dataset result since the number of items is 40,000 the time taken for serial execution is no more approximately 2.2 times that of the parallel execution so it can be stated that the time for parallel execution has increased.

It can be inferred from the results that even though soft computing-based approaches like frequent item set mining using genetic algorithm [12] are better to increase performance when compared to their serial counterparts, when it comes to large datasets, the improvement is not so high. This is where the parallel implementations in OpenMP and CUDA platforms excel and show a large increase in performance. This observation is very resourceful in making a decision when building hardware platforms solely for the purpose of Frequent Item set Mining.

5 Conclusion

In this paper, we have proposed a system that runs the Apriori algorithm on the given input dataset and does it on the CUDA cores of a GPU. The algorithm utilizes the parallelism of a GPU by taking the parallelizable operations and executing them on the GPU. The purpose of using Apriori algorithm to do frequent item set mining in this case with vertical bitmap representation using int4 Boolean array is that this algorithm is executed in the GPU for both candidate generation and frequency counting whereas other algorithm such as Trie-based FP-Growth and Eclat algorithm are CPU/GPU-based algorithms. So not only did the memory size decrease when using this algorithm but also the speed of execution was comparable with other algorithms.

In conclusion, using a vertical bitmap representation made up of int4 integers which is essentially a Boolean array increases the memory efficiency by 8 times and using Apriori algorithm also reduces the execution time by a significant amount.

References

1. https://developer.nvidia.com/cuda-faq
2. Zhang, F., Zhang, Y., Bakos, J.: Gpapriori: Gpu-accelerated frequent itemset mining. In 2011 IEEE International Conference on Cluster Computing, pp. 590–594. IEEE, Sep 2011
3. Chon, K.W., Hwang, S.H., Kim, M.S.: GMiner: a fast GPU-based frequent itemset mining method for large-scale data. Inform. Sci. **439**, 19–38 (2018)
4. Djenouri, Y., Comuzzi, M.: GA-Apriori: combining Apriori heuristic and genetic algorithms for solving the frequent itemsets mining problem. In: Pacific-Asia Conference on Knowledge Discovery and Data Mining, pp. 138–148. Springer, Cham (2017)
5. Li, Y., Xu, J., Chen, L.: A new closed frequent itemsets mining algorithm based on GPU. In: 2015 3rd International Conference on Advanced Cloud and Big Data (CBD 2015) (2016)

6. Li, H.: A GPU-based maximal frequent itemsets mining algorithm over stream. In: 2010 International Conference on Computer and Communication Technologies in Agriculture Engineering, vol. 1, pp. 289–292). IEEE, June 2010

7. Rathi, S., Dhote, C. A., Bangera, V.: Speeding up frequent itemset mining process on XML data using graphic processor. In: 2014 5th International Conference-Confluence the Next Generation Information Technology Summit (Confluence), pp. 206–209. IEEE, Sep 2014

8. Zois, V., Panangadan, A., Prasanna, V.: Accelerating support count for association rule mining on GPUs. In: 2016 IEEE International Parallel and Distributed Processing Symposium Workshops (IPDPSW), pp. 1423–1432. IEEE, May 2016

9. Hryniów, K.: Parallel pattern mining on graphics processing units. In: Proceedings of the 14th International Carpathian Control Conference (ICCC), pp. 134–139. IEEE, May 2013

10. Huang, Y.S., Yu, K.M., Zhou, L.W., Hsu, C.H., Liu, S.H.: Accelerating parallel frequent itemset mining on graphics processors with sorting. In: IFIP International Conference on Network and Parallel Computing, pp. 245–256. Springer, Berlin (2013)

11. http://fimi.ua.ac.be/data

12. Ghosh, S., Biswas, S., Sarkar, D., Sarkar, P.P.: Mining frequent itemsets using genetic algorithm (2010). arXiv preprint arXiv:1011.0328

Analysis of Road Networks Using the Louvian Community Detection Algorithm

R. Rashmi, Shivani Champawat, G. Varun Teja and K. Lavanya

Abstract In today's world, the population is increasing rapidly that make people move in and out of the countries/states for various reasons. With the evolution of technology, there is an advancement in transportation, and traveling across places has become easier than before. Now, we have various means of transport to move around the world. Though the most preferred way to travel is the roadways, there are roads built that connect cities and states together. But we also have a few disadvantages like increase in the vehicle registrations, amount of pollution, and the number of road accidents. This paper addresses an issue related to the traffic congestion caused due to the vehicles and analyzes the traffic at a particular area based on the threshold using the Louvain community detection algorithm. The Louvain algorithm is a graph algorithm used to detect the communities within a particular region and then form clusters. The algorithm is implemented on a US road network in Neo4j to detect the traffic and provide an alternative route for conveyance.

Keywords Clusters · Communities · Louvain algorithm · Neo4j · Road network

1 Introduction

The US street network [1] surpasses 6.58 million kilometers in all out length; it is the world's biggest street arrange pursued by China and India. The US road organize involves 4.3 million kilometers of cleared streets including 76,334 km of turnpikes

R. Rashmi · S. Champawat · G. Varun Teja · K. Lavanya (✉)
School of Computer Science and Engineering, VIT University, Vellore, India
e-mail: lavanya.k@vit.ac.in

R. Rashmi
e-mail: rashmi.r2018@vitstudent.ac.in

S. Champawat
e-mail: shivani.champawat2018@vitstudent.ac.in

G. Varun Teja
e-mail: gvarun.teja2018@vitstudent.ac.in

© Springer Nature Singapore Pte Ltd. 2020
K. N. Das et al. (eds.), *Soft Computing for Problem Solving*,
Advances in Intelligent Systems and Computing 1057,
https://doi.org/10.1007/978-981-15-0184-5_64

and 2.28 million kilometers of unpaved streets. It incorporates a large number of
the world's longest interstates. There are three classes of street: interstate through-
ways, US numbered highways, and state Highways Route 20 (US 20), an east–west
expressway estimating 5415 km long, is the longest single street in the nation and one
of the world's longest parkways. China has the world's second greatest street arrange,
which surpasses 4.24 million kilometers. National interstates and common roadways
separately involve 4 and 7% of the Chinese street arrange. The nation's interstate
system, which reaches out more than 96,000 km, is the world greatest system of this
type. Road network dataset has three attributes start node, end node, and distance.
Start node and end node are the nodes and distance is the relationship. The nodes represent the roads and their relationship between these nodes is established as one road
is connected to the other. Communities are groups of nodes within a network that are
more densely connected to one another than to other nodes. Modularity is a metric
that was proposed by Newman and Girvan that quantifies the quality of a community
assignment by measuring how much denser, and the connections are within communities compared to what they would be in a particular type of random network. The
Louvain method of community detection is an algorithm for detecting communities
in networks that relies upon a heuristic for maximizing the modularity. The method
consists of repeated application of two steps. The first step is a "greedy" assignment
of nodes to communities, favoring local optimizations of modularity. The second step
is the definition of a new coarse-grained network in terms of the communities found
in the first step. These two steps are repeated until no further modularity-increasing
reassignments of communities are possible. The Louvain method achieves modularity comparable to preexisting algorithms, typically in less time, and so, it enables the
study of much larger networks. It also generally reveals a hierarchy of communities
at different scales, and this hierarchical perspective can be useful for understanding
the global functioning of a network.

2 Literature Survey

Generalized Louvain method for community detection in large networks discovers communities possibly large networks using network modularity optimization
approach. Based on the k-paths, the algorithm exploits novel centrality. Ranking of
nodes is computed efficiently using edge ranking technique in near-linear time in
large networks. Central ranking is deliberated; then, the pairwise proximity is calculated between nodes of the network. The community structure is discovered by
efficiently maximizing the network modularity [2]. Graph database: A survey deals
with the evolution of graph database and their significance in current computing
environment. It also essays the details of the current graph databases and their comparisons which give the idea of current scenarios in happening in data modeling
using graphs. This paper gives an overview of the different types of graph databases,
applications, and comparison between their models based on some properties [3].
Tensor-based document retrieval over Neo4j with an application to PubMed mining

explains about the PubMed which is the largest online open-source database which includes document related to life science and biomedical research under the authorization of NIH. It contains near 40 million abstracts so far. For that reason, there need a perfect method which can execute document analysis and extraction. For that, text-mining methodologies are introduced with PubMed which execute traditional document—term matrix representation, an architecture for content-based retrieval implemented with the python and Neo4j [4]. Tensor fusion of social structural and functional analytics over Neo4j provides harmonic centrality with the structural ranking presentation, which can execute easily in sorted time. A directed graph ranking is utilized for vertices to get prioritized. And if the graph contains large number of vertices and nodes, then it is a big challenge to design algorithm and implement for vertices ranking [5]. Storage and parallel topology processing of the power network based on Neo4j illustrates that the traditional technique for data storage and management in power network data system is very hard. With the development of technology, it is getting more complex. The GRAPH-CIM model which is based on Hadoop architecture processes the big data of power network topology processing, which will make data robust and parallel usable [6]. Smart RDF Data storage in graph databases introduces a new methodology to convert RDF to graph databases. Networking enables to form large number of connected datasets. With the growing number of nodes every day, graph databases are effective and efficient solution to represent these datasets. Hence, many complex and connected datasets are converting to graph database to improve performance [7]. Community detection through likelihood optimization: in search of a sound model deals with community detection in network analysis that provides an extensive theoretical and empirical analysis to compare several models. The stochastic block model, which is currently the most widely used model for likelihood optimization, is developed and compared with two likelihood optimization algorithms suitable for the null models under consideration [8].

3 Methodology

3.1 Framework

The clustering roads inside the database are based on the interconnections with other roads. There are two sources of data that can be used when clustering a graph, node properties, and graph structure (i.e., interactions between the nodes). In most clustering algorithms, only one of these sources is used. For example, node properties in a road network could be distance information, while graph structure could be built by following nodes interconnection relationships. Community detector combines the two sources of information as we need to consider both distance information and interconnections between the roads (Fig. 1).

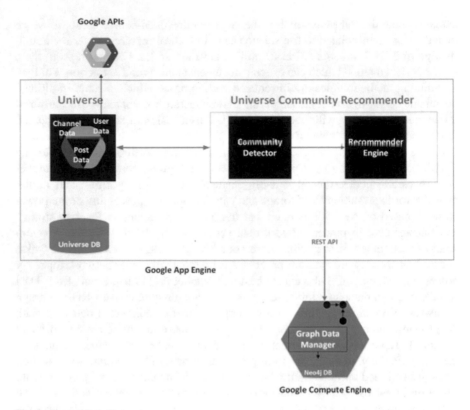

Fig. 1 Framework diagram

Universe Community Recommender is an independent module of universe that needs access to road details (distance) and interconnections with other roads to compute the weights where the clustering of nodes takes place. The UCR's most important components include:

i.   VertexAdapterService, which the two clustering algorithms that transforms road network data into vertices.
ii. Neo4j service, which calls the REST Endpoints exposed by the Graph Data Manager to save or retrieve communities.

The Graph Data Manager exposes a REST API to write and retrieve data from the Neo4j database. It also performs some business logic checks before saving data to the database. The community detection phase will identify the communities.

3.2 Louvain Community Detection Algorithm for Road Networks

The Louvain method of community detection is an algorithm for detecting communities in networks that relies upon a heuristic for maximizing the modularity. The method consists of repeated application of two steps.

i. "Greedy" assignment of nodes to communities that favors local optimizations of modularity.
ii. The definition of a new coarse-grained network in terms of communities formed in first step. Repeat two steps until no further modularity-increasing reassignments of communities are attainable.

Pseudo Code

1. Load the dataset into Neo4j.
2. Merge the values of start node, end node, and distance.
3. Perform algorithm algo.louvian.stream.
4. Load nodes, match road1 return id.
5. Load relationships, match road1, road2, and add weight.
6. If threshold >0.0174, return road1 as source and road2 as target.
7. Display the communities which are greater than threshold value.
8. Alert the displayed communities.

Modularity is the optimized value, defined as a value between -1 and 1 which measures density of inside community links compared to links between communities.
Modularity for a weighted graph is defined as:

$$Q = \frac{1}{2m} \sum_{ij} \left[A_{ij} - \frac{k_i k_j}{2m} \right] \delta(c_i, c_j), \qquad (1)$$

where A_{ij} represents nodes.

k_i and k_j are the sum of the weights of the edges attached to nodes i and j.
m is the sum of all edge weights in graph.
c_i and c_j are the communities of the node.
δ is simple delta function.

First, each node in the network is assigned to its own community. Then, for each node i, the change in modularity is calculated for removing i from its own community and moving it into the community of each of neighbor j of i. This is calculated using the equation.

$$\Delta Q = \left[\frac{\Sigma_{in} + k_{i,in}}{2m} - \left(\frac{\Sigma_{tot} + k_i}{2m} \right)^2 \right] - \left[\frac{\Sigma_{in}}{2m} - \left(\frac{\Sigma_{tot}}{2m} \right)^2 - \left(\frac{k_i}{2m} \right)^2 \right] \qquad (2)$$

where

Σ_{in} $_1$is sum of all the weights of the links inside the community i is moving into.
Σ_{tot} is the sum of all the weights of the links to nodes in the community.
K_i is the weighted degree of i.
$K_{i,in}$ is the sum of the weights of the links between i and other nodes in community.

3.3 Implementation

The dataset has 100 rows with three attributes, namely start node, end node, and distance. Start node and end node are the nodes and distance is the relationship between the nodes. The nodes represent the roads and their relationship is established as connectivity that one road has with the other. The distance density per 10,000 km between the roads is identified as the relationship. Figure 2 given below shows the dataset of the road networks that are chosen.

1. *LOAD CSV FROM "file:///Ldataset.csv" as row*
    //loading CSV file
2. *MERGE (road1:distance {name:start_node})*
    //merge road1 with start node
3. *MERGE (road2:distance {name:end_node})*
    //merge road2 with end node
4. *Merge(road1)-[Density:Density{dist:row.distance}]->(road2)*
    //merge density between road1 and road2.

Figure 3 represents the graph generated when the road network dataset is imported to Neo4j database and the relationships between the road nodes that are autoestablished.

Applying the Louvain community detection algorithm finds out which are the roads that are more densely interconnected and the communities are identified on the basis of the threshold distance that has been previously fixed. The interconnected nodes are identified as communities.

1. *CALL algo.louvain.stream(*
    //load nodes
2. *'MATCH (road1:country) RETURN id(road1) as id',*
    //load relationships
3. *'MATCH (road1:country)-[number]->(road2:country)*
    //similarity threshold
4. *WHERE number.num > "0.0174"*
    RETURN id(road1) as source,id(road2) as target',
    {graph:"cypher"})
    YIELD nodeId,community.

Fig. 2 US road network
dataset

start_node	end_node	distance
0	1	0.002025
0	6	0.005952
1	2	0.01435
2	3	0.012279
3	4	0.011099
5	6	0.006157
5	7	0.001408
5	8	0.012008
7	265	0.003213
8	298	0.005382
9	10	0.01294
9	36	0.018695
9	37	0.002948
10	11	0.013238
11	12	0.028027
12	13	0.016322
13	14	0.03219
14	15	0.009105
15	16	0.012708
16	17	0.026648
17	18	0.009636
18	19	0.014209
19	20	0.008261
20	21	0.011507
21	22	0.008586
22	23	0.011969
23	24	0.011713
24	25	0.011671
25	26	0.007185
26	27	0.010523
27	28	0.00428

Neo4j autogenerates interconnected communities and are represented in node Id column and displays the communities which have heavy traffic in the community column (Fig. 4).

4 Conclusion

The selected algorithm for detecting communities is Louvain; it is a highly scalable and greedy approach for computation. Statistics show that Louvain algorithm can be tested with up to 100 million nodes in two million nodes network takes approximately two minutes on a standard desktop. Here, the dataset which is chosen is road

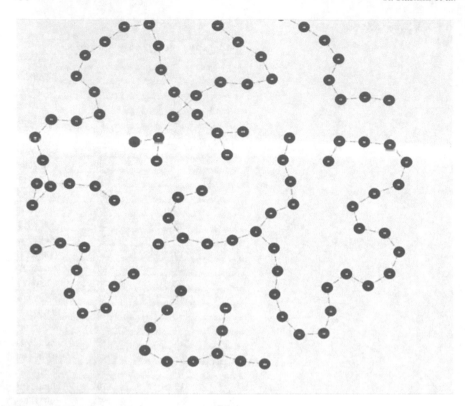

Fig. 3 Graph generated after importing dataset and building relationship

Fig. 4 Communities
identified based on threshold
distance >0.0174

"node Id"	"community"
64	0
65	1
66	2
67	3
68	4
69	5
74	6
75	91
76	8
77	10
78	10
79	12

networks which are expected to grow significantly as time elapses. So, the use of Louvain algorithm for community detection resulted in a larger number of detected communities. Once the community of roads is detected using the algorithm, identification of the communities and providing traffic alerts dynamically in correspondence to the infrastructure are feasible.

References

1. Us Road Network. https://www.roadtraffic-technology.com/features/featurethe-worlds-biggest-road-networks-4159235/
2. Generalized Louvain method for community detection in large networks (Research Article: Pasquale De Meo Published: Nov 2011)
3. Graph Database A Survey (Research Article: Rohit Kumar Kaliyar Published: International Conference on Computing, Communication & Automation ,2015)
4. Tensor-based Document Retrieval over Neo4j with an Application to PubMed Mining (Research Article: G Drakopoulos Published: July 2016)
5. Tensor fusion of social structural and functional analytics over Neo4j (Research Article: G Drakopoulos Published: July 2016)
6. Storage and parallel topology processing of the power network based on Neo4j (Research Article: Xianlong Lv et al Published: 2017)
7. Smart RDF Data storage in Graph Databases: (Research Article: Roberto De Virgilio Published: IEEE/ACM International Symposium on Cluster, Cloud and Grid Computing ,2017)
8. Community detection through likelihood optimization: in search of a sound model (Research Article: Liudmila Prokhorenkova Published: 11 July 20)

Performance Analysis of Single- and Ensemble-Based Classifiers for Intrusion Detection

R. Hariharan, I. Sumaiya Thaseen and G. Usha Devi

Abstract The aim of this paper is to analyze the performance of an intrusion detection model using single- and ensemble-based classifiers. Several tree-based single classifiers were analyzed. The ensemble of tree-based classifiers was also analyzed to differentiate the superiority in their performance. Different proportions of the benchmark KDD dataset are utilized for observing the performance of the model. Classification based on the accuracy, model building time, and kappa statistic is evaluated as the performance measures in this paper. The base and ensemble classifiers resulted in better accuracy are observed in the experiments and only Naive Bayes and random tree resulted in minimum model building time. Most of the classifiers produced better results for kappa statistic. The highest statistic is computed for ADA classifier, and lowest error is computed for the random forest ensemble.

Keywords Accuracy · Classification · Computation time · Ensemble · Error rate · Intrusion · Performance · Traffic

1 Introduction

Intrusion detection system (IDS) is an integral component in the network to discover the malicious activities that emerge endlessly and diversely in the world of the Internet [1]. IDSs are broadly classified as signature-based IDS and anomaly-based IDS. There is a major challenge for IDS to protect the network from denial of service (DoS), unauthorized access to data, and information integrity. Any IDS must be capable to handle system crashes and recover back with minimal usage of resources. Various machine learning methods have been analyzed by researchers for anomaly detection. The widely used techniques are support vector machine (SVM) [2], neural network [3], Naive Bayes (NB) [4], and decision tree [5]. These approaches have resulted in better accuracy in intrusion detection and solved most of the intrusion detection problems like huge data processing, summarization, and visualization of data for

R. Hariharan · I. Sumaiya Thaseen (✉) · G. Usha Devi
School of Information Technology and Engineering, VIT University, Vellore, Tamil Nadu, India
e-mail: sumaiyathaseen@gmail.com

© Springer Nature Singapore Pte Ltd. 2020 759
K. N. Das et al. (eds.), *Soft Computing for Problem Solving*,
Advances in Intelligent Systems and Computing 1057,
https://doi.org/10.1007/978-981-15-0184-5_65

network administrators. However, ensemble models are also utilized to improve the base classifier accuracy [6]. Multiple weak learners are referred to as ensemble. Some of the algorithms that are used are AdaBoost, random forest, random tree, LogitBoost, and bagging. In this paper, a comparative study of different supervised base and ensemble classifiers is based on accuracy, model building time, and error rate. The reason for performing such an analysis is some classifiers result in better accuracy with the trade-off time. Hence, it is very vital to choose a classifier for model building considering the time.

In this paper, various base and ensemble classifiers are analyzed and integrated with chi-square feature selection. Various performance metrics like accuracy, model building time, kappa statistic, and root mean square error. These metrics aid us to evaluate which model performs better on the dataset.

The remaining paper is summarized as follows: The related work is discussed in Sect. 2. The background of the various classifiers used for analysis is outlined in Sect. 3. The model built for analyzing performance is briefed in Sect. 4. The results of the model are analyzed in Sect. 5. The conclusion is specified in Sect. 6.

2 Literature

Many machine learning approaches are deployed for identifying intrusions in recent years. The techniques are an integration of single and ensemble classifiers. Feature selection techniques are also combined along with classifiers for improving accuracy. In the literature below, we discuss the different intrusion detection approaches developed utilizing single- and ensemble-based classifiers.

An intrusion detection system is built combining core vector machine (CVM) and principal component analysis (PCA) [7]. The learning technique is proved to be robust for building intrusion detection systems. Various types of intrusions are identified by deploying a stacking technique in IDS which was developed [8]. The results are good for precision, recall, and ROC; however, the authors have not analyzed the computation time of the system. Different weak learners such as decision tree, Naïve Bayes, random forest, multi-layer perceptron (MLP), LogitBoost, and support vector machine (SVM) classifiers are utilized to create a strong classifier such as AdaBoost for identifying the difficult category of attacks, namely U2R and R2L. The correlation-based technique is also used by the authors for removing the redundant features. Various machine learning algorithms were tested using Aegean WiFi Intrusion Dataset (AWID) public dataset [9]. The best results were given by algorithms namely extra trees, random forests, and bagging. The drawback of this model is the maximum computation time. Rotation forest technique is utilized to enhance the attack detection of an IDS in a wireless network [10]. The authors accessed the performance of 20 machine learning algorithms using the Area Under Curve (AUC) metric. The algorithms utilized were single and ensemble classifiers. Bayesian network and random tree using ensemble approaches [11] were utilized to build an intrusion detection model. The benchmark KDDCUP99 dataset is tested on the model. The model was compared with other base classifiers based on accuracy rate, sensitivity, specificity, AUC, and ROC curve. The comparison resulted in improved performance of the model on sensitivity and specificity. Bayesian network depicted higher accuracy for a limited number of samples, however, lower accuracy rate in comparison to other classifiers. However, random tree resulted in a good performance for larger samples but not for limited samples. The authors utilized the advantages of both bayesian network and random tree for experimenting the model with small and large datasets. A hybrid approach combining genetic programming and K-nearest neighbor is utilized to detect the intrusion, and the approach is the combination of a genetic programming [12]. The model utilized the KDDCUP dataset. The two drawbacks of this approach are as follows: (1) heavy usage of memory to execute; (2) as the dataset increases, the accuracy of the classifier in the detection process decreases.

An IDS [13] using an ensemble classifier based on the distance feature is developed utilizing the UNSW_NB15 dataset. The centroids of the cluster are determined by the k-means algorithm. However, in this approach, an alternative is to utilize a distance-based feature which is one dimensional in nature for every sample data. LogitBoost-based algorithm is used for determining known and unknown Web attack traffic which is developed [14]. Predictive models and classification for an intrusion detection are developed using machine learning classification algorithms [15]. The different algorithms utilized are support vector machine, Gaussian Naive Bayes, logistic regression, and random forest (RF). These algorithms are tested with NSL-KDD dataset. An IDS is built using RF ensemble classifier [16]. RF performs well in comparison to other base classifiers in terms of accuracy of individual attacks. The performance of the model is evaluated by utilizing the NSL-KDD dataset. The four types of attack like U2R, R2L, DOS, and Probe are identified by the random forest (RF) algorithm.

Thus from the literature, it is analyzed that many base and ensemble classifiers are used for identifying intrusions. However, the performance of different base and ensemble classifiers is evaluated in the proposed model to identify the superior classifier based on false alarm rate and accuracy.

3 Background

3.1 Chi-Square Feature Selection (Chi)

Chi-square is a numerical test that calculates deviation from the expected distribution assuming the feature is independent of the class label. Following Eqs. (1) and (2) are chi-square feature selection, respectively.

$$\text{Chi-square-metric} = t(\text{tp}, (\text{tp} + \text{fp})\text{Ppos}) + t(\text{fn}, (\text{fn} + \text{tn})\text{Ppos})$$
$$+ t(\text{fp}, (\text{tp} + \text{fp})\text{Pneg}) + t(\text{tn}, (\text{fn} + \text{tn})\text{Pneg}) \tag{1}$$

where

$$t(\text{count, expect}) = (\text{count} - \text{expect})^2/\text{expect} \tag{2}$$

3.2 Naive Bayes

Naive Bayes [17] is a probabilistic classifier which is computationally fast, simple to implement, and works well with high dimensions. Naive Bayes works well for few categories variables, no distribution requirements, but suffers multicollinearity and compute the multiplication of independent distributions. Naive Bayes classifiers are used for learning to classify text documents [18].

3.3 J48

J48 is an extension of ID3 which is based on decision tree technique [19]. It is also called a statistical classifier. The additional features of J48 are pruning decision trees, continuous attribute value ranges, accounting for missing values and derivation of rules.

3.4 Classification ViaRegression (CVR)

A decision tree which computes linear regression at the leaf nodes is known as CVR [20]. CVR utilizes RF approach. Every tree in the forest classifies an object by building more number of decision trees during training. Any tree which classifies a specific class is considered as a vote for that category. The test sample is designated to the category which results in maximum votes. The accuracy of the approach depends on random inputs and attributes. This technique is widely used for determining the class category according to a large number of samples, and it minimizes overfitting.

3.5 AdaBoost

AdaBoost [21] is an adaptive boosting technique used for building a robust approach which is a linear integration of weak classifiers. This algorithm is one of the most popular machine learning (ML) algorithms. AdaBoost algorithm falls into the ensemble method called boosting. A base learner is called for a particular amount of iteration. The AdaBoost algorithm is very fast, simple, and easy to program. No prior knowledge is needed about the weak learner.

3.6   Bagging

Bagging [22] is a method for improving results of machine learning classification algorithms. Bagging stands for bootstrap aggregating. This method comprises predetermined and parallelized classification trees. This algorithm reduces variance in comparison to regular decision trees. The ensemble can provide variable importance measures like the Gini index for classification and residual sum of squares (RSS) for regression. Bagging algorithm can easily handle qualitative (categorical) features.

3.7   LogitBoost

A logistic model tree (LMT) is a classification model that integrates logistic regression (LR) and decision tree [23]. A binomial log-likelihood is deployed to change the loss function in a linear manner. The LogitBoost algorithm is obtained if it satisfies the following requirements:

- If the AdaBoost technique is casted into a statistical framework and considered as a generalized adaptive model.
- If cost functionality of logistic regression is applied, LogitBoost can be seen as a convex optimization. Following Eqs. (3) and (4) are logistic form and logistic loss, respectively.

It is of the form

$$f = \sum_t \alpha_t h_t \tag{3}$$

The LogitBoost algorithm minimizes the logistic loss:

$$\sum_i \log\left(1 + e^{-y_i f(x_i)}\right) \tag{4}$$

3.8 Random Forest

RF is derived by merging different tree predictors. Every tree is dependent on the sampled random vector. In comparison to existing algorithms, RF results in better accuracy as the algorithm is efficient in executing large datasets also. High-dimensional data can be handled without scaling down to low dimensionality. Random forest is good for parallel or distributed computing. The main advantages of random forest are that it is more robust due to the handling of missing/partial data and also flexible in updating the enhancements to the algorithm.

3.9 Random Tree

Random trees have been introduced [24]. The algorithm can deal with both regression problems and classification. The term forest is obtained as it is a group of tree predictors. Random trees combine two existing algorithms, namely single model trees and random forest. The different kinds of random tree are a random recursive tree, uniform spanning tree, Brownian tree, branching process, random binary tree, rapidly exploring random tree, and random minimal spanning tree. Though distinct training sets are deployed, the same parameters are utilized for training all trees. Bootstrap procedure is utilized for generating the partitions from the original training set. In this technique, an input attribute is analyzed and classified with every tree present in the forest. The final class category is predicted by majority voting technique.

4 Proposed Model

The intrusion detection model is developed as shown in Fig. 1. A preprocessing is performed to normalize the samples in a specific range. The data is divided into training and testing set. The model is then trained on a specific classifier, and the accuracy of the different class labels are predicted from the test set. The model is tested on various base classifiers such as Naïve Bayes, J48, ClassificationViaRegression and ensemble classifiers namely AdaBoost, LogitBoost, random forest, and random tree.

Any IDS can be evaluated based on the accuracy metric which can be calculated by utilizing the following terms: These are true negatives (TN), false positives (FP), true positives (TP), and false negatives (FN) where TN indicates the normal performance that is identified as correct, FP specifies the normal activity wrongly predicted as abnormal, TP specifies the normal behavior which is correctly predicted, and FN specifies the abnormal activity that is misclassified as normal. Following Eqs. (5) and (6) are accuracy and kappa statistic, respectively. Equations (7), (8), and (9) are probability equations, respectively.

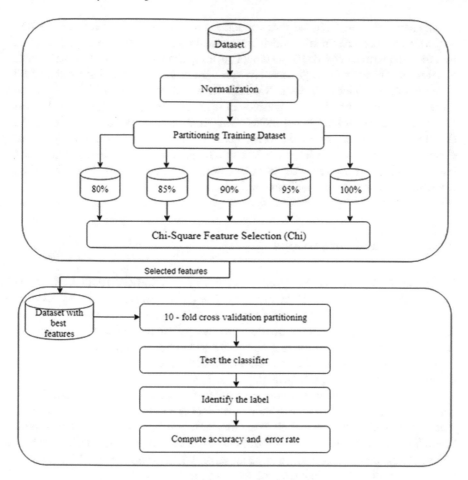

Fig. 1 Intrusion detection model

$$\text{Accuracy} = \frac{\text{TP} + \text{TN}}{\text{TP} + \text{FP} + \text{FN} + \text{TN}} \tag{5}$$

$$\text{Kappa statistic} = \frac{\text{Accuracy} - P_e}{1 - P_e} \tag{6}$$

wherein

$$P_e = P_{yes} + P_{no} \tag{7}$$

$$P_{yes} = \frac{\text{TP} + \text{FP}}{\text{TP} + \text{FP} + \text{TN} + \text{FN}} \cdot \frac{\text{TP} + \text{FN}}{\text{TP} + \text{FP} + \text{TN} + \text{FN}} \tag{8}$$

$$P_{no} = \frac{\text{FN} + \text{TN}}{\text{TP} + \text{FP} + \text{TN} + \text{FN}} \cdot \frac{\text{FP} + \text{TN}}{\text{TP} + \text{FP} + \text{TN} + \text{FN}} \tag{9}$$

The parameters given in the above equations are essential for measuring the IDS performance. The other metrics which are taken into consideration are kappa statistic, root mean square error (RMSE) and model building time. Kappa statistic is a measure to determine the inter-rater agreement for qualitative items. Kappa statistic should be closer to one for an ideal classifier. RMSE is the standard deviation of the prediction errors. It measures how far the regression data points are from the line. RMSE must be closer to zero for an optimal model. The time taken to build the model is the duration of time spent in model development. The different classifiers analyzed for various metrics are Naive Bayes, J48, and ClassificationViaRegression which are base classifiers. The other ensemble classifiers analyzed are AdaBoost, bagging, LogitBoost, random forest, and random tree.

5   Results and Discussions

The experiments were conducted on Weka 3.8 Intel i3 500 GB HDD 4 GB RAM utilizing the benchmark KDD dataset [24]. The dataset consists of nearly 494,025 records out of which 2,96,504 records are taken for training and 197,519 records for testing. The test datasets are divided into two partitions each of 98,000 records, and the average of two partitions is computed as the final measure of every metric. This is done to minimize the data overloading to the model. Table 1 shows features obtained by ranker chi-square attribute evaluation.

Table 2 shows the model building time for different classifiers. The Naïve Bayes classifier results in the least time to build the model as it utilizes a simple probabilistic measure to predict the class label. The random tree which is an ensemble of tree-based learners also requires the least time for model development. AdaBoost which belongs to the family of boosting also requires only 12 s to build the model. However, all other classifiers result in huge time for learning the model.

Table 3 shows the accuracy determined for different base and ensemble classifiers. The best accuracy results are obtained for ClassificationViaRegression and J48 from the family of the decision tree. The other ensemble classifiers such as bagging, LogitBoost, random forest, and random tree also depict higher accuracy. This is due to the reason that ensemble computes the result based on weak decision learners such as a tree. Hence, it is proved that ensemble results in better accuracy as analyzed in the literature.

Table 4 shows the kappa statistic for different classifiers. The base classifiers J48 and ClassificationViaRegression provide best results for kappa as they are closer to 1. The ensembles bagging, LogitBoost, random forest, and random tree also result in a good value. This indicates that there is a direct proportionality between accuracy and kappa. The classifiers which resulted in better accuracy also proved to have a better kappa statistic.

Table 5 shows the RMSE of different classifiers. The results show that J48 and ClassificationViaRegression obtain very low RMSE which is closer to zero. The ensemble classifiers bagging, LogitBoost, random forest, and random tree also result

Table 1 Attributes selected by ranker chi-squared attribute evaluation

Rank	Attribute	Attribute number
1	service	3
2	dst_bytes	6
3	dst_host_diff_srv_rate	35
4	diff_srv_rate	30
5	flag	4
6	dst_host_serror_rate	38
7	dst_host_srv_count	33
8	same_srv_rate	29
9	count	23
10	dst_host_same_srv_rate	34
11	dst_host_srv_serror_rate	39
12	serror_rate	25
13	Src_bytes	5
14	dst_host_srv_diff_host_rate	37
15	srv_serror_rate	26
16	dst_host_same_src_port_rate	36
17	logged in	12
18	dst_host_count	32
19	hot	10
20	dst_host_rerror_rate	40
21	srv_count	24
22	duration	1
23	srv_diff_host_rate	31
24	dst_host_srv_rerror_rate	41
25	rerror_rate	27
26	protocol_type	2
27	srv_rerror_rate	28
28	is_guest_login	22
29	srv_count	24
30	num_compromised	13
31	num_failed_logins	11

Table 2 Time taken for different classifiers

Classifiers	Model building time for test set-1	Model building time for test set-2	Average time (s)
Naive Bayes	0.8	1.31	1.055
J48	11.55	37.2	24.375
ClassificationViaRegression	303.16	179.53	241.345
ADA	13.85	11.89	12.87
Bagging	49.28	42.67	45.975
LogitBoost	253.29	200.39	226.84
Random forest	90.63	72.79	81.71
Random tree	1.9	3.17	2.535

Table 3 Accuracy of different classifiers

Classifiers	Accuracy for test set-1	Accuracy for test set-2	Average accuracy
Naive Bayes	97.4048	93.8551	95.62995
J48	99.9322	99.9088	99.9205
ClassificationViaRegression	99.9149	99.8836	99.89925
ADA	94.9951	96.823	95.90905
Bagging	99.8804	99.9069	99.89365
LogitBoost	99.923	99.8968	99.9099
Random forest	99.9544	99.9595	99.95695
Random tree	99.9068	99.8847	99.89575

Table 4 Kappa statistic for different classifiers

Classifiers	Kappa statistic for test set-1	Kappa statistic for test set-2	Average kappa statistic
Naive Bayes	0.954	0.9027	0.92835
J48	0.9988	0.9985	0.99865
ClassificationViaRegression	0.9986	0.9979	0.99825
ADA	0.9137	0.9429	0.9283
Bagging	0.998	0.9983	0.99815
LogitBoost	0.9987	0.9982	0.99845
Random forest	0.9992	0.9993	0.99925
Random tree	0.9984	0.9979	0.99815

Table 5 Root mean square error of different classifiers

Classifiers	RMSE for test set-1	RMSE for test set-2	Average root mean squared error (RMSE)
Naive Bayes	0.0573	0.077	0.06715
J48	0.0094	0.0093	0.00935
ClassificationViaRegression	0.0089	0.0099	0.0094
ADA	0.1656	0.1544	0.16
Bagging	0.0101	0.0103	0.0102
LogitBoost	0.0077	0.0105	0.0091
Random forest	0.0064	0.0071	0.00675
Random tree	0.0096	0.0124	0.011

in lower RMSE. Hence from Tables 2 and 3, it can be inferred that there is a direct proportion between kappa statistic and RMSE. The classifiers which resulted in better kappa also had better RMSE.

6 Conclusions

This paper presents an analysis of single- and ensemble-based classifiers for intrusion detection. The different base classifiers and ensemble-based classifiers had been validated with results. From the experimental results, it is well observed that the Naïve Bayes and ClassificationViaRegression results in better accuracy in base classifiers and ensembles excludes the ADA method. Furthermore, the RMSE value of the classifiers is observed lesser when compared to other techniques. In addition, Naive Bayes and random tree resulted in reduced time for model construction. Finally, these methods gave superior results in all metrics of performance compared to multiple classifiers even it resulted in better accuracy.

References

1. Denning, D.E.: An intrusion-detection model. IEEE Trans. Software Eng. 222–232 (1987)
2. Khan, L., Awad, M., Thuraisingham, B.: A new intrusion detection system using support vector machines and hierarchical clustering. VLDB J. **16**, 507–521 (2007)
3. Wang, G., Hao, J., Ma, J., Huang, L.: A new approach to intrusion detection using artificial neural networks and fuzzy clustering. Expert Syst. Appl. **37**, 6225–6232 (2010)
4. Amor, N.B., Benferhat, S., Elouedi, Z.: Naive bayes vs decision trees in intrusion detection systems. In: Proceedings of the 2004 ACM Symposium on Applied Computing, pp. 420–424 (2004)

5. Sornsuwit, P., Jaiyen, S.: Intrusion detection model based on ensemble learning for u2r and r2l attacks. In: 7th International Conference on Information Technology and Electrical Engineering (ICITEE), IEEE, pp. 354–359 (2015)
6. Aburomman, A.A., Reaz, M.B.: A survey of intrusion detection systems based on ensemble and hybrid classifiers. Comput. Secur. **65**, 135–152 (2017)
7. Amudha, P., Karthik, S., Sivakumari, S.: Intrusion detection based on core vector machine and ensemble classification methods. In: International Conference on Soft-Computing and Networks Security (ICSNS). IEEE, pp. 1–5 (2015)
8. Roy, S S., Krishna P.V., Yenduri, S.: Analyzing intrusion detection system: an ensemble based stacking approach. In: IEEE International Symposium on Signal Processing and Information Technology (ISSPIT), pp. 307–309 (2014)
9. Alotaibi, B., Elleithy, K.: A majority voting technique for wireless intrusion detection systems. In: IEEE Long Island Systems, Applications and Technology Conference (LISAT), pp. 1–6 (2016)
10. Tama, B.A, Rhee, K.H.: Classifier ensemble design with rotation forest to enhance attack detection of IDS in wireless network. In: 11th Asia Joint Conference on Information Security (Asia JCIS). IEEE, pp. 87–91 (2016)
11. Wang, Y., Shen, Y., Zhang, G.: Research on intrusion detection model using ensemble learning methods. In: 7th IEEE International Conference on Software Engineering and Service Science (ICSESS), pp. 422–425 (2016)
12. Malhotra S., Bali V., Paliwal, K.K.: Genetic programming and K-nearest neighbour classifier based intrusion detection model. In: 7th International Conference Cloud Computing, Data Science & Engineering-Confluence. IEEE, pp. 42–46 (2017)
13. Aravind, M.M, Kalaiselvi, V.K.: Design of an intrusion detection system based on distance feature using ensemble classifier. In: 4th International Conference on Signal Processing, Communication and Networking (ICSCN). IEEE, pp. 1–6 (2017)
14. Kamarudin, M.H., Maple, C., Watson, T., Safa, N.S.: A logitboost-based algorithm for detecting known and unknown web attacks. IEEE Access. **5**, 26190–26200 (2017)
15. Belavagi, M.C., Muniyal, B.: Performance evaluation of supervised machine learning algorithms for intrusion detection. Procedia Comput. Sci. 117–123 (2016)
16. Farnaaz, N., Jabbar, M.A.: Random forest modeling for network intrusion detection system. Procedia Comput. Sci. 213–217 (2016)
17. Quinlan, J.R.: C4. 5: Programs for Machine Learning. Elsevier, Amsterdam (2014)
18. Domingos, P., Pazzani, M.: On the optimality of the simple Bayesian classifier under zero-one loss. Mach. Learn. 103–130 (1997)
19. Arora, T., Dhir, R.: Correlation-based feature selection and classification via regression of segmented chromosomes using geometric features. Med. Biol. Eng. Compu. **55**, 733–745 (2017)
20. Freund, Y., Schapire, R.E.: A decision-theoretic generalization of on-line learning and an application to boosting. J. Comput. Syst. Sci. **55**, 119–139 (1997)
21. Friedman, J., Hastie, T., Tibshirani, R.: Additive logistic regression: a statistical view of boosting (with discussion and a rejoinder by the authors). Ann. Stat. **2**, 337–407 (2000)
22. Breiman, L.: Random forests. Mach. Learn. **4**, 5–32 (2001)
23. Cutler, A., Zhao, G.: Pert-perfect random tree ensembles. Comput. Sci. Stat. **33**, 490–497 (2001)
24. Ruan, Y.X., Lin, H.T., Tsai, M.F.: Improving ranking performance with cost-sensitive ordinal classification via regression. Inf. Retrieval 1–20 (2014)

Timestamp Anomaly Detection Using IBM Watson IoT Platform

Aditi Katiyar, Neha Aktar, Mayank and K. Lavanya

Abstract Anomaly disclosure is an issue of finding startling precedents in a dataset. Amazing precedents can be described as those that do not agree to the general direct of the dataset. Irregularity revelation is basic for a couple of use spaces; for instance, cash related and correspondence organizations, general prosperity, and environment contemplates. In this paper, we base on revelation of irregularities in month-to-month temperature, weight, and significance data on IBM Watson organize for timestamp peculiarity area. IBM Watson features to make chronicled dataset dependent nervous qualities that are gotten from the time plan informational collection. With these principles, we can prepare create informing system for customers IoT devices when a sporadic examining is recognized by the DSX acknowledgment data science experience. In this examination, we took a gander at the results IBM Watson IoT organize and fuzzy rationale abnormality acknowledgment. IBM Watson IoT organize features to deliver alert/caution to the customer. On IBM Watson organize, the z-score is processed to distinguish characteristics in the real-time series data using the IBM Data Science Involvement in direct advances. Also, showed up, how one can deduce the edge a motivating force for the given chronicled data and set the administer as requirements be in IBM Watson IoT Platform to make continuous alerts.

Keywords Anomaly detection in time series data · IBM Watson Platform · Fuzzy logic inference system · Temperature data · Pressure data · Magnitude data

A. Katiyar · N. Aktar · Mayank · K. Lavanya (✉)
School of Computer Science and Engineering, VIT University, Vellore, India
e-mail: lavanya.k@vit.ac.in

A. Katiyar
e-mail: aditi.katiyar2018@vitstudent.ac.in

N. Aktar
e-mail: neha.akthar2018@vitstudent.ac.in

Mayank
e-mail: mayank.2018@vitstudent.ac.in

© Springer Nature Singapore Pte Ltd. 2020   771
K. N. Das et al. (eds.), *Soft Computing for Problem Solving*,
Advances in Intelligent Systems and Computing 1057,
https://doi.org/10.1007/978-981-15-0184-5_66

1   Introduction

Irregularity acknowledgment can be portrayed as 'the issue of finding plans in data that do not change in accordance with expected normal lead.' Moreover, showed up, how one can decide the limit an impetus for the given recorded data and set the oversee suitably in IBM Watson IoT Platform to make continuous cautions. Irregularity disclosure is basic when the surprising behavior in the dataset gives colossal information about the system. Anomalies can be caused by pernicious activities, instrumentation goofs, changes in the earth (i.e., natural change), and human slip-ups. Eccentricity ID is a crucial issue in a couple of use regions; for instance, Mastercard blackmail acknowledgment in cash-related systems, intrusion revelation in correspondence structures, and irresistible ailment area when all is said in done prosperity data. Of late, peculiarity area has also transformed into a basic issue in air considers for recognizing unusual climatic conditions caused by an unnatural climate change. Irregularities in an air data game plan can be illuminated by the going with model. In an occasional game plan of temperature data, one can see changes among high and low temperatures. The normal lead for temperature data is to accomplish high characteristics (e.g., temperature more than 20 °C) in summer months and drop to low characteristics (e.g., temperature underneath 10 °C) in winter months. In a temperature dataset, a winter month with typical temperature more than 20 °C or a pre-summer month with ordinary temperature underneath 10 °C can be recognized as a peculiarity. On a model month-to-month temperature, weight and enormity timestamp values, the edge regard, set apart as (+ve)3 to (−ve)3, can be called a variation from the norm. One of the critical challenges of abnormality acknowledgment is to depict customary and unusual practices. Distinctive challenges can be recorded as to make methodologies to discover weird data and to design computationally capable calculations. The purpose of this examination is to discover irregularities in temperature data, weight data, and size data using IBM Watson features to make controls dependent on decided edge regards and to create alert/caution to the customer. As the bit of complex game plan, this model can be used for finding new segments of peculiarity and making the ready structure while soft method of reasoning determination system gives simply joined graph to portrayal.

2   Literature Survey

In the composition, there are different wide studies looking at anomaly acknowledgment approaches. Landage and Wankhade [1] gave a continuous overview of the variation from the norm area issues, techniques, and application areas. Idika and Mathur [2] gave a review of irregularity area methodologies. The 91 of intrigue and downsides and the motivation of these techniques were discussed in a comparative strategy. Tahir [3] investigated the changed course of action methods for variation from the norm revelation issues, i.e., thickness based, network-based methodologies, and the speculative structure of the irregularity acknowledgment issues. Sun et al. [4] discussed the peculiarity acknowledgment counts, fundamental systems, the focal points, and hindrances of these philosophies and proposed another fluffy-based peculiarity revelation estimation. The idiosyncrasy area systems can be appointed as authentic techniques and partition-based strategies. Quantifiable techniques intend to develop a truthful model of the data and perceive data that do not fit into the model. In separation-based methodologies, the detachments between data are considered in area of eccentricities: the data at a partition more unmistakable than a predefined evacuate is called an anomaly. This examination bases on perceiving special cases using partition-based systems. DBSCAN, which was made by Zang et al. [5], is one of the division-based philosophies, which has been commonly used for handling irregularity disclosure issues. The outline of the methodology used by Saxe and Berlin [6] consists of feature extraction: which extracts four types of features from the static benign and malicious binaries; deep neural network classifier: consisting of input, hidden, and output layers; and lastly, score calibration: which carries out nonparametric score distribution estimation that can be interpreted through Bayesian estimation of a malware. This method is able to achieve a detection rate of 95% with a false positive rate of 0.1% over a dataset of 400,000 software binaries. It requires moderate computation and is able to achieve a good amount of accuracy. Dahl et al. [7] state that automatically generated malware poses as a significant issue for computer users. Feature selection for malware classification prevails but the number of features is still too large for neural networks which are much complex algorithms. Thus, to reduce this, an approach using random projection to even reduce dimensionality of input space further is proposed. To achieve this, firstly, a dataset is constructed. Then to reduce dimensionality of the input space, initial feature selection is carried out. Next, random projections are used to further reduce the dimensionality, while still maintaining the highly important information. Finally, several models such as nonlinear neural networks and linear logistic regression classifiers are trained to classify the files as malwares. This resulted in a 43% reduction in error rate compared to using all the features. For one-layer neural networks with random projections, the two-class error rate came out to be 0.49% and for a group of neural networks, it came out to be 0.42% which proves great performance. According to Golovko et al. [8], neural networks and AIS have been used in many areas for anomaly activity recognition and detection. But the existing methods were quite static and thus posed a problem for detecting new viruses and malwares. Thus, an approach using various intelligent techniques such as fusion of neural networks and AIS was posed which proved to have better potential to recognize novel viruses. Integrating neural networks and AIS increases the performance of security system.

3 Background Study

3.1 IBM Watson Overview

Watson is a computing system that was created for question answering (QA) initially built by IBM in order to apply more advanced processing of natural language, reasoning in automated form, representation of knowledge by using latest machine learning technologies and accessing information to the fields of open source domain having question answering. The vital difference that remains in between QA technology and documentation search is that document search generally takes a word to word query and returns a generated list of finite number of documents, particularly ranked in the order of priority to the executed query (generally based on ranking of page and popularity), while it can be seen QA technology just takes in the given question that is briefed in natural language format, needed to be understood in much meticulous manner, that returns a far more precise or accurate answer to the asked question.

Initially, when IBM was created, it was stated that moreover the 100 highly varied techniques are used in order to analyze different aspects of natural language, find and result in hypotheses, identify and verify the sources and combine them and rank hypotheses accordingly, in the end, score evidence. The capabilities of Watson have been extended and so does the working of Watson that has been changed enormously and enhancing machine learning capabilities and optimizing hardware's available to our developers as well as the researchers.

In social insurance, IBM Watson Platform has joined the Watson's characteristic dialect, speculation age, and proof-based learning capacities so as to add to clinical choice emotionally supportive networks and the expansion in artificial insight in the majority of the human services for use by the best medicinal experts. ENGEO (the building firm) made an online administration with the assistance of the IBM accomplice program named GoFetchCode. GoFetchCode shows assemble codes by utilizing Watson's characteristic dialect handling and question-noting capabilities. It is no longer simply an inquiry replying (QA) processing framework planned from Q&A combines; however, now it can 'read,' 'see.' 'talk,' 'hear,' 'taste,' 'decipher,' 'suggest,' and 'learn.'

Watson has been using the Apache Unstructured Information Management Architecture framework and IBM's Deep Question Answering software implementation. The system was mostly written in various programming languages, which includes Prolog, C++, and java that runs on SUSE Linux Enterprise Server 11 operating system using the Apache Hadoop framework.

3.2 IBM Data Science Experience (DSX)

It not only provides interactive but also a collaborative, cloud-based environment which use the famous open source tools like R and Python. Secondly, a new project named anomaly detection has been created to collect and share notebooks, connect to different data sources, create pipelines, and add datasets all in one place in which new Jupyter Notebook named final anomaly detection has been created.

The Jupyter Notebook is one of the web applications that are open source and more importantly allows you to create documents that not only contain code that is live and has mathematical equations along with the documents that are shareable and contain visualizations and narrative text. It generally includes cleaning of data and transformation provided for simulation of numerical data, implementing statistical model, visualization of data, and using concepts of machine learning.

3.3 Z-Score

A z-score can be mostly defined as the total number of standard deviations from the original mean a data point is. Technically, it can be inferred that it is an absolute measure of how many standard deviation points above or below the population mean that is given a raw score will or will not be. Larger deviation from the mean value can only be seen when z-score value is higher, which can be interpreted as abnormal or outlier.

This is for population z-score:

$$z = (x - \mu)/\sigma \tag{1}$$

In Eq. (1), μ has been taken as the population mean and σ has been taken as the population standard deviation.

This is for sample z-score:

$$Z_i = (X_i - \bar{X})/S \tag{2}$$

In Eq. (2), x bar is taken as the sample mean and s is taken as the sample standard deviation.

3.4 Fuzzy Logic

Anomaly detection is the detection of abnormal behavior. Fuzzy rules are generated using an editor called rule editor for the given sets of data, representing varied behavior of an individual system. The strength of using such rules has been fired by data points. Therefore, instead of using the values that are predicted ones, the fuzzy anomaly detection system will be using the firing strength method for the classification purpose. With the help of MATLAB fuzzy logic toolbox, we will be able to generate rule base and data base. The output produced in MATLAB generally contains the Mamdani-type Fuzzy Rules having version either linguistic or numerical (integer), the types and values of membership function, and the rules with confidence degree.

4 Methodology

The main motive of anomaly detection is to identify cases that are not usually within data that is seemingly homogeneous in nature. To implement anomaly detection, Jupyter Notebook with Data Science Experience in IBM Watson IoT Platform is used. The Jupyter Notebook not only allows us to create documents that contain code that is executable have formulas related to mathematics and share documents that can be visualized graphically (matplotlib) and its text is explanatory in nature.

4.1 Dataset

The paper detects the anomaly by processing the temperature, pressure, and magnitude historical timestamp dataset. Each of the three dataset contains one timestamp column along with the column containing the temperature measured in degree (°F), pressure measured in pascal (Pa), and magnitude measured in seismic moment (Mo), respectively. In total, we have taken 720 target data for this study. The dataset is been taken from GitHub and Kaggle.

4.2 Anomaly Detection Using IBM Watson

In this project firstly, we have created an account on IBM Watson Platform, i.e., Bluemix account for using the Watson IoT Platform services along with the IBM Data Science Experience (DSX). In the DSX menu, we have loaded the temperature, pressure, and magnitude file in CSV formats. In order to show the anomalies, we have calculated the z-score, as shown in Fig. 1.

We have calculated the threshold values from the historical temperature, pressure, and magnitude data using the z-score in order to create the Cloud Rule for each dataset in IBM Watson Platform, and on the basis of these rules, we have created an E-mail action, such that an E-mail will be sent to the concerned person whenever the temperature value, pressure value, or magnitude value crosses the threshold values that we have derived.

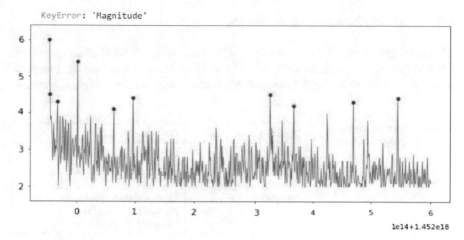

Fig. 1 The red marks are the unexpected spikes and dips are the anomalies in the pressure dataset whose calculated z-score value is greater than 3 or less than −3

4.3 Visualization of Anomaly Detection Using Fuzzy Logic

4.3.1 Defining Input and Output Variables

Firstly in Fuzzy Logic Designer, we will define the two input variables, i.e., temperature and pressure (Table 1) and one output variable, i.e., magnitude (Table 2). The positive point in using fuzzy logic toolbox software is that it will not limit the total number of inputs we are going to use. However, the memory that is available to the machine will limit the inputs. Sometimes, it is difficult to do the analyzation of fuzzy inference system using the tools other than the traditional one in case of large number of inputs or large number of membership functions.

Table 1 Input variables

Input variable	Category	Fuzzification method
Temperature	Numerical	Generalized bell
Pressure	Numerical	Generalized bell

Table 2 Output variable

Output variable	Category	Fuzzification method
Magnitude	Numerical	Triangular

4.3.2 Defining Membership Functions

Membership Function Editor helps us to define the values of all the different membership functions associated with each of the variables, i.e., temperature, pressure, and magnitude (Tables 3 and 4, respectively), it has been shown in Figs. 2, 3 and, 4.

Table 3 Input membership functions along with their ranges

Input linguistic variable	Linguistic value
Temperature	Low (15–28) Normal (28–38) High (38–90)
Pressure	Low (−9999 to 1014) Normal (1014–1017) High (1017–1021)

Table 4 Output membership function with the range

Output linguistic variable	Linguistic value
Magnitude	Low (1.5–2) Normal (2–4) High (4–6)

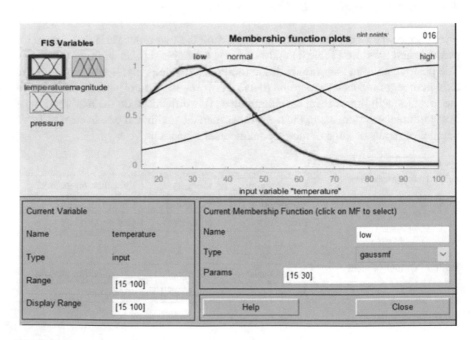

Fig. 2 Temperature membership function

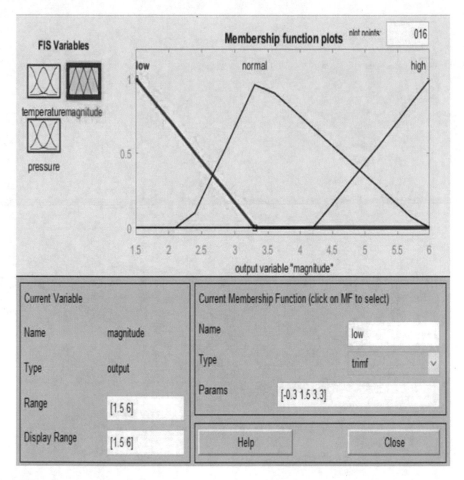

Fig. 3 Pressure membership function

4.3.3 Defining Fuzzy Rules

1. If (temperature is in low range) or (pressure is in low range), then (magnitude will also be in low range).
2. If (temperature is in high range), then (magnitude will also be in high range).
3. If (temperature is in high range) or (pressure is not in high range), then (magnitude is in high range) that defines the behavior of the anomaly detection system.

Fuzzy inference diagram is generally viewed using rule viewer. It views the dependency of output, i.e., magnitude on two of the inputs, i.e., temperature and pressure; it has generated and plot an output surface map for the anomaly detection system (Fig. 5).

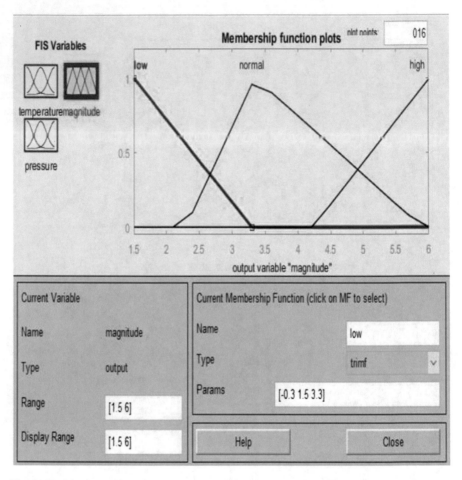

Fig. 4 Magnitude membership function

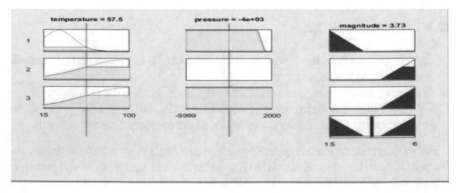

Fig. 5 Rule viewer for temperature, pressure, and magnitude

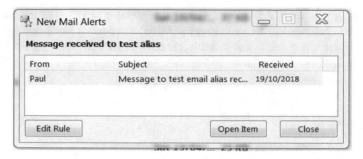

Fig. 6 E-mail alert

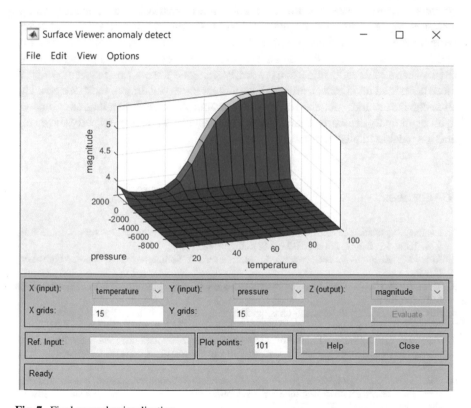

Fig. 7 Final anomaly visualization

5 Results and Discussion

On platform which is open source, alarming system and the anomaly detection are implemented using the IBM Watson IoT Platform using Jupyter Notebook in it and visualization of anomaly detection is done using fuzzy logic toolbox in Matlab. On the basis of z-score value, the anomaly detected in the input variables will generate the alarming system that is the email is sent to the customers shown in Fig. 6. Membership functions in fuzzy inference system are needed to be assigned to the input variables along with their suitable ranges in order to easily detect anomaly though the visualization as shown in Fig. 7.

6 Conclusion

Automatic generation and implementation of such rules are important in complex anomaly detection along with adding new emailing alarm system. Based on the collected historical data, this project modeled the sophisticated relationship between temperature, pressure, and magnitude. Experimental results showcase that fuzzy logic toolbox helps in visualizing the anomaly in given dataset, it is also necessary that the qualities of rules that IF-THEN have are not compromised at any cost and IBM Watson IoT Platform helps us to easily generate the alarming system which is more beneficial for clients in order to have a clear understanding of their devices. The all-inclusive accuracy obtained in detecting anomaly is much better than as compared to other anomaly detection techniques since it provides more user-friendly and easily understandable implementation.

References

1. Landage, J., Wankhade, M.P.: Malware and malware detection techniques: a survey. Int. J. Eng. Res. Technol. (IJERT) 2(12) (2013). ISSN: 2278-0181
2. Idika, N., Mathur, A.P.: A Survey of Malware Detection Techniques, p. 48. Purdue University (2007)
3. Tahir, R.: A study on malware and malware detection techniques. Int. J. Educ. Manage. Eng. 8(2), 20 (2018)
4. Sun, J., Lou, Y., Ye, F: Research on anomaly pattern detection in hydrological time series. In: Web Information Systems and Applications Conference (WISA) IEEE Conferences (2017)
5. Zang, D., Liu, J., Wang, H.: Markov chain-based feature extraction for anomaly detection in time series and its industrial application. In: Chinese Control and Decision Conference (CCDC) (2018)
6. Saxe, J., Berlin, K.: Deep neural network based malware detection using two dimensional binary program features. In: 2015 10th International Conference on Malicious and Unwanted Software (MALWARE). IEEE (2015)
7. Dahl, G.E., et al.: Large-scale malware classification using random projections and neural networks. In: 2013 IEEE International Conference on Acoustics, Speech and Signal Processing. IEEE (2013)
8. Golovko, V., et al.: Neural network and artificial immune systems for malware and network intrusion detection. Advances in Machine Learning II, 485–513. Springer, Berlin, Heidelberg (2010)

Call Churn Prediction with PySpark

Mark Sheridan Nonghuloo, Rangapuram Aravind Reddy, Ganji Manideep, M. R. SarathVamsi and K. Lavanya

Abstract Different markets over the world are ending up progressively more saturated, with an ever-increasing number of customers swapping their enrolled benefits between contending organizations. Consequently, organizations have understood that they should center their promoting endeavors in client maintenance instead of client procurement. It limits client surrender by foreseeing which clients are probably going to cross out a membership to an administration. In spite of the fact that initially utilized inside the telecommunication business, it has turned out to be regular practice crosswise over banks, ISPs, insurance firms and other verticals. In this paper, an end to end churn prediction is done in view of client call information records. We take a gander at what sorts of client information are normally utilized, do some preparatory investigation of the information and create churn prediction models with PySpark. PySpark processes huge datasets at minimal time and when it comes to the synchronization points as well as errors, framework easily handles at the back end. The PySpark API takes advantage of Spark to deliver dramatic improvements in processing speed for large sets of data.

Keywords Client procurement · Churn · PySpark · Call information · Telecommunication

M. S. Nonghuloo · R. Aravind Reddy · G. Manideep · M. R. SarathVamsi · K. Lavanya (✉)
School of Computer Science and Engineering, VIT University, Vellore, India
e-mail: lavanya.k@vit.ac.in

M. S. Nonghuloo
e-mail: marksheridan.n2018@vitstudent.ac.in

R. Aravind Reddy
e-mail: rangapuramaravind.2018@vitstuden.ac.in

G. Manideep
e-mail: ganji.manideep2018@vitstudent.ac.in

M. R. SarathVamsi
e-mail: mr.sarathvamsi2018@vitstudent.ac.in

© Springer Nature Singapore Pte Ltd. 2020
K. N. Das et al. (eds.), *Soft Computing for Problem Solving*,
Advances in Intelligent Systems and Computing 1057,
https://doi.org/10.1007/978-981-15-0184-5_67

1 Introduction

Different markets over the world are winding up progressively more saturated, with an ever-increasing number of clients swapping their enlisted benefits between contending organizations. In this way, organizations have understood that they should center their showcasing endeavors in client maintenance as opposed to client procurement. In reality, examines have demonstrated that the assets a company spends in endeavoring to increase new clients are far more noteworthy than the assets it would spend if it somehow happened to endeavor to hold its clients. Client maintenance techniques can be focused on high hazard clients that are meaning to stop their custom or move their custom to another administration contender. This impact of client misfortune is otherwise called client churn. From a machine learning viewpoint, churn can be planned as a paired characterization issue. In spite of the fact that there are different ways to deal with churn expectation (for instance, survival examination), the most well-known arrangement is to mark 'churners' over a particular timeframe as one class and clients who remain drew in with the item as the correlative class thus precise early distinguishing proof of these clients is basic in limiting the cost of an organization's general maintenance advertising strategy. Churn is characterized distinctively by every association of item. For the most part, the clients who quit utilizing an item or administration for a given timeframe are alluded to as churners. Accordingly, churn is a standout among the most essential components in the Key Performance Indicator (KPI) of an item or administration. A full client lifecycle investigation requires investigating standards for dependability, keeping that in mind the end goal is to readily comprehend the well-being of the business or item. Currently, big companies, government organizations and people at higher positions gradually move toward big data analytics for making important decisions to become more prevailing in this world. However, data analysis is a process for cleaning, modeling and transforming the data on huge datasets to obtain the useful information. There are a lot of big data tools out there for big data analytics. Apache Spark is one among that tools which was introduced in 2010. It has a friendly Scala interface. Spark is compatible with Hadoop and its modules and runs faster than the Hadoop due to its real-time processing. Spark provides a very intelligible application programming interface (API) of great value. In comparison with Scala PySpark is easy to implement and simple to write. Appropriately, PySpark is cast off in this paper.

2 Literature Survey

Jie Lu et al. stated that churn prediction of the customer is one of the main features of the modern telecom customer relationship management systems (CRM). They conducted a real-world study on customer churn and proposed to use boosting to improve a customer churn prediction model [1]. The author Hua Hsu discussed a system which recommends for customer churn by proposing a decision tree algorithm. The data which is used for this analysis has roofed over 4000 members and more than 60,000 transactions, over a period of four months [2]. Malathi and Kamalraj focused their research on understanding customer churn prediction using different data mining techniques. This approach can be used in telecommunication industries for customer retention activities of their customer relationship management efforts. They have used the data mining techniques to predict call churn on the customer details [3]. Author Veronika Effendy has proposed a technique for the imbalanced data handling problem to improve the customer churn prediction model. This proposed technique is a combination of weighted random forest (WRF) and sampling for dataset balancing so that its accuracy of the churn prediction is enhanced [4]. The author G. Ganesh Sundar Kumar proposed one-class SVM-based under-sampling technique for enhancing the insurance fraud detection and churn prediction. Initially, the data is sampled using one-class SVM technique and then classing is performed on that using ML algorithms. Depending on the results, it is concluded that decision tree performance is better than other classification algorithms along with one-class SVM. It reduces system complications and helps in improving the prediction precision [5]. D. Olle et al. proposed hybrid learning model to predict customer churn for the telecommunication industry. Their model utilizes WEKA which is one of the well-known tools of machine learning. They even stated that data mining technique can detect the customers churn with high tendency but lacks in providing the reason of the churn. The primary goal of their study is to prove that models built using data mining techniques can explain the churn behavior with more accuracy than the models using single methods [6]. According to Yen, C et al., customer churn in telecommunication industry refers to the customer shifting from one service provider to other service provider. Customer management systems are the process conducted by telecom company to retain their customers. The evolution of the technology and increase in the service providers made this market more competitive than ever. So, the telecom industry has realized that to sustain their profits and preserve their customers to survive in this competitive world [7]. The authors Yihui et al. proposed a decision tree-based random forest method for characteristic extraction. From the original data with Q characteristics, exactly N samples with $q < Q$ are selected randomly for each tree to form a forest. The number of trees depends on the number of characteristics randomly combined for each decision tree from the whole Q characteristics [8].

3 Background Study

3.1 Apache Spark

Apache Spark is a computation platform designed to be fast and easy to use it. It provides fault tolerance on commodity hardware, fault tolerance and scalability. It contains application programming interfaces (APIs) for Scala, Python, Java and

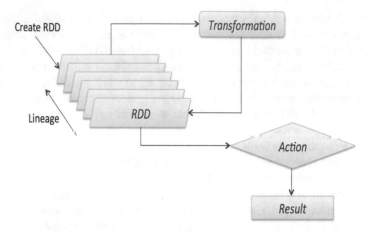

Fig. 1 Example of transformation process in Apache Spark

libraries for SQL, machine learning, streaming and graph processing. It can be performed on Hadoop clusters or as a stand-alone. It is based on the Hadoop MapReduce model and extends it to implement it for more computations that includes stream processing and interactive queries. The main feature of Apache Spark is its in-memory cluster computing which indeed increases the processing speed of an application.

As shown in Fig. 1, Apache Spark uses Resilient Distributed Dataset (RDD) which is a read only collection of objects than can operated parallelly, which are separated across a set of machines which can be restored if a partition is lost. There are two types of RDD operations: transformations and actions. Transformations are the functions that take Resilient Distributed Dataset and convert into RDD of another type, sing functions which are designed by user. Some of the functions of transformations are map, filter, reduceByKey, join, cogroup randomSplit. Thus, actions are RDD operations of Spark which gives non-RDD values. The values generated of action are stored to drivers or to the external storage system. Action is a way of sending data from executer to drivers. Executers are responsible for performing tasks; driver is a JVM process which coordinates execution of a task and workers.

3.2 MongoDB

MongoDB is an open-source document database and one of the leading NoSQL databases. It is a cross-platform database that provides availability, high performance and scalability. It works on the concept of documents and collections. MongoDB server generally consists of multiple databases. Number of fields, size of the document and content differ from one document to another. In MongoDB, data is stored in the form of JSON documents. It is schema less and easy to scale.

3.3 *Boosting*

Boosting algorithm is one of the most powerful learning ideas which are designed for classification problems. Most of the boosting algorithms consist of iteratively learning weak classifiers with respect to a distribution and adding then to a final strong classifier. It converts a set of weak learners to strong ones. There are many types of boosting algorithms such as gradient tree boosting XGBoost and AdaBoost (adaptive boosting).

4 Architecture

Figure 2 shows the call churn prediction architecture that takes customer information as data and predicts the churn. The customer performs several events for example, a stream of billing actions, a stream of call records by the company and the subscriber, SMS, complaints, etc. The Integrator module integrates all these streams and generates a logically unique stream of events. The Data Manager module manages the customer pending predictions queue and customer's information database. The predictions queue contains all predictions which are waiting for refusal or confirmation in the future; whereas, the customer information database contains all the basic information about the subscribers (name, age, type of contract, address, etc.). The Record Processor is the soul of the system. It receives stream of events which are generated by the integrator and uses those events for two purposes. First, it updates the customer's information database according to the event. Second, it generates records from events using information from the customer information database. The Record Processor builds, maintains and applies the predictive models. Therefore, it contains machine learning or data mining algorithms that make prediction possible. Thus, the

Fig. 2 Framework architecture of call churn prediction

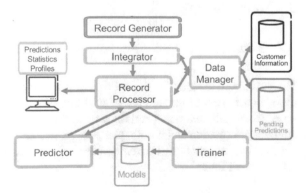

Fig. 3 MongoDB Spark
Connection establishment

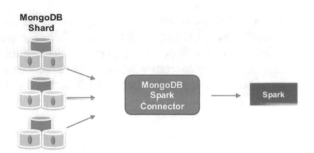

Record Processor module produces predictions and profiles of the predicted churn-ers. The churner's ids and their profiles produced by Record Processor are passed to the user interface or to the other parts of the customer management systems (CMM) so that adequate actions can be assessed and performed.

Figure 3 shows the MongoDB Spark Connection establishment. The MongoDB Spark is an open-source project, to read and write the data from MongoDB using Apache Spark. The connection offers various features like converting a MongoDB collection into a Spark RDD and methods to load collections directly into a Spark Data Frame or Dataset. MongoDB helps us to develop applications faster because it uses tables and stored procedures which are no longer required. It is an advantage to developers because earlier the tables should be translated to object model before they could be used in the application. Now, the stored data and object model have same structure similar to JSON format which is also known as BSON. MongoDB provides various options, supports scalability and maintains data consistency.

5 Methodology

5.1 Dataset

We had chosen to study the dataset from a telecom company which includes data of mobile customers in the number of millions who are active at a particular point of time. We initially extracted variables from the customer's database, which includes contract information and mobile plan, usage, account length, area, billing and prod-uct holding information. It even consists of the customer care inbound/outbound information (Table 1).

Table 1 List of attributes with its type in the dataset

S.no.	Column name	Type
1.	State	String
2.	Account length	Integer
3.	Area code	Integer
4.	International plan	String
5.	Voice mail plan	String
6.	Voicemail messages	Integer
7.	Total day minutes	Double
8.	Total day calls	Integer
9.	Total day charge	Double
10.	Total evening minutes	Double
11.	Total evening calls	Integer
12.	Total evening charge	Double
13.	Total night minutes	Double
14.	Total night calls	Integer
15.	Total night charge	Double
16.	Total international minutes	Double
17.	Total international calls	Integer
18.	Total international charge	Double
19.	Customer service calls	Integer

6 Pseudo Code

1. Loading the data from the MongoDB

$$Data = load()$$

2. Create training and testing data

$$train, test = data.Split(0.8, 0.2)$$

3. Check for correlation among columns

$$train.Correlation()$$

4. Converting from string to numeric values

$$Val = String\ Indexer(Train)$$

5. Acquiring the vectors

$$Vector = vectorAssembler(Train)$$

6. Creating a pipeline

$$Pipeline(val, vector, GBTClassifier)$$

7. Training the model

$$Model = pipeline.fit(train)$$

8. Predicting the test data

$$Predict = model.Transform(test)$$

9. Displaying the predicted data

$$print\ Predict$$

7   Results Analysis

We have evaluated the performance of our churn prediction model by using a train-ing set of customer information which has been collected from the link (https://bigml.com/user/cesareconti89/gallery/dataset/58cfbada49c4a13341003cba). Each customer is classified according to customer churn susceptibility. The testing dataset is monitored frequently and updated with the latest information. In this way, we can replicate the real-world scenario of churn prediction. The input for this model is the customer information database which contains all the information about the customer (state, account length, area code, phone number, etc.). The input contains all information about the customer; the total information is required to predict the churn.

In Fig. 4, churn is predicted for the customers. If the churn value is 1, then it is 'True,' else if it is 0, then it is 'False.' The churns are calculated for each customer, and churn percentage is calculated. In our model, the churning for the 932 customers is 136. The churn rate for these 932 customers is 14%. The bar plot of churn prediction of 932 customers is shown in Fig. 5.

A churn prediction model should be measured by its ability to identifying churners for marketing purpose. We, therefore, used the Receiver Operation Characteristic (ROC) for testing the efficiency of this model.

Fig. 4 Analysis of the
outcomes of churn

```
+-------------+-----+
|phone number|churn|
+-------------+-----+
|    341-9764|    0|
|    389-7073|    0|
|    362-8331|    0|
|    356-5244|    0|
|    364-1134|    0|
|    346-8863|    0|
|    336-6533|    0|
|    366-9781|    0|
|    404-1931|    0|
|    418-9385|    0|
|    359-9454|    0|
|    396-2335|    0|
|    394-4548|    0|
|    333-5295|    0|
|    383-9255|    0|
|    345-8237|    0|
|    402-1381|    0|
|    397-5911|    0|
|    341-4075|    0|
|    349-9060|    0|
|    384-6132|    1|
```

Fig. 5 Bar plot for the
predicted churns

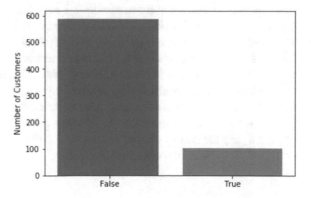

Evaluator = BinaryClassificationEvaluator ()

auroc = evaluator.evaluate(predictions, {evaluator.metricName:

"areaUnderROC"})

print ("Test Area Under ROC "+str(auroc))

Test Area Under ROC 0.741558019577

The churn prediction system is tested by the Receiver Operating Characteristic (ROC). When the testing is performed, it is measured that this model efficiency is 74.1%, for example, if this model generates churn for four customers, three predictions are correct and one prediction of the customer churning might be wrong.

8 Conclusion

The telecommunication industry had suffered huge loss from the high churn rates and immense churning of customers. Though the loss is unavoidable, churn can be managed and kept in an adequate level. This research conducts an experimental investigation of customer churn prediction based on real-world data sets. In this paper, the boosting algorithm is used which has predicted the churn for customers. The evaluation of the model we developed is calculated which has proved that this model is efficient of predicting the churns. There is still a lot of substantial work to do from both business and technical point of view. On the one hand, the performance and efficiency of the model can be further improved as well as other classification methods can be used and compared. On the other hand, churn prediction will only provide a basis for the generation of lists and prioritizes contact customers in spite of identifying the reason of customer's churn behavior and providing the customer needs are also essential for targeted marketing.

References

1. Lu, J., Lin, H., Lu, N., Zhang, G.: A customer churn prediction model in telecom industry using boosting. IEEE Trans. Indu. Inf. **10**(2), 1659–1665 (2014)
2. Chlang, D.A.: A recommender system to avoid customer churn. Expert Syst. Appl. **36**(4), 8071-8075 (2003)
3. Malathi, A., Kamalraj, N.: Applying data mining techniques in telecom churn prediction. Int. J. Adv. Res. Comput. Sci. Softw. Eng. **10** (2013)

4. Baizal, Z.K.A., Effendy, V.: Handling imbalanced data in customer churn prediction using combined sampling and weighted random forest. In: 2014 2nd International Conference (ICoICT), pp. 325–330. IEEE (2014)
5. Siddeshwar, V., Ravi, V., Sundarkumar, G.G.: One-class support vector machine based undersampling: application to churn prediction and insurance fraud detection. In: 2015 IEEE International Conference on Computational Intelligence and Computing Research (ICCIC), pp. 1–7. IEEE (2015)
6. Olle, G.D., Cai, S.: A hybrid churn prediction model in mobile telecommunication industry. Int. J. e-Educ. e-Bus. e-Manage. e-Learn. 4(1), 55 (2014)
7. Hung, S.Y., Yen, D.C., Wang, H.Y.: Applying data mining to telecom churn management. Exp. Syst. Appl. 31, 515–524 (2006)
8. Chiyu, Z., Yihui, Q.: Research of indicator system in customer churn prediction for telecom industry. In: 2016 11th ICCSE, pp. 123–130. IEEE (2016)
9. Ahn, J.-H., Han, S.-P., Lee, Y.-S.: Customer churn analysis: churn determinants and mediation effects of partial defection in the Korean mobile telecommunications service industry. Telecommun. Policy 30(10), 552–568 (2006)
10. Neslin, S.A., Gupta, S., Kamakura, W., Junxiang, L., Mason, C.H.: Defection detection: measuring and understanding the predictive accuracy of customer churn models. J. Mark. Res. 43(2), 204–211 (2006)

Location of UPFC Based on Critical Bus Ranking Index

A. Thamilmaran and P. Vijayapriya

Abstract Unified power flow controller (UPFC) is one among the versatile devices from the FACTS family that has a great potential to enable power system engineers maintain the required security level of a large power systems that are interconnected by controlling the amount of active and/or reactive power flow. Identification of suitable location for installing this costly device is relatively very difficult task. Various criteria need to be satisfied before selecting the location for installation of UPFC. In this paper, a novel method is proposed which identifies the best location for UPFC placement based on the net reduction in the critical bus ranking index under various line outage contingencies. The proposed method was implemented on a 5-bus system. The UPFC was placed at the identified location and it is found that the installation of UPFC has reduced the criticality of the selected bus and it also enhanced the voltage stability margin of the system under selected line outage contingencies.

Keywords UPFC · Voltage stability margin · Minimum singular value · Critical bus · Contingency and severity index

1 Introduction

In the recent power system networks, active and reactive power flow control in a transmission lines are a very crucial aspect as there is possibility of bilateral transactions. Due to new power generation companies entering the power market, there is a serious threat for voltage stability, which may lead to voltage collapse. Hence, there is a need for enhanced secured operation of the entire power systems. In recent days, flexible AC transmission system (FACTS) devises based on power electronic

A. Thamilmaran · P. Vijayapriya (✉)
SELECT, Vellore Institute of Technology, Vellore, India
e-mail: pvijayapriya@vit.ac.in

© Springer Nature Singapore Pte Ltd. 2020
K. N. Das et al. (eds.), *Soft Computing for Problem Solving*,
Advances in Intelligent Systems and Computing 1057,
https://doi.org/10.1007/978-981-15-0184-5_68

components has become an attractive tool that controls power flow, provides the possibility of operating grid at increased efficiency and flexibility. The most promising device that has emanated from the FACTS family is the unified power flow controller (UPFC) [1–3].

In the last decade, the research focus in the FACTS family is on UPFC that had resulted in many beneficial contributions in areas such as modeling [4–6], efficient control strategy [7, 8], and enhancement of stability of system using UPFC [9, 10]. However, in order to obtain good performance, proper placement of UPFC is a more important and is a vital task too. Various methods are reported in literature for finding the locations of UPFC in integrated systems but not much attention has been given to interconnected power system under contingencies [11]. Alamelu et al. [12] have set the location of UPFC based on the sensitivity analysis that is performed on the UPFC transformer model and found that the generation cost will be minimum as compared to other location. Kazemi et al. [13] found the location of UPFC using GA that enhances the loadability of the power system network using injection model. They validated their result only in steady state and the transient effects are left open for further research. Shaheen et al. [14] discuss on optimal placement of UPFC and its parameter setting so that it improves the security of the power system but they have done this study only under single line outage contingencies. In this paper, the location selection for installation of an UPFC is carried out with the objective of reducing the criticality of the most critical bus. Identification of the critical bus based on voltage stability criterion proposed in [15] is used in this paper under various line outage contingencies. A new indicator, net severity index (NetΔSI) is developed to identify the best location for installation of UPFC. The proposed method has been tested under simulated conditions on a standard 5-bus system. Section 2 presents the static model of UPFC incorporated in this paper, Sect. 3 briefly explains the concept of critical bus ranking and the step-wise procedure for bus ranking. Section 4 explains the indicator used in this method and algorithm for placement of UPFC and in Sect. 5 results are presented.

2 Static Model of UPFC

A static model for UPFC, which is referred as UPFC injection model derived in [4], is used in this paper. This model is very useful in understanding the impact of the UPFC on the power system. Furthermore, the UPFC injection model can easily be incorporated into the power flow model.

The UPFC injection model is shown in Fig. 1. The model shows that the net active power interchange of UPFC with the power system is zero, as it is expected for a lossless UPFC.

Fig. 1 UPFC injection
model

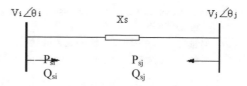

Table 1 Modified Jacobian matrix [4]

$H(i,i) = H^o(i,i) - Q_{sj}$	$N(i,i) = N^o(i,i) - P_{sj}$
$H(i,j) = H^o(i,j) + Q_{sj}$	$N(i,j) = N^o(i,j) - P_{sj}$
$H(j,i) = H^o(j,i) + Q_{sj}$	$N(j,i) = N^o(j,i) + P_{sj}$
$H(j,j) = H^o(j,j) - Q_{sj}$	$N(j,j) = N^o(j,j) + P_{sj}$
$J(i,i) = J^o(i,i)$	$L(i,i) = L^o(i,i) + 2Q_{si}$
$J(i,j) = J^o(i,j)$	$L(i,j) = L^o(i,j)$
$J(j,i) = J^o(j,i) - P_{sj}$	$L(j,i) = L^o(j,i) + Q_{sj}$
$J(j,j) = J^o(j,j) + P_{sj}$	$L(j,j) = L^o(j,j) + Q_{sj}$

$$P_{si} = r_{bs} V_i V_j \sin(\theta_{ij} + \gamma) \quad P_{sj} = -r_{bs} V_i V_j \sin(\theta_{ij} + \gamma)$$
$$Q_{si} = r_{bs} V_i^2 \cos \gamma \qquad Q_{sj} = -r_{bs} V_i V_j \cos(\theta_{ij} + \gamma)$$

The Jacobian matrix is modified as given in Table 1. (The superscript o denotes the Jacobian elements without UPFC.) This modified Jacobian is incorporated in the load flow model used in this work.

3   Critical Bus Ranking

The basic objective of embedding an UPFC into the power system is to obtain an effective control of active power and/or reactive power in the line. However, in order to locate the UPFC in an appropriate location, many criteria are to be considered. In the large interconnected power system, during various line outage contingences the buses are likely to reach to a very critical state and hence reduction of bus criticality is also an important task for power engineers. Literature survey revealed that reduction of criticality of a bus is not being considered as the criterion for suggesting the suitable location for UPFC.

The basic concept of the critical bus ranking used in this paper is to identify the critical buses, which when loaded under a specific contingency case up to their stability margins cause severe impact on overall system voltages. This severity is quantified with an accurate severity Index, which incorporates the overall system impact. This technique involves in computation of severity indices for the following two cases.

(a) System operating in base case load condition and subjected to various contingencies (SIb)
(b) System at a given contingency and at the verge of voltage collapse due to increase in load of any bus up to it's voltage stability margin (SIvc).

The severity index of (a) signifies the impact of a particular contingency on the system compared with the base case condition of the system.

The severity index of (b) signifies the impact of a bus reaching to voltage stability margin under a specific contingency.

The difference between these two severity indices (ΔSI) gives a measure of severity of any bus reaching to its voltage stability margin under a specific contingency. The SI used [15] and the step-wise procedure for its computation is given below.

The general formula for calculating the severity index is

$$SI = \left[\left(\sum (\Delta P_j) \right)^2 + \left(\sum (\Delta Q_j) \right)^2 \right]^{\frac{1}{2}} \tag{1}$$

where

$$\Delta P_j = \frac{P_j 2 \Delta V_j}{V_j^0}, \quad \Delta Q_j = \frac{Q_j 2 \Delta V_j}{V_j^0}$$

$P_j + JQ_j$ Complex load connected at the Jth bus
V_j^0 Pre-contingency voltage at the Jth bus
V_j Post-contingency voltage at the Jth bus (for calculating SIB)
ΔV_j Deviation in Jth bus voltage ($V_j^0 - V_j$).

- Severity index for base case (SIB) is computed using V_j^0 and V_j.
- Severity index at the verge of voltage collapse (SIvc) is computed for the selected contingency condition in which the bus loadings of individual buses (j) are kept at their voltage stability margins (P_{VSM} and Q_{VSM}) while the remaining buses are at their base case loads. Hence, $\Delta V_j = \left(V_j^0 - V_j^1 \right)$ where V_j^1 is the voltage of the jth bus when the load on it is at its stability margin
- For each contingency case, the (ΔSI) difference between SIvc and SIB is computed for all buses.

a. **Step-wise procedure for Bus Ranking**

Step 1: Bus—wise stability margins are computed using load flow divergence technique for all load buses under selected contingency conditions.
Step 2: Severity index (SIB) is computed for base case under selected contingencies.

Step 3: Severity index ($SI_{VC)}$) is computed for selected contingency condition in which the bus loading of Individual buses are kept at their voltage stability margins. (This signifies that the system is at the verge of voltage instability due to increase in load at the bus up to its stability margin.)

Step 4: The bus-wise differences (ΔSI) between SI_{VC} and SI_B is computed for each contingency.

Step 5: Buses are ranked in the decreasing order of ΔSI.

4 Development of NETΔSI

To compute the NETΔSI, the bus ranking indices are computed using the above steps without UPFC and with UPFC in selected lines.

Incorporation of UPFC in the system changes both SI_B and SIvc due to variation in the voltages and hence, ΔSI also changes. For a particular location of UPFC, with one of the bus load at its voltage stability margin, the ΔSI value may increase or decrease for the considered set of contingencies. The net variation in the ΔSI of a bus for all contingencies is calculated as shown below:

$$\text{Net}\Delta SI = \sum_{j=1}^{n} \left(\Delta SI_j^{\text{old}} - \Delta SI_j^{\text{new}} \right) \tag{2}$$

ΔSI^{old} Severity index without UPFC
ΔSI^{new} Severity index with UPFC
j Given set of contingencies.

This NetΔSI is computed for all possible UPFC location.

a. **Flow Chart for Identifying UPFC Location**

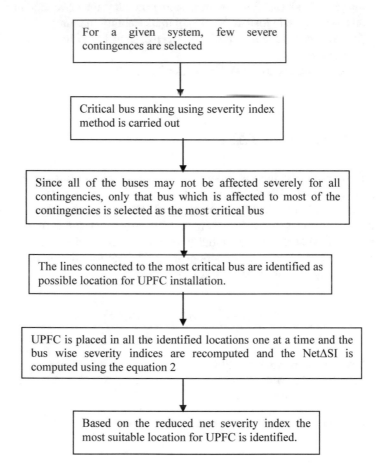

5 Case Study and Results

The proposed method is applied to a standard 5-bus system. Critical Bus ranking results reveals that the bus 3 is more critical for most of the considered contingencies and hence, reducing the criticality of bus 3 is considered as the criterion for identification of location of UPFC. A general MATLAB program was developed to compute the NetΔSI for the selected set of contingencies for a given location of UPFC.

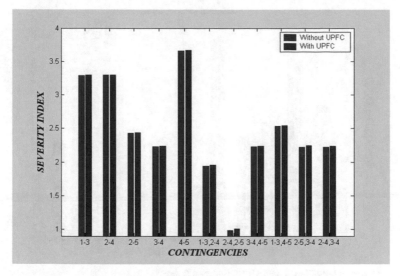

Fig. 2 Severity index of bus 3 with UPFC between 3 and 2

As per the proposed method, to reduce the criticality of bus 3, the UPFC is to placed in any one of transmission lines that are connected to bus 3 [3–4 or 3–1 or 3–2]. NETΔSI is computed for the above-mentioned locations of UPFC.

a. **Case—I: UPFC at line 3–2**

Placement of UPFC in the above line section has only increased the critical bus ranking index of bus 3 considerably for all the considered set of 11 contingencies (5 single line and 6 double line outage) Fig. 2 shows the increase in severity index of bus 3 for various contingencies with UPFC and hence this line section is not suitable for locating UPFC to reduce the criticality of bus 3.

b. **Case—II: UPFC at line 3–1**

Placement of UPFC in the above line section has also increased the critical bus ranking index of bus 3 considerably for all the considered set of converged 11 contingencies (5 single line and 6 double line outage) Fig. 3 shows the increase in severity index of bus 3 for various contingencies with UPFC hence this line section is also not suitable for locating UPFC to reduce the criticality of bus 3.

c. **Case—III: UPFC at line 3–4**

Placement of UPFC in the above line section has reduced the critical bus ranking index of bus 3 considerably for all the considered set of contingencies (5 single line and 4 double line outage) Fig. 4 shows the reduction in severity index of bus 3 for various contingencies with UPFC. It is also seen from the graph that the % reduction in ΔSI for 1–3 contingency is maximum and is around 12%.

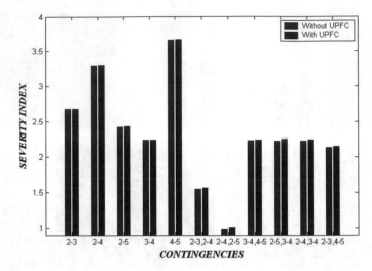

Fig. 3 Severity index of bus 3 with UPFC between 3 and 1

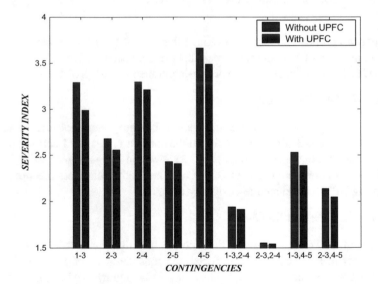

Fig. 4 Severity index of bus 3 with UPFC between 3 and 4

Table 2 NetΔSI for various UPFC location

UPFC Location	NetΔSI		
	Bus 3	Bus 4	Bus 5
3–4	**0.9572**	−1.1401	−0.1372
3–2	−0.1197	−0.1097	0.1097
3–1	−0.1111	−0.1111	−0.1098

Bold value indicates maximum value of NetΔSI at bus 3

d. Results

For the standard 5-bus system considered, the proposed method was implemented and following table gives the value of NetΔSI of a bus for various locations of UPFC (Table 2).

From the above table and from the bar graphs presented (Figs. 2, 3 and 4) it is clear that the criticality of the most critical bus (Bus 3) has reduced by placing an UPFC between 3 and 4 which is indicated by the maximum value of NetΔSI.

Simulations are carried out to verify the possibility of enhancement in the VSM due to installation of UPFC (Table 3).

The minimum singular value for all contingencies when the UPFC is located in line section 3–4 is compared with UPFC placed at other locations and it is evident

Table 3 Enhancement in MSVs with UPFC placed at various locations

Contingency	MSV			
	Without UPFC	With UPFC between lines		
		3–4	3–2	3–1
1–3	2.3941	**2.4108**	2.3958	–
2–3	3.2856	**3.3562**	–	3.2890
2–4	3.1775	**3.2517**	3.1806	3.1806
2–5	1.4368	**1.6567**	1.4436	1.4436
3–4	3.1000	–	3.1016	3.1016
4–5	3.5599	**3.6081**	3.5643	3.5643
1–3, 2–4	1.8110	**1.8589**	1.8168	–
2–3, 2–4	2.2150	**2.4953**	–	2.2235
2–4,2–5	0.8118	Not converged	0.8272	0.8272
3–4, 4–5	3.0151	–	3.0240	3.0240
1–3, 4–5	2.3578	**2.3724**	2.3626	–
2–4, 3–4	1.3156	–	1.3270	1.3270
2–3, 4–5	3.2501	**3.3417**	–	3.2602
2–5, 3–4	0.5225	–	0.5398	0.5398

Bold values indicate maximum value when the UPFC is placed between 3–4

from the table that there is maximum enhancement in the VSM due to installation of UPFC at the identified location.

The results are summarized as follows:

- Identified critical bus: Bus 3
- Best location for UPFC placement: Line 3–4
- Average improvement in MSV for 9 contingencies: 5.03%
- Average reduction in the severity index of bus 3: 7.36%.

It is evident from the results that the proposed UPFC placement strategy helps in achieving the reduction in the criticality of the bus and also in enhancing VSM under all contingencies.

6 Conclusion

A simple yet most effective method of determining the suitable location for installation of UPFC is proposed based on the reduction in the net critical bus ranking index of the most critical bus for the considered set of contingencies. The proposed method is implemented on a standard 5-bus system. For each of the contingency, analyses are carried out by placing UPFC in different transmission lines connected to the most critical bus. It was proved that the installation of UPFC in the identified location has not only reduced the criticality of the most critical bus but also enhanced the voltage stability margin of the system as indicated by enhancement in the minimum singular value of the load flow Jacobian and also improvement in the voltage profile of the system.

References

1. IEEE Power Engineering Society/CIGRE: FACTS overview. Special issue, 95TP108. IEEE Service Centre, Piscataway, NJ (1995)
2. IEEE Power Engineering Society/CIGRE: FACTS applications. Special issue, 96TP116-0. IEEE Service Centre. Piscataway, NJ (1996)
3. Huang, X., Li, Q., Yang, N.: Topology and matching capacity of direct-connected converter for co-phase traction power supply substation. J. Southwest Jiaotong Univ. **52**(2), 379–388 (2017)
4. Noroozian, M., Angquist, L., Ghandhari, M., Andersson, G.: Use of UPFC for optimal power flow control. IEEE Trans. Power Deliv. **12**, 1629–1633 (1997)
5. Fuerte-Esquivel, C.R., Acha, E.: Unified power flow controller: a critical comparison of Newton-Raphson UPFC algorithms in power flow studies. IEE Proc. Gener. Transm. Distrib. **144**(5), 437–444 (1997)
6. Fang, D.Z., Fang, Z., Wang, H.F.: Application of the injection modeling approach to power flow analysis for systems with unified power flow controller. Int. J. Electric Power Energy Syst. **23**, 421–425 (2001)
7. Padiyar, K.R., Kulkarni, A.M.: Control design and simulation of unified power flow controller. IEEE Trans. Power Deliv. 1–7 (1997)

8. Padiyar, K.R., Uma Rao, K.: Modeling and control of unified power flow controller for transient stability. Int. J. Electric Power Energy Syst. **21**, 1–11 (1997)
9. Chen, H., Wang, Y., Zhou, R.: Transient and voltage stability enhancement via coordinated excitation and UPFC control. IEE Proc. Gener. Transm. Distrib. **148**(3), 201–208 (2001)
10. Chen, H., Wang, Y., Zhou, R.: Transient stability enhancement via coordinated excitation and UPFC control. Int. J. Electric Power Energy Syst. **24**, 19–29 (2002)
11. Galiana, F.D.: Bound estimates of the severity of line outages. IEEE Trans. PAS **103**, 2612–2624 (1984)
12. Kazemi, A., Arabkhabori, D., Yari, M., Aghaei, J.: Optimal location of UPFC in power systems for increasing loadability by genetic algorithm. In: Universities Power Engineering Conference, 2006. UPEC'06. Proceedings of the 41st International, vol. 2, pp. 774–779. IEEE (2006)
13. Alamelu, S.M., Kumudhini Devi, R.P.: Novel optimal placement of UPFC based on sensitivity analysis and evolutionary programming. J. Eng.Appl. Sci. **3**(1), 59–63 (2008)
14. Shaheen, H.I., Rashed, G.I., Cheng, S.J.: Optimal location and parameter setting of UPFC for enhancing power system security based on differential evolution algorithm. Int. J. Electr. Power Energy Syst. **33**(1), 94–105 (2011)
15. Kalaivani, R., Dheebika, S.K.: Enhancement of voltage stability and reduction of power loss using genetic algorithm through optimal location of SVC, TCSC and UPFC. J. Eng. Sci. Technol. **11**(10), 1470–1485 (2016)

Personal Assistant for Social Media

Anshul Mathew, Kevin Job Thomas and P. Swarnalatha

Abstract In today's world, with an increase in usage and consumption in social media, it has become paramount as to how we consume and manage our social media. The reason as to why we should do this lies in understanding how social media works and its numerous advantages. Through this paper, we propose a program that culminates data from multiple social networking websites and emails into a single platform from where users can view, edit and update directly into the emails and social networking websites.

Keywords Personal assistant · Social networks · Email

1 Introduction

In today's world, with an increase in usage and consumption in social media, it has become paramount as to how we consume and manage our social media. The reason as to why we should do this lies in understanding how social media works and its numerous advantages. After having chalked that out, we need to understand who would benefit from this by understanding our stakeholder profiles.

2 Literature Review

Social media is a computer-based technology that facilitates the sharing of ideas and information and the building of virtual networks and communities. Using social media, one can share and receive photos, videos as well as other personal information by taking advantage of easy communication systems provided by inter and other

A. Mathew (✉) · K. J. Thomas · P. Swarnalatha
Vellore Institute of Technology, Vellore, Tamil Nadu 632014, India
e-mail: anshulmathew@gmail.com

P. Swarnalatha
e-mail: hgswarnalatha@gmail.com

© Springer Nature Singapore Pte Ltd. 2020
K. N. Das et al. (eds.), *Soft Computing for Problem Solving*,
Advances in Intelligent Systems and Computing 1057,
https://doi.org/10.1007/978-981-15-0184-5_69

telecommunication capabilities present today. It is not only restricted to these activities but can also permeate into spheres of gaming, networking with people, video sharing, virtual worlds, sharing movie reviews of restaurants movies and other items [1]. It also has the capabilities to serve as a market place.

The various platforms are accessed by end of network devices like mobile phones, tablet computers and even regular computer via an application or a web application generally used for messaging in real time.

At an individual level, it was intended for keeping in contact with friends who are in different parts of the world with great ease and simplicity. Finding jobs and various career opportunities through these social media are novel ways that are gaining much traction and popularity these days [2]. There may even be just a common interest on a particular topic which in turn forms forums which then link similar-minded people together [3].

From 'Social media: The new hybrid element of the promotion mix' by Mangold and Faulds highlights the importance of social media in business today. [4]. That paper really talks about how social media really enables a single individual to connect to virtually hundreds and thousands of people with a single click of a button. Using social media, we can shape what exactly is being talked about a product and receive feedback from it. Since there are multiple platforms, it becomes increasingly difficult to manage these discussions without a consolidation of these platforms which is what we have attempted to do with principles of human-computer interaction as our guiding light. In a nutshell, social media provides advertising, personal selling, public relations, publicity, direct marketing and sales promotion to produce a unified customer-focused message and, therefore, achieve various organizational objectives. Without the right tools, taking advantage of this would be very difficult.

By keeping in mind, the eight golden rules enshrined in 'Designing the User Interface: Strategies for Effective Human-Computer Interaction' by Schneiderman [6], we have tried to make the learning curve for the user very gradual.

After understanding all the benefits, we have to consider a design process, and after extensive research, we have concluded that the best way forward is that by adopting a conservative approach as stated by Daniel Fallman in 'Design-oriented Human-Computer Interaction'. It borrows from the school of thought where we progress gradually from abstract SRS to the more concrete implementation [7]. The undertaking of the designer or the group of designers is to find, following the requirements specification, solutions to the described problems that may feasibly be carried out within the boundaries of the design project's constraints, including issues of cost, time, and performance which are what we have learnt from this project [8] .

3 Architecture

Essentially, it is a HTML in Python webserver architecture. The architecture that we followed was one where a Python webserver would run the Python programs to host a site that would be running on the information from the Python files.

3.1   Back End

For this purpose, firstly, a pure Python program was developed for each of the functions in Python on PyCharm. This would then have to be converted to a suitable format understandable by the webserver which was Django in our case. This means that the files of Python must be converted to views of Python. Each module developed in this manner must be installed as an app on the Django webserver.

3.2   Front End

Front end is a website that is capable of displaying the results of the contents retrieved by the back end. The front end is done in HTML/CSS/JS with bootstrap to make the website responsive. Django cannot process static elements just as the JavaScript scripts or the images. Hence, the HTML has to be split into separate static and dynamic components where the dynamic components are identified as templates.

3.3   Interfacing of Back end and Front End

Using Jinja Logic, the Python data structure is embedded into the HTML page which displays the contents of the request.

4   Software Requirements Specifications

This product is being developed for easier management, and use of multiple social media accounts on a single platform to perform publishing of posts, messages, emails, etc. Through this, one can access all notifications through the app/web app. Currently, we go into great depth regarding Gmail module of the app.

4.1   Intended Audience and Reading Suggestions

The intended audience is for this is social media mangers of big celebrities and organizations who want to better reach their customer and add more value to their businesses. The developers of the project would also find this document very handy. Testers can identify what module works with which part and so on and so forth. Future document writers can make the relevant updates based on the insights and inputs from developers. This contains an overall view of the app but more specifically on the Gmail segment of it.

4.2 Product Scope

The product that is being developed has the potential to be used even by single on a day to day basis to remove clutter from their busy lives and be more social media aware and present. This can, in future, learn the preferences of a person and become more personalized reminding them about when to post and suggest topics to post on.

4.3 Product Perspective

With advent of multiple social media platforms and every bodies interest to be present on all of them, we are bound to make mistakes while juggling all these accounts around. That is why we proposed to combine all of these similar functionalities into one seamless application. Using this app, you can identify and connect better with you clients and can avoid carelessly missing out reaching them via multiple platforms just because you were not aware of a notification. Using this app, it becomes easier to make a social media marketing strategy as different app that use different demographics, and so by rolling out each update in more controlled and consciously organized manner, we can handle our reach better. An easier UI for mail can facilitate this.

4.4 Functional Requirements

Currently, we are focusing on Gmail product functionality to go with the first iteration. We should be able to:

- Send mail
- Receive mail
- Search mail
- Sort by labels.

We have incorporated all of this with an improved UI so as to marginally decrease the clutter present otherwise with a minimalistic design.

4.5 User Classes and Characteristics

The business owner—for example, a clothes shop which has many of its products upon sale can display its collection with an appropriate mixture of photos and videos that can be simultaneously posted all from one platform.

The Celebrity—Celebs, these days, have to constantly engage with their fans so as to remain relevant in the pop culture scene and are notoriously famous for posting frequently. An app that manages this will be a blessing in disguise.

4.6 Operating Environment

As of now, we are using a web client because support is mainly available on these platforms by the respective developers. Since it is a web application, it will work beautifully on mobile phones, PC, laptops, tablet devices and any device which supports a browser and a display screen of good resolution.

4.7 Design and Implementation Constraints

Since we are relying on the API clients from Google and the structure of websites from famous social media platforms as long as there are not any major changes in the API calls and UI design of the original sites, the app should not face too many hurdles.

4.8 User Documentation

This project is based on various third-party clients like Google, FB, etc. [9]. Going through their API documentation is paramount for developers if they must uncover any bugs. The appropriate links have been mentioned in references.

4.9 Assumptions and Dependencies

Our mission is not to entirely re-invent the wheel but rather to make a faster and more efficient wheel that is easy to use. In this project, we assume that the users have basic knowledge as to how to operate an email account and its relevant functionalities like composing a mail, searching for mail, viewing different threads and replying to threads. We depend on API given by third-party developers which should not be underestimated. Since we are using Python for this project if the libraries support for Python are not updated or are no longer continued is good to be noted. This project will require a web browser and a working Internet connection for this project as it connects to the Internet. There are several CSS/HTML-based libraries being used, and hence, browser support must be there for all of them.

4.10   External Interface Requirements

4.10.1   User Interfaces

Layout and Structure:

The developed responsive GUI is one that conforms to the traditional left menu format standard of websites, however, due to the constraint that the navigation panel occupies a large space on a mobile device such as a phone or a tablet, and when the same website is viewed through a smaller screen, the layout will be shifted to a top-navigation panel format [10]. This ensures that maximum screen area can be made available to the user to view details from emails and social media.

GUI Standards Followed:

i.   KISS (Keep it Simple Silly): This is a commonly used standard which means that the entirety of the website can be accessed in the most simple and easy to follow manner. By following the title to website approach, we are able to tell the user exactly the format of navigation [11].

ii. Apply UI Elements as they are Originally Defined: Underline to confirm selection: The selected choice of website is underlined in the user interface to remind the user of his/her current selection.

iii. Apply UI Elements as they are Originally Defined: The selected account of the user is highlighted.

iv. Apply UI Elements as they are Originally Defined: Standard Header-Footer-navigator. The elements of the headers, footers and navigators remain same throughout the website; hence, the users know exactly where to look for.

v.   Areal Scroll of the mouse: Just like the standard of Facebook and Twitter scrolling, we too have used the areal scroll concept where there are multiple sections that the user can scroll but the scroll happens where the mouse pointer is placed, and this avoids the hurdle of making a further choice of selective scrolling.

vi. Gestalt's Law of continuation: According to Gestalt, a user interface becomes more acceptable to users if there is continuity in the design.

Application 1: Continual scroll for websites.
For this purpose, we have arranged the social media networks websites one after another in a continuous and efficient manner.
Application 2: Application of a transitional animation rather than a click to present. Older UIs often use a click to view design where as soon as the request is

clicked, the response is loaded. This however causes confusion and often leaves the user questioning whether the response is loaded. Modern UI technique that we used here is a transitional animation that shows the transition of the UI to the new response rather than a sudden load of the same.

vii. Standard of Common Fate: The standard of common fate means that for all belonging to the same family, if a certain action is performed, then the expected result must be similar. We have applied this approach to the individual websites. For instance, when a user clicks Google drive from the navigation bar, the page scrolls up or down to the location of the Google drive module. Similarly, a user can expect the same result for Gmail or YouTube as well.

4.11 Error Handling Within the UI

There are various small or big errors that are possible due to size of the project [12]. Most of the possibility of error is in the back end during the scraping and the storing of data. However, any such occurrence of an error must be noted if necessary, to the user. Some of the possible errors that could occur with their countermeasure are given in Table 1.

Table 1 Error handling within the UI

Error	Details	Countermeasures and Notification to user
Login error	If the user has not logged into the account	The login page of the account is presented to the user, and he/she may enter the credentials. These credentials are hence not saved which means that the user need not worry about the security of password being stolen from the device
Internet error	If there is no internet available for the website	HTTP404 custom page is displayed assuring user that the correct website is reached but there is no internet
Load fail	If any of the individual websites fail to load	The failed to connect page is displayed only in the frame where the website sits. The remainder of the website works with all its functionality

4.12 Hardware Interfaces

The hardware being used here is not at all sophisticated and involves the use of a basic desktop PC and a mouse to access the site along with a working Internet connection.
Software Interfaces:
The software interfaces used include

- Python 3.6
- The Django Framework [13]
- The Http Urllibs in python
- Selenium Drivers
- HTML/CSS
- Jinja.

4.13 Communications Interfaces

The communication interface used include

- SMTP
- HTTP
- TCP/IP.

4.14 Nonfunctional Requirements

4.14.1 Performance Requirements

We must ensure that the time taken to display the information should not exceed 5 s. All the pages must be hyperlinked well and should do what is intended. Sending mail via the API client must be done smoothly without making the back end heavy. The elements in the web page must be responsive without them going out of place. While querying, we should not wait a long time.

4.14.2 Security Requirements

Unwanted access by unauthorized users so as to compromise the persons' email and other accounts must be thwarted. Using tokens to login makes this task more secure. Any changes to login credential could be disastrous and securing that will be paramount.

4.14.3   Scalability

This will have to be hosted on servers that run round the clock so that we access to this service anytime by anyone. Adopting appropriate frameworks like Django makes this possible.

5   Software Design Specification

5.1   Design Considerations

5.1.1   Assumptions and Dependencies

Certain assumptions of the user were made while designing this particular piece of software. They can be described as follows:

1. They have used any one of the major operating systems.
2. They have used an Internet browser of some sort and known how to open a new tab and plug in a URL to use it.
3. They have used popular social media sites like Facebook, Instagram, Gmail and even web WhatsApp.
4. They have a basic understanding of how email works and its functions.

   Certain dependencies of this project include:

1. The structure and functionality of the original third part must not change
2. The APIs used should not become obsolete
3. A working Internet connection
4. A browser which supports flash.

5.2   General Constraints

The success for our project depends on a very good and strong Internet connection without which access to social media let alone our app will not be successful. Another major constraint will depend on the exact structure and functionality provided by the third-party vendors' websites. A dramatic change in their structure and function can render our site buggy and inconvenient. Network connectivity is a major factor that one must consider. Outage of servers by third-party vendor will cause issues.

5.3   Goals and Guidelines

Ultimately, we want a system that is able to manage our social media handles in a fast and efficient manner. For this, we require the apps to be well differentiated but accessible at the same time. We realized that some of the traditional UI has been designed perfectly but their amalgamation has always been the problem which is what our final goal is—'To beautifully amalgamate the platforms'. We desire to keep some of the original UI as we do not want to burden the user with a steeper learning curve and reduce the load on memory m one of Schneiderman's principles of HCI [6].

5.4   Development Methods

For this project, we have adopted an incremental model of software development. While developing, we chunked up the project into several modules which served as milestone points so as to get the project off the ground step by step [14]. For example, we focused on being able to get the Gmail functionality of the ground and subsequently other modules like the WhatsApp module and later the YouTube module up and running. We defined each module as a milestone that we would want to achieve, thereby making the whole project scalable and efficient.

5.5   Architectural Strategies

In our original plan to design and develop the software, we wanted to have a dashboard system wherein every user can view all the notifications on a single page and without any clutter. The UI to be designed should be clean and aesthetic in nature. Hence, we thought having a direct access side panel to different social media apps would be helpful from which one can navigate between the different mediums easily.

We decided to use Django to implement our project as the Python community is strong and very active. Also, Django provides a framework which is scalable and can handle large amounts of data with ease. Several of the API are available in Python, and hence, we decided to build it with Python and Jinja.

To reduce the burden on the user, we have come up with several known icons and kept intact some of the traditional UI of the social media platforms.

To interact with our software, one must be proficient in using a mouse and keyboard. They should also be aware using a web browser.

We designed it as a web app because it can also be accessed from various devices which have browser support, thereby not locking our project into a single platform.

We decide to create our pages dynamically, thereby matching the retrieval times of the original sites and making it lightning fast for the user.

5.6 System Architecture

As has been mentioned earlier, we have used Python along with Django to develop our project. To understand and develop the project, we have adopted a modular approach. We have broken the project into different modules to make our project scalable and easier to handle.

Firstly, we had developed a bare Python script that extracts relevant details from Gmail using the Gmail API. We later imported this script into our Django framework and subsequently used Jinja to connect the front end as well as the back end of our project. Each app within Django has a view, and using this, we map an apps' functionality to the HTML.

An interesting thing to note is that when one adopts the Django framework, we create apps for a particular functionality. What this means is every module that we develop be Facebook, Twitter, etc., is distinct in terms of functionality and features, and each of these is implemented as separate apps which altogether are linked up Django framework.

The Gmail app has the facilities to send email, receive email, view email and even search for mail based on several criteria.

The YouTube app has been designed in such a way that we retain certain features, so we get a familiar experience at the same time providing easy navigation between the different social media app and keeping up to date with your subscriptions. You are also able to view, like and subscribe to more videos as well in our current design.

The Pinterest app functions in a similar way wherein you are updated with notifications and are able to view photos based on your interests.

The architecture has been so designed that in case we wish to expand on more social media platforms, it is quite possible.

5.7 Policies and Tactics

We strongly believed in Schneiderman's golden rules while developing the project [6]. We ensure that we strove for consistency by bundling notifications together which can be accessed by a drop-down menu. We created a side panel that can be used to access/switch from one social media profile to another. We would provide informative feedback. For instance, when a person switches between one social media to another, we show the page either scrolling up or down depending on what their preferences are set as in the side panel.

In order to reduce short-term memory load, we have designed near identical interface within our app to resemble the original interfaces, thereby following Schneiderman's golden rules. We have been able to successfully enable support for an internal locus of control.

To improve speed and reduce delay as a lot of information has to be delayed, we are dynamically creating the data using the third-party vendor sites so as to improve user response time and hence user experience. To see how well we perform, we use the result in validation and testing to fully appreciate this.

6 Verification and Validation

For a project such as the one we have done where there is a HTML/CSS website implemented on a Django framework with Jinja Logic and Python back end, there is much need for both verification and validation.

6.1 Verification

6.1.1 Verification Process

It was through the process of verification that we arrived at the architecture that is presented in the paper. One of the main doubts during the implementation was to decide which one of the following models could make a more powerful platform as compared to the other

1. Python in HTML or
2. HTML in Python.

The verification model used to resolve this dilemma was the V-Model of verification [15].

6.1.2 Verification Objective

The objective of verification was to identify which architectural design better suited the requirements of the proposed project (Table 2).

6.1.3 Results

From the process of verification, it is clear that the number of files to be separately coded for architecture would be enormously high, and hence, it would become unfeasible for a large project such as the personal assistant. Hence, we decided to go for the architecture of the second type which was HTML in Python.

Table 2 Verification process

Analysis	Architecture 1	Architecture 2
Requirements analysis	A suitable way to display messages and updates from multiple websites on a single platform	A suitable way to display messages and updates from multiple websites on a single platform
High-level design	The First Architecture has a Python in HTML approach. This means that the HTML page loads and the information is filled by individual Python function calls. These Python files are run by PHP, and their result obtained is stored and displayed	The Second architecture has a HTML in Python approach. Rather than waiting for the PHP to call the Python functions, we can have run a Python web framework that can build HTML pages from the information of Python files
Detailed specifications	Prototype: For developing and Testing Each section of the webpage for individual websites has one or more Python files attached to it that are run as the HTML loads. These pages are loaded with the information returned from the Python file. The function calls are made by the embedded PHP. For the purpose of testing and developing, the webpage is deployed through a localhost such as Deploying: On Dedicated Server The dedicated server hosts the webpages in the standard manner with support for PHP and Python as additional requirements	With the HTML in Python architecture, we would need a Python webserver such as Django or Flask that runs a Python server that is capable of running webpages and websites without hindering any of its functionalities. This would mean that most of the HTML logic would have to be driven by Python
Coding	Coding for architecture one means that the websites must be developed first with individual Python files for each of its requirements or sections	First, the Python files would have to be prepared which would have to be converted to a format suitable for the webserver, then each module of Python must be installed on the server as an app and the HTML pages must run of the data sent by the Python programs

6.2 Validation

6.2.1 Validation Objective

Through the process of validation, we wished to understand the adaptability of the website to our users, how quickly they were able to understand the purpose, functionality and the features that the website offered.

6.2.2 Validation Process

For the validation process, the website was tested by users, and the users were asked to fill a Google form to understand any shortcomings of the website. The link of the Google form is below: https://goo.gl/forms/57I6zHNvHkJOefe8.

6.2.3 Results of Survey

The users were extremely pleased with the website. To the question whether they would prefer this over the other individual websites, there were no cases where the user did not want to use this website over the individuals. Only one said maybe. And, an overall of 80% felt that the website was either useful or extremely useful (Figs. 1 and 2).

Hence, from the process of validation, it can be made certain that the users took the product well and would use it if it would be made available.

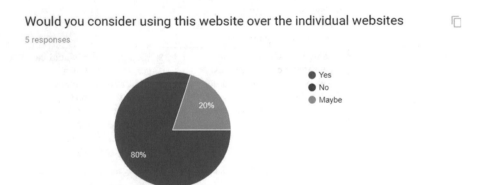

Fig. 1 Feedback of acceptability of the idea

Did you find the application useful

5 responses

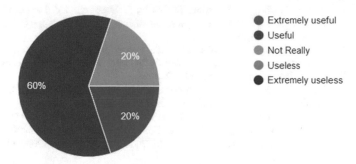

Fig. 2 Survey of usefulness of the application

7 Conclusion and Future Work

With the strong importance of social media today, we need to ensure that we are not overpowered by the information overload. With such tools like these, we can ensure that this shall never happen. However, it does not end here. This project can be further enhanced by introducing voice-controlled mechanisms for input and control. Automated replies for general purpose enquiries can be added, and a more robust reminder capability can be added as well to never miss on posting.

Acknowledgements This project would have been incomplete without the guidance of certain people who motivated us, guided us and supported us. Making of this project was possible only because of teacher who believed in us, encouraged us and pushed us to strive for excellence. We are grateful to Prof. Swarnalatha P. for her constant support which helped us in the preparation of this report.

References

1. Alves, H., Fernandes, C.: Social Media Marketing: A Literature Review and Implications: Implications Social Media Marketing: A Literature, Jan 2018 (2016)
2. Mostafa, M., Morsy, H.: The Impact of Social Media On, Mar 2017
3. Amedie, J.: The Impact of Social Media on Society (2015)
4. Mangold, W.G., Faulds, D.J.: Social media: the new hybrid element of the promotion mix (2009)
5. Fallman, D.: Design-oriented Human—Computer Interaction, vol. 5, pp. 225–232 (2003)
6. Shneiderman, B.: Designing the user interface: strategies for effective human-computer interaction. Pearson Education India (2010)
7. Bannon, L.J.: From human factors to human actors: the role of psychology and human-computer interaction studies in system design. In: Readings in Human–Computer Interaction, pp. 205–214 (1995)

8. Jacko, J.A.: (ed.) Human Computer Interaction Handbook: Fundamentals, Evolving Technologies, and Emerging Applications. CRC Press (2012)
9. https://developers.google.com/gmail/api/
10. www.colorlib.com
11. Abras, C., Maloney-Krichmar, D, Preece, J.: User-centered design. In: Bainbridge, W. (ed.) Encyclopedia of Human-Computer Interaction, vol. 37(4), pp. 445–456. Sage Publications, Thousand Oaks (2004)
12. Kuutti, K.: Activity theory as a potential framework for human-computer interaction research. In: Context and Consciousness: Activity Theory and Human-Computer Interaction, p. 17 (1996)
13. https://docs.djangoproject.com/en/2.1/
14. Vahedha, Jyothi, B.N.: Smart traffic control system using ATMEGA328 micro controller and arduino software. In: International Conference on Signal Processing, Communication, Power and Embedded System (SCOPES). 2016—Proceedings, pp. 1584–1587 (2017)
15. Software Engineering and Standards Committee: IEEE Standard for Software Verification and Validation IEEE Standard for Software Verification and Validation, vol. 1998, July 1998

Mobile Application for Alzheimer's Patients

Kanmuru Vikranth Reddy, Gali Mohan Sreenivas, C. Abhishek
and P. Swarnalatha

Abstract Alzheimer's is a disease similar to dementia that causes issues with memory, considering and conduct. Symptoms grow gradually and deteriorate after some time, finally getting to be extreme enough to meddle with every day assignments. This makes an Alzheimer's patient's life extremely hard to do his/her every day errands. We plan to give an automated app that helps in the above said. The main cause of Alzheimer's is the blocking of the neurons by mainly two proteins Beta-Amyloid and Tau. Beta-Amyloid clumps between two neurons which block the channel between them and the Tau is accumulated in the neuron which blocks the neuron. Maturing is additionally a central point in Alzheimer's which causes dementia. With the drive of having a social effect and removing the deprivation of numerous families, we needed to make something to encourage such patients. Taking in the above components into thought, we have concocted a plan to make Alzheimer's patient's life less demanding. We are making a compact app which can help the Alzheimer's understanding recall vital occasions, recognize close individuals, and make brisk calls to close ones and adjacent cab drivers. The entire app is done by putting a cherry on the best as a consistent talkbot which makes the UI more intuitive and vivid. We have looked into more than eight investigate papers and distinguished the irregularities and holes in the current advances and subsequently concocted the possibility of a predictable app which can cover these irregularities.

Keywords Alzheimer's · Dementia · UI · Mobile application

K. V. Reddy (✉) · G. M. Sreenivas · C. Abhishek · P. Swarnalatha (✉)
Vellore Institute of Technology, Vellore 632014, India
e-mail: vikranth.reddy2016@vitstudent.ac.in

P. Swarnalatha
e-mail: pswarnalatha@vit.ac.in

© Springer Nature Singapore Pte Ltd. 2020
K. N. Das et al. (eds.), *Soft Computing for Problem Solving*,
Advances in Intelligent Systems and Computing 1057,
https://doi.org/10.1007/978-981-15-0184-5_70

1   Introduction

Improve an app for Alzheimer's patients than in the present presence. A convenient app which can help the Alzheimer's understanding recollect essential occasions. Recognize close individuals and make snappy calls to close ones and adjacent cab drivers. The entire app is done by putting a cherry on the best in the shape of a consistent visit bot which makes the UI more intelligent and vivid.

1.1   Motivation

For the most part old and elderly individuals are getting alzheimer infection and it rouses us to make an app that can assist them with remembering their everyday plan [1]. Additionally, their drug time and how frequently they need to utilize the prescription day by day. We need them to recall individuals by utilizing face acknowledgment highlight and after that it gives insights about the individual in the event that he/she is enrolled as of now. This would make life less demanding for the patients and the general population encompassing them.

1.2   Summary of Existing Frameworks and Issues Timeless

In this app, Alzheimer's patients can look through photographs of loved ones, and the app will disclose to them who the individual is and how they are identified with the patient utilizing facial acknowledgment tech. In the event that a patient does not remember somebody in a similar room, they can take an image and the tech will likewise endeavor to naturally distinguish them. The app additionally incorporates a basic update screen that rundowns appointments for the day, alongside a straightforward contacts screen that indicates photographs of relatives alongside names [2]. In the event that a patient endeavors to call a contact repeatedly—something that can now and then happen as a result of the disease—the app will streak a brisk update: "Would you say you are certain you need to call? You just called under five minutes back [3]." A "me" page demonstrates the patient's own name, age, telephone number, and address [4].

2   Literature Review

Demo abstract [5]: Alzimio: It is an app with activity acknowledgment and geofencing, and safety features for dementia patients. Dementia, Autism, and Alzheimer's issue influence a great many individuals around the world. Experiencing absent mindedness influenced patients will in general stray and possibly get into hazardous circumstances. This work builds Alzimio's application, to continuously provide security capacities to patients; including movement based cautions, explore to closest companion, and registration me. Six principle structure objectives progressed the plan of Alzimio, including intermittent well-being checks, client control, adaptability, proficiency cautions, and streamlined activity, among others. Giving dependable persistent detecting and activity without radically influence most IoT human services' gadgets. Novel action acknowledgment and calculations are structured streamlined to meet the objectives and beat the test. This work shows the primary portable app consolidating highlights of adaptable safe-zone geofencing, enhanced action acknowledgment, and route to security focusing on clients with dementia (counting Alzheimer's and mental imbalance), with broad field testing and information investigation and refinement. They intend to demo the versatile well-being app on cell phone and tablet.

The Alzimio application for dementia, autism and Alzheimer's using activity-based recognition algorithms and geofencing [6].

Dementia, autism, and Alzheimer's issue influence a large number of individuals around the world [7]. Experiencing neglect influenced patients' will in general stray and conceivably get into hazardous circumstances [8]. This work presents a very efficient app, which gives notifications and alerts to the patients suffering from the disease; safe-zone geofencing and movement acknowledgment [9]. It also provides supports in terms of giving a range of unsafe areas to the user of the app which is the patient here, and whenever such area or zone is recognized, the patient is notified. At the same time, the application is compatible even on smartphones and other mobile devices. Such difficulties are not one of a kind of Alzimio rather are common to most Internet of Things (IoT) human services application [10]. Their limit-based calculations insightfully channel and process the yield of Android application programming interface for movement acknowledgment and geofencing, at various time scales. The application was assessed utilizing broad situations of utilization for a while. We find that their maximum in-window calculation can accomplish over 95% precision in under 30s in many situations. The ideal edge was observed to be around 65% certainty, to accomplish best exactness and deferral. The Alzimio application running productively on spending plan (below high-end) Android telephones without observably influencing force utilization.

A portable app giving geofencing and movement-based cautions to assist clients with dementia, chemical imbalance, and Alzheimer's, through a few limit-based calculations is introduced in this work (i.e., max-in-window), and it was demonstrated that Alzimio met its structure objectives of exact discovery exercises (greater than 95%) with less postponement (under 30s), while making it running throughout the entire day using a cell phone in a solitary charging, and ideal certainty limit is observed to be around 65% certainty and accomplished an ideal equalization/tradeoff among exactness and deferral. Power utilization expanded by 7.5% for geofencing and 16% for the action acknowledgment work.

A Mobile reminder system for elderly and Alzheimer's patients [11].

Techniques:

Alert leftover portion picture and description of individuals, occasion association, and database management

Disadvantage:

Not valuable for multipurpose just couple of capacities are accessible in this application

Focal points:

Simple to get to, occasion arranging, depiction for pictures, mobile app development and usability research to help dementia and Alzheimer's patients [12]

Techniques utilized:

Sundowning and agitation, music and memory, occupied boards

Disadvantage:

This app has no static warning to recall the patients constantly likewise no face acknowledgment include

Points of interest:

Availability, aged user, assistive technologies, computer science education, software engineering, sundowning are utilized in this venture.

3 Methodology

Improve an app for Alzheimer's patients than in the present presence. A convenient app which can help the Alzheimer understanding recollect essential occasions. Recognize close individuals and make snappy calls to close ones and adjacent cab drivers. The entire app is done by putting a cherry on the best in the shape of a consistent visit bot which makes the UI more intelligent and vivid.

3.1 Overview

The appropriate response is quite basic and includes building up a versatile app that has confronted acknowledgment of their nearby ones whom they can distinguish progressively.

Highlights of the app: Face acknowledgment to identify individuals. Notes for client's reference. Chatbot to robotize the UI. Crisis contacts to rapidly make calls. Static notification to remind the client about the app. Programming interface calls and loads of different highlights!!! Advances utilized = Tensorflow, Android Studio, Python, Agora API, git, and heroku. We are actualizing this app on Android stage for the model

Fig. 1 System architecture

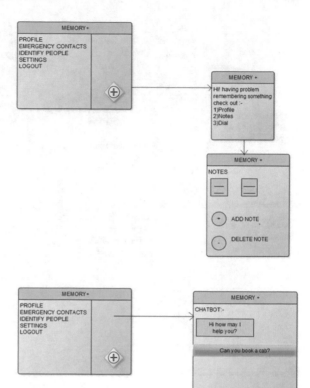

3.2 System Architecture

See Fig. 1.

3.3 Functional Architecture

See Fig. 2.

3.4 Modular Design

See Fig. 3

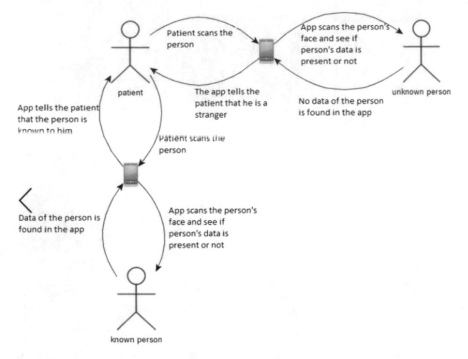

Fig. 2 Functional architecture

3.5 Application of Proposed System

Face acknowledgment: to Identify individuals
Notes: for client's reference
Chatbot: to mechanize the UI
Crisis contacts: to rapidly make calls.

3.6 Novelty of Proposed System

Face identification/acknowledgement: This feature is used to ensure or asset the patient in identifying the people they recently interact on daily basis.

This feature will enable the Alzheimer's patients to have even more profound and certain recognition of their closed ones whom they interact on a regular basis.

Chatbot: A chatbot works similarly as a human keeping an eye on an assistance work area. At the point when a client opens a talk discourse to request help, the chatbot is the medium reacting…. This is likewise a chatbot and, utilizing your area, a reaction can be given, in common, conversational answers.

Fig. 3 Modular
design–activity

Responses: It is likewise fundamental that the chatbot ought to enough react to the sound of the conversationalist and have the capacity to recognize positive and negative implications of the specific situation. It is realized that numerous senior individuals experience the ill effects of discouragement; so, the chatbot will likewise confront the errand to enhance the disposition of the conversationalist and to switch confused and troublesome themes of a discussion, for example, disease or demise to a positive course, so the questioner could appreciate the correspondence.

Update: It is pivotal to make a model for the chatbot, utilizing which it will remind the questioner about the things the individual can overlook, for example, to accept drug or call relatives and companions.

Scan for applicable and intriguing data: As a major aspect of the scholarly improvement of the chatbot, we intend to make it to a greater extent a student. That is to furnish it with a programmed capacity to freely look for answers in Google in the event that it does not know something. This will permit the chatbot to learn consistently. The more it imparts, the more information it will have about its general surroundings.

4 Implementation

4.1 Software Description and Screenshots with Particular Description

Face acknowledgment: This feature is used to ensure or asset the patient in identifying the people they recently interact on daily basis. This feature will enable the Alzheimer's patients to have even more profound and certain recognition of their closed ones whom they interact on a regular basis. We have used OpenCV for this.

Chatbot:- A chatbot works similarly as a human keeping an eye on an assistance work area. At the point when a client opens a visit discourse to request help, the chatbot is the medium reacting…. This is additionally a chatbot and, utilizing your area, a reaction can be given, in regular, conversational answers.

Fig. 4 App functionalities

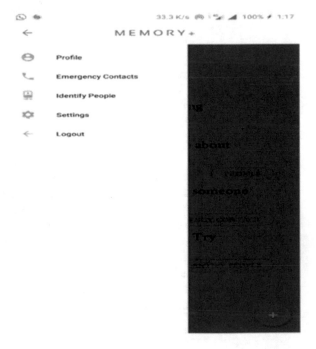

Fig. 5 Patient details

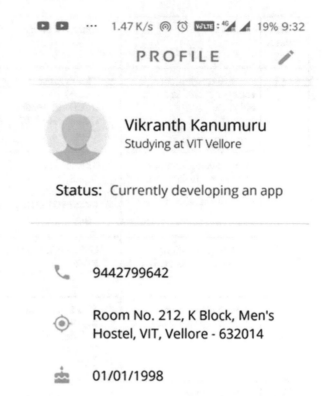

- Notes: for client's reference and used to recollect
- Crisis contacts: to rapidly make calls.

Figure 4 deals with an option which a user can choose according to our need like if we want to open the profile of the ones, we can choose on the profile and there is like emergency contacts, identify people, etc. A user shall save the contact the patient is required to know the number of the ones and can save the contact.

Figure 5 describes the profile of the guy which was saved by the patient. And by this profile, the patient can remember about the one.

5 Test Case and Analysis

See Fig. 6.

Test cases	Input	Expected output	Observed output
T1	Face recognition: azhar(registered in app but old photo)	This is azhar Age:32 Relation:brother Contact:2612102 True	Unidentified person False
T2	Face recognition: akbar(new pic updated and registered)	This is akbar Age:27 Relation:friend Contact:2132102 True	This is akbar Age:27 Relation:friend Contact:2132102 True
T3	Face recognition: Amar(unregistered)	Unidentified person False	Unidentified person False
T4	Face recognition: Anthony(registered)	This is Anthony Age:45 Relation:father Contact 1525126 True	This is Anthony Age:45 Relation:father Contact: 1525126 True
T5	Face recognition: Alex(unregistered)	Unidentified person False	Unidentified person False
T6	Notes: Add notes in the application	Notes added	Notes added
T7	Notes:Delete notes in the application	Notes deleted	Notes deleted
T8	Emergency contact:Mother contact	Call mother	Call mother
T9	Notes:Update the previous note	Alter notes	Alter notes

Fig. 6 Table of graph of the test case analysis

6 Conclusion

This app has new innovative ideas like face recognition and chatbot which are not present in other applications available for patients suffering from Alzheimer's. The existing applications provide only limited features and hence has not been of pivotal support to the patients. But this application's design will solve the existing problems of the patients by providing a service which will keep the patient notified or reminded about their day-to-day happening through the app. The application also comes with additional features which will enable the user/patient to keep track of his/her location by notifying when they cross the predefined boundary location. This is has been overwhelming experience for us in researching and implementing a product complying with the standards of HCI and also we would like to emphasis that for the purpose of study, no Alzheimer's patients were used. But we believe that after further iterations, this prototype can be taken into the market and help a lot of Alzheimer's patients. There is an estimate of over a million Alzheimer's patients in the world and this application is a step toward reaching a wider audience and helping them.

7 Future Work

Some patients have to take medicines regularly and sometimes there might not be people to give the medicines to patients. So, the authors are going to add a feature where the app generates a call when it is time for patients to take medicine and in the call there will be a recorded message reminding about the medicine improving the facial recognition technology

References

1. Hebert, L., Scherr, P., Beckett, L.: Age-specific incidence of Alzheimer's disease in a community population. J. Am. Med. Assoc. **273**, 1354 (1995)
2. Van der Roest, H.G., Meiland, F.J., Comijs, H.C., Derksen, E., Jansen, A.P., van Hout, H.P., Jonker, C., Dröes, R.M.: What do community-dwelling people with dementia need? A survey of those who are known to care and welfare services. Int. Psychogeriatr. **21**(5), 949–965 (2009). https://doi.org/10.1017/S1041610209990147
3. Topo, P.: Technology studies to meet the needs of people with dementia and their caregivers: a literature review. J. Appl. Geront. **28**(1), 5–37 (2008). https://doi.org/10.1177/0733464808324019
4. Miranda-Castillo, C., Woods, B., Orrell, M.: The needs of people with dementia living at home from user, caregiver and professional perspectives: a cross-sectional survey. BMC Health Serv Res. **04**(13), 43 (2013). https://doi.org/10.1186/1472-6963-13-43
5. Astell, A., Alm, N., Gowans, G., Ellis, M., Dye, R., Vaughan, P.: Involving older people with dementia and their careers in designing computer based support systems-some methodological considerations. Univ. Access Inf. Soc. **8**, 49–58 (2009)

6. Gowans, G., Dye, R., Alm, N., Vaughan, P., Astell, A., Ellis, M.: Designing the interface between dementia patients, caregivers and computer-based intervention. Des. J. **20**(1), 12 (2007). (Ashgate Publishing Limited)
7. Gitlin, L.N., et al.: A randomized trial of a web-based platform to help families manage dementia-related behavioral symptoms: the WeCareAdvisor™. Contemp. Clin. Trials **62**, 27–36 (2017)
8. May, Brian H., Feng, Mei, Hyde, Anna J., Hügel, Helmut, Chang, Su-yueh, Dong, Lin, Guo, Xinfeng, Zhang, Anthony L., Chuanjian, Lu, Xue, Charlie C.: Comparisons between traditional medicines and pharmacotherapies for Alzheimer disease: a systematic review and meta-analysis of cognitive outcomes. Int. J. Geriatr. Psychiatry **3**, 449–458 (2017)
9. Woolham, J.: Safe at home: the effectiveness of assistive technology in supporting the independence of people with dementia: the safe at home project. Hawker, London (2005)
10. Gitlin, L.N., Winter, L., Dennis, M.: Assistive devices caregivers use and find helpful to manage problem behaviors of dementia. Gerontechnology **9**(3), 408–414 (2010). https://doi.org/10. 4017/gt.2010.09.03.006.00
11. El Haj, M., Gallouj, K., Antoine, P.: Google calendar enhances prospective memory in Alzheimer's disease: a case report. J. Alzheimers Dis. **57**, 285–291 (2017)
12. Silva, A.R., et al.: It is not only memory: effects of sensecam on improving well-being in patients with mild Alzheimer disease. Int. Psychogeriatr. **29**, 741–754 (2017)

Study of Human–Computer Interaction in Augmented Reality

S. K. Janani and P. Swarnalatha

Abstract In the present day scenario, it can be observed that in majority of the projects coming up, the focus on specific product requirement and their immediate solutions is the approach to the existing and upcoming augmented reality problem statements. Exploring more into the functionalities and the roadmap to it is not taken much interest in. This paper desires to explore a few of such augmented reality applications when overlapped with human and computer interaction concepts for them to be a little more user-centric giving them an advantage of not only being more robust with respect to usability but also be popular in the markets for the same reasons. It is a small-scale attempt to bridge the gap between user convenience and upcoming technology of augmented reality, by studying various augmented reality projects in different fields including medical instruments, augmented dance, augmented architectural study, etc., and assessing them on the basis of the most common human–computer interaction parameters that were identified by literature surveys and further inferring the most important parameter, the problem areas, advantages, and challenges of the system by the results obtained.

Keywords Augmented reality · Human–computer interaction · HCI applications · AR case study–pros and cons

S. K. Janani (✉) · P. Swarnalatha
School of Computer Science and Engineering, Vellore Institute of Technology, Vellore 632014, India
e-mail: sk.janani2016@vitstudent.ac.in

P. Swarnalatha
e-mail: pswarnalatha@vit.ac.in

© Springer Nature Singapore Pte Ltd. 2020
K. N. Das et al. (eds.), *Soft Computing for Problem Solving*,
Advances in Intelligent Systems and Computing 1057,
https://doi.org/10.1007/978-981-15-0184-5_71

835

1   Introduction

1.1   AR as a Technology

Augmented reality (AR) is popular and trending in the technology world today, that focuses on bringing together the virtual world and the real world and to provide the user with additional information of value through innovation. It can provide information at a deeper level to things than the perceived. For example, a street down the lane through normal vision looks like a pathway but through the sight of AR, it is a stack of information about the places, the ways to reach, reviews, all in real time [1].

AR systems can be of different types, monitor-based—that combines an input from the real world and a system searched virtual world parameter, optical—a virtual play on a head-mount screen to give an illusion of real-life experience, and video type—which is an extended version of optical, where visuals are moving graphics of real environment [1].

Although the AR has negligible latency, it takes time to process and load the media that combines various inputs to display. There also might be some distortion due to the difference in vision level and the gadget display level giving rise to misjudgment.

1.2   Why Study HCI in AR

In today's world, the article observed that most of the AR innovations ranging from product to service are all closely clustered around specific problem definitions and their immediate solutions.

Through this paper, the authors' desire is to explore the application of AR overlapping on the basic guidelines of the human and computer interaction (HCI). The authors' aim is at finding out how these rules can go hand in hand with the upcoming field of augmented reality.

The process of AR projects although brilliantly covers even the questionable part of technology, giving us an option of exploring fields like optimization, tracking, visualization, transfer of data, etc. and is turning out to be the driving force of the advancing technology, it quite often misses out on the interface to the customer. To resolve this, the AR systems need to be integrated using a standard design principle that covers all aspects of the product centrally. In addition to this advantage, user-centric products are in higher demand in the markets [2].

2 Literature Survey

In the present technology-dominated world, HCI has been gaining massive popularity in the last few years due to the result of ever-increasing spectrum of real-time innovations and applications in various fields. The major trend is haptic interfaces, that are touch and gesture-based applications and vision-based, that are facial and esthetic interfaces.

Researchers and developers till date have been working on ways to innovate and create new applications, and with the supplementing advancements, the scope to explore the features of HCI aiming at user convenience has opened up to a plethora of opportunities [3].

According to the recent trends the prediction for the future of AR and HCI, the authors come to learn that AR might be on the edge of becoming the most powerful technology that will hold the capability to transform the human and computer relationship completely, giving it a new dimension [1]. There are studies that suggest that in the near future, world will be made of 3D Internet where everything will be a piece of data connected to the other, and it will be a huge workplace for developers, researchers, and analysts [4].

Until now, most of the development that has taken place in the field of AR is technology-centric, the only research phase of it focuses on how to solve immediate concerns regarding the product. It is very new that there is little research that has started to come up on developing and incorporating HCI guidelines to structure the product. From Swan and Gabbard's publication, seeing the positive review about the use of HCI in AR clearly shows us the need to encourage HCI-based research for future applications. Although it might not be any concrete in the near future as AR is still seen as a new technology exploring its boundaries, it cannot be put in constraints of the principles yet to give a standard for its development and operation [2] (Table 1).

3 Comparative Study/Proof of Concept/Technology Demonstration of Applications of AR System

Applications of AR as reflected above include, but are not limited to the field of education (virtual video learning), healthcare (practice, remote surgeries), law enforcement (incident depiction), military (practice weaponry and war), architectural (building demo/prototype), market research (analysis and prediction), and entertainment (3D cinemas and real-world replica games). The study of this article can clearly see how AR as technology plays a major roles in a variety of fields, and further, the authors will see how when combined with HCI it gives rise to very important and promising products [12].

Table 1 Comparative study of different augmented reality applications, their reviews, pros and cons

Paper title	User review	Pros identified	Cons identified
Revealing the shopper experience of using a 'magic-mirror' augmented reality make-up Application [5]	"The makeup moved along with the face when the head turned. The makeup looked continuously existing"	Enable users attempt on a virtual system. The innovations are focused on convenience, authenticity, energy, and surprising new elements	This requires the shop colleague to see how to convey customers to the application and how to urge them to utilize it
Interactive augmented reality for dance [6]	"New innovation to improve esthetic articulation and development. Displays a history of sight and sound to examine and utilize related artistic expressions"	The product gives a stage to the generation of carefully upgraded move execution that is receptive to choreographers with restricted specialized foundation	Not all the moves can be caught. The cost proficiency is a contributing element in actualizing it all over the place
Advanced AR-augmented reality using sixth sense technology [7]	"Denoting the client's fingers with various hues to the webcam can occasionally be extremely bothering, although helpful overall"	Compactness, bolster for multicontact association, connects world and data, straightforward information, delineate, open-source innovation, and financially savvy	To get pictures as RGB, esteems are set in the scope of 0–1 changing over the edge to dark and white, thus esteems other than this range can't be acknowledged
New era of teaching learning: 3D marker-based augmented reality [8]	"Positive outcome to supplant the customary learning framework. Can be extended to e-commerce, medical, military, etc."	Augmented articles can be overlaid. Can be utilized for practical. Indeed, even hypothetical ideas can be made extremely entrancing to learn	Obligatory QR code distinguished document. Event is set to convert some hypothesis point to basis and principle ideas only by the subject teacher.
Applying augmented reality to enable automated and low-cost data capture from medical devices [9]	"To assess the ease of use of the gadget, it's given to specialists and they set the marker that the attendants later take for perusing"	Enables patients to readily track changes in their body and enable doctors to screen and recognize patient needs	Slight but sudden changes in the instrument or the surrounding can influence the exactness of the readings

(continued)

Table 1 (continued)

Paper title	User review	Pros identified	Cons identified
3D survey and AR for cultural heritage: the case study of Aurelian wall at Castra Praetoria in Rome [10]	"Aim is to look at the viewpoint to users. Turning the pictures into tracker images and fit them to 3D mode"	Each SFM procedure can make definite 3D work demonstration effortlessly. Can show information from photogrammetric overview, blended models with virtual remaking	Superimposing the 3D models in the correct position
Sensors for location-based augmented reality—the example of Galileo and Egnos [11]	"Area-based administration uses standard vision-based positions following AR representation of these substantial parts of nature"	Distinctive sensors can estimate close postures. GNS information for area-based administrations	Supplementing the GPS diminishes the even position precision to the meter level
A review on the use of augmented reality to generate safety awareness and enhance emergency response [4]	"It enables the student to take a visual of the patient and assess the clinical state from the skin shading and see the nearness of wounds"	Life bolster background exercise to settle on parameters to prepare and structure test appropriately to secure inside the example space of the framework	Requires camera and a projector over the puppet. They are fairly massive bits of equipment; however, sensibly convenient can be difficult to keep up supplant because of expenses
A real-time augmented reality system to see through cars [12]	"It's productive to diminish the computational time and constrain the information exchange"	Calculation is proficient because of effective design. The large computational load is shared and decreases exchange delay	The picture is hard to produce since it's situated under the vehicle. Just the 3D information is exchanged between the two autos.
Incorporating geo-tagged mobile videos into context-aware augmented reality applications [13]	"With the information-driven methodology, when the problem is recognized, the framework can prescribe to the client"	Three methodologies available for joining video content into AR applications; pre-characterized, on-request, and proposed by hotspots	Given substance is unbending and difficult to be customized even within the sight of rich client setting

4   HCI Principles

The subject and field of human–computer interaction (HCI) mainly focus on the interaction between the system and user, its efficiency, and effectiveness. It builds standards for product development, emphasizing focus on unique and robust designs for hardware and software components to facilitate human interaction with the system in terms of gestures, behaviors, commands, etc. Thereby indicating, implementing HCI can result in more efficient and natural ways of user interaction of any product [12].

Although the composition and selection of rules vary based on developer and customer requirements, the main set of rules accepted largely have been explained in [14].

Today, in a technical aspect, the major subfields under the spectrum of HCI are machine learning (to predict and aid for user requirements), recognition of gesture, voice, behavior, mood/emotions, etc. Its implementation ranges from large-scale laboratories to a small wristwatch and in various fields.

5   Comparative Testing Results and Discussion

As mentioned above, there exist many HCI guidelines for user interfaces. However, they cannot easily and completely be applied to every AR project. The most common attributes that were found in the selected projects are taken as parameters of test of how developers prioritize them. This can be achieved by providing a model describing subject–object relationships in the studies of the projects that are picked.

Top Basic Desirable HCI Attributes: [15]

i.   Safety
ii.   Utility
iii.   Effectiveness
iv.   Efficiency
v.   Usability
vi.   Appeal
vii.   Learnability
viii. Compatibility
ix.   Predictability
x.   Simplicity
xi.   Flexibility
xii.   Responsiveness
xiii. Protection
xiv.   Control
xv.   WYSIWYG
xvi.   Invisible Technology.

While all of the abovementioned principles are the most desirable attributes that a designed product is looked upon for, another way of looking at the scenario is that these HCI methods have no hard and fast set of principles that ought to be applied for a successful design. They majorly depend on the field of work, the developer's opinion, user's requirements, and ease and simplicity of efficient working [2].

Analysing all of these aspects would go beyond the limits of this present work. Therefore, the paper targeted the eight most important of them that are relevant to our scale of study.

i. **Affordability of product by user**: The gap between a user interface and the access of its functional and physical properties. It represents the ratio of customers that the product can be made available to feasibly.

ii. **Low overhead of the product**: The limit of extra backlog load that the system or user can hold for the respective set of features to be put in use. Overhead affects the learnability and ease of usage.

iii. **Less physical effort to user**: Allowing users to complete a task or requirement with least interactions steps or processes reducing the likelihood of fatigue. The number of times of usage should also be taken into consideration.

iv. **Learning to use the product**: The user should not have much trouble learning to operate the system. It impacts the acceptance and usage of the product. Consistency and self-descriptiveness help increase learnability.

v. **User satisfaction**: Focus on user experience gets larger as the closeness with the application increases to perform tasks. Informal subjective surveys can help give direction to the work needed for impressing the customer.

vi. **Flexibility in product use**: Users have different preferences and abilities, and the product has to be able to fit according to the needs. Methods can be incorporating different modes of input and outputs and integrating different modalities to suit user preferences.

vii. **Responsiveness and feedback from the system**: The system is expected to have least lag to keep the user involved or to have a response from the system to know the process status. The feedback from them is very important to be taken into consideration for application updates and improvements.

viii. **Error tolerance of the system**: Identifying and working on problems that might occur include testing them with the stability tracking algorithms that are developed. However, they might not be very accurate as a result of human, technical, calculative, or environmental errors [2].

Data and Graph:

Affordance, learnability, user satisfaction, flexibility in use, responsiveness and feedback and error tolerance to be high are desirable, meanwhile preferring a low cognitive overhead and physical effort scale (Table 2 and Graph 1).

This is a quantised representation of the eight parameters from each AR application that was selected for the study. It can be noticed from the graph, that the user satisfaction, flexibility, and the responsiveness are the most consistent parameters in each case, while the others vary quite noticeably depending on the applications and users.

Table 2 AR applications and their ranking with respect to HCI techniques

	Area of study	Affordance	Cognitive overhead	Physical effort	Learnability	User satisfaction	Flexibility in use	Responsiveness and feedback	Error tolerance	Index
1	AR makeup and shopping technology	2	2	1	3	4	5	4	3	1–Least
2	Interactive AR dance technology	1	2	3	3	4	3	3	3	2–Less
3	Sixth sense AR technology	4	1	1	4	4	4.5	3.5	3	3–Moderate
4	3D marker teaching using AR	5	1	1	5	4.5	5	4.5	4	4–More
5	AR medical devices	3	3	1	3	4.5	5	5	5	5–Most
6	AR architecture survey	3	3.5	2	2.5	4	4	3.5	3.5	–
7	Location-based AR sensors	5	1	1	5	5	4	4	4	–
8	AR see through cars	3	1	1	5	5	5	5	4	–

(continued)

Table 2 (continued)

Area of study	Affordance	Cognitive overhead	Physical effort	Learnability	User satisfaction	Flexibility in use	Responsiveness and feedback	Error tolerance	Index
9 Safety and emergency response using AR	3	4.4	3	1.5	4	5	5	5	–
10 AR incorporation of real-time geographical video	4	1	1	4	5	5	4.5	4.5	–

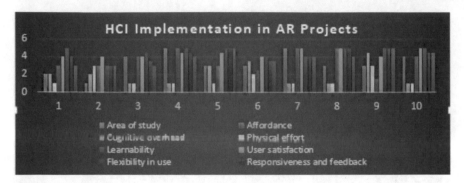

Graph 1 AR applications and their ranking with respect to HCI techniques

It can be safely said that irrespective of the field of work, while implementing HCI in them, the developers prioritize user comfort over all other factors.

Although the HCI framework and its composition majorly depend on the chosen audience/customers and it is aimed at satisfying the general crowd expectations rather than individual preferences over a mass product. It can be proposed that in comparison with the products developed without HCI framework, the ones that are developed with it are more likely to be in demand and have a better future for improvement and innovation [16].

5.1 Advantages of the System

a. The framework of augmented reality is intuitive in nature and works simultane-ously with continuous real-time data.
b. It brings closer or merges the real world and the virtual world.
c. It improves observations and collaborations with this present reality.
d. Due to its utilization in therapeutic industry, life of patients have turned out to be more secure. It helps in productive finding of sicknesses and in early location of them.
e. It can be utilized by anybody according to applications.
f. It can set aside some cash by testing basic circumstances with the end goal to affirm their prosperity without really executing continuously.
g. It can be utilized by military individuals without putting their life at stake by method for war zone reenactment before the real war.

5.2 Challenges in the System

a. As fascinating as AR may seem, some individuals may not prefer wearing a headgear all day in public. "It's bizarre and socially unbalanced." giving it a social rejection.
b. Since in any AR application, the demand for synchronization in real time is high for it to give the user, the correct information limited bandwidth gives latency.
c. The wearable is expected to be small, lightweight, consumes less power, and produces less heat giving scalability issues.
d. Since AR deals with capturing, tracking, and processing, it needs to be accurate, to filter the required and correct data to the user. To keep the virtual and real world in sync, the device needs the quickest data communication and a high bandwidth.
e. In some case, AR can sabotage the privacy of the user and start storing personal preferences and information that facilitate tracking leading to lawful issues.
f. Ordinary purchaser never observes the need to buy most of the AR devices.

6 Conclusion and Future Work

The study of the varied applications revealed the facts that making use of the HCI framework in all of them have resulted in better features and have majorly become an integral part of any development process of the system user-based applications. Also found a few discrepancies related to reading and interpreting the gestures from the user in a few of the applications which made us to proceed that the developers should focus a little more on the accuracy of gestures and commands. For a system to be able to detect gestures with this level of perfection may be a huge deal but for a user to be getting his services on gestures that are sometimes misdirected is not desired. As an overview, our goal was to combine design rules that give importance to the user needs in the field of augmented reality, to identify, understand, and improve the issues found. It is a small-scale attempt to be able to bridge the gap persists in the same. The set of rules and principles are general stated ones, they can be further refined and modified according to the application requirements. For our capabilities at this level, the thorough knowledge of augmented reality and the working is insufficient to be able to produce a set of standard rules that might apply to any AR product in the future. A study and further research in fields that are not specific to certain topics but include variety of disciplines may allow us to understand the spectrum and innovative possibilities of HCI and AR which can further bring about different features, strengths, and expertise to light. The future of this technology and its application depends largely upon the exploitation capabilities of the user.

References

1. Huang, X.: Virtual reality/augmented reality technology: the next chapter of human-computer interaction (2015)
2. Dünser, A., Grasset, R., Seichter, H., Billinghurst, M.: Applying HCI principles to AR systems design (2007)
3. Ali, M.R., Morris, T.: Usability evaluation framework for computer vision based interfaces. World Acad. Sci., Eng. Technol., Int. J. Comput., Electr., Autom., Control. Inf. Eng. 6(10), 1208–1213 (2012)
4. Agrawal, A., Acharya, G., Balasubramanian, K., Agrawal, N., Chaturvedi, R.: A review on the use of augmented reality to generate safety awareness and enhance emergency response. Int. J. Curr. Eng. Technology. Tool Based Augment. R.ity Anat. J. Sci. Educ. Technol. 24(1), 119–124 (2016)
5. Javornik, A., Rogers, Y., Moutinho, A.M., Freeman, R.: Revealing the shopper experience of using a "magic mirror" augmented reality make-up application. In: Conference on Designing Interactive Systems (vol. 2016, pp. 871–882). Association for Computing Machinery (ACM)
6. Brockhoeft, T., Petuch, J., Bach, J., Djerekarov, E., Ackerman, M., Tyson, G.: Interactive augmented reality for dance. In: Proceedings of the Seventh International Conference on Computational Creativity (2016)
7. Udtewar, S., Noronha, J.J., Chheda, Y.A.: Advanced AR-augmented reality using sixth sense technology. Int. J. Eng. Sci. 6153 (2016)
8. Gayathri, D., Om Kumar, S., Sunitha Ram C.: Marker based augmented reality application in education: teaching and learning. Int. J. Res. Appl. Sci. Eng. Technol. (IJRASET) (2016)
9. Chamberlain, D., Jimenez-Galindo, A., Fletcher, R.R., Kodgule, R.: Applying augmented reality to enable automated and low-cost data capture from medical devices. In: Proceedings of the Eighth International Conference on Information and Communication Technologies and Development (p. 42). ACM (2016, June)
10. Canciani, M., Conigliaro, E., Grasso, M.D., Papalini, P., Saccone, M.: 3D survey and augmented reality for cultural heritage. The case study of Aurelian wall at Castra Praetoria in Rome. Int. Arch. Photogramm., Remote. Sens. Spat. Inf. Sci. 41 (2016)
11. Pagani, A., Henriques, J., Stricker, D.: Sensors for location-based augmented reality the example of GALILEO and EGNOS. Int. Arch. Photogramm., Remote. Sens. Spat. Inf. Sci. 41, 1173 (2016)
12. Rameau, F., Ha, H., Joo, K., Choi, J., Park, K., Kweon, I.S.: A real-time augmented reality system to see-through cars. IEEE Trans. Visual Comput. Graphics 22(11), 2395–2404 (2016)
13. To, H., Park, H., Kim, S.H., Shahabi, C.: Incorporating geo-tagged mobile videos into context-aware augmented reality applications. In: 2016 IEEE Second International Conference on Multimedia Big Data (BigMM) (pp. 295–302). IEEE (2016, April)
14. Webb, B.: HCI Lecture 1: principles
15. Demczuk, V.J.: Human-Computer interaction: an overview. (No. DSTO-TR-0260). Defence Science and Technology Organization Canberra (Australia) (1995)
16. Egan, D.E.: Individual differences in human-computer interaction. In: Handbook of Human-Computer Interaction, pp. 543–568 (1988)

Online Real Estate Portal

Nikhil Pavan, Kuppam Sameera, Reddy Ganesh, Kumar Naveen
and Purushotham Swarnalatha

Abstract In this paper, design of a real estate portal has been discussed. Efficiency and ease of using real estate may not be possible with the existing market. This online portal helps clients to sell their property, assisting them in searching their desired property using different filters. Out of this, they can contact the owners for their queries. Also appended with a functionality to help the stakeholders for their constraints posted, attended by the respective users, a feedback characteristic of every seller profile, enabling the users to verify with other profiles for their issues, is also embedded to carry out further communication for their feedbacks. With that of, decision will be made based on the feedback collected from various users. Users generally search property based on area, location, etc., as well as other facilities like lift, furnished and semi-furnished. In spite of that, this paper assists the users by sorting the property based on filters with their results to the users. The authentication feature is also provided to the dealers to advertise their new projects with loan facilities. As a whole, the paper aims to provide an efficient portal and improvement in the sustainability of the real estate business.

Keywords Online real estate portal · Land · Residential properties · Real estate · Online portal · Feedback form

N. Pavan · K. Sameera · R. Ganesh · K. Naveen · P. Swarnalatha (✉)
Vellore Institute of Technology, Vellore 632014, India
e-mail: pswarnalatha@vit.ac.in

N. Pavan
e-mail: pavannikhil99@gmail.com

K. Sameera
e-mail: kuppam.sameera@vit.ac.in

R. Ganesh
e-mail: ganeshreddy881@gmail.com

K. Naveen
e-mail: Naveenkakarla336@gmail.com

© Springer Nature Singapore Pte Ltd. 2020
K. N. Das et al. (eds.), *Soft Computing for Problem Solving*,
Advances in Intelligent Systems and Computing 1057,
https://doi.org/10.1007/978-981-15-0184-5_72

1 Introduction

Online real estate portal helps users to access and manage its information like posting new property, updating or deleting the posted property. The admin has the access to update the status of the property, about selling, buying and cancelation status of the posted property. This is more needed for the entrepreneurs who invest in villas, commercial properties, DDA flats, society flats and builder flats. Users with small investment can also exhibit their properties here. This portal helps users to keep track of data about various properties and agents and their interaction with others. Storing the agents' information also helps by allowing us to update their profiles across the country. Earlier, if people wished to buy a property, they would approach a mediator to fix a deal, but they could not communicate as needed with the seller himself. Such mediators demand a lump sum for helping fix one deal. And if they decided to buy a property, they should take care of a lot of things like finding the agent, fix a time to meet so that everyone is convenient, location and so on. Some real estate portals do not have the security for maintaining the data. This can be achieved by registration form by limiting users. Online banking facility is also included so that the payments can be done online. There is also a scope for the Indian economy to improve the technology based on land conditions and pictures, and it can be done easily using an online real estate portal. We will be creating an online portal which will have the features of uploading the pictures along with your location on the app which will eventually be informed to the nearest three main land dealers, their address and cost of every particular land they sell with the details of the land and condition. Through this online portal, further steps can be taken in order to make the authorization.

2 Related Work

Many research articles have been surveyed in which the authors designed online portals for various purposes. Taking those ideas into consideration and weighing them one another, we have narrowed down our survey into the few articles that follow.

Anirudh et al. proposed a land entryway [1] that enables clients to manage moving and purchasing property on the Web. They designed an online real restate portal that allows users to post property for sale as well as search property and contact its owner online. This entrance holds all the data about the different clients enrolled, the property accessible in spending plan, the area and so on this entryway goes about as an interface between people, merchants and real estate brokers. The creators additionally incorporated an element in this entrance where the clients can apply online for credit.

Dan et al. introduced an article portraying the significance of an online entry in social insurance [2]. They surveyed over 500 patients to assess perceived accessibility and importance of portal-released radiology reports. They had taken a group interview from around 50 doctors to assess the utility of releasing reports. This study demonstrated that the larger part of patients lean toward an online entryway as it is anything but difficult to get to reports from it. The doctors revealed that subsequent messages, telephone calls and office visits had diminished.

Kristy et al. carried out a study within an academic general surgery service compared online and in-person postoperative visits [3]. A few patients had experienced an assortment of treatment amid this period and interfaced with the specialists through an online entryway. The creators had noticed the huge patient-detailed acknowledgment of online visits instead of face-to-face visits. Dominant part of these patients favored online visits notwithstanding for development.

Ajey et al. built up an online entrance which highlights diverse genomic databases and devices [4]. This entry called FisOmics goes about as a stage for sharing fish genomic successions and related data notwithstanding encouraging the entrance of elite computational assets for genome and proteome information examination. It gives the capacity to quarrying, breaking down and picturing genomic successions and related data.

Panagiotis et al. introduced the Inclusive Learning Portal [5] that plans to progress existing arrangements and bolster open access to instructing and learning of individuals with incapacities. This gateway bolsters educators during the time spent creating, sharing and conveying Learning Objects (LOs) that can address the assorted variety of impaired student needs and necessities. It gives instructor preparing chances to upgrading educators' capabilities on comprehensive learning and openness standards.

Harika et al. designed Travel Buddy [6], a carpooling portal which offers cheaper car rides to its users by providing a feature which lets them share the rides with other users. This portal specifically focused on students of Vellore Institute of Technology, Vellore, India. As there is no airport in Vellore, students from this institute need to travel approximately 150 km to catch their respective flights when they go on vacations. As this portal lets them share the fare among the other students of the same institute, the students find it safe and economic.

Vanshaj et al. designed an IoT portal which helps to manage various emergency services like ambulances and other medical services [7]. This portal combines several emergency service departments which helps the service personnel to cooperatively work together. This IoT-enabled portal specifically focuses on ambulance services and other medical services, and so it includes a traffic surveillance model to mark the shortest route that the ambulances can take.

Vamshika et al. designed an interactive portal for alumni [8] whose main objective is to help university faculty to take feedback from the students and make the university better. Through this portal, the students can get career information like internship offers and university news. The primary idea of designing this portal is to keep the alumni updated about the university as well as giving them an opportunity to share their own experiences in the university with their juniors.

3 Proposed System

The proposed system has three models: Home Page, Residential Properties and Administrator. These three models are briefly discussed below:

3.1 Home Page

Home Page comprises details of few trending properties. This page has several modules included in it: The *Login Menu* module has a substantial email id and a legitimate secret key for a secure authorization. This enables only authorized users to log in to the portal [9]. The *Add* module lets the user to add videos or images of the property which makes it easier for the users to get a clear picture of the property and its surroundings without going there in person. The *Counsel* module helps the user to communicate with real estate advisors in real time who can help fix a deal by checking certain constraints like the current demand for a particular area and its price in the market. The *About Us* module contains all the necessary information about the portal, its developers, contact of support team, etc.

3.2 Residential Properties

The Residential Properties model has two modules: The *Search Property* module helps the user to search for property by sorting them out according to certain criteria like budget, location and area. The *Calculator* module lets the users to calculate the price of the property, additional taxes, interest rates, etc., within the portal [10].

3.3 Administrator

The Administrator has the benefit of making changes in the portal contemporarily. The admin is responsible for updating the portal according to the ever-changing scenarios in the real estate market. Users cannot alter the information provided by the admin. The admin is also responsible for the creation of new users as a guest or a client with special features. The admin can also manage the security for access level of the users for usage of the Web site [11]. The most concerned tasks of the admin are given below:

- Existing id and password details
- New login details
- Updating the latest status of the property
- Maintaining the details about the searches done by the users

- Changes in the old account
- Security of the site and data
- Collecting the feedback given by users

4 Entity–Relationship (ER) Model

Figure 1 demonstrates the stream of site. The entity–relationship among various modules of the portal is depicted clearly in the figure. As shown in it, the real estate system comprises two main modules: Admin and Client. Admin has to provide username, password and address in his profile. The admin is responsible for data regarding property and transactions. The client has to provide contact number, user ID, address and budget in his profile which makes it easier for the users to suggest suitable properties to a particular client. Before doing any transaction, it is mandatory for the client to register his profile using his credentials.

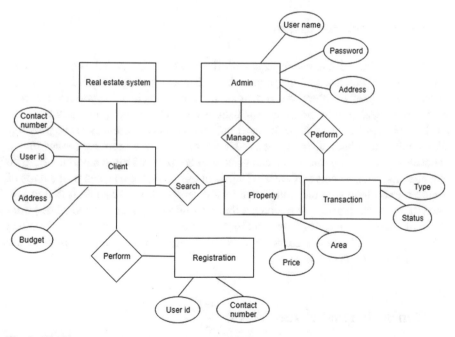

Fig. 1 ER diagram of online real estate portal (OREP)

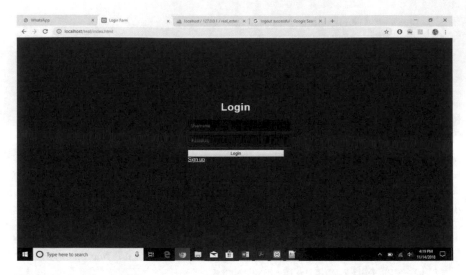

Fig. 2 Login page of OREP

5 Results and Discussion

The fourth section of the paper deals with the results of the portal as given below: Fig. 2 shows a screenshot of the portal which displays the login page. In this stage, users are prompted to enter their credentials in order to log in to the portal in a secure way. After logging in to the portal, the user is redirected to the portal's home page as shown in Fig. 3. In this stage, the user can choose his preferences of the property like location and budget. If any user who wishes to buy a property cannot find it according to his requirements, he can go to another page within the portal where he can post his requirements as shown in Fig. 4. In this page, the user can post details of requirements along with the contact details so that if any other user feels that he can meet these requirements, he can contact that user through phone or email. The portal includes a feedback page as shown in Fig. 5, where the user can share his experience with the portal. All the other users can view this feedback which helps them to decide about the property accordingly.

6 Conclusion and Future Work

The paper concludes with the efficient portal deployment which helps the entrepreneurs to do e-business. This portal acts as a medium between all individuals, agents and real estate brokers. The portal serves as an advertising weapon for all the users to the entire world from home itself. Adding machines helps in making the

Fig. 3 Home page of OREP

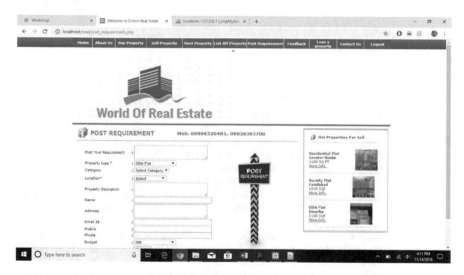

Fig. 4 Requirement posting page of OREP

framework more easy as client can compute the expenses involved to purchase/sell the land in a faster manner. As a future work, the authors focus toward the development of a dynamic incorporation of functionalities with an efficient outcome.

Fig. 5 Feedback page of OREP

References

1. Anirudh, V., Akhil, C. and Kailash, G.: Online real estate portal. Int. Res. J. Eng. Technol. **5**(1) (2018)
2. Dan, H., Grant, O., Karen, C., Terhilda, G., Heather, Q., Joanne, T.: Access to radiology reports via an online patient portal: experiences of referring physicians and patients. J. Am. Coll.E Radiol. **12**(6) (2015)
3. Kristy, K.B., Omobolanle, O.O., Sharon, E.P., Rebeccah, B.B., Michael, D.H., Kenneth, W.S., Richard, A.P., William, H.N., Benjamin, K.P.: Postoperative Care Using a Secure Online Patient Portal: Changing the (Inter)Face of General Surgery. In: Proceedings of American College of Surgeons 101st Annual Clinical Congress, Scientific Forum, Chicago (2015)
4. Ajey, K.P., Iliyas, R., Naresh, S.N., Ravindra, K., Rameshwar, P., Mahender, S., Murali, S., Basdeo, K., Dinesh, K., Anil, R.: FisOmics: A Portal of Fish Genomic Resources. In: Genomics (2019)
5. Panagiotis, Z., Vasilis, K., Demetrios, G.S.: An online educational portal for supporting open access to teaching and learning of people with disabilities. In: Proceedings of IEEE International Conference on Advanced Learning Technologies (2014)
6. Harika, G., Prakhar, L., Akshay, G., Sameera, K. and Swarnalatha, P.: Travel buddy-s carpooling App: a system designed to group students for sharing a cab. J. Comput. Theor. Nanosci. **15** (2018)
7. Vanshaj, G., Suvitha, W., Sameera, K., Swarnalatha, P.: Design of emergency services application using IoT. J. Comput. Theor. Nanosci. **15** (2018)
8. Vamshika, L., Srishti, Deekshitha, N.M., Sophiya, F.D., Swarnalatha, P.: An institutional interactive portal for alumni. J. Comput. Theor. Nanosci. **15** (2018)
9. Chen, H.: Design and implementation of the real estate sales management base on B/S (2000)
10. Chu, T., Pang, Y.: Technological Development of ASP (2007)
11. David, C.H.: Requirements analysis (2004)
12. Jiang, L., Xiang, Y.: Design of the sales management system based On C/S (2005)

Accelerometer Based Home Automation System Using IoT

U. Vanmathi, Hindavi Jadhav, A. Nandhini and M. Rajesh Kumar

Abstract The field of automation technology has been so much improved, and the recent technologies have made the life of people simpler in every aspect. In today's world, automatic systems are being preferred over manual system in all fields that even replaces the remote control mechanisms with automation, such that manpower can be reduced. The incredible increase in the users of internet over recent times has made Internet a much-needed thing in everyday life, with IoT being the latest and emerging Internet technology. Internet of things is a growing field of network which can be used in every industry and machinery that can share information and complete tasks in seconds over Internet which could consume more time if done manually. Accelerometer based home automation system using IoT is a system that uses accelerometer sensor data to control home appliances automatically through Internet based on the tilting actions of the sensor in different directions. It saves electric power and manpower. The home automation system is different from the other manual systems such that it provides the user with the ability to control the system from anywhere around the world provided a Wi-Fi or Internet connection is available.

Keywords Accelerometer sensor ADXL345 · NodeMCU · Internet cloud server · Internet of things

U. Vanmathi · H. Jadhav · A. Nandhini · M. Rajesh Kumar (✉)
School of Electronics Engineering, VIT University, Vellore, India
e-mail: mrajeshkumar@vit.ac.in

U. Vanmathi
e-mail: vanmathiuv@gmail.com

H. Jadhav
e-mail: hindavikj@gmail.com

A. Nandhini
e-mail: 18nandhiniakilan@gmail.com

© Springer Nature Singapore Pte Ltd. 2020
K. N. Das et al. (eds.), *Soft Computing for Problem Solving*,
Advances in Intelligent Systems and Computing 1057,
https://doi.org/10.1007/978-981-15-0184-5_73

855

1 Introduction

Buildings and residences of present years have become more and more self-controlled and automated due to the comfort it provides and the devices or appliances can be switched on or off at any desired time. A home automation system allows users to control electrical appliances without being physically present in the place [1]. The switches need not be turned on or off manually to control the appliances. The models available today are based on wired communication. But this kind of system does not provide a wider range of control and it requires to be installed during the construction of the building. With the advancement of wireless technologies such as Wi-Fi, cloud networks and Internet of things in recent days, wireless systems are used every day and everywhere to make life easier. The control range of appliances is also not limited. They can be controlled from anywhere with the availability of Internet connection. In recent years, wireless systems like cloud computing are becoming more common in networking and industrial automation. In many systems, the usage of wireless techniques such as IoT, Bluetooth and ZigBee gives many advantages that are not achieved through a wired network.

Certain advantages of home automation systems using IoT are as follows:

The installation cost of the set-up is reduced significantly since no wiring is needed. Wired technologies make use of wires which make the system more complicated, and hence, the cost increases. When there is any damage in the wires connected, the system collapses in a wired system. But that is not the case in a wireless system.

Extension of wired system is more difficult task than wireless one. In wireless systems even when the place of installation is changed, quick installation is possible for automation system. It is difficult if an extension to another specified range is required.

Wireless systems can be easily connected to Wi-Fi modems or LAN in that particular organisation; hence, we could get the update status of devices to be controlled. In these wireless systems, connecting smartphones or other Wi-Fi shields with system can be done and it is possible anywhere, at any time.

For the above-mentioned few reasons, the use of wireless technology particularly by IoT has many advantages than a wired connection.

2 Literature Survey

A home automation and home security technique is implemented in which the appliances are controlled through flex sensors using finger gestures [2–4]. The flex sensor is used to detect and measure the deflection or bending, and the Wi-Fi module is used to support cloud storage. The magnetic sensor is used to enhance the door security.

A home automation system using Intel Galileo is implemented that controls the lights, fans and appliances in home and stores the data in the cloud. This design employs the integration of cloud networking, wireless communication, to provide the user with remote based control of appliances. The control action will automatically change on the basis of sensor's data. It is a low cost and expandable system that allows a number of appliances to be controlled [5–11].

Detection of hand gesture by Arduino UNO that is interfaced with accelerometer and gyroscope and the entire set-up is attached to the glove. The movement of the glove, attached with the sensor, senses the tilting and acceleration of the sensor, and the data is transmitted through RF transmitter. The system will act like a remote control for operating the consumer electronic devices, but the control action is confined only to a particular range.

A smart home system using MSP430 microcontroller is designed for controlling the appliances. It also sends SMS using GSM modem, to notify the status of the appliances. The user can give input to the system with the help of touch screen, and special symbols are being assigned to the shape identifier system. Thus, the wireless command sent will control the automation system and sends SMS of the device status.

3 Problem Statement

The currently available systems for home automation suffer from disadvantages such as high cost of installation, less degree of flexibility, lack of easier manageability and confined only to a particular range in cases when ZigBee or Bluetooth is used. Another important threat to system is security. So the ultimate objective of this work is to implement this type of automation system with IoT technique, which is not only capable of controlling but also automating household appliances using web interface with Wi-Fi getting connected to a local server. This system gives more flexibility as it is making use of Wi-Fi.

4 Methodology

Components Used

Accelerometer (ADXL345): ADXL345 is ultralow power accelerometer device, which not only measures the static acceleration but also dynamic acceleration resulting from the tilting action provided and gives value for all the three dimensions of its axis, i.e. it provides the output values for three different directions depending on the 'g' values for that particular direction. SPI or I^2C interface is used for accessing the values.

NodeMCU: It is one among the most integrated Wi-Fi chip in industry and is based on the Arduino module. It is integrated low-power 32-bit processor. It has 17 GPIO (General-Purpose Input/Output) pins that are multiplexed with other functions. It functions as a processor as well as Wi-Fi shield. But the cost of the unit is less than

Fig. 1 System block
diagram

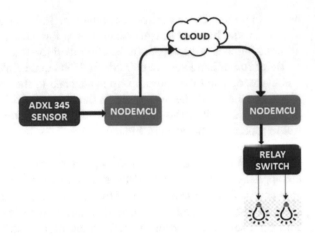

other modules. It supports 802.11b/g/n hence is called Wi-Fi development board. It is very similar to Arduino board, and thus, Arduino IDE can be used for this firmware. The programming for this module is also more similar to Arduino and hence uses Arduino IDE for its software platform.

Relay module: Relay module (4-channel relay switch) is a hardware device used for switching of devices remotely. It can be powered by a 5 V or 12 V source depending on the design. The end devices or appliances are remotely switched ON or OFF through this relay module.

The Proposed Model

This model as shown in Fig. 1 comprises of ADXL345 accelerometer sensor and NodeMCU module. The NodeMCU is also called as the Wi-Fi development board. The NodeMCU is more advantageous than using an Arduino and Wi-Fi shield. The one single module performs the action of both microcontroller and Wi-Fi shield, and also, it is cost-effective. The NodeMCU is enabled when proper connection is established, which is used to read the motion of sensors in all the three directions X, Y, Z, i.e. three-dimensional. The tilting motions of the accelerometer are sensed, and the corresponding control actions are performed on the end device. The threshold levels for the sensor are set in prior to sense and give values for motions in left, right, forward and reverse directions. This data is further sent to web server which is then stored in the cloud, i.e. the data sensed is uploaded to a local web server created through the transmitter side NodeMCU board. A web link is created, also from which the devices can be controlled. This data can be analysed anywhere any time, with Internet connection. If the sensor value matches the value stored in the database, the corresponding action is performed and the appliances are controlled accordingly. In the proposed prototype model, a light and DC fan are used as load. The control action performed on time basis is stored in cloud for analysis. The state of appliances used can also be monitored at different times in the web server with the availability of Internet connectivity. In case if any appliance or light is left switched on, the generated IP address is used for the control action from the NodeMCU module from any computer system nearby. The system block diagram can be shown as follows:

Software Design
Arduino IDE:
Languages such as C and C++ can be supported by Arduino IDE by modifying or by using some different rules for code formation. Many input and output libraries are supplied by this tool which makes the programming easier for all Arduino-based MCU boards. An executable code is converted into a text file with the encoding which is in hexadecimal form. The loader program in the board's firmware is used to load this kind of file. NodeMCU is an open-source IoT platform which can be programmed from Arduino IDE. It acts as Wi-Fi development shield.

Cloud Storage:
Cloud computing is the term that describes anytime and anywhere storage. Rather than using computer systems, this technique makes use of remote servers to manage and perform operations on data. The databases stored in XAMPP file are coded in PHP and are called from Arduino code. XAMPP software provides easier creation of database for an application. A local server is deployed to access the database according to the accelerometer signals and the accelerometer values.

5 System Set-up

The transmitter end:
In the transmitter end, the accelerometer ADXL345 is connected to the NodeMCU [2]. The control action is given by the motions of accelerometer sensor whose directional tilting is sensed through the readings from Arduino IDE software. The threshold for the three-dimensional value is set, and a database is created for each direction of sensor and the control action such as the ON–OFF of the devices used is set in XAMPP software (Fig. 2).

In this work, two end devices such as a DC fan and a DC motor are connected, and the sensor tilting set-up is assigned as follows: (Fig. 3).

DEVICE 1: DC FAN
DEVICE2: DC MOTOR

The directions assigned are as follows:

Forward—Device 1 ON
Reverse—Device 1 OFF
Left—Device 2 ON
Right—Device 2 OFF

Fig. 2 ADXL345 interfaced
with NodeMCU

Fig. 3 Directions assigned
for accelerometer

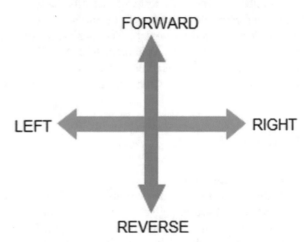

The Receiver end:

The receiver end has another NodeMCU interfaced with the relay unit [12]. The relay unit receives the control action from the accelerometer sensor. Based on the sensor inputs, the relay unit turns the devices ON or OFF. The interfacing of components in the receiver end is shown as follows (Fig. 4).

6 Conclusion

The IoT provides a large number of incredible benefits to society and makes life easier. The work implemented in this paper can support the automation aspects of industrial and smart home concepts, which can be widely deployed on a public platform. This design provides efficient, low power and low-cost way of sensing,

Fig. 4 NodeMCU interfaced with relay unit and connection with end devices

monitoring and controlling the entire system using IoT and implements Industrial automation and smart homes without the use of separate Wi-Fi shield. This way, IoT deployed home automation system can be further deployed to all appliances and smart home or home automation concept is implemented.

References

1. Wadhwani, S., Singh, U., Singh, P., Dwivedi4, S.: Smart home automation and security system using arduino and IOT. Int. Res. J. Eng. Technol. (IRJET), **05**(02) (2018)
2. Abhijit, M., Nair, A., John, J., Basheer, S., Mathai, M.B.: Hand gesture based home automation. Int. J. Adv. Res. Electr., Electron. Instrum. Eng. **6**(3) (2017)
3. Sasikala, K., Ravikumar, D., Vasanth, B.: Wireless touchscreen based assistant for visually impaired. Int. J. Pure Appl. Math. **118**(11), 497–502 (2018)
4. Chandramohan, J., Nagarajan, R., Satheeshkumar, K., Ajithkumar, N., Gopinath, P.A., Ranjithkumar, S.: Intelligent smart home automation and security system using arduino and Wi-fi. IJECS, **6** (2017)
5. Pasha, S.: ThinkSpeak based sensing and monitoring system for IoT with MATLAB Analysis. IJNTR, **2** (2016)
6. Kaur, S., Singh, R., Khairwal, N., Jain, P.: Home automation and security system. ACII, **3** (2016)
7. Vinay Sagar K.N., Kusuma S.M.: Home automation using internet of things. IRJET **2**(3) (2015)
8. Prabhuraj, R., Saravanakumar, B.: Gesture controlled home automation for differently challenged people. IRJE **01**(02) (2014)
9. Bande, V., Pop, S., Loan, C., Pitica, D.: Real time sensor acquisition interfacing using MATLAB. IEEE (2012)
10. Kulkarni, V.S., Lokhande, S.D.: Appearance based recognition of american sign language using gesture segmentation. Int. J. Comput. Sci. Eng. (IJCSE) **2**(3), 56 (2010)
11. Lu, T., Neng, W.: Future internet: the internet of things. In: 3rd International Conference on Computer Theory and Engineering(ICACTE), vol. 5 (2010)
12. Jadhav, K.P., Bari, S.G.: Hand gesture based switching using MATLAB. IJIREEICE, **4** (2016)

Abnormal Notion of Depicting Encrypted Result—Dynamic Usage of Multi-key Encryption Model (DUMKE)

K. Berlin and S. S. Dhenakaran

Abstract Cryptography is a robust security mechanism of converting the secret data into secured format to control unauthorized access, updating, and so on. Due to digitization, all business activities happened through online. So security or trust on transaction is most essential. So it is focused on to increase the strength of security by newly designed encryption mechanisms called dynamic usage of multi-key encryption. Before encryption, input data are compressed by the method of compression technique using matrix mathematical model (MMM). Compressed data are utilized by DUMKE. This mechanism is applied to three public-key algorithms for encryption. The final phase of the system is an outcome of DUMKE which is displayed as image.

Keywords Matrix mathematical model · Multi-key encryption · Dynamic usage · Compression

1 Introduction

1.1 Cryptography

Security of data, in communication spectrum, becomes a serious issue. To ensure the security of data, various algorithms and techniques were evolved under private- and public-key crypto methodologies to retain the authentication of data. Comparing both private-key cryptography and public-key cryptography is more secure and authentic for better communication than the secret-key cryptography [1]. Here, it is decided to deviate work into public crypto models to strengthen the dynamic key selection method. R. L. Rivest et al. implement security algorithm of public-key

K. Berlin (✉) · S. S. Dhenakaran
Department of Computer Science, Alagappa University, Karaikudi, India
e-mail: berlinjenson@gmail.com

S. S. Dhenakaran
e-mail: ssdarvind@yahoo.com

© Springer Nature Singapore Pte Ltd. 2020   863
K. N. Das et al. (eds.), *Soft Computing for Problem Solving*,
Advances in Intelligent Systems and Computing 1057,
https://doi.org/10.1007/978-981-15-0184-5_74

cryptosystem for factoring large numbers [2], and security aspects too prove to be adequate. Through this research, they are proved in which public-key cryptosystem is suitable to provide secured solution of more complex problems too.

1.2 Compression Method

The compression is to decrease the size of the original file through the process of compression and reproduce the file (low redundancy). In this paper, ASCII code is playing the main role to compress the text file because every character having ASCII code for systematic representation. Two types of compression are available such as symmetric compression and asymmetric compression.

The strength of the encryption algorithm is fully based on the data security and how it reacts from cryptographic attacks. Compression-based encryption systems are computationally proved as in the security point of view [3–8]. The existing research results of compression with encryption mainly focused on fixed key length of encryption. Based on the survey of existing methods, some of the mechanisms are designed through the dictionary and index-based encryption [9–11]. Hackers break key while using such fixed key length of encryption mechanisms. To overcome this problem, we are proposed dynamic usage of a crypto algorithm; here, key for encryption is choosing dynamically based on the input data. The need of compression mechanism in proposed method is to reduce the input file size, to reduce file splitting time, to speed up the time of encryption and compare to normal encryption the compressed encryption is very strong while intruder wants to see the secrets.

$$X = \{x_1, x_2, x_3, \ldots, x_n\} \forall x \in \text{Input}_x \tag{1.2.1}$$

$$\overline{X} = \text{Compress Input}_X \tag{1.2.2}$$

$$X_{\text{Compression}} = \{\bar{x}_1, \bar{x}_2, \bar{x}_3, \ldots, \bar{x}_n\} \forall \ \bar{x} \in \text{Compression}_X \tag{1.2.3}$$

$$X_{\text{Compression Size}} = \begin{cases} \overline{X}_{\text{Size}} \ \text{if} \ \overline{X}_{\text{Size}} < X_{\text{size}} \\ X_{\text{Size}} \ \text{Otherwise} \end{cases} \tag{1.2.4}$$

The above formulas (1.2.1), (1.2.2), (1.2.3), (1.2.4) are used to define the compression which is used as the first process in this proposed work

2 Research Fundamental

To provide security and authentication to the data, many algorithms and techniques were evolved. Even though the cryptographic technique remains best, still better secured mechanisms are needed. A lot of crypto techniques exist with some limitations such as fixed key generation techniques, the same number of keys used for encrypting data whether it is public or private. In the situation of the same key usage, there is a chance to break keys easily by the hackers [12]. So it is a problem to every authorized user to maintain their original message. To prevent this kind of problem, the proposed work provides a new mechanism called "Dynamic Usage of Crypto Algorithm." In this approach, every secret message has using more encryption algorithms. This mechanism prevents data from the unauthorized users and also provides perfect privacy to an entire secret. A maximum of three different kinds of encryption methods is utilized to secure the message. The number and selection of an algorithm are depending on the length and contents of an input message. The working process is starting with compression of input text file which is divided/segmented into a maximum of three files. These three files are encrypted dynamically with asymmetric cryptographic algorithms. The encrypted message (ciphertext) is the combination of the outcome of RSA, ElGamal, and ECC crypto algorithms which is converted to gray image for the final output with the help of processing each element in the ciphertext.

3   Initiation of the Proposed Work

Generally, every character is having position to represent a numeric value, which is used for character conversion. The process of the proposed compression method is initiated with alphabet positions.

Table 1 clearly depicts alphabets, numerals, special characters, and their position, which is used in the proposed compression technique. In the proposed scheme, the mathematical mean (average) value is calculated to measure the distance between each element of every row which is placed in array of matrix. The mean value is computed by

$$X_{\text{Div}} = \frac{\sum_{i=0}^{n} X}{N} \tag{3.1}$$

Table 1 Alphabets and positions

Alphabets	a	b	c	d	e	...	y	z
Positions	1	2	3	4	5	...	25	26
Numerals	0	1	2	3	4	...	8	9
Positions	27	28	29	30	31	...	35	36
Special characters	/	?	+	−	*	...	}	~
Positions	37	38	39	40	41	...	54	55

Here, X_{Div} denotes that number of divide parts in input. The symbol $\sum$ represents the summation, i denotes that input characters, X symbolizes the elements, and N denotes number of elements.

4 Compression Mechanism

4.1 Text Compression Using Matrix Mathematical Model [13]

The proposed compression algorithm is designed to compress text files under lossless technique. The text input file has alphabets, numbers, special characters, etc. ASCII values are used to fulfill the process of text compression. Particularly, ASCII values ranging between 32 and 127 printable characters are used for further processing. Generally, every character is given a position for numeric representation. The proposed compression method is initiated with alphabet positions for the conversion process (Table 2).

Example Input text $= \{v, i, r, g, i, n\}$.

Now, the character positions are divided into two partitions as below,

$$A = \begin{cases} a_1 = v \rightarrow 22 \\ a_2 = I \rightarrow 9 \\ a_3 = R \rightarrow 18 \end{cases} \tag{4.1.1}$$

$$B = \begin{cases} b_1 = G \rightarrow 7 \\ b_2 = I \rightarrow 9 \\ b_3 = N \rightarrow 14 \end{cases} \tag{4.1.2}$$

Let A and B be the names of two partitions, whose elements are $a_1, a_2, a_3, \ldots, a_n$ and $b_1, b_2, b_3, \ldots, b_n$.

Table 2 Characters' positions

Character	Position
V	22
I	9
R	18
G	7
I	9
N	14

Table 3 Least values from multiplicative array

*	22	9	18
7	154	**63**	126
9	198	**81**	162
14	308	**126**	252

4.1.1 Multiplicative Array Using Binary Operation

The second phase of proposed compression mechanism is explained below: Let $S = \{a_1, a_2, a_3, ..., a_n\}, \{b_1, b_2, b_3, ..., b_n\}$ be a finite set, and let * be a binary operation on S. The multiplication * is defined below, and the below multiplication table is used to process two partitions' text file.

$$X_{\text{Mull\_Matrix}} = \sum_{i=0}^{n} X_{\text{Div 1}} * \sum_{i=0}^{n} X_{\text{Div 2}} \qquad (4.1.1.1)$$

Here, X_{Div} denotes that number of divided parts using the formula (3.1) (Table 3). The minimum value of multiplication is calculated and added.

$$\text{Min\_Value} = \begin{Bmatrix} b_1 * a_2 \\ b_2 * a_2 \\ b_3 * a_2 \end{Bmatrix} \qquad (4.1.1.2)$$

Hence, the total Min_Value $= 63 + 81 + 126$.

$$\text{Total Min\_Value} = 270 \rightarrow (5)$$

4.1.2 Additive Array Using Binary Operation

The third phase of proposed compression mechanism is using addition operation. Let $S = \{a_1, a_2, a_3, ..., a_n\}, \{b_1, b_2, b_3, ..., b_n\}$ be a finite set, and let $+$ be a binary operation on S. The operation of $+$ is defined here,

$$X_{\text{Add\_Matrix}} = \sum_{i=0}^{n} X_{\text{Div 1}} + \sum_{i=0}^{n} X_{\text{Div 2}} \qquad (4.1.2.1)$$

The above formula (4.1.2.1) is used to create an addition matrix using input value X (Table 4).

Table 4 Highest values from additive array

+	22	9	18
7	**29**	16	25
9	**31**	18	27
14	**36**	23	32

$$\text{Max\_Value} = \begin{Bmatrix} b_1 + a_1 \\ b_2 + a_1 \\ b_3 + a_1 \end{Bmatrix} \qquad (4.1.2.2)$$

The above addition is used to process two partitions text file. Hence, the total Max_Value $= 29 + 31 + 36 = 96$.

$$\text{Total Max\_Value} = 96 \qquad (4.1.2.3)$$

Hence, binary multiplication provides minimum 270 and binary addition provides maximum 96.

$$X_{\text{Diff}} = \sum \text{Total Value}_{\text{MIN}} - \sum \text{Total Value}_{\text{MAX}} \qquad (4.1.2.4)$$

4.1.3 Conversion Process

The final phase of proposed compression methodology is to take difference between two numeric values (TotalMin_value & TotalMax_value). Let D is considered as difference value. That is, final part of proposed compression algorithm fully depends upon the ASCII values. Here, the printable characters of 32–127 are used for further processing. Based on the difference (D) value, mechanism chooses ASCII values to provide final compressed text data. If difference value is less than 32 means, it is needed to add 32 with difference value (D) to maintain D always as greater than or equal to 32. If D value is greater than 32, just write corresponding ASCII values instead of numbers (D). If the difference value is greater than 32 and less than 127, just print equivalent ASCII value as output. Subtract 32 from D, if the value of D is greater than 127 and less than or equal to 160. If D is greater than 160 and less than 256, subtract 127 from D. Output of the above-mentioned procedure, finally, is printed as prefixes of ASCII values.

5　Algorithms for DUMKE Scheme

DUMKE scheme uses three asymmetric algorithms for encryption.

5.1　RSA Cryptosystem

Rivest, Shamir, and Adelman published public-key cryptography algorithm named RSA on 1978. It uses two different keys, public key known to everyone while the private is kept as a secret. The authorized users only know how to open the message. The encryption ratio of RSA algorithm is high, and processing speed is also fast. Key length of this algorithm is more than 1024 bits. The method of timing attack provides security from the cryptanalytic. Block size of RSA algorithm is 446 bytes and 1 round for encryption. RSA is implemented using stream cipher. For encryption, sender transmits public key(n, e) to receiver and kept d(private) as secret. Receiver sends message M to sender in the form of $c = m^e \bmod n$. For decryption d, it is calculated through the form of, $m = cd \bmod n$.

5.2　ElGamal Cryptosystem

The ElGamal algorithm is one of the asymmetric key cryptographic techniques to encrypt and decrypt the message [14]. For key generation of ElGamal cryptosystem, sender generates an efficient description of a cyclic group G of order q with generator g. Sender chooses the value x randomly from $\{1,..., q - 1\}$ and computes $h = g^x$. For encryption, receiver chooses a random y from $\{1,..., q - 1\}$ and calculates $c_1 = g^y$. Shared secret $s = h^y$ calculates by receiver and maps the secret message m with an element m$'$ of G. Finally, receiver calculates $c_2 = m' \cdot s$, and receiver sends the ciphertext.

$$(c_1, c_2) = g^y, m'.h^y \tag{5.2.1}$$

To decrypt ciphertext (c_1, c_2) with private key x, sender calculates shared secret $s = c_1^x$. Then, sender computes $m^! = c_2 \cdot s^{-1}$.

$$c_2 \cdot s' = m' \cdot h^y \cdot (g^{xy})^{-1} = m' \cdot g^{xy} \cdot g^{-xy} = m' \tag{5.2.2}$$

5.3   Elliptic Curve Cryptography

Elliptic curve cryptography is a public-key cryptography in the recent researches [15]. The Elliptic Curve Cryptograhy have the smaller key size. The speed of the ECC is faster than RSA and, works better in handhelds and cell phones. For key generation, sender and receiver choose a finite field fp over an elliptic curve E and choose base point B. Sender chooses a random secret integer e and calculates eB · E. Receiver chooses a random integer d and calculates dB · E. Now, eB and dB are public and e, d are secret.

Finally, sender computes edB $= s = (s_1, s_2)$. Then, compute $k = (s_1 * s_2)$ mod n and compute $c = (k * m)$, and send it to receiver. Receiver receives c and decrypts it as follows: Compute $k = (s_1 * s_2)$ mod N, and also compute k^{-1} mod N, where N is a highest prime number. Finally, get original message as,

$$k^{-1} \times c = k^{-1} \times k \times m = m \tag{5.3.1}$$

5.4   Dynamic Usage of Multi-key Encryption Model

The compressed text file is segmented into minimum of two or a maximum of three files. The splitting procedure of compressed text file is as follows: the first letter of the content of compressed input text file stored in file1, second letter placed on file2, and third letter placed on file3; likewise, the remaining letters are put in the appropriate files. So the contents of compressed text file are a minimum of two or maximum of three files which is created by ratio of length of compressed file divided by number of consonants.

Since a maximum number of three segmented files are created from the compressed input file, a maximum of three encryption algorithm are to be utilized in the encryption process. Now, the compressed input text file is to be encrypted in the dynamic selection of three algorithms named as elliptic curve cryptography (ECC), RSA, ElGamal.

The dynamic use of encryption algorithm with segmented files is discussed below: Find the length of compressed file. Then, length of compressed file divided by number of consonants called f, if $f < 2$ to make segmented file minimum of 2. If $f = 2$, use two asymmetric algorithms for encryption.

The characters (alphabets) of compressed files are counted for the purpose of identifying encryption for dynamic usage.

The counted compressed characters are assigned with three rows. Clearly, the character i occurs 14 times and i placed in third row, which indicates the use of third algorithm to encrypt first segmented file. Similarly, the character e occurs 12 times and e placed in second row, which means second algorithm is to be used with second segmented file and the character a occurs 10 times and a placed in first row, so the

Table 5 Characters in compressed text file

Occurrence of each character in compressed text file		
$a= 10$	$b = 0$	$c = 10$
$d = 1$	$e= 12$	$f = 1$
$g = 4$	$h = 7$	$i= 14$
$j = 0$	$k = 1$	$l = 8$
$m = 4$	$n = 6$	$o = 6$
$p = 6$	$q = 0$	$r = 10$
$s = 7$	$t = 10$	$u = 6$
$v = 2$	$w = 0$	$x = 0$
$y = 7$	$z = 0$	

use of first algorithm used to encrypt third segmented file. So the selection of an algorithm to encrypt the segmented file is in dynamic mode always. Finally, get an encrypted message from the combination of the outcome of three algorithms (ECC, ElGamal, RSA).

Encryption Algorithm Input = Compressed text file
  Output = Encrypted output as image format
  Algorithm:

Step 1: Read compressed text file.
Step 2: Ratio of length of compressed file divided by number of consonants.
Step 3: Based on the ratio value the compressed text file divided/segmented.
Step 4: The splitting procedure: first letter of the content of compressed input text file stored on file1 and second letter placed on file2; likewise, the remaining letters are taken placed.
Step 5: Maximum of three segmented files are created from the compressed text file.
Step 6: So here maximum of three asymmetric encryption algorithms to be utilized for encryption process.
Step 7: Then, length of compressed file divided by number of consonants called f.
Step 8: If $f < 2$ to make segmented file minimum of 2. If $f = 2$, use two asymmetric algorithms for encryption.
Step 9: If $f = 3$, use three asymmetric algorithms for encryption.
Step 10: The characters (alphabets) of compressed files are counted for the purpose of identifying encryption for dynamic usage.
Step 11: Maximum occurrences of each characters are played vital role to choose encryption algorithm (RSA, ElGamal, ECC) to process with segmented file in dynamic mode. That is clearly explained in above Table 5.

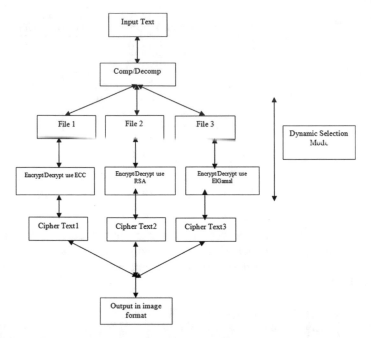

Fig. 1 Proposed encryption model—DUMKE

Step12: Finally, get encrypted message (ciphertext) which is the combination of the outcome of three algorithms.

Figure 1 shows the overall process of DUMKE method.

5.5   Conversion of Ciphertext into Image

In the entire proposed scheme, all the data of cipher (outcome of RSA, ElGamal, ECC) are converted from decimal point into pixel representation to progress the data and provide final output of image. The step-by-step procedure is giving a detailed explanation of how the cipher output is converted into image format.

Step 1: Receive encrypted cipher data from the maximum of three asymmetric encryption algorithms. The outcome of RSA, ElGamal, and ECC is received; for example, the cipher data of RSA are 4202964805184329353621944153173230198648….
ElGamal is 471040.0 417792.0 471040.0 430080.0 430080.0 ….
ECC is 645334876532174643987 ….

Step 2: Divide the encrypted cipher data into multiple blocks, each block having the size of 8 bits. So now got blocks as $B1 = \{42029648\}$ $B2 = \{05184329\}$ $B3 = \{35362194\}$... likewise multiple number of blocks are created from the encrypted data. $B \rightarrow B$ represents block.

Step 3: The entire processes are working with block by block.

Step 4: Make an array of 8×8 matrix from the every block of cipher data. The matrix form is

$$\text{XM}_i = \sum_{j,k=0}^{7} \text{CipherText}_{(\text{SetMatrics})} \tag{5.5.1}$$

The formula (5.5.1) is used to create a 8×8 matrix from the previous step output data.

$$M1 = \begin{pmatrix} 4\,2\,0\,2\,9\,6\,4\,8 \\ 0\,5\,1\,8\,4\,3\,2\,9 \\ 3\,5\,3\,6\,2\,1\,9\,4 \\ 4\,1\,5\,3\,1\,7\,3\,2 \\ 3\,0\,1\,9\,8\,6\,4\,8 \\ 2\,1\,0\,1\,5\,7\,7\,8 \\ 8\,3\,4\,4\,8\,6\,8\,4 \\ 1\,8\,9\,5\,9\,6\,0\,0 \end{pmatrix} \quad M2 = \begin{pmatrix} 7\,5\,0\,7\,1\,0\,1\,4 \\ 1\,4\,7\,3\,6\,9\,8\,0 \\ 2\,0\,3\,0\,3\,3\,5\,9 \\ 3\,4\,9\,0\,4\,7\,1\,0 \\ 7\,1\,1\,9\,7\,4\,3\,6 \\ 8\,1\,7\,5\,7\,5\,6\,5 \\ 5\,3\,9\,4\,8\,9\,8\,5 \\ 2\,7\,1\,1\,6\,2\,0\,0 \end{pmatrix}$$

$$M3 = \begin{pmatrix} 1\,6\,5\,7\,0\,1\,3\,7 \\ 4\,6\,6\,5\,6\,5\,7\,3 \\ 5\,8\,6\,3\,3\,9\,0\,8 \\ 1\,7\,0\,2\,6\,3\,7\,7 \\ 0\,9\,0\,0\,4\,8\,0\,4 \\ 5\,6\,5\,4\,9\,2\,3\,0 \\ 7\,0\,6\,7\,7\,2\,9\,6 \\ 4\,8\,8\,9\,4\,9\,5\,7 \end{pmatrix} \quad M4 = \begin{pmatrix} 6\,5\,5\,9\,4\,3\,6\,6 \\ 4\,7\,3\,7\,4\,0\,7\,6 \\ 1\,4\,4\,3\,2\,5\,3\,0 \\ 0\,5\,2\,1\,3\,9\,8\,1 \\ 1\,9\,9\,0\,0\,0\,8\,4 \\ 6\,8\,8\,7\,9\,5\,3\,5 \\ 8\,5\,9\,4\,8\,4\,6\,5 \\ 6\,3\,4\,4\,3\,7\,6\,2 \end{pmatrix}$$

$M \rightarrow M$ represents matrix, likewise more number of matrix are created based on the input file.

Step 5: The above data of matrix are converted into pixels and get gray image.

Step 6: Finally, the gray image is declared as output. The users give their input as text file and get output as an image formation. The third party could not guess of which kind of data is transferred from source to destination while seen the output.

Table 6 Encrypted output

File name	File size	Output
Java.txt	21 bytes	
Input.txt	6 kb	
File.txt	40 kb	

6   Results and Discussions

The proposed cryptographic algorithm is designed for text files to provide more privacy and authentication. Generally, the existed encryption algorithms are working directly with the input file, but proposed encryption algorithm working differently. The first phase of this encryption algorithm is to compress the input text file; after that, the compressed file is to be encrypted by the dynamic selection of three asymmetric encryption algorithms. The outcome of these three algorithms is processed and displayed the output as image (Table 6).

7   Conclusion

Today's technological world has variety of hacking techniques and hackers to steal the data and earn money through this. Nowadays, this is one of the peak businesses in society. So from the owner of the data point of view, the data will need strong protection. Hence, cryptography remains best among the various research areas. From the survey, many text file encryption techniques provided security in different aspects. For text files, most of the encryption algorithms designed with more than two keys to encrypt data for providing security. The hackers break the keys and collect the secrets. But the proposed encryption algorithm has working differently. The process of compression in encryption algorithm is newly designed to secure the text files. Before the encryption process, the input text file is compressed and segmented. The segmented files are dynamically chosen by an asymmetric algorithm (RSA, ElGamal, ECC) to encrypt it. Finally, the ciphertext is changed to an image showing pixels.

References

1. http://www.dcs.ed.ac.uk/home/adamd/essays/crypto.html
2. Rivest, R.L., Shamir, A., Adleman, L.: A Method for Obtaining Digital Signatures and Public Key Cryptosystems. National Science Foundation Grant MCS76-14294, pp. 1–15. Naval Research grant number N00014-67-A-0204-0063 (1977)
3. Aditya Sundar, N., Pandit Samuel, G., Naidu, Ch.D., Kishore, M.V.: A novel approach for secure communication of text using compression mechanism. Int. J. Comput. Sci. Technol., 243–247 (2015)
4. Elgedawy, I., Srivastava, B., Mittal, S.: Exploring Queriability of Encrypted and Compressed XML Data. CIS Northern Cyprus Campus, pp. 141–146 (2009)
5. Kodabagi, M.M., Jerabandi, M.V., Gadagin, N.: Multilevel Security and Compression of Text Data Using Bit Stuffing Huffman Coding. IEEE, pp. 800–804 (2015)
6. Wong, K.-W., Lin, Q., Chen, J.: Simultaneous arithmetic coding and encryption using chaotic maps. IEEE Trans. Circuits Systems-II, pp. 146–150 (2010)
7. Lakshmana Rao, K., Maringanti, H.B., Maram, B.: A sturdy compression based cryptography algorithm using self-key (ASCCA). Int. J. Eng. Technol., 1236–1245 (2015)
8. Patil, M., Sahu, V., Jain, A.: SMS Text Compression and Encryption on Android OS, IEEE (2014). https://doi.org/10.1109/iccci.2014.6921767
9. Chen, J., Zhou, J., Wong, K.-W.: A modified chaos-based joint compression and encryption scheme. IEEE Trans. Circuits Systems-II, 110–114 (2011)
10. Liu, L., Gai. J.: Bloom filter based index for query over encrypted character strings in database. In: 2009 World Congress on Computer Science and Information Engineering. IEEE, pp. 303–307 (2009)
11. Huang, Y.-M., Liang, Y.-C.: A secure arithmetic coding algorithm based on integer implementation. In: International Symposium on Communications & Information Technologies (ISCIT 2011). IEEE, pp. 518–521 (2011). https://doi.org/10.1109/iscit.2011.6092162
12. Berlin, K., Dhenakaran, S.S.: A novel encryption technique for securing text files. Int. J. Adv. Res. Trends Eng. Technol. 3(20), 68–72 (2016)
13. Berlin, K., Dhenakaran, S.S.: A novel compression technique for text files using matrix mathematical model. In: Proceedings of IETE ICRA-2016. ISBN: 978-93-80609-35-5
14. https://en.wikipedia.org/wiki/ElGamal_encryption
15. http://www.ccs.neu.edu/home/riccardo/courses/cs6750-fa09/talks/Ellis-elliptic-curvecrypto.pdf

Shortwave Infrared-Based Phenology Index Method for Satellite Image Land Cover Classification

KR. Sivabalan and E. Ramaraj

Abstract Recent technology relay upon the satellites. Satellites are having their own technical property to its behavioral methods. Communication methodology and sensor techniques are very important to take earth imagery. Satellites take the images in various combinations of bands to produce the imagery with maximum details. The efficacy of the multispectral imagery is limitless. Multispectral imagery is having the details about a particular region using 11 band combinations. Grouping the pixels into the meaningful class is called as image classification. Satellite image classification is used to define the land surface segmentation and feature extraction. Weather forecasting, agriculture, air force department, and water department are the major departments which rely upon the satellite imagery. Remote sensing input for those real-time applications is endless. Satellite image classification falls with three categories as automatic, semi-automated, and hybrid methods. Medium resolution multi spectral imagery classification algorithms designed with Automatic and Manual classification method to reach the maximum accuracy. High and very high spectral imagery classification algorithms follow the hybrid- or object-based image analysis. Phonology index is designed to classify the multispectral satellite imagery based on the reflectance value of the passive sensor imagery. This paper provides a phenology-based method called shortwave infrared-based phenology index (SIPI) to classify the multispectral imagery with maximum accuracy. Confusion matrix and Kappa coefficient are the quality measures used to justify the classification efficiency.

Keywords Classification · Semi-automated classification · Remote sensing · Phenology index · Multispectral classification

KR. Sivabalan (✉) · E. Ramaraj
Department of Computer Science, Alagappa University, Karaikudi, India
e-mail: sivabalanalu@gmail.com

E. Ramaraj
e-mail: eramaraj@rediffmail.com

© Springer Nature Singapore Pte Ltd. 2020   877
K. N. Das et al. (eds.), *Soft Computing for Problem Solving*,
Advances in Intelligent Systems and Computing 1057,
https://doi.org/10.1007/978-981-15-0184-5_75

1   Introduction

Think of today's nation developments entirely depends on the new creation of technologies. Innovations in recent technological developments are the most interesting and impressive. The moves of technology are extremely fast. New innovations are created for every new day. Among these technology worlds, remote sensing plays a crucial role to achieve impossible requirements as possible from the distance. Remote sensing technology can be used to classify and detect the objects from the distance on the earth. Satellite-based sensors have a main role to execute remote sensing technology.

Sensors are categorized into two types named as active and passive. In the sense of active, internal stimuli used to collect data from the earth surface. Laser beam remote sensing system is the best example of sources used by active sensors. The passive sensors using external stimuli, which means the natural energy of earth surface using as a source of passive sensors. The reflected sunlight energy is a good example of sources used by the passive sensors. The main problem of the passive sensor is it can't observe the earth surface while the sky is covered by clouds. To overcome these kinds of minor problems, active sensors are taken placed. The advantages of the active sensor are to collect imagery at the day as well as night and working with poor weather condition and clouds. The active sensors include LiDAR, laser altimeter, radar, sounder, ranging instrument, and scatterometer, such that passive sensors are spectrometer, sounder, spectroradiometer, imaging radiometer, accelerometer, hyperspectral radiometer.

Image classification techniques play a vital role to categorize the pixels of remotely sensed images into various land cover classes. For example, pixels of remotely captured images are classified into a different field like agriculture, urban, forest, water resources, grassland, industrial zones, and residential zones. This kind of process is also called as land cover mapping. In remote sensing, image classification techniques should need for the purpose of extract information for an application, thematic map creation, field surveys, disaster management, effective decision making, spatial data mining, and visual and digital satellite image interpretation [1].

Generally, there are three different types of classification techniques used to classify the images of remotely sensed data, which includes supervised classification, unsupervised classification, and object-based image analysis. Supervised classification method needs input from the analyst. The process of supervised classification is to select training sets, generate signature files, and classify the training sets. In the supervised image classification, training sets are considered as an important factor to improve accuracy. Two types of training samples are prepared, one for classification and another one for classify the accuracy. Supervised classification methods support the following statistical techniques include artificial neural network, image segmentation, binary decision tree. Data of Landsat imagery and multispectral IKONOS II were classified by fuzzy logic, standard maximum likelihood classifier, and artificial neural networks [2].

The main process of unsupervised classification is to generate clusters and assign classes. The clusters are used to group the pixels of the images into the relevant class. Unsupervised satellite image classification techniques are K-means [3], ISODATA [4] and support vector machine. The process of object-based image analysis is to group the image pixels into objects.

Luo et al. [5] proposed a new mechanism for land cover classification using supervised classification methods; the mechanism was named as "fusion of airborne discrete-return LiDAR and hyperspectral data for land cover classification." Supervised classifiers include support vector machine (SVM) and maximum likelihood classifier (MLC) are used to the classified land cover category into seven classes. Classified seven classes are buildings, road, water bodies, forests, grassland, cropland, and barren land. The data for classification are collected from PCA fusion data, stacking fusion data, LiDAR data, CASI data. Authors got improved overall classification accuracy by 9.1 and 19.6%.

To classify high-resolution satellite images, Wang et al. [6] designed supervised classification methodology in the name of supervised classification high-resolution remote sensing image based on interval type-2 fuzzy membership function. In this research, the authors used the fuzzy membership function to enhance the accuracy of the classification decision. With this proposed algorithm, authors achieved higher classification accuracy.

Routh et al. [7] proposed a new protocol to classify the hyperspectral images using supervised classification technique. The name of research is improving the reliability of mixture tuned matched filtering remote sensing classification results using supervised learning algorithms and cross-validation. The mixture of tuned matched filtering (MTMF) classification expanded by using supervised learning algorithms. After that, the new protocol was designed as a final step of the MTMF classification algorithm. Through their newly designed protocol, authors achieved overall accuracy between 18. 4 and 30.8%.

Ma et al. [8] proposed supervised classification mechanism with the name of parallelizing maximum likelihood classification for supervised image classification by pipelined threat approach through high-level synthesis (HLS) on FPGA cluster. The research focused on the process the big remote sensing images efficiently and effectively. Authors applied maximum likelihood classification (MLC) over big remote sensing images on field-programmable gate array cloud. Comparing CPU cluster, authors achieved the best performance by using FPGA method.

Wu et al. [9] proposed a new supervised classification method that was named as supervised sub-pixel mapping for change detection from remotely sensed images with different resolutions. Authors designed back propagation neural network (BPNN)-based SPM model by using the supervised framework. This method gives more accuracy of high-resolution images and helps to generate sub-pixel change detection map effectively.

Fig. 1 Visualization of
study area [10] details

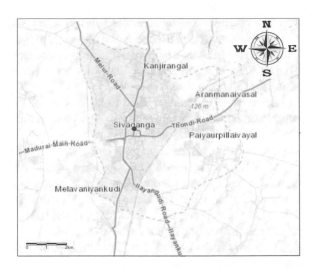

2 Study Area and Data

2.1 Study Area

Sivaganga district of Tamil Nadu extends over 4189 km². The position of study area
is between 9° 43′ and 10° 2′ north latitude and between 77° 47′ and 78° 49′ east
longitude. Part of Sivaganga is taken as the study area. Figure 1 shows the details
about the study area.

Part of Sivaganga is taken as the study area using latitude, longitude [11], and
military grid reference system [12] phenomena.

2.2 Data

Landsat 8 (OLI + TIRS) covers the study area in the 16-day cycle. Landsat-level 1
imagery is used to phenology classification method. It has the 11 bands' combinational multispectral imagery. TIRS bands are covering 100-m-resolution area single
imagery.

Table 1 is used to describe the Landsat 8 band details [13] as per the official
announcement. From the Landsat 8 mission [14], study area data is collected from
the United States Geological Survey is a scientific agency of the US Government's
USGS Earth Explorer official site [15].

The dataset details [15] are tabulated in Table 2. Geo-referencing Tagged Image
File Format (GeoTIFF) [16, 17] is the used format for better classification in multispectral imagery.

Table 1 Landsat 8 (OLI + TIRS) band details

Bands	Name	Wavelength (μm)	Resolution (m)
Band 1	Ultra blue (coastal/aerosol)	0.435 0.451	30
Band 2	Blue	0.452 0.512	30
Band 3	Green	0.533 0.590	30
Band 4	Red	0.636 0.673	30
Band 5	Near infrared (NIR)	0.851 0.879	30
Band 6	Shortwave infrared (SWIR) 1	1.566 1.651	30
Band 7	Shortwave infrared (SWIR) 2	2.107 2.294	30
Band 8	Panchromatic	0.503 0.676	15
Band 9	Cirrus	1.363 1.384	30
Band 10	Thermal infrared (TIRS) 1	10.60 11.19	100 * (30)
Band 11	Thermal infrared (TIRS) 2	11.50 12.51	100 * (30)

Table 2 Details about the Landsat 8 imagery

Property Name	Value
Origin	US Geological Survey
Landsat scene id	LC81430532018158LGN00
Sensor ID	OLI_TIRS
Image format	GeoTIFF
Date acquired	June 12, 2018
Sun elevation	62.61494214
Sun azimuth	59.39342772
Orientation	North Up

3   Methodology

This paper proposes a new methodology called shortwave infrared-based phenology index (SIPI) to classify the multispectral satellite imagery land covers. This SIPI completely works based on the supervised classification aspect (Fig. 2).

Walk 1:

Landsat 8 dataset procured from the USGS earth explorer. Then, the imagery is focused as per the study area selection. Band selection and region of image enrichment are done based on the classification requirement.

Walk 2:

Reflection index worked based on the phenology values of selected band values. Band selection for formulations is done using band specification and band use phenomena.

Walk 3:

Geometric correction performed with the help of radiometric calibration and dark object removal. It helps to enhance the quality of the imagery.

Fig. 2 SIPI architecture

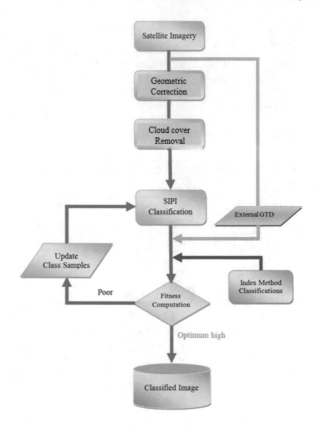

$$\sum_{i=0}^{n} \psi_{DN} \leftrightarrow \sum_{i=0}^{n} \psi_{TOA} \qquad (3.1)$$

where i denotes the band, DN denotes the digital number, and TOA denotes the top of atmosphere

$$\psi_{GC} = \sum_{i=0}^{n} \psi_{TOA} - \sum_{i=0}^{n} \psi_{DO} \qquad (3.2)$$

where GC denotes geometric corrected and DO denotes that dark object. The formulas (3.1) and (3.2) are used to perform the geometric correction on selected band images.

Walk 4:

Cloud cover removal operation is performed on geometric corrected images.

$$\psi_{Cloud\ Detect} = \begin{cases} \psi_i & \text{if } i_{Cloud\ Cover} \geq 0.5 \\ \psi_{null} & \text{Othewise} \end{cases} \qquad (3.3)$$

$$\int_i^n \psi_{\text{Cloud Free}} = \int_i^n \psi_{\text{Cloud Detect}} \pm \int_i^n \psi_{\text{Cirrus}} \tag{3.4}$$

The cloud covers removal performed with the help of above formula (3.3) and (3.4). This cloud removed the images which are used for SIPI formulation for land cover classification.

Walk 5:

Shortwave infrared (SWIR) 1, shortwave infrared (SWIR) 2, coastal, green bands are used to make the classification index.

$$\text{SIPI} = \frac{\sum_{a,b=1}^n (\psi_{\text{SWIR2}} + \psi_{\text{SWIR1}}) - \sum_{a,b=1}^n (\psi_1)}{\sum_{a,b=1}^n (\psi_3)} \tag{3.5}$$

ψ SWIR1—shortwave infrared 1, ψ SWIR2—shortwave infrared 2, ψ 1—coastal band, ψ 3—green band.

Walk 6:

Apply the Following SIPI algorithm for land cover classification.

Algorithm (SIPI)

Step 1: Set the surface class and class hue.

Step 2: Set the surface threshold-based surface class sample using external ground truth data.

Step 3: Get the reflectance values of selected bands.

Step 4: Apply the reflectance values to the SIPI formula (3.5)

Step 5: Group the pixel based on the formula return value and the class threshold value.

$$\text{Class}_i = \left(\text{SIPI}_{x1,x2,x3,...,xn} \right) X_i \in C_i \tag{3.6}$$

$$\text{Min } C_i \leq X_i \leq \text{Max } C_i \tag{3.7}$$

$$\text{Class}_i = \begin{cases} \text{Class}_x & \text{if Min } C_i \leq x_i \leq \text{Max } Ci \\ \text{Class}_{x+1} & \text{Otherwise} \end{cases} \tag{3.8}$$

where i denotes the class number and C denotes the class.

Step 6: Repeat step 3 to step 5 for each pixel in the imagery.

Step 7: Export the land cover classified output image.

Walk 7:

Fitness computation [18] is performed to justify the classification image efficiency. Fitness computation is performed based on the external ground truth data and index methods. The accuracy of the classified image is compared with other phenology index method accuracy.

$$\text{SIPI RES} = \begin{cases} \text{SIPI}_{\text{Output}} & \text{if SIPI}_{\text{Accuracy}} > \text{Existing Index Method}_{\text{Accuracy}} \\ \text{Class Updation and Computation Otherwise} \end{cases} \quad (3.9)$$

If the accuracy is poor than other index methods, class samples are updated and step 3 to step 6 performed for better classification. If the classification produces high accuracy, it stored to the database as the classified image. The above walks 1–7 are used to optimum high accuracy in phenology index-based SIPI classification.

4 Results and Discussion

The SIPI method works on the multispectral imagery which has the shortwave infrared 1, shortwave infrared 2, coastal, and green bands. Five surface classes are created based on the reflectance result. The SIPI method works well in the Landsat 8 imagery. Classification accuracy measured using confusion matrix and Kappa coefficient (Fig. 3).

$$\text{Precision} = \frac{\text{True Positive (TP)}}{(\text{True Positive + False Negative(FN)})} \quad (4.1)$$

$$\text{Recall} = \frac{\text{True Negative(TN)}}{(\text{True Negative + False Positive(FP)})} \quad (4.2)$$

$$\text{Overall Accuracy (OA)} = \frac{\text{Correctly Classified Pixels}}{\text{Total Number of Pixels}} * 100 \quad (4.3)$$

where $C1$—water, $C2$—tree, $C3$—rock and buildings, $C4$—soil and bare land $C5$—vegetated and grassland, Pre.—precision, Re.—recall. Table 3 shows the confusion matrix

$$\text{Overall Accuracy (OA)} = \frac{9286}{9409} * 100 = 98.69,$$

(a)   **(b)**

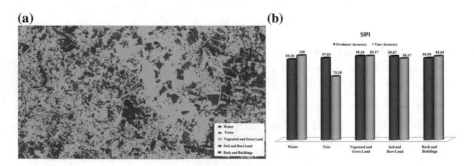

Fig. 3 **a** SIPI classified image, **b** SIPI user and producer accuracy for classification

Table 3 Confusion matrix for SIPI classification

C	C1	C2	C3	C4	C5	Total	TP	FP	TN	Pre.	Re.
C1	**1178**	0	0	0	0	1178	1178	57	8174	0.99	0.99
C2	28	**98**	0	0	4	130	98	3	9276	0.75	0.99
C3	0	0	**407**	4	0	411	407	13	8985	0.99	0.99
C4	0	0	13	**1285**	34	1332	1285	12	8065	0.96	0.99
C5	29	3	0	8	**6318**	6358	6318	38	3013	0.99	0.99

$$\text{kappa}(k) = \frac{N \sum_{i=1}^{n} m_{i,i} - \sum_{i=1}^{n}(G_i C_i)}{N^2 - \sum_{i=1}^{n}(G_i C_i)} \qquad (4.4)$$

where i is the class number, N is the total number of classified values compared to truth values, mi, i is the number of values belonging to the truth class i that have also been classified as class i (i.e., values found along the diagonal of the confusion matrix), Ci is the total number of predicted values belonging to class I, Gi is the total number of truth values belonging to class i [19]

Kappa Coefficient = **0.97**.

5   Comparative Analysis

Same Landsat 8 imagery is classified with standard index methods to improve the SIPI classification accuracy. Normalized difference vegetation index (NDVI) [20], enhanced vegetation index (EVI) [21, 22], soil adjusted vegetation index (SAVI) [21, 23] are the methods used to compare with SIPI method (Fig. 4).

Table 4 and Figs. 5 and 6 show the comparison between SIPI method and existing methods.

Landsat 8 (OLI + TIRS) imagery is taken for the classification process. The same region of an image is classified using four phenology index classification methods. From the results, shortwave infrared-based phenology index (SIPI) method performs well in the Landsat 8 imagery and it produces the maximum classification accuracy among these methods.

6   Conclusion

Real-time applications require the classification result in various aspects for various needs. Classification accuracy relays upon the imagery selection and process methods. This paper provides a new phenology index method for remote sensing satellite image land cover classification. This SIPI method uses the shortwave infrared 1, shortwave infrared 2, coastal, and green band reflectance values for classification. It works well on the medium-resolution multispectral imagery with above-said bands. Sample class update based on fitness computation is helped to optimum the high accuracy result on classification. Five types of land cover classes created and threshold fixed using ground truth data. SIPI classification effectively works on the imagery and produces the maximum accurate result.

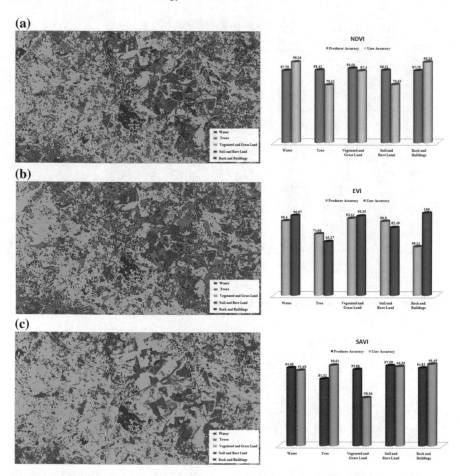

Fig. 4 a NDVI classified image and classification accuracy, **b** EVI classified image and classification accuracy, **c** SAVI classified image and classification accuracy

Table 4 Index methods' classification comparison

Class	NDVI		EVI		SAVI		SIPI	
	P.A	U.A	P.A	U.A	P.A	U.A	P.A	U.A
Water	87.79	98.26	90.4	96.87	94.89	91.69	95.38	100
Tree	88.41	70.11	74.66	65.27	81.32	98.01	97.03	75.38
Veg.	90.56	87.4	93.54	96.35	92.86	58.56	99.40	99.37
Soil	88.21	70.63	89.9	82.49	97.59	96.58	99.07	96.47
Rock	87.79	98.26	59.14	100	94.83	99.49	96.90	99.03
O.A	**88.25**		**91.06**		**95.38**		**98.69**	
Kappa	**0.83**		**0.86**		**0.93**		**0.97**	

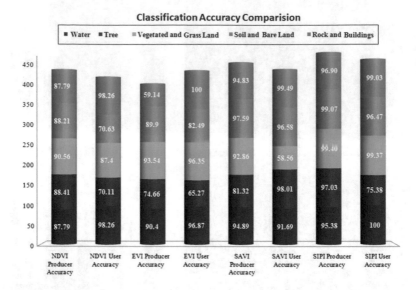

Fig. 5 User and producer accuracy comparison between four phenology index methods

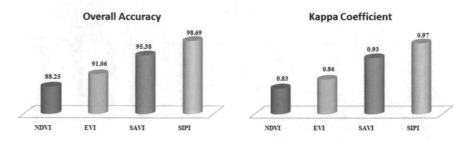

Fig. 6 Overall accuracy and Kappa coefficient comparison between four phenology index methods

Acknowledgment This article has been written with the financial support of RUSA – phase 2.0 grant sanctioned via Letter No F.24-51 / 2014-U, Policy (TNMulti-Gen), Dept. of Edn. Govt. of India, Dt. 09.10.2018.

References

1. Abburu, S., Golla, S.B.: Satellite image classification methods and techniques—a review. Int. J. Comput. Appl. **119**(8), 20–25 (2015)
2. Ayhan, E., Kansu, O.: Analysis of image classification methods for remote sensing. Exp. Tech. **36**(1), 18–25 (2012)
3. Ahmed, R., Mourad, Z., Ahmed, B.H., Mohamed, B.: An optimal unsupervised satellite image segmentation approach based on pearson system and k-Means clustering algorithm initialization. Int. Sci. Index **3**(11), 948–955 (2009)

4. Al-Ahmadi, F.S., Hames, A.S.: Comparison of four classification methods to extract land use and land cover from raw satellite images for some remote arid areas, Kingdom of Saudi Arabia. J. King Abdulaziz Univ.-Earth Sci. **20**(1), 167–191 (2009)
5. Luo, Shezhou, Wang, Cheng, et al.: Fusion of Airborne discrete-return LiDAR and hyperspectral data for land cover classification. Remote Sens. **8**(1), 1–19 (2015)
6. Wang, C., Xu, A., et al.: Supervised classification high-resolution remote-sensing image based on interval type-2 fuzzy membership function. Remote Sens. **10**(5), 1–22 (2018)
7. Routh, D., Seegmiller, L., et al.: Improving the reliability of mixture tuned matched filtering remote sensing classification results using supervised learning algorithms and cross-validation. Remote Sens. **10**(11), 1–19 (2018)
8. Ma, S., Shi, S., et al.: Parallelizing maximum likelihood classification (MLC) for supervised image classification by pipelined thread approach through high-level synthesis (HLS) on FPGA cluster. Big Earth Data **2**(2), 144–158 (2018)
9. Wu, K., Du, Q., et al.: Supervised sub-pixel mapping for change detection from remotely sensed images with different resolutions. Remote Sens. **9**(3), 1–17 (2017)
10. Available at: https://scholar.google.co.in/scholar?hl=en&as_sdt=0%2C5&q=latitude+and+longitude&btnG=&oq=latitude+
11. Voracek, M., Fisher, M.L., Marušiš, A.: The Finno-Ugrian suicide hypothesis: variation in European suicide rates by latitude and longitude. Percept. Motor Skills **97**(2), 401–406 (2003)
12. Available at: http://legallandconverter.com/p50.html
13. Singh, A., Singh, K.K.: Satellite Image classification using genetic algorithm trained radial basis function neural network, application to the detection of flooded areas. J. Vis. Commun. Image Represent. **42** (2016). https://doi.org/10.1016/j.jvcir.2016.11.017
14. Available at: https://landsat.usgs.gov/landsat-8
15. Available at: https://earthexplorer.usgs.gov/
16. Ritter, N., Ruth, M.: The GeoTiff data interchange standard for raster geographic images. Int. J. Remote Sens. **18**(7), 1637–1647 (1997)
17. Available at: https://en.wikipedia.org/wiki/GeoTIFF
18. Story, M., Congalton, R.G.: Accuracy assessment: a user's perspective. Photogram. Eng. Remote Sens. **52**(3), 397–399 (1986)
19. Kappa Coefficient (Available at: https://www.harrisgeospatial.com/docs/CalculatingConfu sionMatrices.html
20. Ehsan, S., Kazem, D.: Analysis of land use-land covers changes using normalized difference vegetation index (NDVI) differencing and classification methods. Afr. J. Agric. Res. **8**(37), 4614–4622 (2013). https://doi.org/10.5897/AJAR11.1825
21. Wang, Cuizhen, Qi, Jiaguo, Cochrane, Mark: Assessment of tropical forest degradation with canopy fractional cover from Landsat ETM+ and IKONOS imagery. Earth Interact. **9**(22), 1–18 (2005)
22. Available at: https://en.wikipedia.org/wiki/Enhanced_vegetation_index
23. Available at: https://en.wikipedia.org/wiki/Soil-adjusted_vegetation_index

Exploration of Various Cloud Security Challenges and Threats

Arun Krishna Chitturi and Purushotham Swarnalatha

Abstract Most of the organizations are stuck in a situation to cloudify due to concerns regarding security of sensitive data. Cloud computing provides many aids to the users and organizations in accordance with expenditure and savings. Other than these benefits, cloud computing has some hurdles that result in restriction of it's usage. Cloud security is the bigger hurdle which is regularly taken into consideration. This paper will give a broad view of major threats challenges in security which are encountered in cloud computing. Cloud computing entities include cloud user, cloud provider, and data owner. The study is carried out based on selection of open-source cloud offerings. This paper can be an exploration tool for any IT person to gain knowledge into security-related risks and challenges which are concerned with cloud computing.

Keywords Security challenges · Security in cloud computing · Major problems in cloud security

1 Introduction

We know that cloud is achieving a lot of pulse, thanks to the mixture of market & automation connected factors. Rapid change in business conditions is leading a modification within the computing infrastructure of the numerous corporations. Quantity of enterprise services and applications is continually increasing, in which new ones being unceasingly adding and older ones being removed. Moreover, one third-quarters of North American and European Corporations obtain the parts of their businesses by contract from outside, which implies that important business linked data is not just spread over the distinctive registering frameworks inside one venture, yet, it is additionally circulated over numerous IT foundations of the organization business arrange. One of the major significant hurdles in cloud while comparing to other IT services is security and privacy [1].

A. K. Chitturi · P. Swarnalatha (✉)
VIT University, Vellore, Tamil Nadu, India
e-mail: hgswarna@gmail.com

© Springer Nature Singapore Pte Ltd. 2020   891
K. N. Das et al. (eds.), *Soft Computing for Problem Solving*,
Advances in Intelligent Systems and Computing 1057,
https://doi.org/10.1007/978-981-15-0184-5_76

Cloud computing builds a network-dependent situation to the users, which gives a way for allocation of resources disregarding the location. In any case, proof is showing that in spite of cloud being viewed as a noteworthy employment road for the following a long time, movement to the cloud worldview is blocked by worries on its security. Let us consider for instance, money-related foundations are pulled in by cloud computing, however, for security issues, they are confined to beginning periods of endorsement (adoption). Ongoing assaults in cloud, as the one of every 2014 where 50 million consumer records of Dropbox were hacked, one demonstrate that cloud information security has turned into an intriguing issue [2]. Proof of the dangers of cloud is uncovered were exhibited by Al Awadhi et al. [3].

Cloud computing in order to become a substitute, it should give minimum security that is given by present traditional IT systems. In order reach this goal, outstanding awareness is required about tools and measures that are feasible to oppose or counter the malicious activities. Because of dynamic scalability features of cloud models, the applications and their data that is present in the cloud models do not have a fixed infrastructure and security [4].

2 Security Challenges

In cloud computing, users do not know accurate location where their important and sensitive data is present because the service providers manage the data centers in different geographical neighborhoods [3]. The ancient security techniques like firewalls and antivirus programs do not provide enough security.

2.1 Major Threats and Security Issues in Cloud Computing

Walker [5] identified the major twelve threats that are concerned with cloud security. They are data breaches, broken authentication, Hacked interface and Application Program Interfaces, exploited system vulnerabilities, account hijacking, malicious insiders, The advanced persistent threat (APT) parasite, permanent data loss, inadequate diligence, cloud service abuses, Denial-of-Service (DoS) attacks, Shared technology, shared dangers. Private keys of the cloud should be maintained privately. They have to protected from attacks like both net attack and also from numerous vulnerabilities of software which have the flaws to give access to authorized information from kernel or hypervisor [6].

[3] The three levels of security challenges faced are classified into three levels as shown in Fig. 1. They are clearly explained with explanation.

Fig. 1 Levels of security issues

3 Communicational Level

3.1 Security Issues at Network Level

The main assets to be discussed regarding security issues at this level are confidentiality and integrity of data. These are the issues concerned with this level:

- Domain Phishing—Domains of customers are hijacked only with the motive to steal their data and collapse the website [7].
- BGP Hijacking—Often addressed as prefix hijacking or route hijacking, which is taking over a group of IP Addresses.
- Sniffer Attacks—Sniffing is capturing the packets which are present in the network.

a. **Security issues at application level**:

All applications require security for stopping or preventing attackers to gain control. The issues concerned with application-level security are [7]

- DDoS—Numerous systems flood the cloud resulting in depletion of the resources
- Dictionary Attack—This is a brute force attack
- Cookie Poisoning—Modifying the cookies of users to gain control

- Hidden Field Manipulation—Modifying the hidden values in the forms
- CAPTCHA Breaking—Bypassing the captcha.

3.2 Security in Host Level

Threats in this level are concerned with operating systems. Major issues concerned with host level are

- Viruses—It is a malicious software that multiplies itself resulting in change of program
- Password cracking—It is a technique to obtain correct password to the system
- Unauthorized access—Gaining access by bypassing authentication
- Foot printing—Gathering information
- Profiling—Getting user profiles and analyzing them for malicious purpose.

4 Computational Level

The biggest challenge that is faced in cloud computing is implementing virtualization.

4.1 Challenges in Virtualization

The main classes in virtualization are desktop, network, server and machine virtualizations. A malignant or malicious virtual machine may try to lure the IaaS (Infrastructure as a Service) VM strategy to execute on same physical machine and then it takes dominance of flaw in hypervisor to deploy a side channel attack [8].

4.1.1 Virtual-Level Security

[3] Virtual machines perform their own cycles in different stages like creation, shutdown, resumed, destroyed, suspended, pending, running, prolong, etc.

VM security-level challenges are categorized as shown in Fig. 2.

(i) **VM Cloning**
Copying existing VM's ID, computer name, IP address, and MAC address is called VM Cloning.

(ii) **VM Isolation**

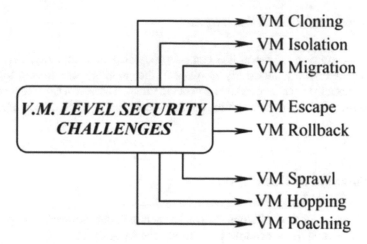

Fig. 2 VM security-level challenges

    Virtual machines are separated for ensuring security. This isolation of virtual machines guarantees security. The break in isolation because of using of IP addresses multiple times among the VMs results in serious challenges.

(iii) **VM migration**:

    Virtual machines have the advantage of easy migration from one server to another. This migration of VMs leads to risks in security for the migrated virtual machine and new-hosting machine.

(iv) **VM Escape**:

    VMs work in isolated conditions inside host. Any trail of direct interaction of VM through intervention in these isolated environments leads to escape of VM.

(v) **Rollback**:

    Virtual machines can return to their previous state from present state. When they are rolled back to former state then they are re-exposed to many vulnerabilities.

(vi) **Sprawl**

    Unchecked deployment of VMs is called sprawl [9]. In this VM sprawl situation, the majority of virtual machines remain inactive. This results in wastage of plenty of resources.

(vii) **Hopping**

    This is nothing but achieving control over other VM by vulnerabilities present in hypervisor. This leads to obscure malicious attacks, etc.

(viii) **Poaching**:

    The vulnerabilities that are existing in the operating system lead to abnormalities of entire system. This leads to entire system failure.

4.2 Hardware-Level Challenges

This layer comprises of assets like central processing unit, memory, networking and storage, etc. These are clearly assigned in the cloud and are divided to virtual machines through virtualization layer. There is a threat of breach in isolated protection if a guest takes care to overpower discretionary access control and mandatory access control.

5 Challenges at Data Level

The most important asset of cloud is data. Breach in data was clearly recognized as the most important threat according to Cloud security Alliance.

Some of the data-level challenges are:

- Data integrity
- Data lineage
- Data recovery
- Data leakage
- Data remanence
- Data backup
- Data isolation
- Data segregation
- Data lockin
- Data provenance
- Data location.

6 Security Open Issues and Threats

ENISA—center for competence of cyber security in the Europe. This works together with member state and private sector divisions to convey advice and solutions [10]. A clear review of research literature was provided. In accordance with [10, 11], it can be said that main challenges in security are mentioned below:

- Shared technologies vulnerabilities
- Data breach
- Account or service traffic hijacking
- Denial-of-Service (DoS)
- Malicious insiders.

6.1 Shared technologies

As Navan et al. [12] demonstrated that hijackers take advantage of vulnerabilities in the hypervisor and get access to the host, where other VMs are present.

6.2 Data Breach

An incident or a situation which involves stealing of sensitive data is called data breach.

6.3 Account or Service Traffic Hijacking

A user looses authority over account and resulting the attacker to get access into sensitive areas and leading to compromise in confidentiality, integrity, and availability of the services.

6.4 Denial-of-Service (DoS)

DOS is more harmful and powerful in cloud than compared to normal context because when work load rises in particular service, then the cloud provides additional power to that particular service [13]. This implies the cloud environment counters DOS attack and at the same time provides additional resources allowing the attacker to continue malicious activity.

6.5 Malicious Insiders

Best example to explain this situation is an employee trying to take advantage of his privileges to access sensitive data.

7 Service Level Agreements

Services that offered to consumers by the providers should follow SLA's resource provision at any instance is based on bandwidth, control processing unit, memory management among others. Security metrics should be included in SLA, these are necessary for mitigation of challenges and also for efficient transmission of responsibilities. Some standards like Secure Provisioning of Cloud Services (SPECS) and European Network and Information Security Agency(ENISA) give security by maintaining Service Level Arguments. Usage of these SLA results in achieving Quality of service. Service level arguments consist of contract definition, negotiation, monitoring and enforcement.

8 Conclusion

The conclusion of the paper consists of various issues regarding communication, computational, and SLA with comparative study. Data-associated security problems are taken as most valuable entities in the computational and virtualization levels, Hardware-level- and virtual-level challenges are discussed in Virtualization layer. Data-level gaps are classified into various levels.

Cloud computing is still not yet completely developed regarding security issues. The author concluded the paper by discussing additional intensive analysis in clouding computing as a key for the comparative purpose. This will be extended by the authors as the future work for deployment in cloud environment.

References

1. Khan, N., Al-Yasiri, A.: Cloud security threats and techniques to strengthen cloud computing adoption framework. In: Cyber Security and Threats: Concepts, Methodologies, Tools, and Applications. IGI Global, pp. 268–285 (2018)
2. Luigi, C., et al.: Cloud security: emerging threats and current solutions. Comput. Electr. Eng. **59**, 126–140 (2017)
3. Al Awadhi, E., Salah, K., Martin, T.: Assessing the security of the cloud environment. In: 2013 7th IEEE GCC Conference and Exhibition (GCC) (2013)
4. Chen, D., Zhao, H.: Data security and privacy protection issues in cloud computing. In: 2012 International Conference on Computer Science and Electronics Engineering (ICCSEE), vol. 1, IEEE (2012)
5. Walker, K.: Cloud security alliance announces software defined perimeter (sdp) initiative (online) (2013). https://cloudsecurityalliance.org/media/news/csa-announcessoftware-defined-perimeter-sdp-initiative/. Accessed October, 2014
6. Yu, W., et al.: Protecting your own private key in cloud: security, scalability and performance. In: 2018 IEEE Conference on Communications and Network Security (CNS), IEEE (2018)
7. Subramanian, N., Jeyaraj, A.: Recent security challenges in cloud computing. Comput. Electr. Eng. **71**, 28–42 (2018)
8. Laniepce, S., et al.: Engineering intrusion prevention services for iaas clouds: the way of the hypervisor. In: 2013 IEEE 7th International Symposium on Service Oriented System Engineering (SOSE), IEEE (2013)
9. Bose, R., Sarddar, D.: A secure hypervisor-based technology create a secure cloud environment. Int. J. Emerg. Res. Manage. Technol. ISSN (2014): 2278-9359
10. Fernandes, D.A.B., et al.: Security issues in cloud environments: a survey. Int. J. Inf. Secur. **13**(2), 113–170 (2014)

11. Jansen, W., Grance, T.: Sp 800-144. Guidelines on Security and Privacy in Public Cloud Computing (2011)
12. Nanavati, M., et al.: Cloud security: a gathering storm. Commun. ACM **57**(5), 70–79 (2014)
13. Deshmukh, R.V., Devadkar, K.K.: Understanding DDoS attack & its effect in cloud environment. Proc. Comput. Sci. **49**, 202–210 (2015)

Smart Traffic Signaling System Using Arduino

Gamidi Vedavasu, K. Vishrutha, G. Janvi Sruthi, S. N. G. S. Karthik
and P. Swarnalatha

Abstract This paper is proposed a system to control traffic based on its density. Traffic congestion is one of the major issues which we face in our daily life. The population is increasing, the resources are decreasing, and as a result of which, the technology is getting better. This technology can be used to control traffic congestion and it can pave its way for the construction of a "smart city". Red light delays are one of the causes of traffic congestion. The main objective of this paper is to propose a sensor-based technique which can act like an automaton. It reduces the workload of the traffic police. The model has four sensors placed at all the sides of a four-way path. These sensors are used for counting the vehicles which use roadways. The sensors used are IR sensors. The sensors send the information to the microcontroller which is mounted on the Arduino UNO board. The controller decides the flashing time of the subsequent lights. Handling overcrowding in urban traffic through coming new generation by artificial intelligence techniques is an important research zone. Various models and approaches have been developed utilizing soft computing techniques to handle this issue. Major soft computing approaches for this object are neural network and genetic algorithms and some more.

Keywords Arduino UNO · IR sensors · RFID tags · RFID reader · Vehicle detection · Traffic management · Intelligent traffic system · Traffic congestion · Roadside infrastructure · LED · Adaptive traffic control system · Soft computing techniques

G. Vedavasu · K. Vishrutha · G. Janvi Sruthi · S. N. G. S. Karthik · P. Swarnalatha (✉)
Vellore Institute of Technology, Vellore, Tamil Nadu 632014, India
e-mail: hgswarna@gmail.com

G. Vedavasu
e-mail: vedavasugamidi@gmail.com

© Springer Nature Singapore Pte Ltd. 2020
K. N. Das et al. (eds.), *Soft Computing for Problem Solving*,
Advances in Intelligent Systems and Computing 1057,
https://doi.org/10.1007/978-981-15-0184-5_77

1 Introduction

TRAFFIC is one of the major issues which we face in our daily life. The population is increasing and so is the technology. There is a drastic increase in the production of cars in the last few years. This can be considered as a milestone for a developing economy. The number of road users is increasing, but the infrastructures are limited. This may lead to traffic congestion, which is not so good to the society. Traffic congestion is a situation on roadways which is caused by slow speeds, increased time intervals to complete the journey and clustering of vehicles. It may lead to accidents and may decrease the efficiency of roadways. Avoiding traffic jams can help us achieve a milestone which is both eco-friendly and socio-friendly. The traditional traffic-based system may not control the problem of traffic congestion. Traffic sign recognition (TSR) helps us to manage the traffic signs, warn the driver, and command or prohibit certain actions (Fig. 1).

2 Literature Review

The paper titled "Advanced Traffic Management System Using Internet of Things" [1] concludes that the growth in population of urban areas has been discussed along with traffic control system. A smart traffic control system is implemented by big data analysis and RFID along with IoT. Supervised learning also helps us to provide a framework for the attributes of effective implementation of a smart traffic management system. The system if implemented will reduce the traffic congestion in big cities and also the system aims at improving the security of the vehicles. The paper titled "Density Based Traffic Control" [2] emphasizes the need of efficient traffic management system in our country as the number of road accidents are increasing. An advance system is designed to deal with the increased chaos in the traffic, and it

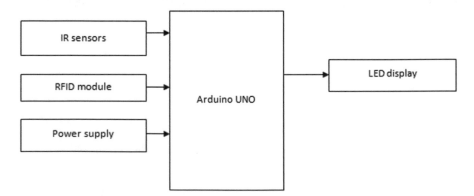

Fig. 1 System design diagram

proposes to effectively distribute the time slots of a vehicle. The prototype has been implemented in the laboratory scale, and the expected outcome was good. The next step is the real-life scenario and for better results, we can implement it on a large scale. It can bring a revolutionary change in traffic management system and its application in the environment. The paper titled "Density Based Traffic Control System Using Arduino UNO" [3] concludes that the proposed method will help us to get into a traffic-free future. The system is of less cost and is efficient because IR sensors are used to maintain the vehicle count. The "Density Based Traffic Signal System" [4] in future work a raspberry pi microcontroller can be used by using OpenCV software which is free. It is used to provide a good view of a traffic controller room. The paper titled "Density Based Intelligent Traffic Signal System Using PIC Microcontroller" [5] describes a model using IR sensors and a microcontroller. By implementing the proposed system, the possibilities of traffic jams have considerably reduced. Time slots are given and hence the density of the traffic is determined by the vehicles using the roadways in that particular time slot. This data can be saved and used for the next interval. The data can be downloaded in the computer using the communication channel between the microcontroller and Arduino. The paper titled "Intelligent Traffic Management for Ambulance and VIP Vehicles" [6] concludes that the model considers not only the priority of the vehicles but also the density of the vehicles on the road and controls the traffic light sequence efficiently and more accurately. The proposed work is of great use since the accuracy of RFID is greater than that of a camera. The paper titled "Intelligent Traffic Signal Control using Image Processing" [7] concludes that a novel system for intelligent traffic control system has been generated. The volume of the traffic has successfully been quantified using various image processing techniques. The paper titled "Design of Emergency Services Application Using IoT" [8] concludes that cameras can be inserted in various parts of the city to monitor the vehicle count. A neural networking system has to be added for reading license plate. The license plates can also be read by using optical character recognition (OCR) technique which is used in image processing. The paper titled "Alarm System to Detect the Location of IOT-based Public Vehicle Accidents" [9] discusses that IoT can also be implemented in transportation. The information regarding the location and accident can also be reported to the police, hospitals, firefighters, related officers, and it can also be used to register all the accidents in the street using SMS and Google map. The paper titled "Smart Traffic Control System Using ATMEGA328 Microcontroller and Arduino Software" [10] concludes that the proposed method can be used for effective management of traffic system. An emergency vehicle is prioritized by using an RFID tag and a reader. The proposed method can be used for effective traffic congestion. The paper titled "Image Processing Based Intelligent Traffic Controlling and Monitoring System Using Arduino" [11] concludes that the proposed method uses image processing and Arduino for effective traffic monitoring system. A Wi-fi module can also be used in particular junctions. It is also used for emergency traveling. The paper titled "An Intelligent Framework for Vehicle Traffic Monitoring System using IoT" [12] concludes that the proposed system can be used for real-time monitoring using IoT platform which can be used to control the traffic signals and hence reliable and hence can be used for designing an effective traffic signaling system to reduce the problem of traffic congestion. The paper titled "An IoT Based Automated Traffic Control System with Real-Time Update Capability" [13] concludes that the proposed method was efficient and safe to give live updates of the traffic system. It also ensures the safety of pedestrians and hence traffic updates can be generated and can be made into a website. The paper titled "Efficient Dynamic Traffic Control System using Wireless Sensor Networks" [14] concludes that RFID tags can be used for emergency vehicles and proximity and IR sensors can be used to check the vehicle count. The paper suggests that capacitive proximity sensors can be used instead of IR sensors to get a better count but they would be costlier to implement. The paper titled "An Improvised Approach for effective precision Farming for Auto Irrigation Using IOT" [15] concludes that any sensors like IR sensors can also be implemented for emergency vehicle detection instead of RFID tags.

3 Methods

Here in this paper, we propose a system in which an IR sensor is used to measure the distance of the vehicle from the signal. This sensor sends the information to the microcontroller mounted on an Arduino. The microcontroller sends the control signals which can make the LED's glow that lights the signal automatically. The system is an "Intelligent" approach for monitoring traffic control. We are inserting RFID tags onto the emergency vehicles and the RFID reader would be inserted near the signal. If the reader detects the tag, then immediately the traffic signals would go red.

Requirements:
Hardware:
The components are:

1. Arduino UNO
2. IR sensors which contain IR transmitter and IR receiver (4)
3. Breadboard
4. LEDs—(red, green, yellow)
5. RFID reader and RFID tags.

Software:
Arduino software.

Arduino UNO: Arduino board has an ATMEGA328 microcontroller mounted on it. It has 14 pins for both input and output, respectively, and out of those 14 pins, 6 can be used as PWM outputs. There are 6 analog inputs pins, a 16 MHz resonator and a USB port and a power pack and a reset button. The board is connected to the computer using a USB cable.

IR sensors: They emit infrared radiations which when come across an object get scattered. The atmosphere acts as the transmission medium for infrared radiations. Infrared sensors have two components.

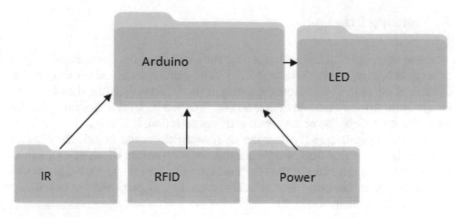

Fig. 2 Module diagram

They are:

1. Transmitter
2. Receiver.

Transmitter is used to generate the IR radiations. Receiver is used to receive the infrared radiations (Fig. 2).

4 Simulation-Based Approaches

PC traffic recreation can be utilized to comprehend traffic stream and create traffic control systems. In [16], traffic simulation with infinitesimal vision is planned which pursues the idea of multi-specialist frameworks. The entire idea is executed in MS C++. A multi-specialist-based simulation approach is proposed in [17]. This methodology utilizes Object-Z and state graph formal dialects. Previously chronicled true information and dynamic mimicked information can be utilized to gauge traffic. In view of this, an online recreation structure is proposed in [18] which coordinates genuine traffic information alongside multi-specialist approach. To control recreated vehicles, responsive operators can be utilized. Utilizing minuscule traffic test system, a methodology is created in [19], wherein vehicles are constrained by driving operators with various conduct settings. In [20], a multi-specialist-based recreation instrument is created which incorporates the components, for example, sharp conduct of an individual causing standard infringement and expectation of basic circumstances by people. A methodology named HUTSIG is gotten from a tiny traffic test system HUTSIM in [21]. It depends on fuzzy inferences utilizing multi-specialists and recreating genuine traffic information. In a multi-specialist-based recreation model is planned utilizing Repast S on java platform. It is used to upgrade the framework limit at the crossing points.

5 Working Principle

When an object gets in the path of the sensor's radiation, the comparator's object either gets low or it gives a voltage 3.3 V. The voltage varies accordingly as the distance of the object varies from the sensor. If the object is detected, then the green led glows for a certain amount of time and then the yellow lights followed by the red light. The main objective of the sensor is object detection.

The working principle is when the IR sensor detects the object, the distance between the object and the sensor is noted down. The sensor which gives the minimum distance is selected and then for that respective lane the green led glows and the countdown is displayed on the LCD screen. In this proposed method, traffic congestion is maintained (Fig. 3).

Operational features:

- Make the circuit on the breadboard and connect it to Arduino UNO.
- Load the program on the board.
- IR sensors are also known as obstacle detecting sensors and these may be used to detect the vehicles (obstacles) on the road and note down the distances of those vehicles from the signal and the vehicle with the least distance is allowed to go first by making the signal of that particular path green. And if that is made green the signals of the rest of the paths become red.
- A countdown is displayed on the LCD screen (Figs. 4 and 5).

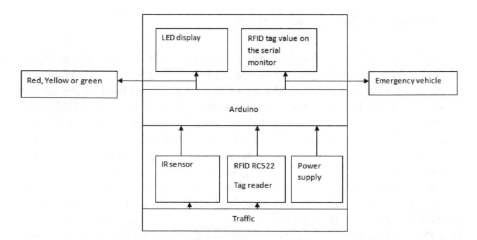

Fig. 3 System architecture

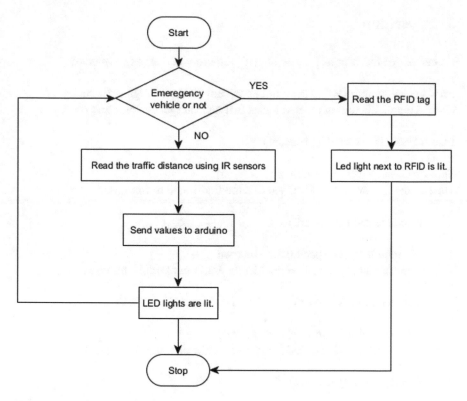

Fig. 4 System flowchart

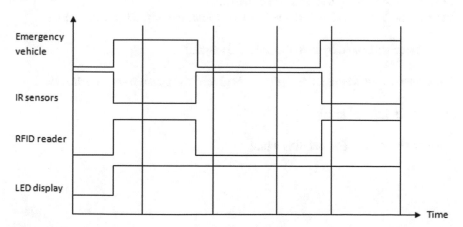

Fig. 5 Timing sequence

6 Algorithm

Procedure Traffic_Control_System() {If (vehicle== emergency vehicle)
{
Print the value of the RFID tag number present on the emergency vehicle;
Glow a blue LED telling other vehicles that an Emergency vehicle is on the way;
}
Else If (vehicle detected by sensor1)
{
Turn on Green led for a specific time interval
Turn on the Yellow led and Red led like the traditional traffic-based system
}
If (vehicle detected by sensor2)
{
Turn on Green led for a specific time interval
Turn on the Yellow led and Red led like the traditional traffic-based system
}
If (vehicle detected by sensor3)
{
Turn on Green led for a specific time interval
Turn on the Yellow led and Red led like the traditional traffic-based system
}
If (vehicle detected by sensor4)
{
Turn on Green led for a specific time interval
Turn on the Yellow led and Red led like the traditional traffic-based system
}
If (distance generated by both the sensors is equal)
{
Follow the cyclic sequence and priority is given based on the sensor's number;
}
If (traffic light is violated)
{
Turn on the camera and get the picture
}
}

7 Implementation

See Figures 6, 7, and 8.

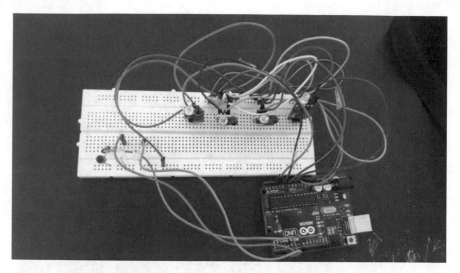

Fig. 6 IR and distance measurement

Fig. 7 RFID module for emergency vehicle detection

8 Conclusion

This paper is very much helpful as it proposes a system that can be used for the control of traffic congestion and also construct an intelligent traffic control system. With the application of this system, the chaos of everyday traffic can be controlled by distributing the time slots based on the density of the traffic. This technology can be first implemented on a smaller scale and results can be observed.

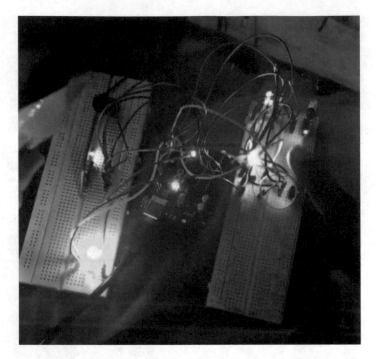

Fig. 8 Implementation of the proposed system

If the results are very much positive, then it can be implemented on a larger scale. We can also use GSM technology to send messages to a phone if there is heavy traffic. This system can be used to provide a smarter traffic control in the modern world. This model when implemented can reduce congestion to a great extent and marks the beginning of the construction of the smart city. There may be many flaws in the model, but these can be rectified when implemented on a large scale. Traffic, one of the major problems in India, also finds solution with Internet of things.

Acknowledgements This project would have been incomplete without the guidance of certain people who motive us, guided us, and supported us. A project which is further beneficial to society was the theme of this paper. Making of this project was possible only because of teacher who believed in us and encouraged us. We are grateful to Prof. Swarnalatha P, for her constant support which helped us in the preparation of this report.

References

1. Lakshminarasimhan, M.: Advanced traffic management system using internet of things (2016)
2. Faruk Bin Poyen, E., Kumar Bhakta, A.K., Manohar, B., Ali, I., Rao, A.: Density based traffic control. Int. J. Adv. Eng. Manag. Sci. Infogain Publ. (Infogainpublication.com), **28**, 2454–1311 (2016)

3. Faruk, E., Poyen, B., Bhakta, A.K., Manohar, B.D., Ali, I.: Density based traffic control. **2016**, 5–7 (2017)
4. Vidhya, K., Banu, A.B.: Density based traffic signal system. Int. J. Innov. Res. Sci. Technol. **3**(3), 1–3 (2014)
5. Kavya*1, B.S.G.: Density based traffic signal system using microcontroller. Electron. HUb, **3** 205–209 (2015)
6. Engineering, C.: Intelligent Traffic Management for Ambulance and VIP Vehicles, pp. 15041–15046 (2016)
7. Mokashi, N.: Intelligent traffic signal control using image processing. Int. J. Adv. Res. Comput. Sci. Manag. Stud. **3**(10), 137–143 (2015)
8. Gupta, V., Wilson, S., Kuppam, S., Swarnalatha, P.: Design of emergency services application using IoT. J. Comput. Theor. Nanosci. **15**, 2635–2639 (2018)
9. Desima, M.A., Ramli, P., Ramdani, D.F., Rahman, S.: Alarm system to detect the location of IOT-based public vehicle accidents, pp. 1–5
10. Vahedha, Jyothi, B.N.: Smart traffic control system using ATMEGA328 micro controller and arduino software. In: Signal Processing, Communication, Power *and* Embedded System, SCOPES 2016—Proceedings, pp. 1584–1587 (2017)
11. Member, I.E., Member, I.E., Member, I.E.: Image processing based intelligent traffic controlling and monitoring system using Arduino, pp. 393–396 (2016)
12. Bhosale, A., Nimbore, P., Shitole, S., Govindwar, O.: Landslides monitoring system Using IoT. **4**, 999–1002 (2017)
13. Talukder, M.Z., Towqir, S.S., Remon, A.R., Zaman, H.U.: An IoT based automated traffic control system with real-time update capability, pp. 1–6 (2017)
14. Bharadwaj, R., Deepak, J., Baranitharan, M., Vaidehi, V.V.: Efficient dynamic traffic control system using wireless sensor networks. In: 2013 International Conference on Recent Trends in Information Technology ICRTIT 2013, pp. 668–673 (2013)
15. Sophiya, D.F., Sameera, K., Siddharth, Swarnalatha, P.: An improvised approach for effective precision farming for auto irrigation using IoT (2018)
16. Improvised approach for effective precision farming for auto irrigation using IoT. J. Comput. Theor. Nanosci. **15**, 2671–2675. Improvised approach for effective precision farming for auto irrigation using IoT. J. Comput. Theor. Nanosci. **15**, 2671–2675
17. Gruer, P., Hilaire, V., Koukarn, A.Q.: Multi agent approach to modelling and simulation of urban transportation system. In: IEEE International Conference on Systems, Man and cybernetics 2001, vol. 4, pp. 2499–2504 (2001)
18. Wahle, J., Schreckenberg M.: A multi agent system for online based simulations based on real world traffic data. In: Proceedings of the 34th Annual Hawaii international Conference on System Sciences, January 2001 (2001)
19. Ahlert, P.A.M., Rothkrantz, L.J.M.: Microscopic traffic simulation with reactive driving agents. In: Proceedings IEEE Intelligent Transportation Systems, pp 860–865 (2001)
20. Doniec, A., Mandiau, R., Piechowiak, S., Espie, S.: A behavioural multi agent model for road traffic simulation. Eng. Appl. Artif. Intell. **21**(8), 1443–1454 (2008)
21. Kang, L., Zhao, Z., Lin, B., Jin, L., The traffic control system at urban intersections during the phase transitions based on VII. In: International Conference on Computer Applications and System Modeling (ICCASM), vol 13, pp 137–141 (2010)

Test Path Identification for Virtual Assistants Based on a Chatbot Flow Specifications

Mani Padmanabhan

Abstract The development of the Internet provides opportunities for new types of communications between virtual assistant and human. The technology which is mainly used in the communications is chatbot. A chatbot is a simulated computer program that enabled human conversation by the Internet. The virtual assistant is currently used for a variety of purposes. The chatbot database flow is the important activity for the development of software for the virtual assistant. The process of chatbot testing is based on the well-formalized test cases. The test cases are based on the chatbot trace in the database. Trace path identification during the development of the chatbot software is the challenging process. This paper presents the methodology to identify the test cases for virtual assistant using chatbot database flow-oriented specification. Chatbot database flow is one of the few specification languages supporting for formal description into an applied specification. The database specification divided into several of specification trace using the proposed algorithm. Finally, the chatbot intent trace has provided the path for software test case generation. The experiments show trace path-based test cases that have yielded the effective coverage criteria in the chatbot software development.

Keywords Validation · Software testing · Internet of things · Real-time systems

1 Introduction

A virtual assistant is a set of a computer program which conducts a communication through auditory or textual. In this artificial intelligence, programs are often designed convincingly simulated how the real human would behave as a conversational partner. Database systems play a major role in the chatbot software development. Artificial intelligence established database management systems (DBMS) provide the efficient organizations to access large amounts of real-time data. The software testing for

M. Padmanabhan (✉)
Faculty of Computer Applications, SSL, Vellore Institute of Technology (VIT), Vellore, Tamil Nadu, India
e-mail: mani.p@vit.ac.in

© Springer Nature Singapore Pte Ltd. 2020
K. N. Das et al. (eds.), *Soft Computing for Problem Solving*,
Advances in Intelligent Systems and Computing 1057,
https://doi.org/10.1007/978-981-15-0184-5_78

chatbot approaches is an important activity in the real-world. The chatbot dialog flow technology for an industry partner to better their internal communication process. Currently, chatbot has been used for online event management, order the food, book the appointment, and interactive voice assistant for business.

The usual technique of quality assurance for the real-time systems is effective software testing. The effective testing to be ensuring with running the program on many test inputs and check if the results conform to the program specification. The success of testing highly depends on the quality of the test inputs [1].

The chatbot is different from normal web applications. The web application has provided the formal input and output for the particular sequence [2]. Model-based test data or sequence has been collected from the software design specification. In the process of software development, software design has provided the essential input for the model-based testing [3]. The testing process is getting the trace from the design engineer. The software specification-based testing has produced the effective software development [4].

In the digital world, product user has got the data from the human through the voice. The quick response for the particular task has been based on the voice call human experience and accessibility of the data from the server. In the real-time example, E-Commerce customer has the query "how to delete the placed order", then human voice assistant has to get the answer from the server, or customer has to search the solution from the frequently asked question databases. In the fast-growing global, this type of searches has spent more time and cost. If any of the assistants have provided the solution, then more time and cost to save for service provider and vendor because the same type of the query repeatedly asked by the customer like how to reset the password and how to cancel the order, order pizza, and book the appointment. The interactive voice response (IVR) or the server database-based hyperlink has determined by the system-based input. The pre-written programs are based on the input from the input device. Applying model-based testing for this computer program is fully based on the input request. In the chatbot, input request has varied from the above-stated process, and the virtual assistant has fully based on the chatbot dialog flow.

A chatbot is typically used in dialog systems for various practical purposes including customer service or information acquisition. A chatbot is a computer program or an artificial intelligence which behaviors a discussion via audio or word-based methods. The chatbot dialog flow has been based on two types that are natural language processing systems and database input keywords [5].

The natural language processing systems have lots of mathematical algorithms and systematic program to understand the process of the chatbot. The database input keywords method more simple system its scan for keywords within the input, then pull in the database for the reply. In the real-time example, Figure 1 describes the types of chatbot. The user asked the virtual assistant to book an appointment, and then, within few seconds, the appointment number has received by the user. The virtual assistant had created the database indent and pull all the related information for the user, and used for appointment booking most matching keywords, or the most similar wording trace from the user database.

Fig. 1 Types of chatbot process

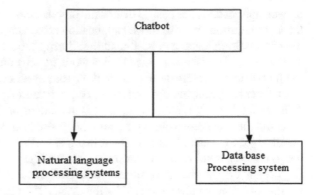

The chatbot testing is an important activity during the real-time software development. The database plays the major role in the chatbot. In this paper proposed the methodology to generate the test cases for chatbot system. Chatbot dialog flow and trace-based test cases are provides the effective software development. This approach has produced the possible path and branch coverage criteria in the chatbot software development. The detailed background for chatbot testing and test case generation approach has described in the following section.

2 Background

Software testing plays a central role in the Software Development Life Cycle (SDLC). The database operations are the major task of almost every IVR modern organization. Software testing has ensured that the IVR system software is defect free through an activity to check whether the actual results match the expected results. More time has to be spent for identifying the actual results of the system software [6]. In the modern virtual assistant to be acknowledgment as a human in the digital world. The effective software testing for the virtual assistant has the major part during the development. The success of the software testing for the particular system software is based on the coverage criteria of generated test cases. Test cases are the art of testing process. According to the ISTQB definition, "test case is a set of input values, execution preconditions, expected results, and execution post conditions, developed for a particular objective or test condition, such as exercising a particular program path or verifying compliance with a specific requirement." If the generated test cases identify the actual results from all possible path for particular IVR system software, then simply ensure the quality of the product. The following division describes the importance of software testing during the SDLC.

Test execution is the process of implementing the system software and comparing the expected and actual results. In the model-based testing, the expected results extracted from the software behavior [7]. The software flow has been produced the systematic input for the test execution. In the example, product order cancel based on web applications has been provided the precondition, data, and post condition for the test execution. The IVR-based applications like virtual assistant have to identify the precondition, data, and post condition during the runtime of the program. The modern virtual assistant has used the keyword search concept. User acceptance testing had done during the development of chatbot systems, but in the runtime, failure has happened, or mismatched results during the runtime have become the user dissatisfaction. The runtime mismatched results called fallback intent in the chatbot. To avoid the occurrence during the runtime, test execution has to be done in all the possible path in chatbot testing. The following section describes the chatbot testing based on the natural language conversations (NLC).

A chatbot is based on the primary blocks called intent. The indent has the solution for the different queries generated by the agent. The agent has done the acceptance testing during the development based on the indent. Figure 2 shows the set of indent for the banking process, balance check indent have multiple input value like how much my balance, tell my balance, what is my balance in SB account then SQL database Administrator called the indent balance check. The testing has to be done during the development based on the user acceptance testing. DBA called the same indent based on the input query. During the runtime, the balance indent triggered in the different forms of the text or voice chatbot testing had the failure. The dialog flow is the major part of the chatbot process. The test execution based on the dialog flow identifies all possible path in the chatbot testing. The maximum path coverage has built the quality software during the runtime process. Natural language conversations-based test path identification had produced the effective software testing for a chatbot.

Chatbot intent is the natural language understanding modules. The intent transform natural user request into actionable data in the human app, product or service. This transformation occurs when a user input matches inside the natural language conversations. The verification or keyword trace matching based on the database. The query-based testing proved the possible coverage in chatbot testing [8].

A testing tool has been defined for data generation incorporating alloy specifications both for the schema and the query. Each table is modeled as a separate n-arity relation over the attribute domains, and the query is specified as a set of constraints

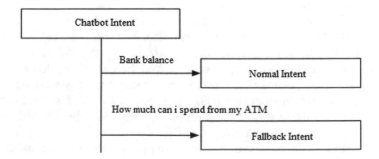

Fig. 2 Types of chatbot intent

that models the condition in the WHERE clause. However, this approach cannot handle tables with a larger number of attributes due to the arity of the table relations. The propose a technique named reverse query processing for generating test databases that take the query and the desired output as input and generates a database instance that could produce that output for the query [9].

A test case in software testing is a set of conditions under which a tester will determine whether features in a software application are working according to the required requirements Software testers might design multiple test cases, called a test suite, to validate or determine that a software program is functioning correctly [10]. The design techniques that the testers use to design test cases vary, but most of the techniques are often based on the feature component of the software requirements. The component for the virtual assistant is the chatbot NLC.

NLC is the human-computer interaction keywords to be used in the chatbot or virtual assistant process. NLC has transferred the human language to JavaScript Object Notation (JSON) file. JavaScript Object Notation is an open-standard file format that uses human interaction keyword to transmit objects with database attribute values. Chatbot dialog was based on the stored database under the particular agent. Database attribute-based testing has provided the effective test execution for real-time process [5]. The proposed methodology has three major parts he tests trace identification from the NLC chatbot, Test path generation based on the proposed approach finally all possible path and branch coverage for the chatbot test cases. Figure 3 describes the internal process of chatbot to be converting as the actionable output object for NLC.

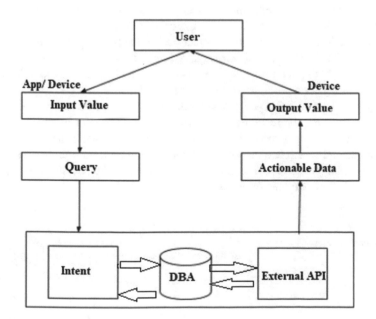

Fig. 3 Chatbot internal process for NLC

3 Proposed Methodology

Significant research in standardized software development practices increasingly improves in building reliable system software for chatbot; errors are still present in the chatbot software than fallback intent during the runtime. The aim of chatbot software testing is to expose or identify such bugs by executing the software on a set of test execution. In the basic form, a test execution has consisted of program inputs and corresponding expected outputs. After the software has successfully passed the testing phase, the developer has a greater degree of confidence in the chatbot-based real-time software [11]. Typically, software testing is a labor intensive, therefore more time and cost expensive. Research has shown that testing can account for 50% of the total cost of software development [12]. Natural language conversations (NLC) is the foundation for chatbot database attributes. NLC-based test execution has reduced the total cost and time of the chatbot execution during the software development. NLC has been divided into two parts: one is trigger, and other is the intent. The trigger part is a single value to start the halt process. Multiple intents have been possible each intent produce the different value if any of the intent unavailable then deadlock or fallback for the particular NLC during the chatbot execution Fig. 4 shows the sample chatbot with trigger and set of intent.

The main idea of the chatbot-based testing approach is to derive from the NLC specification. In the set of specification traces, each of which is used to generate a set of test cases and then to repeatedly execute each test case until all associated possible paths have been covered [13].

The following section explains the trigger and intent identification which are hidden in the NLC. The chatbot working architecture is described in Fig. 5. NLC has to be transformed the actionable object during the passing of the database. The passing flow is called the natural language conversations flow diagram (NLCFD). In this, NLCFD has the valid trace for the test cases. The proposed approach explains the methodology for a possible path from the NLCFD.

Many concurrent chatbot programs contain program nondeterminism. That is, the expected results and the result of a concurrent program are not fixed [14]. The test cases depend on the program path execution [15]. Suppose in the chatbot dialog flow have threads and which can be executed concurrently then each program statement is atomic. An important issue to address is how to derive a specification trace and how to select a set of traces for testing.

Fig. 4 Sample natural language conversation

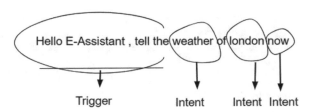

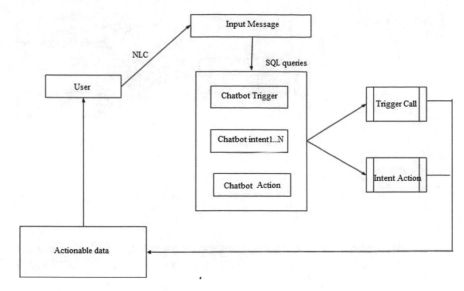

Fig. 5 Natural language conversations flow diagram

A trace from a natural language conversations has to achieve a specification for the testing. If a process in the NLC is nondeterministic, then NLC has to be divided into a set of deterministic sub-processes, each defining a single function of the unique process. Each unique process describes the intent in the chatbot database. In the proposed approach, NLC has to be subtitled and generated the possible test path for the test execution using the proposed algorithm. The pseudocode 1 describes the procedure to identify the trigger and intent form the NLC.

Pseudocode 1: Trace identification for chatbot

```
Pre: Natural language conversations (x)
Data: SQL query for the input
Post: Actionable data with trigger and intent
Public int trace(int x)
string query = " ";
int trace = −1;
SqlConnection tc = new SqlConnection();
tc.ConnectionString = "..";
tc.Open();
query=SELECT * FROM T1 WHERE Intent_a = x;
SqlCommand cmd = new SqlCommand(query, tc);
SqlDataReader results=cmd.ExecuteReader();
while(results.Read())
{
pre=whether;
```

Fig. 6 Chatbot parallel
process

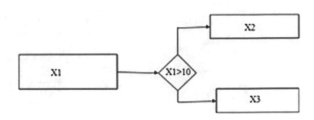

```
post = Intent_a;
data = x;
}
return tc;
}
```

The produced intent trace has the trigger and intent value, each intent trace has selected a set of test inputs on the basis of its input domain, and execution of any of them is expected to satisfy the trace.

For example in the Fig. 6 shows the three types of the intent $x1$, $x2$, $x3$ in the process $x1$ is the constraint indent, $X1$ is the temperature indent, the user ask the virtual assistant for the location temperature the location indent based on the GMT of the location if GMT less value produce the $x2$ or $x3$ if the $x1$ unavailable in the database then called the fallback indent. The following steps describe the trace identification for a parallel process.

Call Process: X1 (L1:int)
pre:L>10
Post:X2=(GMT+12) + (temp(X2))
end_process
Call Proces: X2 (L1:int)
pre:L<10
Post:X2=(GMT+23) + (temp(X2))
end_process
Trigger (trace1) = pre(x1) ^ post(x2)
Trigger (trace2) = pre(x2) ^ post(x3)
Path representation
P1 = trace 1
P2 = Trace 2
P3 = Default path

The proposed algorithm has the input trigger value. The input triggers are the SQL select command and intern value to be the like operation in the SQL command. According to the proposed algorithm, each trigger has to be converted as the precondition, and the expected result has to be converted as the post condition in the trace.

The path representation to combine the predicates on the processes, conditional nodes, and data stores of a trace to derive the domain of the trace. Each path has

Table 1 Test cases for chatbot

Test case ID	Trace	Pre	Data	Post
TS1	P1	$X1$	Trace $1 = (X > 10)$	$X2$
TS2	P2	$X2$	Trace $2 = (X < 10)$	$X3$
TS3	P3	Invalid data	Invalid data	Invalid data

to produce the trace precondition, data, and post condition for the particular trace. Table 1 shows the test cases for the p1. Trace 1 has described the precondition, if the location GMT is less than 10, then the weather is different in the location, and SQL database gets the GMT location and displays the weather with the username from the database.

The generated test cases have to be validated with path and branch coverage of the real-time chatbot. Based on the proposed approach, experiments show chatbot testing achieved effective path and the branch coverage.

4 Experiments and Comparative Results

The five different natural language conversations have to be taken for the experiment of the proposed approach. The value for each indent has depended on the different criteria in a database. In example, NLC1 describes the ticket booking instruction from the user. The SQL query has passed to the database to be converted as the trigger "book the ticket," and location, availability, and user basic information are extracted from the database. Figure 7 shows the chatbot process for the air ticket

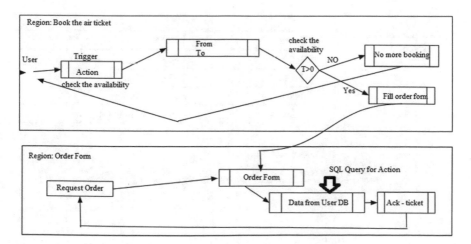

Fig. 7 Chatbot process for booking air ticket

Table 2 List of intent for NLC (Trace 1)

Intent	Data
Intent 1	Source
Intent 2	Destination
Intent 3	User name
Intent 4	Date
Intent 5	Time
Intent 6	Ticket number

booking. The user intent received from the user, the trigger word for the chatbot is book the ticket, and intent 1…*n* are displayed in the table.

Tables 2 and 3 show the list of intent for the trace 1 and 2; if the availability of the tickets is zero, then intent 4 has to be called. In the condition checking, if the data are unavailable, then produce the fallback message to the user.

Figure 8 shows the NLC for setting the alarm on the mobile. The intent has to be divided into the two types: one is the system defined and another is the developer defined.

The system defined task is "set the alarm" but user instructed to the mobile through the virtual assistant "wake up me". The developer defined the task equivalent to the system defined through manual way or the system fallback to happen during the runtime. The test path is to be identified for the NLC based on the proposed approach. The SQL query has to convert as per the pre, post condition. The proposed algorithm provides the effective branch coverage for the chatbot testing. The identified intent has produced the possible path for the test cases identification. Table 4 shows the three different types of the path with trace id. During the chatbot testing, each trace produces the test cases for dialog flow. The coverage criteria are to be calculated

Table 3 List of intent for NLC (Trace 2)

Intent	Data
Intent 1	Source
Intent 2	Destination
Intent 3	Availability
Intent 4	No more booking

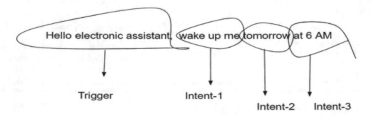

Fig. 8 Chatbot NLC for setting the alarm

Table 4 Test cases for setting the alarm NLC

Test case ID	Trace	Pre	Data	Post
TS1	Trace 1	6 AM	Alarm set—ACK (tomorrow)	TS1
TS2	Trace 2	6	Alarm set—ACK (today)	TS2
TS3	Trace 3	Invalid	Fallback	TS3

based on the path and branch coverage. In the formula 1 has described the number of intent are describe as the NWC. The term N describes the number of generated path. The x describes the trigger value in the chatbot conversation.

$$\text{Brach coverage} = \frac{n(n-1)x^2}{\text{NWC}} \qquad (1)$$

According to the proposed approach, the number of branch coverage percentage and the acceptance testing has to be validated. Figure 9 shows the coverage criteria for the five different NLC.

The chatbot testing has used the user acceptance testing for verification of the test path and the test data. The proposed approach has differed from user acceptance testing. The NLCFD-based test cases are validated with the possible path of chatbot. Based on the propsed approach has identified the less fallback in the chatbot real-time process.

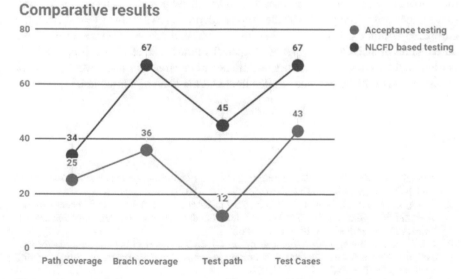

Fig. 9 Comparative results for NLC

5 Comparative Results

The proposed test path identification methodology based on the NLCDF yielded the effective results the comparative analysis of the proposed approach, and the user acceptance testing has to be matched in Fig. 9. In feature research investigating the possibility of automatic test case generation on the basis of NLC specifications.

The NLCDF based test path provide the trigger trace and maximum possible path to be identified based on the algorithm, in acceptance testing fewer test cases and branch coverage so more fallback possible during the runtime. The comparative results of acceptance testing and NLC show the average of path coverage, branch coverage, and the number of test path. The experimental results show that the proposed approach provides the number of test cases higher than the acceptance testing. The coverage criteria for NLC in virtual assistant reduce the fallback intent and transformation of the normal intent during the runtime process.

6 Conclusion

This paper has described an approach to generate the test cases for chatbot interaction for modern virtual assistant using NLCDF specifications to test the corresponding concurrent software systems. By using this approach, a tester can generate a set of test cases covering potential traces derived from the NLC dialog flow diagram and then repeatedly execute the test cases to ensure that all possible paths corresponding to the traces can be traversed. The identified path produced the effective test coverage compared with the acceptance testing. In the future, a detailed approach to achieve all path and branch coverage and other specification-based chatbot testing methods is tried. In the future, research work establishes more strategies and coverage criteria to discover specific defects such as deadlocks in the chatbot interaction.

References

1. Andrade, L., Machado, L.: Generating test cases for real-time systems based on symbolic models. IEEE Trans. Softw. Eng. **39**(9), 1216–1229 (2013)
2. Adam, D., Nojan, N., Marios, F., Marin, L.: Chatbots as assistants: an architectural framework. In: Proceedings of the 27th Annual International Conference on Computer Science and Software Engineering, Nov 2017, pp. 76–86
3. Jee, E., et al.: Automated test case generation for FBD programs implementing reactor protection system software: automated test case generation for FDB programs. Softw. Testing Verification Reliab. **24**(8), 608–628 (2014)
4. Chen, M., Mishra, P., Kalita, D.: Efficient test case generation for validation of UML activity diagrams. Des. Autom. Embed. Syst. **14**(2), 105–130 (2010)
5. API.AI.: What is API.AI. https://dialogflow.com/docs/getting-started (2018)
6. Mani, P., Prasanna, M.: Test case generation for embedded system software using UML interaction diagram. J. Eng. Sci. Technol. **12**(4), 860–874 (2017)

7. Shahbazi, A., Miller, J.: Black-box string test case generation through a multi-objective optimization. IEEE Trans. Softw. Eng. **42**(4), 361–378
8. Chatbot dialogflow. https://dialogflow.com/case-studies/
9. Abdul Khalek, S., Elkarablieh, B., Laleye, Y.O., Khurshid, S.: Query-aware test generation using a relational constraint solver. In: Proceedings in IEEE/ACM International Conference on Automated Software Engineering, May 2008, pp. 238–247
10. Gupta, S., Bhatia, R.: Testing functional requirements using B model specifications. In: ACM SIGSOFT Software Engineering Notes, vol. 35, no. 2, pp. 1–8 (2010)
11. Mani, P., Prasanna, M.: Validation of automated test cases with specification path. J. Stat. Manage. Syst. Issue Mach. Learn. Softw. Syst. **20**(4), 535–542 (2017)
12. Wang, H., Xing, J., Yang, Q., Song, W., Zhang, X.: Generating effective test cases based on satisfiability modulo theory solvers for service-oriented workflow applications: effective test cases for service-oriented workflow applications. Softw. Testing Verification Reliab. **26**(2), 149–169 (2016)
13. Mall, R.: Fundamentals of Software Engineering, 3rd edn. PHI
14. Beizer, B.: Software Testing Techniques. International Thompson Computer Press, London (1990)
15. Ghiduk, A.S.: Automatic generation of basis test paths using variable length genetic algorithm. Inf. Process. Lett. **114**(6), 304–316 (2014)

Feature Extraction from Hyperspectral Image Using Decision Boundary Feature Extraction Technique

R. Venkatesan and S. Prabu

Abstract Hyperspectral pix is captured from satellite TV for pc sensors, which incorporates airborne seen or infrared imaging spectrometer that provides distinct facts approximately spectral and temporal features to categorize the materials than the landscape records. Snapshots contain ridiculous and proper spectral data, and the development of property use/cowl class accuracy is projected from using such photographs. This type of techniques facilitates to enclose efficaciously that is useful to multispectral fact. The essential foundation is with the aim to the extent of educational facts situated while doing no longer parallel toward the augment of ambit of hyperspectral statistics. However, to develop methods to system hyperspectral photograph is quite tough and studies objective. In order to conquer these troubles, the usage is made of functions' extraction with dimensionality reduction. Attribute origin is recognized toward exist a valuable method within each lowering compotation difficulty along with growing precision of hyperspectral picture category. In this paper, an easy and however a pretty influential characteristic removal technique based totally scheduled choice boundary characteristic extraction (DBFE) is proposed. Initially, the hyperspectral picture is partitioned keenly on obstacles of adjoining hyperspectral bands, the bands in every separation are concert by using averaging and that is solitary of the best decision strategies. Finally, the device offers functions with reduced number. Hyperspectral records set turned into are examined to demonstrate the presentation of the novel technique and also compared the present strategies of function extraction.

Keywords Hyperspectral imaging · Features' extraction · Region information · Dimensionality reduction · Hyperspectral bands

R. Venkatesan · S. Prabu (✉)
Vellore Institute of Technology, Vellore 632014, India
e-mail: prabu.sevugan@gmail.com

R. Venkatesan
e-mail: venkatesaneng@gmail.com

© Springer Nature Singapore Pte Ltd. 2020
K. N. Das et al. (eds.), *Soft Computing for Problem Solving*,
Advances in Intelligent Systems and Computing 1057,
https://doi.org/10.1007/978-981-15-0184-5_79

1   Introduction

Imaging spectrometer, a generation, which turned into advance in the 1980s, can obtain loads of spectral bands simultaneously [1, 2]. The image received with spectrometers is known as hyperspectral pix, which no longer best two-dimensional spatial data but additionally contain rich and exceptional spectral records. With those characteristics, they may be used to become aware of surface objects and improve land use/cover class accuracies Hyperspectral imaging sensors gather hyperspectral pictures inside the structure of 3D arrays, amid spatial proportions instead of the picture breadth furthermore top [3], and a spectral size unfolding the supernatural bands, whose quantity is generally a couple of hundred. Owing toward the redundancy of the uncooked illustration, it is far profitable toward the proposed efficient function extractors and toward utilizing the supernatural facts of hyperspectral pictures. Analysis of hyperspectral photographs has a variety of demanding situations. For instance, due to the effect of the sensor's instant discipline of view and the diversity of the land-cover training, the presence of blended pixels is viable. By converting the abundance map right into a better resolution photograph, the sub-pixel mapping technique can specify the spatial distribution of various categories on the sub-pixel scale. In the past three decades, hyperspectral photographs were broadly utilized in fields including mineral identification, flowers mapping, and disaster research. Due to the high dimensionality of hyperspectral statistics, picture processing strategies which encompass have been productively practical to the multispectral facts within the beyond and are not right to hyperspectral information. For instance, it is ineffective while the conventional statistical classification methods are carried out to hyperspectral images with confined schooling samples. In other words, the dimensionality will increase with the range of bands, and the quantity of education samples for type needs to be multiplied as properly (Hsu 2007). This has been describing the "curse of dimensionality" via Bellman (1961). The normally used technique to resolve "curse of dimensionality" is dimensionality reduction, which may be divided into sorts: function choice and characteristic extraction. For hyperspectral pictures, attribute removal is used toward to condense the ambit extra regularly. Furthermore, there are forms of feature extraction strategies. The first kind is based on the statistical assets of information. For example, primary additives transform (PAT) is the most usually used and easy method. Although it worries the distribution of whole data, some beneficial functions for hyperspectral records can be omitted effortlessly [3]. Apparently, the elevated ambit of hyperspectral facts ought to boom the skills and efficiency to allocate property employ/cowl kinds. Though, the categorization methods, which encompass has been efficaciously carried out to multispectral statistics within the precedent are not because powerful towards hyperspectral records. The fundamental reason is to facilitate the level of education records locate does not now become accustomed to the growing dimensionality of hyperspectral facts. The condition in the educational pattern is inadequate for the desires that are not unusual for the hyperspectral container; the assessment of numerical framework turns into imprecise and untrustworthy. As the ambit increases amid the range of bands, the

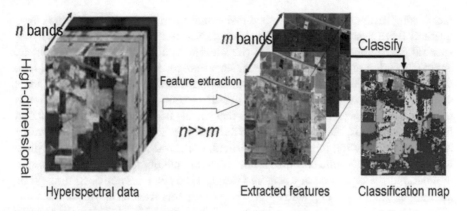

Fig. 1 Hyperspectral image processing

variety of teaching samples wanted intended for preparation are selected classifier that must be multiplied exponentially and properly. The speedy boom in education samples' dimension used for concentration judgment has been termed the "curse of dimensionality" via Bellman [4], which ends up in the "peaking phenomenon" or "Hughes phenomenon" in classifier layout [5]. The outcome is to facilitate the category accurateness that primarily grows after which declines the quantity of supernatural bands and later will increase as schooling pattern is stored the equal. Intended for an agreed classifies, the "curse of dimensionality" can only be averted by supplying an adequately huge illustration length. The greater complicated the classifier, the bigger becomes the relation of pattern length of ambit to keep away from the irritation of ambit. However, in the observation the quantity of instruction pattern is restrained during maximum of the hyperspectral function. Also, the elevated ambit of hyperspectral facts makes it essential to look for novel diagnostic strategies to evade a big enhancement within the compotation point. It is less complicated, however once in a while extremely successful manner of coping amidst excessive spatial statistics is to lessen the quantity of ambit. This can be accomplished by the way of function choice or removal to facilitate a little figure of prominent features is deriving since the hyperspectral records faced is with a restrained set of training samples. The hyperspectral processing is proven in Fig. 1.

2 Literature Review

Kang et al. [6] applied a new approach for hyperspectral photograph characteristic removal, the IFRF, which have been planned. The planned method is established totally as scheduled in the function of IF to lessen the measurement of the facts, using recursive filtering to mix dimensional instruction addicted to the consequential IFRF functions. Attempt has been passed out on three unique real hyperspectral photographs. The outcome of the attempt exhibits the value of the planned approach, which supplied higher consequences than folks of the broadly second-hand pixel-wise classifiers and the spectral–spatial classifiers. Furthermore, the planned approaches have supplied numerous different benefits along with the function to preserve sound conserve of the physical that means the hyperspectral information. In different phrases, the pixel ideals into the characteristic picture unmoving mirror is the spectral reaction of a pixel in a particular spectral variety; it's instance green is due to the fact it's miles is based totally on a completely speedy EPF; and despite the fact that the type accuracy acquired through the IFRF is motivated through the numeral of functions along with the criterion of the recurrent refine, those selections aren't essential. The motive is that there may be huge vicinity across the superior numeral of features for which the planned approach has comparable consequences, which outperforms other classification strategies in conditions of precision. Additional developments of these images encompass an inclusive investigation of the implementation of further EPFs' technique the compound bands have and moreover an examination of the opportunity of bearing in mind the association level of contiguous bands within the band-partitioning procedure.

Zhang et al. [7] preserved the gadget for reading the spectral–spatial in the sequence of the pixel. Because keen exposed via this author, tensor illustration preserves since various while potential the unique spatial constraints of a positive pixel and its neighbors, which enables higher characterization of the pixel's spectral–spatial attribute. Correlated toward the vector-based totally trait illustration, such constitutional records in the tensor characteristic are to be a practical restriction toward lessen the amount of unidentified framework utilized into mastering a function DR representation. The discriminability of lessons intended for class is to be preserved. The adjoining pattern of mutually is the equal instructions, plus characteristic instructions are studied within the planned TDLA escalation, so the discriminative statistics may subsist retain. Distant of the mean aspect, the proposed loom besides preserve address the nonlinear environment of the illustration sharing via considering the district various of the samples, which preserve in addition assist to obtain a enhanced type presentation than the conservative linear DR methods. A comprehensive DR structure for high-order information is provided. And a recommendation of DR structure to just recognize high-order information along with a compilation of third-order facts' cubes or various high-order function records inputs is made, because the inputs are for DR. It may be proven to be the preceding DR [8] as a unique container of our tensor process while the enter facts are first-order vectors.

Kang et al. [9] planned used for attribute removal of hyperspectral pictures for the crucial instance. Compared amid different broadly used spectral–spatial class methods, experiments achieved on three actual hyperspectral statistics showed the superb general presentation of the SVM-IID approach [10] during conditions of category accuracies, especially while the amount of schooling samples is tremendously small. The cause is to facilitate the IID preserve successfully eliminates ineffective spatial records which include shading and consistency to facilitate are not straightly associated towards the substance of various gadgets. It ought to be noted that, even though the planned approach preserve acquires elevated class accuracies amid an undersized quantity of instruction samples, it is not computationally green compared amidst different attribute removal methods. This paper usually targets next to displaying to facilitate IID, which fashions the perceiving role of human revelation, workings well for hyperspectral picture characteristic removal. Based on the scheduled concept of IID, the pixel standards of hyperspectral descriptions are decisive by means of two factors: The spectral reflectance is decided via using the substance of dissimilar gadgets, along with the shading thing, which consists of mild furthermore to form needy residences. Because the subsequent one feature is not always immediately associated with the substance of the entity, IID is taken up in this paper to eliminate inadequate spatial in sequence conserved in the shading factor of the hyperspectral picture. Exclusively, the IID-based totally function removal technique planned in this paper includes the subsequent steps: First, averaging primarily based photograph blending is followed to condense the spectral measurement of the hyperspectral data. Second, an optimization-primarily based IID technique is carried out to molder the size decreased hyperspectral information into inherent additives, and handiest the reflectance mechanism of the photograph are conserved for type.

Li et al. [11] advanced a novel structure for numerous features studying that are based totally happening for the mixing of various form of (linear and nonlinear) functions. A fundamental involvement of the supplied technique is the combined attention of mutually linear and nonlinear element not including at all regularization frameworks to manipulate the burden of both characteristics so that distinctive styles had capabilities that can be mutually oppressed (in a collaborative and bendy technique) used for hyperspectral photograph type. The major purpose is to address not unusual circumstances during observation, in which several lessons can be divided by the use of linearly consequent features at the same time as others could additionally require nonlinearly consequent capabilities. Until now, a chief development while the usage of a couple of characteristic learning is based on using kernels [12], i.e., MKL. Though, extremely only some techniques had been explored with a view to adaptively choose the mainly beneficial sort of characteristic is used for distinctive training within the prospect. In these paintings, that provide an initial step on this path also makes contribution to a structure, which is supplied and gifted to address each linear and nonlinear magnificence obstacles. A foremost improvement of planned method is to facilitate its far extra stretch than MKL, inside the wisdom that it could believe linear and nonlinear capabilities are also no longer handiest kernel features [13]. As an end consequence, MKL can be measured as a unique case of the planned structure. Though the offered structure is to be fashionable as well as appropriate to contain a few forms of contribution features, on this employment, we encompass measured hard and fast consultant capabilities distinctly which include the unique (spectral) in sequence restricted in the prospect, a position of (spatial) morphological features [14] extracted using one of a kind of attributes, in addition to kernel-based transformations of the aforementioned capabilities. The structure consequently permits high-quality resilience within the utilization of the benefits of every form of characteristic, in addition to the incorporation of extra functions in destiny traits.

Jia et al. [8] supplied a top-level view of feature choice and feature extraction techniques were provided with original superior techniques which are especially relevant and intended for type of hyperspectral statistics. This system is easy if sufficient professional understanding or laboratory research is available. However, information-pushed techniques need to regularly carry out to address a huge range of functions, via supervised or unsupervised methods. FS maintains the wavelength information, while FE does no longer, excluding for functions generated commencing organizations of adjacent bands. Linear and nonlinear transforms contain been residential for FE, which commonly squeeze the whole unique dimensions keen on some new capabilities. An admired structure for each linear and nonlinear FEs becomes planned. By choosing the suitable to imply and weight function, various FE methods evolved in premature studies may be expressed via the planned popular shape. A warning is emphasized to facilitate a category reparability measure used in FS and alteration situation era regularly calls for a big quantity of education data in parametric supervised methods, which would not be low priced or physically possible to observe. Nonparametric techniques or unsupervised techniques are favored later. The topical consciousness of studies is on nonlinear FE to discover the nonlinear residences imbedded within the records. Commonly speaking, characteristic discount is extra powerful for binary implementations, wherein best of one magnificent couple is addressed. Multistage binary or hierarchical classifiers are an excellent preference to hold instances amid extra than two training, anywhere distinctive features may be used for character set pairs. Besides, unique classifiers preserve to be engaged as well as to present the possibility used for coping amid without problems divided in instructions with fewer attempts.

Xia et al. [15] advanced novel spectral–spatial classifiers, which comprise two necessary mechanisms, particularly; replacement forests used for the pixel aware class in addition to MRF are used for the spatial regularization, correspondingly. In unique, revolution forests, which produce a spare ridge matrix the usage of function removal and accidentally decided on subsets of the authentic features, are second hand to approximation the elegance chances. Then, spatial appropriate data carried out via using MRF is used to filter the category outcomes access from rotation woodland classifiers. Lastly, the production is shaped by way of solving a maximum a posteriori(MAP) trouble via the α-enlargement diagram cuts optimization algorithm. The foremost involvement of the planned employment is to introduce three various knowledge nearby function elimination techniques, which include neighborhood preserving embedding (NPE), linear local tangent space alignment(LLTSA), and locality preserving projection (LPP), into the rotation forests. The consequences of revolution forest through three nearby characteristic removal strategies are similarly subtle amid the assist of spatial appropriate statistics, in order to be shown to offer an excellent categorization of the satisfied hyperspectral statistics. Experimental effects are provided to schedule three hyperspectral airborne images recorded via the airborne visible/infrared imaging spectrometer (AVIRIS), reflective optics system imaging spectrometer (ROSIS), and digital airborne imaging spectrometer

(DAIS), respectively. They have special spatial–spectral resolutions and communi-cate to exclusive contexts, as a result representative to the toughness of the termi-nation. It needs to be mentioned that we do no longer utilize function removal for dimensionality discount however for revolution of the axes even as maintenance to all the dimensions.

3 Existing Methodologies

Supervised class of hyperspectral snapshots is a complex undertaking. Hughes occur-rence seems as the statistics dimensionality that will increase. This occurrence hap-pens, since the range of education samples intended for the knowledge level of the classifier is normally extremely confined compared amid the amount of spectral bands. To categorize toward the determination of this hassle, some purpose of dis-count techniques have been used which can be capable to perform as it should be within the occurrence of imperfect schooling sets. The function discount of excessive dimensional information is utilized in two popular approaches: function choice and characteristic discount.

3.1 Principal Component Analysis

PCA is an unmanaged characteristic extraction approach that generates the most important additives of facts that are sorted in step with the disagreement standards. In characteristic removal by means of PCA, the mechanism through the bigger author-ity is maintained by furthermore mechanism amid decrease authority is surplus. Consequently, PCA might not work nicely in categorization cause, since it does not keep in mind the facts of division labels and intolerance of training. Solitary of the maximum generally used unsupervised characteristic removal techniques is the most important element evaluation (PCA). For supervised categorization troubles, PCA does not employ the instruction restricted inside the elegance labels [5]. PCA is eigenvalue disintegration of the records besides the instructions of biggest varia-tion in the statistics. PCA represents a preferred characteristic area given by those PCA capabilities, which illustrate elevated amount of electricity compaction. PCA generates uncorrelated production functions. An inferior dimensional illustration of unique information in time period of taking pictures of the statistics course to facil-itate has the biggest difference generated with the aid of PCA. This unsupervised characteristic removal technique tasks a statistics that locate to a new synchronize scheme through formative of the eigen standards with eigenvectors of covariance environment of unique information. Then, following working out of the eigenvalues and eigenvectors, they are sorted in a sliding manner. The important additives may be built as a linear renovation of statistics by way of combining the eigenvector matrix.

The PCA model is described as follows [16]:

(1) Let have N data samples $S_1, S_2, S_3, \ldots, S_n$ in M-dimensional space. Each S_i is a $M \times 1$ vector. Let $\overline{S}$ denotes the mean vector of the contribution data and preserve be represented as

$$\overline{S} = \frac{1}{N} \sum_{i=1}^{N} S_i$$

(2) C, which represents the covariance matrix, is distinct as

$$C = \frac{1}{N} \sum_{i=1}^{N} (S_i - \overline{S})(S_i - \overline{S})^{\mathrm{T}}$$

(3) Let $\Psi_1, \Psi_2, \ldots, \Psi_N$ are the n eigen vectors corresponding n largest eigenvalues of C, appearance the vector $W = [\Psi_1, \Psi_2, \ldots, \Psi_N]$. Find the characteristic vector as

$$Y_i = W^{\mathrm{T}}((S_i - \overline{S}) \, \forall_i = 1, 2, \ldots, N$$

The significant records are specially managed with the aid of the best of a small quantity of most important additives, and in widespread, it is a linear summation of the records from many instructions.

3.2 Linear Discriminant Analysis

LDA strategies are used in information, sample recognition, and device learning to discover a linear mixture of functions. LDA attempts to explicit less one dependent variable as a linear mixture of different features or measurements. LDA is likewise carefully related to PCA and thing analysis in that they both look for linear aggregate of variables, which first rate explains the facts. LDA explicitly attempts to version the difference between the training of statistics. PCA alternatively does not consider any distinction in class, and aspect evaluation builds the characteristic. Combination is based totally on the differences as opposed to similarities. LDA searches for those vectors in the fundamental liberty that an excellent discriminable amid curriculum [17]. Further officially agreed quantity of unbiased features comparative to which the facts is defined, LDA creates a linear aggregate of these which yields the major mean differences among the favored lessons. The inside-class scatter environment is given as,

$$S_{\mathrm{w}} = \sum_{j=1}^{c} \sum_{i=1}^{N_j} (x_i^j - \mu_j)(x_i^j - \mu_j)^{\mathrm{T}}$$

I is the linking set scatter matrix,

$$S_b = \sum_{j=1}^{c} (\mu_j - \mu)(\mu_j - \mu)^T$$

And

$$S_T = S_w + S_b$$

The above equation is the total scatter matrix. In supervised equation, we get to know LDA is more efficient function extraction approach than PCA because its extracted capabilities use the elegance information. However, it is far assumed that the distributions of samples in every elegance are normal and homoscedastic. Therefore, it can be difficult to find a properly and representative characteristic area if this assumption is violated.

4 Modified Decision Boundary Feature Extraction

Attribute removal is usually measured by a statistics mapping method, which determines the correct subspace of dimensionality M from the authentic element area of dimensionality N. The manner of function removal preserve is a linear or nonlinear facts alteration. Despite how the facts alteration is carried out, the function removal set of rules should be intended to maintain the statistics of attention for a unique difficulty which includes density, delousing, or class. For instance, in hyperspectral photograph type, effective capabilities are those that are maximum capable of keeping class reparability. Lee and Landgrebe (1993) confirmed that discriminately information capabilities and redundant data features can be derived from the selection border itself. The technique is called decision boundary characteristic extraction (DBFE). It was shown that everyone had the capabilities wished for type which is ordinary to the effective resolution border. A choice border capabilities matrix (DBFM) was distinct to the expected intrinsic discriminant size [18] and to extract discriminately information capabilities from the decision boundary. In categorize to determine the effective decision boundary, the general public of training samples is first decided on. The quantity of schooling samples required will be much greater for high-dimensional facts. For hyperspectral pictures, the quantity of schooling samples is commonly sufficient to prevent spectacle or defer a great covariance estimate. In addition, DBFE for greater than two instructions is suboptimal.

They additionally showed that the perceptive revealing characteristic vectors contain a factor that is everyday to the judgment border at the slightest at single spot taking place of the selection border. Further, discriminating surplus characteristic vectors are orthogonal to a vector normal to the resolution border at each spot at the choice border. A selection boundary feature matrix (DBFM) changed into distinct with the intention to take out discriminately enlightening functions and discriminately surplus capabilities commencing the choice boundary. It can be exposed to facilitate the status of the DBFM which is the smallest measurement where the identical category precision can be received as in the unique element area. Also, the eigenvectors of the DBFM, similar to nonzero eigenvalues, are the essential feature vectors to reap the same classification precision as inside the unique element space.

The discriminately informative feature is shown in Fig. 2.

The discriminately redundant features are shown in Fig. 3 to predict the classes.

DBFE procedure for neural network can be explained as follows:

Step (1) Instruct the neural network by means of all elements.

Fig. 2 Discriminately information feature

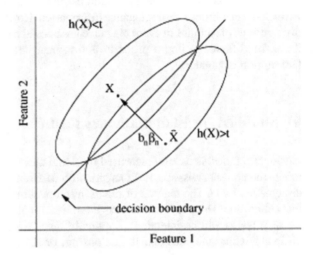

Fig. 3 Discriminately redundant feature

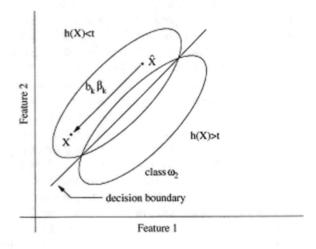

Step (2) Intended for each preparation trial properly confidential as class ω_1, locate the adjoining sample acceptably confidential as class ω_2. Repeat the same process for samples classified as class ω_2.

Step (3) The line linking a couple of samples established in Step 2 that have to exceed throughout the resolution border, while the pair of samples are suitably confidential another way. By moving beside this row, find the point on the resolution border or near the resolution border within a threshold.

Step (4) At each point found in Step (3), estimate the normal vector N. N_i

Step (5) Estimate the decision boundary feature matrix using the normal vectors found in Step (4) by

$$\sum_{\text{DBFM}} = \frac{1}{L} \sum N_i N_i^{\text{T}}$$

where the number of properly confidential samples and T is the reverse operator.

Step (6) Select the eigenvectors of the resolution border element matrix while new aspect vectors according to the magnitude of the consequent eigenvalues. If there are more than two classes, the process preserve to be frequent for every couple of classes behind the network is qualified and intended for all the classes. Then, the total resolution border feature matrix can be calculated by averaging the resolution border feature matrices of every couple of classes.

5 Experimental Results

The real-time dataset was used in this paper over the dataset of PAVIA by means of popular sensor known as airborne imaging spectrometer with spectral location from 0.596 to 1.256. In this segment, we are able to examine the effectiveness of the planned technique DBFE in contrast with PCA, LDA. The common correctness of the lessons is taken into consideration as assessment degree. Accuracy of each set is described as proportion of numeral, which is rightly categorized to the entire set. Take a look at the samples. The neural network classifier is used for category.

The feature extraction results are shown in Fig. 4.

The estimation of DBFE with two different characteristic extraction methods the usage of 10 training samples. The accuracy rate is shown in Fig. 5.

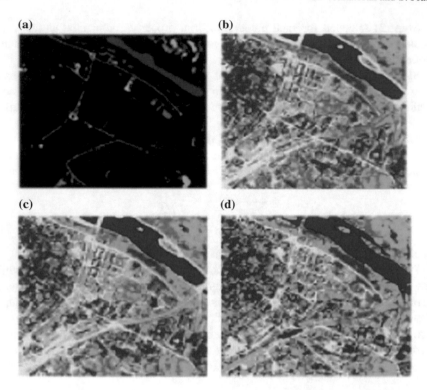

Fig. 4 Feature extraction results

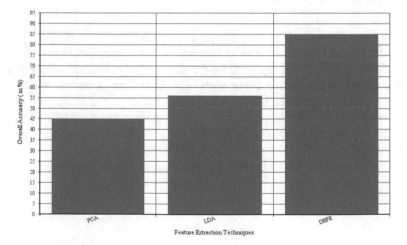

Fig. 5 Overall accuracy

6 Conclusion

In this paper, experiments consist of function extraction techniques with neural network class strategies. The feature extraction strategies together with principal component analysis (PCA), linear discriminative analysis (LDA), and decision boundary feature extraction (DBFE) are used as preprocessing steps to reduce the dimensionality reduction of hyperspectral image. The experimental consequences suggest that the use of DBFE is one in all characteristic extraction technique to prove higher in comparison to PCA and LDA. However, DBFE becomes computationally much less high priced as compared to the alternative feature extraction techniques. In destiny, we used numerous characteristic extraction techniques that are used in numerous areas.

References

1. Huang, X.-M, Hsu, P.-H.: Hyperspectral Image analysis using Hilbert-Huang transform. ACRS. **2011**, 285–290 (2011)
2. Ren, Y., Liao, L., Maybank, S.J., Zhang, Y., Liu, X.: Hyperspectral image spectral-spatial feature extraction via tensor principal component analysis. IEEE Geosci. Remote Sens. Lett. **14**(9), 1431–1435 (2017)
3. Imani, M., Ghassemian, H.: Feature extraction of hyperspectral images using boundary semi-labeled samples and hybrid criterion. J. AI Data Mining. **5**(1), 39–53 (2017)
4. Hsu, P.-H.: Feature extraction of hyperspectral images using Wavelet and matching pursuit. ISPRS J Photogram. Remote Sens. **62**(2), 78–92 (2007)
5. Imani, M., Ghassemian, H.: Principal component discriminant analysis for feature extraction and classification of hyperspectral images. In: 2014 Iranian Conference on Intelligent Systems (ICIS). IEEE (2014)
6. Kang, X., Li, S., Benediktsson, J.A.: Feature extraction of hyperspectral images with image fusion and recursive filtering. IEEE Trans. Geosci. Remote Sens. **52**(6), 3742–3752 (2014)
7. Zhang, L., et al.: Tensor discriminative locality alignment for hyperspectral image spectral-spatial feature extraction. IEEE Trans. Geosci. Remote Sens. **51**(1), 242–256 (2013)
8. Jia, X., Kuo, B.-C., Crawford, M.M.: Feature mining for hyperspectral image classification. Proc. IEEE **101**(3), 676–697 (2013)
9. Kang, X., et al.: Intrinsic image decomposition for feature extraction of hyperspectral images. IEEE Trans. Geosci. Remote Sens. **53**(4), 2241–2253 (2015)
10. Melgani, F., Bruzzone, L.: Classification of hyperspectral remote sensing images with support vector machines. IEEE Trans. Geosci. Remote Sens. **42**(8), 1778–1790 (2004)
11. Li, J., et al.: Multiple feature learning for hyperspectral image classification. IEEE Trans. Geosci. Remote Sens. **53**(3), 1592–1606 (2015)
12. Tuia, D., Matasci, G., Camps-Valls, G., Kanevski, M.: Learning relevant image features with multiple kernel classification. IEEE Trans. Geosci. Remote Sens. **48**(10), 3780–3791 (2010)
13. Gu, Y., Wang, C., You, D., Zhang, Y.: Representative multiple kernel learning for classification in hyperspectral imagery. IEEE Trans. Geosci. Remote Sens. **50**(7), 2852–2865 (2012)
14. Benediktsson, J.A., Palmason, J.A., Sveinsson, J.R.: Classification of hyperspectral data from urban areas based on extended morphological profiles. IEEE Trans. Geosci. Remote Sens. **43**(3), 480–491 (2005)
15. Xia, J., et al.: Spectral-spatial classification for hyperspectral data using rotation forests with local feature extraction and Markov random fields. IEEE Trans. Geosci. Remote Sens. **53**(5), 2532–2546 (2015)

16. Diwaker, M.K., Chaudhary, P.T., et al.: A comparative performance analysis of feature extraction techniques for hyperspectral image classification (2016)
17. Rathi, V.P., Palani, S.: Brain tumor MRI image classification with feature selection and extraction using linear discriminant analysis (2012). arXiv:1208.2128
18. Bandos, T.V., Bruzzone, L., Camps-Valls, G.: Classification of hyperspectral images with regularized linear discriminant analysis. IEEE Trans. Geosci. Remote Sens. **47**(3), 862–873 (2009)

R. Venkatesan received Bachelor of Engineering degree in Computer Science and Engineering from Park College of Engineering and Technology, Coimbatore, and Master of Engineering in Computer Science and Engineering from Sona College of Technology, Salem. He has more than 10 years of experience in teaching and 3 years of experience in research. His current research includes satellite image processing and neural networks. Currently, he is Associate Professor in Department of Computer Science and Engineering, MIET Engineering College, Trichy, and he is Research Scholar in School of Computer Science and Engineering in Vellore Institute of Technology, Vellore.

Dr. Prabu Sevugan received Bachelor of Engineering degree in Computer Science and Engineering from Sona College of Technology (Autonomous) and Master of Technology in Remote Sensing from College of Engineering, Guindy, Anna University, Chennai, and one more Master of Technology in Information Technology at School of Computer Science and Engineering, Bharathidasan University, Trichy. He did his doctoral studies on integration of GIS and artificial neural networks to map the landslide susceptibility from College of Engineering, Guindy, and Anna University, Chennai. He was Postdoctoral Fellow in GISE Advanced Research Lab, Department of Computer Science and Engineering, Indian Institute of Technology Bombay. He has more than 80 publications in national and international journals and conferences. He has organized three international conferences, which include one IEEE Conference as chair and also participated in many workshops and seminars. He is Member of many professional bodies and Senior Member of IACSIT, UACEE, and IEEE. He has more than 14 years of experience in teaching and research. His current research includes remote sensing, satellite image processing, and neural networks. Currently, he is Professor in School of Computer Science and Engineering, Vellore Institute of Technology, Vellore.

Securing the Data in Cloud Environment Using Parallel and Multistage Security Mechanism

Ranjan Goyal, R. Manoov, Prabu Sevugan and P. Swarnalatha

Abstract Cloud computing technology is emerging rapidly due to increasing demand of service required by the different organizations, institutions, and individuals. Nonetheless, the data in the cloud environment is still not completely secure from the outside and inside intrusions and attacks. In this paper, an improvised approach for securing the data in cloud computing environment using parallel and multistage security mechanism (PMSSM) is proposed to provide simultaneous security checks and procedural multistage security using authentication techniques, intrusion detection and encryption. By taking this approach into consideration, the chances of identifying the attack can be increased as the parallelism in the verification and multistage security will enhance the probability of identifying the intrusion or the attack being performed by the attacker. Also, considering the dynamic nature of the intrusion, the parallel and multistage security can help in avoiding such scenario. The discussion of the proposed mechanism suggested that this model can be used to provide enhanced cloud data security to the cloud users.

Keywords Cloud computing · Data security · Intrusion detection · Multistage security · Encryption · Authentication

1 Introduction

The cloud computing environment is being widely used by people for business and personal purposes. It is actually a pool of computers shared to be used over the network in order to overcome the expenses or management efforts. So, instead of having the data locally on the computer of each user, the data is outsourced to a shared or hosted server. The providers of this kind of service are generally seen as the datacenters. The cloud computing is generally used by its users in order to reduce the costs of buying hardware and software. Though cloud services can be accessed globally over the internet, not every cloud service is same and right for

R. Goyal · R. Manoov · P. Sevugan (✉) · P. Swarnalatha
School of Computer Science and Engineering, Vellore Institute of Technology, Vellore, India
e-mail: prabu.sevugan@gmail.com

© Springer Nature Singapore Pte Ltd. 2020   941
K. N. Das et al. (eds.), *Soft Computing for Problem Solving*,
Advances in Intelligent Systems and Computing 1057,
https://doi.org/10.1007/978-981-15-0184-5_80

everyone. There are different models and services offered in the cloud environment. The deployment models for the cloud are generally classified into four types: public, private, community, and hybrid model. The cloud environment can be used to provide different services. These include Software as a Service (SaaS), Performance as a Service (PaaS), and Infrastructure as a Service (IaaS). These can be combined to form Everything as a Service (EaaS). Figure 1 shows the cloud deployment models, and Fig. 2 shows the cloud service models.

As mentioned in the previous paragraph, there are different types of cloud deployment models and cloud services. Nonetheless, the security can be a big issue among all these models. So, to overcome this problem, there should be some mechanisms to provide security to all these models irrespective of the type of the model. Basically, the major requirement is to provide a generic security model that can be integrated with any of the cloud model in order to secure the information or data of the cloud user. For this purpose, the different security issues as well as the security mechanisms

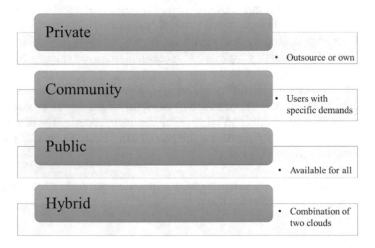

Fig. 1 Cloud deployment models

Fig. 2 Cloud service models

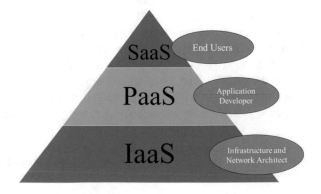

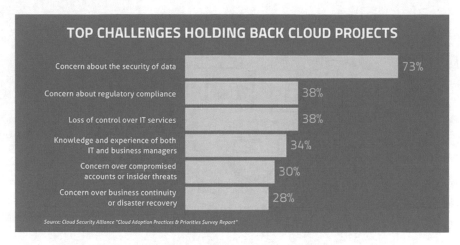

Fig. 3 Top challenges holding back cloud projects

to prevent them are required to be considered. As the goal is to provide a generic security model, the target security issues should also be generic. This means that the security attacks and vulnerabilities will be generalized into the major network attacks if required.

According to the survey report, it is shown that the major challenges that are holding back the cloud projects are due to concern about the security of data. Figure 3 shows the top challenges holding back the cloud projects. Rest of the paper is formulated as follows: Sect. 2 describes the attacks and security issues in cloud computing environment. Section 3 provides the literature survey of related works. Section 4 provides the proposed mechanism for securing the cloud environment from the possible attacks. Section 5 discusses the proposed model based on the possible attack states followed by the conclusion based on the discussion.

2　Attacks and Security Issues

There are several security issues and attacks that still make the cloud environment insecure for data outsourcing [1–3]. The most common attack is distributed denial of service (DDoS) [4]. It is actually based on denial of service (DoS) attack. The DoS attack is a type of cyber-attack in which the attacker attempts to stop the services on the victim's computer by sending a large number of meaningless packets or requests. When the victim system receives these requests, it gets busy in handling these requests making it unavailable to serve the legitimate user. This generally happens when the number of requests received exceeds the number of requests that the system can handle. These also result in wastage of bandwidth of the system and is generally termed as bandwidth depletion. Another type of attack is man-in-the-middle (MITM) attack [5]. There are many types of attack but all are somehow related to the MITM

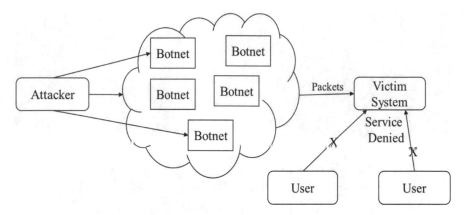

Fig. 4 DDoS attack on victim system

attack considering all these attacks tries to act as a middle man in order to steal the information in some way.

Although the DoS attack seems to be enough to complete the task, this attack is not efficient in accomplishing the task successfully. Thus, this attack is generally performed on a system by a number of systems. These systems are compromised to perform the attack and are termed as zombies or botnets. A botnet or a zombie is a system that is compromised and programmed to perform some malicious activity in order to accomplish a task without the knowledge of its actual user. These botnets are generally controlled by the attacker remotely. These can be formed with the help of a malware that attacker injects into the systems via a number of ways such as the Internet, software, or USB. Therefore, DDoS is performed with the help of these botnets by flooding the victim's system with meaningless packets or requests in order to make the victim's system unavailable for legitimate users. Figure 4 depicts the DDoS attack being performed on victim using botnets network.

3   Related Works

The security of data in the cloud environment is required to be analyzed based on the attacks on the cloud. Jaafar et al. [4] provided a review of detection methods for HTTP-based DDoS attacks. The defense mechanism for DDoS includes prevention, monitoring, detection, and mitigation. In the work proposed by Prabu et al. [5], the work considered the security of data in cloud environment by taking the use of Advanced Encryption Standard (AES) followed by Secure Hash Algorithm (SHA) for hashing the keys in order to prevent the keys to get in hands of the attacker. The work showed that the running time of this mechanism is feasible for implementation in the cloud environment. Nonetheless, the work was only focused toward securing the public cloud as the public cloud can be accessed by anyone on the Internet

[6–9]. In the paper proposed by Himanshu and Afewou [10], the work provided a study on the various security aspects and presented a model based on multiple cloud architectures. The model provided a secure entry to the cloud by using the one-time password (OTP) feature and also enhanced the security of data in transfer using identity anonymization enabling the data to be evaluated and used efficiently without compromising the security of the data.

Zhang et al. [11] proposed an identity-based cloud storage auditing system for shared data, which supports real efficient user revocation with the use of various algorithms like setup algorithm, i.e., generating master secret key, Private Key Generation algorithm, Authenticator Generation algorithm, Proof Generation algorithm, Verification algorithm, User Revocation algorithm to make everything work out. Wani et al. [12] proposed some new security style for cloud data security which uses symmetric cryptographic algorithms for encryption purposes. This style/architecture uses Advanced Encryption Standard (AES), Blowfish and SHA3 and single-use password (OTP) to secure the whole system. Thus, the literature survey of related works suggests to provide a generic model for the security of cloud environment that can be utilized to provide a foolproof mechanism to ensure the security of data in the cloud environment. This paper provides a security model for the same that utilizes the parallel and multistage security in the mechanism to make the cloud secure from major attacks including DDoS and MITM.

4 Proposed Model

In this section, the security model for cloud security is proposed considering the attacks and security issues discussed in the Sect. 2. Figure 5 shows the proposed cloud security model, i.e., PMSSM. Initially, the client is required to provide the details such as user id and password which will be used to check for client account. If the client details are valid, then accept the details and proceed for further verification. It is a notable point that the PAP sends the information in plain text. Nonetheless, an improvement can be made by encrypting the sensitive information like password to prevent MITM attack possibility. In the back end, the Challenge Handshake Authentication Protocol (CHAP) will be executed as soon as the user is asked to provide the details for PAP. The CHAP protocol will start by sending a challenge string to the client and will wait for the response. On arrival of response from the client, the server will verify the response and will accept or reject. In case of acceptance, the mechanism will move to the next step of verification.

The next step is to check for any intrusion. The intrusion can be from inside or from outside. The inside intrusion is avoided by the CHAP, so only outside intrusion is possible at this stage. So, here the login activities and patterns such as DNS and location will be matched and on clearing this stage of verification, the access will be granted to the user for accessing the data such as upload or download the data from the cloud storage. In case the intrusion is detected, the access request is rejected and the details including IP, DNS, and timestamp are recorded. Now, the data in the cloud

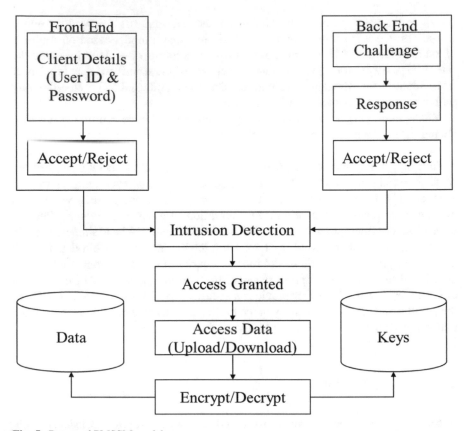

Fig. 5 Proposed PMSSM model

is encrypted with the help of AES 256 bit, so in case of MITM attack, the data cannot be understood by the attacker even though the attacker gets the access to the data. Although it is highly unlikely that the attacker will be able to bypass the security implemented to access the data. Nonetheless, considering a possibility of accessing the data by the attacker, the data is encrypted and stored in the database in order to enhance and maximize the security for the user. Table 1 provides the proposed PMSSM algorithm.

5   Analysis and Discussion

In this section, a discussion is provided to analyze the proposed model based on the possible attack scenarios and its prevention mechanism. Firstly, consider the DDoS attack scenario. In case of DDoS attack, the attacker tries to send a large number of requests to the server in order to make the system unavailable to the legitimate user.

Table 1 Proposed PMSSM algorithm

Algorithm 1. Proposed PMSSM
START PAP and CHAP
IF PAP → Accept & CHAP → Accept
START Intrusion Detection
IF Intrusion Detection → FALSE
Grant Access to environment
FOR Data Upload
ENCRYPT Data using AES
STORE Data in Data Storage
STORE Key in Keys Database
FOR Data Download
GET Data from Storage
GET Key from Keys Database
DECRYPT Data using AES
ELSE   //INTRUSION DETECTION → TRUE
REJECT Access Request
RECORD Details //IP, DNS, Timestamp
ELSE   //PAP → Reject or CHAP → Reject
REJECT Access Request

Here, with the integration of CHAP mechanism, if the attacker tries to send such requests, the server will reject the requests that will be dropped after it crosses the limit. Thus, the model is secure from DDoS attack.

Consider the attack scenarios involving a type of MITM attack. In such case, the attacker can possibly try to get the information from either the PAP mechanism or CHAP mechanism and may bypass any of this mechanism. But it is hard to bypass both of the mechanism at once as it is hard to implement for the attacker that can serve in bypassing both the mechanisms at once. One reason is that there are a limited number of attempts allowed for getting password authentication accepted. As the password is encrypted, so even if the attacker tries to get the password, it is hard to decrypt the password at the same time along with getting the CHAP mechanism authenticated. Nonetheless, even if the attacker bypasses the mechanism, the attacker is most likely going to trigger the intrusion detection mechanism due to some change in behavior or activity. Also, even if the attacker somehow manages to get direct access to the cloud DB containing the data of the user, the attacker will not be able to extract the information from it as the data is encrypted with AES encryption. Thus, the system is secure from issues involving any kind of MITM attack. Table 2 provides the attacks handled at various stages in proposed PMSSM.

Table 2 Attacks handled at various stages in PMSSM

Stage or scenario	Mechanism	Attacks handled
Client login	PAP	MITM
Challenge	CHAP	DDoS, MITM, intrusion
Intrusion check	Intrusion detection	Intrusion
Data access	AES	MITM

6 Conclusion

From the discussion on the possible attack scenarios, it is concluded that the proposed model can be used for providing high security to the cloud users as the model secures the cloud from the major attacks including DDoS and MITM attacks. It is also concluded that with the introduction of this model, the model can be integrated into any type of cloud model, namely public, private, community, and hybrid model as the model is designed for generic use. The public cloud can simply integrate this model by keeping the data of public cloud in data cloud and keys in a private cloud. The private cloud will have only private databases. The community and hybrid will also utilize the databases in a similar fashion. Future work will be based on the actual performance of this model. The model can be further enhanced based on the parallelism in the security mechanism considering the time required to perform individual mechanism by the system.

References

1. Pandith, M.Y.: Data security and privacy concerns in cloud computing. Internet Things Cloud Comput. **2**(2), 6–11 (2014)
2. Singh, N., Kumar, N.: Information security in cloud computing using encryption techniques. Int. J. Sci. Eng. Res. **5**(4), 1111–1113 (2014)
3. Hassan, N., Khalid, A.: A survey of cloud computing security challenges and solutions. Int. J. Comput. Sci. Inf. Secur. **14**(1), 52–56 (2016)
4. Jaafar, G.A., Abdullah, S.M., Ismail, S.: Review of recent detection methods for HTTP DDoS attack. J. Comput. Netw. Commun. **2019**, 1–10 (2019)
5. Prabu, S., Ganapathy, G., Goyal, R.: Enhanced data security for public cloud environment with secured hybrid encryption authentication mechanisms. Scalable Comput.: Pract. Exp. **19**(4), 351–360 (2018)
6. Rajamani, T., Sevugan, P., Purushotham, S.: An investigation on the techniques used for encryption and authentication for data security in cloud computing. IIOABJ **7**(5), 126–138 (2016)
7. Joshi, M., et al.: Secure cloud storage. Int. J. Comput. Sci. Commun. Netw. **1**(2), 171–175 (2011)
8. Surv, N., et al.: Framework for client side AES encryption techniques in cloud computing. Int. Adv. Comput. Conf. (IACC) **6**(1), 525–528 (2015)
9. Moura, J., Hutchison, D.: Review and analysis of networking challenges in cloud computing. J. Netw. Comput. Appl. **60**(6), 113–129 (2016)

10. Himanshu, G., Afewou, K.D.: A trust model for security and privacy in cloud services. In: 6th International Conference on Reliability, Infocom Technologies and Optimization (ICRITO) (Trends and Future Directions), pp. 443–450 (2017)
11. Zhang, Y., Yu, J., Hao, R., Wang, C., Ren, K.: Enabling efficient user revocation in identity-based cloud storage auditing for shared big data. IEEE Trans. Dependable Secure Comput., 1–13 (2017)
12. Wani, A.R., Rana, Q.P., Pandey, N.: Cloud security architecture based on user authentication and symmetric key cryptographic techniques. In: 6th International Conference on Reliability, Infocom Technologies and Optimization (Trends and Future Directions) (ICRITO), pp. 529–534 (2017)

Hybrid Fuzzy Logic-Based MPPT for Wind Energy Conversion System

Vankayalapati Govinda Chowdary, V. Udhay Sankar, Derick Mathew,
CH Hussaian Basha and C. Rani

Abstract Maximum power can be extricated when the turbine keeps running at a consistent and constant speed by using all the vitality present in the wind. The turbine can keep running at a steady speed just when the breeze speed is consistent. The wind vitality being wild in nature, maximum power must be achieved by making the turbine to keep running at the specific breeze speed. To achieve most extreme power, distinctive sorts of maximum power point tracking (MPPT) procedures are utilized. So as to comprehend prudent and proficient power age utilizing wind turbines, modification of fuzzy-based MPPT method is displayed and results are compared with different MPPT techniques for wind energy conversion system have been done and are introduced in subtleties.

Keywords Fuzzy logic · Incremental conductance · Perturb and observe · Wind energy conversion system

Nomenclature

ρ Air density (1.2 kg/m^3)
C_p Power coefficient

V. Govinda Chowdary (✉) · V. Udhay Sankar · D. Mathew · CH Hussaian Basha · C. Rani
School of Electrical Engineering, VIT University, Vellore, India
e-mail: govindachowdaryvankayalapati@gmail.com

V. Udhay Sankar
e-mail: udhaysankarv@gmail.com

D. Mathew
e-mail: derick.mathew@vit.ac.in

CH Hussaian Basha
e-mail: hussaianbasha.ch@vit.ac.in

C. Rani
e-mail: crani@vit.ac.in

K. N. Das et al. (eds.), *Soft Computing for Problem Solving*,
Advances in Intelligent Systems and Computing 1057,
https://doi.org/10.1007/978-981-15-0184-5_81

β Incident angle of the blade
V_{in} Input voltage
V_o Output voltage
P_w Wind power
η_g Generator efficiency
η_m Motor efficiency
I Current
P_e Electric power generated
E Error
DP   Deviation of power over a small time interval
DV   Deviation of voltage over a small time interval
DI   Deviation of current over a small time interval
CE   Deviation in error

1   Introduction

Rapid extinct of fossil fuels and relative increase in the demand of electricity result in the need of different types of renewable sources. Renewable energy resources exist over wide geographical areas, in contrast to other energy sources, which are concentrated in a limited number of countries. Renewable energy systems are rapidly becoming more efficient and cheaper. Their share of total energy consumption is increasing. At the national level, at least 30 nations around the world already have renewable energy contributing more than 20% of energy supply [1]. National renewable energy markets are projected to continue to grow strongly in the coming decade and beyond. Countries like Iceland and Norway generate all their electricity using renewable energy, and many other countries have set a goal to reach 100% renewable energy in the future. Former United Nations Secretary-General Ban Ki-moon has said that renewable energy has the ability to lift the poorest nations to new levels of prosperity [2].

One of the sustainable sources is wind, and it is the most seasoned wellspring of vitality, which was utilized for mechanical purposes in prior days. Wind vitality was utilized for power for the first time in year 1887. The wind energy conversion system (WECS) changes wind vitality into usable type of electrical energy. The extraction of intensity from the inexhaustible assets has been presented from the idea of economic advancement to diminish the weight upon the exhaustible energy sources.

As of 2017, the worldwide total cumulative installed electricity generation capacity from wind power amounted to 486,790 MW, an increase of 12.5% compared to the previous year [3]. There is a straight increment in introduced limit of wind vitality from 2013 to 2016 by 54,642 MW, 63,330 MW, 51,675 MW and 36,023 MW in 2016, 2015, 2014 and 2013 separately. Wind control age limit in India has fundamentally expanded lately. Starting at February 28, 2018, the complete introduced wind control limit was 32.96 GW [4, 5], the fourth biggest introduced wind control limit on the planet.

World's largest turbine generates 8 MW power is Vestas V164 situated at Burbo Bank, off the west shoreline of the United Kingdom in 2016, later refreshed to 9.5 MW. With different kinds of generators accessible, the need to learn about generators is important to comprehend which generator works fine with the MPPT strategy to fulfill criteria, for example, productivity, economy, barometrical condition and future extent of plant extension.

They are generally used to change over the mechanical power yield of wind turbines into electrical power for the lattice. They are known as synchronous generators in light of the fact that the speed of the rotor should dependably coordinate the supply recurrence. Perpetual magnet synchronous generator (PMSG) machine is the most productive of every single synchronous machine since it has a portable attractive source inside itself. Utilization of perpetual magnets for the excitation devours no additional electrical power [6].

2 Wind Turbine Characteristics

The wing energy conversion system changes over input mechanical power into output electrical power. The mechanical power yield of wind turbines is determined utilizing Eq. (1).

$$P_m = \frac{1}{2} C_p(\lambda, \beta) * \rho * A * V^3 \tag{1}$$

where P_m is the mechanical power output, ρ is the air density in kg/m^3, A is the area swept by the rotor, and V is the wind velocity in rad/s. $C_p(\lambda, \beta)$ is the turbine power coefficient which is a function λ, tip speed ratio (TSR) and beta (β) blade pitch angle.

The relation between C_p, TSR and blade pitch angle β is given by

$$C_p(\lambda, \beta) = C_1 \left(\frac{C_2}{\lambda_i} - C_3 * \beta - C_4 \right) e^{\frac{-C_5}{\lambda_i}} + C_6 * \lambda \tag{2}$$

where

$$\frac{1}{\lambda_i} = \frac{1}{\lambda + 0.08\beta} - \frac{0.035}{1 + \beta^3} \tag{3}$$

The coefficients C_1 to C_6 are as follows:

$$C_1 = 0.5176, \quad C_2 = 116, \quad C_3 = 0.4, \quad C_4 = 5, C_5 = 21 \quad \text{and} \quad C_6 = 0.0068$$

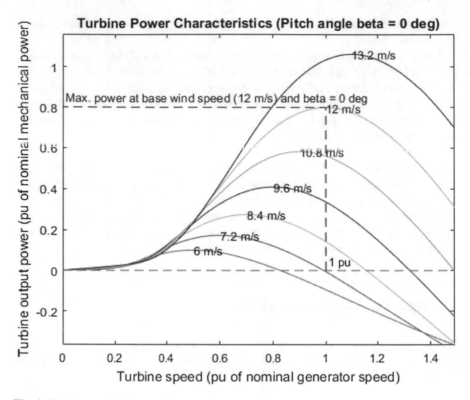

Fig. 1 Turbine power characteristics

The optimal rotor speed at which the turbine operates at maximum power is represented in Eq. (4).

$$\omega_{opt} = \lambda_{opt}(V_w/R) \tag{4}$$

where ω_{opt} is the optimal wind speed in rad/s, λ_{opt} is the optimal TSR in radians, V_w is the wind velocity in m/s, and 'R' is the turbine radius in meters [7]. Figure 1 shows wind turbine power characteristic, and in this paper 12 m/s is taken as base speed and β is $0°$.

3 MPPT Techniques for WECS

a. Requirement of MPPT Techniques for WECS

A typical wind turbine, power–speed trademark, is plotted in Fig. 2. Contingent upon the speed, there exist four working areas. Regions 1 and 4 are unfeasible on the grounds that no solid power yield can be normal before the cut in (NW-Cin) and

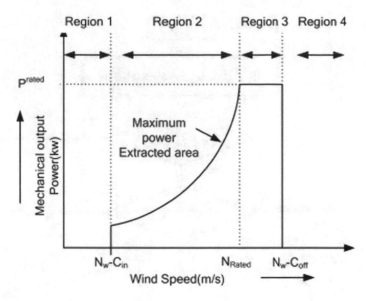

Fig. 2 A typical wind turbine power–speed characteristic

after cutoff (NW-Coff) speed. Subsequently, the wind turbine cannot be connected to grid in this period. In Region 3, MPPT is not required as in this locale the bend as of now achieves most extreme power. Along these lines, the second region between NW-Cin and NRated is the ideal zone for power extraction.

b. **Perturb and Observe (P&O) Method**

Perturb and observe (P&O) is the least difficult method. This is utilized to look for the nearby maxima points of a given capacity. It is generally used in wind energy conversion systems to get the ideal working point that boosts the electric energy. The time unpredictability of this calculation is less yet on achieving near the MPP, and it does not stop at the MPP and continues wavering on both the headings. At the point when this occurs, the calculation has achieved near the MPP and we can set a suitable error limit or can utilize a wait function which ends up increasing time complexity [8–11]. Figure 3 shows the flowchart of P&O algorithm. In this, I_k, V_k, P_k represent current, voltage and power in present iteration, whereas V_{k-1} and P_{k-1} represent the voltage and power of previous iteration.

c. **Incremental Conductance (INC) Method**

The MPP is determined by comparing instant conductance (I/V) to the incremental conductance ($\Delta I/\Delta V$). The INC technique is based on the fact that the slope of $P–V$ curve is zero at MPP. At the point when MPP is a long way from working point, the progression measure is substantial for optimizing while at the same time working point is nearer to MPP, the progression estimate turns out to be little to diminish enduring state oscillation [9, 12–16]. Figure 4 shows the flowchart of incremental

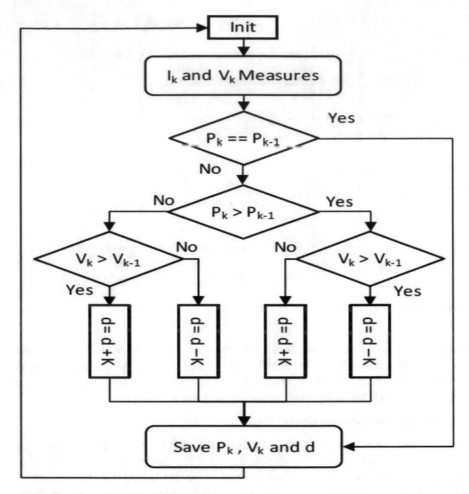

Fig. 3 Flowchart of P&O algorithm

conductance algorithm. In this, dV and dI represent difference of voltage and current, respectively, over a small interval of time.

d. Fuzzy Logic Controller (FLC) Method

An FLC is the artificial decision-making controller that operates in closed loop. FLC generally consists of three stages [17–21].

Fuzzification

Fuzzification is the way toward changing a genuine scalar incentive into a fuzzy esteem. Fuzzy-based controllers have the upsides of working with uncertain sources of info, need not bother with a precise scientific model and handle nonlinearity.

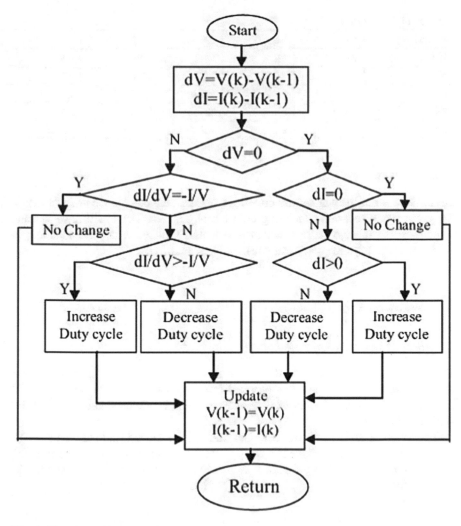

Fig. 4 Flowchart of incremental conductance

Rule base lookup table

The general method for speaking to human learning is by framing natural language articulation given by IF and THEN resulting. This is the second step. The fuzzy rule base incorporates 25 control principles. These principles are actualized by a PC and utilized for the control buck converter with the end goal that greatest power is accomplished at the yield of the breeze turbine at all working conditions.

The rule base is given in Table 1; it represents the set of rules used for modeling FLC. ΔV and ΔI represent the difference of voltage and current over a small interval of time. The rules are framed in five levels, namely negative big (NB), negative small (NS), zero (ZE), positive small (PS) and positive big (PB).

Table 1 FLC set of rules

$\Delta V/\Delta I$	NB	NS	ZE	PS	PB
NB	ZE	PB	ZE	NB	NS
NS	PS	ZE	ZE	NB	NS
ZE	ZE	ZE	ZE	ZE	ZE
PS	PS	PB	ZE	ZE	NS
PB	PS	PB	ZE	NB	ZE

Defuzzification

Defuzzification is the way toward creating a quantifiable outcome in crisp rationale for given fuzzy sets and comparing enrollment degrees. These will have various principles that change various factors into a fluffy outcome, and the outcome is depicted as far as yield enrollment work in fuzzy sets.

4   Proposed Work

FLC-Based Change in Slope Method

The inputs for fuzzy controllers are error signal and variation in error signal. When the signs are determined and etymological factors are acquired, the output of FLC is the obligation proportion for buck converter which is produced dependent on the standards [17, 22–25].

This error signal depends on basic fact that at MPP slope is zero ($\Delta P/\Delta V = 0$). The error signal (E) and change in error signal (CE) are calculated based on Eqs. (5) and (6), respectively. Figure 5 shows the basic structure of FLC-based change in slope method for calculating duty ratio for MPPT.

$$E(k) = [P(k) - P(k-1)]/[V(k) - V(k-1)]   \tag{5}$$

$$\text{CE} = [E(k) - E(k-1)]   \tag{6}$$

Figures 6 and 7 show membership function of input variables E and CE, respectively. The range of E varies from -0.03 to 0.03, whereas CE varies from -1 to 1. These ranges are finalized by trial and error method. Figure 8 shows the output membership function of duty ratio (d), and its range is from 0.4 to 1.

From Fig. 3, the subsystem Simulink block is used to draw the current against voltage curves of PV cell and it is used to measure the difference between the current of photon and diode and reverse saturation current. Similarly, double-diode PV cell operation is the same as a single-diode PV cell. The only difference is in between is reverse saturation current (Table 2).

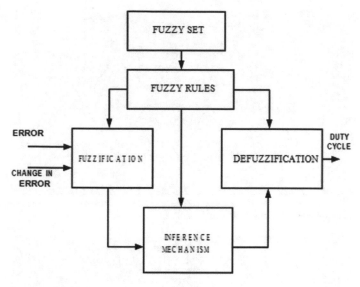

Fig. 5 Basic structure of FLC

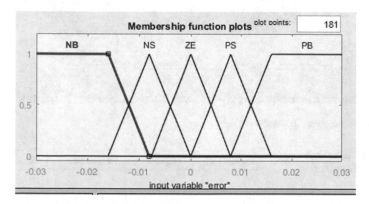

Fig. 6 Input membership function of error

5 Simulation and Analysis

A. *Simulation Results*

For the simulation of various MPPT techniques, a WECS has been created based on block diagram shown in Fig. 9. It consists of wind generator, AC-to-DC rectifier and buck converter with MPPT controller. MATLAB simulation circuit has been created by taking similar working components as in Fig. 9 for comparison purpose. Parameters from Table 3 have been set for creating WECS and are used to analyze various MPPT techniques (Fig. 10).

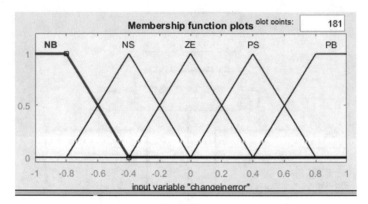

Fig. 7 Input membership function of change in error

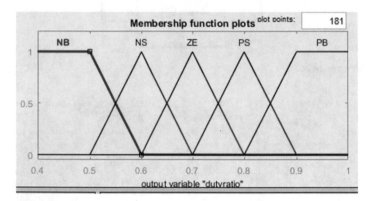

Fig. 8 Input membership function of duty ratio

Table 2 Double-diode operating parameters

Parameters	Values	Parameters	Values
N_p	1	R_p	103.326 Ω
N_s	36	I_{sc_n}	4.2 A
A	1.0	$I_{o1} = I_{o2}$	8.234e^{-10} A
a_1	1.21	V_{oc}	20.359 V
R_s	0.5 Ω	I_{g_STC}	5.432 A
I_{MPPT}	4.78 A	V_{MPPT}	15.10 V

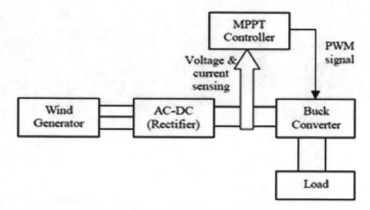

Fig. 9 Block diagram of WECS

Table 3 Parameter specifications

Equipment	Parameter	Value
Wind turbine specifications	Nominal mechanical output power (W)	2500 W
	Base power of the electrical generator (VA)	2500/0.8 VA
	Base wind speed, V (m/s)	12 m/s
Generator specifications	Permanent magnet synchronous generators	
	Number of phases	3
	Rotor type	Round
	Mechanical input	Torque Tm
	Stator phase resistance, R_s (Ω)	0.05 Ω
	Armature inductance, XL (H)	0.000635 H
AC–DC rectifier specifications	Forward voltage, V_f (V)	0.8 V
	Diode type	Universal bridge
DC–DC buck converter specifications	RC branch	$R = 1\ \Omega$
		$C = 1200\ \mu F$
	RL branch	$R = 1\ \Omega$
		$L = 402\ \mu H$
Load type	Resistive load	$R = 18\ \Omega$

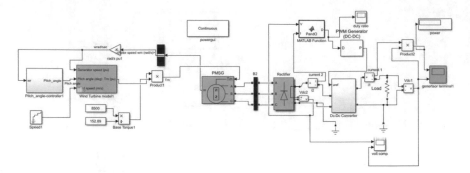

Fig. 10 MATLAB simulation system

Figure 11 shows the rotational speed of wind turbine vs wind speed. The rated wind speed of 12 m/s has been given as input at 3 s.

Figures 12, 13, 14 and 15 demonstrate the duty ratio varieties with respect to wind speed of P&O, INC, FLC, FLC-based 'change in slope' strategies separately. Looking at Figs. 12, 13, 14 and 15 which show obligation proportion regarding wind speed of each MPPT procedure, the accompanying lines have been expressed. P&O MPPT demonstrates no change till 1.2 s, and later on it found a way to reach MPPT as quick as conceivable; this demonstrates the motions of obligation proportion in Fig. 12. INC MPPT demonstrates no variety in Fig. 13, and this demonstrates at

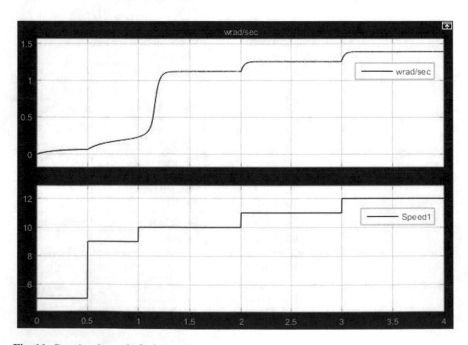

Fig. 11 Rotational speed of wind turbine versus wind speed

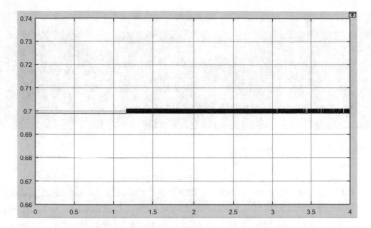

Fig. 12 Duty ratio of P&O MPPT

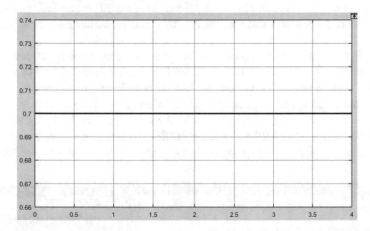

Fig. 13 Duty ratio of INC MPPT

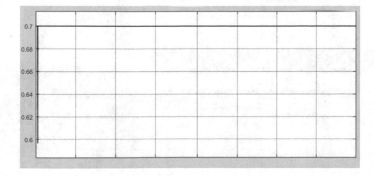

Fig. 14 Duty ratio of FLC-based direct method

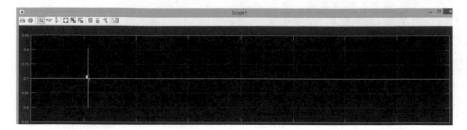

Fig. 15 Duty ratio of FLC proposed work

each moment this strategy is achieving the MPP quickly. Obligation proportion of FLC-based direct input MPPT demonstrates an abrupt low and high in Fig. 14, this is because of unexpected ascent in speed, and this strategy additionally comes to MPP quickly. Obligation proportion of FLC-based change in error MPPT demonstrates an unexpected high or low in Fig. 15 at 0.5 s as till then ΔP is zero in Fig. 19, this is because of ascent in power and voltage, and this strategy additionally comes to MPP quickly.

Figures 16, 17, 18 and 19 demonstrate the yield power and voltage regarding wind speed. Both P&O MPPT and INC MPPT strategies have come to 2912 W at various occasions 3.815 s and 3.57 s separately. FLC-based direct information and 'change in slope' MPPT systems have achieved 2912 W and 2913 W individually. The time taken to achieve greatest power by FLC-based strategies is 3.894 s and 3.699 s individually. Since INC MPPT procedure utilizes conductance incentive to figure MPP and as each framework has conductance esteem, this influences the system to

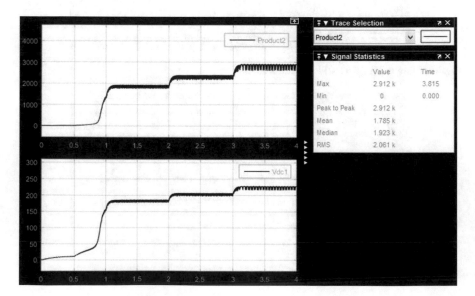

Fig. 16 P&O MPPT power and voltage versus wind speed

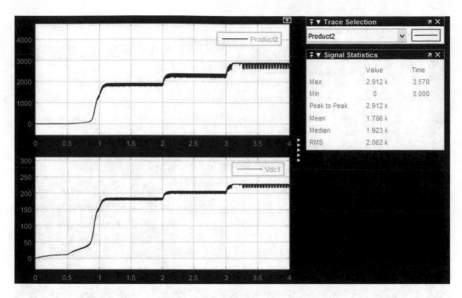

Fig. 17 INC MPPT power and voltage versus wind speed

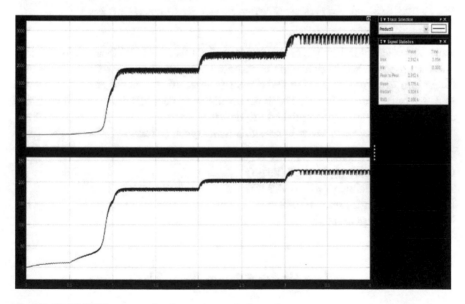

Fig. 18 FLC MPPT power and voltage versus wind speed

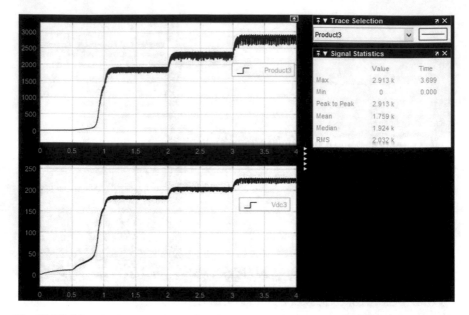

Fig. 19 FLC-based 'change in slope' MPPT power and voltage versus wind speed

work incorrectly for down to earth structure. So, this INC MPPT cannot be utilized and the proposed MPPT defeated the essential disadvantage of P&O, INC and FLC direct info strategy which is the fast ripples in Figs. 16, 17 and 18 at max speed that is 3–4 s time interim. In Fig. 19, those swells have been diminished and an about consistent power was accomplished (Table 4).

Theoretical calculations at base–rated wind speed (12 m/s)

Wind power $P_w = \frac{1}{2}\rho A^3 = 7463.7\,\text{W}$
Generator efficiency $(\eta_g) = 0.9$ (taken)
Transmission efficiency $(\eta_m) = 0.9$ (taken)
Power coefficient C_p max $= 0.48$
Electrical power generated

$$P_e = C_p * \eta_m * \eta_n * P_w = 3049\,\text{W}$$

6 Conclusion

In this paper, different MPPT techniques for WECS have been studied and compared. Even though the P&O, INC, FLC-based MPPT techniques reach maximum power, it is found that P&O and FLC-based direct method take more time when compared to

Table 4 Comparison of various MPPT techniques

MPPT techniques	Complexity	Memory measurement	Performance under varying wind conditions	Maximum power	Time taken to reach max power after reaching max speed (12 m/s)
Perturb and observe	Simple	No	Moderate	2912	3.815
Incremental conductance—IC	Simple	No	Moderate	2912	3.57
Fuzzy logic controller-based direct input	High	Yes	Good	2912	3.894
Fuzzy logic controller-based change in slope method	High	Yes	Good	2913	3.699

INC. While considering oscillations in maximum power, FLC shows better results. So by taking the advantages of FLC direct method and INC, a new FLC-based change in slope method has been proposed. So, the proposed FLC-based 'change in error' method takes less time to reach MPP comparatively. The proposed method works fine and gives an output power of 2913 W nearly same as base electrical power generated.

References

1. Global Trends in Sustainable Energy Investment 2007: Analysis of Trends and Issues in the Financing of Renewable Energy and Energy Efficiency in OECD and Developing Countries (PDF), p. 3. United Nations Environment Programme (2007). unep.org. Archived (PDF) from the original on 13 Oct 2014. Accessed 13 Oct 2014
2. Leone, S.: U.N. Secretary-General: Renewables Can End Energy Poverty. Renewable Energy World (2011)
3. GWEC: Global Wind Report Annual Market Update. Gwec.net. Accessed 20 May 2017
4. Installed Capacity of Wind Power Projects in India. Accessed 7 Apr 2018
5. Global Wind Statistics 2017 (PDF)
6. http://www.alternative-energy-tutorials.com/wind-energy/wind-turbinegenerator.html
7. Tounsi, A., Abid, H., Kharrat, M., Elleuch, K.: MPPT algorithm for wind energy conversion system based on PMSG. In: 2017 18th International Conference on Sciences and Techniques of Automatic Control and Computer Engineering (STA), Monastir, Tunisia, pp. 533–538 (2017)
8. Wafa, H., Aicha, A., Mouna, B.H.: Steps of duty cycle effects in P&O MPPT algorithm for PV system. In: 2017 International Conference on Green Energy Conversion Systems (GECS), Hammamet, pp. 1–4 (2017)

9. Masood, B., Siddique, M.S., Asif, R.M., Zia-ul-Haq, M.: Maximum power point tracking using hybrid perturb & observe and incremental conductance techniques. In: 2014 4th International Conference on Engineering Technology and Technopreneurship (ICE2T), Kuala Lumpur, pp. 354–359 (2014)
10. Khadidja, S., Mountassar, M., M'hamed, B.: Comparative study of incremental conductance and perturb & observe MPPT methods for photovoltaic system. In: 2017 International Conference on Green Energy Conversion Systems (GECS), Hammamet, pp. 1–6 (2017)
11. Lahfaoui, B., Zouggar, S., Elhafyani, M.L., Seddik, M.: Experimental study of P&O MPPT control for wind PMSG turbine. In: 2015 3rd International Renewable and Sustainable Energy Conference (IRSEC), Marrakech, pp. 1 6 (2015)
12. Lee, J.H., Bae, H., Cho, B.H.: Advanced incremental conductance MPPT algorithm with a variable step size. In: 2006 12th International Power Electronics and Motion Control Conference, Portoroz, pp. 603–607 (2006)
13. Heydari, M., Smedley, K.: Comparison of maximum power point tracking methods for medium to high power wind energy systems. In: 2015 20th Conference on Electrical Power Distribution Networks Conference (EPDC), Zahedan, pp. 184–189 (2015)
14. Abdullah, M.A., Yatim, A.H.M., Tan, C.W.: A study of maximum power point tracking algorithms for wind energy system. In: 2011 IEEE Conference on Clean Energy and Technology (CET), Kuala Lumpur, pp. 321–326 (2011)
15. Kumar, D., Chatterjee, K.: A review of conventional and advanced MPPT algorithms for wind energy systems. Renew. Sustain. Energy Rev. 55, 957–970 (2016). ISSN 1364-0321
16. Mehta, G., Dwivedi, M., Yadav, V.K.: Comparison of advance intelligence algorithms for maximum power point tracking. In: 2017 4th IEEE Uttar Pradesh Section International Conference on Electrical, Computer and Electronics (UPCON), Mathura, pp. 262–267 (2017)
17. Heshmatian, S., Kazemi, A., Khosravi, M., Khaburi, D.A.: Fuzzy logic based MPPT for a wind energy conversion system using sliding mode control. In: 2017 8th Power Electronics, Drive Systems & Technologies Conference (PEDSTC), Mashhad, pp. 335–340 (2017)
18. Sarvi, M., et al.: A New Method for Rapid Maximum Power Point Tracking of PMSG Wind Generator Using PSO_Fuzzy Logic (2013)
19. Rajvikram, M., Renuga, P., Swathisriranjani, M.: Fuzzy based MPPT controller's role in extraction of maximum power in wind energy conversion system. In: 2016 International Conference on Control, Instrumentation, Communication and Computational Technologies (ICCICCT), Kumaracoil, pp. 713–719 (2016)
20. Dida, A., Benattous, D.: Fuzzy logic based sensorless MPPT algorithm for wind turbine system driven DFIG. In: 2015 3rd International Conference on Control, Engineering & Information Technology (CEIT), Tlemcen, pp. 1–6 (2015)
21. Soufi, Y., Bechouat, M., Kahla, S., Bouallegue, K.: Maximum power point tracking using fuzzy logic control for photovoltaic system. In: 2014 International Conference on Renewable Energy Research and Application (ICRERA), Milwaukee, WI, pp. 902–906 (2014)
22. Marmouh, S., Boutoubat, M., Mokrani, L.: MPPT fuzzy logic controller of a wind energy conversion system based on a PMSG. In: 2016 8th International Conference on Modelling, Identification and Control (ICMIC), Algiers, pp. 296–302 (2016)
23. Sl-Subhi, A., Alsumiri, M., Alalwani, S.: Novel MPPT algorithm for low cost wind energy conversion systems. In: 2017 International Conference on Advanced Control Circuits Systems (ACCS) Systems & 2017 International Conference on New Paradigms in Electronics & Information Technology (PEIT), Alexandria, Egypt, pp. 144–148 (2017)
24. Harrabi, N., Souissi, M., Aitouche, A., Chaabane, M.: MPPT algorithm for wind energy generation system using T-S fuzzy modeling. In: 2016 5th International Conference on Systems and Control (ICSC), Marrakesh, pp. 157–162 (2016)
25. Patil, S.N., Prasad, R.C.: Design and development of MPPT algorithm for high efficient DC–DC converter for wind energy system connected to grid. In: 2015 International Conference on Computer, Communication and Control (IC4), Indore, pp. 1–7 (2015)

Cyclist Detection Using Tiny YOLO v2

Karattupalayam Chidambaram Saranya, Arunkumar Thangavelu, Ashwin Chidambaram, Sharan Arumugam and Sushant Govindraj

Abstract This paper seeks to evaluate the performance of the state of the art object classification algorithms for the purpose of cyclist detection using the Tsinghua—Daimler Cyclist Benchmark. This model focuses on detecting cyclists on the road for its use in development of autonomous road vehicles and advanced driver-assistance systems for hybrid vehicles. The Tiny YOLO v2 algorithm is used here and requires less computational resources and higher real-time performance than the YOLO method, which is extremely desirable for the convenience of such autonomous vehicles. The model has been trained using the training images in the mentioned benchmark and has been tested for the test images available for the same. The average IoU for all the truth objects is calculated and the precision-recall graph for different thresholds was plotted.

Keywords Tiny Yolo v2 · Tsinghua–Daimler Cyclist Benchmark · Cyclist detection · IoU

1 Introduction

With the pace of development occurring in the deployment of autonomous driving vehicles and the extensive work being done in building advanced driver-assistance systems (ADAS), there is a need to implement state of the art detection models for the purposes of automated and assisted driving. Vehicle manufacturers have moved onto equipping most vehicles with cameras which offer a unique opportunity to design computer vision models which can offer both accurate and fast real-time detection.

Advanced driver-assistance systems (ADAS) have taken advantage of the recent advances in computer vision and have incorporated it in order to analyze and highlight obstacles which hamper driving. One of the challenges that impair the performance of ADAS is the efficient detection of the vulnerable road users (VRU) which consists predominantly of pedestrians and cyclists. While existing approaches have studied

K. C. Saranya (✉) · A. Thangavelu · A. Chidambaram · S. Arumugam · S. Govindraj
Vellore Institute of Technology, Vellore, Tamil Nadu, India
e-mail: saranya.kc@vit.ac.in; saranya.karattupalayam.c@gmail.com

© Springer Nature Singapore Pte Ltd. 2020   969
K. N. Das et al. (eds.), *Soft Computing for Problem Solving*,
Advances in Intelligent Systems and Computing 1057,
https://doi.org/10.1007/978-981-15-0184-5_82

pedestrian systems extensively, systems designed for cyclists have not been worked on as much. Cyclist detection is a very challenging problem to solve due to the diversity in cyclists posture, viewpoints and the vehicle itself. The Tsinghua–Daimler benchmark used in the model allows for the development of systems exclusively toward cyclist detection, addressing the aforementioned gap.

Considering vision-based cyclist detection, one of the earliest image processing models include Li [1] who has adapted methods previously used for pedestrian detection. Their model focusses on detecting crossing cyclists using HOG LP and a linear SVM classifier. The SVM classifier allows it to optimize the time-draining steps of feature extraction involving HOG. The model allows only for detection of crossing cyclists. Approaches to optimize real-time performance include the method described by Tian et al. [2] using cascade detectors to identify cyclists in different perspectives and filters to deal with occlusion. The model uses a geometry-based ROI extraction along with trajectory planning to improve performance but still only achieve a real-time performance of 11 fps.

The variations found while detecting cyclists makes convolutional neural network models an attractive choice for designing systems as they show a higher degree of performance and adaptability in challenging situations compared to conventional image processing techniques.

The KITTI dataset [3] serves as the source for most work done toward cyclist detection using convolutional neural networks. The dataset contains a person class which comprises of both pedestrians and cyclist so the models which have been developed for pedestrian detection also deal with cyclist detection. For instance, Yang in [4] has used convolutional object detector with scale-dependent pooling and cascaded rejection classifiers in order to ensure fast and accurate detection. The scale-dependent pooling allows it to improve detection on small objects and the cascaded rejection classifiers serve to improve speed of detection by eliminating negative detections quickly and increases accuracy. Toward cyclist detection, the model achieves accuracies of 61% in the moderate case. Ren [5] has used single-stage detection by having recurrent rolling convolution (RRC) architecture to work over multi-scale feature maps allowing for bounding boxes showing "deep in context" detection. They were able to overcome previous difficulties facing single-stage detectors such as low accuracy and took advantage of existing benefits such as ease in training and efficiency in performance. Their model achieved a high level of accuracy of 76.47% under a IoU threshold of 0.7 for moderate case in cyclist detection. With the development of Faster RCNN [6], Saleh et al. [7] have proposed a faster RCNN network and have generated training data by using synthetic training using generated image datasets developing a theoretically unlimited batch of images to train on. This has allowed them to outperform the previously outlined HOG-SVM approach by 21% in average precision.

Apart from networks trained on the KITTI dataset, Fast RCNN [8] techniques have been adopted along with stereo techniques as shown by Li et al. [9] but models had deficiencies ranging from execution to difficulty in accuracy for edge cases. These edge cases include cases where the cyclist have a higher degree of occlusion and are less than 40 pixels wide in the detection image. Their work led to the creation of the

Tsinghua–Daimler benchmark dataset which serves as a prime source for the training and testing of focused cyclist detection for the model discussed in this paper as it contains images taken from the perspective of a camera mounted on a car echoing how our intended systems would receive input. It contains a thorough and wide range of images for cyclist detection comprising of 9741 images earmarked for training and another set of 2914 "test" images used to evaluate detector performance.

The models discussed still majorly focus on accuracy and not on real-time performance. As highlighted by Huang in [10] modern networks like YOLO have a slight decrease in accuracy but more than compensate for it in real-time performance [11] when compared to approaches like Fast RCNN which is integral to our aim of developing a model for autonomous vehicles and ADAS systems. With optimization, efficient platforms like YOLO and its developing variants like fast YOLO [12] one can achieve better real-time object detection and also utilize the advantage of the recent advances in graphical processing unit (GPU) technology [13]. These advances enable developers to maximize advantages of the highly parallelized manner of detections, possible in methods like YOLO with GPUs having thousands of efficient cores allowing us to carry out real-time detection with high precision while also reducing computational effort and processing time.

This paper discusses the application of convolutional neural networks through the Tiny YOLO v2 algorithm [11] developed on an open-source platform called "Darkflow" which is written in programming language "Cython" "C" and "Python" which is designed from the ground up for accurate and real-time detection, for the classification and detection of cyclists on roads in diverse scenarios as encountered in the Tsinghua–Daimler benchmark [9] dataset. Tiny Yolo v2 variant [11] is a lightweight model using nine convolutional layers when compared to the full model's 24 and also has better real-time detection with a slight decrease in accuracy. This trade-off is acceptable as both of our prime focus areas of autonomous driving and driving-assistance systems require fast detection. The work carried out has been centered on studying the performance of the above model on the benchmark [9] to determine its efficiency and accuracy for real-time cyclist detection. A system equipped with an Intel 6th gen Core i5 6200u with a Nvidia 940MX GPU has been used for training the Tiny YOLO network, whereas the detection was run on a system equipped with Intel Core i7 8th gen processor on the CPU.

The paper is organized as follows: Sect. 2 sheds details on methodology followed through the paper, which includes the information of dataset, preprocessing included before training, the Tiny YOLO v2 network architecture, and training procedure. Section 3 sheds light on the results during the testing process and discusses the details of the result. Section 4 concludes the entire paper.

Fig. 1 (Clockwise) **a** Non-VRU; **b** test; **c** valid; **d** train

2 Methodology

2.1 Dataset

The Tsinghua–Daimler Cyclist Benchmark [5] has been used as the dataset and it consists of four subsets. The first set of 9741 images has been annotated only for cyclist objects which has occlusion less than 10% and have detection area greater than 60 pixels. The second is a set of 1019 images of valid object of interest belonging to all 7 classes, with the detection area higher than 20 pixels. Thirdly, a set of 2914 annotated images are available which has been used for testing the model. The final set of 1000 images is those with no object of interest.

The dataset is wide, with a range of cyclists, with change in color, hue, saturation, and some of the examples are given below (see Fig. 1). (a) Non-VRU, the subset of 1000 images with no objects, (b) test, subset which is used for testing the metrics required, (c) valid, subset with 10,109 images of objects of interest belonging to all the classes, and (d) train, subset of 9741 images which is used to train the neural network model.

2.2 Preprocessing

All the training images are resized as $416 \times 416 \times 3$ (3 representing RGB color channels) in order to reduce the memory footprint and thus increasing training efficiency and overall performance.

Annotations provided for every training image in the dataset were converted according to the requirements of the Tiny YOLO v2 neural network model. This involved assigning a difficult tag in the derived annotation for objects having "out_of_image" or "unsure_box in the source annotation and an occluded tag for objects tagged with occlusion parameters, if the source annotation had "occluded >40" or "occluded >80", then the occluded tag in the derived annotation was set as 1 else it was set as 0.

2.3   Network Architecture

Tiny YOLO v2 has 9 convolutional layers and 6 Maxpool layers. For every convolutional layer, the filters are chosen initially (see Fig. 2). The number of filters in the final layer of the convolutional layer is calculated through:

$$\text{No. of filters} = num * (classes + 5) \tag{1}$$

The chosen dataset has seven classes, namely "unlabeled," "pedestrian," "cyclist," "motorcyclist," "tricyclist," "wheelchairuser," "mopeduser." As a result, the filters in the final layer sum up to 60.

For each convolutional layer apart from the final detection layer, leaky activation function is chosen, such that it brings nonlinearity to the model. The number of filters used and activation of each layer is mentioned in Table 1.

This is followed by the selection of image padding and stride. "Stride" controls the way the filters convolve around the input. The "stride" and "padding" has been fixed as 2 and 1, respectively, for all convolutional layers.

The selection of the convolutional layers is followed by the selection of pooling layers, which are placed alternatively to every convolutional layer. Tiny YOLO v2 uses Maxpooling which allows for results from convolutional layers to be generalized, allowing for them to be static for changes in scale and orientation. The

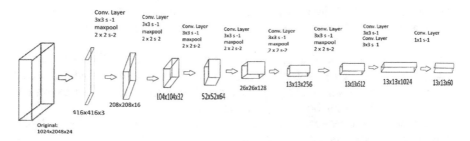

Fig. 2 Network architecture representing the number of convolutional layers, Maxpool layers along with its dimension of the image at the beginning and the end of each layer

Table 1 Network architecture configurations

Type	Activation function	Number of filters	Output
Convolutional	Leaky	16	208×208
Convolutional	Leaky	32	104×104
Convolutional	Leaky	64	52×52
Convolutional	Leaky	128	26×26
Convolutional	Leaky	256	13×13
Convolutional	Leaky	512	13×13
Convolutional	Leaky	1024	13×13
Convolutional	Leaky	1024	13×13
Convolutional	Linear	60	13×13

Maxpool layer serves mainly two purposes, reducing the computation cost and controlling overfitting. Maxpool layers are configured to have 2×2 filters with stride set to 2.

2.4 Training

Training set for the model consisted of 9741 images from the Tsinghua–Daimler benchmark with annotations for only "cyclist" named as the "train" set.

According to the model's architecture, training images are divided into batches of 8, enabling the GPU to process 8 images at a time. With the processing of a batch, the gradients are updated accordingly.

Table 2 gives a summary of all the parameters required for training.

The initial learning rate is set to 0.001 and the decay rate is set to 0.0005 for the leaky activation function. Initial weights for training were acquired from the weights of a Tiny YOLO v2 model trained on VOC dataset.

The model follows a learning rate rescaling schedule as follows: it rescales by a factor of 10 after 100 iterations, 0.1 after 20,000 and again rescales by a factor of 0.1 after 30,000 iterations. One epoch of training requires approximately 1217 iterations.

Each of the preprocessed images is divided into grids of 13×13 cells with each responsible for prediction of 5 different bounding boxes for objects whose center lies in the cell. As initial weights are taken from Tiny YOLO v2 VOC detector, the cells generate 5 bounding boxes for each prediction.

The prediction of every bounding box contains certain information: the label, confidence, bounding box coordinates of the top left and bottom right. These are calculated as [4]:

$$b_x = \sigma(t_x) + C_x \qquad (2)$$

Table 2 Training parameters

Parameter	Value
Dataset	Tsinghua–Daimler Cyclist Benchmark
Training samples	9741
Validation samples	1019
Initial learning rate	0.001
Momentum	0.9
Learning rate decay	0.0005
Pretrained weights	VOC 2007
Batches	8
Epochs	25
Train size (height × width)	416 × 416
GPU	Nvidia 940MX; Intel Core i5-6200u

$$b_y = \sigma(t_y) + C_y \tag{3}$$

$$b_w = p_w e^{tw} \tag{4}$$

$$b_h = p_h e^{th} \tag{5}$$

The model has been trained for 25 epochs, and results obtained are discussed in the following section.

3   Results and Discussion

Apart from the training set of cyclist images, the dataset [6] includes a set of 2914 images. These set of images have been utilized to calculate the required metrics for analyzing the performance of the neural network, after the training of 25 epochs.

Table 3 gives a summary of all the parameters during testing period.

The results obtained through the network included different scenarios of successful detection (see Fig. 3): (a) is the detection of a lone cyclist at a distance farther than the range of vision, (b) and (d) being a side view of the cyclist one of which is blurry, and (c) detects three cyclists who are closer to each other. In all the images, the blue colored bounding box represents the ground truth provided along with the dataset [6] and the green-colored bounding box is the predicted bounding box resulted from the neural network.

The performance of the neural network model has been evaluated by the metrics: Intersection of union (IoU), precision-recall Graph, and F1 score.

Table 3 Test parameters for the experiment

Parameter	Value
Test dataset	Tsinghua–Daimler Cyclist Benchmark
Samples	2914
Detection threshold	0.5 (50% confidence)
IoU threshold	50%
Test image size in pixels (height × width)	1024 × 2048
Image size after conversion	416 × 416

Fig. 3 (Clockwise) **a** Longitudinal cross-section of cyclist; **b** side view of cyclist (blurry); **c** multiple cyclists; **d** side view of cyclist (clear)

The prediction was taken into consideration only when it had a confidence of at least 0.5. The confidence denotes the probability that the neural network of the detected object being a cyclist.

Intersection of Union (IoU)

IoU refers to the percentage of the area that has been detected accurately by the neural network compared to the ground truth. It was calculated as follows:

$$\text{IoU} = \frac{\text{area of overlap}}{\text{area of union}} * 100\% \tag{6}$$

$$\text{area of union} = \text{area of ground truth} + \text{area of prediction} - \text{area of overlap} \quad (7)$$

The area is common to both, the area of ground truth denoted by the green box and the predicted area denoted by the blue box from Fig. 3 was calculated and presented as the area of overlap.

IoU typically ranges from 0 to 1, with 0 indicating no overlap and 1 a perfect overlap of the ground truth and the predicted bounding box, respectively. The average IoU of all the true object was 73.17%, with predicted objects with a confidence factor of 0.5 [] considered true.

Precision-Recall Graph

The efficiency of the system is denoted by the calculation of the accuracy. Accuracy is calculated as below:

$$\text{Accuracy} = \frac{\text{TP} + \text{TN}}{\text{TP} + \text{FP} + \text{FN} + \text{TN}} \quad (8)$$

True negatives, TN, denote the cases where the test image's background was identified appropriately as background by the system. Precision-recall graph is plotted in order to examine this quantity meaningfully as a mere count of the number of successful cases would be devoid of context.

The precision-recall graph as shown in Fig. 3 has been plotted over different for various values of precision and recall. The model sets a confidence factor threshold of 0.5 to determine the validity of its detections.

True positives, TP, require predictions to have an IoU and confidence level greater than the threshold value, 0.5.

False positives, FP, have been considered as those with confidence greater than the threshold but IoU less than 50%.

False negatives, FN, have been considered as those cyclists, which are not detected by the neural net, which was calculated as the difference between the actual number of cyclists present and the number of true positives.

Precision and recall was calculated as below:

$$\text{Precision} = \frac{\text{TP}}{\text{TP} + \text{FP}} \quad (9)$$

$$\text{Recall} = \frac{\text{TP}}{\text{TP} + \text{FN}} \quad (10)$$

The precision value of 0.8 was obtained indicating that it has a high likelihood of finding true objects. The high precision value means that the object detected to be cyclist by the model is very likely to be true, which makes the model more reliable (Fig. 4).

F1 score

To find an optimum mix between precision and recall, the metric, F1 score have been used.

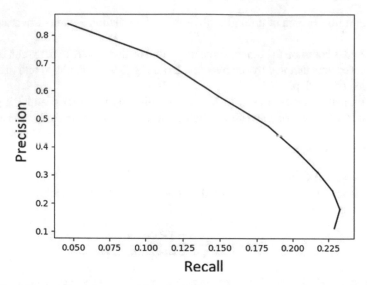

Fig. 4 Precision-recall graph over multiple threshold

F1 score is calculated as the weighted average between precision and recall, thus taking into an account both the false positives and the false negatives. It was calculated as below:

$$\text{F1 score} = 2 * \frac{(\text{Recall} * \text{Precision})}{\text{Recall} + \text{Precision}} \tag{11}$$

The maximum F1 score recorded, through various thresholds, was 0.4. This can be improved by prolonged training and attending to the factors such as occlusion and camera blind spots.

4   Conclusion

The implemented model shows good performance with regards to accuracy and efficiency for the scenarios found in the Tsinghua–Daimler benchmark. The deficiencies in the performance can be attributed to situations where the objects of interest show high degree of occlusion, greater than forty percent, or are limited by the characteristics of the camera used to capture images leading to reduced range of vision causing blind spots and limited field of view. Further studies can be carried out on the above-mentioned difficulties to improve performance. The model prescribed in this paper can also be extended toward overall VRU detection, both pedestrians and cyclists albeit with a few modifications and with more extensive training.

References

1. Li, T., Cao, X., Xu, Y.: An effective crossing cyclist detection on a moving vehicle. In: 2010 8th World Congress on Intelligent Control and Automation, Jinan, pp. 368–372 (2010)
2. Tian, W., Lauer, M.: Fast cyclist detection by cascaded detector and geometric constraint. In: 2015 IEEE 18th International Conference on Intelligent Transportation Systems, Las Palmas, Spain (2015)
3. Geiger, A., Lenz, P., Urtasun, R.: Are we ready for Autonomous Driving? The KITTI Vision Benchmark Suite, CVPR (2012)
4. Yang, F., Choi, W., Lin, Y.: Exploit all the layers: fast and accurate CNN object detector with scale dependent pooling and cascaded rejection classifiers. Proceedings of the IEEE Conference on Computer Vision and Pattern Recognition (2016)
5. Ren, J., Chen, X., Liu, J., Sun, W., Pang, J., Yan, Q., Tai, Y.W., Xu, L.: Accurate single stage detector using recurrent rolling convolution. Proceedings of the IEEE Conference on Computer Vision and Pattern Recognition (2017)
6. Ren, S., He, K., Girshick, R., Sun, J.: Faster R-CNN: towards real-time object detection with region proposal networks (2016). arXiv:1506.01497 [cs.CV]
7. Saleh, K., Hossny, M., Hossny, A., Nahavandi, S.: Cyclist detection in LIDAR scans using faster R-CNN and synthetic depth images. In: 2017 IEEE 20th International Conference on Intelligent Transportation Systems (ITSC), Yokohama, pp. 1–6 (2017)
8. Girshick, R.: Fast R-CNN. In: 2015 IEEE International Conference on Computer Vision (ICCV), Santiago, Chile (2015)
9. Li, X., Flohr, F., Yang, Y., Xiong, H., Braun, M., Pan, S., Li, K., Gavrila, D.M.: A new benchmark for vision-based cyclist detection. In: 2016 IEEE Intelligent Vehicles Symposium (IV), IEEE (2016)
10. Huang, J., et al.: Speed/accuracy trade-offs for modern convolutional object detectors. In: 2017 IEEE Conference on Computer Vision and Pattern Recognition (CVPR) (2017)
11. Redmon, J., Farhadi, A.: YOLO9000: better, faster, stronger (2016). arXiv:1612.08242 [cs.CV]
12. Shafiee , M.J., Chywl, B., Li, F., Wong, A.: Fast YOLO: a fast you only look once system for real-time embedded object detection in video (2017). arXiv:1709.05943 [cs.CV]
13. Kharchenko, V., Chyrka, I.: Detection of airplanes on the ground using YOLO neural network. In: 2018 IEEE 17th International Conference on Mathematical Methods in Electromagnetic Theory (MMET), Kiev, Ukraine (2018)

Automatic Plant Leaf Classification Based on Back Propagation Networks for Medical Applications

Karattupalayam Chidambaram Saranya and Apoorv Goyal

Abstract Recognition of medicinal leaves has been a skill that is passed down ages. Being a skill of great importance, it can help the community if the use of the skill can be generalized. In this paper, a review of an intelligent recognition system is presented to classify different types of leaves (40 classes of leaves) using back propagation neural network and the system presents a very good accuracy. At the end, the portability and ease of use of the system are demonstrated as a GUI making the system user-friendly and rendering it ready to use.

Keywords Plant leaf classification · Back propagation networks · Medical application · Neural net

1 Introduction

Medicinal leaf recognition is a skill that is passed down from one generation to another. Such a skill has great importance and can prove to be really helpful if it can be delivered as a portable, easy to use, and readily available system on devices such as smartphones and computers. With the help of machine learning, a computer can be taught to acquire this skill and with a motive to deliver to the community, it is presented as a GUI which can be easily implemented on any platform.

Şekeroğlu and İnan [1] presented an intelligent recognition system that is able to classify 27 different classes of leaves with an accuracy of 97.2%. The training and testing images used were almost same with the only difference being the addition of noise in the test set. Abdolvahab and Sharath Kumar [2] proposed a plant classification methodology using Gray Level co-occurrence matrix, GLCM and principal component analysis, and PCA for feature extraction that is able to classify 12 different classes of leaves. The disadvantage of using this approach as they have mentioned was that GLCM is very sensitive to changes and slight changes affect the accuracy

K. C. Saranya (✉) · A. Goyal
School of Electronics Engineering, Vellore Institute of Technology, Vellore,
Tamil Nadu, India
e-mail: saranya.kc@vit.ac.in

© Springer Nature Singapore Pte Ltd. 2020
K. N. Das et al. (eds.), *Soft Computing for Problem Solving*,
Advances in Intelligent Systems and Computing 1057,
https://doi.org/10.1007/978-981-15-0184-5_83

drastically. Jyotismita and Ranjan [3] and Wang et al. [4] proposed a methodology based on feature extraction which is able to classify up to three different classes of leaves. Ekshinge et al. [5] devised a multilayer perceptron net for leaf recognition that extracts the leaf features and uses the multilayer perceptron network for classification. They achieved an accuracy of about 94%. Wu et al. [6] in their paper used leaf features such as shape, margin, and vein for leaf classification. The key concern with this research was the accuracy with which the net is able to discriminate uniform or similar features.

In similar researches, Sandeep Kumar [7] proposed an identification method for Indian medicinal plants that uses features such as area, edge, and color of leaf for classification. This method had a constraint on the type of images it can process and it was limited to processing fully mature leaves. Fu and Chi [8] proposed an approach for extraction of leaf vein that comprises of two stages. They used an artificial neural network for classification along with a thresholding method to obtain the results. Zulkifli et al. [9] and Bhardwaj et al. [10] used moment invariant (MI) for feature extraction and different approaches for classification. Prasad et al. [11] used curvelet transform to extract leaf information and classify them. Kulkarni et al. [12] suggested leaf classification using radial basis probabilistic neural network (RBPNN) and Zernike moments. These approaches used a very less number of classes.

The dataset used for training and testing is a standard leaf database from UCI repository consisting of images of leafs and their classes [13]. The number of classes used in many previously proposed methodologies was less and increasing the number of classes would affect the system accuracy drastically. This fact hinders the scalability of a system. That is not the case with our system as demonstrated later we have used 40 different classes of leaves and plan on adding more classes in the future. We use multiple preprocessing techniques and their effect on the results of the net and its accuracy. The entire system is implemented on MATLAB on a dual-core Intel i5 with 4 GB RAM and AMD Radeon graphic card.

The paper discusses the methodology, the training set, the preprocessing of images, back propagation net in Sect. 2 and the results of various experiments for different preprocessing techniques, training and testing sets and the user interface in Sect. 3.

2 Methodology and the Structure of the System

2.1 Training Set

Dataset consists of about 40 classes of different leaves and about 5–13 images for each class of leaf shown in Fig. 1. Five leaves from each class were used for training the network. The images of leaves used for training had a homogenous pink background with varying contrast.

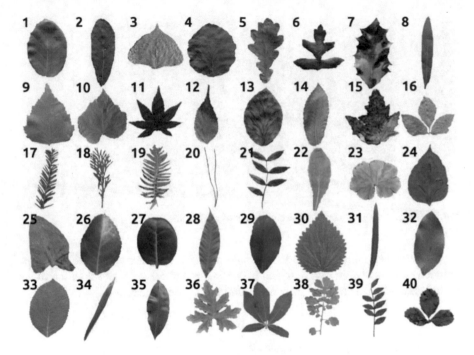

Fig. 1 UCI leaf dataset

2.2 Preprocessing of Images

The first step of the system is the capturing of an image of the leaf for training the network. The original image in RGB that is acquired first converted to gray-scale image and then scaled to 50 × 50 size to minimize the training time of neural network as shown in Fig. 2.

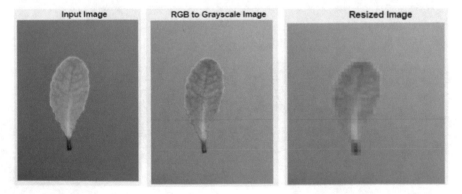

Fig. 2 Preprocessing RGB to gray-scale and resizing

Fig. 3 Preprocessing extracting value component of HSV image

We have also converted the RGB image to HSV and then extracted the value component as shown in Fig. 3, which is then resized to 50×50 and then fed to the neural network. This method provides illumination compensation.

2.3 Back Propagation Network

The network used is a standard back propagation network with two hidden layers each consisting of 200 and 240 units, respectively. The input and output neurons consist of 2500 and 40 units, respectively. The network is run for 1500 epochs, learning rate of 0.05, and has the structure as shown in Fig. 4[]. The gray-scale/HSV images are fed to the neural network shown. The input feeding method is shown in Fig. 5.

For testing, three variants of back propagation are used gradient descent, gradient descent with momentum, and resilient back propagation.

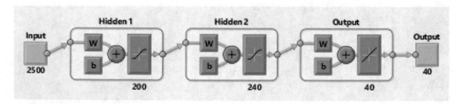

Fig. 4 Network structure

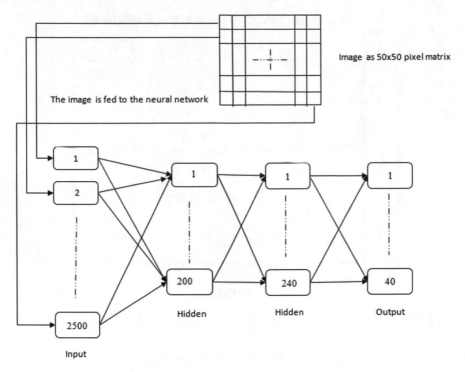

Fig. 5 Back propagation neural network

3 Experiments and Results

Images were used to train the network for all 40 classes.

Training the back propagation network with a momentum of 0.9 the gradient curve is as shown in Fig. 6. The gradient dropped from 37.6 to 0.0353 after 1500 epoch of training. The performance curve is also shown in Fig. 7 and the error changed from 8.29 to 0.0233 after 1500 epochs of training.

Training the same net without the momentum, the gradient curve is as shown in Fig. 8. The gradient dropped from around 51.5 to 0.00654 after 1500 epoch of

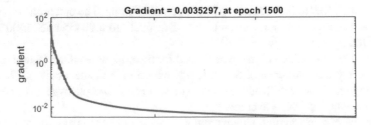

Fig. 6 Gradient curve for back propagation (with momentum of 0.9)

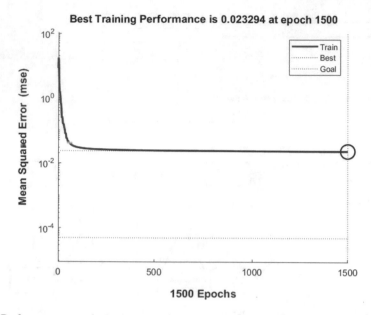

Fig. 7 Performance curve for back propagation (with momentum of 0.9)

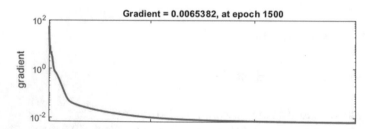

Fig. 8 Gradient curve for back propagation

training. The performance curve is also shown in Fig. 9 and the error changed from 11.6 to 0.02 after 1500 epochs of training.

Training using resilient back propagation with weight increment delta as 1.2 and weight decrement delta as 0.5 the gradient curve is as shown in Fig. 10. The gradient dropped from 40.9 to 0.00171 after 1500 epoch of training. The performance curve is also shown in Fig. 11 and the error changed from 8.10 to 0.00655 after 1500 epochs of training.

The test sets contained of the images used for training with added Gaussian noise. The accuracy was the mean of four test sets where a_i is the accuracy of ith test set containing leaf images with added noise. The recorded accuracy was up to 100% for individual sets and 98.2% average.

After the end of training, another single repetition test was applied to check the ability of trained network to recognize other than the training sequence. The test

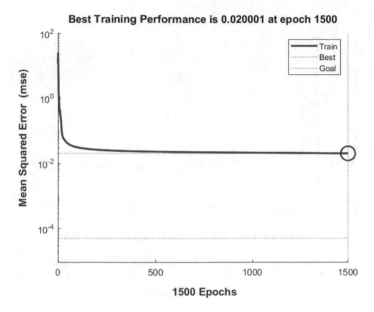

Fig. 9 Performance curve for back propagation

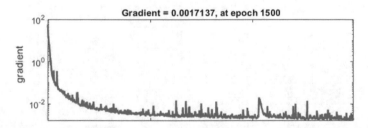

Fig. 10 Gradient curve for resilient back propagation

was performed with images apart from the training vector and with Gaussian noise with images used for training. While the recognition was pretty accurate for training images with added noise, some inaccurate outputs were obtained for images except the training images for some classes but the results were quite consistent for most of the classes.

Figure 12 shows the cumulative heat map for the accuracy of neural network output for RGB images. Figure 13 shows the cumulative heat map for the accuracy of neural network output for HSV images.

To easily visualize the results, the confusion matrix of the tests performed is plotted as a heat map shown in Figs. 12 and 13. Confusion matrix presents the false positives, false negatives, true positives, and true negatives for a class. Rows represent the actual class, columns represent the predicted class, and cells represent the confidence of detection for a particular class. The darker the color of a cell in heat

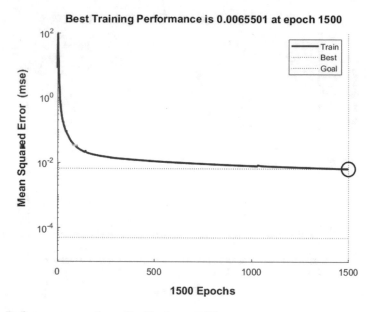

Fig. 11 Performance curve for resilient back propagation

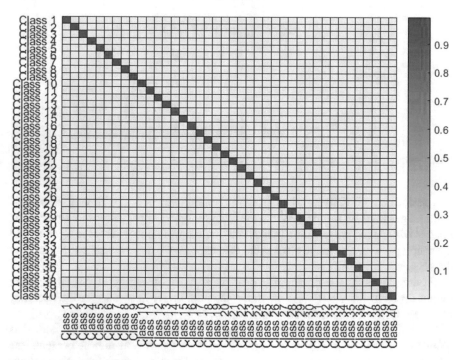

Fig. 12 Confusion matrix–gray-scale preprocessing

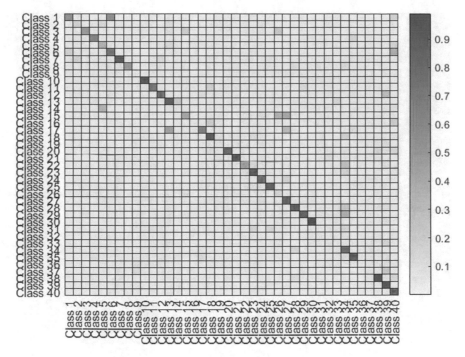

Fig. 13 Confusion matrix–HSV preprocessing

matrix more is the confidence of the neural network in recognition of a particular class. Confusion matrix draws a contrast between the actual class and predicted class.

To demonstrate the portability of the system, it was built with a UI and the classification type so that the user can control the preprocessing technique used depending on the condition of the input image to get the best results and the closest classification output. The GUI is as shown in Fig. 14. The GUI allows the user to select the color space in which the image should be processed so as to compensate for different lighting and other conditions.

The GUI shows the user the predicted class, the botanical name of the leaf, and the closest leaf image to the test image in the database.

4 Conclusions

The leaf recognition system that was proposed in this paper works well in terms of accuracy, fault tolerance, and portability. The GUI makes the system portable and user-friendly. In the future, adding more classes to the system would be prime focus so as to increase the scope of the system. An UGV/UAV can be built around the system so as to make the process of leaf recognition and collection autonomous.

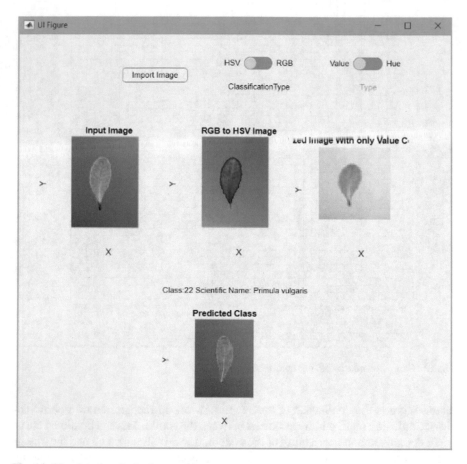

Fig. 14 User interface for leaf recognition

References

1. Şekeroğlu, B., İnan, Y.: Leaves recognition system using a neural network. Procedia Comput. Sci. **102**, 578–582 (2016)
2. Abdolvahab, E., Sharath Kumar, Y.: Leaf recognition for plant classification using GLCM and PCA methods. Orient. J. Comput. Sci. Technol. **3**(1) (2010)
3. Jyotismita, C., Ranjan, P.: Plant leaf recognition using shape based features and neural network classifiers. Int. J. Adv. Comput. Sci. Appl. **2**, 10 (2011)
4. Wang, Z., Chi, Z., Feng, D., Wang, Q.: Leaf image retrieval with shape features. In: Laurini, R. (ed.) VISUAL 2000. LNCS, vol. 1929, pp. 477–487 (2000)
5. Ekshinge, S., Sambhaji, D.B., Andore, M.: Leaf recognition algorithm using neural network based image processing. Asian J. Eng. Technol. Innov. **10**, 16 (2014)
6. Wu, Q., Zhou, C., Wang, C.: Feature extraction and automatic recognition of plant leaf using artificial neural network. Av. Cienc. Comput. **5**, 12 (2006)
7. Sandeep Kumar, E.: Leaf color, area and edge features based approach for identification of indian medicinal plants. Indian J. Comput. Sci. Eng.

8. Fu, H., Chi, Z.: A two-stage approach for leaf vein extraction. In: Proceedings of International Conference on Neural Networks and Signal Processing, vol. 1, pp. 208–211 (2003)

9. Zulkifli, Z., Saad, P., Mohtar, I.A.: Plant leaf identification using moment invariants & General Regression Neural Network. In: 11th International Conference on Hybrid Intelligent Systems (HIS), pp. 430–435 (2011)

10. Bhardwaj, A., Kaur, M., Kumar, A.: Recognition of plants by leaf image using moment invariant and texture analysis. Int. J. Innov. Appl. Stud. **3**(1), 237–248 (2013). ISSN 2028-9324

11. Prasad, S., Kumar, P., Tripathi, R.C.: Plant leaf species identification using Curvelet transform. In: 2nd International Conference on Computer and Communication Technology (ICCCT), pp. 646–652 (2011)

12. Kulkarni, A.H., Rai, H.M., Jahagirdar, K.A., Upparaman, P.S.: A leaf recognition system for classifying plants using RBPNN and pseudo Zernike. Int. J. Latest Trends Eng. Technol. (IJL-TET)

13. Silva, P.F.B., Marcal, A.R.S., Almeida da Silva, R.M.: Evaluation of features for leaf discrimination. Lecture Notes in Computer Science, vol. 7950, pp. 197–204. Springer (2013)

Author Index

© Springer Nature Singapore Pte Ltd. 2020
K. N. Das et al. (eds.), *Soft Computing for Problem Solving*,
Advances in Intelligent Systems and Computing 1057,
https://doi.org/10.1007/978-981-15-0184-5

Printed in the United States
By Bookmasters